COMPLETE SOLUTIONS GUIDE

TO ACCOMPANY

Calculus

LARSON / HOSTETLER

THIRD EDITION

Volume 2
(Chapters 8-13)

Dianna L. Zook
THE PENNSYLVANIA STATE UNIVERSITY
THE BEHREND COLLEGE

D. C. Heath and Company
Lexington, Massachusetts / Toronto

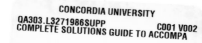

Published simultaneously in Canada.

Printed in the United States of America.

International Standard Book Number: 0-669-10100-1

Preface

This solutions guide is a supplement to *Calculus with Analytic Geometry, Third Edition,* by Roland E. Larson and Robert P. Hostetler. Solutions to every exercise in the text are given with all essential algebraic steps included. There are three volumes in the complete set of solutions guides. Volume I contains Chapters 1–7, Volume II contains Chapters 8–13, and Volume III contains Chapters 14–17. Also available is a one-volume *Study and Solutions Guide to Accompany Calculus, Third Edition,* written by David E. Heyd, which contains worked-out solutions to *selected* representative exercises from the text.

I have made every effort to see that the solutions are correct. However, I would appreciate very much hearing about any errors or other suggestions for improvement.

I would like to thank several people who helped in the production of this guide: David E. Heyd, who assisted the authors of the text and double checked my solutions, Linda L. Matta, who was in charge of the production of the guide, and Timothy R. Larson, who produced the art for the guide. The typing for the guide was done by Nancy K. Stout, of Stout's Secretarial Services, and Linda M. Bollinger. I would also like to thank the students in my mathematics classes. Finally, I would like to thank my husband Ed Schlindwein for his support during the many months I have worked on this project.

Dianna L. Zook
The Pennsylvania State University
The Behrend College
Erie, Pennsylvania 16563

Contents

8 Integration techniques, L'Hopital's Rule, and improper integrals

9 Infinite series

10 Conics

11 Plane curves, parametric equations, and polar coordinates

12 Vectors and the geometry of space

13 Vector-valued functions

8 Integration techniques, L'Hôpital's Rule, and improper integrals

8.1 Review of basic integration formulas

1. Let $u = 3x - 2, \quad u' = 3$

$$\int (3x - 2)^4\,dx = \frac{1}{3}\int (3x - 2)^4(3)\,dx = \frac{1}{15}(3x - 2)^5 + C$$

2. Let $u = t - 9, \quad u' = 1$

$$\int \frac{2}{(t - 9)^2}\,dt = 2\int (t - 9)^{-2}\,dt = \frac{-2}{t - 9} + C$$

3. Let $u = -2x + 5, \quad u' = -2$

$$\int (-2x + 5)^{3/2}\,dx = -\frac{1}{2}\int (-2x + 5)^{3/2}(-2)\,dx = -\frac{1}{5}(-2x + 5)^{5/2} + C$$

4. Let $u = 4 - 2x^2, \quad u' = -4x$

$$\int x\sqrt{4 - 2x^2}\,dx = -\frac{1}{4}\int (4 - 2x^2)^{1/2}(-4x)\,dx = -\frac{1}{6}(4 - 2x^2)^{3/2} + C$$

5. $$\int \left[v + \frac{1}{(3v - 1)^3}\right] dv = \int v\,dv + \frac{1}{3}\int (3v - 1)^{-3}(3)\,dv$$

$$= \frac{1}{2}v^2 - \frac{1}{6(3v - 1)^2} + C$$

6. Let $u = t^2 - t + 2, \quad u' = 2t - 1$

$$\int \frac{2t - 1}{t^2 - t + 2}\,dt = \ln|t^2 - t + 2| + C$$

7. Let $u = -t^3 + 9t + 1, \quad u' = -3t^2 + 9 = -3(t^2 - 3)$

$$\int \frac{t^2 - 3}{-t^3 + 9t + 1}\,dt = -\frac{1}{3}\int \frac{-3(t^2 - 3)}{-t^3 + 9t + 1}\,dt = -\frac{1}{3}\ln|-t^3 + 9t + 1| + C$$

8. $\displaystyle\int \frac{2x}{x-4}\,dx = \int 2\,dx + \int \frac{8}{x-4}\,dx = 2x + 8\ln|x-4| + C$

9. $\displaystyle\int \frac{x^2}{x-1}\,dx = \int (x+1)\,dx + \int \frac{1}{x-1}\,dx = \frac{1}{2}x^2 + x + \ln|x-1| + C$

10. Let $u = x^2 + 2x - 4$, $u' = 2(x+1)$

$$\int \frac{x+1}{\sqrt{x^2+2x-4}}\,dx = \frac{1}{2}\int (x^2+2x-4)^{-1/2}(2)(x+1)\,dx$$

$$= \sqrt{x^2+2x-4} + C$$

11. $\displaystyle\int \left[\frac{1}{3x-1} - \frac{1}{3x+1}\right] dx = \frac{1}{3}\int \frac{1}{3x-1}(3)\,dx - \frac{1}{3}\int \frac{1}{3x+1}(3)\,dx$

$\displaystyle= \frac{1}{3}\ln|3x-1| - \frac{1}{3}\ln|3x+1| + C = \frac{1}{3}\ln\left|\frac{3x-1}{3x+1}\right| + C$

12. Let $u = 5x$, $u' = 5$

$$\int e^{5x}\,dx = \frac{1}{5}\int e^{5x}(5)\,dx = \frac{1}{5}e^{5x} + C$$

13. Let $u = t^2$, $u' = 2t$

$$\int t\sin(t^2)\,dt = \frac{1}{2}\int \sin(t^2)(2t)\,dt = -\frac{1}{2}\cos(t^2) + C$$

14. Let $v = 4u$, $v' = 4$

$$\int \sec(4u)\,du = \frac{1}{4}\int \sec(4u)(4)\,du = \frac{1}{4}\ln|\sec(4u) + \tan(4u)| + C$$

15. $\displaystyle\int \cos x\, e^{\sin x}\,dx = e^{\sin x} + C \qquad (u = \sin x, \quad u' = \cos x)$

16. $\displaystyle\int \frac{e^x}{1+e^x}\,dx = \ln(1+e^x) + C \qquad (u = 1 + e^x, \quad u' = e^x)$

17. $\displaystyle\int \frac{(1+e^t)^2}{e^t}\,dt = \int \frac{1+2e^t+e^{2t}}{e^t}\,dt = \int (e^{-t} + 2 + e^t)\,dt$

$= -e^{-t} + 2t + e^t + C$

18. $\int \frac{1 + \sin x}{\cos x}\,dx = \int (\sec x + \tan x)\,dx = \ln|\sec x + \tan x| + \ln|\sec x| + C$

$= \ln|\sec x(\sec x + \tan x)| + C$

19. Let $u = 3x$, $u' = 3$

$\int \sec 3x \tan 3x\,dx = \frac{1}{3}\int \sec 3x \tan 3x\,(3)\,dx = \frac{1}{3}\sec 3x + C$

20. Let $u = 1 - 2\sqrt{x}$, $u' = -1/\sqrt{x}$

$\int \frac{1}{\sqrt{x}\,(1 - 2\sqrt{x})}\,dx = -\int \frac{1}{1 - 2\sqrt{x}}\left(\frac{-1}{\sqrt{x}}\right)dx = -\ln|1 - 2\sqrt{x}| + C$

21. Let $u = 1 + e^{x}$, $u' = e^{x}$

$\int \frac{2}{e^{-x} + 1}\,dx = 2\int \left(\frac{1}{e^{-x} + 1}\right)\left(\frac{e^{x}}{e^{x}}\right)dx = 2\int \frac{e^{x}}{1 + e^{x}}\,dx = 2\ln|1 + e^{x}| + C$

22. $\int \frac{1}{2e^{x} - 3}\,dx = \int \frac{1}{2e^{x} - 3}\left(\frac{e^{-x}}{e^{-x}}\right)dx = \int \frac{e^{-x}}{2 - 3e^{-x}}\,dx = \frac{1}{3}\int \frac{3e^{-x}}{2 - 3e^{-x}}\,dx$

$= \frac{1}{3}\ln|2 - 3e^{-x}| + C$

23. $\int \frac{1}{1 - \cos x}\,dx = \int \frac{1 + \cos x}{1 - \cos^2 x}\,dx = \int \csc^2 x\,dx + \int \frac{\cos x}{\sin^2 x}\,dx$

$= -\cot x - \frac{1}{\sin x} + C = -\cot x - \csc x + C$

24. $\int \frac{1}{\sec x - 1}\,dx = \int \frac{1}{\sec x - 1}\left(\frac{\sec x + 1}{\sec x + 1}\right)dx = \int \frac{(\sec x + 1)}{\tan^2 x}\,dx$

$= \int \frac{\sec x}{\tan^2 x}\,dx + \int \cot^2 x\,dx = \int \frac{\cos x}{\sin^2 x}\,dx + \int (\csc^2 x - 1)\,dx$

$= -\frac{1}{\sin x} - \cot x - x + C = -(\csc x + \cot x + x) + C$

25. $\int \frac{2t - 1}{t^2 + 4}\,dt = \int \frac{2t}{t^2 + 4}\,dt - \int \frac{1}{4 + t^2}\,dt = \ln(t^2 + 4) - \frac{1}{2}\arctan\frac{t}{2} + C$

26. Let $u = 2t - 1$, $u' = 2$

$\int \frac{2}{(2t - 1)^2 + 4}\,dt = \frac{1}{2}\arctan\frac{2t - 1}{2} + C$

27. $\int \frac{3}{t^2 + 1}\,dt = 3\arctan t + C$

28. $\int \frac{3}{\sqrt{1 - t^2}}\, dt = 3 \arcsin t + C$

29. $\int \frac{1}{x\sqrt{x^2 - 4}}\, dx = \frac{1}{2} \operatorname{arcsec} \frac{|x|}{2} + C$

30. Let $u = x^2 - 4, \quad u' = 2x$

$$\int \frac{-2x}{\sqrt{x^2 - 4}}\, dx = -\int (x^2 - 4)^{-1/2}(2x)\, dx = -2\sqrt{x^2 - 4} + C$$

31. Let $u = 2t - 1, \quad u' = 2$

$$\int \frac{-1}{\sqrt{1 - (2t - 1)^2}}\, dt = -\frac{1}{2} \int \frac{2}{\sqrt{1 - (2t - 1)^2}}\, dt = -\frac{1}{2} \arcsin (2t - 1) + C$$

32. Let $u = \sqrt{3}x, \quad u' = \sqrt{3}$

$$\int \frac{1}{4 + 3x^2}\, dx = \frac{1}{\sqrt{3}} \int \frac{\sqrt{3}}{4 + (\sqrt{3}x)^2}\, dx = \frac{1}{2\sqrt{3}} \arctan \frac{\sqrt{3}x}{2} + C$$

33. $\int \csc (\pi x) \cot (\pi x)\, dx = \frac{1}{\pi} \int \csc (\pi x) \cot (\pi x)\, \pi\, dx = -\frac{1}{\pi} \csc (\pi x) + C$

34. Let $u = 2x, \quad u' = 2$

$$\int \frac{1}{x\sqrt{4x^2 - 1}}\, dx = \int \frac{2}{2x\sqrt{(2x)^2 - 1}}\, dx = \operatorname{arcsec} |2x| + C$$

35. Let $u = t^2, \quad u' = 2t$

$$\int \frac{t}{\sqrt{1 - t^4}}\, dt = \frac{1}{2} \int \frac{2t}{\sqrt{1 - (t^2)^2}}\, dt = \frac{1}{2} \arcsin (t^2) + C$$

36. $\int \frac{\sin x}{\sqrt{\cos x}}\, dx = -\int (\cos x)^{-1/2} (-\sin x)\, dx = -2\sqrt{\cos x} + C$

37. Let $u = \tan x, \quad u' = \sec^2 x$

$$\int \frac{\sec^2 x}{4 + \tan^2 x}\, dx = \frac{1}{2} \arctan \frac{\tan x}{2} + C$$

38. $\int \tan^2 (2x)\, dx = \int [\sec^2 (2x) - 1]\, dx = \frac{1}{2} \int \sec^2 (2x)(2)\, dx - \int dx$

$$= \frac{1}{2} \tan (2x) - x + C = \frac{1}{2}(-2x + \tan 2x) + C$$

39. Let $u = \cos\left(\frac{2}{t}\right)$, $\quad u' = \frac{2 \sin (2/t)}{t^2}$

$$\int \frac{\tan (2/t)}{t^2}\, dt = \frac{1}{2} \int \frac{1}{\cos (2/t)} \left[\frac{2 \sin (2/t)}{t^2}\right] dt = \frac{1}{2} \ln |\cos (2/t)| + C$$

40. Let $u = 1/t$, $\quad u' = -1/t^2$

$$\int \frac{e^{1/t}}{t^2}\, dt = -\int e^{1/t} \left(\frac{-1}{t^2}\right) dt = -e^{1/t} + C$$

41. $$\int (1 + 2x^2)^2\, dx = \int (4x^4 + 4x^2 + 1)\, dx = \frac{4}{5}x^5 + \frac{4}{3}x^3 + x + C$$

$$= \frac{x}{15}(12x^4 + 20x^2 + 15) + C$$

42. $$\int (2 - x^2)^2\, dx = \int (x^4 - 4x^2 + 4)\, dx = \frac{1}{5}x^5 - \frac{4}{3}x^3 + 4x + C$$

$$= \frac{x}{15}(3x^4 - 20x^3 + 60) + C$$

43. $$\int x\left(1 + \frac{1}{x}\right)^3 dx = \int x\left(1 + \frac{3}{x} + \frac{3}{x^2} + \frac{1}{x^3}\right) dx = \int \left(x + 3 + \frac{3}{x} + \frac{1}{x^2}\right) dx$$

$$= \frac{1}{2}x^2 + 3x + 3 \ln |x| - \frac{1}{x} + C$$

44. $$\int (x + x^2)^3\, dx = \int (x^3 + 3x^4 + 3x^5 + x^6)\, dx = \frac{x^4}{4} + \frac{3x^5}{5} + \frac{x^6}{2} + \frac{x^7}{7} + C$$

$$= \frac{x^4}{140}(35 + 84x + 70x^2 + 20x^3) + C$$

45. $$\int (1 + e^x)^2\, dx = \int (e^{2x} + 2e^x + 1)\, dx = \frac{1}{2}e^{2x} + 2e^x + x + C$$

46. $$\int \left(\frac{e^x + e^{-x}}{2}\right)^2 dx = \frac{1}{4} \int (e^{2x} + 2 + e^{-2x})\, dx$$

$$= \frac{1}{4}\left[\frac{1}{2}e^{2x} + 2x - \frac{1}{2}e^{-2x}\right] + C = \frac{1}{8}[e^{2x} + 4x - e^{-2x}] + C$$

47. $$\int \frac{3}{\sqrt{6x - x^2}}\, dx = 3 \int \frac{1}{\sqrt{9 - (x - 3)^2}}\, dx = 3 \arcsin \frac{x - 3}{3} + C$$

48. $\displaystyle\int \frac{1}{(x-1)\sqrt{4x^2-8x+3}}\,dx = \int \frac{2}{[2(x-1)]\sqrt{[2(x-1)]^2-1}}\,dx$

$= \operatorname{arcsec}|2(x-1)| + C$

49. $\displaystyle\int \frac{4}{4x^2+4x+65}\,dx = \int \frac{1}{(x+\frac{1}{2})^2+16}\,dx = \frac{1}{4}\arctan\frac{x+(1/2)}{4} + C$

$\displaystyle= \frac{1}{4}\arctan\frac{2x+1}{8} + C$

50. $\displaystyle\int \frac{1}{\sqrt{2-2x-x^2}}\,dx = \int \frac{1}{\sqrt{3-(x+1)^2}}\,dx = \arcsin\frac{x+1}{\sqrt{3}} + C$

51. Let $u = -x^2$, $u' = -2x$

$\displaystyle\int_0^1 xe^{-x^2}\,dx = -\frac{1}{2}\int_0^1 e^{-x^2}(-2x)\,dx = -\frac{1}{2}e^{-x^2}\Big]_0^1 = \frac{1}{2}(1-e^{-1})$

52. Let $u = \sin t$, $u' = \cos t$

$\displaystyle\int_0^\pi \sin^2 t\cos t\,dt = \frac{1}{3}\sin^3 t\Big]_0^\pi = 0$

53. Let $u = 1 - \ln x$, $u' = \dfrac{-1}{x}$

$\displaystyle\int_1^e \frac{1-\ln x}{x}\,dx = -\int_1^e (1-\ln x)\left(\frac{-1}{x}\right)dx = -\frac{1}{2}(1-\ln x)^2\Big]_1^e = \frac{1}{2}$

54. $\displaystyle\int_1^2 \frac{x-2}{x}\,dx = \int_1^2 \left(1-\frac{2}{x}\right)dx = \Big[x - 2\ln x\Big]_1^2 = 1 - \ln 4$

55. Let $u = x^2 + 9$, $u' = 2x$

$\displaystyle\int_0^4 \frac{2x}{\sqrt{x^2+9}}\,dx = \int_0^4 (x^2+9)^{-1/2}(2x)\,dx = 2\sqrt{x^2+9}\Big]_0^4 = 4$

56. $\displaystyle\int_0^{\pi/4} \cos 2x\,dx = \frac{1}{2}\sin 2x\Big]_0^{\pi/4} = \frac{1}{2}$

57. $\displaystyle\int_{\pi/4}^{\pi/2} \cot x\,dx = \int_{\pi/4}^{\pi/2} \frac{\cos x}{\sin x}\,dx = \ln|\sin x|\Big]_{\pi/4}^{\pi/2} = \ln\sqrt{2} = \frac{1}{2}\ln 2$

58. $\displaystyle\int_0^4 \frac{1}{\sqrt{25-x^2}}\,dx = \arcsin\frac{x}{5}\Big]_0^4 = \arcsin\frac{4}{5}$

59. $\int_1^2 \frac{1}{2x\sqrt{4x^2 - 1}}\,dx = \frac{1}{2}\int \frac{2}{2x\sqrt{(2x)^2 - 1}}\,dx = \frac{1}{2}\operatorname{arcsec}(2x)\Big]_1^2$

$= \frac{1}{2}\left[\operatorname{arcsec} 4 - \frac{\pi}{3}\right]$

60. Let $u = 3x$, $u' = 3$

$\int_0^{2/\sqrt{3}} \frac{1}{4 + 9x^2}\,dx = \frac{1}{3}\int_0^{2/\sqrt{3}} \frac{3}{4 + (3x)^2}\,dx = \frac{1}{6}\arctan\frac{3x}{2}\Big]_0^{2/\sqrt{3}} = \frac{\pi}{18}$

61. Let $u = 1 - x^2$, $u' = -2x$

$A = 4\int_0^1 x\sqrt{1 - x^2}\,dx$

$= -2\int_0^1 (1 - x^2)^{1/2}(-2x)\,dx$

$= -\frac{4}{3}(1 - x^2)^{3/2}\Big]_0^1 = \frac{4}{3}$

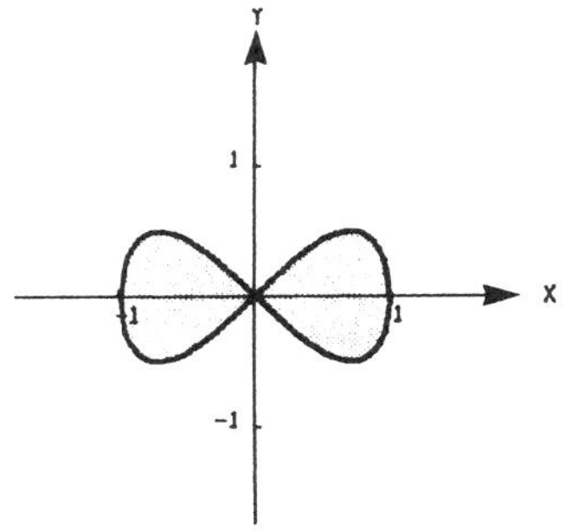

62. $A = \int_0^{\pi/2} \sin 2x\,dx$

$= -\frac{1}{2}\cos 2x\Big]_0^{\pi/2} = 1$

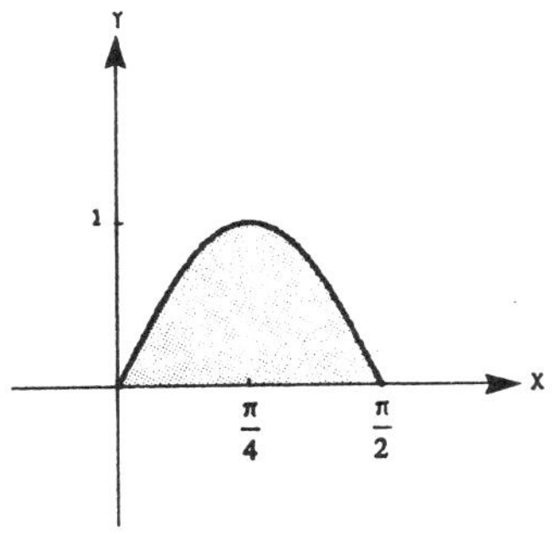

63. $\int_0^{1/a} (x - ax^2)\,dx = \left[\frac{1}{2}x^2 - \frac{a}{3}x^3\right]_0^{1/a} = \frac{1}{6a^2}$

Let $\frac{1}{6a^2} = \frac{2}{3}$

$12a^2 = 3$, $a = \frac{1}{2}$

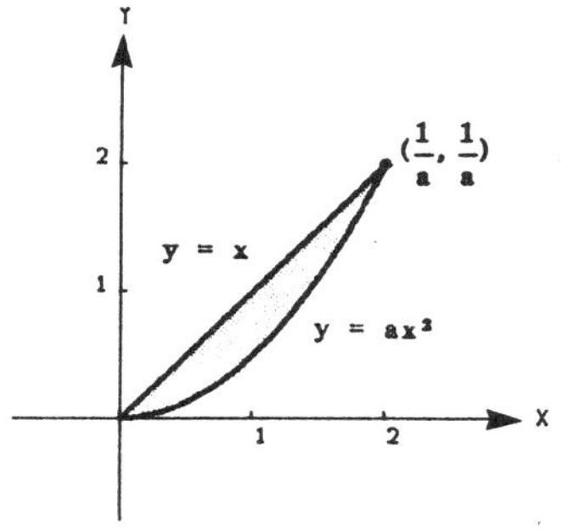

64. Shell

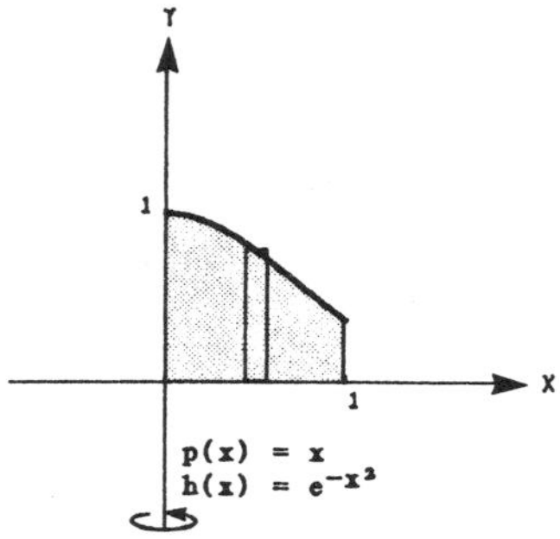

Let $u = -x^2, \qquad u' = -2x$

$$V = 2\pi \int_0^1 xe^{-x^2}\,dx$$

$$= -\pi \int_0^1 e^{-x^2}(-2x)\,dx$$

$$= -\pi e^{-x^2}\Big]_0^1 = \pi(1 - e^{-1})$$

65. Shell (from exercise 64)

$$V = 2\pi \int_0^b xe^{-x^2}\,dx = -\pi e^{-x^2}\Big]_0^b = \pi(1 - e^{-b^2}) = \frac{4}{3}$$

$$e^{-b^2} = \frac{3\pi - 4}{3\pi}, \qquad b = \sqrt{\ln\left(\frac{3\pi}{3\pi - 4}\right)} \approx 0.743$$

66. (a) $$\frac{1}{(\pi/n) - 0}\int_0^{\pi/n} \sin(nx)\,dx = \frac{1}{\pi}\int_0^{\pi/n} \sin(nx)(n)\,dx$$

$$= -\frac{1}{\pi}\cos(nx)\Big]_0^{\pi/n} = \frac{2}{\pi}$$

(b) $$\frac{1}{3 - (-3)}\int_{-3}^{3} \frac{1}{1 + x^2}\,dx = \frac{1}{6}\arctan x\Big]_{-3}^{3} = \frac{1}{3}\arctan 3$$

67. $$A = \int_0^4 \frac{5}{\sqrt{25 - x^2}}\,dx = 5\arcsin\frac{x}{5}\Big]_0^4$$

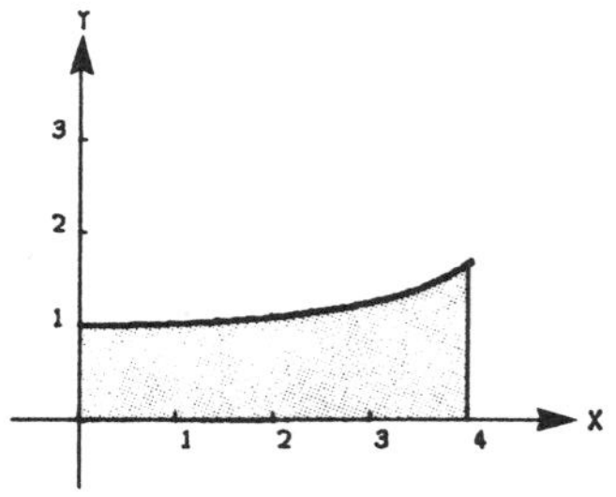

$$= 5\arcsin\frac{4}{5}$$

$$\frac{M_y}{\rho} = \int_0^4 \frac{5x}{\sqrt{25 - x^2}}\,dx$$

$$= -\frac{5}{2}\int (25 - x^2)^{-1/2}(-2x)\,dx$$

$$= -5\sqrt{25 - x^2}\Big]_0^4 = -5[3 - 5] = 10$$

$$\bar{x} = \frac{M_y}{m} = \frac{10}{5\arcsin(4/5)} = \frac{2}{\arcsin(4/5)} \approx 2.157$$

68. $f(x) = 2\sqrt{x}, \qquad f'(x) = \frac{1}{\sqrt{x}}$

$$1 + [f'(x)]^2 = 1 + \frac{1}{x} = \frac{x+1}{x}$$

$$S = 2\pi \int_0^9 2\sqrt{x}\sqrt{\frac{x+1}{x}}\,dx$$

$$= 2\pi \int_0^9 2\sqrt{x+1}\,dx$$

$$= 4\pi\left(\frac{2}{3}\right)(x+1)^{3/2}\Big]_0^9$$

$$= \frac{8\pi}{3}[10\sqrt{10} - 1]$$

$$\approx 256.545$$

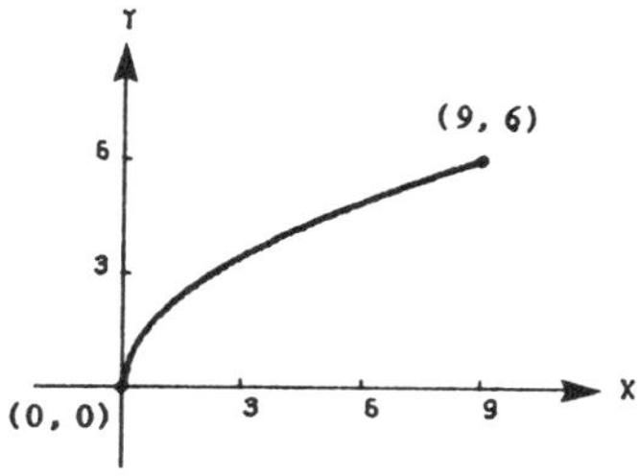

8.2 Integration by parts

1. Let $u = x, \qquad v' = e^{2x}, \qquad u' = 1, \qquad v = \frac{1}{2}e^{2x}$

$$\int xe^{2x}\,dx = \frac{1}{2}xe^{2x} - \int \frac{1}{2}e^{2x}\,dx = \frac{1}{2}xe^{2x} - \frac{1}{4}e^{2x} + C$$

$$= \frac{e^{2x}}{4}(2x - 1) + C$$

2. (1)--> $u = x^2, \qquad v' = e^{2x}, \qquad u' = 2x, \qquad v = \frac{1}{2}e^{2x}$

(2)--> $u = x, \qquad v' = e^{2x}, \qquad u' = 1, \qquad v = \frac{1}{2}e^{2x}$

$$\int x^2e^{2x}\,dx = \frac{1}{2}x^2e^{2x} - \int xe^{2x}\,dx$$

$$= \frac{1}{2}x^2e^{2x} - \left[\frac{1}{2}xe^{2x} - \int \frac{1}{2}e^{2x}\,dx\right]$$

$$= \frac{1}{2}x^2e^{2x} - \frac{1}{2}xe^{2x} + \frac{1}{4}e^{2x} + C = \frac{e^{2x}}{4}(2x^2 - 2x + 1) + C$$

3. $\int xe^{x^2}\,dx = \frac{1}{2}\int e^{x^2}(2x)\,dx = \frac{1}{2}e^{x^2} + C$

4. $\displaystyle\int x^2e^{x^3}\,dx = \frac{1}{3}\int e^{x^3}(3x^2)\,dx = \frac{1}{3}e^{x^3} + C$

5. Let $u = x$, $\quad v' = e^{-2x}$, $\quad u' = 1$, $\quad v = -\frac{1}{2}e^{-2x}$

$$\int xe^{-2x}\,dx = -\frac{1}{2}xe^{-2x} - \int -\frac{1}{2}e^{-2x}\,dx = -\frac{1}{2}xe^{-2x} - \frac{1}{4}e^{-2x} + C$$

$$= \frac{-1}{4e^{2x}}(2x + 1) + C$$

6. Let $u = x$, $\quad v' = e^{-x}$, $\quad u' = 1$, $\quad v = -e^{-x}$

$$\int \frac{x}{e^x}\,dx = \int xe^{-x}\,dx = -xe^{-x} - \int -e^{-x}\,dx = -xe^{-x} - e^{-x} + C$$

$$= -\frac{1}{e^x}(x + 1) + C$$

7. (1)—> $u = x^3$, $\quad v' = e^x$, $\quad u' = 3x^2$, $\quad v = e^x$

(2)—> $u = x^2$, $\quad v' = e^x$, $\quad u' = 2x$, $\quad v = e^x$

(3)—> $u = x$, $\quad v' = e^x$, $\quad u' = 1$, $\quad v = e^x$

$$\int x^3e^x\,dx = x^3e^x - 3\int x^2e^x\,dx = x^3e^x - 3x^2e^x + 6\int xe^x\,dx$$

$$= x^3e^x - 3x^2e^x + 6xe^x - 6e^x + C = e^x(x^3 - 3x^2 + 6x - 6) + C$$

8. $\displaystyle\int \frac{e^{1/t}}{t^2}\,dt = -\int e^{1/t}\left(\frac{-1}{t^2}\right)dt = -e^{1/t} + C$

9. Let $u = \ln x$, $\quad v' = x^3$, $\quad u' = 1/x$, $\quad v = x^4/4$

$$\int x^3\ln x\,dx = \frac{x^4}{4}\ln x - \frac{1}{4}\int x^3\,dx$$

$$= \frac{x^4}{4}\ln x - \frac{x^4}{16} + C = \frac{x^4}{16}(4\ln x - 1) + C$$

10. Let $u = \ln x$, $\quad v' = x^2$, $\quad u' = 1/x$, $\quad v = x^3/3$

$$\int x^2\ln x\,dx = \frac{x^3}{3}\ln x - \frac{1}{3}\int x^2\,dx = \frac{x^3}{3}\ln x - \frac{x^3}{9} + C$$

$$= \frac{x^3}{9}(3\ln x - 1) + C$$

11. Let $u = \ln(t + 1)$, $v' = t$, $u' = 1/(t + 1)$, $v = t^2/2$

$$\int t \ln(t + 1)\, dt = \frac{t^2}{2}\ln(t + 1) - \frac{1}{2}\int \frac{t^2}{t + 1}\, dt$$

$$= \frac{t^2}{2}\ln(t + 1) - \frac{1}{2}\int (t - 1 + \frac{1}{t + 1})\, dt$$

$$= \frac{t^2}{2}\ln(t + 1) - \frac{1}{2}\left[\frac{t^2}{2} - t + \ln(t + 1)\right] + C$$

$$= \frac{1}{4}[2(t^2 - 1)\ln|t + 1| - t^2 + 2t] + C$$

12. Let $u = \ln x$, $u' = 1/x$

$$\int \frac{1}{x(\ln x)^3}\, dx = \int (\ln x)^{-3}(\frac{1}{x})\, dx = \frac{-1}{2(\ln x)^2} + C$$

13. (1)--> $u = (\ln x)^2$, $v' = 1$, $u' = 2(\ln x)(1/x)$, $v = x$

(2)--> $u = \ln x$, $v' = 1$, $u' = 1/x$, $v = x$

$$\int (\ln x)^2\, dx = x(\ln x)^2 - 2\int \ln x\, dx$$

$$= x(\ln x)^2 - 2[x\ln x - \int dx] = x(\ln x)^2 - 2x\ln x + 2x + C$$

14. Let $u = \ln(3x)$, $v' = 1$, $u' = 1/x$, $v = x$

$$\int \ln(3x)\, dx = x\ln(3x) - \int dx = x\ln(3x) - x + C$$

15. $$\int \frac{(\ln x)^2}{x}\, dx = \int (\ln x)^2(\frac{1}{x})\, dx = \frac{(\ln x)^3}{3} + C$$

16. Let $u = \ln x$, $v' = \frac{1}{x^2}$, $u' = \frac{1}{x}$, $v = -\frac{1}{x}$

$$\int \frac{\ln x}{x^2}\, dx = -\frac{\ln x}{x} + \int \frac{1}{x^2}\, dx = -\frac{\ln x}{x} - \frac{1}{x} + C$$

17. Let $u = xe^{2x}$, $v' = \frac{1}{(2x + 1)^2}$, $u' = 2xe^{2x} + e^{2x} = e^{2x}(2x + 1)$,

$v = -\frac{1}{2(2x + 1)}$

$$\int \frac{xe^{2x}}{(2x + 1)^2}\, dx = -\frac{xe^{2x}}{2(2x + 1)} + \int \frac{e^{2x}}{2}\, dx$$

$$= \frac{-xe^{2x}}{2(2x + 1)} + \frac{e^{2x}}{4} + C = \frac{e^{2x}}{4(2x + 1)} + C$$

18. Let $u = x^2e^{x^2}$, $v' = \dfrac{x}{(x^2+1)^2}$

$u' = 2x^3e^{x^2} + 2xe^{x^2} = 2xe^{x^2}(x^2+1)$, $v = -\dfrac{1}{2(x^2+1)}$

$$\int \frac{x^3e^{x^2}}{(x^2+1)^2}\,dx = -\frac{x^2e^{x^2}}{2(x^2+1)} + \int xe^{x^2}\,dx = -\frac{x^2e^{x^2}}{2(x^2+1)} + \frac{e^{x^2}}{2} + C$$

19. Let $u = x$, $v' = \sqrt{x-1}$, $u' = 1$, $v = \frac{2}{3}(x-1)^{3/2}$

$$\int x\sqrt{x-1}\,dx = \frac{2}{3}x(x-1)^{3/2} - \frac{2}{3}\int (x-1)^{3/2}\,dx$$

$$= \frac{2}{3}x(x-1)^{3/2} - \frac{4}{15}(x-1)^{5/2} + C = \frac{2(x-1)^{3/2}}{15}(3x+2) + C$$

20. (1)—> $u = x^2$, $v' = \sqrt{x-1}$, $u' = 2x$, $v = 2/3(x-1)^{3/2}$

(2)—> $u = x$, $v' = (x-1)^{3/2}$, $u' = 1$, $v = 2/5(x-1)^{5/2}$

$$\int x^2\sqrt{x-1}\,dx = \frac{2}{3}x^2(x-1)^{3/2} - \frac{4}{3}\int x(x-1)^{3/2}\,dx$$

$$= \frac{2}{3}x^2(x-1)^{3/2} - \frac{4}{3}\left[\frac{2}{5}x(x-1)^{5/2} - \frac{2}{5}\int (x-1)^{5/2}\,dx\right]$$

$$= \frac{2(x-1)^{3/2}}{105}(15x^2 + 12x + 8) + C$$

21. (1)—> $u = x^2$, $v' = e^x$, $u' = 2x$, $v = e^x$

(2)—> $u = x$, $v' = e^x$, $u' = 1$, $v = e^x$

$$\int (x^2-1)e^x\,dx = \int x^2e^x\,dx - \int e^x\,dx$$

$$= x^2e^x - 2\int xe^x\,dx - e^x$$

$$= x^2e^x - 2\left[xe^x - \int e^x\,dx\right] - e^x$$

$$= x^2e^x - 2xe^x + e^x + C = (x-1)^2e^x + C$$

22. Let $u = \ln(2x)$, $v' = \dfrac{1}{x^2}$, $u' = \dfrac{1}{x}$, $v = -\dfrac{1}{x}$

$$\int \frac{\ln(2x)}{x^2}\,dx = -\frac{\ln(2x)}{x} + \int \frac{1}{x^2}\,dx$$

$$= -\frac{\ln(2x)}{x} - \frac{1}{x} + C = -\frac{\ln(2x)+1}{x} + C$$

23. Let $u = \ln x$, $v' = 1$, $u' = 1/x$, $v = x$

$$\int \ln x \, dx = x \ln x - \int dx = x \ln x - x + C$$

24. Let $u = x$, $v' = \dfrac{1}{\sqrt{2 + 3x}}$, $u' = 1$, $v = \dfrac{2}{3}\sqrt{2 + 3x}$

$$\int \frac{x}{\sqrt{2 + 3x}} \, dx = \frac{2x\sqrt{2 + 3x}}{3} - \frac{2}{3}\int \sqrt{2 + 3x} \, dx$$

$$= \frac{2x\sqrt{2 + 3x}}{3} - \frac{4}{27}(2 + 3x)^{3/2} + C$$

$$= \frac{2\sqrt{2 + 3x}}{27}[9x - 2(2 + 3x)] + C = \frac{2\sqrt{2 + 3x}}{27}(3x - 4) + C$$

25. Let $u = x$, $v' = \cos x$, $u' = 1$, $v = \sin x$

$$\int x \cos x \, dx = x \sin x - \int \sin x \, dx = x \sin x + \cos x + C$$

26. (1)--> $u = x^2$, $v' = \cos x$, $u' = 2x$, $v = \sin x$

(2)--> $u = x$, $v' = \sin x$, $u' = 1$, $v = -\cos x$

$$\int x^2 \cos x \, dx = x^2 \sin x - 2\int x \sin x \, dx$$

$$= x^2 \sin x - 2(-x \cos x + \int \cos x \, dx) = x^2 \sin x + 2x \cos x - 2 \sin x + C$$

27. Let $u = x$, $v' = \sec^2 x$, $u' = 1$, $v = \tan x$

$$\int x \sec^2 x \, dx = x \tan x - \int \tan x \, dx = x \tan x - \int \frac{\sin x}{\cos x} \, dx$$

$$= x \tan x + \ln|\cos x| + C$$

28. Let $u = \theta$, $v' = \sec\theta \tan\theta$, $u' = 1$, $v = \sec\theta$

$$\int \theta \sec\theta \tan\theta \, d\theta = \theta \sec\theta - \int \sec\theta \, d\theta = \theta \sec\theta - \ln|\sec\theta + \tan\theta| + C$$

29. Let $u = \arcsin(2x)$, $v' = 1$, $u' = \dfrac{2}{\sqrt{1 - 4x^2}}$, $v = x$

$$\int \arcsin(2x) \, dx = x \arcsin(2x) - \int \frac{2x}{\sqrt{1 - 4x^2}} \, dx$$

$$= x \arcsin(2x) + (1/2)\sqrt{1 - 4x^2} + C$$

30. Let $u = \arccos x$, $v' = 1$, $u' = \dfrac{-1}{\sqrt{1 - x^2}}$, $v = x$

$$\int \arccos x\, dx = x \arccos x + \int \frac{x}{\sqrt{1 - x^2}}\, dx = x \arccos x - \sqrt{1 - x^2} + C$$

31. Let $u = \arctan x$, $v' = 1$, $u' = \dfrac{1}{1 + x^2}$, $v = x$

$$\int \arctan x\, dx = x \arctan x - \int \frac{x}{1 + x^2}\, dx = x \arctan x - \frac{1}{2} \ln |1 + x^2| + C$$

32. Let $u = \arctan \dfrac{x}{2}$, $v' = 1$, $u' = \dfrac{2}{4 + x^2}$, $v = x$

$$\int \arctan \frac{x}{2}\, dx = x \arctan \frac{x}{2} - \int \frac{2x}{4 + x^2}\, dx$$

$$= x \arctan \frac{x}{2} - \ln (4 + x^2) + C$$

33. (1)—> $u = \sin x$, $v' = e^{2x}$, $u' = \cos x$, $v = \dfrac{1}{2} e^{2x}$

(2)—> $u = \cos x$, $v' = e^{2x}$, $u' = -\sin x$, $v = \dfrac{1}{2} e^{2x}$

$$\int e^{2x} \sin x\, dx = \frac{1}{2} e^{2x} \sin x - \frac{1}{2} \int e^{2x} \cos x\, dx$$

$$= \frac{1}{2} e^{2x} \sin x - \frac{1}{2} \left[\frac{1}{2} e^{2x} \cos x + \frac{1}{2} \int e^{2x} \sin x\, dx\right]$$

$$\frac{5}{4} \int e^{2x} \sin x\, dx = \frac{1}{2} e^{2x} \sin x - \frac{1}{4} e^{2x} \cos x$$

$$\int e^{2x} \sin x\, dx = \frac{1}{5} e^{2x} [2 \sin x - \cos x] + C$$

34. (1)—> $u = \cos 2x$, $v' = e^x$, $u' = -2 \sin 2x$, $v = e^x$

(2)—> $u = \sin 2x$, $v' = e^x$, $u' = 2 \cos 2x$, $v = e^x$

$$\int e^x \cos 2x\, dx = e^x \cos 2x + 2 \int e^x \sin 2x\, dx$$

$$= e^x \cos 2x + 2\left(e^x \sin 2x - 2 \int e^x \cos 2x\, dx\right)$$

$$5 \int e^x \cos 2x\, dx = e^x \cos 2x + 2e^x \sin 2x$$

$$\int e^x \cos 2x\, dx = \frac{e^x}{5} (\cos 2x + 2 \sin 2x) + C$$

35. Let $u = x$, $v' = \sin 2x$, $u' = 1$, $v = 1/2 \cos 2x$

$$\int x \sin 2x \, dx = \frac{-1}{2} x \cos 2x + \frac{1}{2} \int \cos 2x \, dx = \frac{-1}{2} x \cos 2x + \frac{1}{4} \sin 2x + C$$

$$= \frac{1}{4}(\sin 2x - 2x \cos 2x) + C$$

Thus $\displaystyle\int_0^{\pi} x \sin 2x \, dx = \frac{1}{4}(\sin 2x - 2x \cos 2x) \Big]_0^{\pi} = -\frac{\pi}{2}$

36. Let $u = x^2$, $u' = 2x$

$$\int_0^1 x \arcsin x^2 \, dx = \frac{1}{2} \int_0^1 \arcsin (x^2)(2x) \, dx$$

$$= \frac{1}{2} \left[x^2 \arcsin (x^2) + \sqrt{1 - x^4} \right]_0^1 = \frac{1}{4}(\pi - 2)$$

37. $\displaystyle\int_0^1 e^x \sin x \, dx = \frac{e^x}{2}(\sin x - \cos x)\Big]_0^1 = \frac{e[\sin (1) - \cos (1)] + 1}{2}$

38. (1)--> $u = x^2$, $v' = e^x$, $u' = 2x$, $v = e^x$

(2)--> $u = x$, $v' = e^x$, $u' = 1$, $v = e^x$

$$\int_0^1 x^2 e^x \, dx = x^2 e^x \Big]_0^1 - 2 \int_0^1 x e^x \, dx = e - 2 \left[x e^x \Big]_0^1 - \int_0^1 e^x \, dx \right]$$

$$= e - 2e + \left[2e^x \right]_0^1 = e - 2$$

39. $\displaystyle\int_0^{\pi/2} x \cos x \, dx = (x \sin x + \cos x) \Big]_0^{\pi/2} = \frac{\pi}{2} - 1$ (See exercise 25)

40. Let $u = \ln (1 + x^2)$, $v' = 1$, $u' = \dfrac{2x}{1 + x^2}$, $v = x$

$$\int_0^1 \ln (1 + x^2) \, dx = x \ln (1 + x^2) \Big]_0^1 - \int_0^1 \frac{2x^2}{1 + x^2} \, dx$$

$$= \ln 2 - \int_0^1 \left(2 - \frac{2}{1 + x^2}\right) dx$$

$$= \ln 2 - \left[2x - 2 \arctan x \right]_0^1 = \ln 2 + \frac{\pi}{2} - 2$$

41. Let $u = 2x$, $v' = \sqrt{2x - 3}$, $u' = 2$, $v = \frac{1}{3}(2x - 3)^{3/2}$

(a) $$\int 2x\sqrt{2x - 3}\,dx = \frac{2}{3}x(2x - 3)^{3/2} - \frac{2}{3}\int (2x - 3)^{3/2}\,dx$$

$$= \frac{2}{3}x(2x - 3)^{3/2} - \frac{2}{15}(2x - 3)^{5/2} + C$$

$$= \frac{2}{15}(2x - 3)^{3/2}(3x + 3) + C = \frac{2}{5}(2x - 3)^{3/2}(x + 1) + C$$

(b) Let $u = \sqrt{2x - 3}$, $x = \dfrac{u^2 + 3}{2}$, $dx = u\,du$

$$\int 2x\sqrt{x - 3}\,dx = \int (u^2 + 3)u^2\,du = \frac{u^5}{5} + u^3 + C = \frac{u^3}{5}(u^2 + 5) + C$$

$$= \frac{2}{5}(2x - 3)^{3/2}(x + 1) + C$$

42. (a) Let $u = x$, $v' = \sqrt{4 + x}$, $u' = 1$, $v = \frac{2}{3}(4 + x)^{3/2}$

$$\int x\sqrt{4 + x}\,dx = \frac{2}{3}x(4 + x)^{3/2} - \frac{2}{3}\int (4 + x)^{3/2}\,dx$$

$$= \frac{2}{3}x(4 + x)^{3/2} - \frac{4}{15}(4 + x)^{5/2} + C = \frac{2}{15}(4 + x)^{3/2}(3x - 8) + C$$

(b) Let $u = \sqrt{4 + x}$, $x = u^2 - 4$, $dx = 2u\,du$

$$\int x\sqrt{4 + x}\,dx = \int (u^2 - 4)(u)(2u)\,du = 2\int (u^4 - 4u^2)\,du$$

$$= 2\left[\frac{u^5}{5} - \frac{4u^3}{3}\right] + C = \frac{2}{15}u^3[3u^2 - 20] + C$$

$$= \frac{2}{15}(4 + x)^{3/2}[3(4 + x) - 20] + C$$

$$= \frac{2}{15}(4 + x)^{3/2}(3x - 8) + C$$

43. (a) Let $u = x^2$, $v' = \dfrac{x}{\sqrt{4 + x^2}}$, $u' = 2x$, $v = \sqrt{4 + x^2}$

$$\int \frac{x^3}{\sqrt{4 + x^2}}\,dx = x^2\sqrt{4 + x^2} - 2\int x\sqrt{4 + x^2}\,dx$$

$$= x^2\sqrt{4 + x^2} - \frac{2}{3}(4 + x^2)^{3/2} + C = \frac{1}{3}\sqrt{4 + x^2}\,(x^2 - 8) + C$$

43. (b) Let $u = \sqrt{4 + x^2}$, $x^2 = u^2 - 4$, $x\,dx = u\,du$

$$\int \frac{x^3}{\sqrt{4 + x^2}}\,dx = \int \frac{u^2 - 4}{u}(u\,du) = \int (u^2 - 4)\,du$$

$$= \frac{u^3}{3} - 4u + C = \frac{1}{3}u(u^2 - 12) + C$$

$$= \frac{1}{3}\sqrt{4 + x^2}\,(x^2 - 8) + C$$

44. (a) Let $u = x$, $v' = \sqrt{4 - x}$, $u' = 1$, $v = -\frac{2}{3}(4 - x)^{3/2}$

$$\int x\sqrt{4 - x}\,dx = -\frac{2}{3}x(4 - x)^{3/2} + \frac{2}{3}\int (4 - x)^{3/2}\,dx$$

$$= -\frac{2}{3}x(4 - x)^{3/2} - \frac{4}{15}(4 - x)^{5/2} + C$$

$$= -\frac{2}{15}(4 - x)^{3/2}[5x + 2(4 - x)] + C$$

$$= -\frac{2}{15}(4 - x)^{3/2}(3x + 8) + C$$

(b) Let $u = \sqrt{4 - x}$, $x = 4 - u^2$, $dx = -2u\,du$

$$\int x\sqrt{4 - x}\,dx = \int (4 - u^2)\,u\,(-2u)\,du = -2\int (4u^2 - u^4)\,du$$

$$= -2\left[\frac{4u^3}{3} - \frac{u^5}{5}\right] + C = -\frac{2}{15}u^3(20 - 3u^2) + C$$

$$= -\frac{2}{15}(4 - x)^{3/2}[20 - 3(4 - x)] + C$$

$$= -\frac{2}{15}(4 - x)^{3/2}(3x + 8) + C$$

45. Let $u = x^n$, $v' = \sin x$, $u' = nx^{n-1}$, $v = -\cos x$

$$\int x^n \sin x\,dx = -x^n \cos x + n\int x^{n-1}\cos x\,dx$$

46. Let $u = x^n$, $v' = \cos x$, $u' = nx^{n-1}$, $v = \sin x$

$$\int x^n \cos x\,dx = x^n \sin x - n\int x^{n-1}\sin x\,dx$$

47. Let $u = \ln x$, $v' = x^n$, $u' = \frac{1}{x}$, $v = \frac{x^{n+1}}{n+1}$

$$\int x^n \ln x \, dx = \frac{x^{n+1}}{n+1} \ln x - \int \frac{x^n}{n+1} \, dx = \frac{x^{n+1}}{n+1} \ln x - \frac{x^{n+1}}{(n+1)^2} + C$$

$$= \frac{x^{n+1}}{(n+1)^2} [(n+1) \ln x - 1] + C$$

48. Let $u = x^n$, $v' = e^{ax}$, $u' = nx^{n-1}$, $v = \frac{1}{a} e^{ax}$

$$\int x^n e^{ax} \, dx = \frac{x^n e^{ax}}{a} - \frac{n}{a} \int x^{n-1} e^{ax} \, dx$$

49. (1)—> $u = \sin bx$, $v' = e^{ax}$, $u' = b \cos bx$, $v = \frac{1}{a} e^{ax}$

(2)—> $u = \cos bx$, $v' = e^{ax}$, $u' = -b \sin bx$, $v = \frac{1}{a} e^{ax}$

$$\int e^{ax} \sin bx \, dx = \frac{e^{ax} \sin bx}{a} - \frac{b}{a} \int e^{ax} \cos bx \, dx$$

$$= \frac{e^{ax} \sin bx}{a} - \frac{b}{a} \left[\frac{e^{ax} \cos bx}{a} + \frac{b}{a} \int e^{ax} \sin bx \, dx\right]$$

$$= \frac{e^{ax} \sin bx}{a} - \frac{be^{ax} \cos bx}{a^2} - \frac{b^2}{a^2} \int e^{ax} \sin bx \, dx$$

Therefore $(1 + \frac{b^2}{a^2}) \int e^{ax} \sin bx \, dx = \frac{e^{ax} a \sin bx}{a^2} - \frac{be^{ax} \cos bx}{a^2}$

$$\int e^{ax} \sin bx \, dx = \frac{e^{ax}(a \sin bx - b \cos bx)}{a^2 + b^2} + C$$

50. Similar to 49.

51. Let $u = x$, $v' = e^{-x}$, $u' = 1$, $v = -e^{-x}$

$$A = \int_0^4 xe^{-x} \, dx = -xe^{-x}\Big]_0^4 + \int_0^4 e^{-x} \, dx$$

$$= (-xe^{-x} - e^{-x})\Big]_0^4 = 1 - \frac{5}{e^4} \approx 0.908$$

52. Let $u = x$, $v' = e^{-x/3}$, $u' = 1$, $v = -3e^{x/3}$

$$A = \frac{1}{9} \int xe^{-x/3} \, dx = \frac{1}{9} \left[-3xe^{-x/3} + 3 \int e^{-x/3} \, dx\right]_0^3$$

$$= (-\frac{1}{3} xe^{-x/3} - e^{-x/3})\Big]_0^3 = 1 - \frac{2}{e} \approx 0.264$$

53. $$A = \int_0^1 e^{-x} \sin(\pi x)\, dx = \left.\frac{e^{-x}[-\sin \pi x - \pi \cos \pi x]}{1 + \pi^2}\right]_0^1$$

$$= \frac{1}{1 + \pi^2}\left[\frac{\pi}{e} + \pi\right] = \frac{\pi}{1 + \pi^2}\left(\frac{1}{e} + 1\right) \approx 0.395$$

(See Exercise 49)

54. $$A = \int_0^{\pi} x \sin x\, dx = (-x \cos x + \sin x)\Big]_0^{\pi} = \pi$$

(See Exercise 45)

55. (a) $$A = \int_1^e \ln x\, dx = \Big[x \ln x - x\Big]_1^e = 1$$

(See Exercise 23)

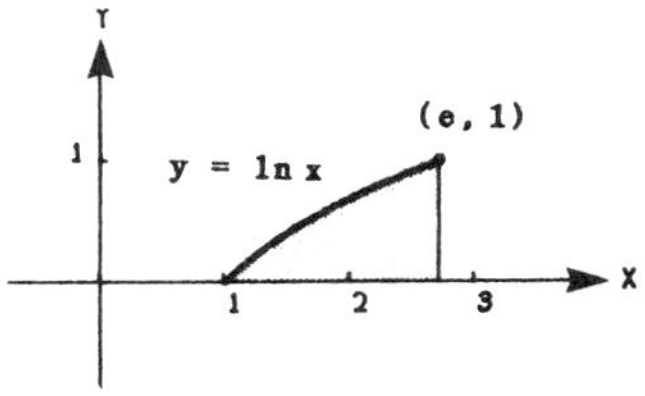

(b) $R(x) = \ln x, \quad r(x) = 0$

$$V = \pi \int_1^e (\ln x)^2\, dx$$

$$= \pi \Big[x(\ln x)^2 - 2x \ln x + 2x\Big]_1^e = \pi(e - 2)$$

(See Exercise 13)

(c) $p(x) = x, \quad h(x) = \ln x$

$$V = 2\pi \int_1^e x \ln x\, dx = 2\pi\left[\frac{x^2}{4}(-1 + 2 \ln x)\right]_1^e = \frac{(e^2 + 1)\pi}{2}$$

(See Exercise 47)

(d) $$\bar{x} = \frac{\int_1^e x \ln x\, dx}{1} = \frac{e^2 + 1}{4} \approx 2.097$$

$$\bar{y} = \frac{\frac{1}{2}\int_1^e (\ln x)^2\, dx}{1} = \frac{e - 2}{2} \approx 0.359$$

56. <u>Shell</u>

Let $u = x, \quad v' = e^x, \quad u' = 1, \quad v = e^x$

$$V = 2\pi \int_0^1 xe^x\, dx = 2\pi\left(\Big[xe^x\Big]_0^1 - \int_0^1 e^x\, dx\right)$$

$$= 2\pi\left(e - \Big[e^x\Big]_0^1\right) = 2\pi[e - (e - 1)] = 2\pi$$

57. (a) Average $= \int_1^2 (1.6t\ln t + 1)\,dt$

$$= \left[0.8t^2\ln t - 0.4t^2 + t\right]_1^2 = 3.2(\ln 2) - 0.2 \approx 2.0181$$

(b) Average $= \int_3^4 (1.6t\ln t + 1)\,dt = \left[0.8t^2\ln t - 0.4t^2 + t\right]_3^4$

$$= 12.8(\ln 4) - 7.2(\ln 3) - 1.8 \approx 8.0346$$

58. (a) $\dfrac{\int_0^{91} [410.5t^2e^{-t/30} + 25{,}000]\,dt}{91}$

$$= \frac{1}{91}\left[410.5\int_0^{91} t^2e^{-t/30}\,dt + \int_0^{91} 25{,}000\,dt\right]$$

$$= \frac{1}{91}\left[410.5\left[\frac{t^2e^{-t/30}}{-1/30} - \frac{2}{-1/30}\int te^{-t/30}\,dt\right] + 25000t\right]_0^{91}$$

$$= \frac{1}{91}\left[410.5\left[-30t^2e^{-t/30} + 60\left[\frac{te^{-t/30}}{-1/30} - \frac{1}{-1/30}\int e^{-t/30}\,dt\right]\right]\right.$$

$$\left. + 25000t\right]_0^{91}$$

$$= \frac{1}{91}\left[410.5[-30t^2e^{-t/30} - 1800te^{-t/30} - 54000e^{-t/30}] + 25000t\right]_0^{91}$$

$$= \frac{1}{91}\left[12315e^{-t/30}[-t^2 - 60t - 1800] + 25000t\right]_0^{91}$$

$$= \$167316.12$$

(b) $\dfrac{1}{365 - 274}\left[12315e^{-t/30}[-t^2 - 60t - 1800] + 25000t\right]_{274}^{365}$

$$= \frac{1}{91}[9114948.975 - 6725882.145] = \$26253.48$$

59. $u = x, \quad v' = \sin(\frac{n\pi}{2}x), \quad u' = 1, \quad v = \frac{2}{n\pi}\cos(\frac{n\pi}{2}x)$

$$I_1 = \int_0^1 x\sin(\frac{n\pi}{2}x)\,dx = \frac{-2x}{n\pi}\cos(\frac{n\pi}{2}x)\Big]_0^1 + \frac{2}{n\pi}\int_0^1 \cos(\frac{n\pi}{2}x)\,dx$$

$$= -\frac{2}{n\pi}\cos(\frac{n\pi}{2}) + \Big[(\frac{2}{n\pi})^2\sin(\frac{n\pi}{2}x)\Big]_0^1$$

$$= -\frac{2}{n\pi}\cos(\frac{n\pi}{2}) + (\frac{2}{n\pi})^2\sin(\frac{n\pi}{2})$$

$$I_2 = \int_1^2 (-x+2)\sin(\frac{n\pi}{2}x)\,dx$$

$$= \frac{-2(-x+2)}{n\pi}\cos(\frac{n\pi}{2}x)\Big]_1^2 - \frac{2}{n\pi}\int_1^2 \cos(\frac{n\pi}{2}x)\,dx$$

$$= \frac{2}{n\pi}\cos(\frac{n\pi}{2}) - \Big[(\frac{2}{n\pi})^2\sin(\frac{n\pi}{2}x)\Big]_1^2$$

$$= \frac{2}{n\pi}\cos(\frac{n\pi}{2}) + (\frac{2}{n\pi})^2\sin(\frac{n\pi}{2})$$

$$h(I_1 + I_2) = b_n = h\Big[(\frac{2}{n\pi})^2\sin(\frac{n\pi}{2}) + (\frac{2}{n\pi})^2\sin(\frac{n\pi}{2})\Big] = \frac{8h}{(n\pi)^2}\sin(\frac{n\pi}{2})$$

60. $\text{Average value} = \frac{1}{\pi}\int_0^\pi e^{-4t}(\cos 2t + 5\sin 2t)\,dt$

$$= \frac{1}{\pi}\Big[e^{-4t}(\frac{-4\cos 2t + 2\sin 2t}{20}) + 5e^{-4t}(\frac{-4\sin 2t - 2\cos 2t}{20})\Big]_0^\pi$$

(from exercises 49 and 50)

$$= \frac{7}{10\pi}(1 - e^{-4t}) \approx 0.223$$

61. For any integrable function, $\int f(x)\,dx = C + \int f(x)\,dx$ but this cannot be used to imply that $C = 0$.

62. The solution is correct.

8.3 Trigonometric integrals

1. Let $u = \cos x$, $u' = -\sin x$

$$\int \cos^3 x \sin x \, dx = -\int \cos^3 x(-\sin x) \, dx = -\frac{1}{4}\cos^4 x + C$$

2. Let $u = \sin x$, $u' = \cos x$

$$\int \cos^3 x \sin^2 x \, dx = \int \cos x(1 - \sin^2 x) \sin^2 x \, dx$$

$$= \int (\sin^2 x - \sin^4 x) \cos x \, dx$$

$$= \frac{1}{3}\sin^3 x - \frac{1}{5}\sin^5 x + C$$

3. Let $u = \sin 2x$, $u' = 2 \cos 2x$

$$\int \sin^5 2x \cos 2x \, dx = \frac{1}{2}\int \sin^5 2x(2 \cos 2x) \, dx = \frac{1}{12}\sin^6 2x + C$$

4. Let $u = \cos x$, $u' = -\sin x$

$$\int \sin^3 x \, dx = \int \sin x(1 - \cos^2 x) \, dx = \int \cos^2 x(-\sin x) \, dx + \int \sin x \, dx$$

$$= \frac{1}{3}\cos^3 x - \cos x + C$$

5. Let $u = \cos x$, $u' = -\sin x$

$$\int \sin^5 x \cos^2 x \, dx = \int \sin x(1 - \cos^2 x)^2 \cos^2 x \, dx$$

$$= -\int (\cos^2 x - 2\cos^4 x + \cos^6 x)(-\sin x) \, dx$$

$$= \frac{-1}{3}\cos^3 x + \frac{2}{5}\cos^5 x - \frac{1}{7}\cos^7 x + C$$

6. Let $u = \sin(x/3)$, $u' = (1/3)\cos(x/3)$

$$\int \cos^3 \frac{x}{3} \, dx = \int (\cos\frac{x}{3})(1 - \sin^2\frac{x}{3}) \, dx = 3\int (1 - \sin^2\frac{x}{3})(\frac{1}{3}\cos\frac{x}{3}) \, dx$$

$$= 3\left[\sin\frac{x}{3} - \frac{1}{3}\sin^3\frac{x}{3}\right] + C = 3\sin\frac{x}{3} - \sin^3\frac{x}{3} + C$$

7. $\int \cos^2 3x\, dx = \int \frac{1 + \cos 6x}{2}\, dx = \frac{1}{2}[x + \frac{1}{6}\sin 6x] + C$

$= \frac{1}{12}[6x + \sin 6x] + C$

8. $\int \sin^4 2x\, dx = \frac{1}{4}\int (1 - \cos 4x)^2\, dx = \frac{1}{4}\int (1 - 2\cos 4x + \cos^2 4x)\, dx$

$= \frac{1}{4}\int (1 - 2\cos 4x + \frac{1 + \cos 8x}{2})\, dx$

$= \frac{1}{4}(x - \frac{1}{2}\sin 4x + \frac{1}{2}x + \frac{1}{16}\sin 8x) + C$

$= \frac{1}{64}(24x - 8\sin 4x + \sin 8x) + C$

9. $\int \sin^2 \pi x \cos^4 \pi x\, dx = \int (\frac{1 - \cos 2\pi x}{2})(\frac{1 + \cos 2\pi x}{2})^2\, dx$

$= \frac{1}{8}\int (1 + \cos 2\pi x - \cos^2 2\pi x - \cos^3 2\pi x)\, dx$

$= \frac{1}{8}\int \left[1 + \cos 2\pi x - (\frac{1 + \cos 4\pi x}{2}) - (1 - \sin^2 2\pi x)\cos 2\pi x\right] dx$

$= \frac{1}{8}\int \left[\frac{1}{2} - \frac{1}{2}\cos 4\pi x + \sin^2 2\pi x \cos 2\pi x\right] dx$

$= \frac{x}{16} - \frac{\sin 4\pi x}{64\pi} + \frac{\sin^3 2\pi x}{48\pi} + C$

10. $\int \frac{\sin^3 3\theta}{\sqrt{\cos 3\theta}}\, d\theta = \int (\cos 3\theta)^{-1/2}(1 - \cos^2 3\theta)\sin 3\theta\, d\theta$

$= -\frac{1}{3}\int [(\cos 3\theta)^{-1/2} - (\cos 3\theta)^{3/2}](-3\sin 3\theta)\, d\theta$

$= -\frac{1}{3}\left[2(\cos 3\theta)^{1/2} - \frac{2}{5}(\cos 3\theta)^{5/2}\right] + C$

$= -\frac{2}{15}(\cos 3\theta)^{1/2}[5 - \cos^2 3\theta] + C = -\frac{2}{15}\sqrt{\cos 3\theta}\,[5 - \cos^2 3\theta] + C$

11. $\int \sin^4 x \cos^2 x\, dx = \int (\frac{1 - \cos 2x}{2})^2(\frac{1 + \cos 2x}{2})\, dx$

$= \frac{1}{8}\int (1 - \cos 2x - \cos^2 2x + \cos^3 2x)\, dx$

$= \frac{1}{8}\int (1 - \cos 2x - \cos^2 2x + \cos 2x - \cos 2x \sin^2 2x)\, dx$

11. (continued)

$$= \frac{1}{8}\int (\frac{1}{2} - \frac{1}{2}\cos 4x - \cos 2x \sin^2 2x)\ dx$$

$$= \frac{1}{16}\left[x - \frac{\sin 4x}{4} - \frac{\sin^3 2x}{3}\right] + C$$

12. $$\int \sin^2 \frac{x}{2} \cos^2 \frac{x}{2}\ dx = \int (\frac{1 - \cos x}{2})(\frac{1 + \cos x}{2})\ dx = \frac{1}{4}\int \sin^2 x\ dx$$

$$= \frac{1}{8}\int (1 - \cos 2x)\ dx = \frac{1}{16}(2x - \sin 2x) + C$$

13. Let $u = x$, $v' = \sin^2 x$, $u' = 1$, $v = 1/4(2x - \sin 2x)$

$$\int x \sin^2 x\ dx = \frac{1}{4}x(2x - \sin 2x) - \frac{1}{4}\int (2x - \sin 2x)\ dx$$

$$= \frac{1}{4}x(2x - \sin 2x) - \frac{1}{4}(x^2 + \frac{1}{2}\cos 2x) + C$$

$$= \frac{1}{8}[2x^2 - 2x \sin 2x - \cos 2x] + C$$

14. Let (1) $u = x^2$, $v' = \sin^2 x$, $u' = 2x$, $v = 1/4(2x - \sin 2x)$

(2) $u = x$, $v' = \sin 2x$, $u' = 1$, $v = -1/2 \cos 2x$

$$\int x^2 \sin^2 x\ dx = \frac{1}{4}x^2(2x - \sin 2x) - \frac{1}{2}\int (2x^2 - x \sin 2x)\ dx$$

$$= \frac{1}{2}x^3 - \frac{1}{4}x^2 \sin 2x - \frac{1}{3}x^3 + \frac{1}{2}\int x \sin 2x\ dx$$

$$= \frac{1}{6}x^3 - \frac{1}{4}x^2 \sin 2x + \frac{1}{2}\left[-\frac{1}{2}x \cos 2x + \frac{1}{2}\int \cos 2x\ dx\right]$$

$$= \frac{1}{6}x^3 - \frac{1}{4}x^2 \sin 2x - \frac{1}{4}x \cos 2x + \frac{1}{8}\sin 2x + C$$

$$= \frac{1}{24}[4x^3 - 6x^2 \sin 2x - 6x \cos 2x + 3 \sin 2x] + C$$

15. $$\int \sec (3x)\ dx = \frac{1}{3}\ln |\sec 3x + \tan 3x| + C$$

16. $$\int \sec^2 (2x - 1)\ dx = \frac{1}{2}\tan (2x - 1) + C$$

17. $\int \sec^4 5x\, dx = \int (1 + \tan^2 5x) \sec^2 5x\, dx$

$$= \frac{1}{5}\left[\tan 5x + \frac{\tan^3 5x}{3}\right] + C = \frac{\tan 5x}{15}[3 + \tan^2 5x] + C$$

18. $\int \sec^6 \frac{x}{2}\, dx = \int \left[1 + \tan^2 \frac{x}{2}\right]^2 \sec^2 \frac{x}{2}\, dx$

$$= \int \left[1 + 2\tan^2 \frac{x}{2} + \tan^4 \frac{x}{2}\right] \sec^2 \frac{x}{2}\, dx$$

$$= 2\tan\frac{x}{2} + \frac{4\tan^3 (x/2)}{3} + \frac{2\tan^5 (x/2)}{5} + C$$

$$= \frac{2}{15}\tan\frac{x}{2}\left[15 + 10\tan^2\frac{x}{2} + 3\tan^4\frac{x}{2}\right] + C$$

19. Let $u = \sec \pi x$, $v' = \sec^2 \pi x$, $v' = \pi \sec \pi x \tan \pi x$, $v = 1/\pi \tan \pi x$

$$\int \sec^3 \pi x\, dx = \frac{1}{\pi}\sec \pi x \tan \pi x - \int \sec \pi x \tan^2 \pi x\, dx$$

$$= \frac{1}{\pi}\sec \pi x \tan \pi x - \int \sec \pi x (\sec^2 \pi x - 1)\, dx$$

$$2\int \sec^3 \pi x\, dx = \frac{1}{\pi}[\sec \pi x \tan \pi x + \ln|\sec \pi x + \tan \pi x|] + C$$

$$\int \sec^3 \pi x\, dx = \frac{1}{2\pi}\left[\sec \pi x \tan \pi x + \ln|\sec \pi x + \tan \pi x|\right] + C$$

20. Let $u = \sec^3 \pi x$, $v' = \sec^2 \pi x$, $u' = 3\pi \sec^3 \pi x \tan \pi x$, $v = \frac{1}{\pi}\tan \pi x$

$$\int \sec^5 \pi x\, dx = \frac{1}{\pi}\sec^3 \pi x \tan \pi x - 3\int \sec^3 \pi x \tan^2 \pi x\, dx$$

$$= \frac{1}{\pi}\sec^3 \pi x \tan \pi x - 3\left[\int \sec^5 \pi x\, dx - \int \sec^3 \pi x\, dx\right]$$

$$4\int \sec^5 \pi x\, dx = \frac{1}{\pi}\sec^3 \pi x \tan \pi x + 3\int \sec^3 \pi x\, dx$$

(See Exercise 19 for $\int \sec^3 \pi x\, dx$)

$$\int \sec^5 \pi x\, dx = \frac{1}{4\pi}\left\{\sec^3 \pi x \tan \pi x + \frac{3}{2}\left[\sec \pi x \tan \pi x + \ln|\sec \pi x + \tan \pi x|\right]\right\} + C$$

21. $\int \tan^3 (1 - x)\, dx = \int [\sec^2 (1 - x) - 1] \tan (1 - x)\, dx$

$= - \int [\tan (1 - x)]^1(-\sec^2 (1 - x))\, dx - \int \tan (1 - x)\, dx$

$= - \dfrac{\tan^2 (1 - x)}{2} - \ln |\cos (1 - x)| + C$

22. $\int \tan^2 x\, dx = \int (\sec^2 x - 1)\, dx = \tan x - x + C$

23. $\int \tan^5 \frac{x}{4}\, dx = \int \left[\sec^2 \frac{x}{4} - 1\right] \tan^3 \frac{x}{4}\, dx$

$= \int \tan^3 \frac{x}{4} \sec^2 \frac{x}{4}\, dx - \int \tan^3 \frac{x}{4}\, dx$

$= \tan^4 \frac{x}{4} - \int \left[\sec^2 \frac{x}{4} - 1\right] \tan \frac{x}{4}\, dx$

$= \tan^4 \frac{x}{4} - 2 \tan^2 \frac{x}{4} - 4 \ln \left|\cos \frac{x}{4}\right| + C$

24. $\int \tan^3 \frac{\pi x}{2} \sec^2 \frac{\pi x}{2}\, dx = \frac{1}{2\pi} \tan^4 \frac{\pi x}{2} + C$

25. $\int \sec^2 x \tan x\, dx = \frac{1}{2} \tan^2 x + C \qquad (u = \tan x, \quad u' = \sec^2 x)$

26. Let $u = \cot 3x, \quad u' = -3 \csc^2 3x$

$\int \csc^2 3x \cot 3x\, dx = -\frac{1}{3} \int \cot 3x(-3 \csc^2 3x)\, dx = -\frac{1}{6} \cot^2 3x + C$

27. $\int \tan^2 x \sec^2 x\, dx = \dfrac{\tan^3 x}{3} + C$

28. $\int \tan^5 2x \sec^2 2x\, dx = \frac{1}{12} \tan^6 2x + C$

29. $\int \sec^5 \pi x \tan \pi x\, dx = \frac{1}{\pi} \int \sec^4 \pi x (\pi \sec \pi x \tan \pi x)\, dx$

$= \frac{1}{5\pi} \sec^5 \pi x + C$

30. $\int \sec^4 (1 - x) \tan (1 - x)\, dx = -\int \sec^3 (1 - x)[-\sec (1 - x) \tan (1 - x)]\, dx$

$= -\dfrac{\sec^4 (1 - x)}{4} + C$

31. $\int \sec^6 4x \tan 4x\, dx = \dfrac{1}{4}\int \sec^5 4x(4 \sec 4x \tan 4x)\, dx = \dfrac{\sec^6 4x}{24} + C$

32. $\int \sec^2 \dfrac{x}{2} \tan \dfrac{x}{2}\, dx$

$= 2 \int \sec \dfrac{x}{2} \left[\dfrac{1}{2} \sec \dfrac{x}{2} \tan \dfrac{x}{2}\right] dx$ or $= 2 \int \tan \dfrac{x}{2} \left[\dfrac{1}{2} \sec^2 \dfrac{x}{2}\right] dx$

$= \sec^2 \dfrac{x}{2} + C$ or $= \tan^2 \dfrac{x}{2} + C$

33. Let $u = \sec x$, $u' = \sec x \tan x$

$\int \sec^3 x \tan x\, dx = \int \sec^2 x(\sec x \tan x)\, dx = \dfrac{1}{3}\sec^3 x + C$

34. $\int \tan^3 3x\, dx = \int (\sec^2 3x - 1) \tan 3x\, dx = \dfrac{1}{3}\int \tan 3x(3 \sec^2 3x)\, dx$

$+ \dfrac{1}{3}\int \dfrac{-3 \sin 3x}{\cos 3x}\, dx = \dfrac{1}{6}\tan^2 3x + \dfrac{1}{3}\ln |\cos 3x| + C$

35. $\int \tan^3 3x \sec 3x\, dx = \int (\sec^2 3x - 1) \sec 3x \tan 3x\, dx$

$= \dfrac{1}{3}\int \sec^2 3x(3 \sec 3x \tan 3x)\, dx - \dfrac{1}{3}\int 3 \sec 3x \tan 3x\, dx$

$= \dfrac{1}{9}\sec^3 3x - \dfrac{1}{3}\sec 3x + C$

36. $\int \sqrt{\tan x} \sec^4 x\, dx = \int \tan^{1/2} x(\tan^2 x + 1) \sec^2 x\, dx$

$= \int (\tan^{5/2} x + \tan^{1/2} x) \sec^2 x\, dx = \dfrac{2}{7}\tan^{7/2} x + \dfrac{2}{3}\tan^{3/2} x + C$

37. $\int \cot^3 2x\, dx = \int (\csc^2 2x - 1)\cot 2x\, dx = -\frac{1}{2}\int \cot 2x(-2\csc^2 2x)\, dx$

$-\frac{1}{2}\int \frac{2\cos 2x}{\sin 2x}\, dx = -\frac{1}{4}\cot^2 2x - \frac{1}{2}\ln|\sin 2x| + C$

$= \frac{1}{4}\left[\ln|\csc^2 2x| - \cot^2 2x\right] + C$

38. Let $u = \tan\frac{x}{2}$, $u' = \frac{1}{2}\sec^2\frac{x}{2}$

$\int \tan^4\frac{x}{2}\sec^4\frac{x}{2}\, dx = \int \tan^4\frac{x}{2}(\tan^2\frac{x}{2} + 1)\sec^2\frac{x}{2}\, dx$

$= 2\int (\tan^6\frac{x}{2} + \tan^4\frac{x}{2})(\frac{1}{2}\sec^2\frac{x}{2})\, dx = \frac{2}{7}\tan^7\frac{x}{2} + \frac{2}{5}\tan^5\frac{x}{2} + C$

39. Let $u = \cot\theta$, $u' = -\csc^2\theta$

$\int \csc^4\theta\, d\theta = \int \csc^2\theta(1 + \cot^2\theta)\, d\theta = \int \csc^2\theta\, d\theta + \int \csc^2\theta\cot^2\theta\, d\theta$

$= -\cot\theta - \frac{1}{3}\cot^3\theta + C$

40. Let $u = \sec t$, $u' = \sec t\tan t$

$\int \tan^3 t\sec^3 t\, dt = \int (\sec^2 t - 1)\sec^3 t\tan t\, dt = \int \sec^4 t(\sec t\tan t)\, dt$

$- \int \sec^2 t(\sec t\tan t)\, dt = \frac{1}{5}\sec^5 t - \frac{1}{3}\sec^3 t + C$

41. $\int \frac{\cot^2 t}{\csc t}\, dt = \int \frac{\csc^2 t - 1}{\csc t}\, dt = \int (\csc t - \sin t)\, dt$

$= \ln|\csc t - \cot t| + \cos t + C$

42. $\int \frac{\cot^3 t}{\csc t}\, dt = \int \frac{\cos^3 t}{\sin^2 t}\, dt = \int \frac{(1 - \sin^2 t)\cos t}{\sin^2 t}\, dt$

$= \int \frac{\cos t}{\sin^2 t}\, dt - \int \cos t\, dt = \frac{-1}{\sin t} - \sin t + C = -\csc t - \sin t + C$

43. $\int \sin 3x\cos 2x\, dx = \frac{1}{2}\int (\sin 5x + \sin x)\, dx = \frac{-1}{2}\left[\frac{1}{5}\cos 5x + \cos x\right] + C$

$= \frac{-1}{10}(\cos 5x + 5\cos x) + C$

44. $\int \cos 3\theta \cos(-2\theta)\, d\theta = \int \cos 3\theta \cos 2\theta\, d\theta = \frac{1}{2}\int (\cos 5\theta + \cos\theta)\, d\theta$

$= \frac{1}{2}\sin\theta + \frac{1}{10}\sin 5\theta + C$

45. $\int \sin\theta \sin 3\theta\, d\theta = \frac{1}{2}\int (\cos 2\theta - \cos 4\theta)\, d\theta = \frac{1}{2}\left[\frac{1}{2}\sin 2\theta - \frac{1}{4}\sin \right] + C$

$= \frac{1}{8}(2\sin 2\theta - \sin 4\theta) + C$

46. $\int x\tan^2 x^2\, dx = \int [x\sec^2 x^2 - x]\, dx = \frac{1}{2}\int \sec^2 x^2(2x)\, dx - \int x\, dx$

$= \frac{1}{2}\tan x^2 - \frac{1}{2}x^2 + C$

47. $\int \frac{1}{\sec x \tan x}\, dx = \int \frac{\cos^2 x}{\sin x}\, dx = \int (\csc x - \sin x)\, dx$

$= \ln|\csc x - \cot x| + \cos x + C$

48. $\int \frac{\sin^2 x - \cos^2 x}{\cos x}\, dx = \int \frac{1 - 2\cos^2 x}{\cos x}\, dx = \int (\sec x - 2\cos x)\, dx$

$= \ln|\sec x + \tan x| - 2\sin x + C$

49. $\int (\tan^4 t - \sec^4 t)\, dt = \int (\tan^2 t + \sec^2 t)(\tan^2 t - \sec^2 t)\, dt$

$= -\int (\tan^2 t + \sec^2 t)\, dt = -\int [2\sec^2 t - 1]\, dt = -2\tan t + t$

50. $\int \frac{1 - \sec t}{\cos t - 1}\, dt = \int \frac{\cos t - 1}{(\cos t - 1)\cos t}\, dt = \int \sec t\, dt$

$= \ln|\sec t + \tan t| + C$

51. $\int_{-\pi}^{\pi} \sin^2 x\, dx = 2\int_0^{\pi} \frac{1 - \cos 2x}{2}\, dx = \left[x - \frac{1}{2}\sin 2x\right]_0^{\pi} = \pi$

52. $\int_0^{\pi/4} \tan^2 x\, dx = \int_0^{\pi/4} (\sec^2 x - 1)\, dx = \left[\tan x - x\right]_0^{\pi/4} = 1 - \frac{\pi}{4}$

53. $\displaystyle\int_0^{\pi/4} \tan^3 x\, dx = \int_0^{\pi/4} (\sec^2 x - 1)\tan x\, dx = \int_0^{\pi/4} \sec^2 x \tan x\, dx$

$\displaystyle - \int_0^{\pi/4} \frac{\sin x}{\cos x}\, dx = \left[\frac{1}{2}\tan^2 x + \ln|\cos x|\right]_0^{\pi/4} = \frac{1}{2}(1 - \ln 2)$

54. $\displaystyle\int_0^{\pi/4} \sec^2 t\sqrt{\tan t}\, dt = \frac{2}{3}\tan^{3/2} t\Big]_0^{\pi/4} = \frac{2}{3}$ $\quad (u = \tan t, \quad u' = \sec^2 t)$

55. $\displaystyle\int_0^{\pi/2} \frac{\cos t}{1 + \sin t}\, dt = \ln|1 + \sin t|\Big]_0^{\pi/2} = \ln 2$ $\quad (u = 1 + \sin t, \quad u' = \cos t)$

56. $\displaystyle\int_{-\pi}^{\pi} \sin 3\theta \cos\theta\, d\theta = \frac{1}{2}\int_{-\pi}^{\pi} (\sin 4\theta + \sin 2\theta)\, d\theta$

$\displaystyle = -\frac{1}{2}\left[\frac{1}{4}\cos 4\theta + \frac{1}{2}\cos 2\theta\right]_{-\pi}^{\pi} = 0$

57. $\displaystyle\int_0^{\pi/4} \sin 2\theta \sin 3\theta\, d\theta = \frac{1}{2}\int_0^{\pi/4} (\cos\theta - \cos 5\theta)\, d\theta$

$\displaystyle = \frac{1}{2}\left[\sin\theta - \frac{1}{5}\sin 5\theta\right]_0^{\pi/4} = \frac{1}{2}\left[\sin\frac{\pi}{4} - \frac{1}{5}\sin\frac{5\pi}{4}\right] = \frac{3\sqrt{2}}{10}$

58. $\displaystyle\int_0^{\pi/2} (1 - \cos\theta)^2\, d\theta = \int_0^{\pi/2} (1 - 2\cos\theta + \cos^2\theta)\, d\theta$

$\displaystyle = \int_0^{\pi/2} \left(1 - 2\cos\theta + \frac{1 + \cos 2\theta}{2}\right) d\theta$

$\displaystyle = \left[\frac{3}{2}\theta - 2\sin\theta + \frac{1}{4}\sin 2\theta\right]_0^{\pi/2} = \frac{3\pi}{4} - 2$

59. Let $u = \sin x, \quad u' = \cos x$

$\displaystyle\int_{-\pi/2}^{\pi/2} \cos^3 x\, dx = 2\int_0^{\pi/2} (1 - \sin^2 x)\cos x\, dx = 2\left[\sin x - \frac{1}{3}\sin^3 x\right]_0^{\pi/2} = \frac{4}{3}$

60. $\displaystyle\int_{-\pi/2}^{\pi/2} (\sin^2 x + 1)\, dx = \int_{-\pi/2}^{\pi/2} \left(\frac{1 - \cos 2x}{2} + 1\right) dx = \int_{-\pi/2}^{\pi/2} \left(\frac{3}{2} - \frac{1}{2}\cos 2x\right) dx$

$\displaystyle = \left[\frac{3}{2}x - \frac{1}{4}\sin 2x\right]_{-\pi/2}^{\pi/2} = \frac{3\pi}{2}$

61. Let $u = \tan 3x$, $u' = 3\sec^2 3x$

$$\int \sec^4 3x \tan^3 3x\, dx = \int \sec^2 3x \tan^3 3x \sec^2 3x\, dx$$

$$= \frac{1}{3}\int (\tan^2 3x + 1)\tan^3 3x(3\sec^2 3x)\, dx$$

$$= \frac{1}{3}\int (\tan^5 3x + \tan^3 3x)(3\sec^2 3x)\, dx = \frac{\tan^6 3x}{18} + \frac{\tan^4 3x}{12} + C$$

OR let $u = \sec 3x$, $u' = 3\sec 3x \tan 3x$

$$\int \sec^4 3x \tan^3 3x\, dx = \int \sec^3 3x \tan^2 3x \sec 3x \tan 3x\, dx$$

$$= \frac{1}{3}\int \sec^3 3x(\sec^2 3x - 1)(3\sec 3x \tan 3x)\, dx = \frac{\sec^6 3x}{18} - \frac{\sec^4 3x}{12} + C$$

$$= \frac{1}{18}\tan^6 3x + \frac{1}{6}\tan^4 3x + \frac{1}{6}\tan^2 3x + \frac{1}{18} - \frac{1}{12}\tan^4 3x - \frac{1}{6}\tan^2 3x - \frac{1}{12} + C$$

$$= \frac{\tan^6 3x}{18} + \frac{\tan^4 3x}{12} + \left(\frac{1}{18} - \frac{1}{12}\right) + C$$

62. Let $u = \tan x$, $u' = \sec^2 x$

$$\int \sec^2 x \tan x\, dx = \frac{1}{2}\tan^2 x + C$$

OR let $u = \sec x$, $u' = \sec x \tan x$

$$\int \sec x(\sec x \tan x)\, dx = \frac{1}{2}\sec^2 x + C = \frac{1}{2}(\tan^2 x + 1) + C$$

$$= \frac{1}{2}\tan^2 x + \left(\frac{1}{2} + C\right)$$

63. (a) $$\int_0^\pi \cos(mx)\cos(nx)\, dx = \frac{1}{2}\int_0^\pi [\cos(m+n)x + \cos(m-n)x]\, dx$$

$$= \frac{1}{2}\left[\frac{\sin(m+n)x}{m+n} + \frac{\sin(m-n)x}{m-n}\right]_0^\pi = 0 \quad (m \neq n)$$

(b) $$\int_{-\pi}^\pi \cos(mx)\cos(nx)\, dx$$

$$= \frac{1}{2}\left[\frac{\sin(m+n)x}{m+n} + \frac{\sin(m-n)x}{m-n}\right]_{-\pi}^\pi = 0 \quad (m \neq n)$$

$$\int_{-\pi}^\pi \sin(mx)\sin(nx)\, dx = \frac{1}{2}\int_{-\pi}^\pi [\cos(m-n)x - \cos(m+n)x]\, dx$$

$$= \frac{1}{2}\left[\frac{\sin(m-n)x}{m-n} - \frac{\sin(m+n)x}{m+n}\right]_{-\pi}^\pi = 0 \quad (m \neq n)$$

63. (continued)

$$\int_{-\pi}^{\pi} \sin(mx)\cos(nx)\,dx = \frac{1}{2}\int_{-\pi}^{\pi} [\sin(m+n)x + \sin(m-n)x]\,dx$$

$$= -\frac{1}{2}\left[\frac{\cos(m+n)x}{m+n} + \frac{\cos(m-n)x}{m-n}\right]_{-\pi}^{\pi} \qquad (m \neq n)$$

$$= -\frac{1}{2}\left[\left(\frac{\cos(m+n)\pi}{m+n} + \frac{\cos(m-n)\pi}{m-n}\right) - \left(\frac{\cos(m+n)(-\pi)}{m+n} + \frac{\cos(m-n)(-\pi)}{m-n}\right)\right]$$

$= 0$ since $\cos(-\theta) = \cos\theta$

64. $$A = \int_{-\pi/4}^{0} \left(\tan^2 x - \frac{4x}{\pi}\right) dx + \int_{0}^{\pi/4} \left(\frac{4x}{\pi} - \tan^2 x\right) dx$$

$$= \int_{-\pi/4}^{0} \left(\sec^2 x - 1 - \frac{4x}{\pi}\right) dx + \int_{0}^{\pi/4} \left(\frac{4x}{\pi} - \sec^2 x + 1\right) dx$$

$$= \left[\tan x - x - \frac{2x^2}{\pi}\right]_{-\pi/4}^{0} + \left[\frac{2x^2}{\pi} - \tan x + x\right]_{0}^{\pi/4}$$

$$= \frac{\pi}{4}$$

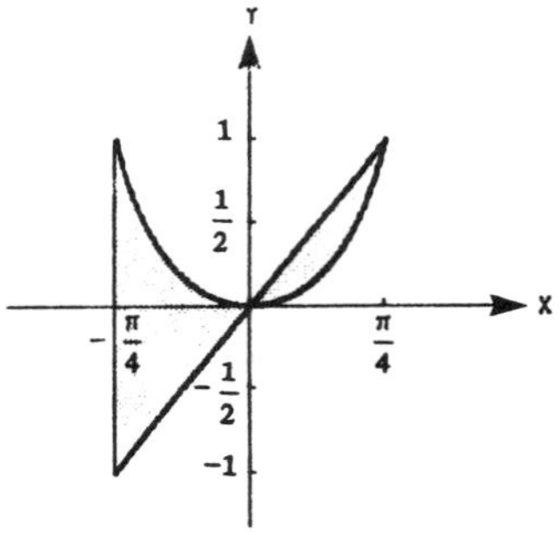

65. $R(x) = \tan x, \qquad r(x) = 0$

$$V = 2\pi \int_{0}^{\pi/4} \tan^2 x\,dx$$

$$= 2\pi \int_{0}^{\pi/4} [\sec^2 x - 1]\,dx$$

$$= 2\pi \left[\tan x - x\right]_{0}^{\pi/4}$$

$$= 2\pi\left(1 - \frac{\pi}{4}\right)$$

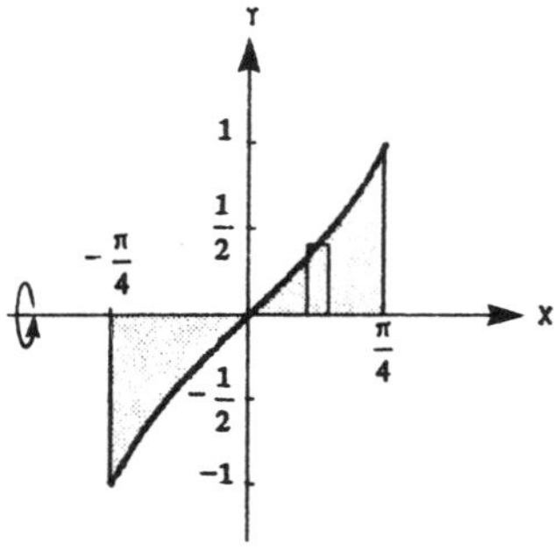

66. $R(x) = \sin x \cos^2 x\, dx, \qquad r(x) = 0$

$$V = \pi \int_0^{\pi/2} \sin^2 x \cos^4 x\, dx = \pi \int_0^{\pi/2} \left(\frac{1 - \cos 2x}{2}\right)\left(\frac{1 + \cos 2x}{2}\right)^2 dx$$

$$= \frac{\pi}{8} \int_0^{\pi/2} [1 + \cos 2x - \cos^2 2x - \cos^3 2x]\, dx$$

$$= \frac{\pi}{8} \int_0^{\pi/2} \left[1 + \cos 2x - \left(\frac{1 + \cos 4x}{2}\right) - (1 - \sin^2 2x) \cos 2x\right] dx$$

$$= \frac{\pi}{8} \int_0^{\pi/2} \left[\frac{1}{2} - \frac{1}{2}\cos 4x + \sin^2 2x \cos 2x\right] dx$$

$$= \left[\frac{\pi x}{16} - \frac{\pi \sin 4x}{64} + \frac{\pi \sin^3 2x}{48}\right]_0^{\pi/2} = \frac{\pi^2}{32}$$

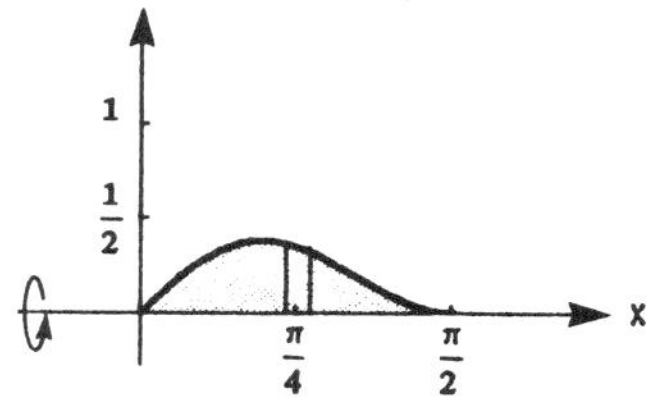

67. (a) $V = \pi \int_0^{\pi} \sin^2 x\, dx = \frac{\pi}{2} \int_0^{\pi} (1 - \cos 2x)\, dx$

$$= \frac{\pi}{2} \left[x - \frac{1}{2}\sin 2x\right]_0^{\pi} = \frac{\pi^2}{2}$$

(b) $A = \int_0^{\pi} \sin x\, dx = -\cos x \Big]_0^{\pi} = 1 + 1 = 2$

Let $u = x, \quad v' = \sin x, \quad u' = 1, \quad v = -\cos x$

$$\overline{x} = \frac{1}{A} \int_0^{\pi} x \sin x\, dx = \frac{1}{2} \left[-x \cos x \Big]_0^{\pi} + \int_0^{\pi} \cos x\, dx\right]$$

$$= \frac{1}{2} \left[-x \cos x + \sin x\right]_0^{\pi} = \frac{\pi}{2}$$

$$\overline{y} = \frac{1}{2A} \int_0^{\pi} \sin^2 x\, dx = \frac{1}{8} \int_0^{\pi} (1 - \cos 2x)\, dx$$

$$= \frac{1}{8} \left[x - \frac{1}{2}\sin 2x\right]_0^{\pi} = \frac{\pi}{8}$$

$$(\overline{x}, \overline{y}) = \left(\frac{\pi}{2}, \frac{\pi}{8}\right)$$

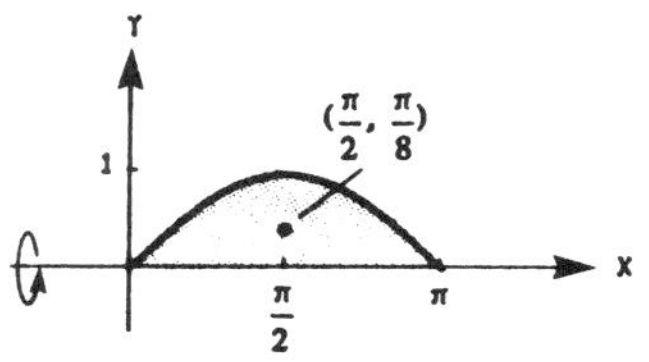

68. (a) $V = \pi \int_0^{\pi/2} \cos^2 x\, dx = \frac{\pi}{2} \int_0^{\pi/2} (1 + \cos 2x)\, dx$

$= \frac{\pi}{2} \left[x + \frac{1}{2} \sin 2x \right]_0^{\pi/2} = \frac{\pi^2}{4}$

(b) $A = \int_0^{\pi/2} \cos x\, dx = \sin x \Big]_0^{\pi/2} = 1$

Let $u = x$, $v' = \cos x$, $u' = 1$, $v = \sin x$

$\overline{x} = \int_0^{\pi/2} x \cos x\, dx = x \sin x \Big]_0^{\pi/2} - \int_0^{\pi/2} \sin x\, dx$

$= \left[x \sin x + \cos x \right]_0^{\pi/2} = \frac{\pi}{2} - 1 = \frac{\pi - 2}{2}$

$\overline{y} = \frac{1}{2} \int_0^{\pi/2} \cos^2 x\, dx = \frac{1}{4} \int_0^{\pi/2} (1 + \cos 2x)\, dx = \frac{1}{4} \left[x + \frac{1}{2} \sin 2x \right]_0^{\pi/2} = \frac{\pi}{8}$

$(\overline{x}, \overline{y}) = \left(\frac{\pi - 2}{2}, \frac{\pi}{8}\right)$

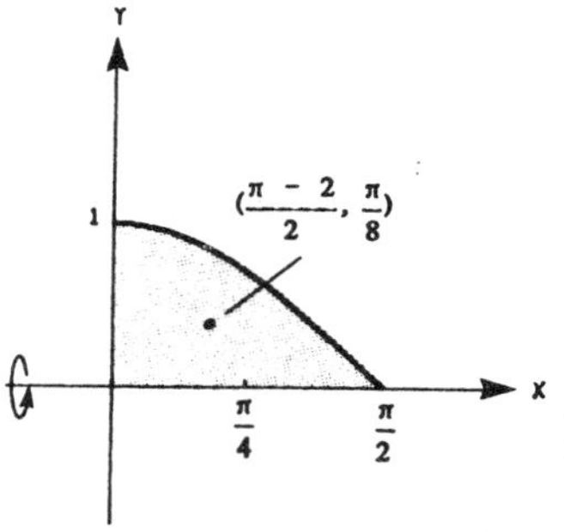

69. Let $u = \sin x$, $u' = \cos x$

$\int_0^{\pi/2} \cos^3 x\, dx = \int_0^{\pi/2} (1 - \sin^2 x) \cos x\, dx = \left[\sin x - \frac{1}{3} \sin^3 x \right]_0^{\pi/2} = \frac{2}{3}$

70. Let $u = \sin x$, $u' = \cos x$

$\int_0^{\pi/2} \cos^5 x\, dx = \int_0^{\pi/2} (1 - \sin^2 x)^2 \cos x\, dx$

$= \int_0^{\pi/2} (1 - 2\sin^2 x + \sin^4 x) \cos x\, dx$

$= \left[\sin x - \frac{2}{3} \sin^3 x + \frac{1}{5} \sin^5 x \right]_0^{\pi/2} = \frac{8}{15}$

71. Let $u = \sin x$, $\quad u' = \cos x$

$$\int_0^{\pi/2} \cos^7 x\, dx = \int_0^{\pi/2} (1 - \sin^2 x)^3 \cos x\, dx$$

$$= \int_0^{\pi/2} (1 - 3\sin^2 x + 3\sin^4 x - \sin^6 x) \cos x\, dx$$

$$= \left[\sin x - \sin^3 x + \frac{3}{5}\sin^5 x - \frac{1}{7}\sin^7 x\right]_0^{\pi/2} = \frac{16}{35}$$

72. $$\int_0^{\pi/2} \sin^2 x\, dx = \frac{1}{2}\int_0^{\pi/2} (1 - \cos 2x)\, dx = \frac{1}{2}\left[x - \frac{1}{2}\sin 2x\right]_0^{\pi/2} = \frac{\pi}{4}$$

73. $$\int_0^{\pi/2} \sin^4 x\, dx = \frac{1}{4}\int_0^{\pi/2} (1 - \cos 2x)^2\, dx$$

$$= \frac{1}{4}\int_0^{\pi/2} (1 - 2\cos 2x + \frac{1 + \cos 4x}{2})\, dx$$

$$= \frac{1}{4}\left[\frac{3x}{2} - \sin 2x + \frac{1}{8}\sin 4x\right]_0^{\pi/2} = \frac{3\pi}{16}$$

74. $$\int_0^{\pi/2} \sin^6 x\, dx = \frac{1}{8}\int_0^{\pi/2} (1 - \cos 2x)^3\, dx$$

$$= \frac{1}{8}\int_0^{\pi/2} (1 - 3\cos 2x + 3\cos^2 2x - \cos^3 2x)\, dx$$

$$= \frac{1}{8}\int_0^{\pi/2} (\frac{5}{2} - 3\cos 2x + \frac{3}{2}\cos 4x - \cos 2x + \sin^2 2x \cos 2x)\, dx$$

$$= \frac{1}{8}\left[\frac{5x}{2} - 2\sin 2x + \frac{3}{8}\sin 4x + \frac{1}{6}\sin^3 2x\right]_0^{\pi/2} = \frac{5\pi}{32}$$

75. Let $u = \sin^{n-1} x$, $\quad v' = \sin x$, $\quad u' = (n-1)\sin^{n-2} x \cos x$, $\quad v = -\cos x$

$$\int \sin^n x\, dx = -\sin^{n-1} x \cos x + (n-1)\int \sin^{n-2} x \cos^2 x\, dx$$

$$= -\sin^{n-1} x \cos x + (n-1)\int \sin^{n-2} x\, dx - (n-1)\int \sin^n x\, dx$$

Therefore $n\int \sin^n x\, dx = -\sin^{n-1} x \cos x + (n-1)\int \sin^{n-2} x\, dx$

$$\int \sin^n x\, dx = \frac{-\sin^{n-1} x \cos x}{n} + \frac{n-1}{n}\int \sin^{n-2} x\, dx$$

76. Let $u = \cos^{n-1} x$, $v' = \cos x$, $u' = -(n-1)\cos^{n-2} x \sin x$, $v = \sin x$

$$\int \cos^n x\,dx = \cos^{n-1} x \sin x + (n-1)\int \cos^{n-2} x \sin^2 x\,dx$$

$$= \cos^{n-1} x \sin x + (n-1)\int \cos^{n-2} x\,dx - (n-1)\int \cos^n x\,dx$$

Therefore $n\int \cos^n x\,dx = \cos^{n-1} x \sin x + (n-1)\int \cos^{n-2} x\,dx$

$$\int \cos^n x\,dx = \frac{\cos^{n-1} x \sin x}{n} + \frac{n-1}{n}\int \cos^{n-2} x\,dx$$

77. Let $u = \sin^{n-1} x$, $v' = \cos^m x \sin x$, $u' = (n-1)\sin^{n-2} x \cos x$,

$$v = \frac{-\cos^{m+1} x}{m+1}$$

$$\int \cos^m x \sin^n x\,dx = \frac{-\sin^{n-1} x \cos^{m+1} x}{m+1} + \frac{n-1}{m+1}\int \sin^{n-2} x \cos^{m+2} x\,dx$$

$$= \frac{-\sin^{n-1} x \cos^{m+1} x}{m+1} + \frac{n-1}{m+1}\int \sin^{n-2} x \cos^m x\,(1 - \sin^2 x)\,dx$$

$$= \frac{-\sin^{n-1} x \cos^{m+1} x}{m+1} + \frac{n-1}{m+1}\int \sin^{n-2} x \cos^m x\,dx$$

$$- \frac{n-1}{m+1}\int \sin^n x \cos^m x\,dx$$

$$\frac{m+n}{m+1}\int \cos^m x \sin^n x\,dx = \frac{-\sin^{n-1} x \cos^{m+1} x}{m+1} + \frac{n-1}{m+1}\int \sin^{n-2} x \cos^m x\,dx$$

$$\int \cos^m x \sin^n x\,dx = \frac{-\cos^{m+1} x \sin^{n-1} x}{m+n} + \frac{n-1}{m+n}\int \cos^m x \sin^{n-2} x\,dx$$

8.4 Trigonometric substitution

1. Let $x = 5\sin\theta$, $dx = 5\cos\theta\, d\theta$, $\sqrt{25 - x^2} = 5\cos\theta$

$$\int \frac{1}{(25 - x^2)^{3/2}}\, dx = \int \frac{5\cos\theta}{(5\cos\theta)^3}\, d\theta = \frac{1}{25}\int \sec^2\theta\, d\theta$$

$$= \frac{1}{25}\tan\theta + C = \frac{x}{25\sqrt{25 - x^2}} + C$$

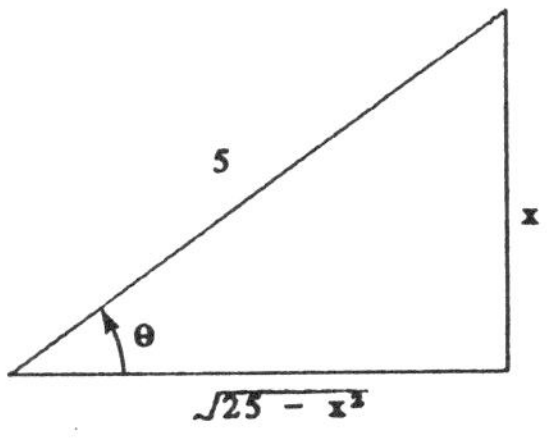

2. Same substitution as in Exercise 1.

$$\int \frac{1}{x^2\sqrt{25 - x^2}}\, dx \int \frac{5\cos\theta\, d\theta}{(25\sin^2\theta)(5\cos\theta)} = \frac{1}{25}\int \csc^2\theta\, d\theta$$

$$= -\frac{1}{25}\cot\theta + C = \frac{-\sqrt{25 - x^2}}{25x} + C$$

3. Same substitution as in Exercise 1.

$$\int \frac{\sqrt{25 - x^2}}{x}\, dx = \int \frac{25\cos^2\theta\, d\theta}{5\sin\theta} = 5\int (\csc\theta - \sin\theta)\, d\theta$$

$$= 5[\ln|\csc\theta - \cot\theta| + \cos\theta] + C$$

$$= 5\ln\left|\frac{5 - \sqrt{25 - x^2}}{x}\right| + \sqrt{25 - x^2} + C$$

4. $$\int \frac{1}{\sqrt{25 - x^2}}\, dx = \arcsin\frac{x}{5} + C$$

5. $$\int \frac{x}{\sqrt{x^2 + 9}}\, dx = \frac{1}{2}\int (x^2 + 9)^{-1/2}(2x)\, dx = \sqrt{x^2 + 9} + C$$

6. Let $2x = 4\tan\theta$, $dx = 2\sec^2\theta\, d\theta$, $\sqrt{4x^2 + 16} = 4\sec\theta$

$$\int \frac{1}{x\sqrt{4x^2 + 16}}\, dx = \int \frac{2\sec^2\theta\, d\theta}{2\tan\theta(4\sec\theta)} = \frac{1}{4}\int \frac{\sec\theta}{\tan\theta}\, d\theta$$

$$\frac{1}{4}\int \csc\theta\, d\theta = \frac{1}{4}\ln|\csc\theta - \cot\theta| + C$$

$$= \frac{1}{4}\ln\left|\frac{\sqrt{x^2 + 4} - 2}{x}\right| + C$$

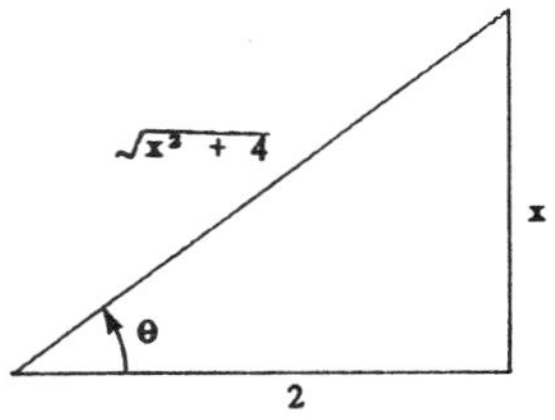

7. Let $x = 2\sin\theta$, $dx = 2\cos\theta\, d\theta$, $\sqrt{4 - x^2} = 2\cos\theta$

$$\int_0^2 \sqrt{16 - 4x^2}\, dx = 2\int_0^2 \sqrt{4 - x^2}\, dx$$

$$= 2\int_0^{\pi/2} 2\cos\theta(2\cos\theta\, d\theta)$$

$$= 8\int_0^{\pi/2} \cos^2\theta\, d\theta$$

$$= 4\int_0^{\pi/2} (1 + \cos 2\theta)\, d\theta$$

$$= 4\left[\theta + \frac{1}{2}\sin 2\theta\right]_0^{\pi/2} = 2\pi$$

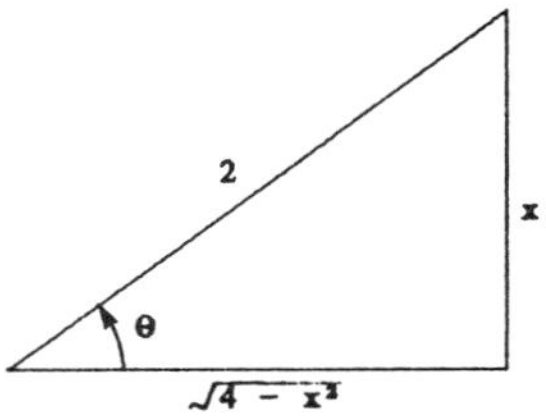

8. Let $u = 16 - 4x^2$, $u' = -8x$

$$\int_0^2 x\sqrt{16 - 4x^2}\, dx = -\frac{1}{8}\int_0^2 (16 - 4x^2)^{1/2}(-8x)\, dx$$

$$= -\frac{1}{12}(16 - 4x^2)^{3/2}\Big]_0^2 = \frac{16}{3}$$

9. Let $x = 3\sec\theta$, $dx = 3\sec\theta\tan\theta\, d\theta$, $\sqrt{x^2 - 9} = 3\tan\theta$

$$\int \frac{1}{\sqrt{x^2 - 9}}\, dx = \int \frac{3\sec\theta\tan\theta\, d\theta}{3\tan\theta}$$

$$= \int \sec\theta\, d\theta = \ln|\sec\theta + \tan\theta| + C_1$$

$$= \ln\left|\frac{x}{3} + \frac{\sqrt{x^2 - 9}}{3}\right| + C_1$$

$$= \ln\left|x + \sqrt{x^2 - 9}\right| + C$$

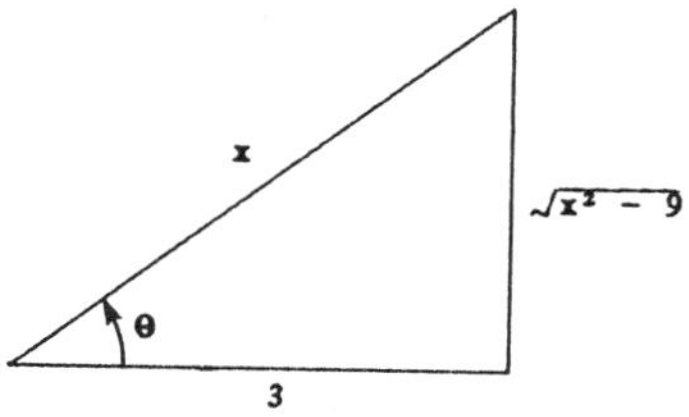

10. Let $u = 1 - t^2$, $u' = -2t$

$$\int \frac{t}{(1 - t^2)^{3/2}}\, dt = -\frac{1}{2}\int (1 - t^2)^{-3/2}(-2t)\, dt = \frac{1}{\sqrt{1 - t^2}} + C$$

11. Let $t = \sin\theta$, $dt = \cos\theta\, d\theta$, $1 - t^2 = \cos^2\theta$

$$\int_0^{\sqrt{3}/2} \frac{t^2}{(1 - t^2)^{3/2}}\, dt$$

$$= \int_0^{\pi/3} \frac{\sin^2\theta\cos\theta\, d\theta}{\cos^3\theta}$$

$$= \int_0^{\pi/3} (\sec^2\theta - 1)\, d\theta$$

$$= \Big[\tan\theta - \theta\Big]_0^{\pi/3} = \sqrt{3} - \frac{\pi}{3}$$

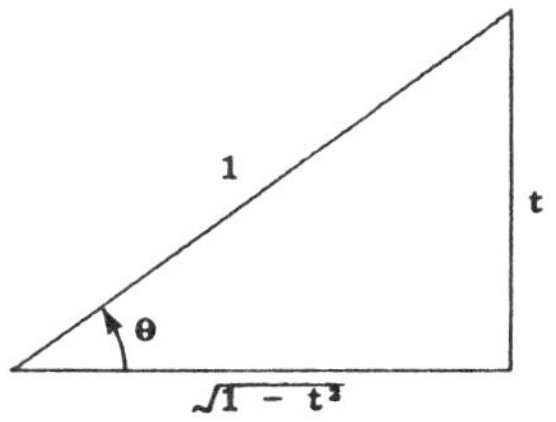

12. Same substitutions as in Exercise 11.

$$\int \frac{1}{(1 - t^2)^{5/2}}\, dt = \int \frac{\cos\theta}{\cos^5\theta}\, d\theta = \int (\tan^2\theta + 1)\sec^2\theta\, d\theta$$

$$= \frac{1}{3}\tan^3\theta + \tan\theta + C = \frac{t^3}{3(1 - t^2)^{3/2}} + \frac{t}{\sqrt{1 - t^2}} + C$$

$$= \frac{3t - 2t^3}{3(1 - t^2)^{3/2}} + C$$

13. Let $x = \sin\theta$, $dx = \cos\theta\, d\theta$, $\sqrt{1 - x^2} = \cos\theta$

$$\int \frac{\sqrt{1 - x^2}}{x^4}\, dx = \int \frac{\cos\theta\,(\cos\theta\, d\theta)}{\sin^4\theta}$$

$$= \int \cot^2\theta\csc^2\theta\, d\theta$$

$$= -\frac{1}{3}\cot^3\theta + C$$

$$= \frac{-(1 - x^2)^{3/2}}{3x^3} + C$$

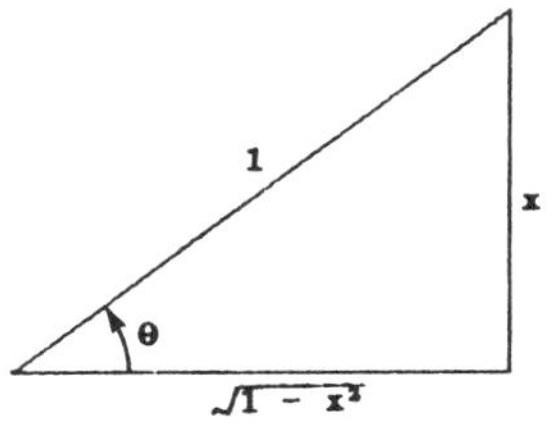

14. Let $2x = 3\tan\theta$, $dx = \dfrac{3}{2}\sec^2\theta\, d\theta$, $\sqrt{4x^2+9} = 3\sec\theta$

$$\int \frac{\sqrt{4x^2+9}}{x^4}\,dx = \int \frac{3\sec\theta[(3/2)\sec^2\theta\, d\theta]}{(3/2)^4\tan^4\theta}$$

$$= \frac{8}{9}\int \frac{\cos\theta}{\sin^4\theta}\,d\theta = \frac{-8}{27\sin^3\theta} + C$$

$$= -\frac{8}{27}\csc^3\theta + C = \frac{-(4x^2+9)^{3/2}}{27x^3} + C$$

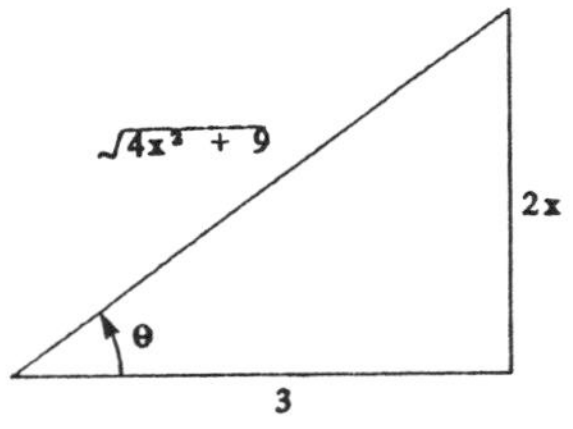

15. Same substitutions as in Exercise 14.

$$\int \frac{1}{x\sqrt{4x^2+9}}\,dx = \int \frac{(3/2)\sec^2\theta\, d\theta}{(3/2)\tan\theta\, 3\sec\theta} = \frac{1}{3}\int \csc\theta\, d\theta$$

$$= \frac{1}{3}\ln\left|\csc\theta - \cot\theta\right| + C$$

$$= \frac{1}{3}\ln\left|\frac{\sqrt{4x^2+9}-3}{2x}\right| + C \qquad \text{(Rationalize)}$$

$$= -\frac{1}{3}\ln\left|\frac{\sqrt{4x^2+9}+3}{2x}\right| + C$$

16. Let $x = \sqrt{3}\tan\theta$, $dx = \sqrt{3}\sec^2\theta\, d\theta$, $x^2 + 3 = 3\sec^2\theta$

$$\int \frac{1}{(x^2+3)^{3/2}}\,dx = \int \frac{\sqrt{3}\sec^2\theta\, d\theta}{3\sqrt{3}\sec^3\theta}$$

$$= \frac{1}{3}\int \cos\theta\, d\theta$$

$$= \frac{1}{3}\sin\theta + C$$

$$= \frac{x}{3\sqrt{x^2+3}} + C$$

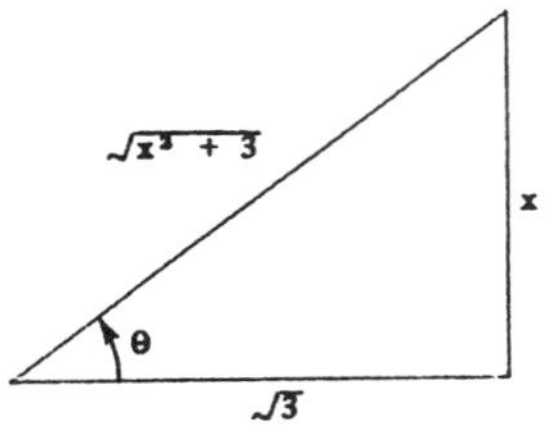

17. Same substitutions as in Exercise 16, or $u = x^2 + 3$, $u' = 2x$

$$\int \frac{x}{(x^2+3)^{3/2}}\,dx = \int \frac{\sqrt{3}\tan\theta(\sqrt{3}\sec^2\theta\, d\theta)}{3\sqrt{3}\sec^3\theta}$$

$$= \frac{\sqrt{3}}{3}\int \sin\theta\, d\theta = -\frac{\sqrt{3}}{3}\cos\theta + C = \frac{-1}{\sqrt{x^2+3}} + C$$

18. Let $x = 2 \sec \theta$, $dx = 2 \sec \theta \tan \theta \, d\theta$, $\sqrt{x^2 - 4} = 2 \tan \theta$

$$\int \frac{x^3}{\sqrt{x^2 - 4}} \, dx = \int \frac{8 \sec^3 \theta (2 \sec \theta \tan \theta \, d\theta)}{2 \tan \theta}$$

$$= 8 \int \sec^4 \theta \, d\theta = 8 \int (\tan^2 \theta + 1) \sec^2 \theta \, d\theta$$

$$= 8 \left[\frac{1}{3} \tan^3 \theta + \tan \theta\right] + C$$

$$= \frac{\sqrt{x^2 - 4}}{3}(x^2 + 8) + C$$

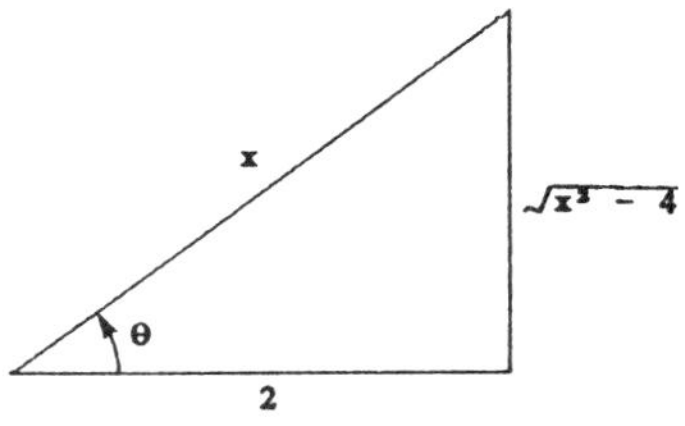

19. Same substitution as in Exercise 18.

$$\int x^3 \sqrt{x^2 - 4} \, dx = \int 8 \sec^3 \theta (2 \tan \theta)(2 \sec \theta \tan \theta) \, d\theta$$

$$= 32 \int \sec^4 \theta \tan^2 \theta \, d\theta = 32 \int (\tan^4 \theta + \tan^2 \theta) \sec^2 \theta \, d\theta$$

$$= 32 \left[\frac{1}{5} \tan^5 \theta + \frac{1}{3} \tan^3 \theta\right] + C$$

$$= \frac{(x^2 - 4)^{3/2}}{15}(3x^2 + 8) + C$$

20. Same substitution as in Exercise 18.

$$\int \frac{\sqrt{x^2 - 4}}{x} \, dx = \int \frac{2 \tan \theta (2 \sec \theta \tan \theta \, d\theta)}{2 \sec \theta} = 2 \int (\sec^2 \theta - 1) \, d\theta$$

$$= 2(\tan \theta - \theta) + C = \sqrt{x^2 - 4} - 2 \operatorname{arcsec} \frac{x}{2} + C$$

21. Let $u = 1 + e^{2x}$, $u' = 2e^{2x}$

$$\int e^{2x} \sqrt{1 + e^{2x}} \, dx = \frac{1}{2} \int (1 + e^{2x})^{1/2} (2e^{2x}) \, dx$$

$$= \frac{1}{3}(1 + e^{2x})^{3/2} + C$$

22. Let $x = \tan\theta$, $dx = \sec^2\theta\, d\theta$, $1 + x^2 = \sec^2\theta$

$$\int_{-1}^{1} \frac{1}{(1+x^2)^3}\,dx = \int_{-\pi/4}^{\pi/4} \frac{\sec^2\theta\, d\theta}{\sec^6\theta}$$

$$= \int_{-\pi/4}^{\pi/4} \cos^4\theta\, d\theta$$

$$= 2\int_{0}^{\pi/4} \left(\frac{1+\cos 2\theta}{2}\right)^2 d\theta$$

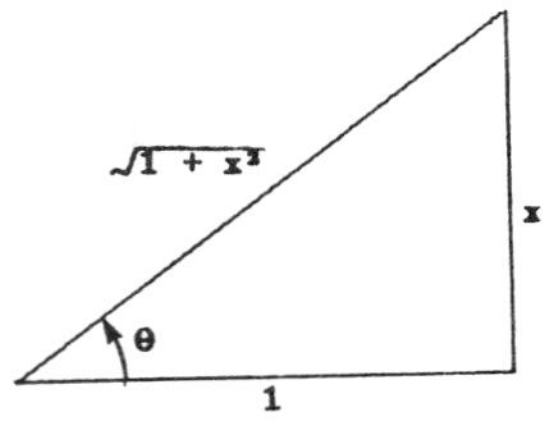

$$= \frac{1}{4}\int_{0}^{\pi/4} (3 + 4\cos 2\theta + \cos 4\theta)\, d\theta$$

$$= \frac{1}{4}\left[3\theta + 2\sin 2\theta + \frac{1}{4}\sin 4\theta\right]_0^{\pi/4}$$

$$= \frac{3\pi}{16} + \frac{1}{2}$$

23. Let $x = 2\tan\theta$, $4 + x^2 = 4\sec^2\theta$, $dx = 2\sec^2\theta\, d\theta$

$$\int_0^1 \frac{x^2}{(4+x^2)^2}\,dx$$

$$= \int_0^{\arctan 1/2} \frac{4\tan^2\theta(2\sec^2\theta\, d\theta)}{16\sec^4\theta}$$

$$= \frac{1}{2}\int_0^{\arctan 1/2} \sin^2\theta\, d\theta$$

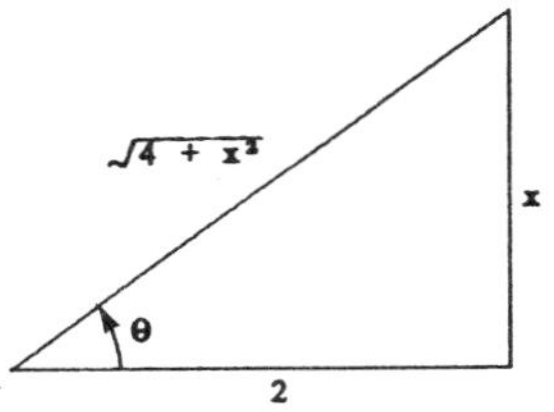

$$= \frac{1}{4}\int_0^{\arctan 1/2} (1 - \cos 2\theta)\, d\theta$$

$$= \frac{1}{4}\left[\theta - \frac{1}{2}\sin 2\theta\right]_0^{\arctan 1/2}$$

$$= \frac{1}{4}\left[\arctan\frac{1}{2} - \sin\left(\arctan\frac{1}{2}\right)\cos\left(\arctan\frac{1}{2}\right)\right]$$

$$= \frac{1}{4}\left[\arctan\frac{1}{2} - \frac{2}{5}\right] = \frac{1}{20}\left[5\arctan\frac{1}{2} - 2\right]$$

24. $$\int \frac{1}{\sqrt{4x - x^2}}\,dx = \int \frac{1}{\sqrt{4 - (x-2)^2}}\,dx = \arcsin\frac{x-2}{2} + C$$

25. Let $u = x^2 + 2x + 2$, $u' = 2x + 2$

$$\int (x+1)\sqrt{x^2 + 2x + 2}\,dx = \frac{1}{2}\int (x^2 + 2x + 2)^{1/2}(2x + 2)\,dx$$

$$= \frac{1}{3}(x^2 + 2x + 2)^{3/2} + C$$

26. Let $x - 1 = \sin\theta$, $dx = \cos\theta\, d\theta$, $\sqrt{1 - (x - 1)^2} = \cos\theta$

$$\int \frac{x^2}{\sqrt{2x - x^2}}\, dx = \int \frac{x^2}{\sqrt{1 - (x - 1)^2}}\, dx = \int \frac{(1 + \sin\theta)^2 (\cos\theta\, d\theta)}{\cos\theta}$$

$$= \int (1 + 2\sin\theta + \sin^2\theta)\, d\theta$$

$$= \int (\frac{3}{2} + 2\sin\theta - \frac{1}{2}\cos 2\theta)\, d\theta$$

$$= \frac{3}{2}\theta - 2\cos\theta - \frac{1}{4}\sin 2\theta + C$$

$$= \frac{3}{2}\theta - 2\cos\theta - \frac{1}{2}\sin\theta\cos\theta + C$$

$$= \frac{3}{2}\arcsin(x - 1) - \frac{1}{2}\sqrt{2x - x^2}\,(x + 3) + C$$

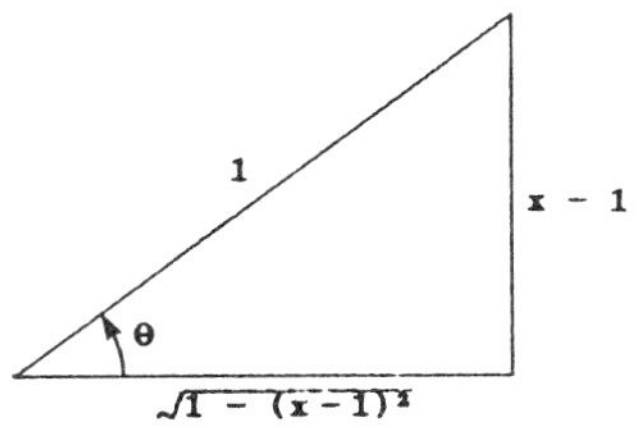

27. Let $e^x = \sin\theta$, $e^x\, dx = \cos\theta\, d\theta$, $\sqrt{1 - e^{2x}} = \cos\theta$

$$\int e^x\sqrt{1 - e^{2x}}\, dx$$

$$= \int \cos^2\theta\, d\theta = \frac{1}{2}\int (1 + \cos 2\theta)\, d\theta$$

$$= \frac{1}{2}(\theta + \frac{1}{2}\sin 2\theta) + C$$

$$= \frac{1}{2}\left[\arcsin e^x + e^x\sqrt{1 - e^{2x}}\right] + C$$

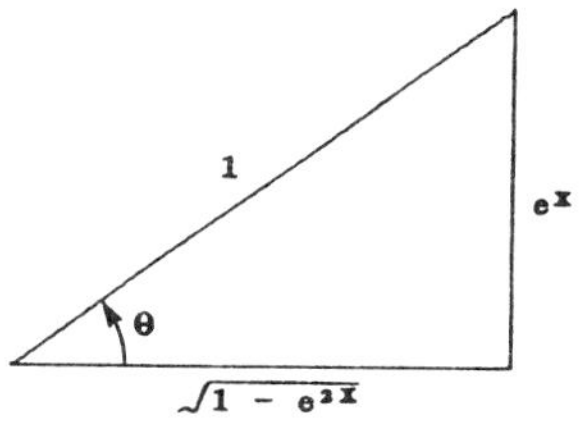

28. Let $\sqrt{x} = \sin\theta$, $x = \sin^2\theta$, $dx = 2\sin\theta\cos\theta\, d\theta$, $\sqrt{1 - x} = \cos\theta$

$$\int \frac{\sqrt{1 - x}}{\sqrt{x}}\, dx$$

$$= \int \frac{\cos\theta(2\sin\theta\cos\theta\, d\theta)}{\sin\theta}$$

$$= 2\int \cos^2\theta\, d\theta = \int (1 + \cos 2\theta)\, d\theta$$

$$= (\theta + \frac{1}{2}\sin 2\theta) + C$$

$$= \arcsin\sqrt{x} + \sqrt{x}\,\sqrt{1 - x} + C$$

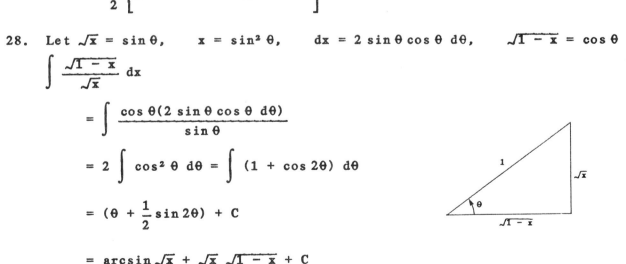

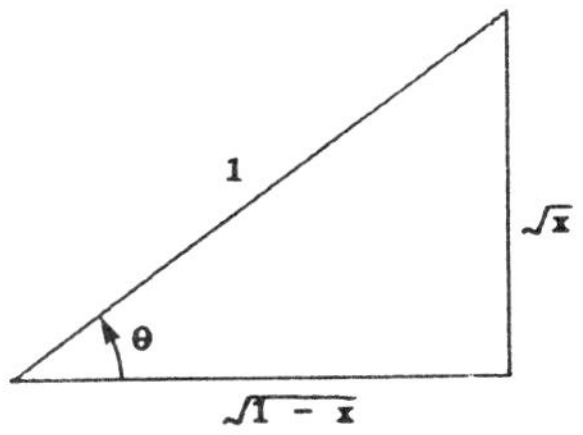

29. Let $x = \sqrt{2}\tan\theta$, $dx = \sqrt{2}\sec^2\theta\, d\theta$, $x^2 + 2 = 2\sec^2\theta$

$$\int \frac{1}{4 + 4x^2 + x^4}\,dx = \int \frac{1}{(x^2 + 2)^2}\,dx$$

$$= \int \frac{\sqrt{2}\sec^2\theta\, d\theta}{4\sec^4\theta}$$

$$= \frac{\sqrt{2}}{4}\int \cos^2\theta\, d\theta$$

$$= \frac{\sqrt{2}}{4}\left(\frac{1}{2}\right)\int (1 + \cos 2\theta)\,d\theta$$

$$= \frac{\sqrt{2}}{8}\left(\theta + \frac{1}{2}\sin 2\theta\right) + C$$

$$= \frac{\sqrt{2}}{8}(\theta + \sin\theta\cos\theta) + C$$

$$= \frac{1}{4}\left[\frac{x}{x^2 + 2} + \frac{1}{\sqrt{2}}\arctan\frac{x}{\sqrt{2}}\right] + C$$

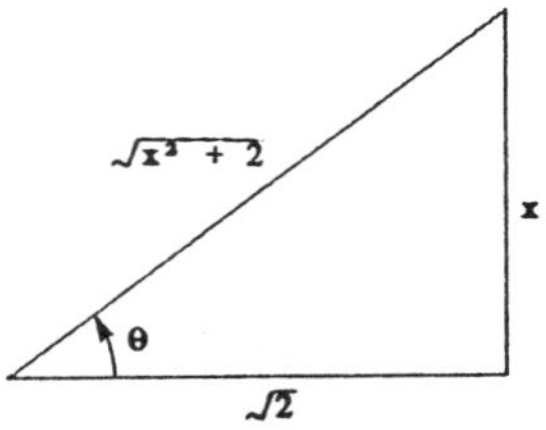

30. Let $x = \tan\theta$, $dx = \sec^2\theta\, d\theta$, $x^2 + 1 = \sec^2\theta$

$$\int \frac{x^3 + x + 1}{x^4 + 2x^2 + 1}\,dx = \frac{1}{4}\int \frac{4x^3 + 4x}{x^4 + 2x^2 + 1}\,dx + \int \frac{1}{(x^2 + 1)^2}\,dx$$

$$= \frac{1}{4}\ln(x^4 + 2x^2 + 1) + \int \frac{\sec^2\theta\, d\theta}{\sec^4\theta}$$

$$= \frac{1}{2}\ln(x^2 + 1) + \frac{1}{2}\int (1 + \cos 2\theta)\,d\theta$$

$$= \frac{1}{2}\ln(x^2 + 1) + \frac{1}{2}(\theta + \sin\theta\cos\theta) + C$$

$$= \frac{1}{2}\left[\ln(x^2 + 1) + \arctan x + \frac{x}{x^2 + 1}\right] + C$$

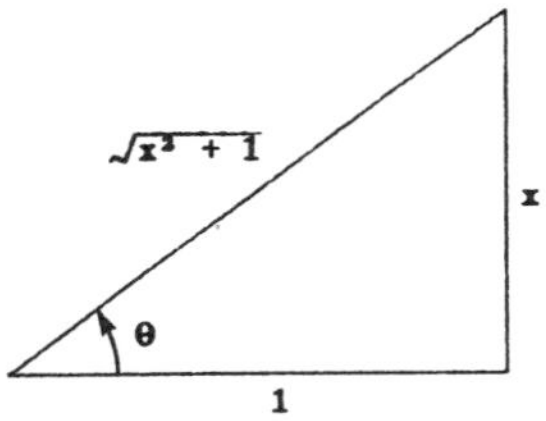

31. Let $a = 2$, $u = 3x$, $du = 3\ dx$ (See Theorem 8.2)

$$\int \sqrt{4 + 9x^2}\ dx = \frac{1}{3}\int \sqrt{9x^2 + 4}\,(3)\ dx$$

$$= \frac{1}{6}\left[3x\sqrt{9x^2 + 4} + 4\ \ln\left|3x + \sqrt{9x^2 + 4}\right|\right] + C$$

32. Let $a = 1$, $u = x$, $du = dx$ (See Theorem 8.2)

$$\int \sqrt{1 + x^2}\ dx = \frac{1}{2}\left[x\sqrt{x^2 + 1} + \ln\left|x + \sqrt{x^2 + 1}\right|\right] + C$$

33. Let $x = \sec\theta$, $\sqrt{x^2 - 1} = \tan\theta$, $dx = \sec\theta\tan\theta\ d\theta$

$$\int \frac{x^2}{\sqrt{x^2 - 1}}\ dx = \int \frac{\sec^2\theta}{\tan\theta}\sec\theta\tan\theta\ d\theta = \int \sec^3\theta\ d\theta$$

$$= \frac{1}{2}\left[\sec\theta\tan\theta + \ln\left|\sec\theta + \tan\theta\right|\right] + C$$

$$= \frac{1}{2}\left[x\sqrt{x^2 - 1} + \ln\left|x + \sqrt{x^2 - 1}\right|\right] + C$$

(See Example 6 in Section 8.2)

34. (1) $x = 2\sec\theta$, $dx = 2\sec\theta\tan\theta\ d\theta$, $\sqrt{x^2 - 4} = 2\tan\theta$

(2) $u = \sec^3\theta$, $v' = \sec^2\theta$, $u' = 3\sec^3\theta\tan\theta$, $v = \tan\theta$

$$\int x^2\sqrt{x^2 - 4}\ dx = \int (2\sec\theta)^2(2\tan\theta)(2\sec\theta\tan\theta)\ d\theta$$

$$= 16\int \sec^3\theta\tan^2\theta\ d\theta = 16\int \sec^3\theta(\sec^2\theta - 1)\ d\theta$$

$$= 16\left[\int \sec^5\theta\ d\theta - \int \sec^3\theta\ d\theta\right]$$

$$= 16\left[\sec^3\theta\tan\theta - 3\int \sec^3\theta\tan^2\theta\ d\theta - \int \sec^3\theta\ d\theta\right]$$

$$64\int \sec^3\theta\tan^2\theta\ d\theta$$

$$= 16\left[\sec^3\theta\tan\theta - \frac{1}{2}\left[\sec\theta\tan\theta + \ln\left|\sec\theta + \tan\theta\right|\right]\right]$$

$$16\int \sec^3\theta\tan^2\theta\ d\theta$$

$$= 4\sec^3\theta\tan\theta - 2[\sec\theta\tan\theta + \ln\left|\sec\theta + \tan\theta\right|] + C_1$$

34. (continued)

$$= 4\left(\frac{x}{2}\right)^3\frac{\sqrt{4 - x^2}}{2} - 2\left[\left(\frac{x}{2}\right)\left(\frac{\sqrt{4 - x^2}}{2}\right) + \ln\left|\frac{x}{2} + \frac{\sqrt{4 - x^2}}{2}\right|\right] + C_1$$

$$= \frac{1}{4}x^3\sqrt{4 - x^2} - \frac{1}{2}x\sqrt{4 - x^2} - 2\ln|x + \sqrt{4 - x^2}| + C$$

35. (1) $u = \operatorname{arcsec} 2x$, $v' = 1$, $u' = \dfrac{1}{x\sqrt{4x^2 - 1}}$, $v = x$

(2) $2x = \sec\theta$, $dx = \dfrac{1}{2}\sec\theta\tan\theta\, d\theta$, $\sqrt{4x^2 - 1} = \tan\theta$

$$\int \operatorname{arcsec} 2x\, dx = x \operatorname{arcsec} 2x - \int \frac{1}{\sqrt{4x^2 - 1}}\, dx$$

$$= x \operatorname{arcsec} 2x - \int \frac{1/2 \sec\theta\tan\theta\, d\theta}{\tan\theta}$$

$$= x \operatorname{arcsec} 2x - \frac{1}{2}\int \sec\theta\, d\theta = x \operatorname{arcsec} 2x - \frac{1}{2}\ln|\sec\theta + \tan\theta| + C$$

$$= x \operatorname{arcsec} 2x - \frac{1}{2}\ln|2x + \sqrt{4x^2 - 1}| + C$$

36. (1) $u = \arcsin x$, $v' = x$, $u' = \dfrac{1}{\sqrt{1 - x^2}}$, $v = \dfrac{x^2}{2}$

(2) $x = \sin\theta$, $dx = \cos\theta\, d\theta$, $\sqrt{1 - x^2} = \cos\theta$

$$\int x \arcsin x\, dx = \frac{x^2}{2}\arcsin x - \frac{1}{2}\int \frac{x^2}{\sqrt{1 - x^2}}\, dx$$

$$= \frac{x^2}{2}\arcsin x - \frac{1}{2}\int \frac{\sin^2\theta}{\cos\theta}\cos\theta\, d\theta$$

$$= \frac{x^2}{2}\arcsin x - \frac{1}{4}\int (1 - \cos 2\theta)\, d\theta$$

$$= \frac{x^2}{2}\arcsin x - \frac{1}{4}\left[\theta - \frac{1}{2}\sin 2\theta\right] + C$$

$$= \frac{x^2}{2}\arcsin x - \frac{1}{4}\left[\theta - \sin\theta\cos\theta\right] + C$$

$$= \frac{x^2}{2}\arcsin x - \frac{1}{4}\left[\arcsin x - x\sqrt{1 - x^2}\right] + C$$

$$= \frac{1}{4}[(2x^2 - 1)\arcsin x + x\sqrt{1 - x^2}] + C$$

37. From Example 4 we have $\int_{\sqrt{3}}^{2} \frac{\sqrt{x^2 - 3}}{x}\,dx$

$$= \sqrt{3}\left[\tan\theta - \theta\right]_0^{\pi/6}$$

$$= \sqrt{3}\left[\frac{\sqrt{x^2 - 3}}{\sqrt{3}} - \operatorname{arcsec}\frac{x}{\sqrt{3}}\right]_{\sqrt{3}}^{2}$$

$$= 1 - \frac{\sqrt{3}\pi}{6}$$

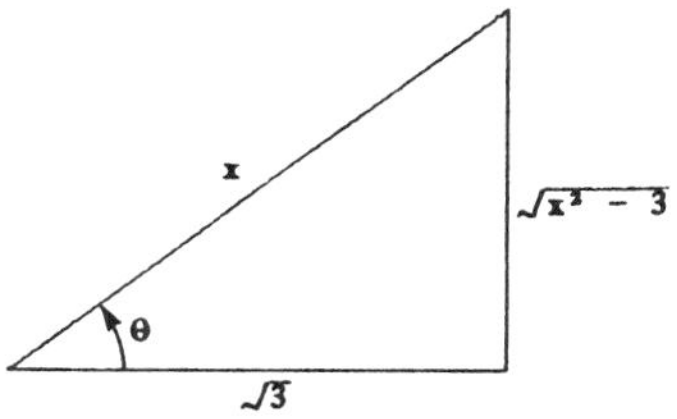

38. Let $x = 3\tan\theta$, $dx = 3\sec^2\theta\,d\theta$, $\sqrt{x^2 + 9} = 3\sec\theta$

$$\int_0^3 \frac{x^3}{\sqrt{x^2 + 9}}\,dx$$

$$= \int_0^{\pi/4} \frac{(27\tan^2\theta)(3\sec^3\theta\,d\theta)}{3\sec\theta}$$

$$= 27\int_0^{\pi/4} (\sec^2\theta - 1)\sec\theta\tan\theta\,d\theta$$

$$= 27\left[\frac{1}{3}\sec^3\theta - \sec\theta\right]_0^{\pi/4} = 9(2 - \sqrt{2})$$

$$= 27\left[\frac{(x^2 + 9)^{3/2}}{81} - \frac{(x^2 + 9)^{1/2}}{3}\right]_0^3$$

$$= 9(2 - \sqrt{2})$$

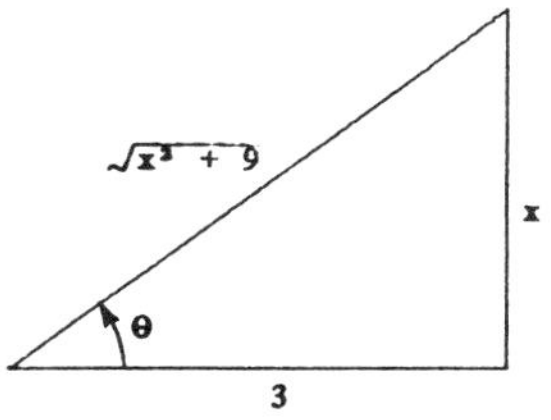

39. Let $3x = 5\sin\theta$, $dx = \frac{5}{3}\cos\theta\,d\theta$, $\sqrt{25 - 9x^2} = 5\cos\theta$

$$\int_0^{5/3} \sqrt{25 - 9x^2}\,dx = \frac{25}{3}\int_0^{\pi/2} \cos^2\theta\,d\theta$$

$$= \frac{25}{6}\int_0^{\pi/2} (1 + \cos 2\theta)\,d\theta$$

$$= \frac{25}{6}\left[\theta + \frac{1}{2}\sin 2\theta\right]_0^{\pi/2} = \frac{25\pi}{12}$$

$$= \frac{25}{6}\left[\arcsin\frac{3x}{5} + \left(\frac{3x}{5}\right)\frac{\sqrt{25 - 9x^2}}{5}\right]_0^{5/3}$$

$$= \frac{25}{6}\arcsin 1 = \frac{25\pi}{12}$$

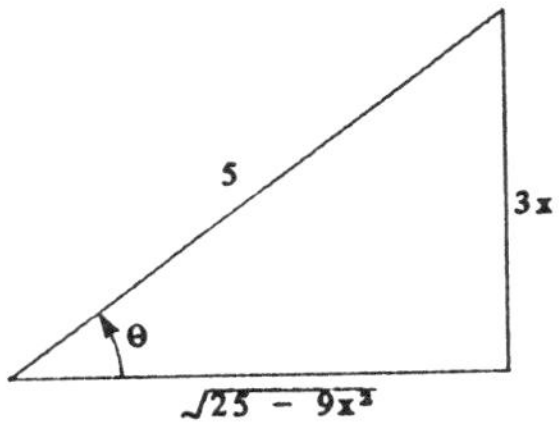

40. Let $r = L\tan\theta$, $dr = L\sec^2\theta\, d\theta$, $r^2 + L^2 = L^2\sec^2\theta$

$$\frac{1}{R}\int_0^R \frac{2mL}{(r^2 + L^2)^{3/2}}\,dr$$

$$= \frac{2mL}{R}\int_a^b \frac{L\sec^2\theta\, d\theta}{L^3\sec^3\theta} = \frac{2m}{RL}\int_a^b \cos\theta\, d\theta$$

$$= \frac{2m}{RL}\sin\theta\Big]_a^b$$

$$= \frac{2m}{RL}\frac{r}{\sqrt{r^2 + L^2}}\Big]_0^R$$

$$= \frac{2m}{L\sqrt{R^2 + L^2}}$$

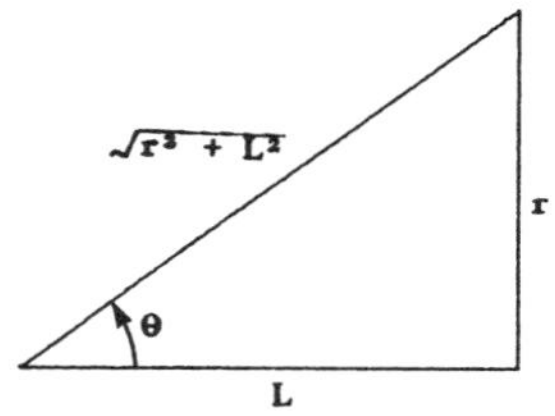

41. Area of representative rectangle $= 2\sqrt{1 - y^2}\,\Delta y$

Pressure $= 2(62.4)(3 - y)\sqrt{1 - y^2}\,\Delta y$

$$F = 124.8\int_{-1}^{1}(3 - y)\sqrt{1 - y^2}\,dy$$

$$= 124.8\left[3\int_{-1}^{1}\sqrt{1 - y^2}\,dy - \int_{-1}^{1} y\sqrt{1 - y^2}\,dy\right]$$

$$= 124.8\left[\frac{3}{2}(\arcsin y + y\sqrt{1 - y^2}) + \frac{1}{2}\left(\frac{2}{3}\right)(1 - y^2)^{3/2}\right]_{-1}^{1}$$

$$= (62.4)3[\arcsin 1 - \arcsin(-1)] = 187.2\pi \text{ lb}$$

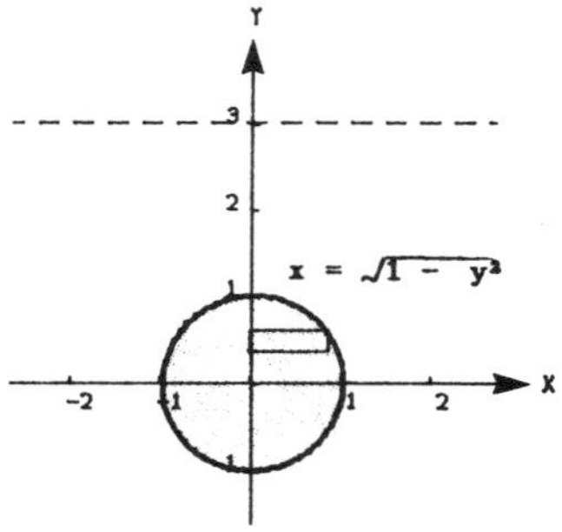

42. $$F = 124.8\int_{-1}^{1}(d - y)\sqrt{1 - y^2}\,dy$$

$$= 124.8d\int_{-1}^{1}\sqrt{1 - y^2}\,dy - 124.8\int_{-1}^{1} y\sqrt{1 - y^2}\,dy$$

$$= 124.8\left(\frac{d}{2}\right)\left[\arcsin y + y\sqrt{1 - y^2}\right]_{-1}^{1} - 124.8(0) = 62.4\pi d \text{ lb}$$

43. Let $x - 3 = \sin\theta$, $dx = \cos\theta$, $\sqrt{1 - (x - 3)^2} = \cos\theta$

<u>Shell</u>

$$V = 4\pi \int_2^4 x\sqrt{1 - (x - 3)^2}\, dx$$

$$= 4\pi \int_{-\pi/2}^{\pi/2} (3 + \sin\theta) \cos^2\theta\, d\theta$$

$$= 4\pi \left[\frac{3}{2} \int_{-\pi/2}^{\pi/2} (1 + \cos 2\theta)\, d\theta + \int_{-\pi/2}^{\pi/2} \cos^2\theta \sin\theta\, d\theta\right]$$

$$= 4\pi \left[\frac{3}{2}(\theta + \frac{1}{2}\sin 2\theta) - \frac{1}{3}\cos^3\theta\right]_{-\pi/2}^{\pi/2}$$

$$= 6\pi^2$$

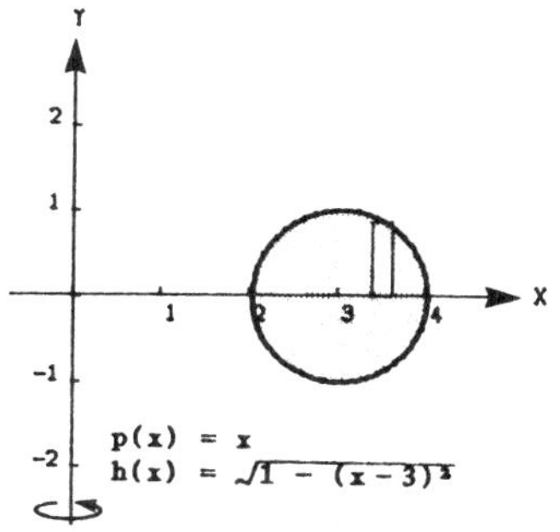

44. Let $x - h = r\sin\theta$, $dx = r\cos\theta\, d\theta$, $\sqrt{r^2 - (x - h)^2} = r\cos\theta$

<u>Shell</u>

$$V = 4\pi \int_{h-r}^{h+r} x\sqrt{r^2 - (x - h)^2}\, dx$$

$$= 4\pi \int_{-\pi/2}^{\pi/2} (h + r\sin\theta) r\cos\theta (r\cos\theta)\, d\theta$$

$$= 4\pi r^2 \int_{-\pi/2}^{\pi/2} (h + r\sin\theta) \frac{1 + \cos 2\theta}{2}\, d\theta$$

$$= 4\pi r^2 \left[\frac{h}{2} \int_{-\pi/2}^{\pi/2} (1 + \cos 2\theta)\, d\theta + r\int_{-\pi/2}^{\pi/2} \sin\theta \cos^2\theta\, d\theta\right]$$

$$= 2\pi r^2 h \left[\theta + \frac{1}{2}\sin 2\theta\right]_{-\pi/2}^{\pi/2} - \left[4\pi r^3 \left(\frac{\cos^3\theta}{3}\right)\right]_{-\pi/2}^{\pi/2}$$

$$= 2\pi^2 r^2 h$$

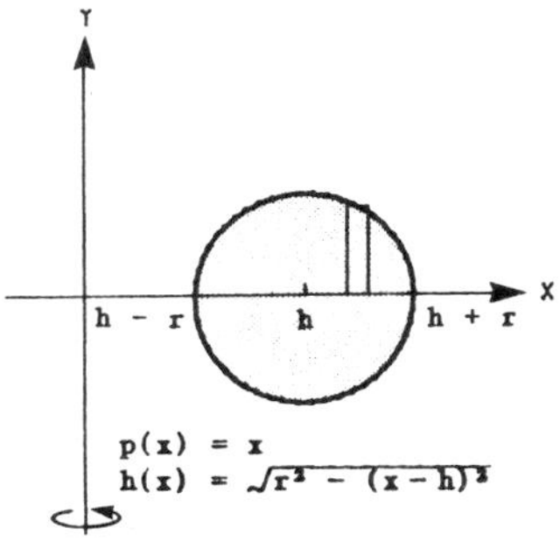

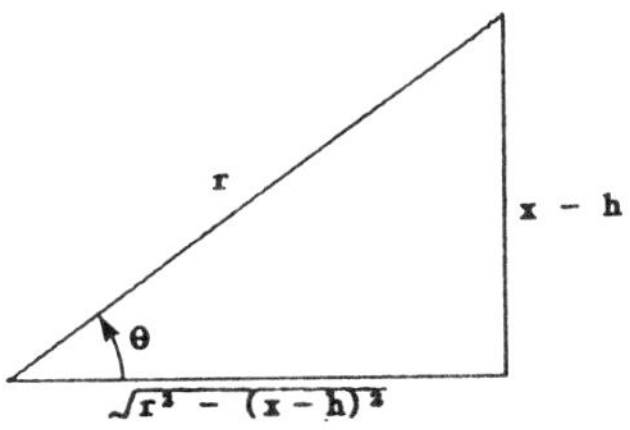

45. $y = \ln x$, $y' = \dfrac{1}{x}$, $1 + (y')^2 = 1 + \dfrac{1}{x^2} = \dfrac{x^2 + 1}{x^2}$

Let $x = \tan\theta$, $dx = \sec^2\theta\, d\theta$, $\sqrt{x^2 + 1} = \sec\theta$

45. (continued)

$$S = \int_1^5 \sqrt{\frac{x^2+1}{x^2}}\,dx = \int_1^5 \frac{\sqrt{x^2+1}}{x}\,dx = \int \frac{\sec\theta}{\tan\theta}\sec^2\theta\,d\theta$$

$$= \int \frac{\sec\theta}{\tan\theta}(1+\tan^2\theta)\,d\theta = \int (\csc\theta + \sec\theta\tan\theta)\,d\theta$$

$$= -\ln|\csc\theta + \cot\theta| + \sec\theta = \left[-\ln\left|\frac{\sqrt{x^2+1}}{x} + \frac{1}{x}\right| + \sqrt{x^2+1}\right]_1^5$$

$$= \left[-\ln\left(\frac{\sqrt{26}+1}{5}\right) + \sqrt{26}\right] - \left[-\ln(\sqrt{2}+1) + \sqrt{2}\right]$$

$$= \ln\left[\frac{5(\sqrt{2}+1)}{\sqrt{26}+1}\right] + \sqrt{26} - \sqrt{2} \approx 4.367$$

$$\left(\text{or } \ln\left[\frac{\sqrt{26}-1}{5(\sqrt{2}-1)}\right] + \sqrt{26} - \sqrt{2}\right)$$

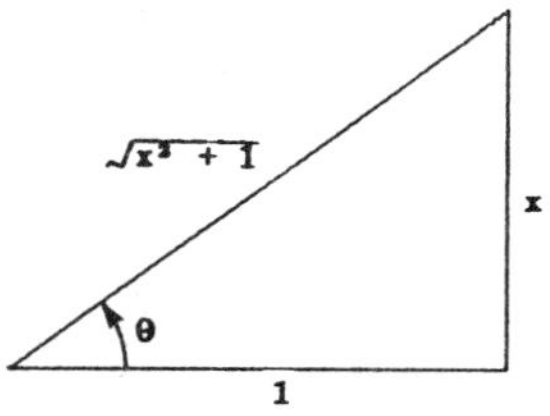

46. Let $u = 2x$, $du = 2\,dx$, $a = 1$ (See Theorem 8.2)

$y = x^2$, $y' = 2x$, $1 + (y')^2 = 1 + 4x^2$

$$S = \int_0^3 \sqrt{1+4x^2}\,dx = \frac{1}{2}\int_0^3 \sqrt{4x^2+1}(2)\,dx$$

$$= \frac{1}{4}\left[2x\sqrt{4x^2+1} + \ln|2x + \sqrt{4x^2+1}|\right]_0^3$$

$$= \frac{1}{4}\left[6\sqrt{37} + \ln(6+\sqrt{37})\right] \approx 9.747$$

47. $y = x - 0.005x^2$, $y' = 1 - 0.01x$, $1 + (y')^2 = 1 + (1 - 0.01x)^2$

Let $u = 1 - 0.01x$, $du = -0.01\,dx$, $a = 1$ (See Theorem 8.2)

$$S = \int_0^{200} \sqrt{1 + (1-0.01x)^2}\,dx$$

$$= -100\int_0^{200} \sqrt{(1-0.01x)^2+1}(-0.01)\,dx$$

$$= -50\Big[(1-0.01x)\sqrt{(1-0.01x)^2+1}$$

$$+ \ln|(1-0.01x) + \sqrt{(1-0.01x)^2+1}|\Big]_0^{200}$$

$$= -50[(-\sqrt{2} + \ln|-1+\sqrt{2}|) - (\sqrt{2} + \ln|1+\sqrt{2}|)]$$

$$= 100\sqrt{2} + 50\ln\left[\frac{\sqrt{2}+1}{\sqrt{2}-1}\right] \approx 229.559$$

48. $y = x - \frac{x^2}{72}, \qquad y' = 1 - \frac{x}{36}, \qquad 1 + (y')^2 = 1 + (1 - \frac{x}{36})^2$

$$S = \int_0^{72}\sqrt{1 + (1 - \frac{x}{36})^2}\,dx = -36\int_0^{72}\sqrt{1 + (1 - \frac{x}{36})^2}\,(-\frac{1}{36})\,dx$$

$$= -\frac{36}{2}\left[(1 - \frac{x}{36})\sqrt{1 + (1 - \frac{x}{36})^2} + \ln\left|(1 - \frac{x}{36}) + \sqrt{1 + (1 - \frac{x}{36})^2}\right|\right]_0^{72}$$

$$= -18[(-\sqrt{2} + \ln|-1 + \sqrt{2}|) - (\sqrt{2} + \ln|1 + \sqrt{2}|]$$

$$= 36\sqrt{2} + 18\ln\left[\frac{\sqrt{2} + 1}{\sqrt{2} - 1}\right] \approx 82.641$$

49. $y = x^2, \qquad y' = 2x, \qquad 1 + (y')^2 = 1 + 4x^2$

$2x = \tan\theta, \qquad dx = \frac{1}{2}\sec^2\theta\,d\theta, \qquad \sqrt{1 + 4x^2} = \sec\theta$

(Use integration by parts on $\int \sec^5\theta\,d\theta$ and $\int \sec^3\theta\,d\theta$.

See problems 19 and 20 in section 8.3.)

$$S = 2\pi\int_0^{\sqrt{2}} x^2\sqrt{1 + 4x^2}\,dx = 2\pi\int (\frac{\tan\theta}{2})^2(\sec\theta)(\frac{1}{2}\sec^2\theta)\,d\theta$$

$$= \frac{\pi}{4}\int \sec^3\theta\tan^2\theta\,d\theta = \frac{\pi}{4}\int \sec^5\theta\,d\theta - \int \sec^3\theta\,d\theta$$

$$= \frac{\pi}{4}\left[\frac{1}{4}\left[\sec^3\theta\tan\theta + \frac{3}{2}(\sec\theta\tan\theta + \ln|\sec\theta + \tan\theta|)\right]\right.$$
$$\left. - \frac{1}{2}(\sec\theta\tan\theta + \ln|\sec\theta + \tan\theta|)\right]$$

$$= \frac{\pi}{4}\left[\frac{1}{4}[(1 + 4x^2)^{3/2}(2x)]\right.$$
$$\left. - \frac{1}{8}[(1 + 4x^2)^{1/2}(2x) + \ln|\sqrt{1 + 4x^2} + 2x|]\right]_0^{\sqrt{2}}$$

$$= \frac{\pi}{4}\left[\frac{54\sqrt{2}}{4} - \frac{6\sqrt{2}}{6} - \frac{1}{8}\ln(3 + 2\sqrt{2})\right]$$

$$= \frac{\pi}{4}(\frac{51\sqrt{2}}{4} - \frac{\ln(3 + 2\sqrt{2})}{8})$$

$$= \frac{\pi}{32}[102\sqrt{2} - \ln(3 + 2\sqrt{2})] \approx 13.989$$

50. Let $x = 3\tan\theta$, $dx = 3\sec^2\theta\, d\theta$, $\sqrt{x^2+9} = 3\sec\theta$

$$A = 2\int_0^4 \frac{3}{\sqrt{x^2+9}}\,dx = 6\int_0^4 \frac{dx}{\sqrt{x^2+9}} = 6\int \frac{3\sec^2\theta\, d\theta}{3\sec\theta}$$

$$= 6\int \sec\theta\, d\theta = 6\ln|\sec\theta + \tan\theta| = 6\ln\left|\frac{\sqrt{x^2+9}+x}{3}\right|\Bigg]_0^4$$

$$= 6\ln 3$$

$\bar{x} = 0$ (by symmetry)

$$\bar{y} = \frac{1}{2}\left(\frac{1}{A}\right)\int_{-4}^4 \left(\frac{3}{\sqrt{x^2+9}}\right)^2 dx$$

$$= \frac{9}{12\ln 3}\int_{-4}^4 \frac{1}{x^2+9}\,dx$$

$$= \frac{3}{4\ln 3}\left[\frac{1}{3}\arctan\frac{x}{3}\right]_{-4}^4$$

$$= \frac{2}{4\ln 3}\arctan\frac{4}{3} \approx 0.422$$

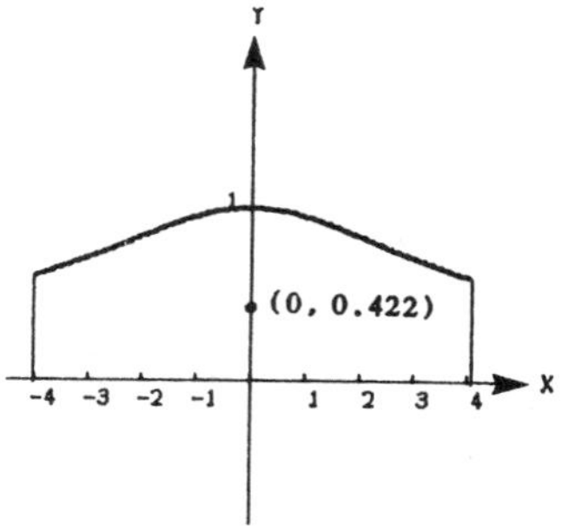

$$(\bar{x}, \bar{y}) = \left(0, \frac{1}{2\ln 3}\arctan\frac{4}{3}\right) \approx (0, 0.422)$$

51. $F = 48\int_{-1}^{0.8} (0.8 - y)(2)\sqrt{1-y^2}\,dy$

$$= 96\left[0.8\int_{-1}^{0.8} \sqrt{1-y^2}\,dy - \int_{-1}^{0.8} y\sqrt{1-y^2}\,dy\right]$$

$$= 96\left[\frac{0.8}{2}(\arcsin y + y\sqrt{1-y^2}) + \frac{1}{3}(1-y^2)^{3/2}\right]_{-1}^{0.8}$$

$$\approx 96[1.263] \approx 121.3 \text{ lbs}$$

52. $F = 64\int_{-1}^{0.4} (0.4 - y)(2)\sqrt{1-y^2}\,dy$

$$= 128\left[0.4\int_{-1}^{0.4} \sqrt{1-y^2}\,dy - \int_{-1}^{0.4} y\sqrt{1-y^2}\,dy\right]$$

$$= 128\left[\frac{0.4}{2}(\arcsin y + y\sqrt{1-y^2}) + \frac{1}{3}(1-y^2)^{3/2}\right]_{-1}^{0.4}$$

$$\approx 128[0.7264] \approx 93 \text{ lbs.}$$

8.5
Partial fractions

1. $\dfrac{1}{x^2 - 1} = \dfrac{1}{(x + 1)(x - 1)} = \dfrac{A}{x + 1} + \dfrac{B}{x - 1}$, $\quad 1 = A(x - 1) + B(x + 1)$

when $x = -1$, $1 = -2A$, $A = -\dfrac{1}{2}$, $\quad$ when $x = 1$, $1 = 2B$, $B = \dfrac{1}{2}$

$$\int \frac{1}{x^2 - 1}\,dx = -\frac{1}{2}\int \frac{1}{x + 1}\,dx + \frac{1}{2}\int \frac{1}{x - 1}\,dx$$

$$= -\frac{1}{2}\ln|x + 1| + \frac{1}{2}\ln|x - 1| + C = \frac{1}{2}\ln\left|\frac{x - 1}{x + 1}\right| + C$$

2. $\dfrac{1}{4x^2 - 9} = \dfrac{1}{(2x - 3)(2x + 3)} = \dfrac{A}{2x - 3} + \dfrac{B}{2x + 3}$, $\quad 1 = A(2x + 3) + B(2x - 3)$

when $x = \dfrac{3}{2}$, $1 = 6A$, $A = \dfrac{1}{6}$, $\quad$ when $x = -\dfrac{3}{2}$, $1 = -6B$, $B = -\dfrac{1}{6}$

$$\int \frac{1}{4x^2 - 9}\,dx = \frac{1}{6}\left[\int \frac{1}{(2x - 3)}\,dx - \int \frac{1}{2x + 3}\,dx\right]$$

$$= \frac{1}{12}[\ln|2x - 3| - \ln|2x + 3|] + C = \frac{1}{12}\ln\left|\frac{2x - 3}{2x + 3}\right| + C$$

3. $\dfrac{3}{x^2 + x - 2} = \dfrac{3}{(x - 1)(x + 2)} = \dfrac{A}{x - 1} + \dfrac{B}{x + 2}$, $\quad 3 = A(x + 2) + B(x - 1)$

when $x = 1$, $3 = 3A$, $A = 1$, $\quad$ when $x = -2$, $3 = -3B$, $B = -1$

$$\int \frac{3}{x^2 + x - 2}\,dx = \int \frac{1}{x - 1}\,dx - \int \frac{1}{x + 2}\,dx$$

$$= \ln|x - 1| - \ln|x + 2| + C = \ln\left|\frac{x - 1}{x + 2}\right| + C$$

4. $\displaystyle\int \frac{x + 1}{x^2 + 4x + 3}\,dx = \int \frac{1}{x + 3}\,dx = \ln|x + 3| + C$

5. $\dfrac{5 - x}{2x^2 + x - 1} = \dfrac{5 - x}{(2x - 1)(x + 1)} = \dfrac{A}{2x - 1} + \dfrac{B}{x + 1}$,

$5 - x = A(x + 1) + B(2x - 1)$

when $x = \dfrac{1}{2}$, $4.5 = 1.5A$, $A = 3$,

when $x = -1$, $6 = -3B$, $B = -2$

5. (continued)

$$\int \frac{5 - x}{2x^2 + x - 1}\,dx = 3\int \frac{1}{2x - 1}\,dx - 2\int \frac{1}{x + 1}\,dx$$

$$= \frac{3}{2}\ln|2x - 1| - 2\ln|x + 1| + C$$

6. $$\frac{3x^2 - 7x - 2}{x(x - 1)(x + 1)} = \frac{A}{x} + \frac{B}{x - 1} + \frac{C}{x + 1}$$

$3x^2 - 7x - 2 = A(x^2 - 1) + Bx(x + 1) + Cx(x - 1)$

when $x = 0$, $-2 = -A$, $A = 2$, when $x = 1$, $-6 = 2B$, $B = -3$

when $x = -1$, $8 = 2C$, $C = 4$

$$\int \frac{3x^2 - 7x - 2}{x^3 - x}\,dx = 2\int \frac{1}{x}\,dx - 3\int \frac{1}{x - 1}\,dx + 4\int \frac{1}{x + 1}\,dx$$

$$= 2\ln|x| - 3\ln|x - 1| + 4\ln|x + 1| + C$$

7. $$\frac{x^2 + 12x + 12}{x(x + 2)(x - 2)} = \frac{A}{x} + \frac{B}{x + 2} + \frac{C}{x - 2}$$

$x^2 + 12x + 12 = A(x + 2)(x - 2) + Bx(x - 2) + Cx(x + 2)$

when $x = 0$, $12 = -4A$, $A = -3$, when $x = -2$, $-8 = 8B$, $B = -1$

when $x = 2$, $40 = 8C$, $C = 5$

$$\int \frac{x^2 + 12x + 12}{x^3 - 4x}\,dx = 5\int \frac{1}{x - 2}\,dx - \int \frac{1}{x + 2}\,dx - 3\int \frac{1}{x}\,dx$$

$$= 5\ln|x - 2| - \ln|x + 2| - 3\ln|x| + C$$

8. $$\frac{x^3 - x + 3}{x^2 + x - 2} = x - 1 + \frac{2x + 1}{(x + 2)(x - 1)} = x - 1 + \frac{A}{x + 2} + \frac{B}{x - 1}$$

$2x + 1 = A(x - 1) + B(x + 2)$

when $x = -2$, $-3 = -3A$, $A = 1$, when $x = 1$, $3 = 3B$, $B = 1$

$$\int \frac{x^3 - x + 3}{x^2 + x - 2}\,dx = \int \left[x - 1 + \frac{1}{x + 2} + \frac{1}{x - 1}\right] dx$$

$$= \frac{x^2}{2} - x + \ln|x + 2| + \ln|x - 1| + C$$

$$= \frac{x^2}{2} - x + \ln|x^2 + x - 2| + C$$

9. $\dfrac{2x^3 - 4x^2 - 15x + 5}{x^2 - 2x - 8} = 2x + \dfrac{x + 5}{(x - 4)(x + 2)} = 2x + \dfrac{A}{x - 4} + \dfrac{B}{x + 2}$

$x + 5 = A(x + 2) + B(x - 4)$

when $x = 4$, $9 = 6A$, $A = \dfrac{3}{2}$, when $x = -2$, $3 = -6B$, $B = -\dfrac{1}{2}$

$$\int \frac{2x^3 - 4x^2 - 15x + 5}{x^2 - 2x - 8}\,dx = \int \left[2x + \frac{3/2}{x - 4} - \frac{1/2}{x + 2}\right] dx$$

$$= x^2 + \frac{3}{2}\ln|x - 4| - \frac{1}{2}\ln|x + 2| + C$$

10. $\dfrac{x + 2}{x(x - 4)} = \dfrac{A}{x - 4} + \dfrac{B}{x}$, $x + 2 = Ax + B(x - 4)$

when $x = 4$, $6 = 4A$, $A = \dfrac{3}{2}$, when $x = 0$, $2 = -4B$, $B = -\dfrac{1}{2}$

$$\int \frac{x + 2}{x^2 - 4x}\,dx = \int \left[\frac{3/2}{x - 4} - \frac{1/2}{x}\right] dx = \frac{3}{2}\ln|x - 4| - \frac{1}{2}\ln|x| + C$$

11. $\dfrac{4x^2 + 2x - 1}{x^2(x + 1)} = \dfrac{A}{x} + \dfrac{B}{x^2} + \dfrac{C}{x + 1}$

$4x^2 + 2x - 1 = Ax(x + 1) + B(x + 1) + Cx^2$

when $x = 0$, $B = -1$, when $x = -1$, $C = 1$, when $x = 1$, $A = 3$

$$\int \frac{4x^2 + 2x - 1}{x^3 + x^2}\,dx = \int \left[\frac{3}{x} - \frac{1}{x^2} + \frac{1}{x + 1}\right] dx$$

$$= 3\ln|x| + \frac{1}{x} + \ln|x + 1| + C = \frac{1}{x} + \ln|x^4 + x^3| + C$$

12. $\dfrac{2x - 3}{(x - 1)^2} = \dfrac{A}{x - 1} + \dfrac{B}{(x - 1)^2}$, $2x - 3 = A(x - 1) + B$

when $x = 1$, $B = -1$, when $x = 0$, $A = 2$

$$\int \frac{2x - 3}{(x - 1)^2}\,dx = \int \left[\frac{2}{x - 1} - \frac{1}{(x - 1)^2}\right] dx$$

$$= 2\ln|x - 1| + \frac{1}{x - 1} + C$$

13. $\dfrac{x^4}{(x - 1)^3} = x + 3 + \dfrac{6x^2 - 8x + 3}{(x - 1)^3} = x + 3 + \dfrac{A}{x - 1} + \dfrac{B}{(x - 1)^2} + \dfrac{C}{(x - 1)^3}$

$6x^2 - 8x + 3 = A(x - 1)^2 + B(x - 1) + C = Ax^2 + (B - 2A)x + (A - B + C)$

13. (continued)

By equating like coefficients:

$6 = A \Rightarrow A = 6, \quad -8 = B - 2A \Rightarrow B = 4, \quad 3 = A - B + C \Rightarrow C = 1$

$$\int \frac{x^4}{(x-1)^3}\,dx = \int \left[x + 3 + \frac{6}{x-1} + \frac{4}{(x-1)^2} + \frac{1}{(x-1)^3}\right] dx$$

$$= \frac{x^2}{2} + 3x + 6\ln|x-1| - \frac{4}{x-1} - \frac{1}{2(x-1)^2} + C$$

14. $\dfrac{4x^2 - 1}{2x(x+1)^2} = \dfrac{A}{2x} + \dfrac{B}{x+1} + \dfrac{C}{(x+1)^2}$

$4x^2 - 1 = A(x+1)^2 + B(2x)(x+1) + C(2x)$

$= (A + 2B)x^2 + (2A + 2B + 2C)x + A$

Therefore $A + 2B = 4, \quad 2A + 2B + 2C = 0, \quad A = -1.$

Solving these equations we get $A = -1, \quad B = \frac{5}{2}, \quad C = -\frac{3}{2}.$

$$\int \frac{4x^2 - 1}{2x(x^2 + 2x + 1)}\,dx = -\frac{1}{2}\int \frac{1}{x}\,dx + \frac{5}{2}\int \frac{1}{x+1}\,dx - \frac{3}{2}\int \frac{1}{(x+1)^2}\,dx$$

$$= \frac{1}{2}\left[5\ln|x+1| - \ln|x| + \frac{3}{x+1}\right] + C$$

15. $\dfrac{3x}{(x-3)^2} = \dfrac{A}{x-3} + \dfrac{B}{(x-3)^2}, \qquad 3x = A(x-3) + B$

when $x = 3$, $B = 9$ and when $x = 0$, $A = 3$

$$\int \frac{3x}{x^2 - 6x + 9}\,dx = 3\int \frac{1}{x-3}\,dx + 9\int \frac{1}{(x-3)^2}\,dx$$

$$= 3\ln|x-3| - \frac{9}{x-3} + C$$

16. $\dfrac{6x^2 + 1}{x^2(x-1)^3} = \dfrac{A}{x} + \dfrac{B}{x^2} + \dfrac{C}{x-1} + \dfrac{D}{(x-1)^2} + \dfrac{E}{(x-1)^3}$

$6x^2 + 1 = Ax(x-1)^3 + B(x-1)^3 + Cx^2(x-1)^2 + Dx^2(x-1) + Ex^2$

when $x = 1$, $E = 7$ and when $x = 0$, $B = -1$

By letting $B = -1$ and $E = 7$ and expanding we get:

$x^3 - 4x^2 + 3x = (A + C)x^4 + (-3A - 2C + D)x^3 + (3A + C - D)x^2 - Ax$

16. (continued)

By equating coefficients of like terms we have the equations:

$A + C = 0, \quad -3A - 2C + D = 1, \quad 3A + C - D = -4, \quad -A = 3$

Solving these equations we obtain $A = -3, \quad C = 3, \quad D = -2.$

$$\int \frac{6x^2 + 1}{x^2(x - 1)^3}\,dx = \int \left[-\frac{3}{x} - \frac{1}{x^2} + \frac{3}{x - 1} - \frac{2}{(x - 1)^2} + \frac{7}{(x - 1)^3}\right] dx$$

$$= 3\ln\left|\frac{x - 1}{x}\right| + \frac{1}{x} + \frac{2}{x - 1} - \frac{7}{2(x - 1)^2} + C$$

17. $\dfrac{x^2 - 1}{x(x^2 + 1)} = \dfrac{A}{x} + \dfrac{Bx + C}{x^2 + 1}, \qquad x^2 - 1 = A(x^2 + 1) + (Bx + C)x$

when $x = 0, \quad A = -1, \qquad$ when $x = 1, \quad 0 = -2 + B + C$

when $x = -1, \quad 0 = -2 + B - C$

Solving these equations we have $A = -1, \quad B = 2, \quad C = 0.$

$$\int \frac{x^2 - 1}{x^3 + x}\,dx = -\int \frac{1}{x}\,dx + \int \frac{2x}{x^2 + 1}\,dx$$

$$= \ln|x^2 + 1| - \ln|x| + C = \ln\left|\frac{x^2 + 1}{x}\right| + C$$

18. $\dfrac{x}{(x - 1)(x^2 + x + 1)} = \dfrac{A}{x - 1} + \dfrac{Bx + C}{x^2 + x + 1}$

$x = A(x^2 + x + 1) + (Bx + C)(x - 1), \qquad$ when $x = 1, \quad 1 = 3A$

when $x = 0, \quad 0 = A - C, \qquad$ when $x = -1, \quad -1 = A + 2B - 2C$

Solving these equations we have $A = \frac{1}{3}, \quad B = -\frac{1}{3}, \quad C = \frac{1}{3}.$

$$\int \frac{x}{x^3 - 1}\,dx = \frac{1}{3}\left[\int \frac{1}{x - 1}\,dx + \int \frac{-x + 1}{x^2 + x + 1}\,dx\right]$$

$$= \frac{1}{3}\left[\int \frac{1}{x - 1}\,dx - \frac{1}{2}\int \frac{2x + 1}{x^2 + x + 1}\,dx + \frac{3}{2}\int \frac{1}{(x + \frac{1}{2})^2 + (3/4)}\,dx\right]$$

$$= \frac{1}{3}\left[\ln|x - 1| - \frac{1}{2}\ln|x^2 + x + 1| + \sqrt{3}\arctan\left(\frac{2x + 1}{\sqrt{3}}\right)\right] + C$$

19. $\dfrac{x^2}{x^4 - 2x^2 - 8} = \dfrac{A}{x - 2} + \dfrac{B}{x + 2} + \dfrac{Cx + D}{x^2 + 2}$

$x^2 = A(x + 2)(x^2 + 2) + B(x - 2)(x^2 + 2) + (Cx + D)(x + 2)(x - 2)$

19. (continued)

when $x = 2$, $4 = 24A$, when $x = -2$, $4 = -24B$

when $x = 0$, $0 = 4A - 4B - 4D$, when $x = 1$, $1 = 9A - 3B - 3C - 3D$

Solving these equations we have $A = \frac{1}{6}$, $B = -\frac{1}{6}$, $C = 0$, $D = \frac{1}{3}$.

$$\int \frac{x^2}{x^4 - 2x^2 - 8}\,dx = \frac{1}{6}\left[\int \frac{1}{x-2}\,dx - \int \frac{1}{x+2}\,dx + 2\int \frac{1}{x^2+2}\,dx\right]$$

$$= \frac{1}{6}\left[\ln\left|\frac{x-2}{x+2}\right| + \sqrt{2}\arctan\frac{x}{\sqrt{2}}\right] + C$$

20. $$\frac{2x^2 + x + 8}{(x^2+4)^2} = \frac{Ax + B}{x^2 + 4} + \frac{Cx + D}{(x^2+4)^2}$$

$$2x^2 + x + 8 = (Ax + B)(x^2 + 4) + Cx + D$$

$$= Ax^3 + Bx^2 + (4A + C)x + (4B + D)$$

By equating the coefficients of like terms we have $A = 0$, $B = 2$,

$4A + C = 1$, $4B + D = 8$.

Solving these equations we have $A = 0$, $B = 2$, $C = 1$, $D = 0$.

$$\int \frac{2x^2 + x + 8}{(x^2+4)^2}\,dx = 2\int \frac{1}{x^2+4}\,dx + \frac{1}{2}\int \frac{2x}{(x^2+4)^2}\,dx$$

$$= \arctan\frac{x}{2} - \frac{1}{2(x^2+4)} + C$$

21. $$\frac{x}{(2x-1)(2x+1)(4x^2+1)} = \frac{A}{2x-1} + \frac{B}{2x+1} + \frac{Cx + D}{4x^2+1}$$

$$x = A(2x+1)(4x^2+1) + B(2x-1)(4x^2+1) + (Cx + D)(2x-1)(2x+1)$$

when $x = \frac{1}{2}$, $\frac{1}{2} = 4A$, when $x = -\frac{1}{2}$, $-\frac{1}{2} = -4B$

when $x = 0$, $0 = A - B - D$, when $x = 1$, $1 = 15A + 5B + 3C + 3D$.

Solving these equations we have $A = \frac{1}{8}$, $B = \frac{1}{8}$, $C = -\frac{1}{2}$, $D = 0$.

$$\int \frac{x}{16x^4 - 1}\,dx = \frac{1}{8}\left[\int \frac{1}{2x-1}\,dx + \int \frac{1}{2x+1}\,dx - 4\int \frac{x}{4x^2+1}\,dx\right]$$

$$= \frac{1}{16}\ln\left|\frac{4x^2-1}{4x^2+1}\right| + C$$

22. $\dfrac{x^2 - 4x + 7}{(x + 1)(x^2 - 2x + 3)} = \dfrac{A}{x + 1} + \dfrac{Bx + C}{x^2 - 2x + 3}$

$x^2 - 4x + 7 = A(x^2 - 2x + 3) + (Bx + C)(x + 1)$

when $x = -1$, $12 = 6A$, when $x = 0$, $7 = 3A + C$

when $x = 1$, $4 = 2A + 2B + 2C$

Solving these equations we have $A = 2$, $B = -1$, $C = 1$.

$$\int \frac{x^2 - 4x + 7}{x^3 - x^2 + x + 3}\,dx = 2\int \frac{1}{x + 1}\,dx + \int \frac{-x + 1}{x^2 - 2x + 3}\,dx$$

$$= 2\ln|x + 1| - \frac{1}{2}\ln|x^2 - 2x + 3| + C$$

23. $\dfrac{x^2 + x + 2}{(x^2 + 2)^2} = \dfrac{Ax + B}{x^2 + 2} + \dfrac{Cx + D}{(x^2 + 2)^2}$

$x^2 + x + 2 = (Ax + B)(x^2 + 2) + Cx + D$

$= Ax^3 + Bx^2 + (2A + C)x + (2B + D)$

By equating the coefficients of like terms we have $A = 0$, $B = 1$,

$2A + C = 1$, $2B + D = 2$.

Solving these equations $A = 0$, $B = 1$, $C = 1$, $D = 0$.

$$\int \frac{x^2 + x + 2}{(x^2 + 2)^2}\,dx = \int \frac{1}{x^2 + 2}\,dx + \int \frac{x}{(x^2 + 2)^2}\,dx$$

$$= \frac{\sqrt{2}}{2}\arctan\frac{x}{\sqrt{2}} - \frac{1}{2(x^2 + 2)} + C$$

24. $\dfrac{x^3}{(x + 2)^2(x - 2)^2} = \dfrac{A}{x - 2} + \dfrac{B}{(x - 2)^2} + \dfrac{C}{(x + 2)} + \dfrac{D}{(x + 2)^2}$

$x^3 = A(x - 2)(x + 2)^2 + B(x + 2)^2 + C(x + 2)(x - 2)^2 + D(x - 2)^2$

when $x = 2$, $8 = 16B$, when $x = -2$, $-8 = 16D$

when $x = 0$, $0 = -8A + 4B + 8C + 4D$, when $x = 1$, $1 = -9A + 9B + 3C + D$

Solving these equations we have $A = \frac{1}{2}$, $B = \frac{1}{2}$, $C = \frac{1}{2}$, $D = -\frac{1}{2}$.

$$\int \frac{x^3}{(x^2 - 4)^2}\,dx$$

$$= \frac{1}{2}\left[\int \frac{1}{x - 2}\,dx + \int \frac{1}{(x - 2)^2}\,dx + \int \frac{1}{x + 2}\,dx - \int \frac{1}{(x + 2)^2}\,dx\right]$$

24. (continued)

$$= \frac{1}{2}\left[\ln|x - 2| - \frac{1}{x - 2} + \ln|x + 2| + \frac{1}{x + 2}\right] + C$$

$$= \frac{1}{2}\ln|x^2 - 4| - \frac{2}{x^2 - 4} + C$$

25. $$\frac{x^2 + 5}{(x + 1)(x^2 - 2x + 3)} = \frac{A}{x + 1} + \frac{Bx + C}{x^2 - 2x + 3}$$

$x^2 + 5 = A(x^2 - 2x + 3) + (Bx + C)(x + 1)$

$= (A + B)x^2 + (-2A + B + C)x + (3A + C)$

when $x = -1$, $A = 1$

By equating coefficients of like terms we have

$A + B = 1$, $-2A + B + C = 0$, $3A + C = 5$.

Solving these equations we have $A = 1$, $B = 0$, $C = 2$.

$$\int \frac{x^2 + 5}{x^3 - x^2 + x + 3}\,dx = \int \frac{1}{x + 1}\,dx + 2\int \frac{1}{(x - 1)^2 + 2}\,dx$$

$$= \ln|x + 1| + \sqrt{2}\arctan\frac{x - 1}{\sqrt{2}} + C$$

26. $$\frac{x^2 + x + 3}{(x^2 + 3)^2} = \frac{Ax + B}{x^2 + 3} + \frac{Cx + D}{(x^2 + 3)^2}$$

$x^2 + x + 3 = (Ax + B)(x^2 + 3) + Cx + D$

$= Ax^3 + Bx^2 + (3A + C)x + (3B + D)$

By equating coefficients of like terms we have $A = 0$, $B = 1$, $3A + C = 1$, $3B + D = 3$. Solving, we have $A = 0$, $B = 1$, $C = 1$, $D = 0$.

$$\int \frac{x^2 + x + 3}{x^4 + 6x^2 + 9}\,dx = \int\left[\frac{1}{x^2 + 3} + \frac{x}{(x^2 + 3)^2}\right]dx$$

$$= \frac{1}{\sqrt{3}}\arctan\frac{x}{\sqrt{3}} - \frac{1}{2(x^2 + 3)} + C$$

27. $$\frac{6x^2 - 3x + 14}{(x - 2)(x^2 + 4)} = \frac{A}{x - 2} + \frac{Bx + C}{x^2 + 4}$$

$6x^2 - 3x + 14 = A(x^2 + 4) + (Bx + C)(x - 2)$

when $x = 2$, $32 = 8A$, $A = 4$, when $x = 0$, $14 = 4A - 2C$, $C = 1$

when $x = 1$, $17 = 5A - B - C$, $B = 2$

27. (continued)

$$\int \frac{6x^2 - 3x + 14}{x^3 - 2x^2 + 4x - 8}\,dx = \int \left[\frac{4}{x - 2} + \frac{2x + 1}{x^2 + 4}\right] dx$$

$$= 4\int \frac{1}{x - 2}\,dx + \int \frac{2x}{x^2 + 4}\,dx + \int \frac{1}{x^2 + 4}\,dx$$

$$= 4\ln|x - 2| + \ln(x^2 + 4) + \frac{1}{2}\arctan\frac{x}{2} + C$$

28. $$\frac{2x^2 - 9x}{(x - 2)^3} = \frac{A}{x - 2} + \frac{B}{(x - 2)^2} + \frac{C}{(x - 2)^3}$$

$2x^2 - 9x = A(x - 2)^2 + B(x - 2) + C,$ when $x = 2,\ -10 = C,\ C = -10$

$2x^2 - 9x + 10 = A(x^2 - 4x + 4) + Bx - 2B = Ax^2 + (B - 4A)x + (4A - 2B)$

$2 = A,\quad -9 = B - 4A \Rightarrow B = -1,\quad 10 = 4A - 2B$

$$\int \frac{x(2x - 9)}{x^3 - 6x^2 + 12x - 8}\,dx = \int \left[\frac{2}{x - 2} - \frac{1}{(x - 2)^2} - \frac{10}{(x - 2)^3}\right] dx$$

$$= 2\ln|x - 2| + \frac{1}{x - 2} + \frac{5}{(x - 2)^2} + C$$

29. $$\frac{2x^2 - 2x + 3}{(x - 2)(x^2 + x + 1)} = \frac{A}{x - 2} + \frac{Bx + C}{x^2 + x + 1}$$

$2x^2 - 2x + 3 = A(x^2 + x + 1) + (Bx + C)(x - 2)$

when $x = 2,\ 7 = 7A,\ A = 1,$ when $x = 0,\ 3 = A - 2C,\ C = -1$

when $x = 1,\ 3 = 3A - B - C,\ B = 1$

$$\int \frac{2x^2 - 2x + 3}{x^3 - x^2 - x - 2}\,dx = \int \left[\frac{1}{x - 2} + \frac{x - 1}{x^2 + x + 1}\right] dx$$

$$= \ln|x - 2| + \frac{1}{2}\int \frac{2x - 2 + 3 - 3}{x^2 + x + 1}\,dx$$

$$= \ln|x - 2| + \frac{1}{2}\int \frac{2x + 1}{x^2 + x + 1}\,dx - \frac{3}{2}\int \frac{1}{x^2 + x + 1}\,dx$$

$$= \ln|x - 2| + \frac{1}{2}\ln|x^2 + x + 1| - \frac{3}{2}\int \frac{1}{(x + \frac{1}{2})^2 + \frac{3}{4}}\,dx$$

$$= \ln|x - 2| + \frac{1}{2}\ln|x^2 + x + 1| - \sqrt{3}\arctan\left[\frac{x + 1/2}{\sqrt{3}/2}\right] + C$$

$$= \ln|x - 2| + \frac{1}{2}\ln|x^2 + x + 1| - \sqrt{3}\arctan\left[\frac{2x + 1}{\sqrt{3}}\right] + C$$

30. $\dfrac{-x^4 + 5x^3 - 9x^2 + 3x - 6}{(x + 1)(x - 1)^2(x^2 + 1)} = \dfrac{A}{x + 1} + \dfrac{B}{x - 1} + \dfrac{C}{(x - 1)^2} + \dfrac{Dx + E}{x^2 + 1}$

$-x^4 + 5x^3 - 9x^2 + 3x - 6 = A(x - 1)^2(x^2 + 1) + B(x + 1)(x - 1)(x^2 + 1)$

$+ C(x + 1)(x^2 + 1) + (Dx + E)(x + 1)(x - 1)^2$

when $x = 1$, $-8 = 4C$, $C = -2$, when $x = -1$, $-24 = 8A$, $A = -3$

when $x = 0$, $-6 = A - B + C + E$, $-1 = -B + E$

when $x = 2$, $-12 = 5A + 15B + 15C + 6D + 3E$, $33 = 15B + 6D + 3E$

when $x = -2$, $-104 = 45A + 15B - 5C + 18D - 9E$, $21 = 15B + 18D - 9E$

$A = -3$, $B = 2$, $C = -2$, $D = 0$, $E = 1$

$$\int \frac{-x^4 + 5x^3 - 9x^2 + 3x - 6}{x^5 - x^4 - x + 1}\,dx = \int \left[\frac{-3}{x + 1} + \frac{2}{x - 1} - \frac{2}{(x - 1)^2} + \frac{1}{x^2 + 1}\right]dx$$

$$= -3\ln|x + 1| + 2\ln|x - 1| + \frac{2}{x - 1} + \arctan x + C$$

31. $\dfrac{3}{(2x + 1)(x + 2)} = \dfrac{A}{2x + 1} + \dfrac{B}{x + 2}$, $3 = A(x + 2) + B(2x + 1)$

when $x = -\dfrac{1}{2}$, $A = 2$, when $x = -2$, $B = -1$

$$\int_0^1 \frac{3}{2x^2 + 5x + 2}\,dx = \int_0^1 \frac{2}{2x + 1}\,dx - \int_0^1 \frac{1}{x + 2}\,dx$$

$$= \Big[\ln|2x + 1| - \ln|x + 2|\Big]_0^1 = \ln 2$$

32. $\dfrac{1}{x^2 - 4} = \dfrac{A}{x - 2} + \dfrac{B}{x + 2}$, $1 = A(x + 2) + B(x - 2)$

when $x = 2$, $A = \dfrac{1}{4}$, when $x = -2$, $B = -\dfrac{1}{4}$

$$\int_3^4 \frac{1}{x^2 - 4}\,dx = \frac{1}{4}\left[\int_3^4 \frac{1}{x - 2}\,dx - \int_3^4 \frac{1}{x + 2}\,dx\right]$$

$$= \frac{1}{4}\ln\left|\frac{x - 2}{x + 2}\right|\Bigg]_3^4 = \frac{1}{4}\ln\frac{5}{3}$$

33. $\dfrac{x + 1}{x(x^2 + 1)} = \dfrac{A}{x} + \dfrac{Bx + C}{x^2 + 1}$, $x + 1 = A(x^2 + 1) + (Bx + C)x$

when $x = 0$, $A = 1$, when $x = 1$, $2 = 2A + B + C$

when $x = -1$, $0 = 2A + B - C$

33. (continued)

Solving these equations we have A = 1, B = −1, C = 1.

$$\int_1^2 \frac{x + 1}{x(x^2 + 1)}\,dx = \int_1^2 \frac{1}{x}\,dx - \int_1^2 \frac{x}{x^2 + 1}\,dx + \int_1^2 \frac{1}{x^2 + 1}\,dx$$

$$= \left[\ln x - \frac{1}{2}\ln(x^2 + 1) + \arctan x\right]_1^2 = \frac{1}{2}\ln\frac{8}{5} - \frac{\pi}{4} + \arctan 2$$

34. $\dfrac{x - 1}{x^2(x + 1)} = \dfrac{A}{x} + \dfrac{B}{x^2} + \dfrac{C}{x + 1}$, $\quad x - 1 = Ax(x + 1) + B(x + 1) + Cx^2$

when $x = 0$, $B = -1$, when $x = -1$, $C = -2$, when $x = 1$, $0 = 2A + 2B + C$

Solving these equations we have A = 2, B = −1, C = −2.

$$\int_1^5 \frac{x - 1}{x^2(x + 1)}\,dx = 2\int_1^5 \frac{1}{x}\,dx - \int_1^5 \frac{1}{x^2}\,dx - 2\int_1^5 \frac{1}{x + 1}\,dx$$

$$= \left[2\ln|x| + \frac{1}{x} - 2\ln|x + 1|\right]_1^5$$

$$= \left[2\ln\left|\frac{x}{x + 1}\right| + \frac{1}{x}\right]_1^5 = 2\ln\frac{5}{3} - \frac{4}{5}$$

35. $\dfrac{x^4 + 4x^3 - x^2 + 5x + 1}{(x - 1)(x^2 + 1)(x^2 + 2x + 2)} = \dfrac{A}{x - 1} + \dfrac{Bx + C}{x^2 + 1} + \dfrac{Dx + E}{x^2 + 2x + 2}$

$x^4 + 4x^3 - x^2 + 5x + 1 = A(x^2 + 1)(x^2 + 2x + 2)$

$\quad + (Bx + C)(x - 1)(x^2 + 2x + 2) + (Dx + E)(x - 1)(x^2 + 1)$

when $x = 1$, $10 = 10A \Rightarrow A = 1$

$x^4 + 4x^3 - x^2 + 5x + 1 = 1(x^4 + 2x^3 + 3x^2 + 2x + 2)$

$\quad + (Bx + C)(x^3 + x^2 - 2) + (Dx + E)(x^3 - x^2 + x - 1)$

$2x^3 - 4x^2 + 3x - 1 = Bx^4 + Bx^3 - 2Bx + Cx^3 + Cx^2 - 2C + Dx^4$

$\quad - Dx^3 + Dx^2 - Dx + Ex^3 - Ex^2 + Ex - E = (B + D)x^4$

$\quad + (B + C - D + E)x^3 + (C + D - E)x^2 + (-2B - D + E)x + (-2C - E)$

$B + D = 0$, $\quad B + C - D + E = 2$, $\quad C + D - E = -4$, $\quad -2B - D + E = 3$,

$-2C - E = -1$

35. (continued)

Solving these equations we have B = 0, C = −1, D = 0, E = 3

$$\int_2^3 \frac{x^4 + 4x^3 - x^2 + 5x + 1}{x^5 + x^4 + x^3 - x^2 - 2}\,dx = \int_2^3 \left[\frac{1}{x - 1} + \frac{-1}{x^2 + 1} + \frac{3}{x^2 + 2x + 2}\right] dx$$

$$= \int_2^3 \frac{1}{x - 1}\,dx - \int_2^3 \frac{1}{x^2 + 1}\,dx + 3\int_2^3 \frac{1}{(x + 1)^2 + 1}\,dx$$

$$= \Big[\ln|x - 1| - \arctan x + 3\arctan(x + 1)\Big]_2^3$$

$$= (\ln 2 - \arctan 3 + 3\arctan 4) - (0 - \arctan 2 + 3\arctan 3)$$

$$= \ln 2 + \arctan 2 - 4\arctan 3 + 3\arctan 4 \approx 0.7816$$

36. $$\int_0^1 \frac{x^2 - x}{x^2 + x + 1}\,dx = \int_0^1 dx - \int_0^1 \frac{2x + 1}{x^2 + x + 1}\,dx$$

$$= \Big[x - \ln|x^2 + x + 1|\Big]_0^1 = 1 - \ln 3$$

37. $$\frac{1}{u(u - 1)} = \frac{A}{u} + \frac{B}{u - 1}, \qquad 1 = A(u - 1) + Bu$$

when $u = 0$, $A = -1$, when $u = 1$, $B = 1$, $u = \cos x$, $du = -\sin x\,dx$

$$\int \frac{\sin x}{\cos x(\cos x - 1)}\,dx = -\int \frac{1}{u(u - 1)}\,du = \int \frac{1}{u}\,du - \int \frac{1}{u - 1}\,du$$

$$= \ln|u| - \ln|u - 1| + C = \ln\left|\frac{u}{u - 1}\right| + C = \ln\left|\frac{\cos x}{\cos x - 1}\right| + C$$

38. $$\frac{1}{u(u + 1)} = \frac{A}{u} + \frac{B}{u + 1}, \qquad 1 = A(u + 1) + Bu$$

when $u = 0$, $A = 1$, when $u = -1$, $B = -1$, $u = \cos x$, $du = -\sin x\,dx$

$$\int \frac{\sin x}{\cos x + \cos^2 x}\,dx = -\int \frac{1}{u(u + 1)}\,du = \int \frac{1}{u + 1}\,du - \int \frac{1}{u}\,du$$

$$= \ln|u + 1| - \ln|u| + C = \ln\left|\frac{u + 1}{u}\right| + C$$

$$= \ln\left|\frac{\cos x + 1}{\cos x}\right| + C = \ln|1 + \sec x| + C$$

39. $$\frac{1}{(u - 1)(u + 4)} = \frac{A}{u - 1} + \frac{B}{u + 4}, \qquad 1 = A(u + 4) + B(u - 1)$$

when $u = 1$, $A = \frac{1}{5}$, when $u = -4$, $B = -\frac{1}{5}$, $u = e^x$, $u' = e^x$

39. (continued)

$$\int \frac{e^x}{(e^x - 1)(e^x + 4)}\,dx = \int \frac{1}{(u - 1)(u + 4)}\,du$$

$$= \frac{1}{5}\left[\int \frac{1}{u - 1}\,du - \int \frac{1}{u + 4}\,du\right]$$

$$= \frac{1}{5}\ln\left|\frac{u - 1}{u + 4}\right| + C = \frac{1}{5}\ln\left|\frac{e^x - 1}{e^x + 4}\right| + C$$

40. $\dfrac{1}{(u^2 + 1)(u - 1)} = \dfrac{A}{u - 1} + \dfrac{Bu + C}{u^2 + 1}$, $\quad 1 = A(u^2 + 1) + (Bu + C)(u - 1)$

when $u = 1$, $A = \dfrac{1}{2}$, when $u = 0$, $1 = A - C$

when $u = -1$, $1 = 2A + 2B - 2C$

Solving these equations we have $A = \dfrac{1}{2}$, $B = -\dfrac{1}{2}$, $C = -\dfrac{1}{2}$, $u = e^x$, $du = e^x\,dx$

$$\int \frac{e^x}{(e^{2x} + 1)(e^x - 1)}\,dx = \int \frac{1}{(u^2 + 1)(u - 1)}\,du$$

$$= \frac{1}{2}\left[\int \frac{1}{u - 1}\,du - \int \frac{u + 1}{u^2 + 1}\,du\right]$$

$$= \frac{1}{2}\left[\ln|u - 1| - \frac{1}{2}\ln|u^2 + 1| - \arctan u\right] + C$$

$$= \frac{1}{4}\left[2\ln|e^x - 1| - \ln|e^{2x} + 1| - 2\arctan e^x\right] + C$$

41. $$\int \frac{3\cos x}{\sin^2 x + \sin x - 2}\,dx = 3\int \frac{1}{u^2 + u - 2}\,du = \ln\left|\frac{u - 1}{u + 2}\right| + C$$

$$= \ln\left|\frac{-1 + \sin x}{2 + \sin x}\right| + C \qquad \text{(from Exercise 3)}$$

42. $\dfrac{1}{a^2 - x^2} = \dfrac{A}{a - x} + \dfrac{B}{a + x}$, $\quad 1 = A(a + x) + B(a - x)$

when $x = a$, $A = \dfrac{1}{2a}$, when $x = -a$, $B = \dfrac{1}{2a}$

$$\int \frac{1}{a^2 - x^2}\,dx = \frac{1}{2a}\int \left[\frac{1}{a - x} + \frac{1}{a + x}\right]dx$$

$$= \frac{1}{2a}\left[-\ln|a - x| + \ln|a + x|\right] + C$$

$$= \frac{1}{2a}\ln\left|\frac{a + x}{a - x}\right| + C$$

43. $\dfrac{x}{(a + bx)^2} = \dfrac{A}{a + bx} + \dfrac{B}{(a + bx)^2}$, $\quad x = A(a + bx) + B$

when $x = -\dfrac{a}{b}$, $B = -\dfrac{a}{b}$, $\quad$ when $x = 0$, $0 = aA + B$, $A = \dfrac{1}{b}$

$$\int \frac{x}{(a + bx)^2}\,dx = \int \left[\frac{1/b}{a + bx} + \frac{-a/b}{(a + bx)^2}\right] dx$$

$$= \frac{1}{b}\int \frac{1}{a + bx}\,dx - \frac{a}{b}\int \frac{1}{(a + bx)^2}\,dx$$

$$= \frac{1}{b^2}\ln|a + bx| + \frac{a}{b^2}\left(\frac{1}{a + bx}\right) + C$$

$$= \frac{1}{b^2}\left[\frac{a}{a + bx} + \ln|a + bx|\right] + C$$

44. $\dfrac{1}{x^2(a + bx)} = \dfrac{A}{x} + \dfrac{B}{x^2} + \dfrac{C}{a + bx}$, $\quad 1 = Ax(a + bx) + B(a + bx) + Cx^2$

when $x = 0$, $1 = Ba$, $B = \dfrac{1}{a}$, $\quad$ when $x = -\dfrac{a}{b}$, $1 = C\left(\dfrac{a^2}{b^2}\right)$, $C = \dfrac{b^2}{a^2}$

when $x = 1$, $1 = (a + b)A + (a + b)B + C$, $A = -\dfrac{b}{a^2}$

$$\int \frac{1}{x^2(a + bx)}\,dx = \int \left[\frac{-b/a^2}{x} + \frac{1/a}{x^2} + \frac{b^2/a^2}{a + bx}\right] dx$$

$$= -\frac{b}{a^2}\ln|x| - \frac{1}{ax} + \frac{b}{a^2}\ln|a + bx| + C$$

$$= -\frac{1}{ax} + \frac{b}{a^2}\ln\left|\frac{a + bx}{x}\right| + C$$

$$= -\frac{1}{ax} - \frac{b}{a^2}\ln\left|\frac{x}{a + bx}\right| + C$$

45. $A = 2\displaystyle\int_0^3 \left(1 - \frac{7}{16 - x^2}\right) dx$

$$= 2\int_0^3 dx - 14\int_0^3 \frac{1}{16 - x^2}\,dx$$

$$= \left[2x - \frac{14}{8}\ln\left|\frac{4 + x}{4 - x}\right|\right]_0^3 \quad \text{(from Exercise 42)}$$

$$= 6 - \frac{7}{4}\ln 7$$

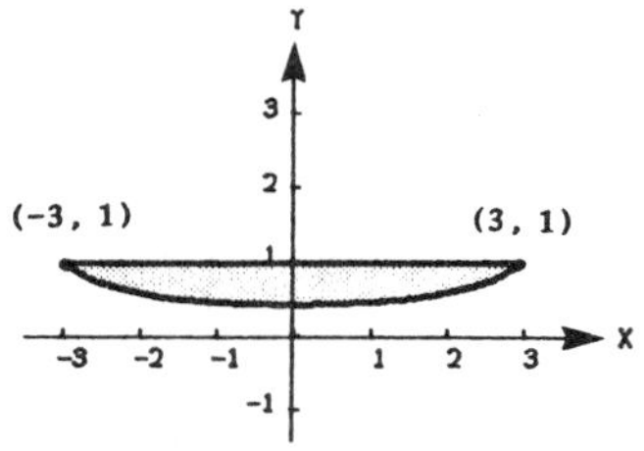

46. $A = \int_0^3 \frac{2x}{x^2+1}\,dx = \ln(x^2+1)\Big]_0^3 = \ln 10$

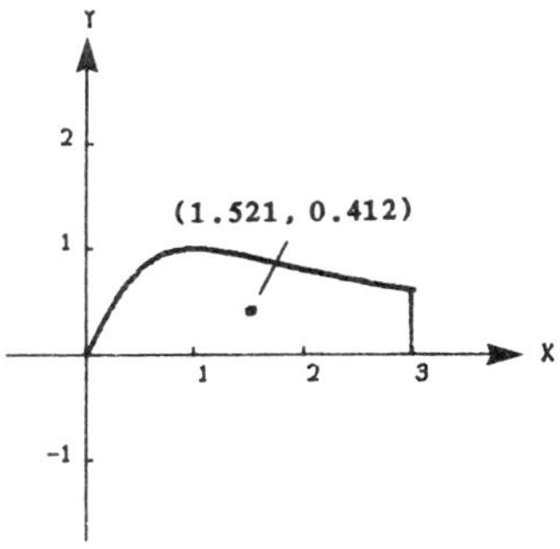

$\bar{x} = \frac{1}{A}\int_0^3 \frac{2x^2}{x^2+1}\,dx = \frac{1}{\ln 10}\int_0^3 \left[2 - \frac{2}{x^2+1}\right] dx$

$= \frac{1}{\ln 10}\left[2x - 2\arctan x\right]_0^3$

$= \frac{2}{\ln 10}[3 - \arctan 3] \approx 1.521$

$\bar{y} = \frac{1}{A}\left(\frac{1}{2}\right)\int_0^3 \left(\frac{2x}{x^2+1}\right)^2 dx = \frac{2}{\ln 10}\int_0^3 \frac{x^2}{(x^2+1)^2}\,dx$

$= \frac{2}{\ln 10}\int_0^3 \left[\frac{1}{x^2+1} - \frac{1}{(x^2+1)^2}\right] dx$ (partial fractions)

$= \frac{2}{\ln 10}\left[\arctan x - \frac{1}{2}\left[\arctan x + \frac{x}{x^2+1}\right]\right]_0^3$ (trig. substitution)

$= \frac{2}{\ln 10}\left[\frac{1}{2}\arctan x - \frac{x}{2(x^2+1)}\right]_0^3$

$= \frac{1}{\ln 10}\left[\arctan x - \frac{x}{x^2+1}\right]_0^3 = \frac{1}{\ln 10}\left[\arctan 3 - \frac{3}{10}\right] \approx 0.412$

$(\bar{x}, \bar{y}) \approx (1.521, 0.412)$

47. $V = \pi\int_0^3 \left(\frac{2x}{x^2+1}\right)^2 dx = 4\pi\int_0^3 \frac{x^2}{(x^2+1)^2}\,dx$

$= 4\pi\int_0^3 \left[\frac{1}{x^2+1} - \frac{1}{(x^2+1)^2}\right] dx$ (partial fractions)

$= 4\pi\left[\arctan x - \frac{1}{2}\left(\arctan x + \frac{x}{x^2+1}\right)\right]_0^3$ (trig. substitution)

$= 2\pi\left[\arctan x - \frac{x}{x^2+1}\right]_0^3 = 2\pi\left[\arctan 3 - \frac{3}{10}\right] \approx 5.963$

48. (a) $\frac{1}{y(L-y)} = \frac{A}{y} + \frac{B}{L-y}$, $\quad 1 = A(L-y) + By$

when $y = 0$, $A = \frac{1}{L}$, $\quad$ when $y = L$, $B = \frac{1}{L}$

$\int \frac{1}{y(L-y)}\,dy = \int k\,dt$, $\quad \frac{1}{L}\left[\int \frac{1}{y}\,dy + \int \frac{1}{L-y}\,dy\right] = \int k\,dt$

48. (continued)

$$\frac{1}{L}[\ln|y| - \ln|L - y|] = kt + C, \qquad \frac{y}{L - y} = C_2e^{kLt}$$

when $t = 0$, $\dfrac{y_0}{L - y_0} = C_2$ and $\dfrac{y}{L - y} = \dfrac{y_0}{L - y_0}e^{kLt}$

Solving for y we have $y = \dfrac{y_0L}{y_0 + (L - y_0)e^{-kLt}}$

(b) From Part (a) $\dfrac{dy}{dt} = ky(L - y)$. This will be maximum when $\dfrac{d^2y}{dt^2} = 0$

or at the point of inflection. $\dfrac{d^2y}{dt^2} = k\left[y(\dfrac{-dy}{dt}) + (L - y)\dfrac{dy}{dt}\right] = 0$

when $y = \dfrac{L}{2}$

49. $$\frac{1}{(x + 1)(n - x)} = \frac{A}{x + 1} + \frac{B}{n - x}, \qquad A = B = \frac{1}{n + 1}$$

$$\frac{1}{n + 1}\int (\frac{1}{x + 1} + \frac{1}{n - x})\,dx = kt + C$$

$$\frac{1}{n + 1}\ln\left|\frac{x + 1}{n - x}\right| = kt + C, \quad \text{when } t = 0, \quad x = 0, \quad C = \frac{1}{n + 1}\ln\frac{1}{n}$$

$$\frac{1}{n + 1}\ln\left|\frac{x + 1}{n - x}\right| = kt + \frac{1}{n + 1}\ln\frac{1}{n}$$

$$\frac{1}{n + 1}\left[\ln\left|\frac{x + 1}{n - x}\right| - \ln\frac{1}{n}\right] = kt$$

$$\ln\frac{nx + n}{n - x} = (n + 1)kt, \qquad \frac{nx + n}{n - x} = e^{(n+1)kt},$$

$$x = \frac{n[e^{(n+1)kt} - 1]}{n + e^{(n+1)kt}}, \qquad \lim_{t \to \infty} x = n$$

8.6
Summary and integration by tables

1. By Formula 6: $\displaystyle\int \frac{x^2}{1 + x}\,dx = -\frac{x}{2}(2 - x) + \ln|1 + x| + C$

2. By Formula 21: $\displaystyle\int \frac{x}{\sqrt{1 + x}}\,dx = -\frac{2}{3}(2 - x)\sqrt{1 + x} + C$

3. By Formula 44: $\int \frac{1}{x^2\sqrt{1-x^2}}\,dx = -\frac{\sqrt{1-x^2}}{x} + C$

4. By Formula 52: $\int x\sin x\,dx = \sin x - x\cos x + C$

5. By Formula 89: $\int x^2 \ln x\,dx = \frac{x^3}{9}[-1 + 3\ln x] + C$

6. By Formula 79: $\int \operatorname{arcsec} 2x\,dx$

$= \frac{1}{2}\left[2x \operatorname{arcsec} 2x - \ln\left|2x + \sqrt{4x^2 - 1}\right|\right] + C$

$u = 2x, \quad u' = 2$

7. By Formula 35: $\int \frac{1}{x^2\sqrt{x^2-4}}\,dx = \frac{\sqrt{x^2-4}}{4x} + C$

8. By Formula 29: $\int \frac{\sqrt{x^2-4}}{x}\,dx = \sqrt{x^2-4} - 2\operatorname{arcsec}\frac{|x|}{2} + C$

9. By Formula 81: $\int xe^{x^2}\,dx = \frac{1}{2}e^{x^2} + C$

10. By Formula 41: $\int \frac{x}{\sqrt{9-x^4}}\,dx = \frac{1}{2}\int \frac{2x}{\sqrt{3^2-(x^2)^2}}\,dx = \arcsin\frac{x^2}{3} + C$

11. By Formula 4: $\int \frac{2x}{(1-3x)^2}\,dx = 2\int \frac{x}{(1-3x)^2}\,dx$

$= \frac{2}{9}\left[\ln|1-3x| + \frac{1}{1-3x}\right] + C$

12. By Formula 14: $\int \frac{1}{2+2x+x^2}\,dx = \frac{2}{\sqrt{4}}\arctan\frac{2x+2}{2} + C$

$= \arctan(x+1) + C$

13. By Formula 76: $\int e^x \arccos e^x\,dx = e^x \arccos e^x - \sqrt{1-e^{2x}} + C$

$u = e^x, \quad u' = e^x$

14. By Formula 56: $\int \frac{\theta^2}{1-\sin\theta^3}\,d\theta = \frac{1}{3}\int \frac{1}{1-\sin\theta^3}3\theta^2\,d\theta$

$= \frac{1}{3}[\tan\theta^3 + \sec\theta^3] + C$

15. By Formula 89: $\int x^3 \ln x \, dx = \frac{x^4}{16}(4 \ln|x| - 1) + C$

16. By Formula 68: $\int \cot^3 \theta \, d\theta = \frac{-\cot^2 \theta}{2} - \int \cot \theta \, d\theta$

$$= -\frac{1}{2}\cot^2 \theta - \ln|\sin \theta| + C$$

17. By Formula 7: $\int \frac{x^2}{(3x - 5)^2} dx = \frac{1}{27}\left[3x - \frac{25}{3x - 5} + 10 \ln|3x - 5|\right] + C$

18. By Formula 13: $\int \frac{1}{2x^2(2x - 1)^2} dx = \frac{1}{2}\int \frac{1}{x^2(2x - 1)^2} dx$

$$= -\frac{1}{2}\left[\frac{4x - 1}{x(2x - 1)} + \frac{4}{-1}\ln\left|\frac{x}{2x - 1}\right|\right] + C$$

$$= \frac{1}{2}\left[4 \ln\left|\frac{x}{2x - 1}\right| - \frac{4x - 1}{x(2x - 1)}\right] + C$$

19. By Formula 73: $\int \frac{x}{1 - \sec x^2} dx = \frac{1}{2}\int \frac{2x}{1 - \sec x^2} dx$

$$= \frac{1}{2}\left[x^2 + \cot x^2 + \csc x^2\right] + C$$

20. By Formula 71: $\int \frac{e^x}{1 - \tan e^x} dx = \frac{1}{2}\left[e^x - \ln|\cos e^x - \sin e^x|\right] + C$

$u = e^x, \quad u' = e^x$

21. By Formula 23: $\int \frac{\cos x}{1 + \sin^2 x} dx = \arctan(\sin x) + C$

$u = \sin x, \quad u' = \cos x$

22. By Formula 23: $\int \frac{1}{t(1 + \ln^2 t)} dt$

$$= \int \frac{1}{1 + (\ln t)^2}\left(\frac{1}{t}\right) dt = \arctan(\ln t) + C$$

$u = \ln t, \quad u' = \frac{1}{t}$

23. By Formula 84: $\int \frac{1}{1 + e^{2x}} dx = x - \frac{1}{2}\ln(1 + e^{2x}) + C$

24. By Formula 2: $\int \frac{1}{\sqrt{x}(1 + 2\sqrt{x})}\,dx$

$= \int \frac{1}{1 + 2\sqrt{x}}\left(\frac{1}{\sqrt{x}}\right)dx = \ln(1 + 2\sqrt{x}) + C$

$u = 1 + 2\sqrt{x}, \quad u' = \frac{1}{\sqrt{x}}$

25. By Formula 14: $\int \frac{\cos\theta}{3 + 2\sin\theta + \sin^2\theta}\,d\theta = \frac{\sqrt{2}}{2}\arctan\frac{1 + \sin\theta}{\sqrt{2}} + C$

$u = \sin\theta, \quad u' = \cos\theta$

26. By Formula 27: $\int x^2\sqrt{2 + (3x)^2}\,dx = \frac{1}{27}\int (3x)^2\sqrt{(\sqrt{2})^2 + (3x)^2}\,dx$

$= \frac{1}{8(27)}\left[3x(18x^2 + 2)\sqrt{2 + 9x^2} - 4\ln\left|3x + \sqrt{2 + 9x^2}\right|\right] + C$

27. By Formula 35: $\int \frac{1}{x^2\sqrt{2 + 9x^2}}\,dx = 3\int \frac{3}{(3x)^2\sqrt{(\sqrt{2})^2 + (3x)^2}}\,dx$

$= -\frac{3\sqrt{2 + 9x^2}}{6x} + C = -\frac{\sqrt{2 + 9x^2}}{2x} + C$

28. By Formula 26: $\int \sqrt{3 + x^2}\,dx = \frac{1}{2}\left[x\sqrt{x^2 + 3} + 3\ln\left|x + \sqrt{x^2 + 3}\right|\right] + C$

29. By Formula 26: $\int e^x\sqrt{1 + e^{2x}}\,dx$

$= \frac{1}{2}\left[e^x\sqrt{e^{2x} + 1} + \ln\left|e^x + \sqrt{e^{2x} + 1}\right|\right] + C$

$u = e^x, \quad u' = e^x$

30. By Formula 36: $\int \frac{1}{\sqrt{x}(x - 4)^{3/2}}\,dx = 2\int \frac{1}{(x - 4)^{3/2}}\left(\frac{1}{2\sqrt{x}}\right)dx$

$= \frac{-\sqrt{x}}{2\sqrt{x - 4}} + C$

$u = \sqrt{x}, \quad u' = \frac{1}{2\sqrt{x}}$

31. By Formulas 50 and 48: $\int \sin^4(2x)\,dx = \frac{1}{2}\int \sin^4(2x)(2)\,dx$

$= \frac{1}{2}\left[\frac{-\sin^3(2x)\cos(2x)}{4} + \frac{3}{4}\int \sin^2(2x)(2)\,dx\right]$

31. (continued)

$$= \frac{1}{2}\left[\frac{-\sin^3(2x)\cos(2x)}{4} + \frac{3}{8}(2x - \sin 2x \cos 2x)\right] + C$$

$$= \frac{1}{16}[6x - 3\sin 2x \cos 2x - 2\sin^3 2x \cos 2x] + C$$

32. By Formulas 51 and 47: $\int \frac{\cos^3\sqrt{x}}{\sqrt{x}}\,dx = 2\int \cos^3\sqrt{x}\left(\frac{1}{2\sqrt{x}}\right)dx$

$$= 2\left[\frac{\cos^2\sqrt{x}\sin\sqrt{x}}{3} + \frac{2}{3}\int \cos\sqrt{x}\left(\frac{1}{2\sqrt{x}}\right)dx\right]$$

$$= \frac{2}{3}\sin\sqrt{x}\,(\cos^2\sqrt{x} + 2) + C$$

$u = \sqrt{x}$, $\quad u' = \frac{1}{2\sqrt{x}}$

33. By Formula 57: $\int \frac{1}{\sqrt{x}(1 - \cos\sqrt{x})}\,dx = 2\int \frac{1}{1 - \cos\sqrt{x}}\left(\frac{1}{2\sqrt{x}}\right)dx$

$$= -2(\cot\sqrt{x} + \csc\sqrt{x}) + C$$

$u = \sqrt{x}$, $\quad u' = \frac{1}{2\sqrt{x}}$

34. By Formula 71: $\int \frac{1}{1 - \tan 5x}\,dx = \frac{1}{5}\int \frac{1}{1 - \tan 5x}(5)\,dx$

$$= \frac{1}{5}\left(\frac{1}{2}\right)(u - \ln|\cos u - \sin u|) + C$$

$$= \frac{1}{10}(5x - \ln|\cos 5x - \sin 5x|) + C$$

$u = 5x$, $\quad u' = 5$

35. By Formulas 55 and 54: $\int t^4\cos t\,dt = t^4\sin t - 4\int t^3\sin t\,dt$

$$= t^4\sin t - 4\left[-t^3\cos t + 3\int t^2\cos t\,dt\right]$$

$$= t^4\sin t + 4t^3\cos t - 12\left[t^2\sin t - 2\int t\sin t\,dt\right]$$

$$= t^4\sin t + 4t^3\cos t - 12t^2\sin t + 24(-t\cos t + \sin t) + C$$

$$= (t^4 - 12t^2 + 24)\sin t + (4t^3 - 24t)\cos t + C$$

36. By Formula 77: $\int \sqrt{x}\arctan(x^{3/2})\,dx = \frac{2}{3}\int \arctan(x^{3/2})(\frac{3}{2}\sqrt{x})\,dx$

$$= \frac{2}{3}\left[x^{3/2}\arctan(x^{3/2}) - \frac{1}{2}\ln|1 + x^3|\right] + C$$

$u = x^{3/2}, \quad u' = \frac{3}{2}x^{1/2}$

37. By Formula 79: $\int x\,\text{arcsec}(x^2 + 1)\,dx = \frac{1}{2}\int \text{arcsec}(x^2 + 1)(2x)\,dx$

$$= \frac{1}{2}\left[(x^2 + 1)\,\text{arcsec}(x^2 + 1) - \ln\left|(x^2 + 1) + \sqrt{x^4 + 2x^2}\right|\right] + C$$

$u = x^2 + 1, \quad u' = 2x$

38. By Formulas 89 and 90: $\int \ln^3 x\,dx = x\ln^3 x - 3\int \ln^2 x\,dx$

$$= x\ln^3 x - 3x(2 - 2\ln x + \ln^2 x) + C$$

$$= x(\ln^3 x - 3\ln^2 x + 6\ln x - 6) + C$$

39. By Formula 3: $\int \frac{\ln x}{x(3 + 2\ln x)}\,dx = \frac{1}{4}\left[2\ln x - 3\ln|3 + 2\ln x|\right] + C$

40. By Formula 45: $\int \frac{e^x}{(1 - e^{2x})^{3/2}}\,dx = \frac{e^x}{\sqrt{1 - e^{2x}}} + C$

$u = e^x, \quad u' = e^x$

41. By Formula 39: $\int \frac{\sqrt{2 - 2x - x^2}}{x + 1}\,dx$

$$= \sqrt{2 - 2x - x^2} - \sqrt{3}\ln\left|\frac{\sqrt{3} + \sqrt{2 - 2x - x^2}}{x + 1}\right| + C$$

$u = x + 1, \quad u' = 1, \quad \sqrt{2 - 2x - x^2} = \sqrt{3 - (x + 1)^2}$

42. By Formulas 25 and 23: $\int \frac{1}{(x^2 - 6x + 10)^2}\,dx = \int \frac{1}{[(x - 3)^2 + 1]^2}\,dx$

$$= \frac{x - 3}{2(x^2 - 6x + 10)} + \frac{1}{2}\int \frac{1}{(x - 3)^2 + 1}\,dx$$

$$= \frac{1}{2}\left[\frac{x - 3}{x^2 - 6x + 10} + \arctan(x - 3)\right] + C$$

$u = x - 3, \quad u' = 1$

43. By Formula 23: $\int \frac{x}{x^4 - 6x^2 + 10}\, dx = \frac{1}{2}\int \frac{2x}{(x^2 - 3)^2 + 1}\, dx$

$= \frac{1}{2}\arctan(x^2 - 3) + C$

$u = x^2 - 3, \quad u' = 2x$

44. By Formula 27: $\int (2x - 3)^2\sqrt{(2x - 3)^2 + 4}\, dx$

$= \frac{1}{2}\int (2x - 3)^2\sqrt{(2x - 3)^2 + 4}\,(2)\, dx$

$= \frac{1}{8}(2x - 3)((2x - 3)^2 + 2)\sqrt{(2x - 3)^2 + 4}$

$- \ln\left|2x - 3 + \sqrt{(2x - 3)^2 + 4}\right| + C$

$u = 2x - 3, \quad u' = 2$

45. By Formula 31: $\int \frac{x}{\sqrt{x^4 - 6x^2 + 5}}\, dx = \frac{1}{2}\int \frac{2x}{\sqrt{(x^2 - 3)^2 - 4}}\, dx$

$= \frac{1}{2}\ln\left|(x^2 - 3) + \sqrt{x^4 - 6x^2 + 5}\right| + C$

$u = x^2 - 3, \quad u' = 2x$

46. By Formula 31: $\int \frac{\cos x}{\sqrt{\sin^2 x + 1}}\, dx = \ln\left|\sin x + \sqrt{\sin^2 x + 1}\right| + C$

$u = \sin x, \quad u' = \cos x$

47. $\int \frac{x^3}{\sqrt{4 - x^2}}\, dx = \int \frac{8\sin^3\theta(2\cos\theta\, d\theta)}{2\cos\theta} = 8\int (1 - \cos^2\theta)\sin\theta\, d\theta$

$= 8\int [\sin\theta - \cos^2\theta(\sin\theta)]\, d\theta = -8\cos\theta + \frac{8\cos^3\theta}{3} + C$

$= \frac{-\sqrt{4 - x^2}}{3}(x^2 + 8) + C$

$x = 2\sin\theta,\ dx = 2\cos\theta\, d\theta,\ \sqrt{4 - x^2} = 2\cos\theta$

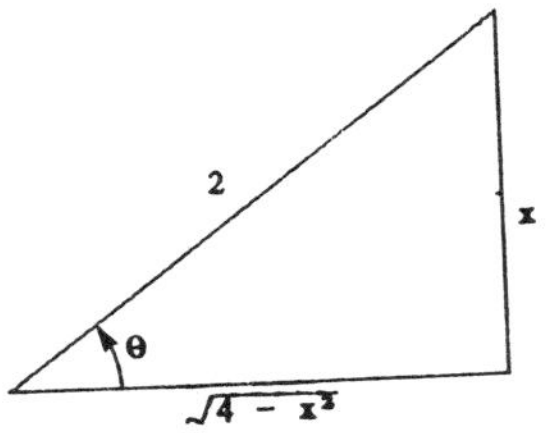

48. $\int \sqrt{\dfrac{3-x}{3+x}}\,dx + \int \dfrac{3-x}{\sqrt{9-x^2}}\,dx = 3\int \dfrac{1}{\sqrt{9-x^2}}\,dx + \int \dfrac{-x}{\sqrt{9-x^2}}\,dx$

$= 3\arcsin\dfrac{x}{3} + \sqrt{9-x^2} + C$

49. By Formula 44: $\int \dfrac{1}{x^{3/2}\sqrt{1-x}}\,dx = 2\int \dfrac{1}{(\sqrt{x})^2\sqrt{1-(\sqrt{x})^2}}\left(\dfrac{1}{2\sqrt{x}}\right)dx$

$= \dfrac{-2\sqrt{1-x}}{\sqrt{x}} + C$

$u = \sqrt{x}, \qquad u' = \dfrac{1}{2\sqrt{x}}$

50. By Formula 26: $\int x\sqrt{x^2+2x}\,dx$

$= \dfrac{1}{2}\int 2(x+1)\sqrt{(x+1)^2-1}\,dx - \int \sqrt{(x+1)^2-1}\,dx$

$= \dfrac{1}{3}(x^2+2x)^{3/2} - \dfrac{1}{2}[(x+1)\sqrt{x^2+2x} - \ln|(x+1)+\sqrt{x^2+2x}|] + C$

$= \dfrac{1}{6}[2(x^2+2x)^{3/2} - 3(x+1)\sqrt{x^2+2x} + 3\ln|x+1+\sqrt{x^2+2x}|] + C$

$u = x+1, \qquad u' = 1$

51. By Formula 8: $\int \dfrac{e^{3x}}{(1+e^x)^3}\,dx = \int \dfrac{(e^x)^2}{(1+e^x)^3}(e^x)\,dx$

$= \dfrac{2}{1+e^x} - \dfrac{1}{2(1+e^x)^2} + \ln|1+e^x| + C$

$u = e^x, \qquad u' = e^x$

52. By Formulas 69 and 61: $\int \sec^5\theta\,d\theta = \dfrac{1}{4}(\sec^3\theta\tan\theta) + \dfrac{3}{4}\int \sec^3\theta\,d\theta$

$= \dfrac{1}{4}(\sec^3\theta\tan\theta) + \dfrac{3}{8}(\sec\theta\tan\theta + \ln|\sec\theta+\tan\theta|) + C$

$= \dfrac{1}{8}[2\sec^3\theta\tan\theta + 3\sec\theta\tan\theta + 3\ln|\sec\theta+\tan\theta|] + C$

53. $\dfrac{u^2}{(a+bu)^2} = \dfrac{1}{b^2} - \dfrac{(2a/b)u + (a^2/b^2)}{(a+bu)^2} = \dfrac{1}{b^2} + \dfrac{A}{a+bu} + \dfrac{B}{(a+bu)^2}$

$-\dfrac{2a}{b}u - \dfrac{a^2}{b^2} = A(a+bu) + B = (aA+B) + bAu$

53. (continued)

Equating the coefficients of like terms we have:

$aA + B = \frac{-a^2}{b^2}$ and $bA = \frac{-2a}{b^2}$. Solving these equations we have:

$A = \frac{-2a}{b^2}$ and $B = \frac{a^2}{b^2}$

$$\int \frac{u^2u'}{(a + bu)^2}\,dx = \frac{1}{b^2}\int u'\,dx - \frac{2a}{b^2}\left(\frac{a}{b}\right)\int \frac{1}{a + bu}(bu')\,dx$$

$$+ \frac{a^2}{b^2}\left(\frac{1}{b}\right)\int \frac{1}{(a + bu)^2}(bu')\,dx = \frac{1}{b^2}u - \frac{2a}{b^3}\ln|a + bu|$$

$$- \frac{a^2}{b^3}\left(\frac{1}{a + bu}\right) + C = \frac{1}{b^3}\left[bu - \frac{a^2}{a + bu} - 2a\ln|a + bu|\right] + C$$

54. $$\int \frac{u^nu'}{\sqrt{a + bu}}\,dx = \frac{2u^n\sqrt{a + bu}}{b} - \frac{2n}{b}\int u^{n-1}\sqrt{a + bu}\;u'\,dx$$

$$= \frac{2u^n\sqrt{a + bu}}{b} - \frac{2na}{b}\int \frac{u^{n-1}u'}{\sqrt{a + bu}}\,dx - 2n\int \frac{u^nu'}{\sqrt{a + bu}}\,dx$$

$w = u^n$, $\quad v' = \frac{u'}{\sqrt{a + bu}}$, $\quad w' = nu^{n-1}u'$, $\quad v = \frac{2}{b}\sqrt{a + bu}$

Therefore $(2n + 1)\int \frac{u^nu'\,dx}{\sqrt{a + bu}} = \frac{2}{b}\left[u^n\sqrt{a + bu} - na\int \frac{u^{n-1}u'}{\sqrt{a + bu}}\,dx\right]$

and $\int \frac{u^nu'\,dx}{\sqrt{a + bu}} = \frac{2}{(2n + 1)b}\left[u^n\sqrt{a + bu} - na\int \frac{u^{n-1}u'}{\sqrt{a + bu}}\,dx\right]$

$$= \frac{2}{(2n + 1)b}\left[u^n\sqrt{a + bu} - na\int \frac{u^{n-1}}{\sqrt{a + bu}}\,du\right]$$

55. When we have $u^2 + a^2$:

$$\int \frac{u'}{(u^2 + a^2)^{3/2}}\,dx = \int \frac{a\sec^2\theta\,d\theta}{a^3\sec^3\theta}$$

$$= \frac{1}{a^2}\int \cos\theta\,d\theta = \frac{1}{a^2}\sin\theta + C = \frac{u}{a^2\sqrt{u^2 + a^2}} + C$$

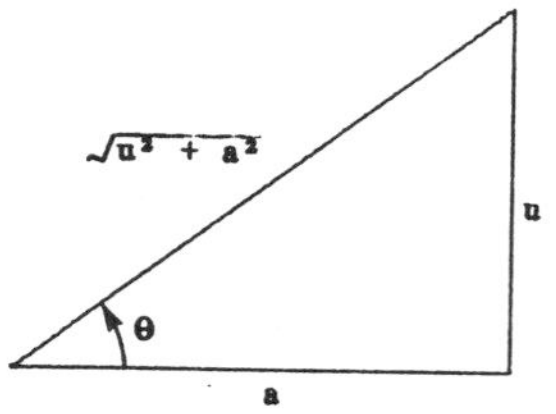

$u = a\tan\theta$, $\quad du = a\sec^2\theta\,d\theta$,

$u^2 + a^2 = a^2\sec^2\theta$

55. (continued)

When we have $u^2 - a^2$:

$$\int \frac{u'}{(u^2 - a^2)^{3/2}}\,dx = \int \frac{a \sec\theta \tan\theta \, d\theta}{a^3 \tan^3\theta}$$

$$= \frac{1}{a^2}\int \frac{\cos\theta}{\sin^2\theta}\,d\theta$$

$$= -\frac{1}{a^2}\csc\theta + C = \frac{-u}{a^2\sqrt{u^2 - a^2}} + C$$

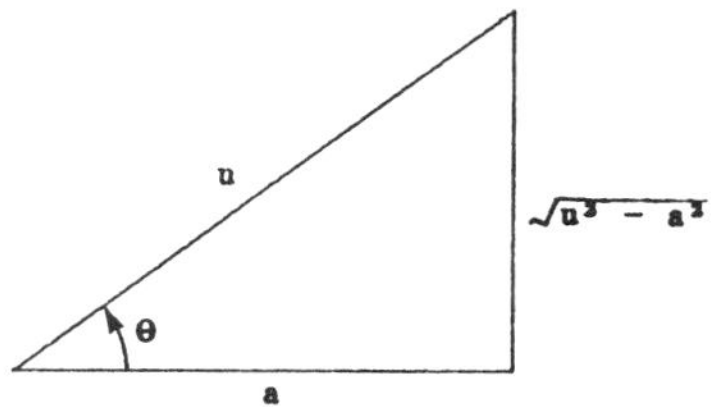

$u = a\sec\theta, \quad du = a\sec\theta\tan\theta\,d\theta,$

$u^2 - a^2 = a^2\tan^2\theta$

56. $\displaystyle\int u^n(\cos u)u'\,dx = u^n \sin u - n\int u^{n-1}(\sin u)u'\,dx$

$$= u^n \sin u - n\int u^{n-1}(\sin u)\,du$$

$w = u^n, \quad v' = \cos u\ u', \quad w' = nu^{n-1}u', \quad v = \sin u$

57. $\displaystyle\int (\arctan u)u'\,dx = u\arctan u - \frac{1}{2}\int \frac{2uu'}{1 + u^2}\,du$

$$= u\arctan u - \frac{1}{2}\ln(1 + u^2) + C = u\arctan u - \ln\sqrt{1 + u^2} + C$$

$w = \arctan u, \quad v' = u', \quad w' = \dfrac{1}{1 + u^2}, \quad v = u$

58. $\displaystyle\int (\ln u)^n\,du = \int (\ln u)^n u'\,dx = u(\ln u)^n - \int n(\ln u)^{n-1}\left(\frac{u'}{u}\right)u\,dx$

$$= u(\ln u)^n - n\int (\ln u)^{n-1}u'dx = u(\ln u)^n - n\int (\ln u)^{n-1}\,du$$

$w = (\ln u)^n, \quad v' = u', \quad w' = n(\ln u)^{n-1}\left(\dfrac{u'}{u}\right), \quad v = u$

59. $\displaystyle\int \frac{1}{2 - 3\sin\theta}\,d\theta = \int \left[\frac{\dfrac{2\,du}{1 + u^2}}{2 - 3\left(\dfrac{2u}{1 + u^2}\right)}\right] = \int \frac{2}{(1 + u^2) - 6u}\,du$

$$= \int \frac{1}{u^2 - 3u + 1}\,du = \int \frac{1}{(u - 3/2)^2 - 5/4}\,du$$

$$= \frac{1}{\sqrt{5}}\ln\left|\frac{(u - \frac{3}{2}) - \frac{\sqrt{5}}{2}}{(u - \frac{3}{2}) + \frac{\sqrt{5}}{2}}\right| + C$$

59. (continued)

$$= \frac{1}{\sqrt{5}} \ln \left| \frac{2u - 3 - \sqrt{5}}{2u - 3 + \sqrt{5}} \right| + C$$

$$= \frac{1}{\sqrt{5}} \ln \left| \frac{2(\tan \theta/2) - 3 - \sqrt{5}}{2(\tan \theta/2) - 3 + \sqrt{5}} \right| + C$$

60. $\displaystyle\int \frac{\sin\theta}{1 + \cos^2\theta}\, d\theta = -\int \frac{-\sin\theta}{1 + (\cos\theta)^2}\, d\theta = -\arctan(\cos\theta) + C$

61. $\displaystyle\int_0^{\pi/2} \frac{1}{1 + \sin\theta + \cos\theta}\, d\theta = \int_0^1 \left[\frac{\frac{2\, du}{1 + u^2}}{1 + \frac{2u}{1 + u^2} + \frac{1 - u^2}{1 + u^2}} \right]$

$$= \int_0^1 \frac{1}{1 + u}\, du = \ln|1 + u| \Big]_0^1 = \ln 2$$

$u = \tan\dfrac{\theta}{2}$

62. $\displaystyle\int_0^{\pi/2} \frac{1}{3 - 2\cos\theta}\, d\theta = \int_0^1 \left[\frac{\frac{2\, du}{1 + u^2}}{3 - \frac{2(1 - u^2)}{1 + u^2}} \right]$

$$= 2 \int_0^1 \frac{1}{5u^2 + 1}\, du = \frac{2}{\sqrt{5}} \arctan(\sqrt{5}\, u) \Big]_0^1 = \frac{2}{\sqrt{5}} \arctan\sqrt{5}$$

$u = \tan\dfrac{\theta}{2}$

63. $\displaystyle\int \frac{\sin\theta}{3 - 2\cos\theta}\, d\theta = \frac{1}{2}\int \frac{1}{u}(u'\, d\theta) = \frac{1}{2}\ln|u| + C = \frac{1}{2}\ln(3 - 2\cos\theta) + C$

$u = 3 - 2\cos\theta, \quad u' = 2\sin\theta$

64. $\displaystyle\int \frac{\sin\theta}{1 + \sin\theta}\, d\theta = \int \frac{\sin\theta - \sin^2\theta}{\cos^2\theta}\, d\theta$

$$= \int (\sec\theta\tan\theta - \sec^2\theta + 1)\, d\theta = \sec\theta - \tan\theta + \theta + C$$

65. $\displaystyle\int \frac{\cos\sqrt{\theta}}{\sqrt{\theta}}\, d\theta = 2\int \cos u(u'\, d\theta) = 2\sin u + C = 2\sin\sqrt{\theta} + C$

$u = \sqrt{\theta}, \quad u' = \dfrac{1}{2\sqrt{\theta}}$

66. $\int \frac{\sin\theta}{\cos\theta(1+\sin\theta)}\,d\theta = \int \frac{\sin\theta - \sin^2\theta}{\cos^3\theta}\,d\theta$

$= \int (\tan\theta\sec^2\theta - \tan^2\theta\sec\theta)\,d\theta = \int (\tan\theta\sec^2\theta - \sec^3\theta + \sec\theta)\,d\theta$

$= \frac{1}{2}\tan^2\theta - \frac{1}{2}\sec\theta\tan\theta + \frac{1}{2}\ln|\sec\theta + \tan\theta| + C$

$= \frac{1}{2}\left[\frac{-\sin\theta}{1+\sin\theta} + \ln\left|\frac{1+\sin\theta}{\cos\theta}\right|\right] + C$

67. $\int \frac{1}{\sin\theta\tan\theta}\,d\theta = \int \frac{\cos\theta}{\sin^2\theta}\,d\theta = -\frac{1}{\sin\theta} + C = -\csc\theta + C$

$u = \sin\theta$

68. $\int \frac{1}{\sec\theta - \tan\theta}\,d\theta = \int \frac{1}{(1/\cos\theta) - (\sin\theta/\cos\theta)}\,d\theta$

$= -\int \frac{-\cos\theta}{1-\sin\theta}\,d\theta = -\ln|1-\sin\theta| + C$

$u = 1 - \sin\theta, \quad u' = -\cos\theta$

69. $W = \int_0^5 2000xe^{-x}\,dx = -2000\int_0^5 -xe^{-x}\,dx$

$= 2000\int_0^5 (-x)e^{-x}(-1)\,dx = 2000\Big[(-x)e^{-x} - e^{-x}\Big]_0^5$

$= 2000(-\frac{6}{e^5} + 1) \approx 1919.145$ ft·lbs

70. $W = \int_0^5 \frac{500x}{\sqrt{26-x^2}}\,dx = -250\int_0^5 (26-x^2)^{-1/2}(-2x)\,dx$

$= -500\sqrt{26-x^2}\Big]_0^5 = 500(\sqrt{26} - 1) \approx 2049.51$ ft·lbs

8.7 Numerical integration

1. **Exact:** $\int_0^2 x^2\,dx = \frac{1}{3}x^3\Big]_0^2 = \frac{8}{3} \approx 2.6667$

Trap.: $\int_0^2 x^2\,dx \approx \frac{1}{4}\left[0 + 2(\frac{1}{2})^2 + 2(1)^2 + 2(\frac{3}{2})^2 + (2)^2\right] = \frac{11}{4} \approx 2.7500$

Simp.: $\int_0^2 x^2\,dx \approx \frac{1}{6}\left[0 + 4(\frac{1}{2})^2 + 2(1)^2 + 4(\frac{3}{2})^2 + (2)^2\right] = \frac{8}{3} \approx 2.6667$

2. **Exact:** $\int_0^1 (\frac{x^2}{2} + 1)\,dx = \left[\frac{x^3}{6} + x\right]_0^1 = \frac{7}{6} = 1.1667$

Trap.: $\int_0^1 (\frac{x^2}{2} + 1)\,dx \approx \frac{1}{8}\left[1 + 2(\frac{(1/4)^2}{2} + 1) + 2(\frac{(1/2)^2}{2} + 1)\right.$

$\left. + 2(\frac{(3/4)^2}{2} + 1) + (\frac{1^2}{2} + 1)\right] = \frac{75}{64} \approx 1.1719$

Simp.: $\int_0^1 (\frac{x^2}{2} + 1)\,dx \approx \frac{1}{12}\left[1 + 4(\frac{(1/4)^2}{2} + 1) + 2(\frac{(1/2)^2}{2} + 1)\right.$

$\left. + 4(\frac{(3/4)^2}{2} + 1) + (\frac{1^2}{2} + 1)\right] = \frac{7}{6} \approx 1.1667$

3. **Exact:** $\int_0^2 x^3\,dx = \frac{x^4}{4}\Big]_0^2 = 4.000$

Trap.: $\int_0^2 x^3\,dx \approx \frac{1}{4}\left[0 + 2(\frac{1}{2})^3 + 2(1)^3 + 2(\frac{3}{2})^3 + (2)^3\right] = \frac{17}{4} = 4.2500$

Simp.: $\int_0^2 x^3\,dx \approx \frac{1}{6}\left[0 + 4(\frac{1}{2})^3 + 2(1)^3 + 4(\frac{3}{2})^3 + (2)^3\right] = \frac{24}{6} = 4.0000$

4. **Exact:** $\int_1^2 \frac{1}{x}\,dx = \ln x\Big]_1^2 \approx 0.6931$

Trap.: $\int_1^2 \frac{1}{x}\,dx \approx \frac{1}{8}\left[1 + 2(\frac{4}{5}) + 2(\frac{4}{6}) + 2(\frac{4}{7}) + \frac{1}{2}\right] \approx 0.6970$

Simp.: $\int_1^2 \frac{1}{x}\,dx \approx \frac{1}{12}\left[1 + 4(\frac{4}{5}) + 2(\frac{4}{6}) + 4(\frac{4}{7}) + \frac{1}{2}\right] \approx 0.6933$

5. **Exact:** $\displaystyle\int_0^2 x^3\,dx = \frac{1}{4}x^4\Big]_0^2 = 4.000$

Trap.: $\displaystyle\int_0^2 x^3\,dx \approx \frac{1}{8}\Big[0 + 2(\frac{1}{4})^3 + 2(\frac{2}{4})^3 + 2(\frac{3}{4})^3 + 2(1)^3 + 2(\frac{5}{4})^3 + 2(\frac{6}{4})^3$

$\displaystyle + 2(\frac{7}{4})^3 + 8\Big] = 4.0625$

Simp.: $\displaystyle\int_0^2 x^3\,dx \approx \frac{1}{12}\Big[0 + 4(\frac{1}{4})^3 + 2(\frac{2}{4})^3 + 4(\frac{3}{4})^3 + 2(1)^3 + 4(\frac{5}{4})^3 + 2(\frac{6}{4})^3$

$\displaystyle + 4(\frac{7}{4})^3 + 8\Big] = 4.000$

6. **Exact:** $\displaystyle\int_1^2 \frac{1}{x}\,dx = \ln x\Big]_1^2 \approx 0.6931$

Trap.: $\displaystyle\int_1^2 \frac{1}{x}\,dx \approx \frac{1}{16}\Big[1 + 2(\frac{8}{9}) + 2(\frac{8}{10}) + 2(\frac{8}{11}) + 2(\frac{8}{12}) + 2(\frac{8}{13}) + 2(\frac{8}{14})$

$\displaystyle + 2(\frac{8}{15}) + \frac{1}{2}\Big] \approx 0.6941$

Simp.: $\displaystyle\int_1^2 \frac{1}{x}\,dx \approx \frac{1}{24}\Big[1 + 4(\frac{8}{9}) + 2(\frac{8}{10}) + 4(\frac{8}{11}) + 2(\frac{8}{12}) + 4(\frac{8}{13}) + 2(\frac{8}{14})$

$\displaystyle + 4(\frac{8}{15}) + \frac{1}{2}\Big] \approx 0.6932$

7. **Exact:** $\displaystyle\int_1^2 \frac{1}{x^2}\,dx = \frac{-1}{x}\Big]_1^2 = 0.5000$

Trap.: $\displaystyle\int_1^2 \frac{1}{x^2}\,dx \approx \frac{1}{8}\Big[1 + 2(\frac{4}{5})^2 + 2(\frac{4}{6})^2 + 2(\frac{4}{7})^2 + \frac{1}{4}\Big] \approx 0.5090$

Simp.: $\displaystyle\int_1^2 \frac{1}{x^2}\,dx \approx \frac{1}{12}\Big[1 + 4(\frac{4}{5})^2 + 2(\frac{4}{6})^2 + 4(\frac{4}{7})^2 + \frac{1}{4}\Big] \approx 0.5004$

8. **Exact:** $\displaystyle\int_0^4 \sqrt{x}\,dx = \frac{2}{3}x^{3/2}\Big]_0^4 = \frac{16}{3} \approx 5.3333$

Trap.: $\displaystyle\int_0^4 \sqrt{x}\,dx \approx \frac{1}{4}\Big[0 + 2\sqrt{1/2} + 2\sqrt{1} + 2\sqrt{3/2} + 2\sqrt{2} + 2\sqrt{5/2} + 2\sqrt{3}$

$\displaystyle + 2\sqrt{7/2} + \sqrt{4}\Big] \approx 5.2650$

8. (continued)

Simp.: $\int_0^4 \sqrt{x}\,dx \approx \frac{1}{6}\Big[0 + 4\sqrt{1/2} + 2\sqrt{1} + 4\sqrt{3/2} + 2\sqrt{2} + 4\sqrt{5/2} + 2\sqrt{3}$

$+ 4\sqrt{7/2} + 2\Big] \approx 5.3046$

9. Exact: $\int_0^1 \frac{1}{1+x^2}\,dx = \arctan x\Big]_0^1 = \frac{\pi}{4} \approx 0.7854$

Trap.: $\int_0^1 \frac{1}{1+x^2}\,dx \approx \frac{1}{8}\Big[1 + \frac{2}{1+(1/4)^2} + \frac{2}{1+(1/2)^2} + \frac{2}{1+(3/4)^2}$

$+ \frac{1}{2}\Big] \approx 0.7828$

Simp.: $\int_0^1 \frac{1}{1+x^2}\,dx \approx \frac{1}{12}\Big[1 + \frac{4}{1+(1/4)^2} + \frac{2}{1+(1/2)^2} + \frac{4}{1+(3/4)^2}$

$+ \frac{1}{2}\Big] \approx 0.7854$

10. Exact: $\int_0^2 x\sqrt{x^2+1}\,dx = \frac{1}{3}(x^2+1)^{3/2}\Big]_0^2 = \frac{1}{3}(5^{3/2} - 1) \approx 3.3934$

Trap.: $\int_0^2 x\sqrt{x^2+1}\,dx \approx \frac{1}{4}\Big[0 + 2(\frac{1}{2})\sqrt{(1/2)^2+1} + 2(1)\sqrt{1^2+1}$

$+ 2(\frac{3}{2})\sqrt{(3/2)^2+1} + 2\sqrt{2^2+1}\Big] \approx 3.4567$

Simp.: $\int_0^2 x\sqrt{x^2+1}\,dx \approx \frac{1}{6}\Big[0 + 4(\frac{1}{2})\sqrt{(1/2)^2+1} + 2(1)\sqrt{1^2+1}$

$+ 4(\frac{3}{2})\sqrt{(3/2)^2+1} + 2\sqrt{2^2+1}\Big] \approx 3.3922$

11. Trap.: $\int_0^{\sqrt{\pi/2}} \cos(x^2)\,dx \approx \frac{\sqrt{\pi/2}}{8}\Big[\cos 0 + 2\cos(\frac{\sqrt{\pi/2}}{4})^2 + 2\cos(\frac{\sqrt{\pi/2}}{2})^2$

$+ 2\cos(\frac{3\sqrt{\pi/2}}{4})^2 + \cos(\sqrt{\pi/2})^2\Big] \approx 0.957$

Simp.: $\int_0^{\sqrt{\pi/2}} \cos(x^2)\,dx \approx \frac{\sqrt{\pi/2}}{12}\Big[\cos 0 + 4\cos(\frac{\sqrt{\pi/2}}{4})^2 + 2\cos(\frac{\sqrt{\pi/2}}{2})^2$

$+ 4\cos(\frac{3\sqrt{\pi/2}}{4})^2 + \cos(\sqrt{\pi/2})^2\Big] \approx 0.978$

12. Trap.: $\int_0^{\sqrt{\pi/4}} \tan(x^2)\,dx \approx \frac{\sqrt{\pi/4}}{8}\left[\tan 0 + 2\tan\left(\frac{\sqrt{\pi/4}}{4}\right)^2 + 2\tan\left(\frac{\sqrt{\pi/4}}{2}\right)^2\right.$

$\left. + 2\tan\left(\frac{3\sqrt{\pi/4}}{4}\right)^2 + \tan(\sqrt{\pi/4})^2\right] \approx 0.2705$

Simp.: $\int_0^{\sqrt{\pi/4}} \tan(x^2)\,dx \approx \frac{\sqrt{\pi/4}}{12}\left[\tan 0 + 4\tan\left(\frac{\sqrt{\pi/4}}{4}\right)^2 + 2\tan\left(\frac{\sqrt{\pi/4}}{2}\right)^2\right.$

$\left. + 4\tan\left(\frac{3\sqrt{\pi/4}}{4}\right)^2 + \tan(\sqrt{\pi/4})^2\right] \approx 0.2575$

13. Trap.: $\int_0^2 \sqrt{1 + x^3}\,dx \approx \frac{1}{2}(1 + 2\sqrt{2} + 3) = 2 + \sqrt{2} \approx 3.41$

Simp.: $\int_0^2 \sqrt{1 + x^3}\,dx \approx \frac{1}{3}(1 + 4\sqrt{2} + 3) = \frac{4}{3}(1 + \sqrt{2}) \approx 3.22$

14. Trap.: $\int_0^2 \frac{1}{\sqrt{1 + x^3}}\,dx \approx \frac{1}{4}\left[1 + 2\left(\frac{1}{\sqrt{1 + (1/2)^3}}\right) + 2\left(\frac{1}{\sqrt{1 + 1^3}}\right)\right.$

$\left. + 2\left(\frac{1}{\sqrt{1 + (3/2)^3}}\right) + \frac{1}{3}\right] \approx 1.3973$

Simp.: $\int_0^2 \frac{1}{\sqrt{1 + x^3}}\,dx \approx \frac{1}{6}\left[1 + 4\left(\frac{1}{\sqrt{1 + (1/2)^3}}\right) + 2\left(\frac{1}{\sqrt{1 + 1^3}}\right)\right.$

$\left. + 4\left(\frac{1}{\sqrt{1 + (3/2)^3}}\right) + \frac{1}{3}\right] \approx 1.4052$

15. $\int_0^1 \sqrt{x}\,\sqrt{1 - x}\,dx = \int_0^1 \sqrt{x(1 - x)}\,dx$

Trap.: $\int_0^1 \sqrt{x(1 - x)}\,dx \approx \frac{1}{8}\left[0 + 2\sqrt{\frac{1}{4}\left(1 - \frac{1}{4}\right)} + 2\sqrt{\frac{1}{2}\left(1 - \frac{1}{2}\right)}\right.$

$\left. + 2\sqrt{\frac{3}{4}\left(1 - \frac{3}{4}\right)}\right] \approx 0.342$

Simp.: $\int_0^1 \sqrt{x(1 - x)}\,dx \approx \frac{1}{12}\left[0 + 4\sqrt{\frac{1}{4}\left(1 - \frac{1}{4}\right)} + 2\sqrt{\frac{1}{2}\left(1 - \frac{1}{2}\right)}\right.$

$\left. + 4\sqrt{\frac{3}{4}\left(1 - \frac{3}{4}\right)}\right] \approx 0.372$

16. Trap.: $\int_0^\pi \frac{\sin x}{x}\,dx \approx \frac{\pi}{8}\left[1 + \frac{2\sin(\pi/4)}{\pi/4} + \frac{2\sin(\pi/2)}{\pi/2} + \frac{2\sin(3\pi/4)}{3\pi/4} + 0\right]$

≈ 1.8355

16. (continued)

Simp.: $\int_0^{\pi} \frac{\sin x}{x}\,dx \approx \frac{\pi}{12}\left[1 + \frac{4\sin(\pi/4)}{\pi/4} + \frac{2\sin(\pi/2)}{\pi/2} + \frac{4\sin(3\pi/4)}{3\pi/4} + 0\right]$

≈ 1.8522

17. Trap.: $\int_0^1 \sin(x^2)\,dx \approx \frac{1}{4}\left[0 + 2\sin\frac{1}{4} + \sin 1\right] \approx 0.334$

Simp.: $\int_0^1 \sin(x^2)\,dx \approx \frac{1}{6}\left[0 + 4\sin\frac{1}{4} + \sin 1\right] \approx 0.305$

18. Trap.: $\int_0^{\pi} \sqrt{x}\sin x\,dx \approx \frac{\pi}{8}\left[0 + 2\sqrt{\pi/4}\sin(\frac{\pi}{4}) + 2\sqrt{\pi/2}\sin(\frac{\pi}{2})\right.$

$\left. + 2\sqrt{3\pi/4}\sin(\frac{3\pi}{4}) + 0\right] \approx 2.3290$

Simp.: $\int_0^{\pi} \sqrt{x}\sin x\,dx \approx \frac{\pi}{12}\left[0 + 4\sqrt{\pi/4}\sin(\frac{\pi}{4}) + 2\sqrt{\pi/2}\sin(\frac{\pi}{2})\right.$

$\left. + 4\sqrt{3\pi/4}\sin(\frac{3\pi}{4}) + 0\right] \approx 2.4491$

19. Trap.: $\int_0^{\pi/4} x\tan x\,dx \approx \frac{\pi}{32}\left[0 + 2(\frac{\pi}{16})\tan(\frac{\pi}{16}) + 2(\frac{2\pi}{16})\tan(\frac{2\pi}{16})\right.$

$\left. + 2(\frac{3\pi}{16})\tan(\frac{3\pi}{16}) + \frac{\pi}{4}\right] \approx 0.194$

Simp.: $\int_0^{\pi/4} x\tan x\,dx \approx \frac{\pi}{48}\left[0 + 4(\frac{\pi}{16})\tan(\frac{\pi}{16}) + 2(\frac{2\pi}{16})\tan(\frac{2\pi}{16})\right.$

$\left. + 4(\frac{3\pi}{16})\tan(\frac{3\pi}{16}) + \frac{\pi}{4}\right] \approx 0.186$

20. Trap.: $\int_0^{\pi/2} \sqrt{1 + \cos^2 x}\,dx \approx \frac{\pi}{8}\left[\sqrt{2} + 2\sqrt{1 + \cos^2(\pi/4)} + 1\right] \approx 1.9100$

Simp.: $\int_0^{\pi/2} \sqrt{1 + \cos^2 x}\,dx \approx \frac{\pi}{12}\left[\sqrt{2} + 4\sqrt{1 + \cos^2(\pi/4)} + 1\right] \approx 1.9146$

21. $f(x) = x^3$, $f'(x) = 3x^2$, $f''(x) = 6x$, $f'''(x) = 6$, $f^{(4)}(x) = 0$

Trap.: Error $\leq (\frac{(2 - 0)^3}{12(4^2)})(12) = 0.5$

(since $f''(x)$ is maximum in $[0, 2]$ when $x = 2$)

Simp.: Error $\leq (\frac{(2 - 0)^5}{180(4^4)})(0) = 0$ (since $f^{(4)}(x) = 0$)

22. $f(x) = \dfrac{1}{x+1}$, $f'(x) = \dfrac{-1}{(x+1)^2}$, $f''(x) = \dfrac{2}{(x+1)^3}$

$f'''(x) = \dfrac{-6}{(x+1)^4}$, $f^{(4)}(x) = \dfrac{24}{(x+1)^5}$

Trap.: Error $\leq \left(\dfrac{(1-0)^3}{12(4^2)}\right)(2) = \dfrac{1}{96} \approx 0.01$

(since $f''(x)$ is maximum in [0, 1] when $x = 0$)

Simp.: Error $\leq \left(\dfrac{(1-0)^5}{180(4^4)}\right)(24) = \dfrac{1}{1920} \approx 0.0005$

(since $f^{(4)}(x)$ is maximum in [0, 1] when $x = 0$)

23. $f(x) = e^{x^3}$

$f''(x) = 3(3x^4 + 2x)e^{x^3}$ which in [0, 1] is maximum when

$x = 1$ and $f''(1) = 15e$

$f^{(4)}(x) = 3(27x^8 + 128x^5 + 60x^2)e^{x^3}$ which in [0, 1] is maximum when

$x = 1$ and $f^{(4)}(1) = 585e$

Trap.: Error $\leq \left(\dfrac{(1-0)^3}{12(4^2)}\right)(15e) = \dfrac{5e}{64} \approx 0.212$

Simp.: Error $\leq \left(\dfrac{(1-0)^5}{180(4^4)}\right)(585e) = \dfrac{13e}{1024} \approx 0.035$

24. $f(x) = e^{-x^2}$

$f''(x) = -2(-2x^2 + 1)e^{-x^2}$ and in [0, 1], $|f''(x)|$ is maximum when

$x = 0$ and $|f''(0)| = 2$

$f^{(4)}(x) = -2(-8x^4 + 24x^2 - 6)e^{-x^2}$ and in [0, 1], $|f^{(4)}(x)|$ is maximum

when $x = 0$ and $|f^{(4)}(0)| = 12$

Trap.: Error $\leq \left(\dfrac{(1-0)^3}{12(4^2)}\right)(2) = \dfrac{1}{96} \approx 0.01$

Simp.: Error $\leq \left(\dfrac{(1-0)^5}{180(4^4)}\right)(12) = \dfrac{1}{3840} \approx 0.00026$

25. $f(x) = \tan(x^2)$

$f''(x) = 2\sec^2(x^2)[1 + 4x^2\tan(x^2)]$ and in [0, 1], $|f''(x)|$ is maximum

when $x = 1$ and $|f''(1)| \approx 49.5305$

25. (continued)

$f^{(4)}(x) = 8 \sec^2 (x^2)[12x^2 + (3 + 32x^4) \tan (x^2) + 36x^2 \tan^2 (x^2)$

$+ 48x^4 \tan^3 (x^2)]$ and in [0, 1], $|f^{(4)}(x)|$ is maximum when

$x = 1$ and $|f^{(4)}(1)| \approx 9184.4734$

Trap.: Error $\leq \dfrac{(1 - 0)^3}{12(4)^2}(49.5305) \approx 0.258$

Simp.: Error $\leq \dfrac{(1 - 0)^5}{180(4)^4}(9184.4734) \approx 0.199$

26. $f(x) = \sin (x^2)$

$f''(x) = 2[-2x^2 \sin (x^2) + \cos (x^2)]$ and in [0, 1], $|f''(x)|$ is maximum

when $x = 1$ and $|f''(1)| = 2.2853$

$f^{(4)}(x) = (16x^4 - 4) \sin (x^2) - 32x^2 \cos (x^2)$ and in [0, 1], $|f^{(4)}(x)|$ is

maximum when $x = 0.795$ and $|f^{(4)}(0.795)| = 14.9$

Trap.: Error $\leq \dfrac{(1 - 0)^3}{12(2^2)}(2.2853) = 0.05$

Simp.: Error $\leq \dfrac{(1 - 0)^5}{180(2^4)}(15) = 0.005$

27. $f''(x) = -2(-2x^2 + 1)e^{-x^2}$ and in [0, 1], $|f''(x)|$ is maximum when

$x = 0$ and $|f''(0)| = 2$

Error $\leq \dfrac{2}{12n^2} < 0.00001$, $n^2 > 16666.67$, $n > 129.1$, let $n = 130$ (Trap.)

$f^{(4)}(x) = -2(-8x^4 + 24x^2 - 6)e^{-x^2}$ and in [0, 1], $|f^{(4)}(x)|$ is maximum when

$x = 0$ and $|f^{(4)}(0)| = 12$

Error $\leq (\dfrac{12}{180n^4}) < 0.00001$, $n^4 > 6666.67$, $n > 9.04$, let $n = 10$ (Simp.)

28. $f''(x) = -2(-2x^2 + 1)e^{-x^2}$ and in [0, 2], $|f''(x)|$ is maximum when

$x = 0$ and $|f''(0)| = 2$

Error $\leq \dfrac{2^3}{12n^2}(2) < 0.00001$, $n^2 > 133333.33$, $n > 365.15$,

let $n = 366$

28. (continued)

$f^{(4)}(x) = -2(-8x^4 + 24x^2 - 6)e^{-x^2}$ and in [0, 2], $|f^{(4)}(x)|$ is maximum when

$x = 0$ and $|f^{(4)}(0)| = 12$

Error $\leq \dfrac{2^5}{180n^4}(12) < 0.00001, \quad n^4 > 213333.33, \quad n > 21.49$

let $n = 22$

29. $f''(x) = \dfrac{2}{x^3}$ and in [1, 3], $|f''(x)|$ is maximum when

$x = 1$ and $|f''(1)| = 2$

Trap.: Error $\leq \dfrac{2^3}{12n^2}(2) < 0.00001, \quad n^2 > 133333.33, \quad n > 365.15,$

let $n = 366$

$f^{(4)}(x) = \dfrac{24}{x^5}$ and in [1, 3], $|f^{(4)}(x)|$ is maximum when

$x = 1$ and $|f^{(4)}(1)| = 24$

Simp.: Error $\leq \dfrac{2^5}{180n^4}(24) < 0.00001, \quad n^4 > 426666.67, \quad n > 25.55$

let $n = 26$

30. $f''(x) = \dfrac{e^x(x^2 - 2x + 2)}{x^3}$ and in [1, 2], $|f''(x)|$ is maximum when

$x = 1$ and $|f''(1)| \approx 2.7183$

Error $\leq \dfrac{1}{12n^2}(2.7183) < 0.00001, \quad n^2 > 22652.34, \quad n > 150.5$

let $n = 151$

$f^{(4)}(x) = \dfrac{e^x(x^4 - 4x^3 + 12x^2 - 24x + 24)}{x^5}$ and in [1, 2], $|f^{(4)}(x)|$ is

maximum when $x = 1$ and $|f^{(4)}(1)| \approx 24.4645$

Error $\leq \dfrac{1}{180n^4}(24.4645) < 0.00001, \quad n^4 > 13591.39, \quad n > 10.797$

let $n = 12$ (for Simpson's Rule n must be even)

31. Simp.: $(n = 6)$

$$\pi = 4\int_0^1 \frac{1}{1 + x^2}\,dx = \frac{4}{3(6)}\left[1 + \frac{4}{1 + (1/6)^2} + \frac{2}{1 + (2/6)^2} + \frac{4}{1 + (3/6)^2} + \frac{2}{1 + (4/6)^2} + \frac{4}{1 + (5/6)^2} + \frac{1}{2}\right] = 3.14159$$

32. Simp.: (n = 6)

(a) $P_r\ (0 \le z \le 1) = \dfrac{1}{\sqrt{2\pi}} \displaystyle\int_0^1 e^{-z^2/2}\, dx$

$$= \frac{1}{\sqrt{2\pi}}\left(\frac{1}{18}\right)\left[1 + 4e^{-(1/6)^2/2} + 2e^{-(2/6)^2/2} + 4e^{-(3/6)^2/2} + 2e^{-(4/6)^2/2} + 4e^{-(5/6)^2/2} + e^{-1/2}\right] = 0.3414$$

(b) $P(0 \le z \le 2) = \dfrac{1}{\sqrt{2\pi}}\left(\dfrac{1}{9}\right)\left[1 + 4e^{-(1/3)^2/2} + 2e^{-(2/3)^2/2} + 4e^{-(3/3)^2/2} + 2e^{-(4/3)^2/2} + 4e^{-(5/3)^2/2} + e^{-2}\right] = 0.4772$

33. $A = \displaystyle\int_0^{\pi/2} \sqrt{x}\cos x\, dx$

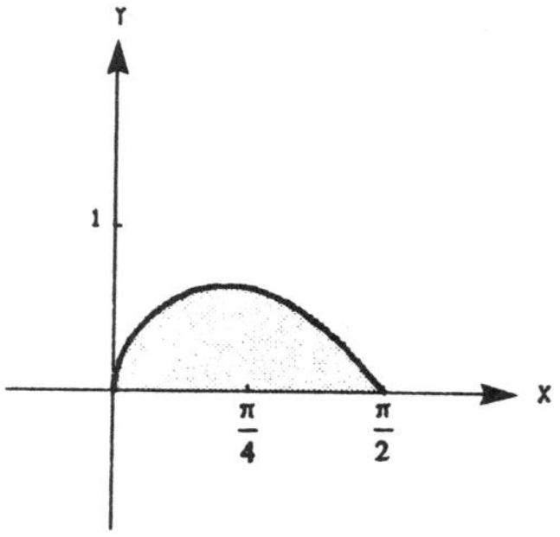

Simp.: n = 14

$$\int_0^{\pi/2} \sqrt{x}\cos x\, dx \approx \frac{\pi}{84}\left[\sqrt{0}\cos 0 + 4\sqrt{\pi/28}\cos\frac{\pi}{28} + 2\sqrt{\pi/14}\cos\frac{\pi}{14} + 4\sqrt{3\pi/28}\cos\frac{3\pi}{28} + \cdots + \sqrt{\pi/2}\cos\frac{\pi}{2}\right] \approx 0.70$$

34. $y = \cos x, \qquad y' = -\sin x, \qquad 1 + (y')^2 = 1 + \sin^2 x$

$$s = \int_0^{\pi/2} \sqrt{1 + \sin^2 x}\, dx$$

Simp.: n = 4

$$s \approx \frac{\pi}{24}\left[\sqrt{1 + \sin^2 0} + 4\sqrt{1 + \sin^2\frac{\pi}{8}} + 2\sqrt{1 + \sin^2\frac{\pi}{4}} + 4\sqrt{1 + \sin^2\frac{3\pi}{8}} + \sqrt{1 + \sin^2\frac{\pi}{2}}\right] \approx 1.910$$

35. Simp.: n = 8

$$8\sqrt{3}\int_0^{\pi/2} \sqrt{1 - \frac{2}{3}\sin^2\theta}\, d\theta \approx \frac{\sqrt{3}\pi}{6}\left[\sqrt{1 - \frac{2}{3}\sin^2 0} + 4\sqrt{1 - \frac{2}{3}\sin^2\frac{\pi}{16}} + 2\sqrt{1 - \frac{2}{3}\sin^2\frac{\pi}{8}} + \cdots + \sqrt{1 - \frac{2}{3}\sin^2\frac{\pi}{2}}\right] \approx 17.476$$

36. Simp.: $|\text{Error}| = \frac{(b-a)^5}{180n^4} f^{(4)}(c)$ for $a \le c \le b$

If $f(x)$ is third degree then $f^{(4)}(x) = 0$. Therefore Error = 0.

$\int_0^1 x^3\,dx = \frac{x^4}{4}\Big]_0^1 = \frac{1}{4}$, By Simpson's Rule: $\frac{1}{6}\left[0 + 4(\frac{1}{2})^3 + 1\right] = \frac{1}{6}(\frac{3}{2}) = \frac{1}{4}$

37. Area $\approx \frac{1000}{3(10)}[125 + 4(125) + 2(120) + 4(112) + 2(90) + 4(90)$
$+ 2(95) + 4(88) + 2(75) + 4(35)] = 89500$ sq. ft.

38. Area $\approx \frac{120}{3(12)}[75 + 4(81) + 2(84) + 4(76) + 2(67) + 4(68) + 2(69)$
$+ 4(72) + 2(68) + 4(56) + 2(42) + 4(23) + 0] \approx 7463.33$ sq. ft.

8.8
Indeterminate forms and L'Hôpital's Rule

1. $\lim_{x\to 2} \frac{x^2 - x - 2}{x - 2} = \lim_{x\to 2} \frac{(x-2)(x+1)}{x-2} = \lim_{x\to 2} (x+1) = 3$

2. $\lim_{x\to -1} \frac{x^2 - x - 2}{x + 1} = \lim_{x\to -1} \frac{(x-2)(x+1)}{x+1} = \lim_{x\to -1} (x-2) = -3$

3. $\lim_{x\to 0} \frac{\sqrt{4 - x^2} - 2}{x} = \lim_{x\to 0} \frac{-x^2}{x(\sqrt{4-x^2} + 2)} = \lim_{x\to 0} \frac{-x}{\sqrt{4-x^2} + 2} = 0$

4. $\lim_{x\to 2^-} \frac{\sqrt{4 - x^2}}{x - 2} = \lim_{x\to 2^-} \frac{-x/\sqrt{4-x^2}}{1} = \lim_{x\to 2^-} \frac{-x}{\sqrt{4-x^2}} = -\infty$

5. $\lim_{x\to 0} \frac{e^x - (1 - x)}{x} = \lim_{x\to 0} \frac{e^x + 1}{1} = 2$

6. $\lim_{x\to 0^+} \frac{e^x - (1 + x)}{x^3} = \lim_{x\to 0^+} \frac{e^x - 1}{3x^2} = \lim_{x\to 0^+} \frac{e^x}{6x} = \infty$

7. Case 1: $n = 1$

$\lim_{x\to 0^+} \frac{e^x - (1 + x)}{x} = \lim_{x\to 0^+} \frac{e^x - 1}{1} = 0$

7. (continued)

Case 2: n = 2

$$\lim_{x\to 0^+} \frac{e^x - (1 + x)}{x^2} = \lim_{x\to 0^+} \frac{e^x - 1}{2x} = \lim_{x\to 0^+} \frac{e^x}{2} = \frac{1}{2}$$

Case 3: $n \geq 3$

$$\lim_{x\to 0^+} \frac{e^x - (1 + x)}{x^n} = \lim_{x\to 0^+} \frac{e^x - 1}{nx^{n-1}} = \lim_{x\to 0^+} \frac{e^x}{n(n - 1)x^{n-2}} = \frac{1}{0} = \infty$$

8. $$\lim_{x\to 1} \frac{\ln x}{x^2 - 1} = \lim_{x\to 1} \frac{(1/x)}{2x} = \lim_{x\to 1} \frac{1}{2x^2} = \frac{1}{2}$$

9. $$\lim_{x\to \infty} \frac{\ln x}{x} = \lim_{x\to \infty} \frac{(1/x)}{1} = 0$$

10. $$\lim_{x\to \infty} \frac{e^x}{x} = \lim_{x\to \infty} \frac{e^x}{1} = \infty$$

11. $$\lim_{x\to \infty} \frac{3x^2 - 2x + 1}{2x^2 + 3} = \lim_{x\to \infty} \frac{3 - (2/x) + (1/x^2)}{2 + (3/x^2)} = \frac{3}{2}$$

12. $$\lim_{x\to \infty} \frac{x - 1}{x^2 + 2x + 3} = \lim_{x\to \infty} \frac{(1/x) - (1/x^2)}{1 + (2/x) + (3/x^2)} = 0$$

13. $$\lim_{x\to \infty} \frac{x^2 + 2x + 3}{x - 1} = \lim_{x\to \infty} \frac{1 + (2/x) + (3/x^2)}{(1/x) - (1/x^2)} = \infty$$

14. $$\lim_{x\to \infty} \frac{x^2}{e^x} = \lim_{x\to \infty} \frac{2x}{e^x} = \lim_{x\to \infty} \frac{2}{e^x} = 0$$

15. $$\lim_{x\to 0^+} x^2 \ln x = \lim_{x\to 0^+} \frac{\ln x}{(1/x^2)} = \lim_{x\to 0^+} \frac{(1/x)}{(-2/x^3)} = \lim_{x\to 0^+} \frac{-x^2}{2} = 0$$

16. $$\lim_{x\to 0} \left[\frac{1}{x} - \frac{1}{x^2}\right] = \lim_{x\to 0} \frac{x - 1}{x^2} = -\infty$$

17. $$\lim_{x\to 2} \left[\frac{8}{x^2 - 4} - \frac{x}{x - 2}\right] = \lim_{x\to 2} \frac{8 - x(x + 2)}{x^2 - 4} = \lim_{x\to 2} \frac{(2 - x)(4 + x)}{(x + 2)(x - 2)}$$

$$= \lim_{x\to 2} \frac{-(x + 4)}{x + 2} = \frac{-3}{2}$$

18. $\lim_{x \to 2} \left[\dfrac{1}{x^2 - 4} - \dfrac{\sqrt{x - 1}}{x^2 - 4}\right] = \lim_{x \to 2} \dfrac{1 - \sqrt{x - 1}}{x^2 - 4} = \lim_{x \to 2} \dfrac{-1/(2\sqrt{x - 1})}{2x}$

$= \lim_{x \to 2} \dfrac{-1}{4x\sqrt{x - 1}} = \dfrac{-1}{8}$

19. $\lim_{x \to \infty} \dfrac{x}{\sqrt{x^2 - 1}} = \lim_{x \to \infty} \dfrac{1}{\sqrt{1 - (1/x^2)}} = 1$

20. $\lim_{x \to 1^+} \left[\dfrac{3}{\ln x} - \dfrac{2}{x - 1}\right] = \lim_{x \to 1^+} \dfrac{3x - 3 - 2 \ln x}{(x - 1) \ln x}$

$= \lim_{x \to 1^+} \dfrac{3 - (2/x)}{[(x - 1)/x] + \ln x} = \infty$

21. $\lim_{x \to 0^+} x^{1/x} = 0^\infty = 0$

22. $\lim_{x \to 0} (e^x + x)^{1/x}$, $\quad y = (e^x + x)^{1/x}$, $\quad \ln y = \dfrac{1}{x} \ln (e^x + x)$

$\lim_{x \to 0} \dfrac{1}{x} \ln (e^x + x) = \lim_{x \to 0} \dfrac{\ln (e^x + x)}{x} = \lim_{x \to 0} \dfrac{e^x + 1}{e^x + x} = 2$

As $x \to 0$ $\ln y = 2$ and therefore $y = e^2$, $\lim_{x \to 0} (e^x + x)^{1/x} = e^2$

23. $\lim_{x \to \infty} x^{1/x}$, $\quad y = x^{1/x}$, $\quad \ln y = \dfrac{1}{x} \ln x \qquad \lim_{x \to \infty} \dfrac{\ln x}{x} = \lim_{x \to \infty} \dfrac{(1/x)}{1} = 0$

As $x \to \infty$ $\ln y = 0$ and $y = 1$, $\lim_{x \to \infty} x^{1/x} = 1$

24. $\lim_{x \to \infty} (1 + \dfrac{1}{x})^x$, $\quad$ let $y = \lim_{x \to \infty} (1 + \dfrac{1}{x})^x$, $\quad \ln y = \lim_{x \to \infty} \left[x \ln (1 + \dfrac{1}{x})\right]$

$\ln y = \lim_{x \to \infty} \dfrac{\ln (1 + (1/x))}{(1/x)} = \lim_{x \to \infty} \dfrac{\left[\dfrac{(-1/x^2)}{1 + (1/x)}\right]}{(-1/x^2)} = \lim_{x \to \infty} \dfrac{1}{1 + (1/x)} = 1$

As $x \to \infty$ $\ln y = 1$ and $y = e$, $\lim_{x \to \infty} (1 + \dfrac{1}{x})^x = e$

25. $\lim_{x \to \infty} (1 + x)^{1/x} \qquad y = (1 + x)^{1/x} \qquad \ln y = \dfrac{1}{x} \ln (1 + x)$

$\lim_{x \to \infty} \dfrac{1}{x} \ln (1 + x) = \lim_{x \to \infty} \dfrac{1/(1 + x)}{1} = 0$, As $x \to \infty$ $\ln y = 0$ and $y = 1$

$\lim_{x \to \infty} (1 + x)^{1/x} = 1$

26. $\lim_{x \to \pi} \frac{\sin x}{x - \pi} = \lim_{x \to \pi} \frac{\cos x}{1} = -1$

27. $\lim_{x \to 0} \frac{\sin 2x}{\sin 3x} = \lim_{x \to 0} \frac{2 \cos 2x}{3 \cos 3x} = \frac{2}{3}$

28. $\lim_{x \to 0} \frac{\sin ax}{\sin bx} = \lim_{x \to 0} \frac{a \cos ax}{b \cos bx} = \frac{a}{b}$

29. $\lim_{x \to 0} x \csc x = \lim_{x \to 0} \frac{x}{\sin x} = \lim_{x \to 0} \frac{1}{\cos x} = 1$

30. $\lim_{x \to 0} x^2 \cot x = \lim_{x \to 0} \frac{x^2}{\tan x} = \lim_{x \to 0} \frac{2x}{\sec^2 x} = 0$

31. $\lim_{x \to \infty} x \sin (\frac{1}{x}) = \lim_{x \to \infty} \frac{\sin (1/x)}{(1/x)}$

$= \lim_{x \to \infty} \frac{(-1/x^2) \cos (1/x)}{(-1/x^2)} = \lim_{x \to \infty} \cos (\frac{1}{x}) = 1$

32. $\lim_{x \to \infty} x \tan \frac{1}{x} = \lim_{x \to \infty} \frac{\tan (1/x)}{(1/x)} = \lim_{x \to \infty} \frac{-(1/x^2) \sec^2 (1/x)}{-(1/x^2)}$

$= \lim_{x \to \infty} \sec^2 (\frac{1}{x}) = 1$

33. $\lim_{x \to 0} \frac{\arcsin x}{x} = \lim_{x \to 0} \frac{(1/\sqrt{1 - x^2})}{1} = 1$

34. $\lim_{x \to 1} \frac{\arctan x - (\pi/4)}{x - 1} = \lim_{x \to 1} \frac{1/(1 + x^2)}{1} = \frac{1}{2}$

35. $\lim_{x \to \infty} \frac{x^2}{e^{5x}} = \lim_{x \to \infty} \frac{2x}{5e^{5x}} = \lim_{x \to \infty} \frac{2}{25e^{5x}} = 0$

36. $\lim_{x \to \infty} \frac{x^3}{e^{2x}} = \lim_{x \to \infty} \frac{3x^2}{2e^{2x}} = \lim_{x \to \infty} \frac{6x}{4e^{2x}} = \lim_{x \to \infty} \frac{6}{8e^{2x}} = 0$

37. $\lim_{x \to \infty} \frac{(\ln x)^3}{x} = \lim_{x \to \infty} \frac{3(\ln x)^2(1/x)}{1} = \lim_{x \to \infty} \frac{3(\ln x)^2}{x}$

$= \lim_{x \to \infty} \frac{6 (\ln x)(1/x)}{1} = \lim_{x \to \infty} \frac{6 (\ln x)}{x} = \lim_{x \to \infty} \frac{6}{x} = 0$

38. $\lim_{x\to\infty} \frac{(\ln x)^2}{x^3} = \lim_{x\to\infty} \frac{(2\ln x)/x}{3x^2} = \lim_{x\to\infty} \frac{2\ln x}{3x^3}$

$= \lim_{x\to\infty} \frac{(2/x)}{9x^2} = \lim_{x\to\infty} \frac{2}{9x^3} = 0$

39. $\lim_{x\to\infty} \frac{(\ln x)^n}{x^m} = \lim_{x\to\infty} \frac{n(\ln x)^{n-1}/x}{mx^{m-1}} = \lim_{x\to\infty} \frac{n(\ln x)^{n-1}}{mx^m}$

$= \lim_{x\to\infty} \frac{n(n-1)(\ln x)^{n-2}}{m^2x^m} = \cdots = \lim_{x\to\infty} \frac{n!}{m^nx^m} = 0$

40. $\lim_{x\to\infty} \frac{x^m}{e^{nx}} = \lim_{x\to\infty} \frac{mx^{m-1}}{ne^{nx}} = \lim_{x\to\infty} \frac{m(m-1)x^{m-2}}{n^2e^{nx}} = \cdots = \lim_{x\to\infty} \frac{m!}{n^me^{nx}} = 0$

41. $\lim_{x\to 0} \frac{e^{3x}-1}{e^x} = \frac{0}{1} = 0$ L'Hôpital's Rule does not apply

42. $\lim_{x\to 0} \frac{\sin \pi x - 1}{x} = \frac{-1}{0}$ L'Hôpital's Rule does not apply

43. $\lim_{x\to\infty} x\cos\frac{1}{x} = \infty(1)\to\infty$ L'Hôpital's Rule does not apply

44. $\lim_{x\to\infty} \frac{e^{-x}}{1+e^{-x}} = \frac{0}{1+0} = 0$ L'Hôpital's Rule does not apply

45.

x	10	10^2	10^4	10^6	10^8	10^{10}
$\frac{(\ln x)^4}{x}$	2.811	4.498	0.720	0.036	0.001	0.000

46.

x	1	5	10	20	30	100
$\frac{e^x}{x^5}$	2.718	0.047	0.220	151.614	4.39×10^5	2.69×10^{23}

47. $\lim_{x\to a} f(x)^{g(x)}$, $y = f(x)^{g(x)}$, $\ln y = g(x)\ln f(x)$

$\lim_{x\to a} g(x)\ln f(x) = (\infty)(-\infty) = -\infty$, As $x\to a$ $\ln y = -\infty$ and $y = 0$

$\lim_{x\to a} f(x)^{g(x)} = 0$

48. $\lim_{x \to a} f(x)^{g(x)}$, $\quad y = f(x)^{g(x)}$, $\quad \ln y = g(x) \ln f(x)$

$\lim_{x \to a} g(x) \ln f(x) = (-\infty)(-\infty) = \infty$

As $x \to a$ $\ln y = \infty$ and $y = \infty$, $\quad \lim_{x \to a} f(x)^{g(x)} = \infty$

49. $\lim_{x \to \infty} \frac{x}{\ln x} = \lim_{x \to \infty} \frac{1}{(1/x)} = \lim_{x \to \infty} x = \infty$

50. Let N be a fixed value for n. Then

$$\lim_{x \to \infty} \frac{x^{N-1}}{e^x} = \lim_{x \to \infty} \frac{(N-1)\, x^{N-2}}{e^x} = \lim_{x \to \infty} \frac{(N-1)(N-2)x^{N-3}}{e^x}$$

$$= \cdots = \lim_{x \to \infty} \left[\frac{(N-1)!}{e^x}\right] = 0$$

51. $$\lim_{k \to 0} \frac{32(1 - e^{-kt} + \frac{v_0 k e^{-kt}}{32})}{k} = \lim_{k \to 0} \frac{32(1 - e^{-kt})}{k} + \lim_{k \to 0} (v_0 e^{-kt})$$

$$= \lim_{k \to 0} \frac{32(0 + te^{-kt})}{1} + \lim_{k \to 0} (\frac{v_0}{e^{kt}}) = 32t + v_0$$

52. $A = P(1 + \frac{r}{n})^{nt}$

$\ln A = \ln P + nt \ln (1 + \frac{r}{n}) = \ln P + \frac{\ln (1 + (r/n))}{(1/nt)}$

$$\lim_{n \to \infty} \left[\frac{\ln (1 + \frac{r}{n})}{\frac{1}{nt}}\right] = \lim_{n \to \infty} \left[\frac{-\frac{r}{n^2}(\frac{1}{1 + (r/n)})}{-(\frac{1}{n^2 t})}\right] = \lim_{n \to \infty} \left[rt(\frac{1}{1 + (r/n)})\right] = rt$$

Since $\lim_{n \to \infty} \ln A = \ln P + rt$, we have

$$\lim_{n \to \infty} A = e^{(\ln P + rt)} = e^{\ln P} e^{rt} = Pe^{rt}$$

8.9 Improper integrals

1. $\displaystyle\int_0^4 \frac{1}{\sqrt{x}}\,dx = 2\sqrt{x}\Big]_0^4 = 4$

2. $\displaystyle\int_3^4 \frac{1}{\sqrt{x-3}}\,dx = 2\sqrt{x-3}\Big]_3^4 = 2$

3. $\displaystyle\int_0^2 \frac{1}{(x-1)^{2/3}}\,dx = \int_0^1 \frac{1}{(x-1)^{2/3}}\,dx + \int_1^2 \frac{1}{(x-1)^{2/3}}\,dx$

$\displaystyle = \Big[3(x-1)^{1/3}\Big]_0^1 + \Big[3(x-1)^{1/3}\Big]_1^2 = 6$

4. $\displaystyle\int_0^2 \frac{1}{(x-1)^2}\,dx = \int_0^1 \frac{1}{(x-1)^2}\,dx + \int_1^2 \frac{1}{(x-1)^2}\,dx = \frac{-1}{x-1}\Big]_0^1 + \frac{-1}{x-1}\Big]_1^2$

$\displaystyle = \left(\frac{-1}{0^-} - 1\right) + \left(-1 + \frac{1}{0^+}\right) = (\infty - 1) + (-1 + \infty) = \infty$ diverges

5. $\displaystyle\int_0^\infty e^{-x}\,dx = -e^{-x}\Big]_0^\infty = 0 + 1 = 1$

6. $\displaystyle\int_{-\infty}^0 e^{2x}\,dx = \frac{1}{2}e^{2x}\Big]_{-\infty}^0 = \frac{1}{2} - 0 = \frac{1}{2}$

7. $\displaystyle\int_{-\infty}^0 xe^{-2x}\,dx = \frac{1}{4}\int_{-\infty}^0 (-2x)e^{-2x}(-2)\,dx = \frac{1}{4}\Big[(-2x-1)e^{-2x}\Big]_{-\infty}^0 = -\infty$ diverges

8. $\displaystyle\int_0^\infty xe^{-x}\,dx = \lim_{b\to\infty}\Big[-e^{-x}(x+1)\Big]_0^b = \lim_{b\to\infty}[1 - e^{-b}(b+1)] = 1$

since $\displaystyle\lim_{b\to\infty}\left[\frac{b+1}{e^x}\right] = 0$ by L'Hôpital's Rule

9. $\displaystyle\int_0^\infty x^2e^{-x}\,dx = \lim_{b\to\infty}\Big[-e^{-x}(x^2+2x+2)\Big]_0^b = \lim_{b\to\infty}\left[-\frac{b^2+2b+2}{e^b} + 2\right] = 2$

since $\displaystyle\lim_{b\to\infty}\left(-\frac{b^2+2b+2}{e^b}\right) = 0$ by L'Hôpital's Rule

10. $\int_0^{\infty} (x - 1)e^{-x}\, dx = \lim_{b \to \infty} \Big[-xe^{-x}\Big]_0^b = \lim_{b \to \infty} \left(\frac{-b}{e^b} + 0\right) = 0$

by L'Hôpital's Rule

11. $\int_1^{\infty} \frac{1}{x^2}\, dx = \frac{-1}{x}\Big]_1^{\infty} = 1$

12. $\int_1^{\infty} \frac{1}{\sqrt{x}}\, dx = 2\sqrt{x}\Big]_1^{\infty} = \infty - 2$ diverges

13. $\int_0^{\infty} e^{-x}\cos x\, dx = \frac{1}{2}\Big[e^{-x}(-\cos x + \sin x)\Big]_0^{\infty} = \frac{1}{2}[0 - (-1)] = \frac{1}{2}$

14. $\int_0^{\infty} e^{-ax}\sin bx\, dx = \frac{e^{-ax}(-a \sin bx - b \cos bx)}{a^2 + b^2}\Big]_0^{\infty}$

$= 0 - \frac{-b}{a^2 + b^2} = \frac{b}{a^2 + b^2}$

15. $\int_{-\infty}^{\infty} \frac{1}{1 + x^2}\, dx = 2\int_0^{\infty} \frac{1}{1 + x^2}\, dx = 2 \arctan x\Big]_0^{\infty} = 2\left(\frac{\pi}{2}\right) = \pi$

16. $\int_0^{\infty} \frac{x^3}{(x^2 + 1)^2}\, dx = \int_0^{\infty} \frac{x}{x^2 + 1}\, dx - \int_0^{\infty} \frac{x}{(x^2 + 1)^2}\, dx$

$= \Big[\frac{1}{2}\ln(x^2 + 1) + \frac{1}{2(x^2 + 1)}\Big]_0^{\infty} = \infty - \frac{1}{2}$ diverges

17. $\int_0^{\infty} \frac{1}{e^x + e^{-x}}\, dx = \int_0^{\infty} \frac{e^x}{1 + e^{2x}}\, dx = \arctan(e^x)\Big]_0^{\infty} = \frac{\pi}{2} - \frac{\pi}{4} = \frac{\pi}{4}$

18. $\int_0^{\infty} \frac{e^x}{1 + e^x}\, dx = \ln(1 + e^x)\Big]_0^{\infty} = \infty - \ln 2$ diverges

19. $\int_0^{\infty} \cos \pi x\, dx = \frac{1}{\pi}\sin \pi x\Big]_0^{\infty}$ diverges

Since $\sin \pi x$ does not approach a limit as $x \to \infty$

20. $\int_0^{\infty} \sin\frac{x}{2}\, dx = -2\cos\frac{x}{2}\Big]_0^{\infty}$ diverges

Since $\cos\frac{x}{2}$ does not approach a limit as $x \to \infty$

21. $\displaystyle\int_0^1 \frac{1}{x^2}\,dx = \frac{-1}{x}\Big]_0^1 = -1 + \frac{1}{0^+} = -1 + \infty$ diverges

22. $\displaystyle\int_0^1 \frac{1}{x}\,dx = \ln|x|\Big]_0^1 = 0 - (-\infty)$ diverges

23. $\displaystyle\int_0^8 \frac{1}{\sqrt[3]{8-x}}\,dx = \frac{-3}{2}(8-x)^{2/3}\Big]_0^8 = 6$

24. $\displaystyle\int_0^e \ln x\,dx = \Big[x\ln x - x\Big]_0^e = (e - e) - (0 - 0) = 0$

Since $\displaystyle\lim_{b\to 0^+} b(\ln b) = \lim_{b\to 0^+} \frac{\ln b}{(1/b)} = 0$ by L'Hôpital's Rule

25. $\displaystyle\int_0^1 x\ln x\,dx = \lim_{a\to 0^+}\left[\frac{x^2}{2}\ln|x| - \frac{x^2}{4}\right]_a^1 = \lim_{a\to 0^+}\left[\frac{-1}{4} - \frac{a^2\ln a}{2} + \frac{a^2}{4}\right] = \frac{-1}{4}$

Since $\displaystyle\lim_{a\to 0^+}(a^2\ln a) = 0$ by L'Hôpital's Rule

26. $\displaystyle\int_0^{\pi/2} \sec\theta\,d\theta = \ln|\sec\theta + \tan\theta|\Big]_0^{\pi/2} = \infty$ diverges

27. $\displaystyle\int_0^{\pi/2} \tan\theta\,d\theta = \ln|\sec\theta|\Big]_0^{\pi/2} = \infty$ diverges

28. $\displaystyle\int_0^2 \frac{1}{\sqrt{4-x^2}}\,dx = \arcsin\left(\frac{x}{2}\right)\Big]_0^2 = \frac{\pi}{2}$

29. $\displaystyle\int_2^4 \frac{1}{\sqrt{x^2-4}}\,dx = \ln|x + \sqrt{x^2-4}|\Big]_2^4 = \ln(4 + 2\sqrt{3}) - \ln 2 = \ln(2 + \sqrt{3})$

30. $\displaystyle\int_0^2 \frac{1}{4-x^2}\,dx = \frac{1}{4}\ln\left|\frac{2+x}{2-x}\right|\Big]_0^2 = \infty - 0$ diverges

31. $\displaystyle\int_0^2 \frac{1}{\sqrt[3]{x-1}}\,dx = \int_0^1 \frac{1}{\sqrt[3]{x-1}}\,dx + \int_1^2 \frac{1}{\sqrt[3]{x-1}}\,dx$

$\displaystyle = \left[\frac{3}{2}(x-1)^{2/3}\right]_0^1 + \left[\frac{3}{2}(x-1)^{2/3}\right]_1^2 = \frac{-3}{2} + \frac{3}{2} = 0$

32. $\int_0^2 \frac{1}{(x-1)^{4/3}}\,dx = \int_0^1 \frac{1}{(x-1)^{4/3}}\,dx + \int_1^2 \frac{1}{(x-1)^{4/3}}\,dx$

$= \left[\frac{-3}{(x-1)^{1/3}}\right]_0^1 + \left[\frac{-3}{(x-1)^{1/3}}\right]_1^2 = \frac{-3}{0^-} - 3 + 3 + \frac{3}{0^+}$

$= \infty + \infty$ diverges

33. $\int_1^\infty \frac{1}{x^p}\,dx = \lim_{b\to\infty}\left[\frac{x^{1-p}}{1-p}\right]_1^b = \lim_{b\to\infty}\left[\frac{b^{1-p}}{1-p} - \frac{1}{1-p}\right]$

This converges to $\frac{1}{p-1}$ if $1 - p \leq 0$ or $p \geq 1$.

34. (a) Assume $\int_a^\infty g(x)\,dx = L$. $0 \leq \int_a^\infty f(x)\,dx < \int_a^\infty g(x)\,dx = L$

Therefore $\int_a^\infty f(x)\,dx$ converges.

(b) Assume $\int_a^\infty f(x)\,dx = \infty$. $\int_a^\infty g(x)\,dx \geq \int_a^\infty f(x)\,dx = \infty$

Therefore $\int_a^\infty g(x)\,dx$ diverges.

35. Since $\frac{1}{x^2+5} \leq \frac{1}{x^2}$ and $\int_1^\infty \frac{1}{x^2}\,dx$ converges by Exercise 33,

$\int_1^\infty \frac{1}{x^2+5}\,dx$ converges.

36. Since $\frac{1}{\sqrt{x}-1} \geq \frac{1}{x}$ and $\int_2^\infty \frac{1}{x}\,dx$ diverges, $\int_2^\infty \frac{1}{\sqrt{x}-1}\,dx$ diverges.

37. Since $\frac{1}{\sqrt[3]{x}\,(x-1)} \geq \frac{1}{\sqrt[3]{x}}$ and $\int_2^\infty \frac{1}{\sqrt[3]{x}}\,dx$ diverges,

$\int_2^\infty \frac{1}{\sqrt[3]{x}(x-1)}\,dx$ diverges.

38. Since $\frac{1}{\sqrt{x}(1+x)} \leq \frac{1}{x^{3/2}}$ and $\int_1^\infty \frac{1}{x^{3/2}}\,dx$ converges by Exercise 33,

$\int_1^\infty \frac{1}{\sqrt{x}(1+x)}\,dx$ converges.

39. Since $e^{-x^2} \leq e^{-x}$ on $[0, \infty)$ and $\int_0^\infty e^{-x}\,dx$ converges (see Exercise 5)

$$\int_0^\infty e^{-x^2}\,dx \text{ converges.}$$

40. $\frac{1}{\sqrt{x}\ln x} \geq \frac{1}{x}$ since $\sqrt{x} \geq \ln x$ on $[2, \infty)$.

Since $\int_2^\infty \frac{1}{x}\,dx$ diverges, $\int_2^\infty \frac{1}{\sqrt{x}\ln x}\,dx$ diverges.

41. (a) $A = \int_1^\infty \frac{1}{x^2}\,dx = -\frac{1}{x}\Big]_1^\infty = 1$

(b) Disc

$$V = \pi \int_1^\infty \frac{1}{x^4}\,dx = \frac{\pi}{3}$$

(c) Shell

$$V = 2\pi \int_1^\infty x\left(\frac{1}{x^2}\right)\,dx = 2\pi(\ln x)\Big]_1^\infty = \infty \text{ diverges}$$

42. (a) $A = \int_0^\infty e^{-x}\,dx = -e^{-x}\Big]_0^\infty = 0 - (-1) = 1$

(b) Shell

$$V = 2\pi \int_0^\infty xe^{-x}\,dx = 2\pi\left[-e^{-x}(x+1)\right]_0^\infty = 2\pi$$

43. $x^{2/3} + y^{2/3} = 1$

$$\frac{2}{3}x^{-1/3} + \frac{2}{3}y^{-1/3}y' = 0 \quad \text{or} \quad y' = \frac{-y^{1/3}}{x^{1/3}}$$

$$\sqrt{1 + (y')^2} = \sqrt{1 + \frac{y^{2/3}}{x^{2/3}}} = \sqrt{(1/x^{2/3})} = \frac{1}{x^{1/3}}$$

$$s = 4\int_0^1 \frac{1}{x^{1/3}}\,dx = 4\left(\frac{3}{2}x^{2/3}\right)\Big]_0^1 = 6$$

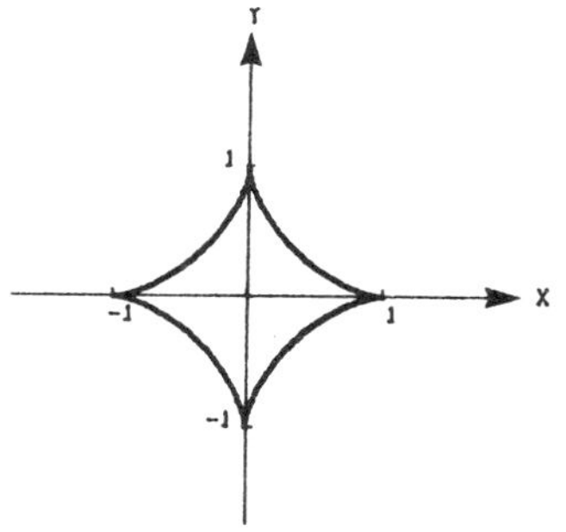

44. $(x - 2)^2 + y^2 = 1, \quad 2(x - 2) + 2yy' = 0, \quad y' = \dfrac{-(x - 2)}{y}$

$$\sqrt{1 + (y')^2} = \sqrt{1 + \frac{(x - 2)^2}{y^2}} = \frac{1}{y}$$

$$S = 4\pi \int_1^3 \frac{x}{y}\, dx = 4\pi \int_1^3 \frac{x}{\sqrt{1 - (x - 2)^2}}\, dx$$

$$= 4\pi \int_1^3 \left[\frac{x - 2}{\sqrt{1 - (x - 2)^2}} + \frac{2}{\sqrt{1 - (x - 2)^2}}\right] dx$$

$$= 4\pi \left[-\sqrt{1 - (x - 2)^2} + 2 \arcsin(x - 2)\right]_1^3$$

$$= 4\pi\, [0 + 2 \arcsin(1) - 2 \arcsin(-1)] = 8\pi^2$$

45. $\Gamma(n) = \displaystyle\int_0^\infty x^{n-1}e^{-x}\, dx$

(a) $\Gamma(1) = \displaystyle\int_0^\infty e^{-x}\, dx = -e^{-x}\Big]_0^\infty = 1$

(b) $\Gamma(2) = \displaystyle\int_0^\infty xe^{-x}\, dx = -e^{-x}(x + 1)\Big]_0^\infty = 1$

(c) $\Gamma(3) = \displaystyle\int_0^\infty x^2e^{-x}\, dx = \Big[-x^2e^{-x} + 2(-e^{-x}(x + 1))\Big]_0^\infty = 2$

(d) $\Gamma(n + 1) = \displaystyle\int_0^\infty x^ne^{-x}\, dx = -x^ne^{-x}\Big]_0^\infty + n\int_0^\infty x^{n-1}e^{-x}\, dx = 0 + n\Gamma(n)$

46. $F(x) = \dfrac{K}{x^2}, \quad 5 = \dfrac{K}{(4000)^2}, \quad K = 80{,}000{,}000$

$$W = \int_{4000}^\infty \frac{80{,}000{,}000}{x^2}\, dx = \frac{-80{,}000{,}000}{x}\Big]_{4000}^\infty = 20{,}000 \text{ mi-ton}$$

$$\frac{W}{2} = 10{,}000 = \frac{-80{,}000{,}000}{x}\Big]_{4000}^b = \frac{-80{,}000{,}000}{b} + 20{,}000,$$

$$\frac{80{,}000{,}000}{b} = 10{,}000, \quad b = 8{,}000,$$

therefore 4000 miles above the earth's surface

47. (a) $\displaystyle\int_{-\infty}^\infty \frac{1}{7}e^{-t/7}\, dt = \int_0^\infty \frac{1}{7}e^{-t/7}\, dt = -e^{-t/7}\Big]_0^\infty = 1$

(b) $\displaystyle\int_0^5 \frac{1}{7}e^{-t/7}\, dt = -e^{-t/7}\Big]_0^5 = -e^{-5/7} + 1 \approx 0.5105$

(c) $\displaystyle\int_0^\infty t\left[\frac{1}{7}e^{-t/7}\right]dt = \Big[-te^{-t/7} - 7e^{-t/7}\Big]_0^\infty = 0 + 7 = 7$

48. (a) $\displaystyle\int_{-\infty}^{\infty} \frac{2}{5}e^{-2t/5}\,dt = \int_{0}^{\infty} \frac{2}{5}e^{-2t/5}\,dt = -e^{-2t/5}\Big]_{0}^{\infty} = 1$

(b) $\displaystyle\int_{0}^{3} \frac{2}{5}e^{-2t/5}\,dt = -e^{-2t/5}\Big]_{0}^{3} = -e^{-6/5} + 1 \approx 0.6988$

(c) $\displaystyle\int_{0}^{\infty} t\left[\frac{2}{5}e^{-2t/5}\right]dt = \left[-te^{2t/5} - \frac{5}{2}e^{-2t/5}\right]_{0}^{\infty} = \frac{5}{2}$

49. (a) $\displaystyle C = 650{,}000 + \int_{0}^{5} 25{,}000e^{-0.12t}\,dt = 650{,}000 - \left[\frac{25{,}000}{0.12}e^{-0.12t}\right]_{0}^{5}$

$\approx \$743{,}997.58$

(b) $\displaystyle C = 650{,}000 + \int_{0}^{10} 25{,}000e^{-0.12t}\,dt = 650{,}000 - \left[\frac{25{,}000}{0.12}e^{-0.12t}\right]_{0}^{10}$

$\approx \$795{,}584.54$

(c) $\displaystyle C = 650{,}000 + \int_{0}^{\infty} 25{,}000e^{-0.12t}\,dt = 650{,}000 - \left[\frac{25{,}000}{0.12}e^{-0.12t}\right]_{0}^{\infty}$

$\approx \$858{,}333.33$

50. (a) $\displaystyle C = 650{,}000 + \int_{0}^{5} 25{,}000(1 + 0.08t)e^{-0.12t}\,dt = 650{,}000$

$\displaystyle + 25{,}000\left\{-\frac{1}{0.12}e^{-0.12t} - 0.08\left[\frac{t}{0.12}e^{-0.12t} + \frac{1}{(0.12)^2}e^{-0.12t}\right]\right\}\Bigg]_{0}^{5}$

$\approx \$760{,}928.32$

(b) $\displaystyle C = 650{,}000 + \int_{0}^{10} 25{,}000(1 + 0.08t)e^{-0.12t}\,dt = 650{,}000$

$\displaystyle + 25{,}000\left\{-\frac{1}{0.12}e^{-0.12t} - 0.08\left[\frac{t}{0.12}e^{-0.12t} + \frac{1}{(0.12)^2}e^{-0.12t}\right]\right\}\Bigg]_{0}^{10}$

$\approx \$842{,}441.86$

(c) $\displaystyle C = 650{,}000 + \int_{0}^{\infty} 25{,}000(1 + 0.08t)e^{-0.12t}\,dt = 650{,}000$

$\displaystyle + 25{,}000\left\{-\frac{1}{0.12}e^{-0.12t} - 0.08\left[\frac{t}{0.12}e^{-0.12t} + \frac{1}{(0.12)^2}e^{-0.12t}\right]\right\}\Bigg]_{0}^{\infty}$

$\approx \$997{,}222.22$

Review Exercises for Chapter 8

1. $\int \frac{x^2}{x^2 + 2x - 15}\,dx = \int dx + \frac{9}{8}\int \frac{1}{x-3}\,dx - \frac{25}{8}\int \frac{1}{x+5}\,dx$

$= x + \frac{9}{8}\ln|x-3| - \frac{25}{8}\ln|x+5| + C$

2. $\int \frac{\sqrt{x^2-9}}{x}\,dx = 3\int \tan^2\theta\,d\theta = 3\int (\sec^2\theta - 1)\,d\theta$

$= 3(\tan\theta - \theta) + C$

$= \sqrt{x^2-9} - 3\,\text{arcsec}\,(\frac{x}{3}) + C$

$x = 3\sec\theta,\qquad dx = 3\sec\theta\tan\theta\,d\theta,$

$\sqrt{x^2-9} = 3\tan\theta$

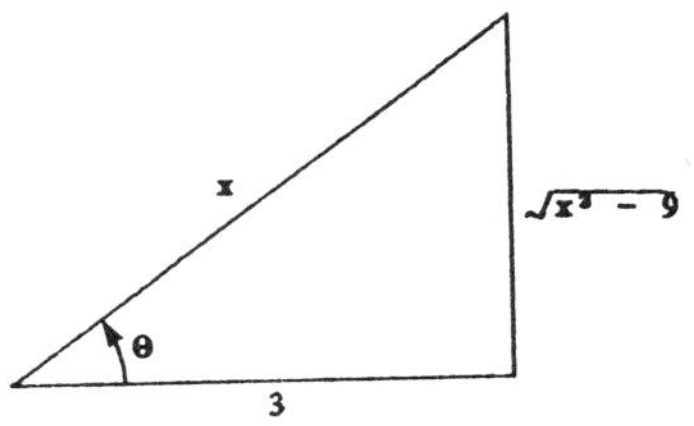

3. $\int \frac{1}{1-\sin\theta}\,d\theta = \int \frac{1+\sin\theta}{\cos^2\theta} = \int (\sec^2\theta + \sec\theta\tan\theta)\,d\theta$

$= \tan\theta + \sec\theta + C$

4. $\int x^2\sin 2x\,dx = -\frac{1}{2}x^2\cos 2x + \int x\cos 2x\,dx$

$= -\frac{1}{2}x^2\cos 2x + \frac{1}{2}x\sin 2x - \frac{1}{2}\int \sin 2x\,dx$

$= -\frac{1}{2}x^2\cos 2x + \frac{x}{2}\sin 2x + \frac{1}{4}\cos 2x + C$

(1)--> $u = x^2,\qquad v' = \sin 2x,\qquad u' = 2x,\qquad v = -\frac{1}{2}\cos 2x$

(2)--> $u = x,\qquad v' = \cos 2x,\qquad u' = 1,\qquad v = \frac{1}{2}\sin 2x$

5. Let $I = \int e^{2x}\sin 3x\,dx = -\frac{1}{3}e^{2x}\cos 3x + \frac{2}{3}\int e^{2x}\cos 3x\,dx$

$= -\frac{1}{3}e^{2x}\cos 3x + \frac{2}{3}\left[\frac{1}{3}e^{2x}\sin 3x - \frac{2}{3}I\right]$

$\frac{13}{9}I = -\frac{1}{3}e^{2x}\cos 3x + \frac{2}{9}e^{2x}\sin 3x,\quad I = \frac{e^{2x}}{13}(2\sin 3x - 3\cos 3x) + C$

5. (continued)

(1)--> $u = e^{2x}$, $v' = \sin 3x$, $u' = 2e^{2x}$, $v = -\frac{1}{3}\cos 3x$

(2)--> $u = e^{2x}$, $v' = \cos 3x$, $u' = 2e^{2x}$, $v = \frac{1}{3}\sin 3x$

6. $\int (x^2 - 1)e^x\,dx = (x^2 - 1)e^x - 2\int xe^x\,dx$

$= (x^2 - 1)e^x - 2xe^x + 2\int e^x\,dx = e^x(x^2 - 2x + 1) + C$

(1)--> $u = x^2 - 1$, $v' = e^x$, $u' = 2x$, $v = e^x$

(2)--> $u = x$, $v' = e^x$, $u' = 1$, $v = e^x$

7. $\int \frac{\ln 2x}{x^2}\,dx = \frac{-\ln 2x}{x} + \int \frac{1}{x^2}\,dx = \frac{-\ln 2x}{x} - \frac{1}{x} + C = -\frac{1}{x}(1 + \ln 2x) + C$

$u = \ln 2x$, $v' = \frac{1}{x^2}$, $u' = \frac{1}{x}$, $v = -\frac{1}{x}$

8. $\int 2x\sqrt{2x - 3}\,dx = \int (u^4 + 3u^2)\,du = \frac{u^5}{5} + u^3 + C = \frac{2(2x - 3)^{3/2}}{5}(x + 1) + C$

$u = \sqrt{2x - 3}$, $x = \frac{u^2 + 3}{2}$, $dx = u\,du$

9. $\int \sqrt{4 - x^2}\,dx = \int (2\cos\theta)(2\cos\theta)\,d\theta = 2\int (1 + \cos 2\theta)\,d\theta$

$= 2\left[\theta + \frac{1}{2}\sin 2\theta\right] + C = 2[\theta + \sin\theta\cos\theta] + C$

$= 2\left[\arcsin\left(\frac{x}{2}\right) + \frac{x}{2}\left(\frac{\sqrt{4 - x^2}}{2}\right)\right] + C$

$= \frac{1}{2}\left[4\arcsin\left(\frac{x}{2}\right) + x\sqrt{4 - x^2}\right] + C$

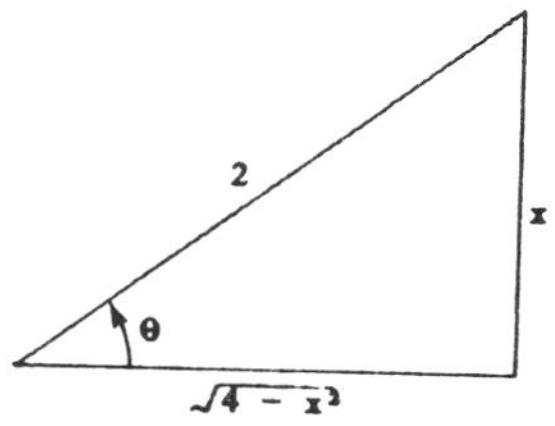

$x = 2\sin\theta$, $dx = 2\cos\theta\,d\theta$,

$\sqrt{4 - x^2} = 2\cos\theta$

10. $\int \frac{\sqrt{4 - x^2}}{x}\,dx = \int \frac{2\cos\theta(2\cos\theta)\,d\theta}{2\sin\theta} = 2\int (\csc\theta - \sin\theta)\,d\theta$

$= 2[\ln|\csc\theta - \cot\theta| + \cos\theta] + C = 2\ln\left|\frac{2 - \sqrt{4 - x^2}}{x}\right| + \sqrt{4 - x^2} + C$

Same substitution as in Exercise 9.

11. $\int \frac{-12}{x^2\sqrt{4-x^2}}\,dx = \int \frac{-24\cos\theta\,d\theta}{(4\sin^2\theta)(2\cos\theta)} = -3\int \csc^2\theta\,d\theta$

$= 3\cot\theta + C = \frac{3\sqrt{4-x^2}}{x} + C$

Same substitution as in Exercise 9.

12. $\int \tan\theta\sec^4\theta\,d\theta = \int (\tan^3\theta + \tan\theta)\sec^2\theta\,d\theta = \frac{1}{4}\tan^4\theta + \frac{1}{2}\tan^2\theta + C$

13. $\int \sec^4\left(\frac{x}{2}\right)dx = \int \left[\tan^2\left(\frac{x}{2}\right) + 1\right]\sec^2\left(\frac{x}{2}\right)dx$

$= \int \tan^2\left(\frac{x}{2}\right)\sec^2\left(\frac{x}{2}\right)dx + \int \sec^2\left(\frac{x}{2}\right)dx$

$= \frac{2}{3}\tan^3\left(\frac{x}{2}\right) + 2\tan\left(\frac{x}{2}\right) + C = \frac{2}{3}\left[\tan^3\left(\frac{x}{2}\right) + 3\tan\left(\frac{x}{2}\right)\right] + C$

14. $\int \sec\theta\cos 2\theta\,d\theta = \int \sec\theta(2\cos^2\theta - 1)\,d\theta = \int (2\cos\theta - \sec\theta)\,d\theta$

$= 2\sin\theta - \ln|\sec\theta + \tan\theta| + C$

15. $\int \frac{9}{x^2-9}\,dx = \frac{3}{2}\left[\int \frac{1}{x-3}\,dx - \int \frac{1}{x+3}\,dx\right] = \frac{3}{2}\ln\left|\frac{x-3}{x+3}\right| + C$

16. $\int \frac{\sec^2\theta}{\tan\theta(\tan\theta - 1)}\,d\theta = \int \frac{1}{u(u-1)}\,du = \int \frac{1}{u-1}\,du - \int \frac{1}{u}\,du$

$= \ln|u-1| - \ln|u| + C = \ln\left|\frac{\tan\theta - 1}{\tan\theta}\right| + C$

$u = \tan\theta, \quad du = \sec^2\theta\,d\theta$

17. $\int \frac{x^2+2x}{x^3-x^2+x-1}\,dx = \frac{3}{2}\int \frac{1}{x-1}\,dx - \frac{1}{2}\int \frac{x-3}{x^2+1}\,dx = \frac{3}{2}\int \frac{1}{x-1}\,dx$

$- \frac{1}{4}\int \frac{2x}{x^2+1}\,dx + \frac{3}{2}\int \frac{1}{x^2+1}\,dx = \frac{3}{2}\ln|x-1| - \frac{1}{4}\ln|x^2+1|$

$+ \frac{3}{2}\arctan x + C = \frac{1}{4}\left[6\ln|x-1| - \ln(x^2+1) + 6\arctan x\right] + C$

18. $\displaystyle\int \frac{4x - 2}{3(x - 1)^2}\,dx = \frac{4}{3}\int \frac{1}{x - 1}\,dx + \frac{2}{3}\int \frac{1}{(x - 1)^2}\,dx$

$\displaystyle = \frac{4}{3}\ln|x - 1| - \frac{2}{3(x - 1)} + C = \frac{2}{3}\left[2\ln|x - 1| - \frac{1}{x - 1}\right] + C$

19. $\displaystyle\int \frac{3x^3 + 4x}{(x^2 + 1)^2}\,dx = 3\int \frac{x}{x^2 + 1}\,dx + \int \frac{x}{(x^2 + 1)^2}\,dx$

$\displaystyle = \frac{3}{2}\ln(x^2 + 1) - \frac{1}{2(x^2 + 1)} + C = \frac{1}{2}\left[3\ln(x^2 + 1) - \frac{1}{x^2 + 1}\right] + C$

20. $\displaystyle\int \sqrt{\frac{x - 2}{x + 2}}\,dx = \int \frac{x - 2}{\sqrt{x^2 - 4}}\,dx = \int \frac{x}{\sqrt{x^2 - 4}}\,dx - 2\int \frac{1}{\sqrt{x^2 - 4}}\,dx$

$\displaystyle = \sqrt{x^2 - 4} - 2\ln|x + \sqrt{x^2 - 4}| + C$

21. $\displaystyle\int \frac{16}{\sqrt{16 - x^2}}\,dx = 16\arcsin\left(\frac{x}{4}\right) + C$

22. $\displaystyle\int \frac{\sin\theta}{1 + 2\cos^2\theta}\,d\theta = \frac{-1}{\sqrt{2}}\int \frac{1}{1 + 2\cos^2\theta}(-\sqrt{2}\sin\theta)\,d\theta$

$\displaystyle = \frac{-1}{\sqrt{2}}\arctan(\sqrt{2}\cos\theta) + C$

$u = \sqrt{2}\cos\theta, \quad u' = -\sqrt{2}\sin\theta$

23. $\displaystyle\int \frac{e^x}{4 + e^{2x}}\,dx = \frac{1}{2}\arctan\left(\frac{e^x}{2}\right) + C$

$u = e^x, \quad u' = e^x$

24. $\displaystyle\int \frac{x}{x^2 - 4x + 8}\,dx = \frac{1}{2}\int \frac{2x - 4 + 4}{x^2 - 4x + 8}\,dx = \frac{1}{2}\int \frac{2x - 4}{x^2 - 4x + 8}\,dx$

$\displaystyle + 2\int \frac{1}{(x - 2)^2 + 4}\,dx = \frac{1}{2}\ln|x^2 - 4x + 8| + \arctan\left(\frac{x - 2}{2}\right) + C$

25. $\displaystyle\int \frac{x}{x^2 + 4x + 8}\,dx = \frac{1}{2}\int \frac{2x + 4}{x^2 + 4x + 8}\,dx - 2\int \frac{1}{(x + 2)^2 + 4}\,dx$

$\displaystyle = \frac{1}{2}\ln|x^2 + 4x + 8| - \arctan\left(\frac{x + 2}{2}\right) + C$

26. $\displaystyle\int \frac{3}{2x\sqrt{9x^2 - 1}}\,dx = \frac{3}{2}\int \frac{3}{3x\sqrt{9x^2 - 1}}\,dx = \frac{3}{2}\,\mathrm{arcsec}\,(3x) + C$

27. $\int \theta \sin\theta \cos\theta \, d\theta = \frac{1}{2}\int \theta \sin 2\theta \, d\theta = -\frac{1}{4}\theta \cos 2\theta + \frac{1}{4}\int \cos 2\theta \, d\theta$

$= -\frac{1}{4}\theta \cos 2\theta + \frac{1}{8}\sin 2\theta + C = \frac{1}{8}(\sin 2\theta - 2\theta \cos 2\theta) + C$

$u = \theta, \quad v' = \sin 2\theta, \quad u' = 1, \quad v = -\frac{1}{2}\cos 2\theta$

28. $\int \frac{\csc\sqrt{2x}}{\sqrt{x}} \, dx = \sqrt{2}\int \csc\sqrt{2x}\left(\frac{1}{\sqrt{2x}}\right) dx = \sqrt{2}\ln|\csc\sqrt{2x} - \cot\sqrt{2x}| + C$

$u = \sqrt{2x}, \quad u' = \frac{1}{\sqrt{2x}}$

29. $\int (\sin\theta + \cos\theta)^2 \, d\theta = \int (\sin^2\theta + 2\sin\theta\cos\theta + \cos^2\theta) \, d\theta$

$= \int (1 + \sin 2\theta) \, d\theta = \theta + \left(-\frac{1}{2}\right)\cos 2\theta + C = \frac{1}{2}(2\theta - \cos 2\theta) + C$

30. $\int \cos 2\theta(\sin\theta + \cos\theta)^2 \, d\theta = \int (\cos^2\theta - \sin^2\theta)(\sin\theta + \cos\theta)^2 \, d\theta$

$= \int (\sin\theta + \cos\theta)^3(\cos\theta - \sin\theta) \, d\theta = \frac{1}{4}(\sin\theta + \cos\theta)^4 + C$

31. $\int \frac{x^{1/4}}{1 + x^{1/2}} \, dx = 4\int \frac{u(u^3)}{1 + u^2} \, du = 4\int \left(u^2 - 1 + \frac{1}{u^2 + 1}\right) du$

$= 4\left[\frac{1}{3}u^3 - u + \arctan u\right] + C = \frac{4}{3}\left[x^{3/4} - 3x^{1/4} + 3\arctan(x^{1/4})\right] + C$

$u = \sqrt[4]{x}, \quad x = u^4, \quad dx = 4u^3 \, du$

32. $\int \sqrt{1 - \cos x} \, dx = \int \frac{\sin x}{\sqrt{1 + \cos x}} \, dx = -\int (1 + \cos x)^{-1/2}(-\sin x) \, dx$

$= -2\sqrt{1 + \cos x} + C, \quad u = 1 + \cos x, \quad u' = -\sin x$

33. $\int \sqrt{1 + \cos x} \, dx = \int \frac{\sin x}{\sqrt{1 - \cos x}} \, dx = \int (1 - \cos x)^{-1/2}(\sin x) \, dx$

$= 2\sqrt{1 - \cos x} + C, \quad u = 1 - \cos x, \quad u' = \sin x$

34. $\int \ln\sqrt{x^2 - 1} \, dx = \frac{1}{2}\int \ln(x^2 - 1) \, dx = \frac{1}{2}x\ln|x^2 - 1| - \int \frac{x^2}{x^2 - 1} \, dx$

$= \frac{1}{2}x\ln|x^2 - 1| - \int dx - \int \frac{1}{x^2 - 1} \, dx = \frac{1}{2}x\ln|x^2 - 1|$

34. (continued)

$$- x - \frac{1}{2}\ln\left|\frac{x - 1}{x + 1}\right| + C = \frac{1}{2}\left[x \ln|x^2 - 1| - 2x - \ln\left|\frac{x - 1}{x + 1}\right|\right] + C$$

$u = \ln|x^2 - 1|, \quad v' = 1, \quad u' = \dfrac{2x}{x^2 - 1}, \quad v = x$

35. $$\int \ln(x^2 + x)\,dx = x\ln|x^2 + x| - \int \frac{2x^2 + x}{x^2 + x}\,dx = x\ln|x^2 + x|$$

$$- \int \frac{2x + 1}{x + 1}\,dx = x\ln|x^2 + x| - \int 2\,dx + \int \frac{1}{x + 1}\,dx$$

$$= x\ln|x^2 + x| - 2x + \ln|x + 1| + C$$

$u = \ln(x^2 + x), \quad v' = 1, \quad u' = \dfrac{2x + 1}{x^2 + x}, \quad v = x$

36. $$\int x \arcsin 2x\,dx = \frac{x^2}{2}\arcsin 2x - \int \frac{x^2}{\sqrt{1 - 4x^2}}\,dx = \frac{x^2}{2}\arcsin 2x$$

$$- \frac{1}{8}\int \frac{2(2x)^2}{\sqrt{1 - (2x)^2}}\,dx = \frac{x^2}{2}\arcsin 2x - \frac{1}{8}\left(\frac{1}{2}\right)\left[-(2x)\sqrt{1 - 4x^2}\right.$$

$$\left. + \arcsin 2x\right] + C \quad \text{(from Formula 43 of Section 8.6)}$$

$$= \frac{1}{16}\left[(8x^2 - 1)\arcsin 2x + 2x\sqrt{1 - 4x^2}\right] + C$$

$u = \arcsin 2x, \quad v' = x, \quad u' = \dfrac{2}{\sqrt{1 - 4x^2}}, \quad v = \dfrac{x^2}{2}$

37. $$\int \cos x \ln(\sin x)\,dx = \sin x \ln(\sin x) - \int \cos x\,dx$$

$$= \sin x \ln(\sin x) - \sin x + C$$

$u = \ln(\sin x), \quad v' = \cos x, \quad u' = \dfrac{\cos x}{\sin x}, \quad v = \sin x$

38. $$\int e^x \arctan(e^x)\,dx = e^x \arctan(e^x) - \int \frac{e^{2x}}{1 + e^{2x}}\,dx = e^x \arctan(e^x)$$

$$- \frac{1}{2}\ln(1 + e^{2x}) + C$$

$u = \arctan(e^x), \quad v' = e^x, \quad u' = \dfrac{e^x}{1 + e^{2x}}, \quad v = e^x$

39. $$\int \frac{x^4 + 2x^2 + x + 1}{(x^2 + 1)^2}\,dx = \int dx + \frac{1}{2}\int \frac{2x}{(x^2 + 1)^2}\,dx = x - \frac{1}{2(x^2 + 1)} + C$$

40. $\int \sqrt{1 + \sqrt{x}}\, dx = \int u(4u^3 - 4u)\, du = \int (4u^4 - 4u^2)\, du = \frac{4u^5}{5} - \frac{4u^3}{3} + C$

$= \frac{4}{15}(1 + \sqrt{x})^{3/2}(3\sqrt{x} - 2) + C$

$u = \sqrt{1 + \sqrt{x}}, \quad x = u^4 - 2u^2 + 1, \quad dx = (4u^3 - 4u)\, du$

41. (a) $\int \frac{1}{x^2\sqrt{4 + x^2}}\, dx = \frac{1}{4}\int \frac{\cos\theta}{\sin^2\theta}\, d\theta = \frac{-1}{4\sin\theta} + C$

$= -\frac{1}{4}\csc\theta + C = \frac{-\sqrt{4 + x^2}}{4x} + C$

$x = 2\tan\theta, \quad dx = 2\sec^2\theta\, d\theta, \quad \sqrt{4 + x^2} = 2\sec\theta$

(b) $\int \frac{1}{x^2\sqrt{4 + x^2}}\, dx = -\frac{1}{4}\int \frac{u}{\sqrt{1 + u^2}}\, du = -\frac{1}{4}\sqrt{1 + u^2} + C$

$= -\frac{1}{4}\sqrt{1 + \frac{4}{x^2}} = \frac{-\sqrt{4 + x^2}}{4x} + C$

$x = \frac{2}{u}, \quad dx = -\frac{2\, du}{u^2}$

42. (a) $\int \frac{1}{x\sqrt{4 + x^2}}\, dx = \frac{1}{2}\int \csc\theta\, d\theta = \frac{1}{2}\ln|\csc\theta - \cot\theta| + C$

$= \frac{1}{2}\ln\left|\frac{\sqrt{x^2 + 4} - 2}{x}\right| + C = -\frac{1}{2}\ln\left|\frac{2 + \sqrt{x^2 + 4}}{x}\right| + C$

$x = 2\tan\theta, \quad dx = 2\sec^2\theta\, d\theta$

(b) $\int \frac{1}{x\sqrt{4 + x^2}}\, dx = \int \frac{du}{u^2 - 4} = \frac{1}{4}\ln\left|\frac{u - 2}{u + 2}\right| + C$

$= \frac{1}{4}\ln\left|\frac{\sqrt{4 + x^2} - 2}{\sqrt{4 + x^2} + 2}\right| + C = \frac{1}{4}\ln\left|\frac{(\sqrt{4 + x^2} - 2)^2}{x^2}\right| + C$

$= \frac{1}{2}\ln\left|\frac{\sqrt{4 + x^2} - 2}{x}\right| + C$

43. (a) $\int \frac{x^3}{\sqrt{4 + x^2}}\, dx = 8\int \frac{\sin^3\theta}{\cos^4\theta}\, d\theta = 8\int (\cos^{-4}\theta - \cos^{-2}\theta)\sin\theta\, d\theta$

$= \frac{8}{3}\sec\theta(\sec^2\theta - 3) + C = \frac{\sqrt{4 + x^2}}{3}(x^2 - 8) + C$

(b) $\int \frac{x^3}{\sqrt{4 + x^2}}\, dx = \int (u^2 - 4)\, du = \frac{1}{3}u^3 - 4u + C = \frac{u}{3}(u^2 - 12) + C$

$= \frac{\sqrt{4 + x^2}}{3}(x^2 - 8) + C$

$u^2 = 4 + x^2, \quad 2u\, du = 2x\, dx$

43. (continued)

(c) $\displaystyle\int \frac{x^3}{\sqrt{4 + x^2}}\,dx = x^2\sqrt{4 + x^2} - \int 2x\sqrt{4 + x^2}\,dx$

$$= x^2\sqrt{4 + x^2} - \frac{2}{3}(4 + x^2)^{3/2} + C = \frac{\sqrt{4 + x^2}}{3}(x^2 - 8) + C$$

$u = x^2, \quad v' = \dfrac{x}{\sqrt{4 + x^2}}, \quad u' = 2x, \quad v = \sqrt{4 + x^2}$

44. (a) $\displaystyle\int x\sqrt{4 + x}\,dx = 64\int \tan^3\theta \sec^3\theta\,d\theta$

$$= 64\int (\sec^4\theta - \sec^2\theta)\sec\theta\tan\theta\,d\theta$$

$$= \frac{64\sec^3\theta}{15}(3\sec^2\theta - 5) + C = \frac{2(4 + x)^{3/2}}{15}(3x - 8) + C$$

$x = 4\tan^2\theta, \quad dx = 8\tan\theta\sec^2\theta\,d\theta, \quad \sqrt{4 + x} = 2\sec\theta$

(b) $\displaystyle\int x\sqrt{4 + x}\,dx = 2\int (u^4 - 4u^2)\,du = \frac{2u^3}{15}(3u^2 - 20) + C$

$$= \frac{2(4 + x)^{3/2}}{15}(3x - 8) + C$$

$u^2 = 4 + x, \quad dx = 2u\,du$

(c) $\displaystyle\int x\sqrt{4 + x}\,dx = \int (u^{3/2} - 4u^{1/2})\,du = \frac{2u^{3/2}}{15}(3u - 20) + C$

$$= \frac{2(4 + x)^{3/2}}{15}(3x - 8) + C$$

$u = 4 + x, \quad du = dx$

(d) $\displaystyle\int x\sqrt{4 + x}\,dx = \frac{2x}{3}(4 + x)^{3/2} - \frac{2}{3}\int (4 + x)^{3/2}\,dx$

$$= \frac{2x}{3}(4 + x)^{3/2} - \frac{4}{15}(4 + x)^{5/2} + C = \frac{2(4 + x)^{3/2}}{15}(3x - 8) + C$$

$u = x, \quad v' = \sqrt{4 + x}, \quad u' = 1, \quad v = \dfrac{2}{3}(4 + x)^{3/2}$

45. $\displaystyle\int_0^{\pi/2} \frac{1}{2 - \cos\theta}\,d\theta = 2\int_0^1 \frac{1}{1 + (\sqrt{3}u)^2}\,du = \frac{2}{\sqrt{3}}\arctan(\sqrt{3}u)\Big]_0^1$

$$= \frac{2\pi}{3\sqrt{3}} \approx 1.21$$

$u = \tan\left(\dfrac{\theta}{2}\right), \quad d\theta = \dfrac{2\,du}{1 + u^2}, \quad \cos\theta = \dfrac{1 - u^2}{1 + u^2}$

46. Simpson's Rule (n = 4)

$$\int_0^1 \frac{x^{3/2}}{2 - x^2}\,dx \approx \frac{1}{12}\left[0 + \frac{4(1/4)^{3/2}}{2 - (1/4)^2} + \frac{2(1/2)^{3/2}}{2 - (1/2)^2} + \frac{4(3/4)^{3/2}}{2 - (3/4)^2} + 1\right] \approx 0.29$$

47. Simpson's Rule (n = 8)

$$\int_0^2 \frac{1}{\sqrt{1 + x^3}}\,dx$$

$$\approx \frac{1}{12}\left[1 + \frac{4}{\sqrt{1 + (1/4)^3}} + \frac{2}{\sqrt{1 + (1/2)^3}} + \frac{4}{\sqrt{1 + (3/4)^3}} + \frac{2}{\sqrt{1 + 1^3}} + \frac{4}{\sqrt{1 + (5/4)^3}} + \frac{2}{\sqrt{1 + (6/4)^3}} + \frac{4}{\sqrt{1 + (7/4)^3}} + \frac{1}{\sqrt{1 + 2^3}}\right] \approx 1.40$$

48. Simpson's Rule (n = 4)

$$\int_0^{\pi/4} \theta \tan\theta \, d\theta \approx \frac{\pi/4}{3(4)}\left[0 + 4\left(\frac{\pi}{16}\right)\tan\left(\frac{\pi}{16}\right) + 2\left(\frac{2\pi}{16}\right)\tan\left(\frac{2\pi}{16}\right) + 4\left(\frac{3\pi}{16}\right)\tan\left(\frac{3\pi}{16}\right) + \frac{\pi}{4}\right] \approx 0.186$$

49. Simpson's Rule (n = 4)

$$\int_1^2 \frac{1}{1 + \ln x}\,dx \approx \frac{1}{12}\left[1 + \frac{4}{1 + \ln(1.25)} + \frac{2}{1 + \ln(1.5)} + \frac{4}{1 + \ln(1.75)} + \frac{1}{1 + \ln 2}\right] \approx 0.74$$

50. Simpson's Rule (n = 8)

$$\int_0^2 \sqrt{1 + x^3}\,dx \approx \frac{1}{12}\left[1 + 4\sqrt{1 + (1/4)^3} + 2\sqrt{1 + (1/2)^3} + 4\sqrt{1 + (3/4)^3} + 2\sqrt{1 + 1^3} + 4\sqrt{1 + (5/4)^3} + 2\sqrt{1 + (6/4)^3} + 4\sqrt{1 + (7/4)^3} + 3\right]$$

$$\approx 3.24$$

51. (a) $\displaystyle\int_0^1 e^x\,dx = e^x\Big]_0^1 = e - 1 \approx 1.72$

(b) $\displaystyle\int_0^1 xe^x\,dx = e^x(x - 1)\Big]_0^1 = 1.00$

(c) $\displaystyle\int_0^1 xe^{x^2}\,dx = \frac{1}{2}e^{x^2}\Big]_0^1 = \frac{1}{2}(e - 1) \approx 0.86$

51. (continued)

(d) Simpson's Rule (n = 8)

$$\int_0^1 e^{x^2}\,dx = \frac{1}{24}\Big[1 + 4e^{1/8^2} + 2e^{1/4^2} + 4e^{(3/8)^2} + 2e^{1/2^2} + 4e^{(5/8)^2}$$

$$+ 2e^{(3/4)^2} + 4e^{(7/8)^2} + e\Big] \approx 1.46$$

52. (a) $\displaystyle\int_0^{\pi/2} \cos x\,dx = \sin x\Big]_0^{\pi/2} = 1$

(b) $\displaystyle\int_0^{\pi/2} \cos^2 x\,dx = \frac{1}{2}\int_0^{\pi/2} (1 + \cos 2x)\,dx = \frac{1}{2}\Big[x + \sin x\cos x\Big]_0^{\pi/2}$

$$= \frac{\pi}{4} \approx 0.78$$

(c) Simpson's Rule (n = 4)

$$\int_0^{\pi/2} \cos(x^2)\,dx = \frac{\pi}{24}\Big[1 + 4\cos\left(\frac{\pi}{8}\right)^2 + 2\cos\left(\frac{\pi}{4}\right)^2 + 4\cos\left(\frac{3\pi}{8}\right)^2$$

$$+ \cos\left(\frac{\pi}{2}\right)^2\Big] \approx 0.85$$

(d) Simpson's Rule (n = 4)

$$\int_0^{\pi/2} \cos\sqrt{x}\,dx = \frac{\pi}{24}\Big[1 + 4\cos\sqrt{\pi/8} + 2\cos\sqrt{\pi/4}$$

$$+ 4\cos\sqrt{3\pi/8} + \cos\sqrt{\pi/2}\Big] \approx 1.01$$

53. $\displaystyle\lim_{x\to 1}\left[\frac{(\ln x)^2}{x - 1}\right] = \lim_{x\to 1}\left[\frac{2(1/x)\ln x}{1}\right] = 0$

54. $\displaystyle\lim_{x\to k}\left(\frac{x^{1/3} - k^{1/3}}{x - k}\right) = \lim_{x\to k}\left(\frac{1/3\,x^{-2/3}}{1}\right) = \frac{1}{3\sqrt[3]{k^2}}$

55. $\displaystyle\lim_{x\to\infty}\frac{e^{2x}}{x^2} = \lim_{x\to\infty}\frac{2e^{2x}}{2x} = \lim_{x\to\infty}\frac{4e^{2x}}{2} = \infty$

56. $\displaystyle\lim_{x\to 1^+}\left(\frac{2}{\ln x} - \frac{3}{x - 1}\right) = \lim_{x\to 1^+}\left[\frac{2x - 2 - 3\ln x}{(\ln x)(x - 1)}\right]$

$$= \lim_{x\to 1^+}\left[\frac{2 - (3/x)}{(x - 1)(1/x) + \ln x}\right] = -\infty$$

57. $y = \lim_{x \to \infty} (\ln x)^{2/x}$, $\quad \ln y = \lim_{x \to \infty} \dfrac{2 \ln(\ln x)}{x}$

$$= \lim_{x \to \infty} \left[\frac{\frac{2}{x \ln x}}{1} \right] = 0$$

Since $\ln y = 0$, $y = 1$

58. $y = \lim_{x \to 1} (x - 1)^{\ln x}$, $\quad \ln y = \lim_{x \to 1} \left[(\ln x) \ln(x - 1) \right]$

$$= \lim_{x \to 1} \left[\frac{\ln(x - 1)}{\frac{1}{\ln x}} \right] = \lim_{x \to 1} \left[\frac{\frac{1}{x - 1}}{(\frac{1}{x})\frac{-1}{\ln^2 x}} \right]$$

$$= \lim_{x \to 1} \left[\frac{-\ln^2 x}{(\frac{x - 1}{x})} \right] = \lim_{x \to 1} \left[\frac{-2(1/x)(\ln x)}{(1/x^2)} \right] = \lim_{x \to 1} 2x(\ln x) = 0$$

Since $\ln y = 0$, $y = 1$

59. $\lim_{n \to \infty} 1000(1 + \frac{0.09}{n})^n = 1000e^{0.09} \approx 1094.17$

60. $\lim_{x \to \infty} xe^{-x^2} = \lim_{x \to \infty} \dfrac{x}{e^{x^2}} = \lim_{x \to \infty} \dfrac{1}{2xe^{x^2}} = 0$

61. $\lim_{x \to 0} \csc 3x \tan \pi x = \lim_{x \to 0} \dfrac{\tan \pi x}{\sin 3x} = \lim_{x \to 0} \dfrac{\pi \sec^2 \pi x}{3 \cos 3x} = \dfrac{\pi}{3}$

62. $\lim_{x \to 0} \dfrac{\sin \pi x}{\sin 2\pi x} = \lim_{x \to 0} \dfrac{\pi \cos \pi x}{2\pi \cos 2\pi x} = \dfrac{\pi}{2\pi} = \dfrac{1}{2}$

63. $s = \int_0^{\pi} \sqrt{1 + \cos^2 x}\, dx \approx 3.82$

64. $s = \int_0^{\pi} \sqrt{1 + \sin^2 2x}\, dx \approx 3.82$

65. By symmetry $\bar{x} = 0$, $A = \dfrac{1}{2}\pi$

$$\bar{y} = \frac{2}{\pi}(\frac{1}{2}) \int_{-1}^{1} (\sqrt{1 - x^2})^2\, dx = \frac{1}{\pi}\left[x - \frac{1}{3}x^3 \right]_{-1}^{1} = \frac{4}{3\pi}$$

$(\bar{x}, \bar{y}) = (0, \frac{4}{3\pi})$

66. By symmetry $\overline{y} = 0$

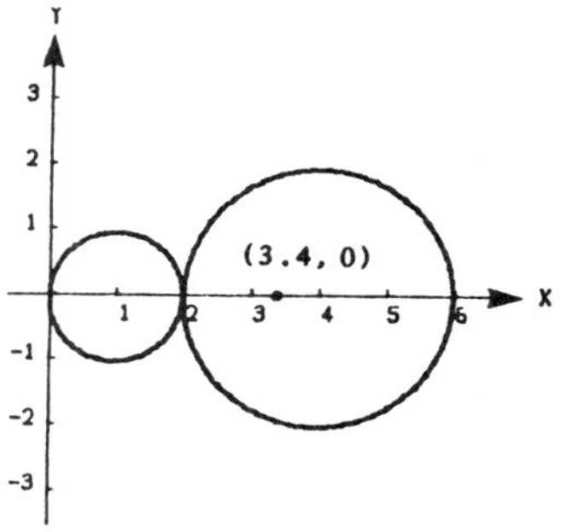

$A = \pi + 4\pi = 5\pi$

$$\overline{x} = \frac{1}{5\pi}\left[\int_0^2 x\sqrt{1-(x-1)^2}\,dx + \int_2^6 x\sqrt{4-(x-4)^2}\,dx\right]$$

$$= \frac{1}{5\pi}\left[\int_{-\pi/2}^{\pi/2} (1+\sin\theta)\cos^2\theta\,d\theta + \int_{-\pi/2}^{\pi/2} (4+2\sin\theta)(4\cos^2\theta)\,d\theta\right]$$

By the trig substitutions $x - 1 = \sin\theta$, and $x - 4 = 2\sin\theta$

$$= \frac{1}{5\pi}\left[\frac{1}{2}\theta + \frac{1}{4}\sin 2\theta - \frac{1}{3}\cos^3\theta + 8\left(\theta + \frac{1}{2}\sin 2\theta\right) + \frac{8}{3}\cos^3\theta\right]_{-\pi/2}^{\pi/2}$$

$$= \frac{17\pi}{5\pi} = 3.4$$

67. $$\int_2^\infty \left[\frac{1}{x^5} + \frac{1}{x^{10}} + \frac{1}{x^{15}}\right]dx < \int_2^\infty \frac{1}{x^5 - 1}\,dx < \int_0^\infty \left[\frac{1}{x^5} + \frac{1}{x^{10}} + \frac{2}{x^{15}}\right]dx$$

$$\left[-\frac{1}{4x^4} - \frac{1}{9x^9} - \frac{1}{14x^{14}}\right]_2^\infty < \int_2^\infty \frac{1}{x^5 - 1}\,dx < \left[-\frac{1}{4x^4} - \frac{1}{9x^9} - \frac{1}{7x^{14}}\right]_2^\infty$$

$$0.015846 < \int_2^\infty \frac{1}{x^5 - 1}\,dx < 0.015851$$

68. $$I_n = \int_0^\infty \frac{x^{2n-1}}{(x^2+1)^{n+3}} = \left[\frac{-x^{2n-2}}{2(n+2)(x^2+1)^{n+2}}\right]_0^\infty$$

$$+ \frac{n-1}{n+2}\int_0^\infty \frac{x^{2n-3}}{(x^2+1)^{n+2}} = 0 + \frac{n-1}{n+2}(I_{n-1}) = \frac{n-1}{n+2}(I_{n-1})$$

$u = x^{n-2}$, $v' = \dfrac{x}{(x^2+1)^{n+3}}$, $u' = (2n-2)x^{2n-3}$, $v = \dfrac{-1}{2(n+2)(x^2+1)^{n+2}}$

(a) $$\int_0^\infty \frac{x^3}{(x^2+1)^5}\,dx = \frac{1}{4}\int_0^\infty \frac{x}{(x^2+1)^4}\,dx = \left[\frac{-1}{24(x^2+1)^3}\right]_0^\infty = \frac{1}{24}$$

(b) $$\int_0^\infty \frac{x^5}{(x^2+1)^6}\,dx = \frac{2}{5}\int_0^\infty \frac{x^3}{(x^2+1)^5}\,dx = \frac{1}{60}$$

(c) $$\int_0^\infty \frac{x^7}{(x^2+1)^7}\,dx = \frac{1}{2}\int_0^\infty \frac{x^5}{(x^2+1)^6}\,dx = \frac{1}{120}$$

9 Infinite series

9.1 Introduction: Taylor polynomials and approximations

1. $f(x) = e^{-x}$, $f'(x) = -e^{-x}$, $f''(x) = e^{-x}$, $f'''(x) = -e^{-x}$

$$P_3(x) = f(0) + f'(0)x + \frac{f''(0)}{2!}x^2 + \frac{f'''(0)}{3!}x^3 = 1 - x + \frac{x^2}{2} - \frac{x^3}{6}$$

2. $f(x) = e^{-x}$, $f'(x) = -e^{-x}$, $f''(x) = e^{-x}$, $f'''(x) = -e^{-x}$,

$f^{(4)}(x) = e^{-x}$, $f^{(5)}(x) = -e^{-x}$

$$P_5(x) = f(0) + f'(0)x + \frac{f''(0)}{2!}x^2 + \frac{f'''(0)}{3!}x^3 + \frac{f^{(4)}(0)}{4!}x^4 + \frac{f^{(5)}(0)}{5!}x^5$$

$$= 1 - x + \frac{x^2}{2} - \frac{x^3}{6} + \frac{x^4}{24} - \frac{x^5}{120}$$

3. $f(x) = e^{2x}$, $f'(x) = 2e^{2x}$, $f''(x) = 4e^{2x}$, $f'''(x) = 8e^{2x}$,

$f^{(4)}(x) = 16e^{2x}$

$$P_4(x) = 1 + 2x + \frac{4}{2!}x^2 + \frac{8}{3!}x^3 + \frac{16}{4!}x^4$$

$$= 1 + 2x + 2x^2 + \frac{4}{3}x^3 + \frac{2}{3}x^4$$

4. $f(x) = e^{3x}$, $f'(x) = 3e^{3x}$, $f''(x) = 9e^{3x}$, $f'''(x) = 27e^{3x}$,

$f^{(4)}(x) = 81e^{3x}$

$$P_4(x) = 1 + 3x + \frac{9}{2!}x^2 + \frac{27}{3!}x^3 + \frac{81}{4!}x^4$$

$$= 1 + 3x + \frac{9}{2}x^2 + \frac{9}{2}x^3 + \frac{27}{8}x^4$$

5. $f(x) = \sin x, \quad f'(x) = \cos x, \quad f''(x) = -\sin x, \quad f'''(x) = -\cos x$

$f^{(4)}(x) = \sin x, \quad f^{(5)}(x) = \cos x$

$$P_5(x) = 0 + (1)x + \frac{0}{2!}x^2 + \frac{-1}{3!}x^3 + \frac{0}{4!}x^4 + \frac{1}{5!}x^5$$

$$= x - \frac{1}{6}x^3 + \frac{1}{120}x^5$$

6. $f(x) = \sin \pi x, \quad f'(x) = \pi \cos \pi x, \quad f''(x) = -\pi^2 \sin \pi x,$

$f'''(x) = -\pi^3 \cos \pi x$

$$P_3(x) = 0 + \pi x + \frac{0}{2!}x^2 + \frac{-\pi^3}{3!}x^3 = \pi x - \frac{\pi^3}{6}x^3$$

7. $f(x) = xe^x, \quad f'(x) = xe^x + e^x, \quad f''(x) = xe^x + 2e^x$

$f'''(x) = xe^x + 3e^x, \quad f^{(4)}(x) = xe^x + 4e^x$

$$P_4(x) = 0 + x + \frac{2}{2!}x^2 + \frac{3}{3!}x^3 + \frac{4}{4!}x^4$$

$$= x + x^2 + \frac{1}{2}x^3 + \frac{1}{6}x^4$$

8. $f(x) = x^2e^{-x}, \quad f'(x) = 2xe^{-x} - x^2e^{-x}, \quad f''(x) = 2e^{-x} - 4xe^{-x} + x^2e^{-x}$

$f'''(x) = -6e^{-x} + 6xe^{-x} - x^2e^{-x}, \quad f^{(4)}(x) = 12e^{-x} - 8xe^{-x} + x^2e^{-x}$

$$P_4(x) = 0 + 0x + \frac{2}{2!}x^2 + \frac{-6}{3!}x^3 + \frac{12}{4!}x^4 = x^2 - x^3 + \frac{1}{2}x^4$$

9. $f(x) = \dfrac{1}{x + 1}, \quad f'(x) = -\dfrac{1}{(x + 1)^2}, \quad f''(x) = \dfrac{2}{(x + 1)^3}$

$f'''(x) = \dfrac{-6}{(x + 1)^4}, \quad f^{(4)}(x) = \dfrac{24}{(x + 1)^5}$

$$P_4(x) = 1 - x + \frac{2}{2!}x^2 + \frac{-6}{3!}x^3 + \frac{24}{4!}x^4 = 1 - x + x^2 - x^3 + x^4$$

10. $f(x) = \dfrac{1}{x^2 + 1}, \quad f'(x) = \dfrac{-2x}{(x^2 + 1)^2}, \quad f''(x) = \dfrac{2(3x^2 - 1)}{(x^2 + 1)^3}$

$f'''(x) = \dfrac{24x(1 - x^2)}{(x^2 + 1)^4}, \quad f^{(4)}(x) = \dfrac{24(5x^4 - 10x^2 + 1)}{(x^2 + 1)^5}$

$$P_4(x) = 1 + 0x + \frac{-2}{2!}x^2 + \frac{0}{3!}x^3 + \frac{24}{4!}x^4 = 1 - x^2 + x^4$$

11. $f(x) = \sec x, \quad f'(x) = \sec x \tan x, \quad f''(x) = \sec^3 x + \sec x \tan^2 x$

$P_2(x) = 1 + 0x + \frac{1}{2!}x^2 = 1 + \frac{1}{2}x^2$

12. $f(x) = \tan x, \quad f'(x) = \sec^2 x, \quad f''(x) = 2\sec^2 x \tan x$

$f'''(x) = 4\sec^2 x \tan^2 x + 2\sec^4 x$

$P_3(x) = 0 + x + \frac{0}{2!}x^2 + \frac{2}{3!}x^3 = x + \frac{1}{3}x^3$

13. $f(x) = 2 - 3x^3 + x^4, \quad f'(x) = -9x^2 + 4x^3, \quad f''(x) = -18x + 12x^2$

$f'''(x) = -18 + 24x, \quad f^{(4)}(x) = 24$

$P_4(x) = 2 + 0x + \frac{0}{2!}x^2 + \frac{-18}{3!}x^3 + \frac{24}{4!}x^4 = 2 - 3x^3 + x^4$

14. $f(x) = 3 - 2x^2 + x^3, \quad f'(x) = -4x + 3x^2, \quad f''(x) = -4 + 6x, \quad f'''(x) = 6$

$P_3(x) = 3 + 0x + \frac{-4}{2!}x^2 + \frac{6}{3!}x^3 = 3 - 2x^2 + x^3$

15. $f(x) = \frac{1}{x}, \quad f'(x) = -\frac{1}{x^2}, \quad f''(x) = \frac{2}{x^3}, \quad f'''(x) = -\frac{6}{x^4}, \quad f^{(4)}(x) = \frac{24}{x^5}$

$f(1) = 1, \quad f'(1) = -1, \quad f''(1) = 2, \quad f'''(1) = -6, \quad f^{(4)}(1) = 24$

$$P_4(x) = 1 - (x - 1) + \frac{2}{2!}(x - 1)^2 + \frac{-6}{3!}(x - 1)^3 + \frac{24}{4!}(x - 1)^4$$
$$= 1 - (x - 1) + (x - 1)^2 - (x - 1)^3 + (x - 1)^4$$

16. $f(x) = \sqrt{x}, \quad f'(x) = \frac{1}{2\sqrt{x}}, \quad f''(x) = -\frac{1}{4x\sqrt{x}}, \quad f'''(x) = \frac{3}{8x^2\sqrt{x}}$

$f^{(4)}(x) = -\frac{15}{16x^3\sqrt{x}}$

$f(4) = 2, \quad f'(4) = \frac{1}{4}, \quad f''(4) = -\frac{1}{32}, \quad f'''(4) = \frac{3}{256}, \quad f^{(4)}(4) = -\frac{15}{2048}$

$$P_4(x) = 2 + \frac{1}{4}(x - 4) + \frac{-1/32}{2!}(x - 4)^2 + \frac{3/256}{3!}(x - 4)^3 + \frac{-15/2048}{4!}(x - 4)^4$$
$$= 2 + \frac{1}{4}(x - 4) - \frac{1}{64}(x - 4)^2 + \frac{1}{512}(x - 4)^3 - \frac{5}{16384}(x - 4)^4$$

17. $f(x) = x^2 \cos x, \quad f'(x) = 2x \cos x - x^2 \sin x$

$f''(x) = 2 \cos x - 4x \sin x - x^2 \cos x$

$f(\pi) = -\pi^2, \quad f'(\pi) = -2\pi, \quad f''(\pi) = -2 + \pi^2$

$P_2(x) = -\pi^2 - 2\pi(x - \pi) + \dfrac{(\pi^2 - 2)}{2}(x - \pi)^2$

18. $f(x) = \ln x, \quad f'(x) = \dfrac{1}{x}, \quad f''(x) = -\dfrac{1}{x^2}, \quad f'''(x) = \dfrac{2}{x^3}, \quad f^{(4)}(x) = -\dfrac{6}{x^4}$

$f(1) = 0, \quad f'(1) = 1, \quad f''(1) = -1, \quad f'''(1) = 2, \quad f^{(4)}(1) = -6$

$P_4(x) = 0 + (x - 1) - \dfrac{1}{2}(x - 1)^2 + \dfrac{1}{3}(x - 1)^3 - \dfrac{1}{4}(x - 1)^4$

19. $f(x) = \sin x, \quad P_1(x) = x, \quad P_3(x) = x - \dfrac{1}{6}x^3,$

$P_5(x) = x - \dfrac{1}{6}x^3 + \dfrac{1}{120}x^5$

x	$\sin x$	$P_1(x)$	$P_3(x)$	$P_5(x)$
0.00	0.0000	0.00	0.0000	0.0000
0.25	0.2474	0.25	0.2474	0.2474
0.50	0.4794	0.50	0.4792	0.4794
0.75	0.6816	0.75	0.6797	0.6817
1.00	0.8415	1.00	0.8333	0.8417

20. $f(x) = \ln x, \quad P_1(x) = x - 1$

$P_4(x) = (x - 1) - \dfrac{1}{2}(x - 1)^2 + \dfrac{1}{3}(x - 1)^3 - \dfrac{1}{4}(x - 1)^4$

x	$\ln x$	$P_1(x)$	$P_4(x)$
1.00	0.0000	0.00	0.0000
1.25	0.2231	0.25	0.2230
1.50	0.4055	0.50	0.4010
1.75	0.5596	0.75	0.5303
2.00	0.6931	1.00	0.5833

21. $f(x) = e^{-x} \approx 1 - x + \dfrac{x^2}{2} - \dfrac{x^3}{6}, \quad f(\tfrac{1}{2}) \approx 0.6042$

22. $f(x) = x^2e^{-x} \approx x^2 - x^3 + \dfrac{1}{2}x^4, \quad f(\tfrac{1}{4}) \approx 0.0488$

23. $f(x) = x^2 \cos x \approx -\pi^2 - 2\pi(x - \pi) + (\frac{\pi^2 - 2}{2})(x - \pi)^2$, $f(\frac{7\pi}{8}) \approx -6.7954$

24. $f(x) = \sqrt{x} \approx 2 + \frac{1}{4}(x - 4) - \frac{1}{64}(x - 4)^2 + \frac{1}{512}(x - 4)^3 - \frac{5}{16384}(x - 4)^4$

$f(5) \approx 2.2360$

25. $f(x) = e^x \approx 1 + x + \frac{x^2}{2} + \frac{x^3}{6} + \frac{x^4}{24} = P_4(x)$

$g(x) = xe^x \approx x + x^2 + \frac{x^3}{2} + \frac{x^4}{6} = Q_4(x)$

$Q_4(x) = xP_4(x) - \frac{x^5}{24}$

26. $f(x) = \sin x$, $P_5(x) = x - \frac{1}{6}x^3 + \frac{1}{120}x^5$, $P_5'(x) = 1 - \frac{1}{2}x^2 + \frac{1}{24}x^4$

$g(x) = \cos x$, $Q_4(x) = 1 - \frac{1}{2}x^2 + \frac{1}{24}x^4$, $P_5'(x) = Q_4(x)$

27. $f(x) = e^x \approx 1 + x + \frac{x^2}{2} + \frac{x^3}{6}$, $x < 0$, $R_4(x) = \frac{e^z}{4!}x^4 < 0.001$

$e^z x^4 < 0.024$, $xe^{z/4} < 0.3936$, $x < \frac{0.3936}{e^{z/4}} < 0.3936$ for $z < 0$

$-0.3936 < x < 0$

28. $f(x) = \sin x \approx x - \frac{x^3}{3!}$, $|R_4(x)| = \left|\frac{\sin z}{4!}x^4\right| \le \frac{|x^4|}{4!} < 0.001$

$x^4 < 0.024$, $|x| < 0.3936$, $-0.3936 < x < 0.3936$

29. $f(x) = e^x \approx 1 + x + \frac{x^2}{2} + \frac{x^3}{6} + \frac{x^4}{24} + \frac{x^5}{120}$

$R_6(x) = \frac{e^z}{6!}x^6$ where $0 < x < 0.6$

$R_6(x) = \frac{e^z}{6!}(0.6)^6 < \frac{e^{0.6}}{6!}(0.6)^6 \approx 0.00012$

30. $f(x) = \ln(x + 1)$, $f^{(n+1)}(x) = \frac{(-1)^{n+1}n!}{(x + 1)^{n+1}}$

$R_n(0.5) = \frac{n!}{(z + 1)^{n+1}} \cdot \frac{1}{(n + 1)!}(0.5)^{n+1}$ where $0 < z < 0.5$

$\frac{(0.5)^{n+1}}{(1)^{n+1}(n + 1)} < 0.0001$, By trial and error, $n = 9$

31.(a) $f(x) = \arcsin x$, $P_3(x) = x + \frac{x^3}{6}$

(b)

x	-1	-0.75	-0.50	-0.25
$f(x)$ $P_3(x)$	-1.5708 -1.1667	-0.8481 -0.8203	-0.5236 -0.5208	-0.2527 -0.2526

x	0	0.25	0.50	0.75	1
$f(x)$ $P_3(x)$	0 0	0.2527 0.2526	0.5236 0.5208	0.8481 0.8203	1.5708 1.1667

(c)

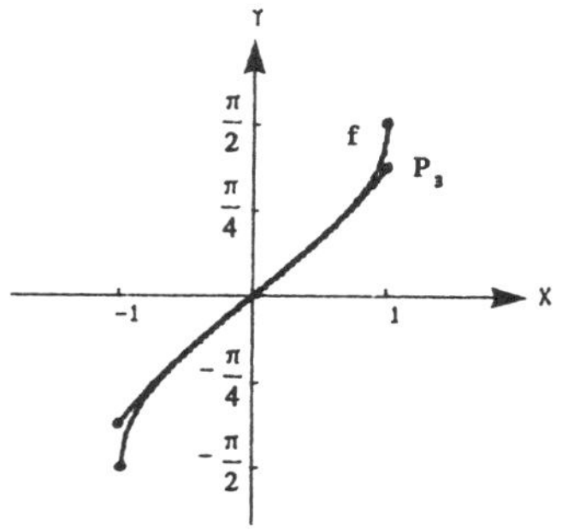

32.(a) $f(x) = \arctan x$, $P_3(x) = x - \frac{x^3}{3}$

(b)

x	-1	-0.75	-0.50	-0.25
$f(x)$ $P_3(x)$	-0.7854 -0.6667	-0.6435 -0.6094	-0.4636 -0.4583	-0.2450 -0.2448

x	0	0.25	0.50	0.75	1
$f(x)$ $P_3(x)$	0 0	0.2450 0.2448	0.4636 0.4583	0.6435 0.6094	0.7854 0.6667

(c)

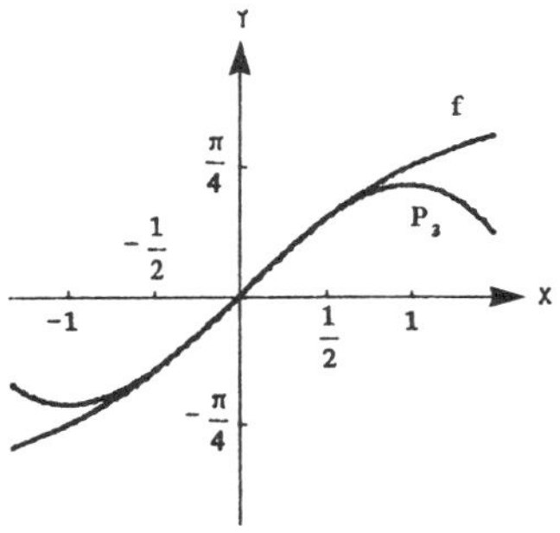

9.2 Sequences

1. $\{2, 4, 8, 16, 32, \ldots\}$

2. $\{\frac{1}{2}, \frac{2}{3}, \frac{3}{4}, \frac{4}{5}, \frac{5}{6}, \ldots\}$

3. $\{-\frac{1}{2}, \frac{1}{4}, -\frac{1}{8}, \frac{1}{16}, -\frac{1}{32}, \ldots\}$

4. $\{1, 0, -1, 0, 1, \ldots\}$

5. $\{\frac{3}{1!}, \frac{3^2}{2!}, \frac{3^3}{3!}, \frac{3^4}{4!}, \frac{3^5}{5!}, \ldots\}$

6. $\{5, \frac{19}{4}, \frac{43}{9}, \frac{77}{16}, \frac{121}{25}, \ldots\}$

7. $\{-1, -\frac{1}{4}, \frac{1}{9}, \frac{1}{16}, -\frac{1}{25}, \ldots\}$

8. $\{3, 6, 9, 12, 15, \ldots\}$

9. $a_n = 3n - 2$

10. $a_n = 4n - 1$

11. $a_n = n^2 - 2$

12. $a_n = \frac{1}{n^2}$

13. $a_n = \frac{n + 1}{n + 2}$

14. $a_n = \frac{n + 1}{2n - 1}$

15. $a_n = \frac{(-1)^{n-1}}{2^{n-2}}$

16. $a_n = \frac{1}{2}(\frac{2}{3})^{n-1}$

17. $a_n = 2 - \frac{n - 1}{n} = \frac{n + 1}{n}$

18. $a_n = 1 + \frac{2^n - 1}{2^n} = \frac{2^{n+1} - 1}{2^n}$

19. $a_n = \frac{n}{(n + 1)(n + 2)}$

20. $a_n = \frac{1}{n!}$

21. $a_n = \frac{1}{1\cdot 3\cdot 5\cdots(2n - 1)} = \frac{2^n n!}{(2n)!}$

22. $a_n = (-1)^{n+1}(2n)$

23. $a_n = (-1)^{n(n-1)/2}$

24. $a_n = \frac{x^{n-1}}{(n - 1)!}$

25. $\lim_{n\to\infty} \frac{n + 1}{n} = 1$, converges

26. $\lim_{n\to\infty} \frac{1}{n^{3/2}} = 0$, converges

27. $\lim_{n\to\infty} (-1)^n(\frac{n}{n + 1})$ does not exist, diverges

28. $\lim_{n\to\infty} (\frac{n - 1}{n} - \frac{n}{n - 1}) = \lim_{n\to\infty} \frac{(n - 1)^2 - n^2}{n(n - 1)}$

$= \lim_{n\to\infty} \frac{1 - 2n}{n^2 - n} = 0$, converges

29. $\lim_{n\to\infty} \frac{3n^2 - n + 4}{2n^2 + 1} = \frac{3}{2}$, converges

30. $\lim_{n\to\infty} \frac{\sqrt{n}}{\sqrt{n} + 1} = 1$, converges

31. $\lim_{n\to\infty} \frac{n^2 - 1}{n + 1} = \infty$, diverges

32. $\lim_{n\to\infty} [1 + (-1)^n]$ does not exist, diverges

33. $\lim_{n\to\infty} \frac{1 + (-1)^n}{n} = 0$, converges

34. $\lim_{n\to\infty} \frac{\ln(n^2)}{n} = \lim_{n\to\infty} \frac{2/n}{1} = 0$, converges

35. $\lim_{n \to \infty} \cos\left(\frac{n\pi}{2}\right)$ does not exist, diverges

36. $\lim_{n \to \infty} \frac{n}{\sqrt{n^2 + 1}} = 1$, converges

37. $\lim_{n \to \infty} \left(\frac{3}{4}\right)^n = 0$, converges

38. $\lim_{n \to \infty} \frac{(n - 2)!}{n!} = \lim_{n \to \infty} \frac{1}{n(n - 1)} = 0$, converges

39.

n	1	2	3	4	5
$f^{(n-1)}(2)$	ln(2)	1/2	-1/4	2/8	-6/16

$a_n = \frac{(-1)^n(n - 2)!}{2^{n-1}}$, for $n > 1$, $\lim_{n \to \infty} \frac{(-1)^n(n - 2)!}{2^{n-1}} = \infty$, diverges

40. $\lim_{n \to \infty} \left(\frac{n^2}{2n + 1} - \frac{n^2}{2n - 1}\right) = \lim_{n \to \infty} \frac{-2n^2}{4n^2 - 1} = -\frac{1}{2}$, converges

41. $\lim_{n \to \infty} \left(3 - \frac{1}{2^n}\right) = 3$, converges

42. $0 < \frac{n!}{n^n} = \frac{n}{n} \cdot \frac{n - 1}{n} \cdot \frac{n - 2}{n} \cdot \frac{n - 3}{n} \cdots \frac{3}{n} \cdot \frac{2}{n} \cdot \frac{1}{n} < \frac{1}{n}$ for $n > 1$

Since $\lim_{n \to \infty} \frac{1}{n} = 0$, we have $\lim_{n \to \infty} \frac{n!}{n^n} = 0$, converges

43. $a_n = \left(1 + \frac{k}{n}\right)^n$

$\lim_{n \to \infty} \left(1 + \frac{k}{n}\right)^n = \lim_{u \to 0} [(1 + u)^{1/u}]^k = e^k$, where $u = \frac{k}{n}$, converges

44. $\lim_{n \to \infty} 2^{1/n} = 1$, converges

45. $\lim_{n \to \infty} \frac{n^p}{e^n} = 0$, converges

46. $a_n = n\sin\frac{1}{n}$, Let $f(x) = x\sin\frac{1}{x}$

$$\lim_{x\to\infty} x\sin\frac{1}{x} = \lim_{x\to\infty}\frac{\sin(1/x)}{1/x} = \lim_{x\to\infty}\frac{(-1/x^2)\cos(1/x)}{-1/x^2}$$

$$= \lim_{x\to\infty}\cos\frac{1}{x} = \cos 0 = 1 \qquad \text{(L'Hopital's Rule)}$$

Therefore $\lim_{n\to\infty} n\sin\frac{1}{n} = 1$

47. $a_n = 4 - \frac{1}{n} < 4 - \frac{1}{n+1} = a_{n+1}$, Monotonic

48. Let $f(x) = \frac{4x}{x+1}$, then $f'(x) = \frac{4}{(x+1)^2}$ thus f is increasing which implies $\{a_n\}$ is increasing. Monotonic

49. $a_n = \frac{\cos n}{n}$, $a_1 = 0.5403$, $a_2 = -0.2081$, $a_3 = -0.3230$, $a_4 = -0.1634$

Not monotonic

50. $a_n = ne^{-n/2}$, $a_1 = 0.6065$, $a_2 = 0.7358$, $a_3 = 0.6694$

Not monotonic

51. $a_n = (-1)^n(\frac{1}{n})$, $a_1 = -1$, $a_2 = \frac{1}{2}$, $a_3 = -\frac{1}{3}$

Not monotonic

52. $a_n = (-\frac{2}{3})^n$, $a_1 = -\frac{2}{3}$, $a_2 = \frac{4}{9}$, $a_3 = -\frac{8}{27}$, Not monotonic

53. $a_n = (\frac{2}{3})^n > (\frac{2}{3})^{n+1} = a_{n+1}$, Monotonic

54. $a_n = (\frac{3}{2})^n < (\frac{3}{2})^{n+1} = a_{n+1}$, Monotonic

55. $a_n = \sin(\frac{n\pi}{6})$, $a_1 = 0.500$, $a_2 = 0.8660$, $a_3 = 1.000$, $a_4 = 0.500$

Not monotonic

56. $\dfrac{n}{2^{n+2}} \overset{?}{\geq} \dfrac{n+1}{2^{(n+1)+2}}$, $\quad 2^{n+3}n \overset{?}{\geq} 2^{n+2}(n+1)$

$2n \overset{?}{\geq} n + 1$, $\quad n \geq 1$, True, Monotonic

57. $a_n = 5 + \dfrac{1}{n}$, $\quad \left|5 + \dfrac{1}{n}\right| \leq 6 \Longrightarrow \{a_n\}$ is bounded

$5 + \dfrac{1}{n} > 5 + \dfrac{1}{n+1} \Longrightarrow \{a_n\}$ is monotonic, $\quad \lim_{n \to \infty} (5 + \dfrac{1}{n}) = 5$

58. $a_n = 3 - \dfrac{4}{n}$, $\quad \left|3 - \dfrac{4}{n}\right| < 3 \Longrightarrow \{a_n\}$ is bounded

$3 - \dfrac{4}{n} < 3 - \dfrac{4}{n+1} \Longrightarrow \{a_n\}$ is monotonic, $\quad \lim_{n \to \infty} (3 - \dfrac{4}{n}) = 3$

59. $a_n = \dfrac{1}{3}(1 - \dfrac{1}{3^n})$, $\quad \left|\dfrac{1}{3}(1 - \dfrac{1}{3^n})\right| < \dfrac{1}{3} \Longrightarrow \{a_n\}$ is bounded

$\dfrac{1}{3}(1 - \dfrac{1}{3^n}) < \dfrac{1}{3}(1 - \dfrac{1}{3^{n+1}}) \Longrightarrow \{a_n\}$ is monotonic

$\lim_{n \to \infty} \left[\dfrac{1}{3}(1 - \dfrac{1}{3^n})\right] = \dfrac{1}{3}$

60. $a_n = 4 + \dfrac{1}{2^n}$, $\quad \left|4 + \dfrac{1}{2^n}\right| \leq 4.5 \Longrightarrow \{a_n\}$ is bounded

$4 + \dfrac{1}{2^n} > 4 + \dfrac{1}{2^{n+1}} \Longrightarrow \{a_n\}$ is monotonic, $\quad \lim_{n \to \infty} (4 + \dfrac{1}{2^n}) = 4$

61. $A_n = P\left[1 + \dfrac{r}{12}\right]^n$

(a) $\lim_{n \to \infty} A_n = \infty$, Divergent

(b) $A_n = 9000\left[1 + \dfrac{0.115}{12}\right]^n$

A_1 = \$9086.25	A_6 = \$9530.06
A_2 = \$9173.33	A_7 = \$9621.39
A_3 = \$9261.24	A_8 = \$9713.59
A_4 = \$9349.99	A_9 = \$9806.68
A_5 = \$9439.60	A_{10} = \$9900.66

62. $A_n = 100(101)[(1.01)^n - 1]$

(a) $A_1 = \$101.00$
$A_2 = \$203.01$
$A_3 = \$306.04$
$A_4 = \$410.10$
$A_5 = \$515.20$
$A_6 = \$621.35$

(b) $A_{60} = \$8248.64$

(c) $A_{240} = \$99{,}914.79$

63. (a) $A_n = (0.8)^n(2.5)$ billion

(b) $A_1 = \$2$ billion
$A_2 = \$1.6$ billion
$A_3 = \$1.28$ billion
$A_4 = \$1.024$ billion

(c) $\lim_{n \to \infty} (0.8)^n(2.5) = 0$

64. $P_n = \$11{,}000(1.055)^n$
$P_1 = \$11{,}605.00$
$P_2 = \$12{,}243.28$
$P_3 = \$12{,}916.66$
$P_4 = \$13{,}627.07$
$P_5 = \$14{,}376.56$

65. $S_6 = 130 + 70 + 40 = 240$
$S_7 = 240 + 130 + 70 = 440$
$S_8 = 440 + 240 + 130 = 810$
$S_9 = 810 + 440 + 240 = 1490$
$S_{10} = 1490 + 810 + 440 = 2740$

66. $x_n = x_{n-1} - \dfrac{f(x_{n-1})}{f'(x_{n-1})}$, $\quad f(x) = x^2 + x - 1$, $\quad f'(x) = 2x + 1$

$$x_1 = 0.5, \quad x_2 = 0.5 - \frac{(0.5)^2 + 0.5 - 1}{2(0.5) + 1} = 0.62500$$

$$x_3 = 0.6250 - \frac{(0.6250)^2 + 0.6250 - 1}{2(0.6250) + 1} = 0.61806$$

$$x_4 = 0.61806 - \frac{(0.61806)^2 + 0.61806 - 1}{2(0.61806) + 1} = 0.61803$$

67. $$\lim_{n \to \infty} \frac{ar^n}{1 - r} = \frac{a}{1 - r} \lim_{n \to \infty} r^n = \begin{cases} \infty, & \text{if } r > 1 \\ \text{does not exist}, & \text{if } r < -1 \end{cases}$$

Therefore, the sequence diverges if $|r| > 1$.

If $|r| < 1$, we have $\displaystyle\lim_{n \to \infty} \frac{ar^n}{1 - r} = \frac{a}{1 - r} \lim_{n \to \infty} r^n = 0$ and the sequence converges.

68. Since $\lim_{n\to\infty} s_n = L > 0$, there exists, for each $\varepsilon > 0$, an integer N such that $|s_n - L| < \varepsilon$ for every $n > N$. Let $\varepsilon = L > 0$ and we have

$|s_n - L| < L$, $-L < s_n - L < L$, or $0 < s_n < 2L$ for each $n > N$.

69. Since $\{s_n\}$ is a convergent sequence, $\lim_{n\to\infty} s_n = L$. For each $\varepsilon > 0$, there exists an integer N such that $|s_n - L| < \frac{\varepsilon}{2}$ for $n > N$ and

$|s_{n-1} - L| < \frac{\varepsilon}{2}$ for $n > N + 1$. For $n > N + 1$, we have

$$|s_n - s_{n-1}| = |(s_n - L) - (s_{n-1} - L)| \le |s_n - L| + |s_{n-1} - L| < \frac{\varepsilon}{2} + \frac{\varepsilon}{2} = \varepsilon$$

Therefore, $0 = \lim_{n\to\infty} (s_n - s_{n-1}) = \lim_{n\to\infty} s_n - \lim_{n\to\infty} s_{n-1}$

and $\lim_{n\to\infty} s_n = \lim_{n\to\infty} s_{n-1}$

9.3 Series and convergence

1. $s_1 = 1$

$s_2 = 1 + \frac{1}{4} = 1.2500$

$s_3 = 1 + \frac{1}{4} + \frac{1}{9} = 1.3611$

$s_4 = 1 + \frac{1}{4} + \frac{1}{9} + \frac{1}{16} = 1.4236$

$s_5 = 1 + \frac{1}{4} + \frac{1}{9} + \frac{1}{16} + \frac{1}{25} = 1.4636$

2. $s_1 = \frac{1}{6} = 0.1667$

$s_2 = \frac{1}{6} + \frac{1}{6} = 0.3333$

$s_3 = \frac{1}{6} + \frac{1}{6} + \frac{3}{20} = 0.4833$

$s_4 = \frac{1}{6} + \frac{1}{6} + \frac{3}{20} + \frac{2}{15} = 0.6167$

$s_5 = \frac{1}{6} + \frac{1}{6} + \frac{3}{20} + \frac{2}{15} + \frac{5}{42} = 0.7357$

3. $s_1 = 3$

$s_2 = 3 - \frac{9}{2} = -1.5$

$s_3 = 3 - \frac{9}{2} + \frac{27}{4} = 5.25$

$s_4 = 3 - \frac{9}{2} + \frac{27}{4} - \frac{81}{8} = -4.875$

$s_5 = 3 - \frac{9}{2} + \frac{27}{4} - \frac{81}{8} + \frac{243}{16} = 10.3125$

4. $s_1 = 1$

$s_2 = 1 + \frac{1}{3} = 1.3333$

$s_3 = 1 + \frac{1}{3} + \frac{1}{5} = 1.5333$

$s_4 = 1 + \frac{1}{3} + \frac{1}{5} + \frac{1}{9} = 1.6444$

$s_5 = 1 + \frac{1}{3} + \frac{1}{5} + \frac{1}{9} + \frac{1}{11} = 1.7354$

5. $s_1 = 3$

$s_2 = 3 + \frac{3}{2} = 4.5$

$s_3 = 3 + \frac{3}{2} + \frac{3}{4} = 5.250$

$s_4 = 3 + \frac{3}{2} + \frac{3}{4} + \frac{3}{8} = 5.625$

$s_5 = 3 + \frac{3}{2} + \frac{3}{4} + \frac{3}{8} + \frac{3}{16} = 5.8125$

6. $s_1 = 1$

$s_2 = 1 - \frac{1}{2} = 0.5$

$s_3 = 1 - \frac{1}{2} + \frac{1}{6} = 0.6667$

$s_4 = 1 - \frac{1}{2} + \frac{1}{6} - \frac{1}{24} = 0.6250$

$s_5 = 1 - \frac{1}{2} + \frac{1}{6} - \frac{1}{24} + \frac{1}{120} = 0.6333$

7. $\sum_{n=1}^{\infty} \frac{n}{n+1}$, $\lim_{n \to \infty} \frac{n}{n+1} = 1 \neq 0$, diverges by Theorem 9.8

8. $\sum_{n=1}^{\infty} (-1)^{n+1}\left(\frac{3^n}{2^{n-1}}\right)$, $\lim_{n \to \infty} \left[(-1)^{n+1}\left(\frac{3^n}{2^{n-1}}\right)\right] \neq 0$, diverges by Theorem 9.8

9. $\sum_{n=1}^{\infty} \frac{3n}{n+1}$, $\lim_{n \to \infty} \frac{3n}{n+1} = 3 \neq 0$, diverges by Theorem 9.8

10. $\sum_{n=1}^{\infty} \frac{n}{2n+3}$, $\lim_{n \to \infty} \frac{n}{2n+3} = \frac{1}{2} \neq 0$, diverges by Theorem 9.8

11. $\sum_{n=1}^{\infty} \frac{n^2}{n^2+1}$, $\lim_{n \to \infty} \frac{n^2}{n^2+1} = 1 \neq 0$, diverges by Theorem 9.8

12. $\sum_{n=1}^{\infty} \frac{n}{\sqrt{n^2+1}}$, $\lim_{n \to \infty} \frac{n}{\sqrt{n^2+1}} = \lim_{n \to \infty} \frac{1}{\sqrt{1 + 1/n^2}} = 1 \neq 0$,
diverges by Theorem 9.8

13. $\sum_{n=0}^{\infty} 3(\frac{3}{2})^n$, $r = \frac{3}{2} > 1$, diverges by Theorem 9.9

14. $\sum_{n=0}^{\infty} (\frac{4}{3})^n$, $r = \frac{4}{3} > 1$, diverges by Theorem 9.9

15. $\sum_{n=0}^{\infty} 1000(1.055)^n$, $r = 1.055 > 1$, diverges by Theorem 9.9

16. $\sum_{n=0}^{\infty} 2(-1.03)^n$, $|r| = 1.03 > 1$, diverges by Theorem 9.9

17. $\sum_{n=1}^{\infty} \frac{2^n + 1}{2^{n+1}}$, $\lim_{n \to \infty} \frac{2^n + 1}{2^{n+1}} = \lim_{n \to \infty} \frac{1 + 2^{-n}}{2} = \frac{1}{2} \neq 0$,
diverges by Theorem 9.8

18. $\sum_{n=1}^{\infty} \frac{n!}{2^n}$, $\lim_{n \to \infty} \frac{n!}{2^n} = \infty$, diverges by Theorem 9.8

19. $\sum_{n=0}^{\infty} 2(\frac{3}{4})^n$ geometric series with $r = \frac{3}{4}$, converges by Theorem 9.9

20. $\sum_{n=0}^{\infty} 2(-\frac{1}{2})^n$ geometric series with $r = -\frac{1}{2}$, converges by Theorem 9.9

21. $\sum_{n=0}^{\infty} (0.9)^n$ geometric series with $r = 0.9$, converges by Theorem 9.9

22. $\sum_{n=0}^{\infty} (-0.6)^n$ geometric series with $r = -0.6$, converges by Theorem 9.9

23. $\sum_{n=1}^{\infty} \frac{1}{n(n + 1)} = \sum_{n=1}^{\infty} (\frac{1}{n} - \frac{1}{n + 1}) = (1 - \frac{1}{2}) + (\frac{1}{2} - \frac{1}{3})$

$+ (\frac{1}{3} - \frac{1}{4}) + (\frac{1}{4} - \frac{1}{5}) + \ldots = 1$

24. $\sum_{n=1}^{\infty} \frac{1}{n(n + 2)} = \sum_{n=1}^{\infty} (\frac{1}{2n} - \frac{1}{2(n + 2)}) = (\frac{1}{2} - \frac{1}{6}) + (\frac{1}{4} - \frac{1}{8}) + (\frac{1}{6} - \frac{1}{10})$

$+ (\frac{1}{8} - \frac{1}{12}) + (\frac{1}{10} - \frac{1}{14}) + \ldots = \frac{1}{2} + \frac{1}{4} = \frac{3}{4}$

25. $\sum_{n=0}^{\infty} (\frac{1}{2})^n = \frac{1}{1 - (1/2)} = 2$

26. $\sum_{n=0}^{\infty} 2(\frac{2}{3})^n = \frac{2}{1 - (2/3)} = 6$

27. $\sum_{n=0}^{\infty} (-\frac{1}{2})^n = \frac{1}{1 - (-1/2)} = \frac{2}{3}$

28. $\sum_{n=0}^{\infty} 2(-\frac{2}{3})^n = \frac{2}{1 - (-2/3)} = \frac{6}{5}$

29. $\sum_{n=0}^{\infty} 2(\frac{1}{\sqrt{2}})^n = \frac{2}{1 - (1/\sqrt{2})} = \frac{2\sqrt{2}}{\sqrt{2} - 1} = 2\sqrt{2}(\sqrt{2} + 1) = 4 + 2\sqrt{2}$

30. $\sum_{n=0}^{\infty} 4(\frac{1}{4})^n = \frac{4}{1 - (1/4)} = \frac{16}{3}$

31. $\sum_{n=0}^{\infty} (\frac{1}{10})^n = \frac{1}{1 - (1/10)} = \frac{10}{9}$

32. $\sum_{n=0}^{\infty} 8(\frac{3}{4})^n = \frac{8}{1 - (3/4)} = 32$

33. $\sum_{n=0}^{\infty} 3(-\frac{1}{3})^n = \frac{3}{1 - (-1/3)} = \frac{9}{4}$

34. $\sum_{n=0}^{\infty} 4(-\frac{1}{2})^n = \frac{4}{1 - (-1/2)} = \frac{8}{3}$

35. $\sum_{n=2}^{\infty} \frac{1}{n^2 - 1} = \sum_{n=2}^{\infty} (\frac{1/2}{n - 1} - \frac{1/2}{n + 1}) = \frac{1}{2}\sum_{n=2}^{\infty} (\frac{1}{n - 1} - \frac{1}{n + 1})$

$= \frac{1}{2}\left[(1 - \frac{1}{3}) + (\frac{1}{2} - \frac{1}{4}) + (\frac{1}{3} - \frac{1}{5}) + (\frac{1}{4} - \frac{1}{6}) + \ldots\right]$

$= \frac{1}{2}\left[1 + \frac{1}{2}\right] = \frac{3}{4}$

36. $\sum_{n=1}^{\infty} \frac{1}{n(n + 1)} = \sum_{n=1}^{\infty} (\frac{1}{n} - \frac{1}{n + 1})$

$= (1 - \frac{1}{2}) + (\frac{1}{2} - \frac{1}{3}) + (\frac{1}{3} - \frac{1}{4}) + \ldots = 1$

37. $\sum_{n=1}^{\infty} \frac{4}{n(n + 2)} = 2\sum_{n=1}^{\infty} (\frac{1}{n} - \frac{1}{n + 2})$

$= 2\left[(1 - \frac{1}{3}) + (\frac{1}{2} - \frac{1}{4}) + (\frac{1}{3} - \frac{1}{5}) + \ldots\right] = 2(1 + \frac{1}{2}) = 3$

38. $\displaystyle\sum_{n=1}^{\infty} \frac{1}{(2n+1)(2n+3)} = \frac{1}{2}\sum_{n=1}^{\infty}\left(\frac{1}{2n+1} - \frac{1}{2n+3}\right)$

$\displaystyle = \frac{1}{2}\left[\left(\frac{1}{3} - \frac{1}{5}\right) + \left(\frac{1}{5} - \frac{1}{7}\right) + \left(\frac{1}{7} - \frac{1}{9}\right) + \ldots\right] = \frac{1}{2}\left(\frac{1}{3}\right) = \frac{1}{6}$

39. $\displaystyle\sum_{n=0}^{\infty}\left(\frac{1}{2^n} - \frac{1}{3^n}\right) = \sum_{n=0}^{\infty}\left(\frac{1}{2}\right)^n - \sum_{n=0}^{\infty}\left(\frac{1}{3}\right)^n = \frac{1}{1-(1/2)} - \frac{1}{1-(1/3)}$

$\displaystyle = 2 - \frac{3}{2} = \frac{1}{2}$

40. $\displaystyle\sum_{n=1}^{\infty}\left[(0.7)^n + (0.9)^n\right] = \sum_{n=0}^{\infty}\left(\frac{7}{10}\right)^n + \sum_{n=0}^{\infty}\left(\frac{9}{10}\right)^n - 2$

$\displaystyle = \frac{1}{1-(7/10)} + \frac{1}{1-(9/10)} - 2 = \frac{10}{3} + 10 - 2 = \frac{34}{3}$

41. $\displaystyle 0.666\overline{6} = \sum_{n=0}^{\infty}\frac{3}{5}(0.1)^n$, geometric series with $a = \frac{3}{5}$ and $r = \frac{1}{10}$

$\displaystyle S = \frac{a}{1-r} = \frac{3/5}{9/10} = \frac{2}{3}$

42. $\displaystyle 0.2323\overline{23} = \sum_{n=0}^{\infty}\frac{23}{100}\left(\frac{1}{100}\right)^n$, geometric series with $a = \frac{23}{100}$

and $r = \frac{1}{100}$, $\qquad S = \frac{a}{1-r} = \frac{23/100}{99/100} = \frac{23}{99}$

43. $\displaystyle 0.075\overline{75} = \sum_{n=0}^{\infty}\frac{3}{40}\left(\frac{1}{100}\right)^n$, geometric series with $a = \frac{3}{40}$ and $r = \frac{1}{100}$

$\displaystyle S = \frac{a}{1-r} = \frac{3/40}{99/100} = \frac{5}{66}$

44. $\displaystyle 0.215\overline{15} = \frac{1}{5} + \sum_{n=0}^{\infty}\frac{3}{200}\left(\frac{1}{100}\right)^n$, geometric series with $a = \frac{3}{200}$

and $r = \frac{1}{100}$, $\qquad S = \frac{1}{5} + \frac{a}{1-r} = \frac{1}{5} + \frac{3/200}{99/100} = \frac{71}{330}$

45. $\displaystyle\sum_{n=1}^{\infty}\frac{n+10}{10n+1}$, $\displaystyle\lim_{n\to\infty}\frac{n+10}{10n+1} = \frac{1}{10} \neq 0$, diverges by Theorem 9.8

46. $\displaystyle\sum_{n=0}^{\infty}\frac{4}{2^n} = 4\sum_{n=0}^{\infty}\left(\frac{1}{2}\right)^n$, geometric series with $r = \frac{1}{2}$

converges by Theorem 9.9

47. $\sum_{n=1}^{\infty} \left(\frac{1}{n} - \frac{1}{n+2}\right) = \left(1 - \frac{1}{3}\right) + \left(\frac{1}{2} - \frac{1}{4}\right) + \left(\frac{1}{3} - \frac{1}{5}\right) + \left(\frac{1}{4} - \frac{1}{6}\right) + \ldots$

$= 1 + \frac{1}{2} = \frac{3}{2}$, converges

48. $\sum_{n=1}^{\infty} \frac{n+1}{2n-1}$, $\lim_{n\to\infty} \frac{n+1}{2n-1} = \frac{1}{2} \neq 0$, diverges by Theorem 9.8

49. $\sum_{n=1}^{\infty} \frac{3n-1}{2n+1}$, $\lim_{n\to\infty} \frac{3n-1}{2n+1} = \frac{3}{2} \neq 0$, diverges by Theorem 9.8

50. $\sum_{n=0}^{\infty} \frac{1}{4^n}$ geometric series with $r = \frac{1}{4}$, converges by Theorem 9.9

51. $\sum_{n=0}^{\infty} (1.075)^n$ geometric series with $r = 1.075$, diverges by Theorem 9.9

52. $\sum_{n=1}^{\infty} \frac{2^n}{100}$ geometric series with $r = 2$, diverges by Theorem 9.9

53. $\sum_{n=2}^{\infty} \frac{n}{\ln n}$, $\lim_{n\to\infty} \frac{n}{\ln n} = \lim_{n\to\infty} \frac{1}{1/n} = \infty$ (by L'Hopital's Rule)
diverges by Theorem 9.8

54. $\sum_{n=1}^{\infty} \frac{2^n}{n^2}$, $\lim_{n\to\infty} \frac{2^n}{n^2} = \lim_{n\to\infty} \frac{(\ln 2)2^n}{2n} = \lim_{n\to\infty} \frac{(\ln 2)^2 2^n}{2} = \infty$,
(by L'Hopital's Rule) diverges by Theorem 9.8

55. $\sum_{n=1}^{\infty} \left(1 + \frac{k}{n}\right)^n$, $\lim_{n\to\infty} \left(1 + \frac{k}{n}\right)^n = e^k \neq 0$, diverges by Theorem 9.8

56. $\sum_{n=1}^{\infty} \frac{1}{n(n+3)} = \frac{1}{3} \sum_{n=1}^{\infty} \left(\frac{1}{n} - \frac{1}{n+3}\right) = \frac{1}{3}\left[\left(1 - \frac{1}{4}\right) + \left(\frac{1}{2} - \frac{1}{5}\right) + \left(\frac{1}{3} - \frac{1}{6}\right)\right.$

$\left. + \left(\frac{1}{4} - \frac{1}{7}\right) + \left(\frac{1}{5} - \frac{1}{8}\right) + \left(\frac{1}{6} - \frac{1}{9}\right) + \ldots\right]$

$= \frac{1}{3}\left(1 + \frac{1}{2} + \frac{1}{3}\right) = \frac{11}{18}$, converges

57. $\sum_{i=0}^{n-1} 8000(0.9)^i = \frac{8000[1 - (0.9)^{(n-1)+1}]}{1 - 0.9} = 80{,}000(1 - 0.9^n)$

58. $\sum_{i=0}^{n-1} 8000(0.75)^i = 8000 \left[\frac{1 - (0.75)^{(n-1)+1}}{1 - 0.75}\right] = 32{,}000(1 - 0.75^n)$

59. $D_1 = 16$

$D_2 = \underbrace{0.81(16)}_{\text{up}} + \underbrace{0.81(16)}_{\text{down}} = 32(0.81)$

$D_3 = 16(0.81)^2 + 16(0.81)^2 = 32(0.81)^2$

$\vdots$

$D = 16 + 32(0.81) + 32(0.81)^2 + \ldots = -16 + \sum_{n=0}^{\infty} 32(0.81)^n$

$= -16 + \dfrac{32}{1 - 0.81} = 152.42$ feet

60. The ball in exercise 59 takes the following times for each fall:

$S_1 = -16t^2 + 16$	$S_1 = 0$ if $t = 1$
$S_2 = -16t^2 + 16(0.81)$	$S_2 = 0$ if $t = 0.9$
$S_3 = -16t^2 + 16(0.81)^2$	$S_3 = 0$ if $t = (0.9)^2$
$\vdots$	$\vdots$
$S_n = -16t^2 + 16(0.81)^{n-1}$	$S_n = 0$ if $t = (0.9)^{n-1}$

Beginning with S_2, the ball takes the same amount of time to bounce up as it takes to fall. The total time elapsed before the ball comes to rest is

$t = 1 + 2 \sum_{n=1}^{\infty} (0.9)^n = -1 + 2 \sum_{n=0}^{\infty} (0.9)^n = -1 + \dfrac{2}{1 - 0.9} = 19$ sec

61. $\sum_{n=0}^{\infty} \frac{1}{4}\left(\frac{1}{4}\right)^n = \dfrac{1/4}{1 - (1/4)} = \dfrac{1}{3}$ of the square is shaded

62. $a = 100, \quad r = 1 + \dfrac{0.10}{12}, \quad n = 61$

$$A + 100 = \frac{100\left[1 - \left(1 + \frac{0.10}{12}\right)^{61}\right]}{1 - \left(1 + \frac{0.10}{12}\right)} \implies A \approx \$7808.24$$

63. $a = 50, \quad r = 1 + \dfrac{0.12}{12} = 1.01, \quad n = 121$

$$A + 50 = \frac{50[1 - (1.01)^{121}]}{1 - 1.01} \implies A \approx \$11{,}616.95$$

64. $A + P = \dfrac{P(1 - [1 + (r/12)]^{N+1})}{1 - [1 + (r/12)]} = -\dfrac{12}{r}P\left[1 - (1 + \dfrac{r}{12})(1 + \dfrac{r}{12})^N\right]$

$$A = -P + \frac{12}{r}P\left[(1 + \frac{r}{12})(1 + \frac{r}{12})^N - 1\right]$$

$$= P\left[-1 + \frac{12}{r}(1 + \frac{r}{12})(1 + \frac{r}{12})^N - \frac{12}{r}\right]$$

$$= P\left[-1 - \frac{12}{r} + (\frac{12}{r} + 1)(1 + \frac{r}{12})^N\right]$$

$$= P\left[-(1 + \frac{12}{r}) + (1 + \frac{12}{r})(1 + \frac{r}{12})^N\right]$$

$$= P\left[-1 + (1 + \frac{r}{12})^N\right](1 + \frac{12}{r}) = P\left[(1 + \frac{r}{12})^N - 1\right](1 + \frac{12}{r})$$

65. $A = 50\left[(1 + \dfrac{0.09}{12})^{480} - 1\right](1 + \dfrac{12}{0.09}) \approx \$235,821.51$

66. $\displaystyle\sum_{n=0}^{66} 2^n = \frac{1 - 2^{67}}{1 - 2} \approx 1.476 \times 10^{20}$, No

67. Let $x = 0.a_1a_2a_3 \ldots a_k999 \ldots$, then we have

$$x = 0.a_1a_2a_3 \ldots a_k + \frac{9}{10^{k+1}}\sum_{n=0}^{\infty}(\frac{1}{10})^n$$

$$= 0.a_1a_2a_3 \ldots a_k + \frac{9}{10^{k+1}}(\frac{1}{1 - 0.1})$$

$$= 0.a_1a_2a_3 \ldots a_k + \frac{1}{10^k}$$

$$= 0.a_1a_2a_3 \ldots (a_k + 1)000 \ldots$$

68. $x = 0.a_1a_2a_3 \ldots a_k\overline{a_1a_2a_3 \ldots a_k}$

$$= 0.a_1a_2a_3 \ldots a_k\left[1 + \frac{1}{10^k} + (\frac{1}{10^k})^2 + (\frac{1}{10^k})^3 + \ldots\right]$$

$$= 0.a_1a_2a_3 \ldots a_k\sum_{n=0}^{\infty}(\frac{1}{10^k})^n$$

$$= 0.a_1a_2a_3 \ldots a_k\left[\frac{1}{1 - (1/10^k)}\right] = \text{a rational number}$$

69. By letting $S_0 = 0$, we have $a_n = \sum_{k=1}^{n} a_k - \sum_{k=1}^{n-1} a_k = S_n - S_{n-1}$

Thus, $\sum_{n=1}^{\infty} a_n = \sum_{n=1}^{\infty} (S_n - S_{n-1}) = \sum_{n=1}^{\infty} (S_n - S_{n-1} + C - C)$

$= \sum_{n=1}^{\infty} \left[(C - S_{n-1}) - (C - S_n)\right]$

70. Let $\{S_n\}$ be the sequence of partial sums for the convergent series $\sum_{n=1}^{\infty} a_n = L$. Then $\lim_{n \to \infty} S_n = L$ and since $R_n = \sum_{k=n+1}^{\infty} a_k = L - S_n$,

we have $\lim_{n \to \infty} R_n = \lim_{n \to \infty} (L - S_n) = \lim_{n \to \infty} L - \lim_{n \to \infty} S_n = L - L = 0$

9.4 The Integral Test and *p*-Series

1. $\sum_{n=1}^{\infty} \frac{1}{n^3}$ Convergent p-series, $p = 3 > 1$

2. $\sum_{n=1}^{\infty} \frac{1}{n^{1/3}}$ Divergent p-series, $p = \frac{1}{3} < 1$

3. $\sum_{n=1}^{\infty} \frac{1}{\sqrt[5]{n}} = \sum_{n=1}^{\infty} \frac{1}{n^{1/5}}$ Divergent p-series, $p = \frac{1}{5} < 1$

4. $\sum_{n=1}^{\infty} \frac{1}{n^{4/3}}$ Convergent p-series, $p = \frac{4}{3} > 1$

5. $\sum_{n=1}^{\infty} \frac{1}{n^{1/2}}$ p-series with $p = \frac{1}{2}$, diverges by Theorem 9.11

6. $\sum_{n=1}^{\infty} \frac{1}{n^2}$ p-series with $p = 2$, converges by Theorem 9.11

7. $\sum_{n=1}^{\infty} \frac{1}{n^{3/2}}$ p-series with $p = \frac{3}{2}$, converges by Theorem 9.11

8. $\displaystyle\sum_{n=1}^{\infty} \frac{1}{n^{2/3}}$ p-series with $p = \frac{2}{3}$, diverges by Theorem 9.11

9. $\displaystyle\sum_{n=1}^{\infty} \frac{1}{n^{1.04}}$ p-series with $p = 1.04$, converges by Theorem 9.11

10. $\displaystyle\sum_{n=1}^{\infty} \frac{1}{n^{\pi}}$ p-series with $p = \pi$, converges by Theorem 9.11

11. $\displaystyle\sum_{n=1}^{\infty} \frac{1}{n+1}, \quad \int_1^{\infty} \frac{1}{x+1}\,dx = \ln(x+1)\Big]_1^{\infty} = \infty,$ diverges by Theorem 9.10

12. $\displaystyle\sum_{n=1}^{\infty} e^{-n}, \quad \int_1^{\infty} e^{-x}\,dx = -e^{-x}\Big]_1^{\infty} = \frac{1}{e},$ converges by Theorem 9.10

13. $\displaystyle\sum_{n=1}^{\infty} ne^{-n}, \quad \int_1^{\infty} xe^{-x}\,dx = -e^{-x}(x+1)\Big]_1^{\infty} = \frac{2}{e},$ converges by Theorem 9.10

14. $\displaystyle\sum_{n=1}^{\infty} e^{-n}\cos n, \quad \int_1^{\infty} e^{-x}\cos x\,dx = \frac{e^{-x}}{2}(\sin x - \cos x)\Big]_1^{\infty}$

$\displaystyle = \frac{\cos 1 - \sin 1}{2e} \approx -0.0554,$ converges by Theorem 9.10

15. $\displaystyle\sum_{n=1}^{\infty} \frac{1}{n^2+1}, \quad \int_1^{\infty} \frac{1}{x^2+1}\,dx = \arctan x\Big]_1^{\infty} = \frac{\pi}{4},$ converges by Theorem 9.10

16. $\displaystyle\sum_{n=1}^{\infty} \frac{1}{2n+1}, \quad \int_1^{\infty} \frac{1}{2x+1}\,dx = \ln\sqrt{2x+1}\Big]_1^{\infty} = \infty,$ diverges by Theorem 9.10

17. $\displaystyle\sum_{n=1}^{\infty} \frac{\ln(n+1)}{n+1}, \quad \int_1^{\infty} \frac{\ln(x+1)}{x+1}\,dx = \ln^2\sqrt{x+1}\Big]_1^{\infty} = \infty$
diverges by Theorem 9.10

18. $\displaystyle\sum_{n=1}^{\infty} \frac{n}{n^2+3}, \quad \int_1^{\infty} \frac{x}{x^2+3}\,dx = \ln\sqrt{x^2+3}\Big]_1^{\infty} = \infty,$ diverges by Theorem 9. 10

19. $\displaystyle\sum_{n=1}^{\infty} \frac{n^{k-1}}{n^k+c}, \quad \int_1^{\infty} \frac{x^{k-1}}{x^k+c}\,dx = \frac{1}{k}\ln(x^k+c)\Big]_1^{\infty} = \infty,$
diverges by Theorem 9.10

20. $\sum_{n=1}^{\infty} n^k e^{-n}$, $\int_1^{\infty} x^k e^{-x}\,dx = -x^k e^{-x}\Big]_1^{\infty} + k\int_1^{\infty} x^{k-1}e^{-x}\,dx$

$= \frac{1}{e} + \frac{k}{e} + \frac{k(k-1)}{e} + \ldots + \frac{k!}{e}$, converges by Theorem 9.10

21. $\sum_{n=1}^{\infty} \frac{1}{2n-1}$, $\int_1^{\infty} \frac{1}{2x-1}\,dx = \ln\sqrt{2x+1}\Big]_1^{\infty} = \infty$, diverges by Theorem 9.10

22. $\sum_{n=2}^{\infty} \frac{1}{n\sqrt{n^2-1}}$, $\int_2^{\infty} \frac{1}{x\sqrt{x^2-1}}\,dx = \text{arcsec}\,x\Big]_2^{\infty} = \frac{\pi}{2} - \frac{\pi}{3}$,
converges by Theorem 9.10

23. $\sum_{n=1}^{\infty} \frac{1}{n^{4/3}}$ p-series with $p = \frac{4}{3}$, converges by Theorem 9.11

24. $3\sum_{n=1}^{\infty} \frac{1}{n^{0.95}}$ p-series with $p = 0.95$, diverges by Theorem 9.11

25. $\sum_{n=0}^{\infty} (\frac{2}{3})^n$ geometric series with $r = \frac{2}{3}$, converges by Theorem 9.9

26. $\sum_{n=1}^{\infty} \frac{n}{\sqrt{n^2+1}}$, $\int_1^{\infty} \frac{x}{\sqrt{x^2+1}}\,dx = \sqrt{x^2+1}\Big]_1^{\infty} = \infty$, diverges by Theorem 9.10

27. $\sum_{n=0}^{\infty} (1.075)^n$ geometric series with $r = 1.075$, diverges by Theorem 9.9

28. $\sum_{n=1}^{\infty} (\frac{1}{n^2} - \frac{1}{n^3}) = \sum_{n=1}^{\infty} \frac{1}{n^2} - \sum_{n=1}^{\infty} \frac{1}{n^3}$
Since these are both convergent p-series, the difference is convergent.

29. $\sum_{n=1}^{\infty} (1 + \frac{1}{n})^n$, $\lim_{n \to \infty} (1 + \frac{1}{n})^n = e \neq 0$

Fails n-th term test, diverges by Theorem 9.8

30. $\sum_{n=2}^{\infty} \ln(n)$, $\lim_{n \to \infty} \ln(n) = \infty$, diverges by Theorem 9.8

31. $S_6 = (0.075) + (0.075)^2 + (0.075)^3 + (0.075)^4 + (0.075)^5 + (0.075)^6$

$\approx 0.0810810666, \quad \sum_{n=1}^{\infty} (0.075)^n = \dfrac{1}{1 - 0.075} - 1 = \dfrac{40}{37} - 1 = \dfrac{3}{37} \approx 0.08\overline{108}$

$R_6 \le S - S_6 \approx 1.4 \times 10^{-8}, \quad \sum_{n=1}^{\infty} (0.075)^n \approx 0.081081066$

32. $S_4 = 1 + \dfrac{1}{2^5} + \dfrac{1}{3^5} + \dfrac{1}{4^5} \approx 1.0353, \qquad R_4 \le \dfrac{4^{1-5}}{5 - 1} = \dfrac{4^{-4}}{4} \approx 0.0010$

$1.0353 \le \sum_{n=1}^{\infty} \dfrac{1}{n^5} \le 1.0373$

33. $S_{10} = \dfrac{1}{2} + \dfrac{1}{5} + \dfrac{1}{10} + \dfrac{1}{17} + \dfrac{1}{26} + \dfrac{1}{37} + \dfrac{1}{50} + \dfrac{1}{65} + \dfrac{1}{82} + \dfrac{1}{101} \approx 0.9818$

$R_{10} \le \int_{10}^{\infty} \dfrac{1}{x^2 + 1}\, dx = \arctan x \Big]_{10}^{\infty} = \dfrac{\pi}{2} - \arctan 10 \approx 0.0997$

$0.9818 \le \sum_{n=1}^{\infty} \dfrac{1}{n^5} \le 1.0815$

34. $S_{10} = \dfrac{1}{2(\ln 2)^3} + \dfrac{1}{3(\ln 3)^3} + \dfrac{1}{4(\ln 4)^3} + \dots + \dfrac{1}{11(\ln 11)^3} \approx 1.9821$

$R_{10} \le \int_{10}^{\infty} \dfrac{1}{(x + 1)[\ln (x + 1)]^3}\, dx = -\dfrac{1}{2[\ln (x + 1)]^2}\Big]_{10}^{\infty}$

$= \dfrac{1}{2(\ln 11)^2} \approx 0.0870$

$1.9821 \le \sum_{n=1}^{\infty} \dfrac{1}{(n + 1)[\ln(n + 1)]^3} \le 2.0691$

35. $S_4 = \dfrac{1}{e} + \dfrac{2}{e^4} + \dfrac{3}{e^9} + \dfrac{4}{e^{16}} \approx 0.404881398$

$R_4 \le \int_{4}^{\infty} xe^{-x^2}\, dx = -\dfrac{1}{2}e^{-x^2}\Big]_{4}^{\infty} \approx 5.6 \times 10^{-8}$

$\sum_{n=1}^{\infty} ne^{-n^2} \le 0.404881398$

36. $S_4 = \dfrac{1}{e} + \dfrac{1}{e^2} + \dfrac{1}{e^3} + \dfrac{1}{e^4} \approx 0.5713$

$R_4 \le \int_{4}^{\infty} e^{-x}\, dx = -e^{-x}\Big]_{4}^{\infty} \approx 0.0183, \qquad 0.5713 \le \sum_{n=1}^{\infty} e^{-n} \le 0.5896$

37. $0 < R_n < \dfrac{N^{1-4}}{4-1} < 0.001, \quad \dfrac{1}{N^3} < 0.003, \quad N^3 > 333.33, \quad N > 6.93, \quad N > 7$

38. $0 < R_n < \dfrac{N^{1-3/2}}{3/2-1} < 0.001, \quad N^{-1/2} < 0.0005, \quad \sqrt{N} > 2000, \quad N > 4{,}000{,}000$

39. $R_n < \displaystyle\int_N^\infty e^{-5x}\,dx = -\frac{1}{5}e^{-5x}\Big]_N^\infty = \frac{e^{-5N}}{5} < 0.001, \quad \frac{1}{e^{5N}} < 0.005$

$e^{5N} > 200, \quad 5N > \ln 200, \quad N > \dfrac{\ln 200}{5}, \quad N > 1.0597, \quad N > 2$

40. $R_N < \displaystyle\int_N^\infty \frac{1}{x^2+1}\,dx = \arctan x\Big]_N^\infty = \frac{\pi}{2} - \arctan N < 0.001$

$-\arctan N < -1.5698, \quad \arctan N > 1.5698, \quad N > \tan 1.5698, \quad N > 1004$

9.5 Comparisons of series

1. $\dfrac{1}{n^2+1} < \dfrac{1}{n^2}$ Therefore, $\displaystyle\sum_{n=1}^\infty \frac{1}{n^2+1}$ converges by comparison with the convergent p-series $\displaystyle\sum_{n=1}^\infty \frac{1}{n^2}$

2. $\dfrac{1}{3n^2+2} < \dfrac{1}{3n^2}$ Therefore, $\displaystyle\sum_{n=1}^\infty \frac{1}{3n^2+2}$ converges by comparison with the convergent p-series $\displaystyle\frac{1}{3}\sum_{n=1}^\infty \frac{1}{n^2}$

3. $\dfrac{1}{n-1} > \dfrac{1}{n}$ for $n \geq 2$ Therefore, $\displaystyle\sum_{n=2}^\infty \frac{1}{n-1}$ diverges by comparison with the divergent p-series $\displaystyle\sum_{n=2}^\infty \frac{1}{n}$

4. $\dfrac{1}{\sqrt{n}-1} > \dfrac{1}{\sqrt{n}}$ for $n \geq 2$ Therefore, $\displaystyle\sum_{n=2}^\infty \frac{1}{\sqrt{n}-1}$ diverges by comparison with the divergent p-series $\displaystyle\sum_{n=2}^\infty \frac{1}{\sqrt{n}}$

5. $\frac{1}{3^n + 1} < \frac{1}{3^n}$ Therefore, $\sum_{n=0}^{\infty} \frac{1}{3^n + 1}$ converges by comparison with the convergent geometric series $\sum_{n=0}^{\infty} (\frac{1}{3})^n$

6. $\frac{2^n}{3^n + 5} < (\frac{2}{3})^n$ Therefore, $\sum_{n=0}^{\infty} \frac{2^n}{3^n + 5}$ converges by comparison with the convergent geometric series $\sum_{n=0}^{\infty} (\frac{2}{3})^n$

7. For $n \geq 3$, $\frac{\ln n}{n + 1} > \frac{1}{n + 1}$ Therefore, $\sum_{n=1}^{\infty} \frac{\ln n}{n + 1}$ diverges by comparison with the divergent series $\sum_{n=1}^{\infty} \frac{1}{n + 1}$

(Note that $\sum_{n=1}^{\infty} \frac{1}{n + 1}$ diverges by a limit comparison with $\sum_{n=1}^{\infty} \frac{1}{n}$)

8. $\frac{1}{\sqrt{n^3 + 1}} < \frac{1}{n^{3/2}}$ Therefore, $\sum_{n=0}^{\infty} \frac{1}{\sqrt{n^3 + 1}}$ converges by comparison with the convergent p-series $\sum_{n=1}^{\infty} \frac{1}{n^{3/2}}$

9. For $n > 3$, $\frac{1}{n^2} > \frac{1}{n!}$ Therefore, $\sum_{n=0}^{\infty} \frac{1}{n!}$ converges by comparison with the convergent p-series $\sum_{n=1}^{\infty} \frac{1}{n^2}$

10. $\frac{1}{3\sqrt[4]{n} - 1} > \frac{1}{3\sqrt[4]{n}}$ Therefore, $\sum_{n=1}^{\infty} \frac{1}{3\sqrt[4]{n} - 1}$ diverges by comparison with the divergent p-series $\frac{1}{3} \sum_{n=1}^{\infty} \frac{1}{\sqrt[4]{n}}$

11. $\frac{1}{e^{n^2}} \leq \frac{1}{e^n}$ Therefore, $\sum_{n=0}^{\infty} \frac{1}{e^{n^2}}$ converges by comparison with the convergent geometric series $\sum_{n=0}^{\infty} (\frac{1}{e})^n$

12. $\frac{4^n}{3^n - 1} > \frac{4^n}{3^n}$ Therefore, $\sum_{n=0}^{\infty} \frac{4^n}{3^n - 1}$ diverges by comparison with the divergent geometric series $\sum_{n=0}^{\infty} (\frac{4}{3})^n$

13. $\lim_{n \to \infty} \frac{n/(n^2 + 1)}{1/n} = \lim_{n \to \infty} \frac{n^2}{n^2 + 1} = 1$ Therefore, $\sum_{n=1}^{\infty} \frac{n}{n^2 + 1}$ diverges by a limit comparison with the divergent p-series $\sum_{n=1}^{\infty} \frac{1}{n}$

14. $\lim_{n \to \infty} \frac{1/\sqrt{n^2 - 1}}{1/n} = \lim_{n \to \infty} \frac{n}{\sqrt{n^2 - 1}} = 1$ Therefore, $\sum_{n=2}^{\infty} \frac{1}{\sqrt{n^2 - 1}}$ diverges by a limit comparison with the divergent p-series $\sum_{n=1}^{\infty} \frac{1}{n}$

15. $\lim_{n \to \infty} \frac{1/\sqrt{n^2 + 1}}{1/n} = \lim_{n \to \infty} \frac{n}{\sqrt{n^2 + 1}} = 1$ Therefore, $\sum_{n=0}^{\infty} \frac{1}{\sqrt{n^2 + 1}}$ diverges by a limit comparison with the divergent p-series $\sum_{n=1}^{\infty} \frac{1}{n}$

16. $\lim_{n \to \infty} \frac{1/(2^n - 5)}{1/2^n} = \lim_{n \to \infty} \frac{2^n}{2^n - 5} = 1$ Therefore, $\sum_{n=0}^{\infty} \frac{1}{2^n - 5}$ converges by a limit comparison with the convergent geometric series $\sum_{n=0}^{\infty} (\frac{1}{2})^n$

17. $\lim_{n \to \infty} \frac{\frac{2n^2 - 1}{3n^5 + 2n + 1}}{1/n^3} = \lim_{n \to \infty} \frac{2n^5 - 1}{3n^5 + 2n + 1} = \frac{2}{3}$

Therefore, $\sum_{n=1}^{\infty} \frac{2n^2 - 1}{3n^5 + 2n + 1}$ converges by a limit comparison with the convergent p-series $\sum_{n=1}^{\infty} \frac{1}{n^3}$

18. $$\lim_{n\to\infty} \frac{\dfrac{5n - 3}{n^2 - 2n + 5}}{1/n} = \lim_{n\to\infty} \frac{5n^2 - 3}{n^2 - 2n + 5} = 5$$

Therefore, $\sum_{n=1}^{\infty} \frac{5n - 3}{n^2 - 2n + 5}$ diverges by a limit comparison with the divergent p-series $\sum_{n=1}^{\infty} \frac{1}{n}$

19. $$\lim_{n\to\infty} \frac{(n + 3)/n(n + 2)}{1/n} = \lim_{n\to\infty} \frac{n^2 + 3n}{n^2 + 2n} = 1$$ Therefore, $\sum_{n=1}^{\infty} \frac{n + 3}{n(n + 2)}$ diverges by a limit comparison with the divergent p-series $\sum_{n=1}^{\infty} \frac{1}{n}$

20. $$\lim_{n\to\infty} \frac{1/n(n^2 + 1)}{1/n^3} = \lim_{n\to\infty} \frac{n^3}{n^3 + n} = 1$$ Therefore, $\sum_{n=1}^{\infty} \frac{1}{n(n^2 + 1)}$ converges by a limit comparison with the convergent p-series $\sum_{n=1}^{\infty} \frac{1}{n^3}$

21. $$\lim_{n\to\infty} \frac{1/n\sqrt{n^2 + 1}}{1/n^2} = \lim_{n\to\infty} \frac{n^2}{n\sqrt{n^2 + 1}} = 1$$ Therefore, $\sum_{n=1}^{\infty} \frac{1}{n\sqrt{n^2 + 1}}$ converges by a limit comparison with the convergent p-series $\sum_{n=1}^{\infty} \frac{1}{n^2}$

22. $$\lim_{n\to\infty} \frac{n/(n + 1)2^{n-1}}{1/2^{n-1}} = \lim_{n\to\infty} \frac{n}{n + 1} = 1$$ Therefore, $\sum_{n=1}^{\infty} \frac{n}{(n + 1)2^{n-1}}$ converges by a limit comparison with the convergent geometric series $\sum_{n=1}^{\infty} (\frac{1}{2})^{n-1}$

23. $$\lim_{n\to\infty} \frac{\dfrac{n^{k-1}}{n^k + 1}}{1/n} = \lim_{n\to\infty} \frac{n^k}{n^k + 1} = 1$$ Therefore, $\sum_{n=1}^{\infty} \frac{n^{k-1}}{n^k + 1}$ diverges by a limit comparison with the divergent p-series $\sum_{n=1}^{\infty} \frac{1}{n}$

24. $$\lim_{n \to \infty} \frac{1/(n + \sqrt{n^2 + 1})}{1/n} = \lim_{n \to \infty} \frac{n}{n + \sqrt{n^2 + 1}} = \frac{1}{2}$$

Therefore, $\sum_{n=1}^{\infty} \frac{1}{n + \sqrt{n^2 + 1}}$ diverges by a limit comparison with the divergent p-series $\sum_{n=1}^{\infty} \frac{1}{n}$

25. $$\lim_{n \to \infty} \frac{\sin(1/n)}{1/n} = \lim_{n \to \infty} \frac{(-1/n^2)\cos(1/n)}{-1/n^2} = \lim_{n \to \infty} \cos\left(\frac{1}{n}\right) = 1$$

Therefore, $\sum_{n=1}^{\infty} \sin\left(\frac{1}{n}\right)$ diverges by a limit comparison with the divergent p-series $\sum_{n=1}^{\infty} \frac{1}{n}$

26. $$\lim_{n \to \infty} \frac{\tan(1/n)}{1/n} = \lim_{n \to \infty} \frac{(-1/n^2)\sec^2(1/n)}{-1/n^2} = \lim_{n \to \infty} \sec^2\left(\frac{1}{n}\right) = 1$$

Therefore, $\sum_{n=1}^{\infty} \tan\left(\frac{1}{n}\right)$ diverges by a limit comparison with the divergent p-series $\sum_{n=1}^{\infty} \frac{1}{n}$

27. $\sum_{n=1}^{\infty} \frac{\sqrt{n}}{n}$ Diverges, p-series with $p = \frac{1}{2}$

28. $\sum_{n=0}^{\infty} 5\left(-\frac{1}{5}\right)^n$ Converges, geometric series with $r = -\frac{1}{5}$

29. $\sum_{n=1}^{\infty} \frac{1}{3^n + 2}$ Converges, comparison with $\sum_{n=1}^{\infty} \left(\frac{1}{3}\right)^n$

30. $\sum_{n=4}^{\infty} \frac{1}{3n^2 - 2n - 15}$ Converges, limit comparison with $\sum_{n=1}^{\infty} \frac{1}{n^2}$

31. $\sum_{n=1}^{\infty} \frac{n}{2n + 3}$ Diverges, nth term test, $\lim_{n \to \infty} \frac{n}{2n + 3} = \frac{1}{2} \neq 0$

32. $\sum_{n=1}^{\infty} \left(\frac{1}{n + 1} - \frac{1}{n + 2}\right) = \left(\frac{1}{2} - \frac{1}{3}\right) + \left(\frac{1}{3} - \frac{1}{4}\right) + \left(\frac{1}{4} - \frac{1}{5}\right) + \ldots = \frac{1}{2}$, converges

33. $\displaystyle\sum_{n=1}^{\infty} \frac{n}{(n^2 + 1)^2}$ Converges, integral test

34. $\displaystyle\sum_{n=1}^{\infty} \frac{3}{n(n + 3)}$ Converges, telescoping series $\displaystyle\sum_{n=1}^{\infty} (\frac{1}{n} - \frac{1}{n + 3})$

35. $\displaystyle\sum_{n=2}^{\infty} \frac{\ln n}{n^p}$, $\displaystyle\int_2^{\infty} \frac{\ln x}{x^p}\,dx = \left[\frac{x^{1-p} \ln x}{1 - p} - \frac{x^{1-p}}{(1 - p)^2}\right]_2^{\infty}$

Converges only if $1 < p$

36. $\displaystyle\sum_{n=0}^{\infty} 2(\frac{1}{3})^n$ The sum of the first n terms of this geometric series is

$S_n = \dfrac{a(1 - r^n)}{1 - r}$. Thus, $S_{10} = \dfrac{2(1 - (1/3)^{10})}{1 - (1/3)} = 3 - (\frac{1}{3})^9$

Since $\displaystyle\sum_{n=0}^{\infty} 2(\frac{1}{3})^n = \frac{2}{1 - (1/3)} = 3$, the error is $(\frac{1}{3})^9$

37. For $n \geq 6$, $\dfrac{1}{6^n} \geq \dfrac{1}{n^n}$ Therefore we have

$$\sum_{n=1}^{5} \frac{1}{n^n} < \sum_{n=1}^{\infty} \frac{1}{n^n} < \sum_{n=1}^{5} \frac{1}{n^n} + \sum_{n=6}^{\infty} (\frac{1}{6})^n$$

Since $\displaystyle\sum_{n=6}^{\infty} (\frac{1}{6})^n = \sum_{n=0}^{\infty} (\frac{1}{6})^n - \sum_{n=0}^{5} (\frac{1}{6})^n = \frac{1}{1 - (1/6)} - \frac{1 - (1/6)^6}{1 - (1/6)}$

$= \dfrac{(1/6)^6}{5/6} = \dfrac{1}{5(6^5)}$, the error in approximating $\displaystyle\sum_{n=1}^{\infty} \frac{1}{n^n}$ by

the sum of its first-five terms is approximately $\dfrac{1}{5(6^5)} = 0.000026$

38. If $j < k - 1$, then $k - j > 1$. The p-series with $p = k - j$ converges and since $\displaystyle\lim_{n \to \infty} \frac{P(n)/Q(n)}{1/n^{k-j}} = L > 0$, the series $\displaystyle\sum_{n=1}^{\infty} \frac{P(n)}{Q(n)}$ converges by the limit comparison test. Similarly, if $j \geq k - 1$, then $k - j \leq 1$ which implies that $\displaystyle\sum_{n=1}^{\infty} \frac{P(n)}{Q(n)}$ diverges by the limit comparison test.

39. $\frac{1}{2} + \frac{2}{5} + \frac{3}{10} + \frac{4}{17} + \frac{5}{26} + \ldots = \sum_{n=1}^{\infty} \frac{n}{n^2 + 1}$, which diverges since the degree of the numerator is only one less than the degree of the denominator.

40. $\frac{1}{3} + \frac{1}{8} + \frac{1}{15} + \frac{1}{24} + \frac{1}{35} + \ldots = \sum_{n=2}^{\infty} \frac{1}{n^2 - 1}$, which converges since the degree of the numerator is two less than the degree of the denominator.

41. $\sum_{n=1}^{\infty} \frac{1}{n^3 + 1}$ converges since the degree of the numerator is three less than the degree of the denominator.

42. $\sum_{n=1}^{\infty} \frac{n^2}{n^3 + 1}$ diverges since the degree of the numerator is only one less than the degree of the denominator.

43. $\lim_{n \to \infty} \frac{a_n}{1/n} = \lim_{n \to \infty} na_n \neq 0$ Therefore, $\sum_{n=1}^{\infty} a_n$ diverges by a limit comparison with the p-series $\sum_{n=1}^{\infty} \frac{1}{n}$

44. $\lim_{n \to \infty} n\left(\frac{n^3}{5n^4 + 3}\right) = \lim_{n \to \infty} \frac{n^4}{5n^4 + 3} = \frac{1}{5} \neq 0$

Therefore, $\sum_{n=1}^{\infty} \frac{n^3}{5n^4 + 3}$ diverges.

45. $\lim_{n \to \infty} \frac{n}{\ln n} = \lim_{n \to \infty} \frac{1}{1/n} = \lim_{n \to \infty} n = \infty$

Therefore, $\sum_{n=2}^{\infty} \frac{1}{\ln n}$ diverges.

9.6 Alternating series

1. $\sum_{n=1}^{\infty} \frac{(-1)^{n+1}}{n}$, $a_{n+1} = \frac{1}{n + 1} < \frac{1}{n} = a_n$ and $\lim_{n \to \infty} \frac{1}{n} = 0$

Therefore, by Theorem 9.14, the series converges.

2. $\sum_{n=1}^{\infty} \frac{(-1)^{n+1}n}{2n - 1}$, $\lim_{n \to \infty} \frac{n}{2n - 1} = \frac{1}{2}$ Therefore, by Theorem 9.14, the series diverges.

3. $\sum_{n=1}^{\infty} \frac{(-1)^{n+1}}{2n-1}$, $a_{n+1} = \frac{1}{2(n+1)-1} < \frac{1}{2n-1} = a_n$ and $\lim_{n \to \infty} \frac{1}{2n-1} = 0$
Therefore, by Theorem 9.14, the series converges.

4. $\sum_{n=2}^{\infty} \frac{(-1)^{n}}{\ln n}$, $a_{n+1} = \frac{1}{\ln(n+1)} < \frac{1}{\ln n} = a_n$ and $\lim_{n \to \infty} \frac{1}{\ln n} = 0$
Therefore, by Theorem 9.14, the series converges.

5. $\sum_{n=1}^{\infty} \frac{(-1)^{n} n^2}{n^2+1}$, $\lim_{n \to \infty} \frac{n^2}{n^2+1} = 1$ Therefore, by Theorem 9.14, the series diverges.

6. $\sum_{n=1}^{\infty} \frac{(-1)^{n+1} n}{n^2+1}$, $a_{n+1} = \frac{n+1}{(n+1)^2+1} < \frac{n}{n^2+1} = a_n$ and $\lim_{n \to \infty} \frac{n}{n^2+1} = 0$
Therefore, by Theorem 9.14, the series converges.

7. $\sum_{n=1}^{\infty} \frac{(-1)^{n}}{\sqrt{n}}$, $a_{n+1} = \frac{1}{\sqrt{n+1}} < \frac{1}{\sqrt{n}} = a_n$ and $\lim_{n \to \infty} \frac{1}{\sqrt{n}} = 0$
Therefore, by Theorem 9.14, the series converges.

8. $\sum_{n=1}^{\infty} \frac{(-1)^{n+1} n^3}{n^3+6}$, $\lim_{n \to \infty} \frac{n^3}{n^3+6} = 1$
Therefore, by Theorem 9.14, the series diverges.

9. $\sum_{n=1}^{\infty} \frac{(-1)^{n+1}(n+1)}{\ln(n+1)}$, $\lim_{n \to \infty} \frac{n+1}{\ln(n+1)} = \lim_{n \to \infty} \frac{1}{1/(n+1)}$

$= \lim_{n \to \infty} (n+1) = \infty$ Therefore, by Theorem 9.14, the series diverges.

10. $\sum_{n=1}^{\infty} \frac{(-1)^{n+1} \ln(n+1)}{n+1}$, $a_{n+1} = \frac{\ln[(n+1)+1]}{(n+1)+1} < \frac{\ln(n+1)}{n+1}$

for $n \geq 2$ and $\lim_{n \to \infty} \frac{\ln(n+1)}{n+1} = \lim_{n \to \infty} \frac{1/(n+1)}{1} = 0$
Therefore, by Theorem 9.14, the series converges.

11. $\sum_{n=0}^{\infty} (-1)^n e^{-n}$, $a_{n+1} = \frac{1}{e^{n+1}} < \frac{1}{e^n} = a_n$ and $\lim_{n \to \infty} \frac{1}{e^n} = 0$
Therefore, by Theorem 9.14, the series converges.

12. $\sum_{n=1}^{\infty} \frac{(-1)^{n+1}\sqrt{n+2}}{\sqrt{n(n+2)}} = \sum_{n=1}^{\infty} \frac{(-1)^{n+1}}{\sqrt{n}}$ Converges (See Exercise 7)

13. $\sum_{n=1}^{\infty} \sin\left[\frac{(2n - 1)\pi}{2}\right] = \sum_{n=1}^{\infty} (-1)^{n+1}$ Diverges

14. $\sum_{n=1}^{\infty} \cos n\pi = \sum_{n=1}^{\infty} (-1)^{n}$ Diverges

15. $\sum_{n=1}^{\infty} \frac{1}{n} \sin\left[\frac{(2n - 1)\pi}{2}\right] = \sum_{n=1}^{\infty} \frac{(-1)^{n+1}}{n}$ Converges (See Exercise 1)

16. $\sum_{n=1}^{\infty} \frac{1}{n} \cos n\pi = \sum_{n=1}^{\infty} \frac{(-1)^{n}}{n}$ Converges (See Exercise 1)

17. $\sum_{n=0}^{\infty} \frac{(-1)^n}{n!}$, $a_{n+1} = \frac{1}{(n + 1)!} < \frac{1}{n!} = a_n$ and $\lim_{n \to \infty} \frac{1}{n!} = 0$
Therefore by Theorem 9.14 the series converges.

18. $\sum_{n=0}^{\infty} \frac{(-1)^n}{(2n)!}$, $a_{n+1} = \frac{1}{(2n + 2)!} < \frac{1}{(2n)!} = a_n$ and $\lim_{n \to \infty} \frac{1}{(2n)!} = 0$
Therefore by Theorem 9.14 the series converges.

19. $\sum_{n=1}^{\infty} \frac{(-1)^{n+1}\sqrt{n}}{n + 2}$, $a_{n+1} = \frac{\sqrt{n + 1}}{(n + 1) + 2} < \frac{\sqrt{n}}{n + 2}$ for $n \geq 2$ and

$\lim_{n \to \infty} \frac{\sqrt{n}}{n + 2} = 0$ Therefore, by Theorem 9.14, the series converges.

20. $\sum_{n=1}^{\infty} \frac{(-1)^{n+1}\sqrt{n}}{\sqrt[3]{n}}$, $\lim_{n \to \infty} \frac{n^{1/2}}{n^{1/3}} = \lim_{n \to \infty} n^{1/6} = \infty$
Therefore, the series diverges by Theorem 9.14.

21. $\sum_{n=1}^{\infty} \frac{(-1)^{n+1}(2)}{e^n - e^{-n}} = \sum_{n=1}^{\infty} \frac{(-1)^{n+1}(2e^n)}{e^{2n} - 1}$, Let $f(x) = \frac{2e^x}{e^{2x} - 1}$

then $f'(x) = \frac{-2e^x(e^{2x} + 1)}{(e^{2x} - 1)^2} < 0$ thus $f(x)$ is decreasing.

Therefore $a_{n+1} < a_n$, $\lim_{n \to \infty} \frac{2e^n}{e^{2n} - 1} = \lim_{n \to \infty} \frac{2e^n}{2e^{2n}} = \lim_{n \to \infty} \frac{1}{e^n} = 0$

Therefore, by Theorem 9.14, the series converges.

22. $\sum_{n=1}^{\infty} \frac{2(-1)^{n+1}}{e^n + e^{-n}} = \sum_{n=1}^{\infty} \frac{(-1)^{n+1}(2e^n)}{e^{2n} + 1}$

Let $f(x) = \frac{2e^x}{e^{2x} + 1}$ then $f'(x) = \frac{2e^{2x}(1 - e^{2x})}{(e^{2x} + 1)^2} < 0$ for $x > 0$

thus $f(x)$ is decreasing for $x > 0$ which implies $a_{n+1} < a_n$.

$\lim_{n \to \infty} \frac{2e^n}{e^{2n} + 1} = \lim_{n \to \infty} \frac{2e^n}{2e^{2n}} = \lim_{n \to \infty} \frac{1}{e^n} = 0$

Therefore, by Theorem 9.14, the series converges.

23. $\sum_{n=1}^{\infty} \frac{(-1)^{n+1}}{(n + 1)^2}$, $\sum_{n=1}^{\infty} \frac{1}{(n + 1)^2}$ converges by comparison to the

p-series $\sum_{n=1}^{\infty} \frac{1}{n^2}$ Therefore, the given series converges absolutely.

24. $\sum_{n=1}^{\infty} \frac{(-1)^{n+1}}{n + 1}$, The given series converges by the alternating series test, but does not converge absolutely since the series $\sum_{n=1}^{\infty} \frac{1}{n + 1}$ diverges by the integral test.

25. $\sum_{n=1}^{\infty} \frac{(-1)^{n+1}}{\sqrt{n}}$ The given series converges by the alternating series test, but does not converge absolutely since $\sum_{n=1}^{\infty} \frac{1}{\sqrt{n}}$ is a divergent p-series. Therefore, the series converges conditionally.

26. $\sum_{n=1}^{\infty} \frac{(-1)^{n+1}}{n\sqrt{n}}$, $\sum_{n=1}^{\infty} \frac{1}{n\sqrt{n}} = \sum_{n=1}^{\infty} \frac{1}{n^{3/2}}$ which is a convergent p-series. Therefore, the given series converges absolutely.

27. $\sum_{n=2}^{\infty} \frac{(-1)^n}{\ln(n)}$ The given series converges by the alternating series test, but does not converge absolutely since the series $\sum_{n=2}^{\infty} \frac{1}{\ln(n)}$ diverges by comparison to the harmonic series $\sum_{n=1}^{\infty} \frac{1}{n}$. Therefore, the series converses conditionally.

28. $\sum_{n=0}^{\infty} \frac{(-1)^n}{e^{n^2}}$, $\sum_{n=0}^{\infty} \frac{1}{e^{n^2}}$ converges by a comparison to the convergent geometric series $\sum_{n=0}^{\infty} (\frac{1}{e})^n$. Therefore, the given series converges absolutely.

29. $\sum_{n=2}^{\infty} \frac{(-1)^n n}{n^3 - 1}$, $\sum_{n=2}^{\infty} \frac{n}{n^3 - 1}$ converges by a limit comparison to the convergent p-series $\sum_{n=2}^{\infty} \frac{1}{n^2}$. Therefore, the given series converges absolutely.

30. $\sum_{n=1}^{\infty} \frac{(-1)^{n+1}}{n^{1.5}}$, $\sum_{n=1}^{\infty} \frac{1}{n^{1.5}}$ is a convergent p-series. Therefore, the given series converges absolutely.

31. $\sum_{n=0}^{\infty} \frac{(-1)^n}{(2n + 1)!}$, $\sum_{n=0}^{\infty} \frac{1}{(2n + 1)!}$ is convergent by comparison to the convergent geometric series $\sum_{n=0}^{\infty} (\frac{1}{2})^n$ since $\frac{1}{(2n + 1)!} < \frac{1}{2^n}$ for $n > 0$. Therefore, the given series converges absolutely.

32. $\sum_{n=0}^{\infty} \frac{(-1)^n}{\sqrt[3]{n + 1}}$ The given series converges by the alternating series test, but $\sum_{n=0}^{\infty} \frac{1}{\sqrt[3]{n + 1}}$ diverges by a limit comparison to the divergent p-series $\sum_{n=1}^{\infty} \frac{1}{\sqrt[3]{n}}$. Therefore, the series converges conditionally.

33. $\sum_{n=0}^{\infty} \frac{\cos n\pi}{n + 1} = \sum_{n=0}^{\infty} \frac{(-1)^n}{n + 1}$ The given series converges by the alternating series test, but $\sum_{n=0}^{\infty} \frac{|\cos n\pi|}{n + 1}$ diverges by a limit comparison to the divergent harmonic series.

$\sum_{n=1}^{\infty} \frac{1}{n}$, $\lim_{n \to \infty} \frac{|\cos n\pi|/(n + 1)}{1/n} = 1$.

Therefore, the series converges conditionally.

34. $\sum_{n=1}^{\infty} (-1)^{n+1} \arctan n$, $\lim_{n \to \infty} \arctan n = \frac{\pi}{2} \neq 0$. The given series diverges.

35. $\sum_{n=1}^{\infty} \frac{\cos n}{n^2}$, $\frac{|\cos n|}{n^2} \leq \frac{1}{n^2}$ so $\sum_{n=1}^{\infty} \left|\frac{\cos n}{n^2}\right|$ converges by comparison to the convergent p-series $\sum_{n=1}^{\infty} \frac{1}{n^2}$ Therefore, the given series converges absolutely.

36. $\sum_{n=1}^{\infty} \frac{\cos n}{n\sqrt{n}}$, $\frac{|\cos n|}{n\sqrt{n}} \leq \frac{1}{n\sqrt{n}}$ so $\sum_{n=1}^{\infty} \left|\frac{\cos n}{n\sqrt{n}}\right|$ converges by comparison to the convergent p-series $\sum_{n=1}^{\infty} \frac{1}{n^{3/2}}$ Therefore, the given series converges absolutely.

37. $\sum_{n=1}^{\infty} \frac{\sin[(2n - 1)\pi/2]}{\sqrt{n}} = \sum_{n=1}^{\infty} \frac{(-1)^n}{\sqrt{n}}$ The given series converges by the alternating series test, but $\sum_{n=1}^{\infty} \frac{|\sin[(2n - 1)\pi/2]|}{\sqrt{n}}$ diverges by a limit comparison to the divergent p-series $\sum_{n=1}^{\infty} \frac{1}{\sqrt{n}}$

38. $\sum_{n=1}^{\infty} \frac{\sin[(2n - 1)\pi/2]}{n} = \sum_{n=1}^{\infty} \frac{(-1)^{n+1}}{n}$ The given series converges by the alternating series test but $\sum_{n=1}^{\infty} \left|\frac{\sin[(2n - 1)\pi/2]}{n}\right|$ diverges by a limit comparison to the divergent p-series $\sum_{n=1}^{\infty} \frac{1}{n}$. Therefore, the series converges conditionally.

39. $\sum_{n=1}^{\infty} \frac{(-1)^{n+1}}{2n^3 - 1}$ By Theorem 9.15 $|R_N| \leq a_{N+1} = \frac{1}{2(N + 1)^3 - 1} \leq 0.001$
This inequality is valid when N = 7, and we may approximate the series by $\sum_{n=1}^{7} \frac{(-1)^{n+1}}{2n^3 - 1} = 1 - \frac{1}{15} + \frac{1}{53} - \frac{1}{127} + \frac{1}{249} - \frac{1}{431} + \frac{1}{685} = 0.947$

40. $\sum_{n=1}^{\infty} \frac{(-1)^{n+1}}{n^4}$ By Theorem 9.15, $|R_N| \leq a_{N+1} = \frac{1}{(N + 1)^4} \leq 0.001$
This inequality is valid when N = 5, and we may approximate the series by $\sum_{n=1}^{5} \frac{(-1)^{n+1}}{n^4} = 1 - \frac{1}{16} + \frac{1}{81} - \frac{1}{256} + \frac{1}{625} \approx 0.9475$

41. $\sum_{n=0}^{\infty} \frac{(-1)^n}{n!}$ By Theorem 9.15, $|R_N| \leq a_{N+1} = \frac{1}{(N + 1)!} \leq 0.001$
This inequality is valid when N = 6, and we may approximate the series by $\sum_{n=0}^{6} \frac{(-1)^n}{n!} = 1 - 1 + \frac{1}{2} - \frac{1}{6} + \frac{1}{24} - \frac{1}{120} + \frac{1}{720} = 0.368$

42. $\sum_{n=0}^{\infty} \frac{(-1)^n}{2^n n!}$ By Theorem 9.15, $|R_N| \leq a_{N+1} = \frac{1}{2^{N+1}(N + 1)!} \leq 0.001$

This inequality is valid when N = 4, and we may approximate the series by $\sum_{n=0}^{4} \frac{(-1)^n}{2^n n!} = 1 - \frac{1}{2} + \frac{1}{8} - \frac{1}{48} + \frac{1}{384} \approx 0.607$

43. $\sum_{n=0}^{\infty} \frac{(-1)^n}{(2n + 1)!}$ By Theorem 9.15, $|R_N| \leq a_{N+1} = \frac{1}{[2(N + 1) + 1]!} \leq 0.001$

This inequality is valid when N = 2, and we may approximate the series by $\sum_{n=0}^{2} \frac{(-1)^n}{(2n + 1)!} = 1 - \frac{1}{6} + \frac{1}{120} \approx 0.842$

44. $\sum_{n=0}^{\infty} \frac{(-1)^n}{(2n)!}$ By Theorem 9.15, $|R_N| \leq a_{N+1} = \frac{1}{(2N + 2)!} \leq 0.001$

This inequality is valid when N = 3, and we may approximate the series by $\sum_{n=0}^{3} \frac{(-1)^n}{(2n)!} = 1 - \frac{1}{2} + \frac{1}{24} - \frac{1}{720} = 0.540$

45. $\sum_{n=1}^{\infty} \frac{(-1)^{n+1}}{n2^n}$ By Theorem 9.15, $|R_N| \leq a_{N+1} = \frac{1}{(N + 1)2^{N+1}} \leq 0.001$

This inequality is valid when N = 7, and we may approximate the series by $\sum_{n=1}^{7} \frac{(-1)^{n+1}}{n2^n} = \frac{1}{2} - \frac{1}{8} + \frac{1}{24} - \frac{1}{64} + \frac{1}{160} - \frac{1}{384} + \frac{1}{896} = 0.406$

46. $\sum_{n=1}^{\infty} \frac{(-1)^{n+1}}{4^n n}$ By Theorem 9.15 $|R_N| \leq a_{N+1} = \frac{1}{4^{N+1}(N + 1)} \leq 0.001$

This inequality is valid when N = 3, and we may approximate the series by $\sum_{n=1}^{3} \frac{(-1)^{n+1}}{4^n n} = \frac{1}{4} - \frac{1}{32} + \frac{1}{192} \approx 0.224$

47. $\sum_{n=0}^{\infty} \frac{(-1)^n}{2n + 1}$ By Theorem 9.15 $|R_N| \leq a_{N+1} = \frac{1}{2(N + 1) + 1} \leq 0.001$

This inequality is valid when $1000 \leq 2N + 3$, or when $N \geq 499$

48. $\sum_{n=1}^{\infty} \frac{(-1)^{n+1}}{n^2}$ By Theorem 9.15 $|R_N| \leq a_{N+1} = \frac{1}{(N + 1)^2} \leq 0.001$

This inequality is valid when $1000 \leq (N + 1)^2$, or when $N \geq 31$

49. $\sum_{n=1}^{\infty} \frac{(-1)^n}{n^p}$ If $p < 0$ then $\lim_{n \to \infty} \frac{1}{n^p} = \lim_{n \to \infty} n^{-p} = \infty$ and the series diverges. If $p = 0$ then $\lim_{n \to \infty} \frac{1}{n^p} = 1$ and the series diverges.

If $p > 0$ then $\lim_{n \to \infty} \frac{1}{n^p} = 0$ and $\frac{1}{(n+1)^p} < \frac{1}{n^p}$.
Therefore, the series converges by the alternating series test.

50. Since $\sum_{n=1}^{\infty} |a_n|$ converges we have $\lim_{n \to \infty} |a_n| = 0$ Thus, there must exist an $N > 0$ such that $|a_N| < 1$ for all $n > N$ and it follows that $a_n^2 \leq |a_n|$ for all $n > N$. Hence, by the comparison test, $\sum_{n=1}^{\infty} a_n^2$ converges. To see that the converse is false, let $a_n = \frac{1}{n}$

9.7 The Ratio and Root Tests

1. $\sum_{n=0}^{\infty} \frac{n!}{3^n}$, $\lim_{n \to \infty} \left|\frac{a_{n+1}}{a_n}\right| = \lim_{n \to \infty} \left|\frac{(n+1)!}{3^{n+1}} \cdot \frac{3^n}{n!}\right| = \lim_{n \to \infty} \frac{n+1}{3} = \infty$
Therefore, by the Ratio Test, the series diverges.

2. $\sum_{n=1}^{\infty} n(\frac{2}{3})^n$, $\lim_{n \to \infty} \left|\frac{a_{n+1}}{a_n}\right| = \lim_{n \to \infty} \left|\frac{(n+1)(2/3)^{n+1}}{n(2/3)^n}\right|$

$= \lim_{n \to \infty} \frac{2(n+1)}{3n} = \frac{2}{3}$
Therefore, by the Ratio Test, the series converges.

3. $\sum_{n=0}^{\infty} \frac{3^n}{n!}$, $\lim_{n \to \infty} \left|\frac{a_{n+1}}{a_n}\right| = \lim_{n \to \infty} \left|\frac{3^{n+1}}{(n+1)!} \cdot \frac{n!}{3^n}\right| = \lim_{n \to \infty} \frac{3}{n+1} = 0$
Therefore, by the Ratio Test, the series converges.

4. $\sum_{n=1}^{\infty} n(\frac{3}{2})^n$, $\lim_{n \to \infty} \left|\frac{a_{n+1}}{a_n}\right| = \lim_{n \to \infty} \left|\frac{(n+1)3^{n+1}}{2^{n+1}} \cdot \frac{2^n}{n\, 3^n}\right|$

$= \lim_{n \to \infty} \frac{3(n+1)}{2n} = \frac{3}{2}$
Therefore, by the Ratio Test, the series diverges.

5. $\sum_{n=1}^{\infty} \frac{n}{2^n}$, $\lim_{n \to \infty} \left|\frac{a_{n+1}}{a_n}\right| = \lim_{n \to \infty} \left|\frac{n+1}{2^{n+1}} \cdot \frac{2^n}{n}\right| = \lim_{n \to \infty} \frac{n+1}{2n} = \frac{1}{2}$
Therefore, by the Ratio Test, the series converges.

6. $\sum_{n=1}^{\infty} \frac{n^2}{2^n}$, $\lim_{n\to\infty} \left|\frac{a_{n+1}}{a_n}\right| = \lim_{n\to\infty} \left|\frac{n^2 + 2n + 1}{2^{n+1}} \cdot \frac{2^n}{n^2}\right|$

$= \lim_{n\to\infty} \frac{n^2 + 2n + 1}{2n^2} = \frac{1}{2}$

Therefore, by the Ratio Test, the series converges.

7. $\sum_{n=0}^{\infty} \frac{(-1)^{n+1} n!}{1\cdot 3\cdot 5\cdots(2n + 1)}$

$\lim_{n\to\infty} \left|\frac{a_{n+1}}{a_n}\right| = \lim_{n\to\infty} \left|\frac{(n + 1)!}{1\cdot 3\cdot 5\cdots(2n + 1)(2n + 3)} \cdot \frac{1\cdot 3\cdot 5\cdots(2n + 1)}{n!}\right|$

$= \lim_{n\to\infty} \frac{n + 1}{2n + 3} = \frac{1}{2}$ Therefore, by the Ratio Test, the series converges.

8. $\sum_{n=1}^{\infty} \frac{(-1)^n 2\cdot 4\cdot 6\cdots 2n}{2\cdot 5\cdot 8\cdots(3n - 1)}$

$\lim_{n\to\infty} \left|\frac{a_{n+1}}{a_n}\right| = \lim_{n\to\infty} \left|\frac{2\cdot 4\cdots 2n(2n + 2)}{2\cdot 5\cdots(3n - 1)(3n + 2)} \cdot \frac{2\cdot 5\cdots(3n - 1)}{2\cdot 4\cdots 2n}\right|$

$= \lim_{n\to\infty} \frac{2n + 2}{3n + 2} = \frac{2}{3}$ Therefore, by the Ratio Test, the series converges.

9. $\sum_{n=1}^{\infty} \frac{2^n}{n^2}$, $\lim_{n\to\infty} \left|\frac{a_{n+1}}{a_n}\right| = \lim_{n\to\infty} \left|\frac{2^{n+1}}{(n + 1)^2} \cdot \frac{n^2}{2^n}\right| = \lim_{n\to\infty} \frac{2n^2}{(n + 1)^2} = 2$

Therefore, by the Ratio Test, the series diverges.

10. $\sum_{n=1}^{\infty} \frac{(-1)^{n+1}(n + 2)}{n(n + 1)}$, $a_{n+1} = \frac{n + 3}{(n + 1)(n + 2)} \le \frac{n + 2}{n(n + 1)} = a_n$

and $\lim_{n\to\infty} \frac{n + 2}{n(n + 1)} = 0$ Therefore, by Theorem 9.14, the series converges.

11. $\sum_{n=1}^{\infty} \frac{(-1)^n 2^n}{n!}$, $\lim_{n\to\infty} \left|\frac{a_{n+1}}{a_n}\right| = \lim_{n\to\infty} \left|\frac{2^{n+1}}{(n + 1)!} \cdot \frac{n!}{2^n}\right| = \lim_{n\to\infty} \frac{2}{n + 1} = 0$

Therefore, by the Ratio Test, the series converges.

12. $\sum_{n=1}^{\infty} \frac{(-1)^{n-1}(3/2)^n}{n^2}$, $\lim_{n\to\infty} \left|\frac{a_{n+1}}{a_n}\right| = \lim_{n\to\infty} \left|\frac{(3/2)^{n+1}}{n^2 + 2n + 1} \cdot \frac{n^2}{(3/2)^n}\right|$

$= \lim_{n\to\infty} \frac{3n^2}{2(n^2 + 2n + 1)} = \frac{3}{2} > 1$

Therefore, by the Ratio Test, the series diverges.

13. $\sum_{n=0}^{\infty} \frac{n!}{n3^n}$, $\lim_{n \to \infty} \left|\frac{a_{n+1}}{a_n}\right| = \lim_{n \to \infty} \left|\frac{(n+1)!}{(n+1)3^{n+1}} \cdot \frac{n\,3^n}{n!}\right| = \lim_{n \to \infty} \frac{n}{3} = \infty$

Therefore, by the Ratio Test, the series diverges.

14. $\sum_{n=1}^{\infty} \frac{(2n)!}{n^5}$, $\lim_{n \to \infty} \left|\frac{a_{n+1}}{a_n}\right| = \lim_{n \to \infty} \left|\frac{(2n+2)!}{(n+1)^5} \cdot \frac{n^5}{(2n)!}\right|$

$= \lim_{n \to \infty} \frac{(2n+2)(2n+1)n^5}{(n+1)^5} = \infty$

Therefore, by the Ratio Test, the series diverges.

15. $\sum_{n=0}^{\infty} \frac{4^n}{n!}$, $\lim_{n \to \infty} \left|\frac{a_{n+1}}{a_n}\right| = \lim_{n \to \infty} \left|\frac{4^{n+1}}{(n+1)!} \cdot \frac{n!}{4^n}\right| = \lim_{n \to \infty} \frac{4}{n+1} = 0$

Therefore, by the Ratio Test, the series converges.

16. $\sum_{n=1}^{\infty} \frac{n^n}{n!}$, $\lim_{n \to \infty} \left|\frac{a_{n+1}}{a_n}\right| = \lim_{n \to \infty} \left|\frac{(n+1)^{n+1}}{(n+1)!} \cdot \frac{n!}{n^n}\right|$

$= \lim_{n \to \infty} \left(\frac{n+1}{n}\right)^n = e > 1$

Therefore, by the Ratio Test, the series diverges.

17. $\sum_{n=0}^{\infty} \frac{3^n}{(n+1)^n}$, $\lim_{n \to \infty} \left|\frac{a_{n+1}}{a_n}\right| = \lim_{n \to \infty} \left|\frac{3^{n+1}}{(n+2)^{n+1}} \cdot \frac{(n+1)^n}{3^n}\right|$

$= \lim_{n \to \infty} \frac{3(n+1)^n}{(n+2)^{n+1}} = \lim_{n \to \infty} \frac{3}{n+2}\left(\frac{n+1}{n+2}\right)^n = (0)\left(\frac{1}{e}\right) = 0$

To find $\lim_{n \to \infty} \left(\frac{n+1}{n+2}\right)^n$, let $y = \lim_{n \to \infty} \left(\frac{n+1}{n+2}\right)^n$.

Then $\ln y = \lim_{n \to \infty} n \ln\left(\frac{n+1}{n+2}\right) = \lim_{n \to \infty} \frac{\ln[(n+1)/(n+2)]}{1/n} = \frac{0}{0}$,

$\ln y = \lim_{n \to \infty} \frac{[(1)/(n+1)] - [(1)/(n+2)]}{-(1/n^2)} = -1$. So $y = e^{-1} = \frac{1}{e}$.

Therefore, by the Ratio Test, the series converges.

18. $\sum_{n=1}^{\infty} \frac{(n!)^2}{(3n)!}$, $\lim_{n \to \infty} \left|\frac{a_{n+1}}{a_n}\right| = \lim_{n \to \infty} \left|\frac{[(n+1)!]^2}{(3n+3)!} \cdot \frac{(3n)!}{(n!)^2}\right|$

$= \lim_{n \to \infty} \frac{(n+1)^2}{(3n+3)(3n+2)(3n+1)} = 0$

Therefore, by the Ratio Test, the series converges.

19. $\sum_{n=0}^{\infty} \frac{4^n}{3^n + 1}$, $\lim_{n\to\infty} \left|\frac{a_{n+1}}{a_n}\right| = \lim_{n\to\infty} \left|\frac{4^{n+1}}{3^{n+1} + 1} \cdot \frac{3^n + 1}{4^n}\right|$

$= \lim_{n\to\infty} \frac{4(3^n + 1)}{3^{n+1} + 1} = \lim_{n\to\infty} \frac{4(1 + 1/3^n)}{3 + 1/3^n} = \frac{4}{3}$

Therefore, by the Ratio Test, the series diverges.

20. $\sum_{n=0}^{\infty} \frac{(-1)^n 2^{4n}}{(2n + 1)!}$, $\lim_{n\to\infty} \left|\frac{a_{n+1}}{a_n}\right| = \lim_{n\to\infty} \left|\frac{2^{4n+4}}{(2n + 3)!} \cdot \frac{(2n + 1)!}{2^{4n}}\right|$

$= \lim_{n\to\infty} \frac{2^4}{(2n + 3)(2n + 2)} = 0$

Therefore, by the Ratio Test, the series converges.

21. $\sum_{n=1}^{\infty} \left(\frac{n}{2n + 1}\right)^n$, $\lim_{n\to\infty} \sqrt[n]{|a_n|} = \lim_{n\to\infty} \sqrt[n]{\left(\frac{n}{2n + 1}\right)^n} = \lim_{n\to\infty} \frac{n}{2n + 1} = \frac{1}{2}$

Therefore, by the Root Test, the series converges.

22. $\sum_{n=1}^{\infty} \left(\frac{2n}{n + 1}\right)^n$, $\lim_{n\to\infty} \sqrt[n]{|a_n|} = \lim_{n\to\infty} \sqrt[n]{\left(\frac{2n}{n + 1}\right)^n} = \lim_{n\to\infty} \frac{2n}{n + 1} = 2$

Therefore, by the Root Test, the series diverges.

23. $\sum_{n=1}^{\infty} (2\sqrt[n]{n} + 1)^n$, $\lim_{n\to\infty} \sqrt[n]{|a_n|} = \lim_{n\to\infty} \sqrt[n]{(2\sqrt[n]{n} + 1)^n} = \lim_{n\to\infty} (2\sqrt[n]{n} + 1)$

To find $\lim_{n\to\infty} \sqrt[n]{n}$ let $y = \lim_{x\to\infty} \sqrt[x]{x}$ then $\ln y = \lim_{x\to\infty} (\ln \sqrt[x]{x})$

$= \lim_{x\to\infty} \frac{1}{x} \ln x = \lim_{x\to\infty} \frac{\ln x}{x} = \lim_{x\to\infty} \frac{1/x}{1} = 0$

Thus $\ln y = 0$ so $y = e^0 = 1$ and $\lim_{n\to\infty} (2\sqrt[n]{n} + 1) = 2(1) + 1 = 3$

Therefore, by the Root Test, the series diverges.

24. $\sum_{n=1}^{\infty} (\sqrt[n]{n} - 1)^n$, $\lim_{n\to\infty} \sqrt[n]{|a_n|} = \lim_{n\to\infty} \sqrt[n]{(\sqrt[n]{n} - 1)^n} = \lim_{n\to\infty} (\sqrt[n]{n} - 1)$

To find $\lim_{n\to\infty} \sqrt[n]{n}$ let $y = \lim_{x\to\infty} \sqrt[x]{x}$ then $\ln y = \lim_{x\to\infty} \ln \sqrt[x]{x}$

$= \lim_{x\to\infty} \frac{1}{x} \ln x = \lim_{x\to\infty} \frac{\ln x}{x} = \lim_{x\to\infty} \frac{1/x}{1} = 0$

Thus $\ln y = 0$ so $y = e^0 = 1$ and $\lim_{n\to\infty} (\sqrt[n]{n} - 1) = 1 - 1 = 0$

Therefore, by the Root Test, the series converges.

25. $\sum_{n=2}^{\infty} (\frac{\ln n}{n})^n$, $\lim_{n\to\infty} \sqrt[n]{|a_n|} = \lim_{n\to\infty} \sqrt[n]{(\frac{\ln n}{n})^n} = \lim_{n\to\infty} \frac{\ln n}{n} = \lim_{n\to\infty} \frac{1/n}{1} = 0$

Therefore, by the Root Test, the series converges.

26. $\sum_{n=0}^{\infty} e^{-n}$, $\lim_{n\to\infty} \sqrt[n]{|a_n|} = \lim_{n\to\infty} \sqrt[n]{\frac{1}{e^n}} = \frac{1}{e}$

Therefore, by the Root Test, the series converges.

27. $\sum_{n=3}^{\infty} \frac{1}{(\ln n)^n}$, $\lim_{n\to\infty} \sqrt[n]{|a_n|} = \lim_{n\to\infty} \sqrt[n]{\frac{1}{(\ln n)^n}} = \lim_{n\to\infty} \frac{1}{\ln n} = 0$

Therefore, by the Root Test, the series converges.

28. $\sum_{n=0}^{\infty} \frac{n+1}{3^n}$, $\lim_{n\to\infty} \sqrt[n]{|a_n|} = \lim_{n\to\infty} \sqrt[n]{\frac{n+1}{3^n}} = \lim_{n\to\infty} \frac{\sqrt[n]{n+1}}{3} = \frac{1}{3}$

Therefore, by the Root Test, the series converges.

29. $\sum_{n=1}^{\infty} \frac{(-1)^n 3^{n-2}}{2^n}$, $\lim_{n\to\infty} \left|\frac{a_{n+1}}{a_n}\right| = \lim_{n\to\infty} (\frac{3^{n-1}}{2^{n+1}} \cdot \frac{2^n}{3^{n-2}}) = \lim_{n\to\infty} \frac{3}{2} = \frac{3}{2}$

Therefore, by the Ratio Test, the series diverges.

30. $\sum_{n=1}^{\infty} \frac{10}{3\sqrt{n^3}}$, $\lim_{n\to\infty} \frac{10/3n^{3/2}}{1/n^{3/2}} = \frac{10}{3}$ Therefore, the series converges by a limit comparison test with the p-series $\sum_{n=1}^{\infty} \frac{1}{n^{3/2}}$

31. $\sum_{n=1}^{\infty} \frac{10n+3}{n2^n}$, $\lim_{n\to\infty} \frac{(10n+3)/n2^n}{1/2^n} = \lim_{n\to\infty} \frac{10n+3}{n} = 10$

Therefore, the series converges by a limit comparison test with the geometric series $\sum_{n=0}^{\infty} (\frac{1}{2})^n$

32. $\sum_{n=1}^{\infty} \frac{2^n}{4n^2-1}$, $\lim_{n\to\infty} \frac{2^n}{4n^2-1} \neq 0$ Therefore, the series diverges.

33. $\sum_{n=1}^{\infty} (-1)^n \ln(\frac{n+2}{n})$, $a_{n+1} = \ln(\frac{n+3}{n+1}) \le \ln(\frac{n+2}{n}) = a_n$ and

$\lim_{n\to\infty} \ln(\frac{n+2}{n}) = 0$

Therefore, by the alternating series test, the series converges.

34. $\sum_{n=1}^{\infty} \frac{1}{\sqrt{n}+2}$, $\lim_{n\to\infty} \frac{1/(\sqrt{n}+2)}{1/\sqrt{n}} = 1$ Therefore, the series diverges by a limit comparison test with the p-series $\sum_{n=1}^{\infty} \frac{1}{n^{1/2}}$

35. $\sum_{n=1}^{\infty} \frac{\cos(n)}{2^n}$, $\left|\frac{\cos(n)}{2^n}\right| \le \frac{1}{2^n}$ Therefore, the series $\sum_{n=1}^{\infty} \left|\frac{\cos(n)}{2^n}\right|$ converges by comparison with the geometric series $\sum_{n=0}^{\infty} (\frac{1}{2})^n$ which in turn implies that the given series converges.

36. $\sum_{n=2}^{\infty} \frac{(-1)^n}{n \ln(n)}$, $a_{n+1} = \frac{1}{(n+1)\ln(n+1)} \le \frac{1}{n \ln(n)} = a_n$

and $\lim_{n\to\infty} \frac{1}{n \ln(n)} = 0$

Therefore, by the alternating series test, the series converges.

37. $\sum_{n=1}^{\infty} \frac{n7^n}{n!}$, $\lim_{n\to\infty} \left|\frac{a_{n+1}}{a_n}\right| = \lim_{n\to\infty} \left|\frac{(n+1)7^{n+1}}{(n+1)!} \cdot \frac{n!}{n\,7^n}\right| = \lim_{n\to\infty} \frac{7}{n} = 0$

Therefore, by the Ratio Test, the series converges.

38. $\sum_{n=1}^{\infty} \frac{\ln(n)}{n^2}$, $\frac{\ln(n)}{n^2} \le \frac{1}{n^{3/2}}$ Therefore, the series converges by comparison with the p-series $\sum_{n=1}^{\infty} \frac{1}{n^{3/2}}$

39. $\sum_{n=1}^{\infty} \frac{(-1)^n 3^{n-1}}{n!}$, $\lim_{n\to\infty} \left|\frac{a_{n+1}}{a_n}\right| = \lim_{n\to\infty} \left|\frac{3^n}{(n+1)!} \cdot \frac{n!}{3^{n-1}}\right|$

$= \lim_{n\to\infty} \frac{3}{n+1} = 0$

Therefore, by the Ratio Test, the series converges.

40. $\sum_{n=1}^{\infty} \frac{(-1)^n 3^n}{n2^n}$, $\lim_{n\to\infty} \left|\frac{a_{n+1}}{a_n}\right| = \lim_{n\to\infty} \left|\frac{3^{n+1}}{(n+1)2^{n+1}} \cdot \frac{n2^n}{3^n}\right|$

$= \lim_{n\to\infty} \frac{3n}{2(n+1)} = \frac{3}{2}$

Therefore, by the Ratio Test, the series diverges.

41. First, let $\lim_{n\to\infty} \sqrt[n]{|a_n|} = r < 1$ and choose R such that $0 \leq r < R < 1$

There must exist some $N > 0$ such that $\sqrt[n]{|a_n|} < R$ for all $n > N$

Thus, for $n > N$, we have $|a_n| < R^n$ and since the geometric series $\sum_{n=0}^{\infty} R^n$ converges, we can apply the comparison test to conclude that $\sum_{n=1}^{\infty} |a_n|$ converges which in turn implies that $\sum_{n=1}^{\infty} a_n$ converges.

Second, let $\lim_{n\to\infty} \sqrt[n]{|a_n|} = r > R > 1$. Then there must exist some $M > 0$ such that $\sqrt[n]{|a_n|} > R$ for all $n > M$. Thus, for $n > M$, we have $|a_n| > R^n > 1$ which implies that $\lim_{n\to\infty} a_n \neq 0$ which in turn implies that $\sum_{n=1}^{\infty} a_n$ diverges.

42. $\sum_{n=1}^{\infty} \frac{1}{n^p}$, $\lim_{n\to\infty} \left|\frac{a_{n+1}}{a_n}\right| = \lim_{n\to\infty} \left|\frac{1}{(n+1)^p} \cdot \frac{n^p}{1}\right|$

$$= \lim_{n\to\infty} \frac{n^p}{n^p + pn^{p-1} + \ldots + 1} = 1$$

9.8 Power series

1. $\sum_{n=0}^{\infty} (\frac{x}{2})^n$ Since the series is geometric, it converges only if

$\left|\frac{x}{2}\right| < 1$ or $-2 < x < 2$

2. $\sum_{n=0}^{\infty} (\frac{x}{k})^n$ Since the series is geometric, it converges only if

$\left|\frac{x}{k}\right| < 1$ or $-k < x < k$

3. $\sum_{n=1}^{\infty} \frac{(-1)^n x^n}{n}$, $\lim_{n\to\infty} \left|\frac{a_{n+1}}{a_n}\right| = \lim_{n\to\infty} \left|\frac{x^{n+1}}{n+1} \cdot \frac{n}{x^n}\right|$

$= |x| \lim_{n\to\infty} \frac{n}{n+1} = |x| < 1$ or $-1 < x < 1$

When $x = 1$, the alternating series $\sum_{n=1}^{\infty} \frac{(-1)^n}{n}$ converges

When $x = -1$, the p-series $\sum_{n=1}^{\infty} \frac{1}{n}$ diverges

Therefore, the interval of convergence is $-1 < x \leq 1$

4. $\sum_{n=0}^{\infty} (-1)^{n+1} n x^n$, $\lim_{n\to\infty} \left|\frac{a_{n+1}}{a_n}\right| = \lim_{n\to\infty} \left|\frac{(n+1)x^{n+1}}{nx^n}\right|$

$= |x| \lim_{n\to\infty} \frac{n+1}{n} = |x| < 1$ or $-1 < x < 1$

When $x = 1$, the alternating series $\sum_{n=0}^{\infty} (-1)^{n+1} n$ diverges

When $x = -1$, the series $\sum_{n=0}^{\infty} (-n)$ diverges

Therefore, the interval of convergence is $-1 < x < 1$

5. $\sum_{n=0}^{\infty} \frac{x^n}{n!}$, $\lim_{n\to\infty} \left|\frac{a_{n+1}}{a_n}\right| = \lim_{n\to\infty} \left|\frac{x^{n+1}}{(n+1)!} \cdot \frac{n!}{x^n}\right| = |x| \lim_{n\to\infty} \frac{1}{n+1} = 0$

Therefore, the interval of convergence is $-\infty < x < \infty$

6. $\sum_{n=0}^{\infty} \frac{(3x)^n}{(2n)!}$, $\lim_{n\to\infty} \left|\frac{a_{n+1}}{a_n}\right| = \lim_{n\to\infty} \left|\frac{(3x)^{n+1}}{(2n+2)!} \cdot \frac{(2n)!}{(3x)^n}\right|$

$= |x| \lim_{n\to\infty} \frac{3}{(2n+2)(2n+1)} = 0$

Therefore, the interval of convergence is $-\infty < x < \infty$

7. $\sum_{n=0}^{\infty} (2n)!(\frac{x}{2})^n$, $\lim_{n\to\infty} \left|\frac{a_{n+1}}{a_n}\right| = \lim_{n\to\infty} \left|\frac{(2n+2)!(x/2)^{n+1}}{(2n)!(x/2)^n}\right|$

$= |x| \lim_{n\to\infty} \frac{(2n+2)(2n+1)}{2} = \infty$

Therefore, the series converges only for $x = 0$

8. $\sum_{n=0}^{\infty} \frac{(-1)^n x^n}{(n + 1)(n + 2)}$, $\lim_{n\to\infty} \left|\frac{a_{n+1}}{a_n}\right| = \lim_{n\to\infty} \left|\frac{x^{n+1}/(n + 2)(n + 3)}{x^n/(n + 1)(n + 2)}\right|$

$= |x| \lim_{n\to\infty} \frac{n + 1}{n + 3} = |x| < 1$ or $-1 < x < 1$

When $x = 1$, the series $\sum_{n=0}^{\infty} \frac{(-1)^n}{(n + 1)(n + 2)}$ converges

When $x = -1$, the series $\sum_{n=0}^{\infty} \frac{1}{(n + 1)(n + 2)}$ converges

Therefore, the interval of convergence is $-1 \le x \le 1$

9. $\sum_{n=1}^{\infty} \frac{(-1)^{n+1} x^n}{4^n}$ Since the series is geometric, it converges only if

$\left|\frac{x}{4}\right| < 1$ or $-4 < x < 4$

10. $\sum_{n=0}^{\infty} \frac{(-1)^n n!(x - 4)^n}{3^n}$, $\lim_{n\to\infty} \left|\frac{a_{n+1}}{a_n}\right| = \lim_{n\to\infty} \left|\frac{(n + 1)!(x - 4)^{n+1}/3^{n+1}}{n!(x - 4)^n/3^n}\right|$

$= |x - 4| \lim_{n\to\infty} \frac{n + 1}{3} = \infty$

Therefore, the series converges only for $x = 4$

11. $\sum_{n=1}^{\infty} \frac{(-1)^{n+1}(x - 5)^n}{n5^n}$, $\lim_{n\to\infty} \left|\frac{a_{n+1}}{a_n}\right| = \lim_{n\to\infty} \left|\frac{(x - 5)^{n+1}}{(n + 1)5^{n+1}} \cdot \frac{n5^n}{(x - 5)^n}\right|$

$= |x - 5| \lim_{n\to\infty} \frac{n}{5(n + 1)} = \frac{|x - 5|}{5} < 1$ or $|x - 5| < 5$ or $0 < x < 10$

When $x = 0$, the series $\sum_{n=1}^{\infty} \frac{-1}{n}$ diverges

When $x = 10$, the series $\sum_{n=1}^{\infty} \frac{(-1)^{n+1}}{n}$ converges

Therefore, the interval of convergence is $0 < x \le 10$

12. $\displaystyle\sum_{n=0}^{\infty} \frac{(x-2)^{n+1}}{(n+1)3^{n+1}}$, $\displaystyle\lim_{n\to\infty} \left|\frac{a_{n+1}}{a_n}\right| = \lim_{n\to\infty} \left|\frac{(x-2)^{n+2}}{(n+2)3^{n+2}} \cdot \frac{(n+1)3^{n+1}}{(x-2)^{n+1}}\right|$

$\displaystyle = |x-2| \lim_{n\to\infty} \frac{n+1}{3(n+2)} = \frac{|x-2|}{3} < 1$ or $|x-2| < 3$ or $-1 < x < 5$

When $x = -1$, the series $\displaystyle\sum_{n=0}^{\infty} \frac{(-1)^{n+1}}{n+1}$ converges

When $x = 5$, the series $\displaystyle\sum_{n=0}^{\infty} \frac{1}{n+1}$ diverges

Therefore, the interval of convergence is $-1 \le x < 5$

13. $\displaystyle\sum_{n=0}^{\infty} \frac{(-1)^{n+1}(x-1)^{n+1}}{n+1}$, $\displaystyle\lim_{n\to\infty} \left|\frac{a_{n+1}}{a_n}\right| = \lim_{n\to\infty} \left|\frac{(x-1)^{n+2}}{n+2} \cdot \frac{n+1}{(x-1)^{n+1}}\right|$

$\displaystyle = |x-1| \lim_{n\to\infty} \frac{n+1}{n+2} = |x-1| < 1$ or $0 < x < 2$

When $x = 0$, the series $\displaystyle\sum_{n=0}^{\infty} \frac{1}{n+1}$ diverges

When $x = 2$, the series $\displaystyle\sum_{n=0}^{\infty} \frac{(-1)^{n+1}}{n+1}$ converges

Therefore, the interval of convergence is $0 < x \le 2$

14. $\displaystyle\sum_{n=1}^{\infty} \frac{(-1)^{n+1}(x-c)^n}{nc^n}$, $\displaystyle\lim_{n\to\infty} \left|\frac{a_{n+1}}{a_n}\right| = \lim_{n\to\infty} \left|\frac{(x-c)^{n+1}}{(n+1)c^{n+1}} \cdot \frac{nc^n}{(x-c)^n}\right|$

$\displaystyle = |x-c| \lim_{n\to\infty} \frac{n}{c(n+1)} = \frac{|x-c|}{c} < 1$

or $|x-c| < c$ or $-c < x-c < c$ or $0 < x < 2c$

When $x = 0$, the series $\displaystyle\sum_{n=1}^{\infty} \frac{-1}{n}$ diverges

When $x = 2c$, the series $\displaystyle\sum_{n=1}^{\infty} \frac{(-1)^{n+1}}{n}$ converges

Therefore, the interval of convergence is $0 < x \le 2c$

15. $\displaystyle\sum_{n=1}^{\infty} \frac{(x-c)^{n-1}}{c^{n-1}}$, $\displaystyle\lim_{n\to\infty} \left|\frac{a_{n+1}}{a_n}\right| = \lim_{n\to\infty} \left|\frac{(x-c)^n}{c^n} \cdot \frac{c^{n-1}}{(x-c)^{n-1}}\right|$

$\displaystyle = |x-c| \lim_{n\to\infty} \frac{1}{c} = \frac{|x-c|}{c} < 1$

or $|x-c| < c$ or $-c < x-c < c$ or $0 < x < 2c$

When $x = 0$ or $x = 2c$, the series diverges and the interval of convergence is $0 < x < 2c$

16. $\displaystyle\sum_{n=1}^{\infty} \frac{2\cdot4\cdot6\cdots(2n)}{3\cdot5\cdot7\cdots(2n+1)}(x^{2n+1}), \quad \lim_{n\to\infty}\left|\frac{a_{n+1}}{a_n}\right|$

$$= \lim_{n\to\infty}\left|\frac{2\cdot4\cdots(2n)(2n+2)}{3\cdot5\cdots(2n+1)(2n+3)}(x^{2n+3})\frac{3\cdot5\cdots(2n+1)}{2\cdot4\cdots(2n)}\left(\frac{1}{x^{2n+1}}\right)\right|$$

$$= x^2 \lim_{n\to\infty}\frac{(2n+2)}{(2n+3)} = x^2 < 1$$

When $x = \pm1$, the series diverges by comparing it to $\displaystyle\sum_{n=1}^{\infty}\frac{1}{2n+1}$ which diverges. Therefore the interval of convergence is $-1 < x < 1$

17. $\displaystyle\sum_{n=1}^{\infty}\frac{(-1)^{n+1}x^{2n-1}}{2n-1}, \quad \lim_{n\to\infty}\left|\frac{a_{n+1}}{a_n}\right| = \lim_{n\to\infty}\left|\frac{x^{2n+1}}{2n+1}\cdot\frac{2n-1}{x^{2n-1}}\right|$

$$= x^2 \lim_{n\to\infty}\frac{2n-1}{2n+1} = x^2 < 1$$

When $x = -1$, the series $\displaystyle\sum_{n=1}^{\infty}\frac{(-1)^n}{2n-1}$ converges

When $x = 1$, the series $\displaystyle\sum_{n=1}^{\infty}\frac{(-1)^{n+1}}{2n-1}$ converges

Therefore, the interval of convergence is $-1 \le x \le 1$

18. $\displaystyle\sum_{n=1}^{\infty}\frac{n!(x-c)^n}{1\cdot3\cdot5\cdots(2n+1)}, \quad \lim_{n\to\infty}\left|\frac{a_{n+1}}{a_n}\right| = \lim_{n\to\infty}\left|\frac{(n+1)!(x-c)^{n+1}}{1\cdot3\cdots(2n-1)(2n+1)}\cdot\right.$

$$\left.\frac{1\cdot3\cdots(2n-1)}{n!(x-c)^n}\right| = |x-c| \lim_{n\to\infty}\frac{n+1}{2n+1} = \frac{|x-c|}{2} < 1$$

or $|x - c| < 2$ or $-2 < x - c < 2$ or $c - 2 < x < c + 2$

When $x = c \pm 2$, the series does not converge and the interval of convergence is $c - 2 < x < c + 2$

19. $\displaystyle\sum_{n=1}^{\infty}\frac{n}{n+1}(-2x)^{n-1}, \quad \lim_{n\to\infty}\left|\frac{a_{n+1}}{a_n}\right| = \lim_{n\to\infty}\left|\frac{(n+1)(-2x)^n}{n+2}\cdot\frac{n+1}{n(-2x)^{n-1}}\right|$

$$= |2x| \lim_{n\to\infty}\frac{(n+1)^2}{n(n+2)} = |2x| < 1 \text{ or } -1 < 2x < 1 \text{ or } -\frac{1}{2} < x < \frac{1}{2}$$

When $x = \pm\frac{1}{2}$, the series diverges and the interval of convergence is $-\frac{1}{2} < x < \frac{1}{2}$

20. $\sum_{n=0}^{\infty} \frac{(-1)^n x^{2n}}{n!}$, $\lim_{n\to\infty} \left|\frac{a_{n+1}}{a_n}\right| = \lim_{n\to\infty} \left|\frac{x^{2n+2}}{(n+1)!} \cdot \frac{n!}{x^{2n}}\right|$

$= x^2 \lim_{n\to\infty} \frac{1}{n+1} = 0$

Therefore, the interval of convergence is $-\infty < x < \infty$

21. $\sum_{n=0}^{\infty} \frac{x^{2n+1}}{(2n+1)!}$, $\lim_{n\to\infty} \left|\frac{a_{n+1}}{a_n}\right| = \lim_{n\to\infty} \left|\frac{x^{2n+3}}{(2n+3)!} \cdot \frac{(2n+1)!}{x^{2n+1}}\right|$

$= x^2 \lim_{n\to\infty} \frac{1}{(2n+2)(2n+3)} = 0$

Therefore, the interval of convergence is $-\infty < x < \infty$

22. $\sum_{n=1}^{\infty} \frac{n!x^n}{(2n)!}$, $\lim_{n\to\infty} \left|\frac{a_{n+1}}{a_n}\right| = \lim_{n\to\infty} \left|\frac{(n+1)!x^{n+1}}{(2n+2)!} \cdot \frac{(2n)!}{n!x^n}\right|$

$= |x| \lim_{n\to\infty} \frac{n+1}{(2n+2)(2n+1)} = 0$

Therefore, the interval of convergence is $-\infty < x < \infty$

23. $\sum_{n=1}^{\infty} \frac{k(k+1)\cdots(k+n-1)x^n}{n!}$, $\lim_{n\to\infty} \left|\frac{a_{n+1}}{a_n}\right|$

$= \lim_{n\to\infty} \left|\frac{k(k+1)\cdots(k+n-1)(k+n)x^{n+1}}{(n+1)!} \cdot \frac{n!}{k(k+1)\cdots(k+n-1)x^n}\right|$

$= |x| \lim_{n\to\infty} \frac{k+n}{n+1} = |x| < 1$ or $-1 < x < 1$

When $x = \pm 1$, the series diverges and the interval of convergence is $-1 < x < 1$

24. $\sum_{n=1}^{\infty} \frac{(-1)^n 2^{2n-1} x^{2n}}{(2n)!}$, $\lim_{n\to\infty} \left|\frac{a_{n+1}}{a_n}\right| = \lim_{n\to\infty} \left|\frac{2^{2n+1}x^{2n+2}}{(2n+2)!} \cdot \frac{(2n)!}{2^{2n-1}x^{2n}}\right|$

$= x^2 \lim_{n\to\infty} \frac{4}{(2n+2)(2n+1)} = 0$

Therefore, the interval of convergence is $-\infty < x < \infty$

25.(a) $f(x) = \sum_{n=0}^{\infty} (\frac{x}{2})^n, \quad -2 < x < 2$

(b) $f'(x) = \sum_{n=0}^{\infty} (\frac{n}{2})(\frac{x}{2})^{n-1}, \quad -2 < x < 2$

(c) $f''(x) = \sum_{n=0}^{\infty} (\frac{n}{2})(\frac{n-1}{2})(\frac{x}{2})^{n-2}, \quad -2 < x < 2$

(d) $\int f(x)\,dx = \sum_{n=0}^{\infty} \frac{2}{n+1}(\frac{x}{2})^{n+1}, \quad -2 \leq x < 2$

26.(a) $f(x) = \sum_{n=1}^{\infty} \frac{(-1)^{n+1}(x-5)^n}{n5^n}, \quad 0 < x \leq 10$

(b) $f'(x) = \sum_{n=1}^{\infty} \frac{(-1)^{n+1}(x-5)^{n-1}}{5^n}, \quad 0 < x < 10$

(c) $f''(x) = \sum_{n=1}^{\infty} \frac{(-1)^{n+1}(n-1)(x-5)^{n-2}}{5^n}, \quad 0 < x < 10$

(d) $\int f(x)\,dx = \sum_{n=1}^{\infty} \frac{(-1)^{n+1}(x-5)^{n+1}}{n(n+1)5^n}, \quad 0 \leq x \leq 10$

27.(a) $f(x) = \sum_{n=0}^{\infty} \frac{(-1)^{n+1}(x-1)^{n+1}}{n+1}, \quad 0 < x \leq 2$

(b) $f'(x) = \sum_{n=0}^{\infty} (-1)^{n+1}(x-1)^n, \quad 0 < x < 2$

(c) $f''(x) = \sum_{n=1}^{\infty} (-1)^{n+1}n(x-1)^{n-1}, \quad 0 < x < 2$

(d) $\int f(x)\,dx = \sum_{n=1}^{\infty} \frac{(-1)^{n+1}(x-1)^{n+2}}{(n+1)(n+2)}, \quad 0 \leq x \leq 2$

28.(a) $f(x) = \sum_{n=1}^{\infty} \frac{(-1)^{n+1}(x-1)^n}{n}, \quad 0 < x \leq 2$

(b) $f'(x) = \sum_{n=1}^{\infty} (-1)^{n+1}(x-1)^{n-1}, \quad 0 < x < 2$

(c) $f''(x) = \sum_{n=2}^{\infty} (-1)^{n+1}(n-1)(x-1)^{n-2}, \quad 0 < x < 2$

(d) $\int f(x)dx = \sum_{n=1}^{\infty} \frac{(-1)^{n+1}(x-1)^{n+1}}{n(n+1)}, \quad 0 \leq x \leq 2$

29. $f(x) = \sum_{n=0}^{\infty} \frac{(-1)^n x^{2n}}{(2n+1)!}, \quad -\infty < x < \infty$

$$\int f(x)\,dx = \sum_{n=0}^{\infty} \frac{(-1)^n x^{2n+1}}{(2n+1)(2n+1)!}, \quad -\infty < x < \infty$$

$$\int_0^1 f(x)\,dx = \sum_{n=0}^{\infty} \frac{(-1)^n}{(2n+1)(2n+1)!} = 1 - \frac{1}{18} + \frac{1}{600} - \frac{1}{35280} + \cdots$$

Since $\frac{1}{35280} < 0.001$, we can approximate the alternating series with an error of less than 0.001 by its first three terms:

$$\int_0^1 f(x)\,dx \approx 1 - \frac{1}{18} + \frac{1}{600} \approx 0.946$$

30. $f(x) = \sum_{n=0}^{\infty} \frac{(-1)^n x^n}{(2n)!}, \quad -\infty < x < \infty$

$$\int f(x)\,dx = \sum_{n=0}^{\infty} \frac{(-1)^n x^{n+1}}{(n+1)(2n)!}, \quad -\infty < x < \infty$$

$$\int_0^1 f(x)dx = \sum_{n=0}^{\infty} \frac{(-1)^n}{(n+1)(2n)!} = 1 - \frac{1}{4} + \frac{1}{72} - \frac{1}{2880} + \frac{1}{201{,}600} - \cdots$$

Since $\frac{1}{2880} < 0.001$, we can approximate the alternating series with an error of less than 0.001 by its first three terms:

$$\int_0^1 f(x)\,dx \approx 1 - \frac{1}{4} + \frac{1}{72} \approx 0.764$$

31. $f(x) = \sum_{n=1}^{\infty} \frac{(-1)^{n+1} n\, x^{n-1}}{3^n\, n!}, \quad -\infty < x < \infty$

$$\int f(x)\,dx = \sum_{n=1}^{\infty} \frac{(-1)^{n+1}\, x^n}{3^n\, n!}, \quad -\infty < x < \infty$$

$$\int_0^1 f(x)\,dx = \sum_{n=1}^{\infty} \frac{(-1)^{n+1}}{3^n\, n!} = \frac{1}{3} - \frac{1}{18} + \frac{1}{162} - \frac{1}{1944} + \cdots$$

Since $\frac{1}{1944} < 0.001$, we can approximate the alternating series with an error of less than 0.001 by its first three terms:

$$\int_0^1 f(x)\,dx \approx \frac{1}{3} - \frac{1}{18} + \frac{1}{162} \approx 0.284$$

32. $f(x) = \sum_{n=0}^{\infty} \frac{(-1)^n x^{2n}}{2^n\, n!}, \quad -\infty < x < \infty$

$$\int f(x)\,dx = \sum_{n=0}^{\infty} \frac{(-1)^n x^{2n+1}}{2^n(2n+1)n!}, \quad -\infty < x < \infty$$

$$\int_0^1 f(x)\,dx = \sum_{n=0}^{\infty} \frac{(-1)^n}{2^n(2n+1)n!} = 1 - \frac{1}{6} + \frac{1}{40} - \frac{1}{336} + \frac{1}{3456} - \cdots$$

Since $\frac{1}{3456} < 0.001$, we can approximate the alternating series with an error of less than 0.001 by its first four terms:

$$\int_0^1 f(x)dx \approx 1 - \frac{1}{6} + \frac{1}{40} - \frac{1}{336} \approx 0.855$$

33.(a) $f(x) = \sum_{n=0}^{\infty} \frac{(-1)^n x^{2n+1}}{(2n+1)!}, \quad -\infty < x < \infty$

$g(x) = \sum_{n=0}^{\infty} \frac{(-1)^n x^{2n}}{(2n)!}, \quad -\infty < x < \infty$

(b) $f'(x) = \sum_{n=0}^{\infty} \frac{(-1)^n x^{2n}}{(2n)!} = g(x)$

(c) $g'(x) = \sum_{n=1}^{\infty} \frac{(-1)^n x^{2n-1}}{(2n-1)!} = \sum_{n=0}^{\infty} \frac{(-1)^{n+1} x^{2n+1}}{(2n+1)!}$

$= -\sum_{n=0}^{\infty} \frac{(-1)^n x^{2n+1}}{(2n+1)!} = -f(x)$

(d) $f(x) = \sin x$ and $g(x) = \cos x$

34. Let $f'(x) = \sum_{n=0}^{\infty} a_n x^n$ and $f(x) = \sum_{n=0}^{\infty} \frac{a_n x^{n+1}}{n+1} = x \sum_{n=0}^{\infty} \frac{a_n x^n}{n+1}$

Since $\left|\frac{a_n x^n}{n+1}\right| < |a_n x^n|$, the convergence of $\sum_{n=0}^{\infty} a_n x^n$ implies the convergence of $x \sum_{n=0}^{\infty} \frac{a_n x^n}{n+1}$

9.9
Representation of functions by power series

1. $\dfrac{a}{x-b} = \dfrac{1}{2-x} = \dfrac{-1}{x-2}$ Therefore, we have a = −1, b = 2, and by Theorem 9.21, $\dfrac{1}{2-x} = \dfrac{1}{2}\sum_{n=0}^{\infty}\left(\dfrac{x}{2}\right)^n = \sum_{n=0}^{\infty}\dfrac{x^n}{2^{n+1}}$, $|x| < 2$ or $-2 < x < 2$

2. $\dfrac{a}{x-b} = \dfrac{3}{4-x} = \dfrac{-3}{x-4}$ Therefore, a = −3, b = 4, and by Theorem 9.21

$\dfrac{3}{4-x} = \dfrac{3}{4}\sum_{n=0}^{\infty}\left(\dfrac{x}{4}\right)^n = \sum_{n=0}^{\infty}\dfrac{3x^n}{4^{n+1}}$, $|x| < 4$ or $-4 < x < 4$

3. $\dfrac{a}{x-b} = \dfrac{1}{2-x} = \dfrac{-1}{x-2}$ Therefore, a = −1, b = 2, c = 5, and by Theorem 9.21,

$\dfrac{1}{2-x} = \dfrac{1}{-3}\sum_{n=0}^{\infty}\left(\dfrac{x-5}{-3}\right)^n = \sum_{n=0}^{\infty}\dfrac{(x-5)^n}{(-3)^{n+1}}$, $|x-5| < 3$ or $2 < x < 8$

4. $\dfrac{a}{x-b} = \dfrac{3}{4-x} = \dfrac{-3}{x-4}$ Therefore, a = −3, b = 4, c = −2 and by Theorem 9.21,

$\dfrac{3}{4-x} = \dfrac{3}{6}\sum_{n=0}^{\infty}\left(\dfrac{x+2}{6}\right)^n = \sum_{n=0}^{\infty}\dfrac{3(x+2)^n}{6^{n+1}}$, $|x+2| < 6$ or $-8 < x < 4$

5. $\dfrac{a}{x-b} = \dfrac{3}{2x-1} = \dfrac{3/2}{x-(1/2)}$, $a = \dfrac{3}{2}$, $b = \dfrac{1}{2}$, and by Theorem 9.21,

$\dfrac{3}{2x-1} = \dfrac{-3/2}{1/2}\sum_{n=0}^{\infty}\left(\dfrac{x}{1/2}\right)^n = -3\sum_{n=0}^{\infty}(2x)^n$, $|2x| < 1$ or $-\dfrac{1}{2} < x < \dfrac{1}{2}$

6. $\dfrac{a}{x-b} = \dfrac{3}{2x-1} = \dfrac{3/2}{x-(1/2)}$, $a = \dfrac{3}{2}$, $b = \dfrac{1}{2}$, c = 2, and by Theorem 9.21,

$\dfrac{3}{2x-1} = \dfrac{-3/2}{-3/2}\sum_{n=0}^{\infty}\left(\dfrac{x-2}{-3/2}\right)^n = \sum_{n=0}^{\infty}\dfrac{(-2)^n(x-2)^n}{3^n}$,

$|x-2| < \dfrac{3}{2}$ or $\dfrac{1}{2} < x < \dfrac{7}{2}$

7. $\dfrac{a}{x - b} = \dfrac{1}{2x - 5} = \dfrac{1/2}{x - (5/2)}$, $a = \dfrac{1}{2}$, $b = \dfrac{5}{2}$, $c = -3$, and by Theorem 9.21,

$$\frac{1}{2x - 5} = \frac{-1/2}{11/2}\sum_{n=0}^{\infty}\left(\frac{x + 3}{11/2}\right)^n = -\sum_{n=0}^{\infty}\frac{2^n(x + 3)^n}{11^{n+1}},$$

$|x + 3| < \dfrac{11}{2}$ or $-\dfrac{17}{2} < x < \dfrac{5}{2}$

8. $\dfrac{a}{x - b} = \dfrac{1}{2x - 5} = \dfrac{1/2}{x - (5/2)}$, $a = \dfrac{1}{2}$, $b = \dfrac{5}{2}$, and by Theorem 9.21,

$$\frac{1}{2x - 5} = \frac{-1/2}{5/2}\sum_{n=0}^{\infty}\left(\frac{x}{5/2}\right)^n = -\sum_{n=0}^{\infty}\frac{2^n x^n}{5^{n+1}}, \quad |x| < \frac{5}{2} \text{ or } -\frac{5}{2} < x < \frac{5}{2}$$

9. $\dfrac{a}{x - b} = \dfrac{3}{x + 2} = \dfrac{3}{x - (-2)}$, $a = 3$, $b = -2$, and by Theorem 9.21,

$$\frac{3}{x + 2} = \frac{3}{2}\sum_{n=0}^{\infty}\left(\frac{x}{-2}\right)^n = 3\sum_{n=0}^{\infty}\frac{(-1)^n x^n}{2^{n+1}}, \quad |x| < 2 \text{ or } -2 < x < 2$$

10. $\dfrac{a}{x - b} = \dfrac{4}{3x + 2} = \dfrac{4/3}{x - (-2/3)}$, $a = \dfrac{4}{3}$, $b = -\dfrac{2}{3}$, $c = 2$,
and by Theorem 9.21,

$$\frac{4}{3x + 2} = \frac{-4/3}{-8/3}\sum_{n=0}^{\infty}\left(\frac{x - 2}{-8/3}\right)^n = \frac{1}{2}\sum_{n=0}^{\infty}\frac{(-3)^n(x - 2)^n}{8^n},$$

$|x - 2| < \dfrac{8}{3}$ or $-\dfrac{2}{3} < x < \dfrac{14}{3}$

11. $\dfrac{3x}{x^2 + x - 2} = \dfrac{2}{x + 2} + \dfrac{1}{x - 1} = \dfrac{2}{x - (-2)} + \dfrac{1}{x - 1}$

$$= \frac{-2}{-2}\sum_{n=0}^{\infty}\left(\frac{x}{-2}\right)^n + \frac{-1}{1}\sum_{n=0}^{\infty}\left(\frac{x}{1}\right)^n = \sum_{n=0}^{\infty}\left(\frac{1}{(-2)^n} - 1\right)x^n$$

The interval of convergence is $-1 < x < 1$ since

$$\lim_{n \to \infty}\left|\frac{a_{n+1}}{a_n}\right| = \lim_{n \to \infty}\left|\frac{1 - (-2)^{n+1}}{(-2)^{n+1}} \cdot \frac{(-2)^n}{1 - (-2)^n}\right|$$

$$= \lim_{n \to \infty}\left|\frac{1 - (-2)^{n+1}}{-2 - (-2)^{n+1}}\right| = 1$$

12. $\dfrac{4x - 7}{2x^2 + 3x - 2} = \dfrac{3}{x + 2} - \dfrac{2}{2x - 1} = \dfrac{3}{x - (-2)} + \dfrac{-1}{x - (1/2)}$

$$= \frac{-3}{-2}\sum_{n=0}^{\infty}\left(\frac{x}{-2}\right)^n + \frac{1}{1/2}\sum_{n=0}^{\infty}\left(\frac{x}{1/2}\right)^n = \sum_{n=0}^{\infty}\left(\frac{(-1)^n 3}{2^{n+1}} + 2^{n+1}\right)x^n$$

The interval of convergence is $-\frac{1}{2} < x < \frac{1}{2}$

13. $\dfrac{2}{1 - x^2} = \dfrac{1}{1 - x} + \dfrac{1}{1 + x} = \dfrac{-1}{x - 1} + \dfrac{1}{x - (-1)}$

$$= \sum_{n=0}^{\infty} x^n + \sum_{n=0}^{\infty}(-1)^n x^n = \sum_{n=0}^{\infty}(1 + (-1)^n)x^n = \sum_{n=0}^{\infty} 2x^{2n}$$

The interval of convergence is $|x^2| < 1$ or $-1 < x < 1$ since

$$\lim_{n \to \infty}\left|\frac{a_{n+1}}{a_n}\right| = \lim_{n \to \infty}\left|\frac{2}{2}\right| = 1$$

14. $\dfrac{4}{4 + x} = \dfrac{4}{x - (-4)} = \sum_{n=0}^{\infty}\left(\dfrac{x}{-4}\right)^n, \quad |x| < 4$

$$\frac{4}{4 + x^2} = \sum_{n=0}^{\infty}\left(\frac{x^2}{-4}\right)^n = \sum_{n=0}^{\infty}\frac{(-1)^n x^{2n}}{4^n}$$

The interval of convergence is $|x^2| < 4$ or $-2 < x < 2$ since

$$\lim_{n \to \infty}\left|\frac{a_{n+1}}{a_n}\right| = \lim_{n \to \infty}\left|\frac{(-1)^{n+1}}{4^{n+1}} \cdot \frac{4^n}{(-1)^n}\right| = \lim_{n \to \infty}\left|-\frac{1}{4}\right| = \frac{1}{4}$$

15. By taking the first derivative, we have $\dfrac{d}{dx}\left[\dfrac{1}{x + 1}\right] = \dfrac{-1}{(x + 1)^2}$

Therefore, $\dfrac{-1}{(x + 1)^2} = \dfrac{d}{dx}\left[\sum_{n=0}^{\infty}(-1)^n x^n\right]$

$$= \sum_{n=1}^{\infty}(-1)^n n x^{n-1} = \sum_{n=0}^{\infty}(-1)^{n+1}(n + 1)x^n$$

16. By taking the second derivative, we have $\dfrac{d^2}{dx^2}\left[\dfrac{1}{x + 1}\right] = \dfrac{2}{(x + 1)^3}$

Therefore, $\dfrac{2}{(x + 1)^3} = \dfrac{d^2}{dx^2}\left[\sum_{n=0}^{\infty}(-1)^n x^n\right] = \dfrac{d}{dx}\left[\sum_{n=1}^{\infty}(-1)^n n x^{n-1}\right]$

$$= \sum_{n=2}^{\infty}(-1)^n\, n(n - 1)x^{n-2} = \sum_{n=2}^{\infty}(-1)^n(n + 2)(n + 1)x^n, \quad -1 < x < 1$$

17. By integrating, we have $\int \frac{1}{x+1}\,dx = \ln(x+1)$

Therefore, $\ln(x+1) = \int \left[\sum_{n=1}^{\infty} (-1)^n x^n\right] dx = \sum_{n=0}^{\infty} \frac{(-1)^n x^{n+1}}{n+1}, \quad -1 < x \leq 1$

18. $\frac{1}{x^2+1} = \sum_{n=0}^{\infty} (-1)^n (x^2)^n = \sum_{n=0}^{\infty} (-1)^n x^{2n}$ and $\frac{2x}{x^2+1} = 2\sum_{n=0}^{\infty} (-1)^n x^{2n+1}$

Since $\frac{d}{dx}\,|\ln(x^2+1)| = \frac{2x}{x^2+1}$, we have

$\ln(x^2+1) = \int \left[\sum_{n=0}^{\infty} (-1)^n 2x^{2n+1}\right] dx = \sum_{n=0}^{\infty} \frac{(-1)^n x^{2n+2}}{n+1}, \quad -1 \leq x \leq 1$

19. Since $\frac{1}{x+1} = \sum_{n=0}^{\infty} (-1)^n x^n$, we have $\frac{1}{4x^2+1} = \sum_{n=0}^{\infty} (-1)^n (4x^2)^n$

$= \sum_{n=0}^{\infty} (-1)^n 4^n x^{2n}, \quad -\frac{1}{2} < x < \frac{1}{2}$

20. Since $\int \frac{1}{4x^2+1}\,dx = \frac{1}{2}\arctan(2x)$, we can use the result of Exercise 19

to obtain $\arctan(2x) = 2\int \frac{1}{4x^2+1}\,dx = 2\int \left[\sum_{n=0}^{\infty} (-1)^n 4^n x^{2n}\right] dx$

$= 2\sum_{n=0}^{\infty} \frac{(-1)^n 4^n x^{2n+1}}{2n+1}, \quad -\frac{1}{2} < x \leq \frac{1}{2}$

21. $\ln\sqrt{\frac{1+x}{1-x}} = \frac{1}{2}[\ln(x+1) - \ln(-x+1)]$

$= \frac{1}{2}\left[\sum_{n=0}^{\infty} \frac{(-1)^n x^{n+1}}{n+1} - \sum_{n=0}^{\infty} \frac{(-1)^n (-x)^{n+1}}{n+1}\right]$

$= \frac{1}{2}\sum_{n=0}^{\infty} [(-1)^n + 1]\frac{x^{n+1}}{n+1} = \sum_{n=0}^{\infty} \frac{x^{2n+1}}{2n+1}, \quad -1 < x < 1$

22. $\frac{1}{x^2-x+1} = \frac{x+1}{x^3+1}$, since $\frac{1}{x^3+1} = \sum_{n=0}^{\infty} (-1)^n (x^3)^n = \sum_{n=0}^{\infty} (-1)^n x^{3n}$,

$\frac{x+1}{x^3+1} = \sum_{n=0}^{\infty} (-1)^n x^{3n}(x+1)$

$= 1 + x - x^3 - x^4 + x^6 + x^7 - x^9 - x^{10} + \ldots \quad -1 \leq x < 1$

23.

x	$x - \frac{x^2}{2}$	$\ln(x + 1)$	$x - \frac{x^2}{2} + \frac{x^3}{3}$
0.0	0.000	0.000	0.000
0.2	0.180	0.182	0.183
0.4	0.320	0.336	0.341
0.6	0.420	0.470	0.492
0.8	0.480	0.588	0.651
1.0	0.500	0.693	0.833

24.

x	$x - \frac{x^2}{2} + \frac{x^3}{3} - \frac{x^4}{4}$	$\ln(x + 1)$	$x - \frac{x^2}{2} + \frac{x^3}{3} - \frac{x^4}{4} + \frac{x^5}{5}$
0.0	0.0	0.0	0.0
0.2	0.18227	0.18232	0.18233
0.4	0.33493	0.33647	0.33698
0.6	0.45960	0.47000	0.47515
0.8	0.54827	0.58779	0.61380
1.0	0.58333	0.69315	0.78333

In Exercises 25–28, $\arctan x = \sum_{n=0}^{\infty} (-1)^n \frac{x^{2n+1}}{2n + 1}$

25. $\arctan \frac{1}{4} = \sum_{n=0}^{\infty} (-1)^n \frac{(1/4)^{2n+1}}{2n + 1} = \sum_{n=0}^{\infty} \frac{(-1)^n}{(2n + 1)4^{2n+1}} = \frac{1}{4} - \frac{1}{192} + \frac{1}{5120} - \cdots$

Since $\frac{1}{5120} < 0.001$, we can approximate the series by its first two terms:

$\arctan \frac{1}{4} \approx \frac{1}{4} - \frac{1}{192} \approx 0.245$

26. $\arctan x^2 = \sum_{n=0}^{\infty} (-1)^n \dfrac{x^{4n+2}}{2n+1}$

$$\int \arctan x^2\, dx = \sum_{n=0}^{\infty} (-1)^n \frac{x^{4n+3}}{(4n+3)(2n+1)} + C \quad (C = 0)$$

$$\int_0^{3/4} \arctan x^2\, dx = \sum_{n=0}^{\infty} (-1)^n \frac{(3/4)^{4n+3}}{(4n+3)(2n+1)}$$

$$= \sum_{n=0}^{\infty} (-1)^n \frac{3^{4n+3}}{(4n+3)(2n+1)4^{4n+3}} = \frac{27}{192} - \frac{2187}{344{,}064} + \frac{177{,}147}{230{,}686{,}720}$$

Since $\dfrac{177{,}147}{230{,}686{,}720} < 0.001$, we can approximate the series

by its first two terms: $\displaystyle\int_0^{3/4} \arctan x^2\, dx \approx \frac{27}{192} - \frac{2187}{344{,}064} \approx 0.134$

27. $\dfrac{\arctan x^2}{x} = \sum_{n=0}^{\infty} (-1)^n \dfrac{x^{4n+1}}{2n+1}$

$$\int \frac{\arctan x^2}{x}\, dx = \sum_{n=0}^{\infty} (-1)^n \frac{x^{4n+2}}{(4n+2)(2n+1)}$$

$$\int_0^{1/2} \frac{\arctan x^2}{x}\, dx = \sum_{n=0}^{\infty} (-1)^n \frac{1}{(4n+2)(2n+1)2^{4n+2}} = \frac{1}{8} - \frac{1}{1152} + \cdots$$

Since $\dfrac{1}{1152} < 0.001$, we can approximate the series by its first term:

$$\int_0^{1/2} \frac{\arctan x^2}{x}\, dx \approx 0.125$$

28. $x^2 \arctan x = \sum_{n=0}^{\infty} (-1)^n \dfrac{x^{2n+3}}{2n+1}$

$$\int x^2 \arctan x\, dx = \sum_{n=0}^{\infty} (-1)^n \frac{x^{2n+4}}{(2n+4)(2n+1)}$$

$$\int_0^{1/2} x^2 \arctan x\, dx = \sum_{n=0}^{\infty} (-1)^n \frac{1}{(2n+4)(2n+1)2^{2n+4}} = \frac{1}{64} - \frac{1}{1152} + \cdots$$

Since $\dfrac{1}{1152} < 0.001$, we can approximate the series by its first term:

$$\int_0^{1/2} x^2 \arctan x\, dx \approx 0.016$$

29.
$$
\begin{array}{r|l}
 & 1 - x + x^2 - x^3 + \cdots \\
1 + x & 1 \\
 & \underline{1 + x} \\
 & -x \\
 & \underline{-x - x^2} \\
 & x^2 \\
 & \underline{x^2 + x^3} \\
 & -x^3 \\
 & \underline{-x^3 - x^4} \\
 & x^4
\end{array}
$$

$$\frac{1}{1 + x} = \sum_{n=0}^{\infty} (-1)^n x^n$$

9.10 Taylor and Maclaurin series

1. Since $e^x = \sum_{n=0}^{\infty} \frac{x^n}{n!} = 1 + x + \frac{x^2}{2!} + \frac{x^3}{3!} + \frac{x^4}{4!} + \ldots$, we have

$$e^{2x} = \sum_{n=0}^{\infty} \frac{(2x)^n}{n!} = 1 + 2x + \frac{4x^2}{2!} + \frac{8x^3}{3!} + \frac{16x^4}{4!} + \ldots$$

2. Since $e^x = \sum_{n=0}^{\infty} \frac{x^n}{n!} = 1 + x + \frac{x^2}{2!} + \frac{x^3}{3!} + \frac{x^4}{4!} + \ldots$, we have

$$e^{-2x} = \sum_{n=0}^{\infty} \frac{(-2x)^n}{n!} = 1 - 2x + \frac{4x^2}{2!} - \frac{8x^3}{3!} + \frac{16x^4}{4!} - \ldots$$

3. For $c = \frac{\pi}{4}$, we have $f(x) = \cos(x)$, $f'(x) = -\sin(x)$, $f''(x) = -\cos(x)$, $f'''(x) = \sin(x)$, $f^{(4)}(x) = \cos(x)$, and so on. Therefore, we have

$$\cos x = \sum_{n=0}^{\infty} \frac{f^{(n)}(\pi/4)[x - (\pi/4)]^n}{n!}$$

$$= \frac{\sqrt{2}}{2}\left[1 - \left(x - \frac{\pi}{4}\right) - \frac{[x - (\pi/4)]^2}{2!} + \frac{[x - (\pi/4)]^3}{3!} + \frac{[x - (\pi/4)]^4}{4!} - \ldots\right]$$

$$= \frac{\sqrt{2}}{2} \sum_{n=0}^{\infty} \frac{(-1)^{n(n+1)/2}[x - (\pi/4)]^n}{n!}$$

4. Since $\cos x = \frac{\sqrt{2}}{2} \sum_{n=0}^{\infty} \frac{(-1)^{n(n+1)/2}[x - (\pi/4)]^n}{n!}$ (Exercise 3)

$$\sin x = \int \cos x \, dx = \frac{\sqrt{2}}{2} \sum_{n=0}^{\infty} \frac{(-1)^{n(n+1)/2}[x - (\pi/4)]^{n+1}}{(n + 1)n!}$$

5. Since $\frac{1}{x} = \sum_{n=0}^{\infty} (-1)^n (x - 1)^n$, we have $\ln x = \int \frac{1}{x} \, dx = \sum_{n=0}^{\infty} (-1)^n \frac{(x - 1)^{n+1}}{n + 1}$

6. For $c = 1$, we have $f(x) = e^x = \sum_{n=0}^{\infty} \frac{f^{(n)}(1)(x-1)^n}{n!}$

and since $f^{(n)}(1) = e$, we have $e^x = \sum_{n=0}^{\infty} \frac{e(x-1)^n}{n!}$

$= e\left[1 + (x-1) + \frac{(x-1)^2}{2!} + \frac{(x-1)^3}{3!} + \frac{(x-1)^4}{4!} + \ldots\right]$

7. Since $\sin(x) = \sum_{n=0}^{\infty} \frac{(-1)^n x^{2n+1}}{(2n+1)!} = x - \frac{x^3}{3!} + \frac{x^5}{5!} - \frac{x^7}{7!} + \frac{x^9}{9!} - \ldots$, we have

$\sin(2x) = \sum_{n=0}^{\infty} \frac{(-1)^n (2x)^{2n+1}}{(2n+1)!} = 2x - \frac{8x^3}{3!} + \frac{32x^5}{5!} - \frac{128x^7}{7!} + \ldots$

8. $f(x) = \tan(x)$, $f'(x) = \sec^2(x)$, $f''(x) = 2\sec^2(x)\tan(x)$,

$f'''(x) = 2\left[\sec^4(x) + 2\sec^2(x)\tan^2(x)\right]$

$f^{(4)}(x) = 8\left[2\sec^4(x)\tan(x) + \sec^2(x)\tan^3(x)\right]$

$f^{(5)}(x) = 8\left[2\sec^6(x) + 11\sec^4(x)\tan^2(x) + 2\sec^2(x)\tan^4(x)\right]$

$\tan(x) = \sum_{n=0}^{\infty} \frac{f^{(n)}(0)x^n}{n!} = x + \frac{2x^3}{3!} + \frac{16x^5}{5!} + \ldots$

9. $f(x) = \sec(x)$, $f'(x) = \sec(x)\tan(x)$, $f''(x) = \sec^3(x) + \sec(x)\tan^2(x)$

$f'''(x) = 5\sec^3(x)\tan(x) + \sec(x)\tan^3(x)$

$f^{(4)}(x) = 5\sec^5(x) + 18\sec^3(x)\tan^2(x) + \sec(x)\tan^4(x)$

$\sec(x) = \sum_{n=0}^{\infty} \frac{f^{(n)}(0)x^n}{n!} = 1 + \frac{x^2}{2!} + \frac{5x^4}{4!} + \ldots$

10. Since $\frac{1}{x+1} = \sum_{n=0}^{\infty} (-1)^n x^n = 1 - x + x^2 - x^3 + x^4 - x^5 + \ldots$, we have

$\ln(x+1) = \int \frac{1}{x+1}\,dx = \sum_{n=0}^{\infty} \frac{(-1)^n x^{n+1}}{n+1} = x - \frac{x^2}{2} + \frac{x^3}{3} - \frac{x^4}{4} + \frac{x^5}{5} - \frac{x^6}{6} + \ldots$

and $\ln(x^2+1) = \sum_{n=0}^{\infty} \frac{(-1)^n (x^2)^{n+1}}{n+1} = x^2 - \frac{x^4}{2} + \frac{x^6}{3} - \frac{x^8}{4} + \frac{x^{10}}{5} - \frac{x^{12}}{6} + \ldots$

11. Since $(1 + x)^{-k} = 1 - kx + \frac{k(k + 1)x^2}{2!} - \frac{k(k + 1)(k + 2)x^3}{3!} + \ldots,$

we have $(1 + x)^{-2} = 1 - 2x + \frac{2(3)x^2}{2!} - \frac{2(3)(4)x^3}{3!} + \frac{2(3)(4)(5)x^4}{5!} - \ldots$

$= 1 - 2x + 3x^2 - 4x^3 + 5x^4 - \ldots = \sum_{n=0}^{\infty} (-1)^n(n + 1)x^n$

12. Since $(1 + x)^{-k} = 1 - kx + \frac{k(k + 1)x^2}{2!} - \frac{k(k + 1)(k + 2)x^3}{3!} + \ldots,$

we have $\left[1 + (-x)\right]^{-1/2}$

$= 1 + (\frac{1}{2})x + \frac{(1/2)(3/2)x^2}{2!} + \frac{(1/2)(3/2)(5/2)x^3}{3!} + \ldots$

$= 1 + \frac{x}{2} + \frac{(1)(3)x^2}{2^2 2!} + \frac{(1)(3)(5)x^3}{2^3 3!} + \ldots$

$= 1 + \sum_{n=1}^{\infty} \frac{1 \cdot 3 \cdot 5 \cdots (2n - 1)x^n}{2^n n!}$

13. $\frac{1}{\sqrt{4 + x^2}} = (\frac{1}{2}) \left[1 + (\frac{x}{2})^2\right]^{-1/2}$ and since $(1 + x)^{-1/2}$

$= 1 + \sum_{n=1}^{\infty} \frac{(-1)^n \, 1 \cdot 3 \cdot 5 \cdots (2n - 1)x^n}{2^n n!}$

we have $\frac{1}{\sqrt{4 + x^2}} = \frac{1}{2} \left[1 + \sum_{n=1}^{\infty} \frac{(-1)^n 1 \cdot 3 \cdot 5 \cdots (2n - 1)(x/2)^{2n}}{2^n n!}\right]$

$= \frac{1}{2} + \sum_{n=1}^{\infty} \frac{(-1)^n \, 1 \cdot 3 \cdot 5 \cdots (2n - 1)x^{2n}}{2^{3n+1} n!}$

14. Since $(1 + x)^k = 1 + kx + \frac{k(k - 1)x^2}{2!} + \frac{k(k - 1)(k - 2)x^3}{3!} + \ldots,$ we have

$(1 + x)^{1/2} = 1 + (\frac{1}{2})x + \frac{(1/2)(-1/2)x^2}{2!} + \frac{(1/2)(-1/2)(-3/2)x^3}{3!} + \ldots$

$= 1 + \frac{x}{2} - \frac{x^2}{2^2 2!} + \frac{1 \cdot 3x^3}{2^3 3!} - \frac{1 \cdot 3 \cdot 5x^4}{2^4 5!} + \ldots$

$= 1 + \frac{x}{2} + \sum_{n=2}^{\infty} \frac{(-1)^{n-1} \, 1 \cdot 3 \cdot 5 \cdots (2n - 3)x^n}{2^n n!}$

15. Since $(1 + x)^{1/2} = 1 + \frac{x}{2} + \sum_{n=2}^{\infty} \frac{(-1)^{n-1}\, 1\cdot 3\cdot 5\cdots(2n - 3)x^n}{2^n n!}$ (Exercise 14)

we have $(1 + x^2)^{1/2} = 1 + \frac{x^2}{2} + \sum_{n=2}^{\infty} \frac{(-1)^{n-1}\, 1\cdot 3\cdot 5\cdots(2n - 3)x^{2n}}{2^n n!}$

16. Since $(1 + x)^{1/2} = 1 + \frac{x}{2} + \sum_{n=2}^{\infty} \frac{(-1)^{n-1}\, 1\cdot 3\cdot 5\cdots(2n - 3)x^n}{2^n n!}$ (Exercise 14)

we have $(1 + x^3)^{1/2} = 1 + \frac{x^3}{2} + \sum_{n=2}^{\infty} \frac{(-1)^{n-1}\, 1\cdot 3\cdot 5\cdots(2n - 3)x^{3n}}{2^n n!}$

17. $\cos(x) = \sum_{n=0}^{\infty} \frac{(-1)^n x^{2n}}{(2n)!} = 1 - \frac{x^2}{2!} + \frac{x^4}{4!} - \frac{x^6}{6!} + \ldots$

$-\sin(x) = \frac{d}{dx}[\cos(x)] = \sum_{n=1}^{\infty} \frac{(-1)^n 2n x^{2n-1}}{(2n)!} = \sum_{n=1}^{\infty} \frac{(-1)^n x^{2n-1}}{(2n - 1)!}$

$\sin(x) = \sum_{n=1}^{\infty} \frac{(-1)^{n+1} x^{2n-1}}{(2n - 1)!} = x - \frac{x^3}{3!} + \frac{x^5}{5!} - \frac{x^7}{7!} + \ldots$

18. $\cos^2(x) = \frac{1}{2}[1 + \cos(2x)]$

$= \frac{1}{2}\left[1 + 1 - \frac{(2x)^2}{2!} + \frac{(2x)^4}{4!} - \frac{(2x)^6}{6!} - \ldots\right] = \frac{1}{2}\left[1 + \sum_{n=0}^{\infty} \frac{(-1)^n (2x)^{2n}}{(2n)!}\right]$

19. $e^x = \sum_{n=0}^{\infty} \frac{x^n}{n!} = 1 + x + \frac{x^2}{2!} + \frac{x^3}{3!} + \frac{x^4}{4!} + \frac{x^5}{5!} + \ldots$

$e^{x^2/2} = \sum_{n=0}^{\infty} \frac{(x^2/2)^n}{n!} = \sum_{n=0}^{\infty} \frac{x^{2n}}{2^n n!} = 1 + \frac{x^2}{2} + \frac{x^4}{2^2 2!} + \frac{x^6}{2^3 3!} + \frac{x^8}{2^4 4!} + \ldots$

20. From Exercise 8, we have $\tan(x) = x + \frac{2x^3}{3!} + \frac{16x^5}{5!} + \ldots$, and by differentiating, we have $\sec^2(x) = 1 + x^2 + \frac{2x^4}{3} + \ldots$

21. $e^x = 1 + x + \frac{x^2}{2!} + \frac{x^3}{3!} + \frac{x^4}{4!} + \frac{x^5}{5!} + \ldots$

$e^{-x} = 1 - x + \frac{x^2}{2!} - \frac{x^3}{3!} + \frac{x^4}{4!} - \frac{x^5}{5!} + \ldots$

$e^x - e^{-x} = 2x + \frac{2x^3}{3!} + \frac{2x^5}{5!} + \frac{2x^7}{7!} + \ldots$

$\sinh(x) = \frac{1}{2}(e^x - e^{-x}) = x + \frac{x^3}{3!} + \frac{x^5}{5!} + \frac{x^7}{7!} + \ldots = \sum_{n=0}^{\infty} \frac{x^{2n+1}}{(2n + 1)!}$

22. From Exercise 21, we have $\sinh(x) = \sum_{n=0}^{\infty} \frac{x^{2n+1}}{(2n+1)!}$

$= x + \frac{x^3}{3!} + \frac{x^5}{5!} + \frac{x^7}{7!} + \ldots$ Therefore, $\cosh(x) = \frac{d}{dx}[\sinh(x)]$

$= \sum_{n=0}^{\infty} \frac{x^{2n}}{(2n)!} = 1 + \frac{x^2}{2!} + \frac{x^4}{4!} + \frac{x^6}{6!} + \ldots$

23. $e^{ix} = 1 + ix + \frac{(ix)^2}{2!} + \frac{(ix)^3}{3!} + \frac{(ix)^4}{4!} + \ldots$

$= 1 + ix - \frac{x^2}{2!} - \frac{ix^3}{3!} + \frac{x^4}{4!} + \frac{ix^5}{5!} - \frac{x^6}{6!} - \ldots$

$e^{-ix} = 1 - ix + \frac{(-ix)^2}{2!} + \frac{(-ix)^3}{3!} + \frac{(-ix)^4}{4!} + \ldots$

$= 1 - ix - \frac{x^2}{2!} + \frac{ix^3}{3!} + \frac{x^4}{4!} - \frac{ix^5}{5!} - \frac{x^6}{6!} + \ldots$

(a) $e^{ix} - e^{-ix} = 2ix - \frac{2ix^3}{3!} + \frac{2ix^5}{5!} - \frac{2ix^7}{7!} + \ldots$

$\frac{e^{ix} - e^{-ix}}{2i} = x - \frac{x^3}{3!} + \frac{x^5}{5!} - \frac{x^7}{7!} + \ldots = \sin(x)$

(b) $e^{ix} + e^{-ix} = 2 - \frac{2x^2}{2!} + \frac{2x^4}{4!} - \frac{2x^6}{6!} + \ldots$

$\frac{e^{ix} + e^{-ix}}{2} = 1 - \frac{x^2}{2!} + \frac{x^4}{4!} - \frac{x^6}{6!} + \ldots = \cos(x)$

24. From Exercise 13, we have $(1 + x)^{-1/2} = 1 + \sum_{n=1}^{\infty} \frac{(-1)^n 1\cdot3\cdot5\cdots(2n-1)x^n}{2^n n!}$

which implies that $(1 + x^2)^{-1/2} = 1 + \sum_{n=1}^{\infty} \frac{(-1)^n 1\cdot3\cdot5\cdots(2n-1)x^{2n}}{2^n n!}$

$\ln(x + \sqrt{x^2 + 1}) = \int \frac{1}{\sqrt{x^2 + 1}}\,dx = x + \sum_{n=1}^{\infty} \frac{(-1)^n 1\cdot3\cdot5\cdots(2n-1)x^{2n+1}}{2^n(2n+1)n!}$

$= x - \frac{x^3}{2\cdot3} + \frac{1\cdot3x^5}{2\cdot4\cdot5} - \frac{1\cdot3\cdot5x^7}{2\cdot4\cdot6\cdot7} + \ldots$

25. $\frac{\sin(x)}{x} = (\frac{1}{x}) \sum_{n=0}^{\infty} \frac{(-1)^n x^{2n+1}}{(2n+1)!} = \sum_{n=0}^{\infty} \frac{(-1)^n x^{2n}}{(2n+1)!}$

$= 1 - \frac{x^2}{3!} + \frac{x^4}{5!} - \frac{x^6}{7!} + \frac{x^8}{9!} - \ldots$

26. $\lim_{x \to \infty} \frac{\sin(x)}{x} = \lim_{x \to \infty} \left[1 - \frac{x}{3!} + \frac{x}{5!} - \frac{x}{7!} + \ldots\right]$

$= [1 - 0 + 0 - 0 + \ldots] = 1$

27. $\int_0^{\pi/2} \frac{\sin(x)}{x}\, dx = \int_0^{\pi/2} \sum_{n=0}^{\infty} \frac{(-1)^n x^{2n}}{(2n+1)!}\, dx$

$= \left[\sum_{n=0}^{\infty} \frac{(-1)^n x^{2n+1}}{(2n+1)(2n+1)!}\right]_0^{\pi/2} = \sum_{n=0}^{\infty} \frac{(-1)^n (\pi/2)^{2n+1}}{(2n+1)(2n+1)!}$

$= \frac{\pi}{2} - \frac{\pi^3}{3 \cdot 3! \cdot 2^3} + \frac{\pi^5}{5 \cdot 5! \cdot 2^5} - \frac{\pi^7}{7 \cdot 7! \cdot 2^7} + \ldots \approx 1.3708$

28. Since $\cos(x) = \sum_{n=0}^{\infty} \frac{(-1)^n x^{2n}}{(2n)!} = 1 - \frac{x^2}{2!} + \frac{x^4}{4!} - \frac{x^6}{6!} + \frac{x^8}{8!} - \ldots$, we have

$\cos(x^2) = \sum_{n=0}^{\infty} \frac{(-1)^n (x^2)^{2n}}{(2n)!} = 1 - \frac{x^4}{2!} + \frac{x^8}{4!} - \frac{x^{12}}{6!} + \frac{x^{16}}{8!} - \ldots$

Thus, $\int_0^1 \cos(x^2)\, dx = \int_0^1 \sum_{n=0}^{\infty} \frac{(-1)^n x^{4n}}{(2n)!}\, dx$

$= \left[\sum_{n=0}^{\infty} \frac{(-1)^n x^{4n+1}}{(2n)!(4n+1)}\right]_0^1 = \sum_{n=0}^{\infty} \frac{(-1)^n}{(2n)!(4n+1)}$

$\approx 1 - \frac{1}{2!(5)} + \frac{1}{4!(9)} - \frac{1}{6!(13)} \approx 0.9045$

29. $\int_0^{\pi/2} \sqrt{x} \cos x\, dx = \int_0^{\pi/2} \left(\sqrt{x} - \frac{\sqrt{x}x^2}{2!} + \frac{\sqrt{x}x^4}{4!} - \ldots\right) dx$

$= \left[\frac{2x^{3/2}}{3} - \frac{2x^{7/2}}{7 \cdot 2!} + \frac{2x^{11/2}}{11 \cdot 4!} - \ldots\right]_0^{\pi/2}$

Since $\frac{2(\pi/2)^{n/2}}{n[(n-3)/2]!} < 0.0001$ when $n = 23$, we have

$\int_0^{\pi/2} \sqrt{x} \cos x\, dx \approx \left(\frac{2(\pi/2)^{3/2}}{3} - \frac{2(\pi/2)^{7/2}}{7 \cdot 2!} + \frac{2(\pi/2)^{11/2}}{11 \cdot 4!} - \frac{2(\pi/2)^{15/2}}{15 \cdot 6!}\right.$

$\left. + \frac{2(\pi/2)^{19/2}}{19 \cdot 8!} - \frac{2(\pi/2)^{23/2}}{23 \cdot 10!}\right) \approx 0.7040$

30. $\int_0^{1/2} \frac{\ln(x+1)}{x}\,dx = \int_0^{1/2} \left(1 - \frac{x}{2} + \frac{x^2}{3} - \frac{x^3}{4} + \ldots\right) dx$

$= \left[x - \frac{x^2}{2^2} + \frac{x^3}{3^2} - \frac{x^4}{4^2} + \ldots\right]_0^{1/2}$

Since $\frac{1}{n^2 2^n} < 0.0001$ when $n = 8$, we have $\int_0^{1/2} \frac{\ln(x+1)}{x}\,dx$

$\approx \left(\frac{1}{2} - \frac{1}{2^2 2^2} + \frac{1}{3^2 2^3} - \frac{1}{4^2 2^4} + \frac{1}{5^2 2^5} - \frac{1}{6^2 2^6} + \frac{1}{7^2 2^7} - \frac{1}{8^2 2^8}\right) \approx 0.4484$

$\int_0^{1/2} \frac{\arctan x}{x}\,dx = \int_0^{1/2} \left(1 - \frac{x^2}{3} + \frac{x^4}{5} - \frac{x^6}{7} + \ldots\right) dx$

$= \left[x - \frac{x^3}{3^2} + \frac{x^5}{5^2} - \frac{x^7}{7^2} + \ldots\right]_0^{1/2}$, $\quad \frac{1}{n^2 2^n} < 0.0001$ when odd $n = 9$

31. $\int_0^{1/2} \frac{\arctan x}{x}\,dx \approx \left(\frac{1}{2} - \frac{1}{3^2 2^3} + \frac{1}{5^2 2^5} - \frac{1}{7^2 2^7} + \frac{1}{9^2 2^9}\right) \approx 0.4872$

32. $\int_1^2 e^{-x^2}\,dx = \int_1^2 \left(1 - x + \frac{x^2}{2!} - \frac{x^3}{3!} + \ldots\right)dx$

$= \left[x - \frac{x^3}{3} + \frac{x^5}{5 \cdot 2!} - \frac{x^7}{7 \cdot 3!} + \frac{x^9}{9 \cdot 4!} - \ldots\right]_1^2$

Since $\frac{2^n - 1}{n[(n-1)/2]!} < 0.001$ when odd $n = 33$, we have

$\int_1^2 e^{-x^2}\,dx \approx \left((2 - 1) - \frac{2^3 - 1}{3} + \frac{2^5 - 1}{5 \cdot 2!} - \ldots - \frac{2^{31} - 1}{31 \cdot 15!} + \frac{2^{33} - 1}{33 \cdot 16!}\right)$

≈ 0.1353

33. $\int_{0.1}^{0.3} \sqrt{1 + x^3}\,dx = \int_{0.1}^{0.3} \left(1 + \frac{x^3}{2} - \frac{x^6}{8} + \frac{x^9}{16} - \frac{5x^{12}}{128} + \ldots\right)dx$

$= \left[x + \frac{x^4}{8} - \frac{x^7}{56} + \frac{x^{10}}{160} - \frac{5x^{13}}{1664} + \ldots\right]_{0.1}^{0.3}$

$\int_{0.1}^{0.3} \sqrt{1 + x^3}\,dx = \left[(0.3 - 0.1) + \frac{1}{8}(0.3^4 - 0.1^4) - \frac{1}{56}(0.3^7 - 0.1^7) + \ldots\right]$

≈ 0.2010

34. $\int_{0.5}^{1} \cos\sqrt{x}\,dx = \int_{0.5}^{1} \left(1 - \frac{x}{2!} + \frac{x^2}{4!} - \frac{x^3}{6!} + \frac{x^4}{8!} - \ldots\right) dx$

$= \left[x - \frac{x^2}{2(2!)} + \frac{x^3}{3(4!)} - \frac{x^4}{4(6!)} + \frac{x^5}{5(8!)} - \ldots\right]_{0.5}^{1}$

$\approx \left[(1 - 0.5) - \frac{1}{4}(1 - 0.5^2) + \frac{1}{72}(1 - 0.5^3) - \frac{1}{2880}(1 - 0.5^4) + \ldots\right]$

≈ 0.3243

35. $\int_0^x (e^{-t^2} - 1)\,dt = \int_0^x \left[\left(\sum_{n=0}^{\infty} \frac{(-1)^n t^{2n}}{n!}\right) - 1\right] dt$

$= \int_0^x \left[\sum_{n=0}^{\infty} \frac{(-1)^{n+1} t^{2n+2}}{(n+1)!}\right] dt = \sum_{n=0}^{\infty} \frac{(-1)^{n+1} t^{2n+3}}{(2n+3)(n+1)!}\Bigg]_0^x$

$= \sum_{n=0}^{\infty} \frac{(-1)^{n+1} x^{2n+3}}{(2n+3)(n+1)!}$

36. $\int_0^x \sqrt{1+t^3}\,dt = \int_0^x \left(1 + \frac{t^3}{2} + \sum_{n=2}^{\infty} \frac{(-1)^{n-1}\, 1\cdot3\cdot5\cdots(2n-3)t^{3n}}{2^n n!}\right) dt$

$= \left[t + \frac{t^4}{8} + \sum_{n=2}^{\infty} \frac{(-1)^{n-1}\, 1\cdot3\cdot5\cdots(2n-3)t^{3n+1}}{(3n+1)2^n n!}\right]_0^x$

$= x + \frac{x^4}{8} + \sum_{n=2}^{\infty} \frac{(-1)^{n-1}\, 1\cdot3\cdot5\cdots(2n-3)x^{3n+1}}{(3n+1)2^n n!}$

37. From Exercise 19, we have $\frac{1}{\sqrt{2\pi}} \int_0^1 e^{-x^2/2}\,dx = \frac{1}{\sqrt{2\pi}} \int_0^1 \sum_{n=0}^{\infty} \frac{(-1)^n x^{2n}}{2^n n!}\,dx$

$= \frac{1}{\sqrt{2\pi}} \left[\sum_{n=0}^{\infty} \frac{(-1)^n x^{2n+1}}{2^n n!(2n+1)}\right]_0^1 = \frac{1}{\sqrt{2\pi}} \sum_{n=0}^{\infty} \frac{(-1)^n}{2^n n!(2n+1)}$

$= \frac{1}{\sqrt{2\pi}} \left[1 - \frac{1}{2\cdot1\cdot3} + \frac{1}{2^2\cdot2!\cdot5} - \frac{1}{2^3\cdot3!\cdot7}\right] = 0.3413$

38. $\frac{1}{\sqrt{2\pi}} \int_1^2 e^{-x^2/2}\,dx = \frac{1}{\sqrt{2\pi}} \int_1^2 \sum_{n=0}^{\infty} \frac{(-1)^n x^{2n}}{2^n n!}\,dx$

$= \frac{1}{\sqrt{2\pi}} \left[\sum_{n=0}^{\infty} \frac{(-1)^n x^{2n+1}}{2^n n!(2n+1)}\right]_1^2 = \frac{1}{\sqrt{2\pi}} \sum_{n=0}^{\infty} \frac{(-1)^n[2^{2n+1} - 1]}{2^n n!(2n+1)}$

$= \frac{1}{\sqrt{2\pi}} \left[1 - \frac{7}{2\cdot1\cdot3} + \frac{31}{2^2\cdot2!\cdot5} - \frac{127}{2^3\cdot3!\cdot7} + \frac{511}{2^4\cdot4!\cdot9} - \frac{2047}{2^5\cdot5!\cdot11}\right.$

$\left. + \frac{8191}{2^6\cdot6!\cdot13} - \frac{32767}{2^7\cdot7!\cdot15} + \frac{131071}{2^8\cdot8!\cdot17} - \frac{524287}{2^9\cdot9!\cdot19}\right] \approx 0.1359$

39. Since $\sin(x) = \sum_{n=0}^{\infty} \frac{(-1)^n x^{2n+1}}{(2n+1)!} = x - \frac{x^3}{3!} + \frac{x^5}{5!} - \frac{x^7}{7!} + \ldots$

we have $\sin(1) = \sum_{n=0}^{\infty} \frac{(-1)^n}{(2n+1)!} = 1 - \frac{1}{3!} + \frac{1}{5!} - \frac{1}{7!} + \ldots$

40. Since $e^x = \sum_{n=0}^{\infty} \frac{x^n}{n!} = 1 + x + \frac{x^2}{2!} + \frac{x^3}{3!} + \frac{x^4}{4!} + \frac{x^5}{5!} + \ldots$

we have $e^{-1} = 1 - 1 + \frac{1}{2!} - \frac{1}{3!} + \frac{1}{4!} - \frac{1}{5!} + \ldots$ and

$\frac{e-1}{e} = 1 - e^{-1} = 1 - \frac{1}{2!} + \frac{1}{3!} - \frac{1}{4!} + \frac{1}{5!} - \ldots$

$= \sum_{n=1}^{\infty} \frac{(-1)^{n-1}}{n!}$

Review Exercises for Chapter 9

1. $a_n = \frac{1}{n!}$

2. $a_n = \frac{n}{n^2 + 1}$

3. $\lim_{n \to \infty} \frac{n+1}{n^2} = 0$, converges

4. $\lim_{n \to \infty} \frac{1}{\sqrt{n}} = 0$, converges

5. $\lim_{n \to \infty} \frac{n^3}{n^2 + 1} = \infty$, diverges

6. $\lim_{n \to \infty} \frac{n}{\ln(n)} = \lim_{n \to \infty} \frac{1}{1/n} = \infty$, diverges

7. $\lim_{n \to \infty} (\sqrt{n+1} - \sqrt{n}) = \lim_{n \to \infty} (\sqrt{n+1} - \sqrt{n}) \frac{\sqrt{n+1} + \sqrt{n}}{\sqrt{n+1} + \sqrt{n}}$

$= \lim_{n \to \infty} \frac{1}{\sqrt{n+1} + \sqrt{n}} = 0$, converges

8. $\lim_{n \to \infty} \left(1 + \frac{1}{2n}\right)^n = \lim_{k \to \infty} \left[\left(1 + \frac{1}{k}\right)^k\right]^{1/2} = e^{1/2}$, converges, $k = 2n$

9. $\lim_{n \to \infty} \frac{\sin(n)}{\sqrt{n}} = 0$, converges

10. Since b and c are positive real numbers, we have $(b^{n+1} + c^{n+1})^{1/(n+1)} \leq (b^n + c^n)^{1/n}$ which implies that the sequence is non-increasing and bounded below by zero and thus, converges.

11. $S_1 = 1$, $S_2 = 1 + \frac{3}{2} = 2.5$, $S_3 = 1 + \frac{3}{2} + \frac{9}{4} = 4.75$

$S_4 = 1 + \frac{3}{2} + \frac{9}{4} + \frac{27}{8} = 8.125$, $S_5 = 1 + \frac{3}{2} + \frac{9}{4} + \frac{27}{8} + \frac{81}{16} = 13.3875$

12. $S_1 = \frac{1}{2}$, $S_2 = \frac{1}{2} - \frac{1}{4} = 0.25$, $S_3 = \frac{1}{2} - \frac{1}{4} + \frac{1}{6} = 0.41667$

$S_4 = \frac{1}{2} - \frac{1}{4} + \frac{1}{6} - \frac{1}{8} = 0.29167$, $S_5 = \frac{1}{2} - \frac{1}{4} + \frac{1}{6} - \frac{1}{8} + \frac{1}{10} = 0.39167$

13. $S_1 = \frac{1}{2!} = 0.5$, $S_2 = \frac{1}{2!} - \frac{1}{4!} = 0.45833$, $S_3 = \frac{1}{2!} - \frac{1}{4!} + \frac{1}{6!} = 0.45972$

$S_4 = \frac{1}{2!} - \frac{1}{4!} + \frac{1}{6!} - \frac{1}{8!} = 0.45970$, $S_5 = \frac{1}{2!} - \frac{1}{4!} + \frac{1}{6!} - \frac{1}{8!} + \frac{1}{10!} = 0.45970$

14. $S_1 = \frac{1}{2}$, $S_2 = \frac{1}{2} + \frac{1}{6} = 0.66667$, $S_3 = \frac{1}{2} + \frac{1}{6} + \frac{1}{12} = 0.75000$

$S_4 = \frac{1}{2} + \frac{1}{6} + \frac{1}{12} + \frac{1}{20} = 0.80000$, $S_5 = \frac{1}{2} + \frac{1}{6} + \frac{1}{12} + \frac{1}{20} + \frac{1}{30} = 0.83333$

15. $\sum_{n=0}^{\infty} (\frac{2}{3})^n$, geometric series with $a = 1$ and $r = \frac{2}{3}$

$$S = \frac{a}{1 - r} = \frac{1}{1 - (2/3)} = \frac{1}{1/3} = 3$$

16. $\sum_{n=0}^{\infty} \frac{2^{n+2}}{3^n} = 4 \sum_{n=0}^{\infty} (\frac{2}{3})^n = 4(3) = 12$ (See Exercise 15)

17. $\sum_{n=0}^{\infty} (\frac{1}{2^n} - \frac{1}{3^n}) = \sum_{n=0}^{\infty} (\frac{1}{2})^n - \sum_{n=0}^{\infty} (\frac{1}{3})^n$

$$= \frac{1}{1 - (1/2)} - \frac{1}{1 - (1/3)} = 2 - \frac{3}{2} = \frac{1}{2}$$

18. $\sum_{n=0}^{\infty} \left[(\frac{2}{3})^n - \frac{1}{(n + 1)(n + 2)}\right] = \sum_{n=0}^{\infty} (\frac{2}{3})^n - \sum_{n=0}^{\infty} (\frac{1}{n + 1} - \frac{1}{n + 2})$

$$= \frac{1}{1 - (2/3)} - \left[(1 - \frac{1}{2}) + (\frac{1}{2} - \frac{1}{3}) + (\frac{1}{3} - \frac{1}{4}) + \ldots\right] = 3 - 1 = 2$$

19. $0.090909\ldots = 0.09 + 0.0009 + 0.000009 + \ldots$

$$= 0.09(1 + 0.01 + 0.0001 + \ldots) = \sum_{n=0}^{\infty} (0.09)(0.01)^n = \frac{0.09}{1 - 0.01} = \frac{1}{11}$$

20. $0.923076\overline{923076} = 0.923076\,[1 + 0.000001 + (0.000001)^2 + \ldots]$

$$= \sum_{n=0}^{\infty} (0.923076)(0.000001)^n = \frac{0.923076}{1 - 0.000001} = \frac{923076}{999999} = \frac{12(76923)}{13(76923)} = \frac{12}{13}$$

21. $\displaystyle\sum_{n=1}^{\infty} \frac{2^n}{n^3},\quad \lim_{n \to \infty} \left|\frac{a_{n+1}}{a_n}\right| = \lim_{n \to \infty} \left|\frac{2^{n+1}}{(n+1)^3} \cdot \frac{n^3}{2^n}\right| = \lim_{n \to \infty} \frac{2n^3}{(n+1)^3} = 2$

Therefore, by the Ratio Test, the series diverges.

22. $\displaystyle\sum_{n=1}^{\infty} \frac{1}{\sqrt{n^2 + 2n}},\quad \lim_{n \to \infty} \frac{1/\sqrt{n^2 + 2n}}{1/n} = \lim_{n \to \infty} \frac{n}{\sqrt{n^2 + 2n}} = 1$ Therefore,

by a limit comparison test with $\displaystyle\sum_{n=1}^{\infty} \frac{1}{n}$, the series diverges.

23. $\displaystyle\sum_{n=1}^{\infty} \frac{1}{\sqrt{n^3 + 2n}},\quad \lim_{n \to \infty} \frac{1/\sqrt{n^3 + 2n}}{1/n^{3/2}} = \lim_{n \to \infty} \frac{n^{3/2}}{\sqrt{n^3 + 2n}} = 1$ Therefore,

by a limit comparison test with $\displaystyle\sum_{n=1}^{\infty} \frac{1}{n^{3/2}}$, the series converges.

24. $\displaystyle\sum_{n=1}^{\infty} \frac{n + 1}{n(n + 2)},\quad \lim_{n \to \infty} \frac{(n + 1)/n(n + 2)}{1/n} = \lim_{n \to \infty} \frac{n + 1}{n + 2} = 1$ Therefore,

by a limit comparison test with $\displaystyle\sum_{n=1}^{\infty} \frac{1}{n}$, the series diverges.

25. $\displaystyle\sum_{n=1}^{\infty} \frac{1}{\sqrt[3]{n^3 + 2n}},\quad \lim_{n \to \infty} \frac{1/\sqrt[3]{n^3 + 2n}}{1/n} = \lim_{n \to \infty} \frac{n}{\sqrt[3]{n^3 + 2n}} = 1$ Therefore,

by a limit comparison test with $\displaystyle\sum_{n=1}^{\infty} \frac{1}{n}$, the series diverges.

26. $\displaystyle\sum_{n=1}^{\infty} \frac{n!}{e^n},\quad \lim_{n \to \infty} \left|\frac{a_{n+1}}{a_n}\right| = \lim_{n \to \infty} \left|\frac{(n + 1)!}{e^{n+1}} \quad \frac{e^n}{n!}\right| = \lim_{n \to \infty} \frac{n + 1}{e} = \infty$

Therefore, by the Ratio Test, the series diverges.

27. $\displaystyle\sum_{n=1}^{\infty} \frac{(-1)^n n}{\ln(n)},\quad \lim_{n \to \infty} \frac{n}{\ln(n)} = \infty \neq 0$ Therefore the series diverges.

28. $\sum_{n=1}^{\infty} \frac{(-1)^n\sqrt{n}}{n+1}$, $a_{n+1} = \frac{\sqrt{n+1}}{n+2} \leq \frac{\sqrt{n}}{n+1} = a_n$ and $\lim_{n \to \infty} \frac{\sqrt{n}}{n+1} = 0$

Therefore, by the alternating series test, the series converges.

29. $\sum_{n=1}^{\infty} \frac{(-1)^n 1\cdot 3\cdot 5\cdots(2n-1)}{2\cdot 4\cdot 6\cdots(2n)(2n+1)}$, $\lim_{n \to \infty} \frac{1\cdot 3\cdot 5\cdots(2n-1)}{2\cdot 4\cdot 6\cdots(2n)(2n+1)} = 0$

and $a_{n+1} = \frac{1\cdot 3\cdot 5\cdots(2n-1)(2n+1)}{2\cdot 4\cdot 6\cdots(2n)(2n+2)(2n+3)} \leq \frac{1\cdot 3\cdot 5\cdots(2n-1)}{2\cdot 4\cdot 6\cdots(2n)(2n+1)} = a_n$

Therefore, by the alternating series test, the series converges.

30. $\sum_{n=1}^{\infty} \frac{1\cdot 3\cdot 5\cdots(2n-1)}{2\cdot 5\cdot 8\cdots(3n-1)}$, $\lim_{n \to \infty} \left|\frac{a_{n+1}}{a_n}\right|$

$= \lim_{n \to \infty} \left|\frac{1\cdot 3\cdots(2n-1)(2n+1)}{2\cdot 5\cdots(3n-1)(3n+2)} \cdot \frac{2\cdot 5\cdots(3n-1)}{1\cdot 3\cdots(2n-1)}\right|$

$= \lim_{n \to \infty} \frac{2n+1}{3n+2} = \frac{2}{3}$

Therefore, by the Ratio Test, the series converges.

31. $\sum_{n=1}^{\infty} \left(\frac{1}{n^2} - \frac{1}{n}\right) = \sum_{n=1}^{\infty} \frac{1}{n^2} - \sum_{n=1}^{\infty} \frac{1}{n}$

Since the second series is a divergent p-series while the first series is a convergent p-series, the difference diverges.

32. $\sum_{n=1}^{\infty} \left(\frac{1}{n^2} - \frac{1}{2^n}\right) = \sum_{n=1}^{\infty} \frac{1}{n^2} - \sum_{n=1}^{\infty} \frac{1}{2^n}$

The first series is a convergent p-series and the second series is a convergent geometric series. Therefore their difference converges.

33. $\sum_{n=0}^{\infty} \frac{(-1)^n(x-2)^n}{(n+1)^2}$, $\lim_{n \to \infty} \left|\frac{a_{n+1}}{a_n}\right| = \lim_{n \to \infty} \left|\frac{(x-2)^{n+1}}{(n+2)^2} \cdot \frac{(n+1)^2}{(x-2)^n}\right|$

$= |x-2| < 1$

When $x = 1$ and when $x = 3$, the series converges and the interval of convergence is $1 \leq x \leq 3$

34. $\sum_{n=0}^{\infty} (2x)^n$ Geometric series which converges only if $|2x| < 1$

or $-\frac{1}{2} < x < \frac{1}{2}$

35. $\sum_{n=0}^{\infty} n!(x - 2)^n, \quad \lim_{n\to\infty} \left|\frac{a_{n+1}}{a_n}\right| = \lim_{n\to\infty} \left|\frac{(n + 1)!(x - 2)^{n+1}}{n!(x - 2)^n}\right| = \infty$

which implies that the series converges only for $x = 2$

36. $\sum_{n=0}^{\infty} \frac{(x - 2)^n}{2^n} = \sum_{n=0}^{\infty} \left(\frac{x - 2}{2}\right)^n$ Geometric series which converges only if

$\left|\frac{x - 2}{2}\right| < 1$ or $0 < x < 4$

37. $f(x) = \sin(x)$, $f'(x) = \cos(x)$, $f''(x) = -\sin(x)$, $f'''(x) = -\cos(x)$, ...

$$\sin(x) = \sum_{n=0}^{\infty} \frac{f^{(n)}(x)[x - (3\pi/4)]^n}{n!}$$

$$= \frac{\sqrt{2}}{2} - \frac{\sqrt{2}}{2}\left(x - \frac{3\pi}{4}\right) - \frac{\sqrt{2}}{2\cdot 2!}\left(x - \frac{3\pi}{4}\right)^2 + \ldots$$

$$= \frac{\sqrt{2}}{2} \sum_{n=0}^{\infty} \frac{(-1)^{n(n+1)/2}[x - (3\pi/4)]^n}{n!}$$

38. $f(x) = x^{1/2}$, $f'(x) = \frac{1}{2}x^{-1/2}$, $f''(x) = -\left(\frac{1}{2}\right)\left(\frac{1}{2}\right)x^{-3/2}$,

$f'''(x) = \left(\frac{1}{2}\right)\left(\frac{1}{2}\right)\left(\frac{3}{2}\right)x^{-5/2}$, $f^{(4)}(x) = -\left(\frac{1}{2}\right)\left(\frac{1}{2}\right)\left(\frac{3}{2}\right)\left(\frac{5}{2}\right)x^{-7/2}$, ...

$$\sqrt{x} = \sum_{n=0}^{\infty} \frac{f^{(n)}(4)(x - 4)^n}{n!}$$

$$= 2 + \frac{(x - 4)}{2^2} - \frac{(x - 4)^2}{2^5 2!} + \frac{1\cdot 3(x - 4)^3}{2^8 3!} - \frac{1\cdot 3\cdot 5(x - 4)^4}{2^{11} 4!} + \ldots$$

$$= 1 + \frac{x}{4} + \sum_{n=0}^{\infty} \frac{(-1)^{n+1}\, 1\cdot 3\cdot 5\cdots(2n - 3)(x - 4)^n}{2^{3n-1} n!}$$

The interval of convergence is given by $0 < x < 8$

39. $3^x = (e^{\ln(3)})^x = e^{x\ln(3)}$ and since $e^x = \sum_{n=0}^{\infty} \frac{x^n}{n!}$, we have

$$3^x = \sum_{n=0}^{\infty} \frac{(x \ln 3)^n}{n!} = 1 + x\ln 3 + \frac{x^2 \ln^2 3}{2!} + \frac{x^3 \ln^3 3}{3!} + \frac{x^4 \ln^4 3}{4!} + \ldots$$

40. $f(x) = \csc(x)$, $f'(x) = -\csc(x)\cot(x)$,

$f''(x) = \csc^3(x) + \csc(x)\cot^2(x)$,

$f'''(x) = -5\csc^3(x)\cot(x) - \csc(x)\cot^3(x)$,

$f^{(4)}(x) = 5\csc^5(x) + 15\csc^3(x)\cot^2(x) + \csc(x)\cot^4(x)$

$$\csc(x) = \sum_{n=0}^{\infty} \frac{f^{(n)}(\pi/2)(x - (\pi/2))^n}{n!}$$

$$= 1 + \frac{1}{2!}\left(x - \frac{\pi}{2}\right)^2 + \frac{5}{4!}\left(x - \frac{\pi}{2}\right)^4 + \ldots$$

41. $f(x) = \frac{1}{x}$, $f'(x) = -\frac{1}{x^2}$, $f''(x) = \frac{2}{x^3}$, $f'''(x) = -\frac{6}{x^4}$, ...

$$\frac{1}{x} = \sum_{n=0}^{\infty} \frac{f^{(n)}(-1)(x+1)^n}{n!} = \sum_{n=0}^{\infty} \frac{-n!(x+1)^n}{n!} = -\sum_{n=0}^{\infty}(x+1)^n$$

42. $f(x) = \cos x$, $f'(x) = -\sin x$, $f''(x) = -\cos x$, $f'''(x) = \sin x$

$$\cos x = \sum_{n=0}^{\infty} \frac{f^{(n)}(-\pi/4)(x + \pi/4)^n}{n!}$$

$$= \frac{\sqrt{2}}{2} + \frac{\sqrt{2}}{2}\left(x + \frac{\pi}{4}\right) - \frac{\sqrt{2}}{2\cdot 2!}\left(x + \frac{\pi}{4}\right)^2 - \frac{\sqrt{2}}{2\cdot 3!}\left(x + \frac{\pi}{4}\right)^3 + \frac{\sqrt{2}}{2\cdot 4!}\left(x + \frac{\pi}{4}\right)^4 + \ldots$$

$$= \frac{\sqrt{2}}{2}\left[1 + \left(x + \frac{\pi}{4}\right) + \sum_{n=1}^{\infty} \frac{(-1)^{[n(n+1)]/2}[x + (\pi/4)]^{n+1}}{(n+1)!}\right]$$

43. $$\sin t = \sum_{n=0}^{\infty} \frac{(-1)^n t^{2n+1}}{(2n+1)!}, \qquad \frac{\sin t}{t} = \sum_{n=0}^{\infty} \frac{(-1)^n t^{2n}}{(2n+1)!}$$

$$\int_0^x \frac{\sin t}{t}\,dt = \sum_{n=0}^{\infty} \frac{(-1)^n t^{2n+1}}{(2n+1)(2n+1)!}\Bigg]_0^x = \sum_{n=0}^{\infty} \frac{(-1)^n x^{2n+1}}{(2n+1)(2n+1)!}$$

44. $$\cos t = \sum_{n=0}^{\infty} \frac{(-1)^n t^{2n}}{(2n)!}, \qquad \cos\frac{\sqrt{t}}{2} = \sum_{n=0}^{\infty} \frac{(-1)^n t^n}{2^{2n}(2n)!}$$

$$\int_0^x \cos\frac{\sqrt{t}}{2}\,dt = \sum_{n=0}^{\infty} \frac{(-1)^n t^{n+1}}{2^{2n}(2n)!(n+1)}\Bigg]_0^x = \sum_{n=0}^{\infty} \frac{(-1)^n x^{n+1}}{2^{2n}(2n)!(n+1)}$$

45. $\dfrac{1}{1+t} = \sum_{n=0}^{\infty} (-1)^n t^n, \qquad \ln(1+t) = \int \dfrac{1}{1+t}\,dt = \sum_{n=0}^{\infty} \dfrac{(-1)^n t^{n+1}}{n+1}$

$\dfrac{\ln(t+1)}{t} = \sum_{n=0}^{\infty} \dfrac{(-1)^n t^n}{n+1}$

$\int_0^x \dfrac{\ln(t+1)}{t}\,dt = \sum_{n=0}^{\infty} \dfrac{(-1)^n t^{n+1}}{(n+1)^2}\Bigg]_0^x = \sum_{n=0}^{\infty} \dfrac{(-1)^n x^{n+1}}{(n+1)^2}$

46. $e^t = \sum_{n=0}^{\infty} \dfrac{t^n}{n!}, \qquad e^t - 1 = \sum_{n=1}^{\infty} \dfrac{t^n}{n!}, \qquad \dfrac{e^t - 1}{t} = \sum_{n=1}^{\infty} \dfrac{t^{n-1}}{n!}$

$\int_0^x \dfrac{e^t - 1}{t}\,dt = \sum_{n=1}^{\infty} \dfrac{t^n}{n \cdot n!}\Bigg]_0^x = \sum_{n=1}^{\infty} \dfrac{x^n}{n \cdot n!}$

47. $\sin(95°) = \sin\left(\dfrac{95\pi}{180}\right) = \dfrac{95\pi}{180} - \dfrac{(95\pi)^3}{180^3 3!} + \dfrac{(95\pi)^5}{180^5 5!} - \dfrac{(95\pi)^7}{180^7 7!} = 0.996$

48. $\cos(0.75) = 1 - \dfrac{(0.75)^2}{2!} + \dfrac{(0.75)^4}{4!} - \dfrac{(0.75)^6}{6!} \approx 0.7317$

49. $\ln(1.75) = (0.75) - \dfrac{(0.75)^2}{2} + \dfrac{(0.75)^3}{3} - \dfrac{(0.75)^4}{4} + \dfrac{(0.75)^5}{5}$

$- \dfrac{(0.75)^6}{6} + \ldots + \dfrac{(0.75)^{15}}{15} \approx 0.560$

50. $e^{-0.25} = 1 - 0.25 + \dfrac{(0.25)^2}{2!} - \dfrac{(0.25)^3}{3!} + \dfrac{(0.25)^4}{4!} - \dfrac{(0.25)^5}{5!} = 0.779$

51. If $f(x) = e^x$, $P_5(x) = 1 + x + \dfrac{x^2}{2!} + \dfrac{x^3}{3!} + \dfrac{x^4}{4!} + \dfrac{x^5}{5!}$

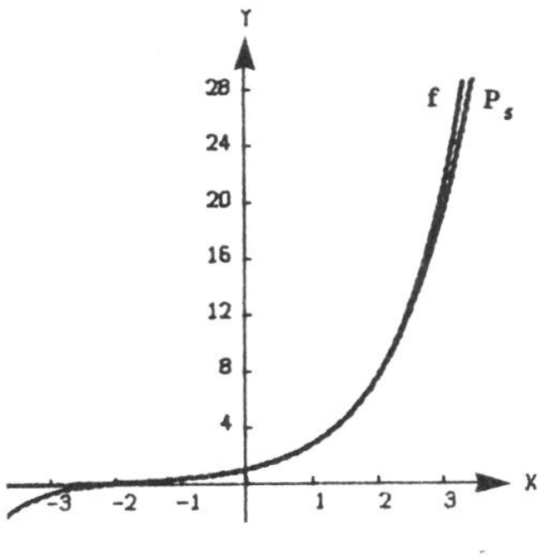

52. $D_1 = 8$

$D_2 = 0.7(8) + 0.7(8) = 16(0.7)$

$\vdots$

$D = 8 + 16(0.7) + 16(0.7)^2 + \ldots + 16(0.7)^n + \ldots$

$= -8 + \sum_{n=0}^{\infty} 16(0.7)^n = -8 + \dfrac{16}{1 - 0.7} = 45\tfrac{1}{3}$ feet

10 Conics

10.1 Parabolas

1. $y^2 = 4x$, Vertex: $(0, 0)$

$p = 1 > 0$, opens to the right

Matches graph (e)

2. $x^2 = -2y$, Vertex: $(0, 0)$,

$p = -\frac{1}{2} < 0$, opens downward

Matches graph (f)

3. $x^2 = 8y$, Vertex: $(0, 0)$

$p = 2 > 0$, opens upward

Matches graph (a)

4. $y^2 = -12x$, Vertex: $(0, 0)$

$p = -3 < 0$, opens to the left

Matches graph (c)

5. $(y - 1)^2 = 4(x - 2)$

Vertex: $(2, 1)$

$p = 1 > 0$, opens to the right

Matches graph (d)

6. $(x + 3)^2 = -2(y - 2)$

Vertex: $(-3, 2)$

$p = -\frac{1}{2} < 0$, opens downward

Matches graph (b)

7. $y = 4x^2$, $x^2 = \frac{1}{4}y = 4(\frac{1}{16})y$

Vertex: $(0, 0)$, Focus: $(0, \frac{1}{16})$

Directrix: $y = -\frac{1}{16}$

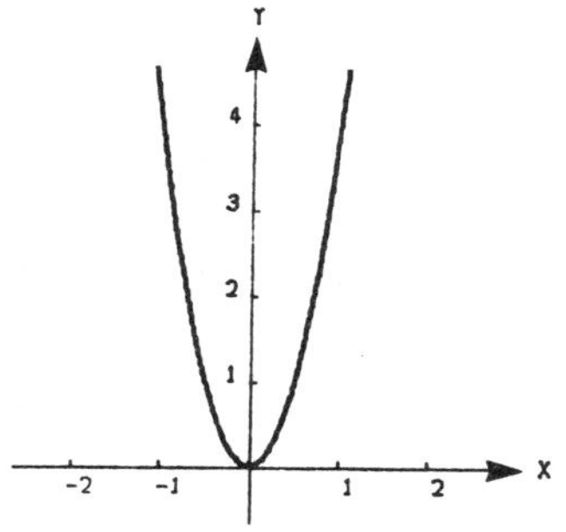

8. $y = 2x^2$, $x^2 = \frac{1}{2}y = 4(\frac{1}{8})y$

Vertex: $(0, 0)$, Focus: $(0, \frac{1}{8})$

Directrix: $y = -\frac{1}{8}$

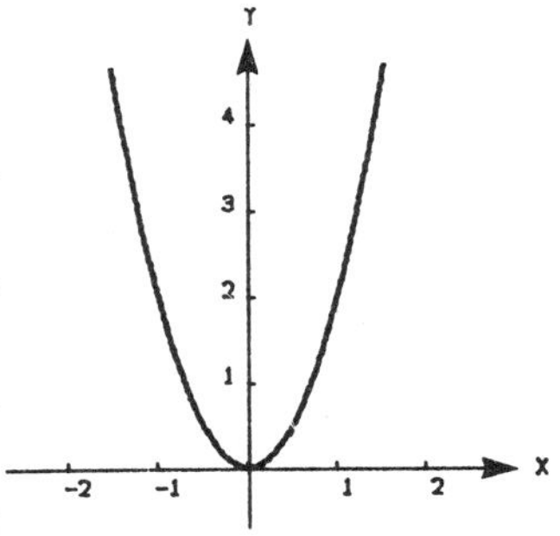

9. $y^2 = -6x = 4(-\frac{3}{2})x$

Vertex: $(0, 0)$, Focus: $(-\frac{3}{2}, 0)$

Directrix: $x = \frac{3}{2}$

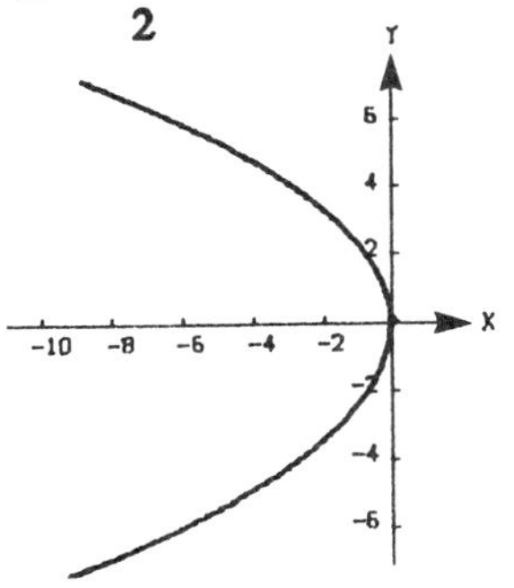

10. $y^2 = 3x = 4(\frac{3}{4})x$

Vertex: $(0, 0)$, Focus: $(\frac{3}{4}, 0)$

Directrix: $x = -\frac{3}{4}$

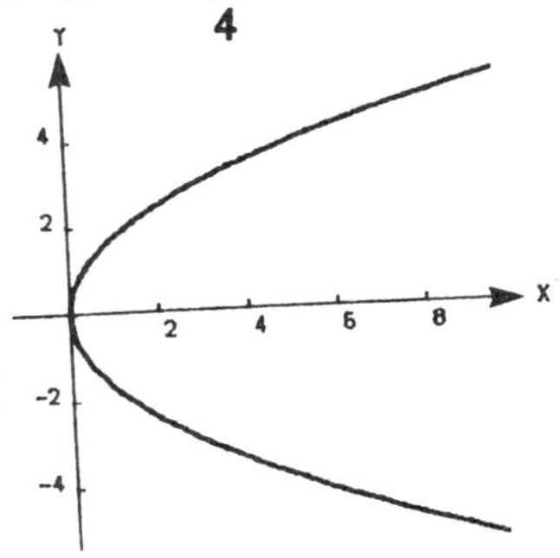

11. $x^2 + 8y = 0$, $x^2 = 4(-2)y$

Vertex: $(0, 0)$, Focus: $(0, -2)$

Directrix: $y = 2$

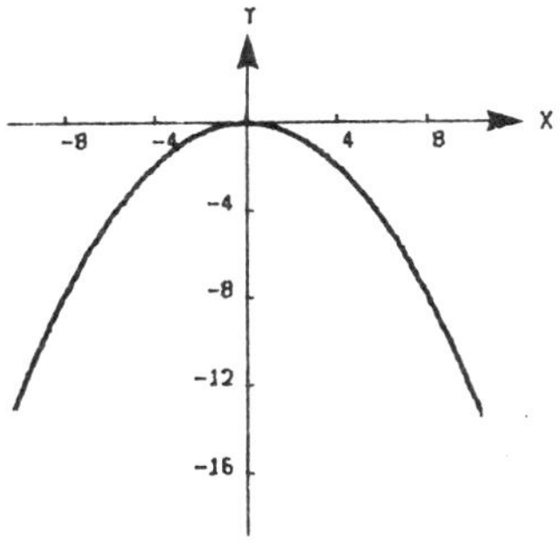

12. $x + y^2 = 0$, $y^2 = -x = 4(-\frac{1}{4})x$

Vertex: $(0, 0)$, Focus: $(-\frac{1}{4}, 0)$

Directrix: $x = \frac{1}{4}$

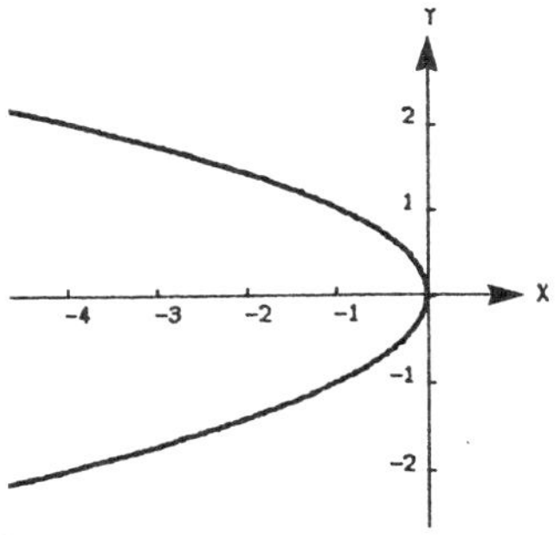

13. $(x - 1)^2 + 8(y + 2) = 0$

$(x - 1)^2 = 4(-2)(y + 2)$

Vertex: $(1, -2)$, Focus: $(1, -4)$

Directrix: $y = 0$

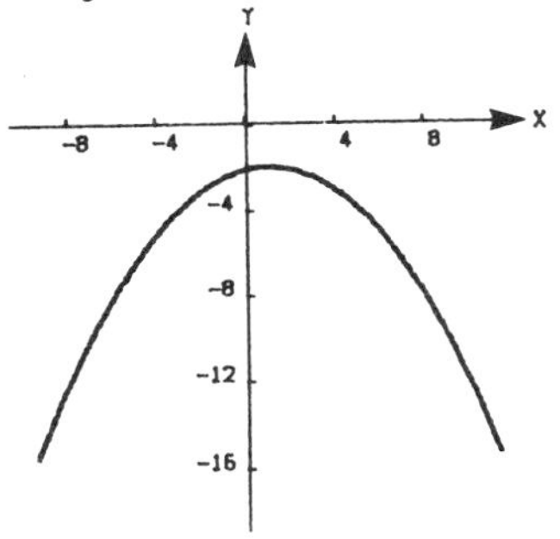

14. $(x + 3) + (y - 2)^2 = 0$

$(y - 2)^2 = 4(-\frac{1}{4})(x + 3)$

Vertex: $(-3, 2)$, Focus: $(-3.25, 2)$

Directrix: $x = -2.75$

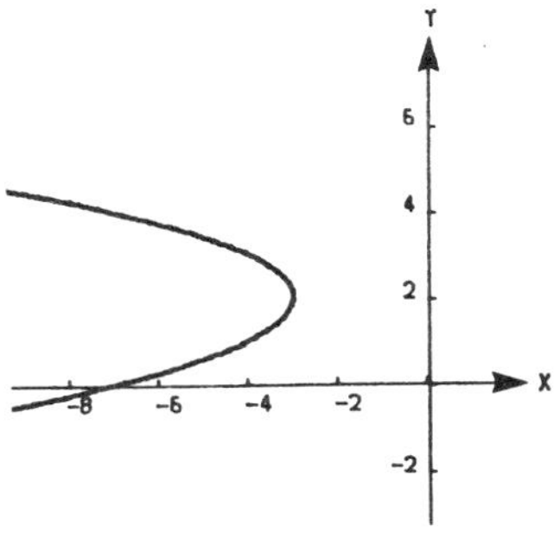

15. $(y + \frac{1}{2})^2 = 2(x - 5) = 4(\frac{1}{2})(x - 5)$

Vertex: $(5, -\frac{1}{2})$

Focus: $(\frac{11}{2}, -\frac{1}{2})$

Directrix: $x = \frac{9}{2}$

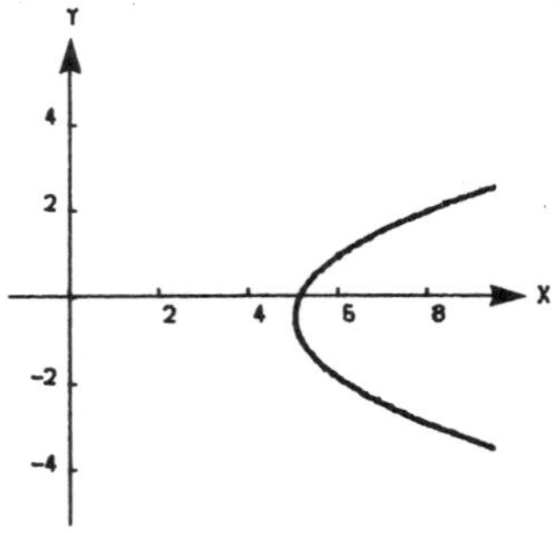

16. $(x + \frac{1}{2})^2 - 4(y - 3) = 0$

$(x + \frac{1}{2})^2 = 4(1)(y - 3)$

Vertex: $(-\frac{1}{2}, 3)$

Focus: $(-\frac{1}{2}, 4)$

Directrix: $y = 2$

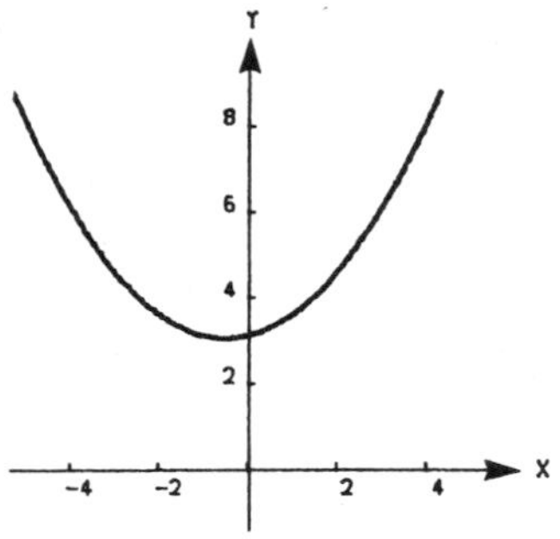

17. $y = \frac{1}{4}(x^2 - 2x + 5)$

$4y - 4 = (x - 1)^2$

$(x - 1)^2 = 4(1)(y - 1)$

Vertex: $(1, 1)$
Focus: $(1, 2)$
Directrix: $y = 0$

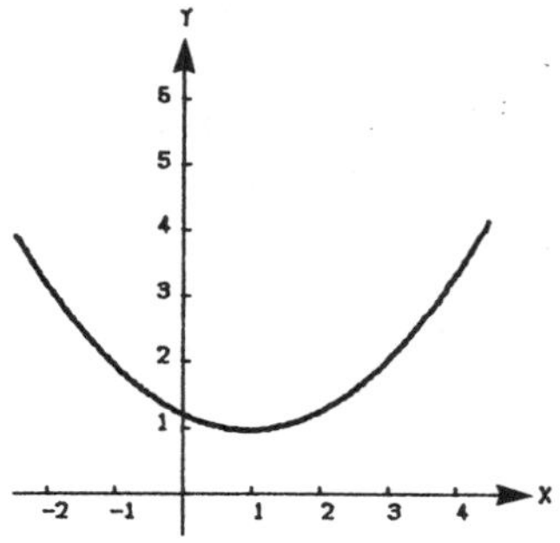

18. $y = -\frac{1}{6}(x^2 + 4x - 2)$

$-6y + 6 = (x + 2)^2$

$(x + 2)^2 = 4(-\frac{3}{2})(y - 1)$

Vertex: $(-2, 1)$

Focus: $(-2, -\frac{1}{2})$

Directrix: $y = \frac{5}{2}$

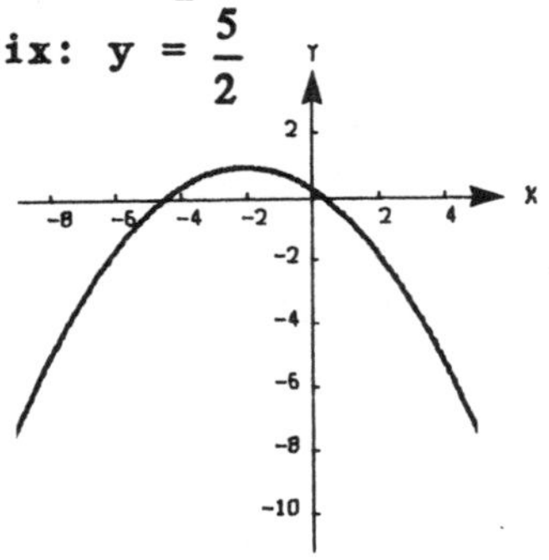

19. $4x - y^2 - 2y - 33 = 0$

$y^2 + 2y + 1 = 4x - 33 + 1$

$(y + 1)^2 = 4(1)(x - 8)$

Vertex: $(8, -1)$
Focus: $(9, -1)$
Directrix: $x = 7$

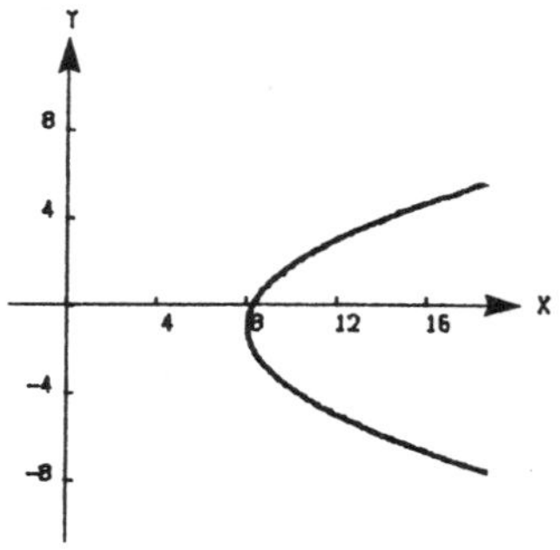

20. $y^2 + x + y = 0$

$y^2 + y + \frac{1}{4} = -x + \frac{1}{4}$

$(y + \frac{1}{2})^2 = 4(-\frac{1}{4})(x - \frac{1}{4})$

Vertex: $(\frac{1}{4}, -\frac{1}{2})$

Focus: $(0, -\frac{1}{2})$

Directrix: $x = \frac{1}{2}$

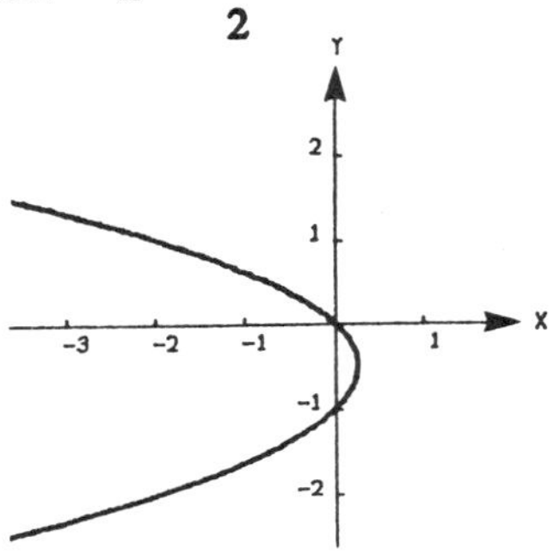

21. $y^2 + 6y + 8x + 25 = 0$

$y^2 + 6y + 9 = -8x - 25 + 9$

$(y + 3)^2 = 4(-2)(x + 2)$

Vertex: $(-2, -3)$

Focus: $(-4, -3)$

Directrix: $x = 0$

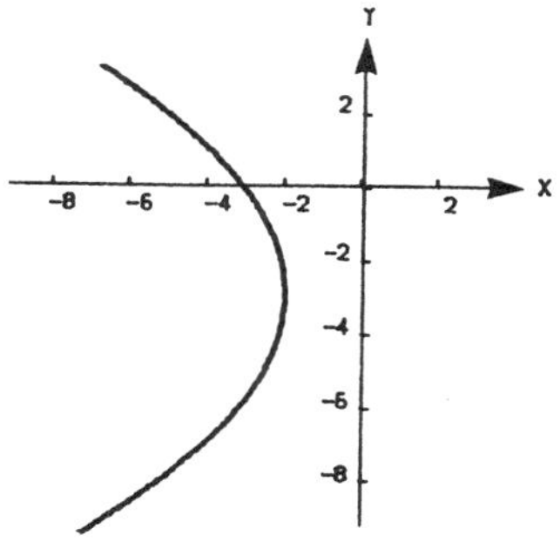

22. $x^2 - 2x + 8y + 9 = 0$

$x^2 - 2x + 1 = -8y - 9 + 1$

$(x - 1)^2 = 4(-2)(y + 1)$

Vertex: $(1, -1)$

Focus: $(1, -3)$

Directrix: $y = 1$

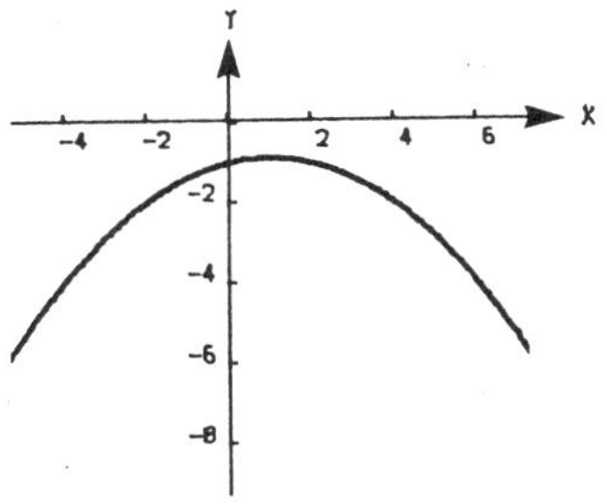

23. $y^2 - 4y - 4x = 0$

$y^2 - 4y + 4 = 4x + 4$

$(y - 2)^2 = 4(1)(x + 1)$

Vertex: $(-1, 2)$

Focus: $(0, 2)$

Directrix: $x = -2$

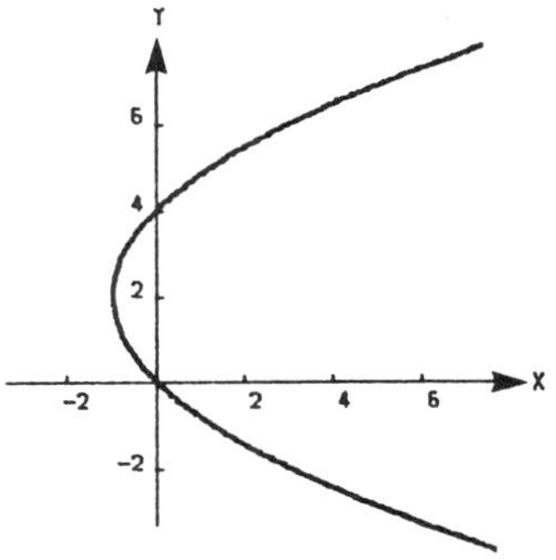

24. $y^2 - 4x - 4 = 0$

$y^2 = 4x + 4 = 4(1)(x + 1)$

Vertex: $(-1, 0)$
Focus: $(0, 0)$
Directrix: $x = -2$

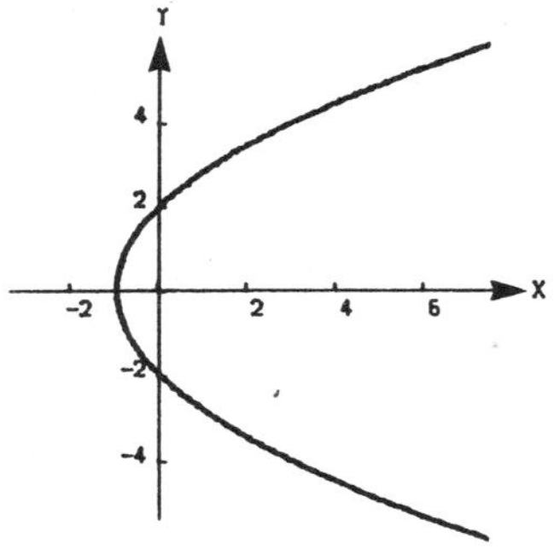

25. $x^2 + 4x + 4y - 4 = 0$

$x^2 + 4x + 4 = -4y + 4 + 4$

$(x + 2)^2 = 4(-1)(y - 2)$

Vertex: $(-2, 2)$
Focus: $(-2, 1)$
Directrix: $y = 3$

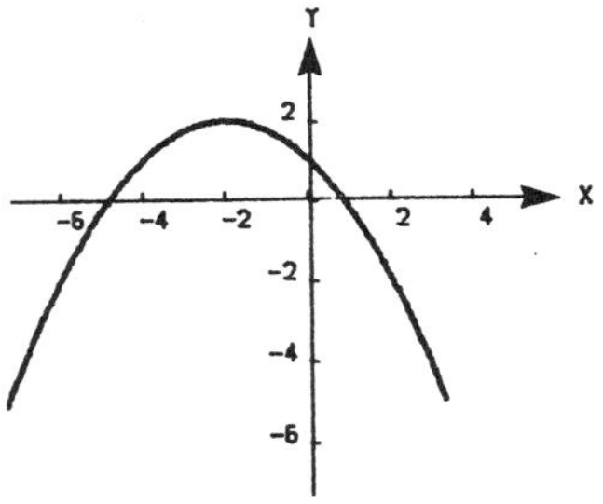

26. $y^2 + 4y + 8x - 12 = 0$

$y^2 + 4y + 4 = -8x + 12 + 4$

$(y + 2)^2 = -4(2)(x - 2)$

Vertex: $(2, -2)$
Focus: $(0, -2)$
Directrix: $x = 4$

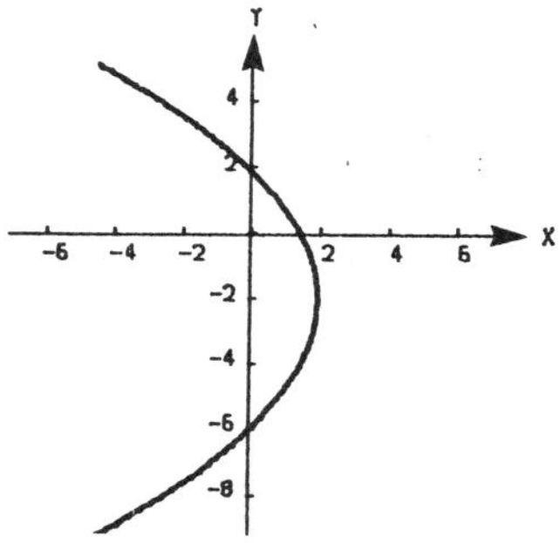

27. $x^2 = -4(\frac{3}{2})y$

$x^2 = -6y$

$x^2 + 6y = 0$

28. $y^2 = 4(2)x$

$y^2 = 8x$

29. $(y - 2)^2 = -4(2)(x - 3)$

$y^2 - 4y + 8x - 20 = 0$

30. $(x + 1)^2 = -4(2)(y - 2)$

$x^2 + 2x + 8y - 15 = 0$

31. $x^2 = 4(-6)(y + 4)$

$x^2 + 24y + 96 = 0$

32. $(y - 1)^2 = -4(3)(x + 2)$

$y^2 - 2y + 12x + 25 = 0$

33. $x^2 = -4(2)(y - 2)$

$x^2 + 8y - 16 = 0$

34. $(y - 2)^2 = 4(2)(x - 0)$

$y^2 - 4y - 8x + 4 = 0$

35. Since the axis of the parabola is vertical, the form of the equation is: $y = ax^2 + bx + c$. Now, substituting the values of the given coordinates into this equation, we obtain: $3 = c$, $4 = 9a + 3b + c$, $11 = 16a + 4b + c$. Solving this system, we have $a = \frac{5}{3}$, $b = -\frac{14}{3}$, $c = 3$. Therefore, $y = \frac{5}{3}x^2 - \frac{14}{3}x + 3$ or $5x^2 - 14x - 3y + 9 = 0$.

36. Since the axis of the parabola is horizontal, the form of the equation is $x = ay^2 + by + c$. Now, substituting the values of the given coordinates into this equation, we obtain: $0 = c$, $4 = 4a - 2b + c$, $3 = 9a - 3b + c$. Solving this system, we have $a = -1$, $b = -4$, $c = 0$. Therefore, $x = -y^2 - 4y$ or $y^2 + 4y + x = 0$.

37. $y = 4 - x^2$

38. $y = 4 - (x - 1)^2 = 4x - x^2$

39. From Example 4:

$4p = 8$, or $p = 2$

Vertex: $(h, -1)$

$(x - h)^2 = -8(y + 1)$

40. From Example 4:

$4p = 8$, or $p = 2$

Vertex: $(4, 0)$

$(x - 4)^2 = 8(y - 0)$ or

$x^2 - 8x - 8y + 16 = 0$

41. Vertex: $(0, 0)$

$p = \frac{3}{8}$

$y^2 = 4(\frac{3}{8})x$

$y^2 = \frac{3}{2}x$

$3x - 2y^2 = 0$

42. Vertex: $(0, 0)$

$p = 3$

$x^2 = 4(3)y$

$x^2 - 12y = 0$

43. $y = -\frac{1}{8}x^2, \qquad y' = -\frac{1}{4}x, \qquad 1 + (y')^2 = 1 + \frac{x^2}{16}$

$$s = \int_0^4 \sqrt{1 + (x^2/16)}\, dx = \frac{1}{4}\int_0^4 \sqrt{16 + x^2}\, dx$$

$$= \frac{1}{8}\left[x\sqrt{16 + x^2} + 16 \ln\left|x + \sqrt{16 + x^2}\right|\right]_0^4$$

$$= \frac{1}{8}\left[4\sqrt{32} + 16 \ln(4 + \sqrt{32}) - 16 \ln 4\right]$$

$$= \frac{1}{8}\left[16\sqrt{2} + 16 \ln(1 + \sqrt{2})\right] = 2\left[\sqrt{2} + \ln(1 + \sqrt{2})\right]$$

44. $x = -y^2, \qquad x' = -2y, \qquad 1 + (x')^2 = 1 + 4y^2$

$$s = \int_0^2 \sqrt{1 + 4y^2}\, dy = \frac{1}{2}\int_0^2 \sqrt{1 + 4y^2}\, 2\, dy$$

$$= \frac{1}{4}\left[2y\sqrt{1 + 4y^2} + \ln\left|2y + \sqrt{1 + 4y^2}\right|\right]_0^2$$

$$= \frac{1}{4}\left[4\sqrt{17} + \ln(4 + \sqrt{17}) - \ln 1\right] = \sqrt{17} + \frac{1}{4}\ln(4 + \sqrt{17})$$

45. $x = \frac{1}{4}y^2, \qquad x' = \frac{1}{2}y, \qquad 1 + (x')^2 = 1 + \frac{y^2}{4}$

$$s = \int_0^4 \sqrt{1 + (y^2/4)}\, dy = \frac{1}{2}\int_0^4 \sqrt{4 + y^2}\, dy$$

$$= \frac{1}{4}\left[y\sqrt{4 + y^2} + 4 \ln\left|y + \sqrt{4 + y^2}\right|\right]_0^4$$

$$= \frac{1}{4}\left[4\sqrt{20} + 4 \ln\left|4 + \sqrt{20}\right| - 4 \ln 2\right] = 2\sqrt{5} + \ln(2 + \sqrt{5})$$

46. $y = \frac{1}{2}x^2, \qquad y' = x, \qquad 1 + (y')^2 = 1 + x^2$

$$s = \int_0^1 \sqrt{1 + x^2}\, dx = \frac{1}{2}\left[x\sqrt{1 + x^2} + \ln\left|x + \sqrt{1 + x^2}\right|\right]_0^1$$

$$= \frac{1}{2}[\sqrt{2} + \ln(1 + \sqrt{2})]$$

47. We place the coordinate axes so that the form of the equation of the parabola is $y = ax^2$.

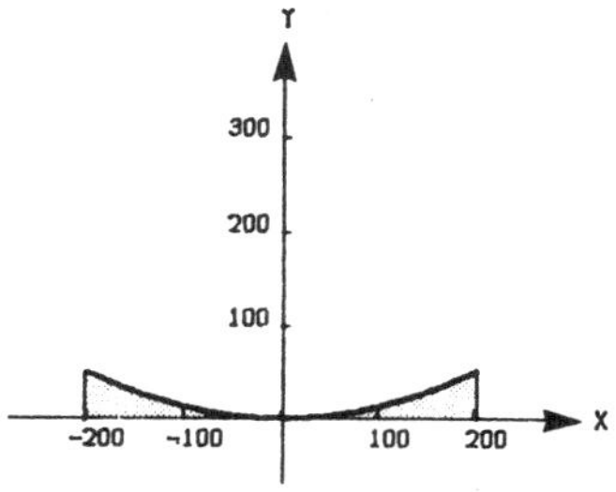

$50 = 40{,}000a$ or $a = \dfrac{1}{800}$

$y = \dfrac{1}{800}x^2$

$$s = 2\int_0^{200} \sqrt{1 + (x/400)^2}\,dx$$

$$= \frac{1}{200}\int_0^{200} \sqrt{(400)^2 + x^2}\,dx$$

$$= \frac{1}{400}\left[x\sqrt{(400)^2 + x^2} + 400^2 \ln(x + \sqrt{400^2 + x^2})\right]_0^{200}$$

$$= 100\left[\sqrt{5} + 4\ln\left(\frac{1 + \sqrt{5}}{2}\right)\right] \approx 416.1 \text{ ft}$$

48. If we place the coordinate system so that the origin is at the center of the earth, then

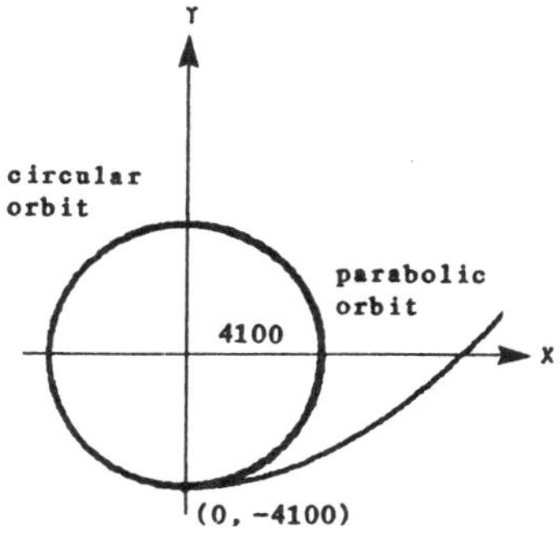

$x^2 = 4p(y - k)$

$x^2 = 4(4100)(y + 4100)$

$x^2 = 16{,}400(y + 4100)$

49. Height of the water: $s = -16t^2 + v_0t + s_0 = -16t^2 + 48$

When $s = 0$: $16t^2 = 48$, $t = \sqrt{3}$

Since the horizontal velocity is constant, hits ground $10\sqrt{3}$ feet away

50. The area enclosed by the line $y = k$ parallel to the directrix and the parabola is:

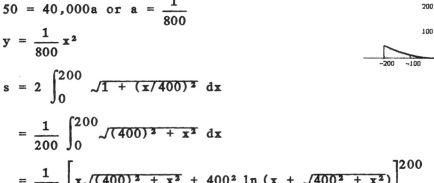

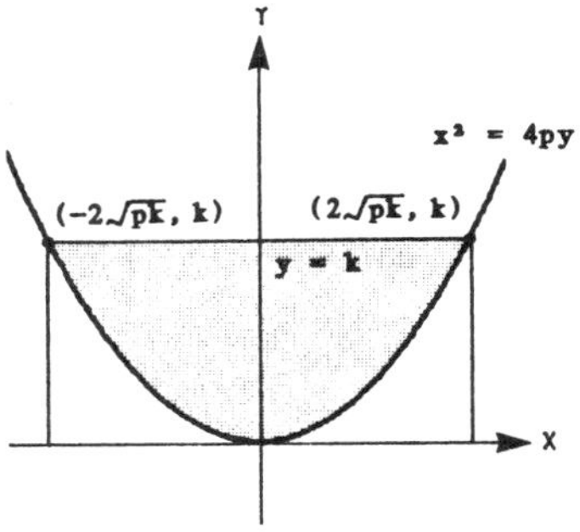

$$A_1 = 2\int_0^{2\sqrt{pk}} \left(k - \frac{1}{4p}x^2\right)dx = \frac{8k\sqrt{pk}}{3}$$

The area by the circumscribed rectangle is: $A_2 = 2(2\sqrt{pk})k = 4k\sqrt{pk}$

hence, $A_1 = \dfrac{2}{3}A_2$

51. $y = ax^2, \quad y' = 2ax$

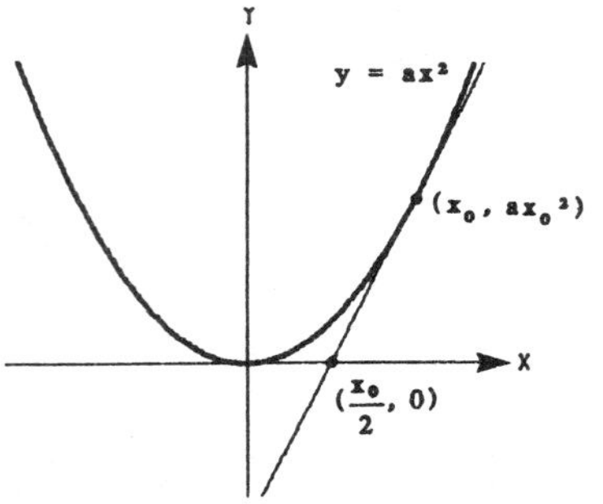

The equation of the tangent line is

$y - ax_0^2 = 2ax_0(x - x_0)$ or $y = 2ax_0x - ax_0^2$

Let $y = 0$: $-ax_0^2 = 2ax_0x - 2ax_0^2$,

$ax_0^2 = 2ax_0x$

Therefore, $\frac{x_0}{2} = x$ is the x-intercept.

52. Without loss of generality, place the coordinate system so that the equation of the parabola is $x^2 = 4py$ and hence $y' = \frac{1}{2p}x$. Therefore, for distinct tangent lines, the slopes are unequal and the lines intersect.

53. Let m_0 be the slope of the one tangent line and therefore $\frac{-1}{m_0}$ is the slope of the second. From the derivative given in Exercise 52, we have

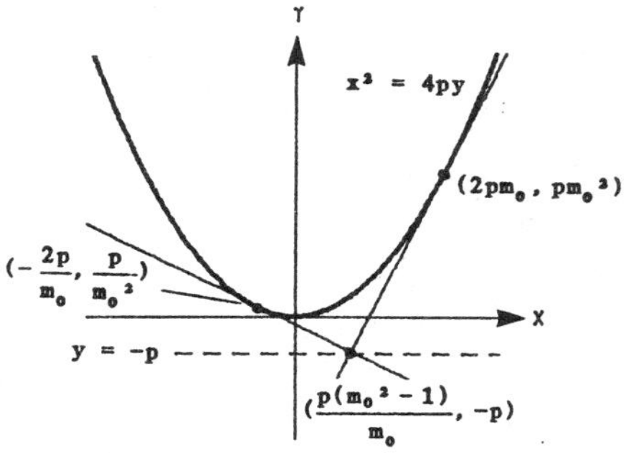

$m_0 = \frac{1}{2p}x_1$ or $x_1 = 2pm_0$

$\frac{-1}{m_0} = \frac{1}{2p}x_2$ or $x_2 = \frac{-2p}{m_0}$

Substituting these values of x into the equation $x^2 = 4py$, we have the coordinates of the points of tangency: $(2pm_0, pm_0^2)$ and $(\frac{-2p}{m_0}, \frac{p}{m_0^2})$ and the equations of the tangent lines are

$(y - pm_0^2) = m_0(x - 2pm_0)$, $(y - \frac{p}{m_0^2}) = \frac{-1}{m_0}(x + \frac{2p}{m_0})$

The point of intersection of these lines is $(\frac{p(m_0^2 - 1)}{m_0}, -p)$ and is on the directrix.

54. The two points of tangency are $(2pm_0, pm_0^2)$ and $\left(\frac{-2p}{m_0}, \frac{p}{m_0^2}\right)$ (from solutions to Exercises 52 and 53). The coordinates of the focus are (0, p). The slope of the line segment from the focus to each point of tangency is $\frac{m_0^2 - 1}{2m_0}$. Therefore, the three points are collinear.

55. Results are dependent on the tangent lines chosen.

10.2 Ellipses

1. $\frac{x^2}{1} + \frac{y^2}{9} = 1$, Center: (0, 0), a = 3, b = 1

   vertical major axis, matches graph (e)

2. $\frac{x^2}{9} + \frac{y^2}{1} = 1$, Center: (0, 0), a = 3, b = 1

   horizontal major axis, matches graph (a)

3. $\frac{x^2}{9} + \frac{y^2}{4} = 1$, Center: (0, 0), a = 3, b = 2

   horizontal major axis, matches graph (c)

4. $\frac{x^2}{9} + \frac{y^2}{9} = 1$, Center: (0, 0), a = b = 3

   circle, matches graph (b)

5. $\frac{(x - 2)^2}{16} + \frac{(y + 1)^2}{4} = 1$, Center: (2, −1), a = 4, b = 2

   horizontal major axis, matches graph (f)

6. $\frac{(x + 2)^2}{4} + \frac{(y + 2)^2}{25} = 1$, Center: (−2, −2), a = 5, b = 2

   vertical major axis, matches graph (d)

7. $\frac{x^2}{25} + \frac{y^2}{16} = 1$

$a^2 = 25, \quad b^2 = 16, \quad c^2 = 9$

Center: (0, 0), Foci: $(\pm 3, 0)$

Vertices: $(\pm 5, 0)$, $e = \frac{3}{5}$

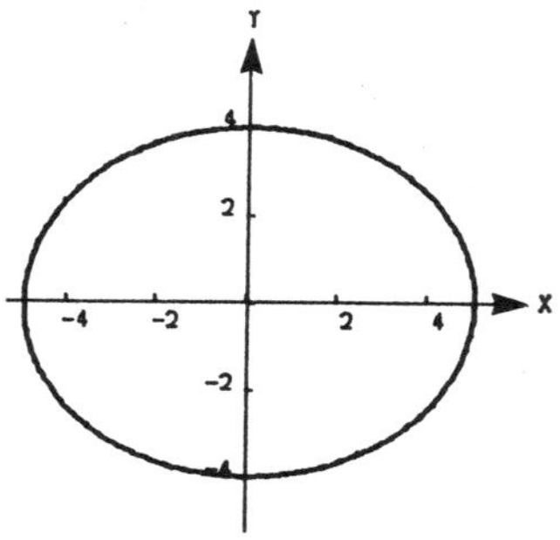

8. $\frac{x^2}{144} + \frac{y^2}{169} = 1$

$a^2 = 169, \quad b^2 = 144, \quad c^2 = 25$

Center: (0, 0), Foci: $(0, \pm 5)$

Vertices: $(0, \pm 13)$, $e = \frac{5}{13}$

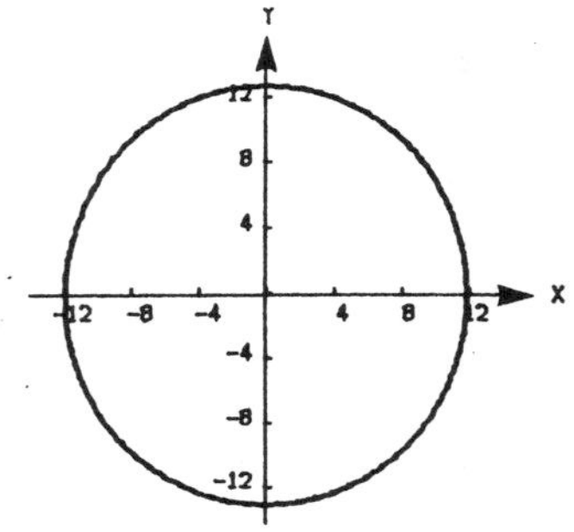

9. $\frac{x^2}{16} + \frac{y^2}{25} = 1$

$a^2 = 25, \quad b^2 = 16, \quad c^2 = 9$

Center: (0, 0), Foci: $(0, \pm 3)$

Vertices: $(0, \pm 5)$, $e = \frac{3}{5}$

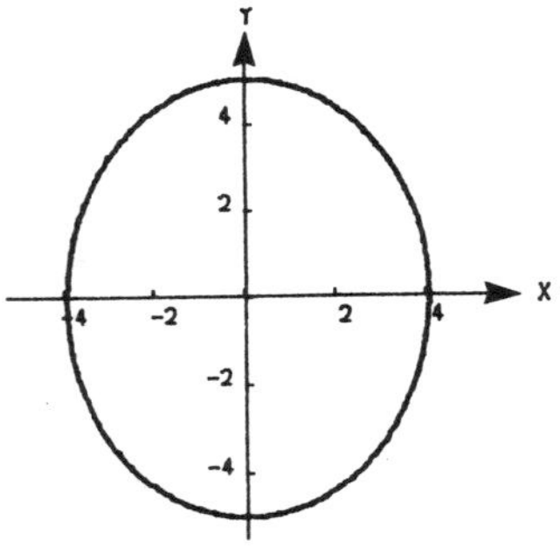

10. $\frac{x^2}{169} + \frac{y^2}{144} = 1$

$a^2 = 169, \quad b^2 = 144, \quad c^2 = 25$

Center: (0, 0), Foci: $(\pm 5, 0)$

Vertices: $(\pm 13, 0)$, $e = \frac{5}{13}$

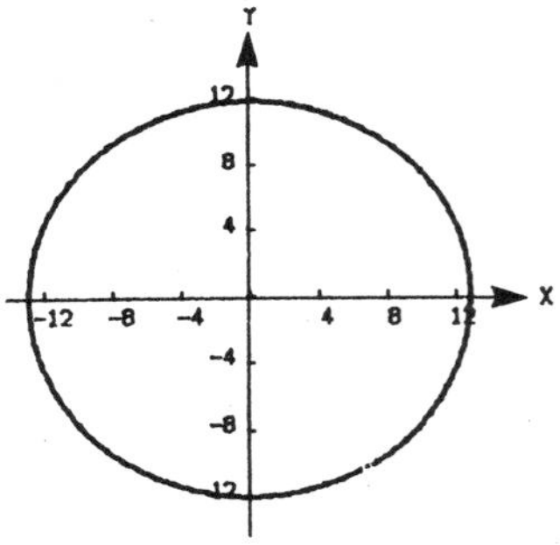

11. $\dfrac{x^2}{9} + \dfrac{y^2}{5} = 1$

$a^2 = 9, \quad b^2 = 5, \quad c^2 = 4$

Center: $(0, 0)$, Foci: $(\pm 2, 0)$

Vertices: $(\pm 3, 0)$, $e = \dfrac{2}{3}$

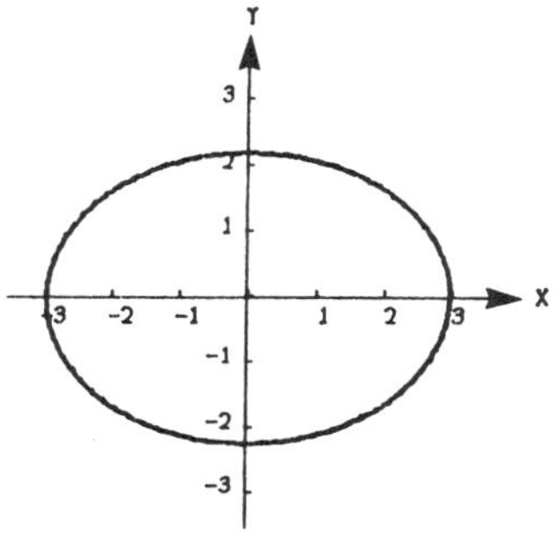

12. $\dfrac{x^2}{28} + \dfrac{y^2}{64} = 1$

$a^2 = 64, \quad b^2 = 28, \quad c^2 = 36$

Center: $(0, 0)$, Foci: $(0, \pm 6)$

Vertices: $(0, \pm 8)$, $e = \dfrac{3}{4}$

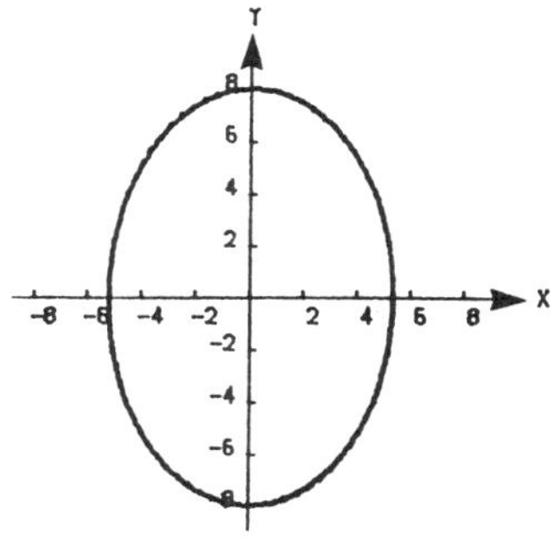

13. $x^2 + 4y^2 = 4, \quad \dfrac{x^2}{4} + \dfrac{y^2}{1} = 1$

$a^2 = 4, \quad b^2 = 1, \quad c^2 = 3$

Center: $(0, 0)$, Foci: $(\pm\sqrt{3}, 0)$

Vertices: $(\pm 2, 0)$, $e = \dfrac{\sqrt{3}}{2}$

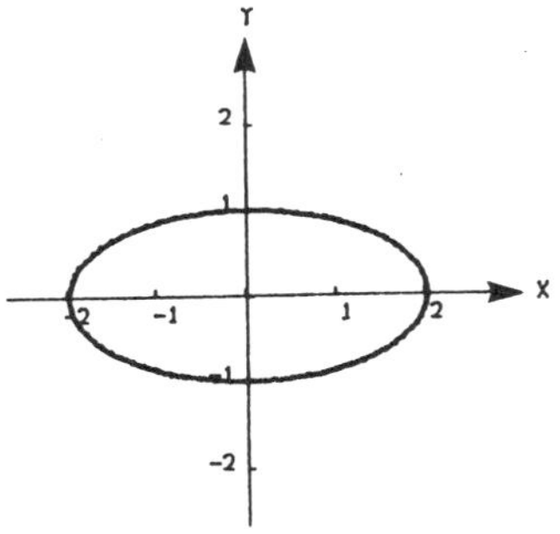

14. $5x^2 + 3y^2 = 15, \quad \dfrac{x^2}{3} + \dfrac{y^2}{5} = 1$

$a^2 = 5, \quad b^2 = 3, \quad c^2 = 2$

Center: $(0, 0)$, Foci: $(0, \pm\sqrt{2})$

Vertices: $(0, \pm\sqrt{5})$, $e = \sqrt{2/5}$

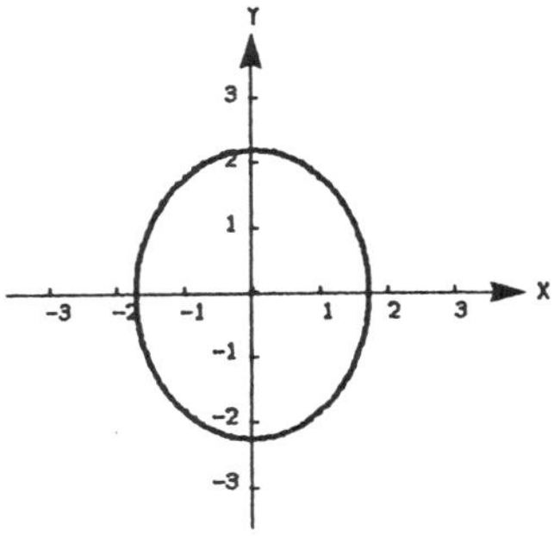

15. $3x^2 + 2y^2 = 6, \quad \frac{x^2}{2} + \frac{y^2}{3} = 1$

$a^2 = 3, \quad b^2 = 2, \quad c^2 = 1$

Center: $(0, 0)$, Foci: $(0, \pm 1)$

Vertices: $(0, \pm\sqrt{3})$

$e = \frac{1}{\sqrt{3}} = \frac{\sqrt{3}}{3}$

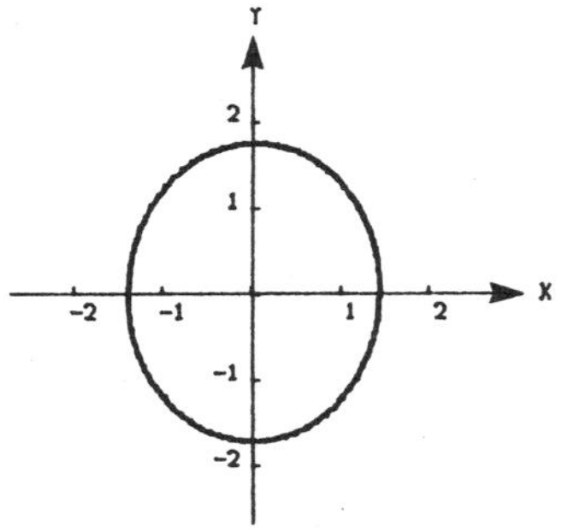

16. $5x^2 + 7y^2 = 70, \quad \frac{x^2}{14} + \frac{y^2}{10} = 1$

$a^2 = 14, \quad b^2 = 10, \quad c^2 = 4$

Center: $(0, 0)$, Foci: $(\pm 2, 0)$

Vertices: $(\pm\sqrt{14}, 0)$, $e = \frac{2}{\sqrt{14}}$

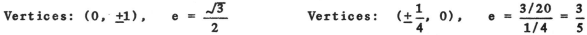

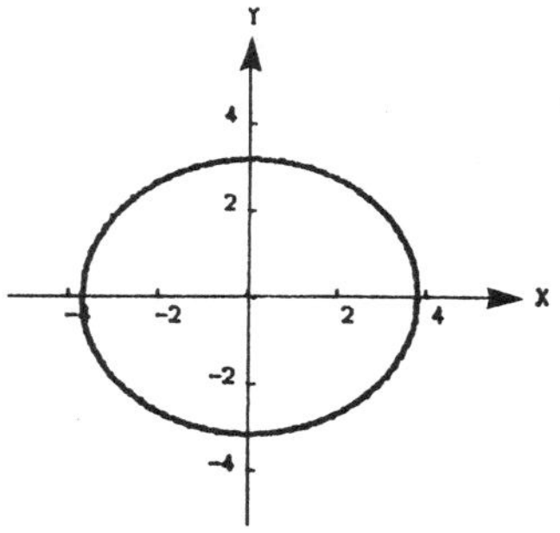

17. $4x^2 + y^2 = 1, \quad \frac{x^2}{1/4} + \frac{y^2}{1} = 1$

$a^2 = 1, \quad b^2 = \frac{1}{4}, \quad c^2 = \frac{3}{4}$

Center: $(0, 0)$, Foci: $(0, \frac{\pm\sqrt{3}}{2})$

Vertices: $(0, \pm 1)$, $e = \frac{\sqrt{3}}{2}$

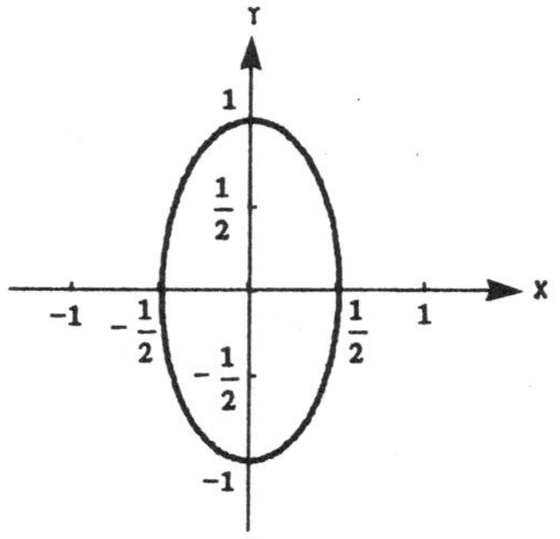

18. $16x^2 + 25y^2 = 1, \quad \frac{x^2}{1/16} + \frac{y^2}{1/25} = 1$

$a^2 = \frac{1}{16}, \quad b^2 = \frac{1}{25}, \quad c^2 = \frac{9}{400}$

Center: $(0, 0)$, Foci: $(\pm\frac{3}{20}, 0)$

Vertices: $(\pm\frac{1}{4}, 0)$, $e = \frac{3/20}{1/4} = \frac{3}{5}$

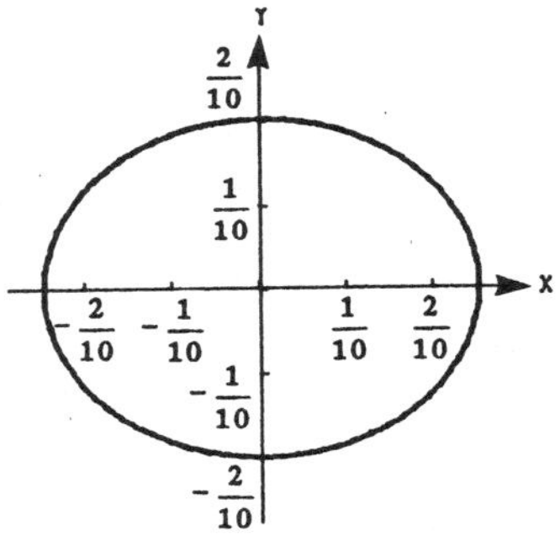

19. $\dfrac{(x-1)^2}{9} + \dfrac{(y-5)^2}{25} = 1$

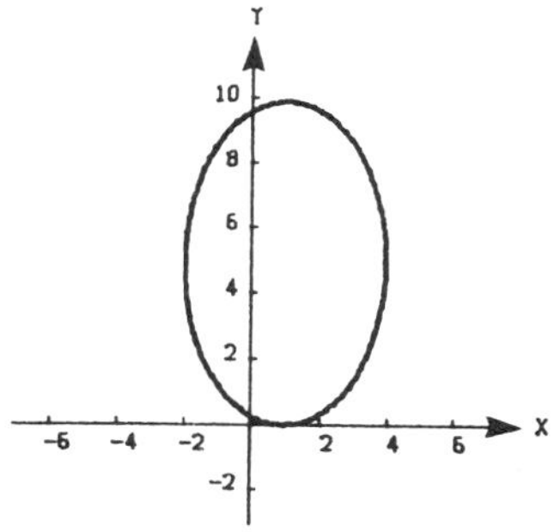

$a^2 = 25, \quad b^2 = 9, \quad c^2 = 16$

Center: $(1, 5)$

Foci: $(1, 9), (1, 1)$

Vertices: $(1, 10), (1, 0)$

$e = \dfrac{4}{5}$

20. $\dfrac{(x+2)^2}{1} + \dfrac{(y+4)^2}{1/4} = 1$

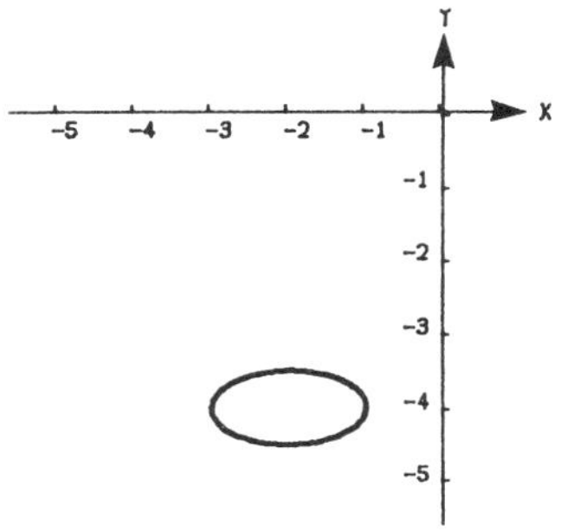

$a^2 = 1, \quad b^2 = \dfrac{1}{4}, \quad c^2 = \dfrac{3}{4}$

Center: $(-2, -4)$

Foci: $\left(-2 \pm \dfrac{\sqrt{3}}{2}, -4\right)$

Vertices: $(-1, -4), (-3, -4)$

$e = \dfrac{\sqrt{3}}{2}$

21. $9x^2 + 4y^2 + 36x - 24y + 36 = 0$

$9(x^2 + 4x + 4) + 4(y^2 - 6y + 9)$

$= -36 + 36 + 36$

$\dfrac{(x+2)^2}{4} + \dfrac{(y-3)^2}{9} = 1$

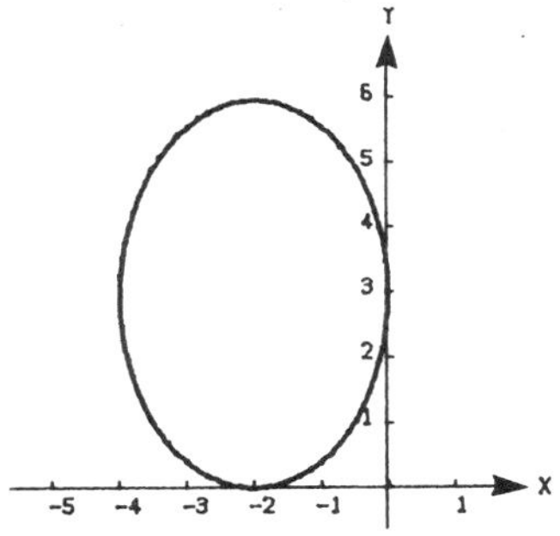

$a^2 = 9, \quad b^2 = 4, \quad c^2 = 5$

Center: $(-2, 3)$, Foci: $(-2, 3 \pm \sqrt{5})$

Vertices: $(-2, 6), (-2, 0)$, $e = \dfrac{\sqrt{5}}{3}$

22. $9x^2 + 4y^2 - 36x + 8y + 31 = 0$

$9(x^2 - 4x + 4) + 4(y^2 + 2y + 1) = -31 + 36 + 4$

$\frac{(x - 2)^2}{1} + \frac{(y + 1)^2}{9/4} = 1$

$a^2 = \frac{9}{4}, \quad b^2 = 1, \quad c^2 = \frac{5}{4}$

Center: $(2, -1)$, Foci: $(2, -1 \pm \frac{\sqrt{5}}{2})$

Vertices: $(2, -1 \pm \frac{3}{2})$, $e = \frac{\sqrt{5}}{3}$

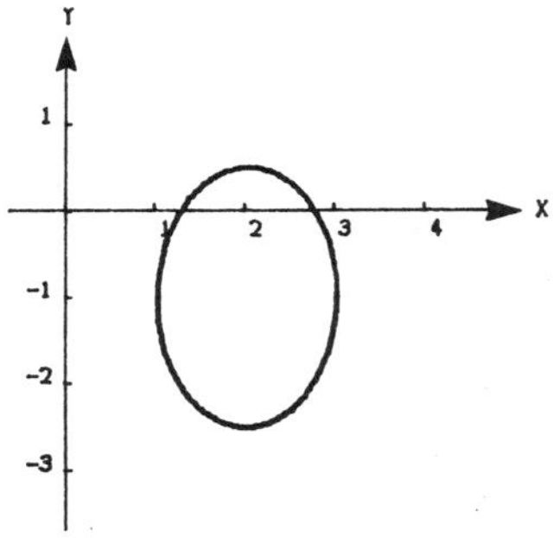

23. $16x^2 + 25y^2 - 32x + 50y + 31 = 0$

$16(x^2 - 2x + 1) + 25(y^2 + 2y + 1)$

$= -31 + 16 + 25$

$\frac{(x - 1)^2}{5/8} + \frac{(y + 1)^2}{2/5} = 1$

$a^2 = \frac{5}{8}, \quad b^2 = \frac{2}{5}, \quad c^2 = \frac{9}{40}$

Center: $(1, -1)$, Foci: $(1 \pm \frac{3\sqrt{10}}{20}, -1)$

Vertices: $(1 \pm \frac{\sqrt{10}}{4}, -1)$, $e = \frac{3}{5}$

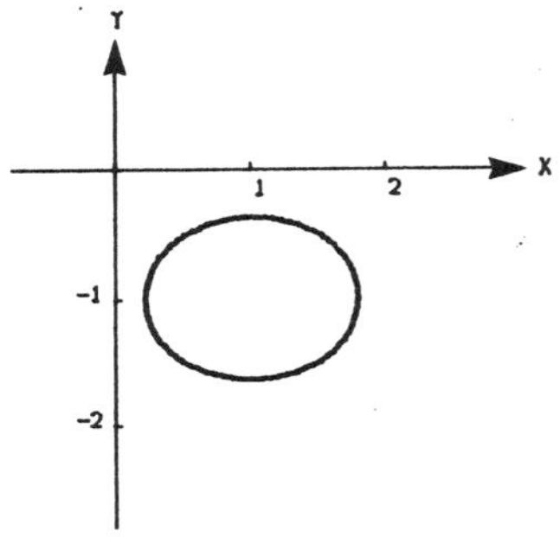

24. $9x^2 + 25y^2 - 36x - 50y + 61 = 0$

$9(x^2 - 4x + 4) + 25(y^2 - 2y + 1)$

$= -61 + 36 + 25$

$9(x - 2)^2 + 25(y - 1)^2 = 0$

Degenerate ellipse with center $(2, 1)$

as the only point

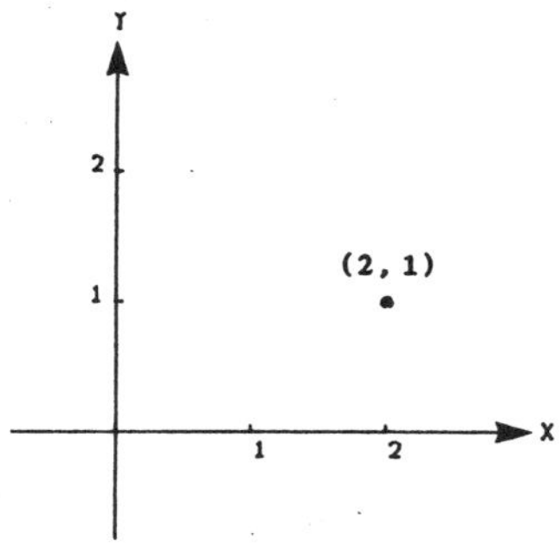

25. $12x^2 + 20y^2 - 12x + 40y - 37 = 0$

$12(x^2 - x + \frac{1}{4}) + 20(y^2 + 2y + 1)$

$= 37 + 3 + 20$

$\frac{[x - (1/2)]^2}{5} + \frac{(y + 1)^2}{3} = 1$

$a^2 = 5, \quad b^2 = 3, \quad c^2 = 2$

Center: $(\frac{1}{2}, -1)$, Foci: $(\frac{1}{2} \pm \sqrt{2}, -1)$

Vertices: $(\frac{1}{2} \pm \sqrt{5}, -1)$, $e = \frac{\sqrt{10}}{5}$

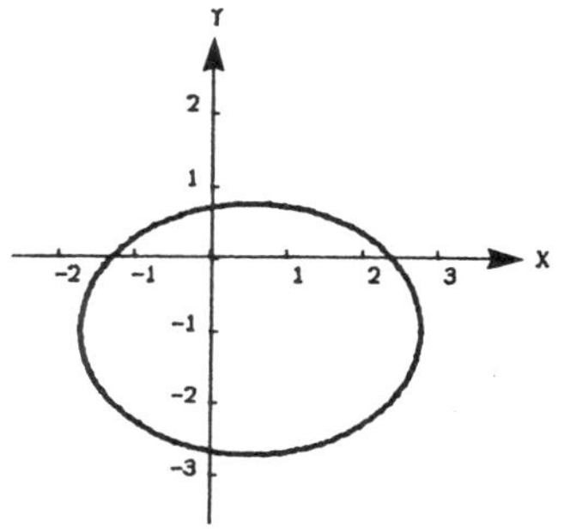

26. $36x^2 + 9y^2 + 48x - 36y + 43 = 0$

$36(x^2 + \frac{4}{3}x + \frac{4}{9}) + 9(y^2 - 4y + 4)$

$= -43 + 16 + 36$

$\frac{[x + (2/3)]^2}{1/4} + \frac{(y - 2)^2}{1} = 1$

$a^2 = 1, \quad b^2 = \frac{1}{4}, \quad c^2 = \frac{3}{4}$

Center: $(-\frac{2}{3}, 2)$, Foci: $(-\frac{2}{3}, 2 \pm \frac{\sqrt{3}}{2})$

Vertices: $(-\frac{2}{3}, 2 \pm 1)$, $e = \frac{\sqrt{3}}{2}$

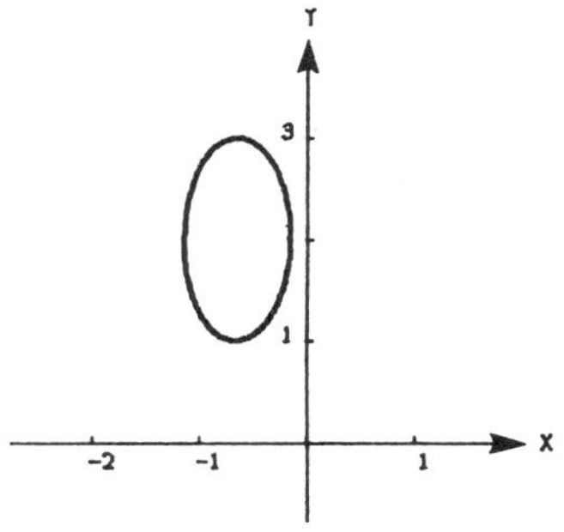

27. $\frac{x^2}{9} + \frac{y^2}{5} = 1$, since $a^2 = 9$, $b^2 = 5$, $c^2 = 4$

horizontal major axis

28. $\frac{x^2}{4} + \frac{y^2}{9/4} = 1$, $\frac{x^2}{4} + \frac{4y^2}{9} = 1$, since $a^2 = 4$, $b^2 = \frac{9}{4}$

horizontal major axis

29. $\frac{x^2}{25} + \frac{y^2}{16} = 1$, since $a^2 = 25$, $\frac{c}{5} = \frac{3}{5}$, $c^2 = 9$, $b^2 = 16$

horizontal major axis

30. $\frac{x^2}{48} + \frac{y^2}{64} = 1$, since $a^2 = 64$, $\frac{c}{8} = \frac{1}{2}$, $c^2 = 16$, $b^2 = 48$

vertical major axis

31. $\frac{(x-2)^2}{4} + \frac{(y-2)^2}{1} = 1$, since center = (2, 2), $a^2 = 4$, $b^2 = 1$

horizontal major axis

32. $\frac{x^2}{16} + \frac{y^2}{12} = 1$, since $a^2 = 16$, $c^2 = 4$, $b^2 = 12$

horizontal major axis

33. $\frac{(x-3)^2}{9} + \frac{(y-5)^2}{16} = 1$, since center = (3, 5), $a^2 = 16$, $b^2 = 9$

vertical major axis

34. Since the major axis is horizontal, $\frac{x^2}{a^2} + \frac{y^2}{b^2} = 1$

Substituting the values of the coordinates of the given points

into this equation, we have: $\frac{9}{a^2} + \frac{1}{b^2} = 1$, $\frac{16}{a^2} = 1$

The solution to this system is $a^2 = 16$, $b^2 = \frac{16}{7}$

Therefore, $\frac{x^2}{16} + \frac{y^2}{16/7} = 1$, $x^2 + 7y^2 = 16$

35. Since the sum of the distances from the foci is 14, we have:

$2a = 14$, $a = 7$, $a^2 = 49$, $c^2 = 25$, $b^2 = 24$, $\frac{x^2}{24} + \frac{y^2}{49} = 1$

36. From the sketch, we can see that

$h = 1$, $k = 2$, $a = 4$, $b = 2$

$\frac{(x-1)^2}{4} + \frac{(y-2)^2}{16} = 1$

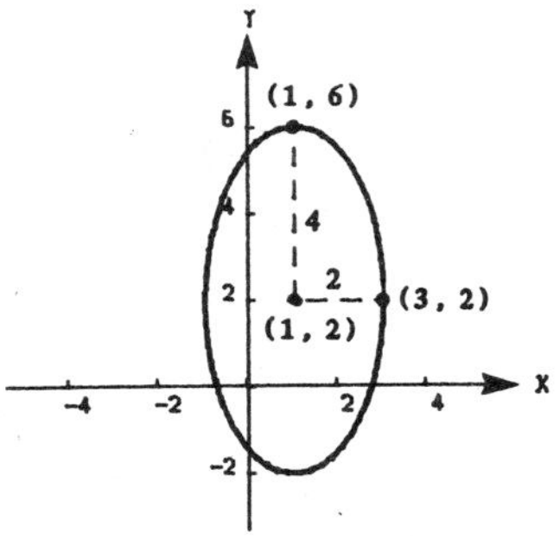

37. Suppose the ellipse is centered at the origin with its major axis horizontal. Then $\frac{x^2}{a^2} + \frac{y^2}{b^2} = 1$. Solving for y when x = c, we have

$\frac{c^2}{a^2} + \frac{y^2}{b^2} = 1$, $y^2 = \frac{b^2}{a^2}(a^2 - c^2) = \frac{b^4}{a^2}$ (since $a^2 - c^2 = b^2$)

Therefore, the endpoints of the latus rectum are $(c, \frac{b^2}{a})$, $(c, -\frac{b^2}{a})$ and its length is $\frac{2b^2}{a}$.

38.(a) $\frac{x^2}{4} + \frac{y^2}{1} = 1$

$a^2 = 4$, $b^2 = 1$, $c^2 = 3$

Endpoints of latus recta:

$(\pm\sqrt{3}, \pm\frac{1}{2})$

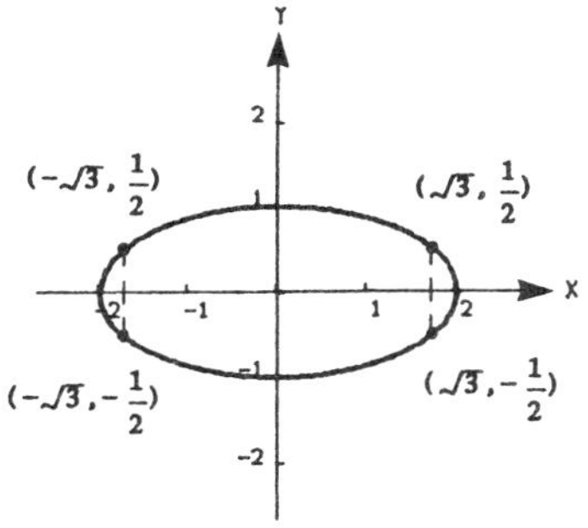

(b) $5x^2 + 3y^2 = 15$

$a^2 = 5$, $b^2 = 3$, $c^2 = 2$

Endpoints of latus recta:

$(\frac{\pm 3\sqrt{5}}{5}, \pm\sqrt{2})$

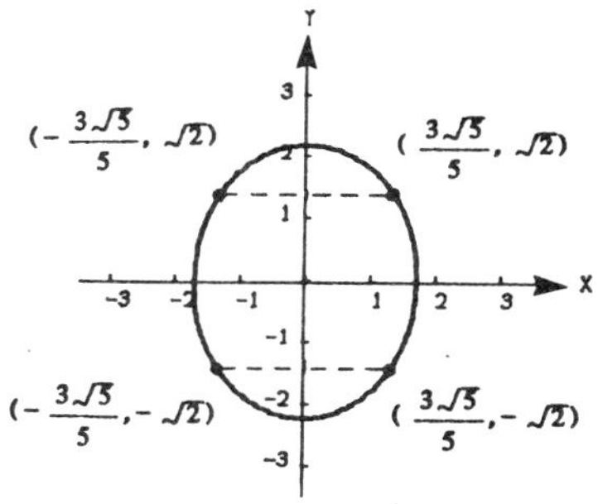

39. $16x^2 + 9y^2 + 96x + 36y + 36 = 0$, $32x + 18yy' + 96 + 36y' = 0$

$y'(18y + 36) = -(32x + 96)$, $y' = \frac{-(32x + 96)}{18y + 36}$

$y' = 0$ when $x = -3$, y' is undefined when $y = -2$

At $x = -3$: $y = 2$ or -6, Endpoints of major axis: $(-3, 2)$, $(-3, -6)$

At $y = -2$: $x = 0$ or -6, Endpoints of minor axis: $(0, -2)$, $(-6, -2)$

40. $9x^2 + 4y^2 + 36x - 24y + 36 = 0$, $18x + 8yy' + 36 - 24y' = 0$

$(8y - 24)y' = -(18x + 36)$, $y' = \frac{-(18x + 36)}{8y - 24}$

$y' = 0$ when $x = -2$, y' undefined when $y = 3$

At $x = -2$: $y = 0$ or 6, Endpoints of major axis: $(-2, 0)$, $(-2, 6)$

At $y = 3$, $x = 0$ or -4, Endpoints of minor axis: $(0, 3)$, $(-4, 3)$

41. $\frac{x^2}{10^2} + \frac{y^2}{5^2} = 1$, $\frac{2x}{10^2} + \frac{2yy'}{5^2} = 0$, $y' = \frac{-5^2x}{10^2y}$. At $(-8, 3)$,

$y' = \frac{5^2(8)}{10^2(3)} = \frac{2}{3}$. The equation of the tangent line is $y - 3 = \frac{2}{3}(x + 8)$

and it will cross the y-axis when $x = 0$ and $y = \frac{2}{3}(8) + 3 = \frac{25}{3}$

42. $\frac{x^2}{16} + \frac{y^2}{25} = 1$, $\frac{2x}{16} + \frac{2yy'}{25} = 0$, $y' = -\frac{25x}{16y}$

At $(3, \frac{5\sqrt{7}}{4})$, $y' = -\frac{75}{20\sqrt{7}} = -\frac{15}{4\sqrt{7}}$

The equation of the tangent line is $y - \frac{5\sqrt{7}}{4} = -\frac{15}{4\sqrt{7}}(x - 3)$

and it will cross the y-axis when $x = 0$ and $y = \frac{45}{4\sqrt{7}} + \frac{5\sqrt{7}}{4} \approx 7.559$

43. $C = 4a \int_0^{\pi/2} \sqrt{1 - e^2 \sin^2\theta}\, d\theta$

For $\frac{x^2}{9} + \frac{y^2}{16} = 1$, $a = 4$, $b = 3$, $c = \sqrt{7}$ and $e = \frac{\sqrt{7}}{4}$

$C = 4(4) \int_0^{\pi/2} \sqrt{1 - (\sqrt{7}/4)^2 \sin^2\theta}\, d\theta = 16 \int_0^{\pi/2} \sqrt{1 - 7/16 \sin^2\theta}\, d\theta$

$C \approx 16(\frac{\pi}{48})[1 + 4(0.9916) + 2(0.9674) + 4(0.9300) + 2(0.8839)$
$+ 4(0.8352) + 2(0.7916) + 4(0.7610) + (0.7500)] \approx 22.10$

44. $C = 4a \int_0^{\pi/2} \sqrt{1 - e^2 \sin^2\theta}\, d\theta$

For $\frac{x^2}{9} + \frac{y^2}{1} = 1$, $a = 3$, $b = 1$, $c = 2\sqrt{2}$ and $e = \frac{2\sqrt{2}}{3}$

$C = 4(3) \int_0^{\pi/2} \sqrt{1 - (\frac{2\sqrt{2}}{3})^2 \sin^2\theta}\, d\theta = 12 \int_0^{\pi/2} \sqrt{1 - (8/9) \sin^2\theta}\, d\theta$

$C \approx 12(\frac{\pi}{48})[1 + 4(0.9829) + 2(0.9326) + 4(0.8518) + 2(0.7454)$
$+ 4(0.6209) + 2(0.4912) + 4(0.3807) + 0.3333] \approx 13.37$

45. $C = 4a \int_0^{\pi/2} \sqrt{1 - e^2 \sin^2 \theta}\, d\theta$

For $\frac{x^2}{3} + \frac{y^2}{2} = 1$, $a = \sqrt{3}$, $b = \sqrt{2}$, $c = 1$ and $e = \frac{1}{\sqrt{3}}$

$C = 4(\sqrt{3}) \int_0^{\pi/2} \sqrt{1 - (1/3) \sin^2 \theta}\, d\theta$

$C \approx (4\sqrt{3})(\frac{\pi}{48})[1 + 4(0.9936) + 2(0.9753) + 4(0.9472) + 2(0.9129)$
$+ 4(0.8772) + 2(0.8459) + 4(0.8242) + 0.8165] \approx 9.91$

46. $C = 4a \int_0^{\pi/2} \sqrt{1 - e^2 \sin^2 \theta}\, d\theta$, $\frac{(x - 3)^2}{4} + \frac{(y + 1)^2}{3} = 1$ has the same circumference as $\frac{x^2}{4} + \frac{y^2}{3} = 1$, $a = 2$, $b = \sqrt{3}$, $c = 1$ and $e = \frac{1}{2}$

$C = 4(2) \int_0^{\pi/2} \sqrt{1 - (1/4) \sin^2 \theta}\, d\theta$

$C \approx 8(\frac{\pi}{48})[1 + 4(0.9952) + 2(0.9815) + 4(0.9606) + 2(0.9354)$
$+ 4(0.9095) + 2(0.8869) + 4(0.8715) + 0.8660] \approx 11.74$

47. (a) $A = 4 \int_0^2 \frac{1}{2}\sqrt{4 - x^2}\, dx = \left[x\sqrt{4 - x^2} + 4 \arcsin (\frac{x}{2}) \right]_0^2 = 2\pi$

(b) <u>Disc</u>

$V = 2\pi \int_0^2 \frac{1}{4}(4 - x^2)\, dx = \frac{1}{2}\pi \left[4x - \frac{1}{3}x^3 \right]_0^2 = \frac{8\pi}{3}$

$y = \frac{1}{2}\sqrt{4 - x^2}$, $y' = \frac{-x}{2\sqrt{4 - x^2}}$

$\sqrt{1 + (y')^2} = \sqrt{1 + \frac{x^2}{16 - 4x^2}} = \frac{\sqrt{16 - 3x^2}}{4y}$

$S = 2(2\pi) \int_0^2 y(\frac{\sqrt{16 - 3x^2}}{4y})\, dx$

$= \frac{\pi}{2\sqrt{3}} \left[\sqrt{3}x\sqrt{16 - 3x^2} + 16 \arcsin (\frac{\sqrt{3}x}{4}) \right]_0^2 = \frac{2\pi}{9}(9 + 4\sqrt{3}\pi)$

47. (c) Shell

$$V = 2\pi \int_0^2 x\sqrt{4 - x^2}\, dx = -\pi \int_0^2 -2x(4 - x^2)^{1/2}\, dx$$

$$= -\frac{2\pi}{3}(4 - x^2)^{3/2}\Big]_0^2 = \frac{16\pi}{3}$$

$$x = 2\sqrt{1 - y^2}, \qquad x' = \frac{-2y}{\sqrt{1 - y^2}}$$

$$\sqrt{1 + (x')^2} = \sqrt{1 + \frac{4y^2}{1 - y^2}} = \frac{\sqrt{1 + 3y^2}}{\sqrt{1 - y^2}}$$

$$S = 2(2\pi) \int_0^1 2\sqrt{1 - y^2}\, \frac{\sqrt{1 + 3y^2}}{\sqrt{1 - y^2}}\, dy = 8\pi \int_0^1 \sqrt{1 + 3y^2}\, dy$$

$$= \frac{8\pi}{2\sqrt{3}} \Big[\sqrt{3}y\sqrt{1 + 3y^2} + \ln\left|\sqrt{3}y + \sqrt{1 + 3y^2}\right|\Big]_0^1$$

$$= \frac{4\pi}{3}[6 + \sqrt{3}\ln(2 + \sqrt{3})] \approx 34.69$$

48. (a) $$A = 4\int_0^4 \frac{3}{4}\sqrt{16 - x^2}\, dx = \frac{3}{2}\Big[x\sqrt{16 - x^2} + 16 \arcsin\frac{x}{4}\Big]_0^4 = 12\pi$$

(b) Disc

$$V = 2\pi \int_0^4 \frac{9}{16}(16 - x^2)\, dx = \frac{9\pi}{8}(16x - \frac{1}{3}x^3)\Big]_0^4 = 48\pi$$

$$y = \frac{3}{4}\sqrt{16 - x^2}, \quad y' = \frac{-3x}{4\sqrt{16 - x^2}}, \quad \sqrt{1 + (y')^2} = \sqrt{1 + \frac{9x^2}{16(16 - x^2)}}$$

$$S = 2(2\pi)\int_0^4 \frac{3}{4}\sqrt{16 - x^2}\sqrt{\frac{16(16 - x^2) + 9x^2}{16(16 - x^2)}}\, dx$$

$$= 4\pi \int_0^4 \frac{3}{4}\sqrt{16 - x^2}\, \frac{\sqrt{256 - 7x^2}}{4\sqrt{16 - x^2}}\, dx = \frac{3\pi}{4}\int_0^4 \sqrt{256 - 7x^2}\, dx$$

$$= \frac{3\pi}{8\sqrt{7}}\Big[\sqrt{7}x\sqrt{256 - 7x^2} + 256 \arcsin\frac{\sqrt{7}x}{16}\Big]_0^4$$

$$= \frac{3\pi}{8\sqrt{7}}(48\sqrt{7} + 256 \arcsin\frac{\sqrt{7}}{4}) \approx 138.93$$

48. (c) <u>Shell</u>

$$V = 4\pi \int_0^4 x\left[\frac{3}{4}\sqrt{16 - x^2}\right]dx = 3\pi(-\frac{1}{2})(\frac{2}{3})(16 - x^2)^{3/2}\Big]_0^4 = 64\pi$$

$$x = \frac{4}{3}\sqrt{9 - y^2}, \quad x' = \frac{-4y}{3\sqrt{9 - y^2}}, \quad \sqrt{1 + (x')^2} = \sqrt{1 + \frac{16y^2}{9(9 - y^2)}}$$

$$S = 2(2\pi)\int_0^3 \frac{4}{3}\sqrt{9 - y^2}\sqrt{\frac{9(9 - y^2) + 16y^2}{9(9 - y^2)}}\,dy$$

$$= 4\pi \int_0^3 \frac{4}{9}\sqrt{81 + 7y^2}\,dy$$

$$= \frac{16}{9}(\frac{\pi}{2\sqrt{7}})\left[\sqrt{7}\,y\sqrt{81 + 7y^2} + 81\ln\left|\sqrt{7}\,y + \sqrt{81 + 7y^2}\right|\right]_0^3$$

$$= \frac{8\pi}{9\sqrt{7}}[3\sqrt{7}(12) + 81\ln(3\sqrt{7} + 12) - 81\ln 9] \approx 168.53$$

49. For $\frac{x^2}{a^2} + \frac{y^2}{b^2} = 1$, we have $y = \frac{b}{a}\sqrt{a^2 - x^2}$

Let (x, y) be the point in Quadrant I where the rectangle meets the ellipse. The dimensions of the rectangle are:

Length = 2x, Width = $2y = \frac{2b}{a}\sqrt{a^2 - x^2}$

Then $A = lw = 2x\left[2(\frac{b}{a})\sqrt{a^2 - x^2}\right]$

$$= \frac{4b}{a}[x\sqrt{a^2 - x^2}]$$

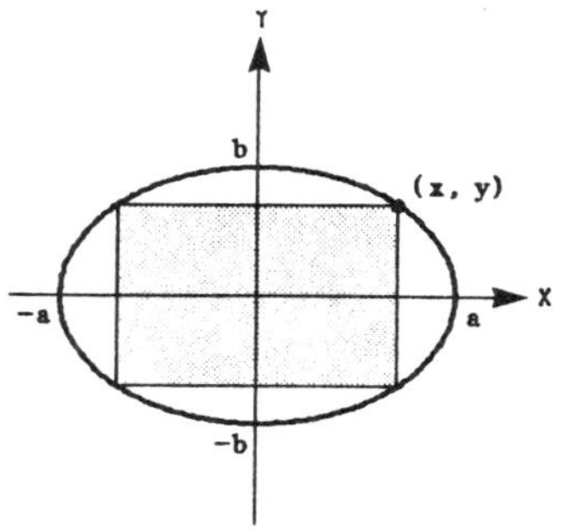

$$A' = \frac{4b}{a}\left[\frac{-x^2}{\sqrt{a^2 - x^2}} + \sqrt{a^2 - x^2}\right]$$

$$= \frac{4b}{a}\left[\frac{a^2 - 2x^2}{\sqrt{a^2 - x^2}}\right]$$

$A' = 0$ when $x = \frac{a}{\sqrt{2}}$

Thus the dimensions of the rectangle of maximum area are:

Length = $2x = \sqrt{2}\,a$, Width = $\frac{2b}{a}\sqrt{a^2 - x^2} = \sqrt{2}\,b$

50. Area of a cross section:

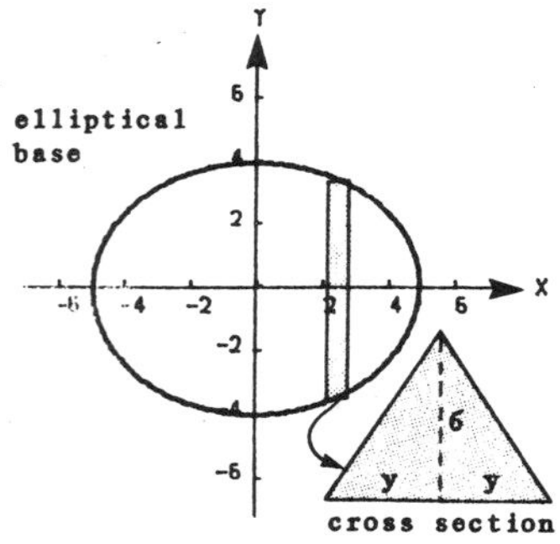

$\frac{1}{2}bh = \frac{1}{2}(2y)6 = 6y$

$A(x) = 6y = 6\left[\frac{4}{5}\sqrt{25 - x^2}\right] dx$

$$V = \int_{-5}^{5} \frac{24}{5}\sqrt{25 - x^2}\, dx$$

$$= \frac{48}{5}\int_{0}^{5} \sqrt{25 - x^2}\, dx$$

$$= \frac{24}{5}\left[x\sqrt{25 - x^2} + 25 \arcsin \frac{x}{5}\right]_0^5 = 60\pi$$

51. $\frac{x^2}{a^2} + \frac{y^2}{b^2} = 1, \quad \frac{2x}{a^2} + \frac{2y}{b^2}y' = 0, \quad y' = \frac{-b^2x}{a^2y}$

The slope of the tangent line at (x_0, y_0) is $\frac{-b^2x_0}{a^2y_0}$

The equation of the tangent line is: $y - y_0 = \frac{-b^2x_0}{a^2y_0}(x - x_0)$

$$a^2y_0y - a^2y_0^2 = -b^2x_0x + b^2x_0^2, \quad b^2x_0x + a^2y_0y = b^2x_0^2 + a^2y_0^2$$

$$\frac{x_0}{a^2}x + \frac{y_0}{b^2}y = \frac{b^2x_0^2 + a^2y_0^2}{a^2b^2} = \frac{x_0^2}{a^2} + \frac{y_0^2}{b^2} = 1$$

52. $\frac{x^2}{a^2} + \frac{y^2}{b^2} = 1, \quad \frac{x^2}{a^2} + \frac{y^2}{a^2(b^2/a^2)} = 1, \quad \frac{x^2}{a^2} + \frac{y^2}{a^2(a^2 - c^2)/a^2} = 1$

$\frac{x^2}{a^2} + \frac{y^2}{a^2(1 - e^2)} = 1$

As $e \to 0$, $1 - e^2 \to 1$ and we have $\frac{x^2}{a^2} + \frac{y^2}{a^2} = 1$ or the circle

$x^2 + y^2 = a^2$.

53. $\frac{x^2}{a^2} + \frac{y^2}{b^2} = 1, \quad \frac{(tx)^2}{a^2} + \frac{(ty)^2}{b^2} = 1, \quad \frac{x^2}{a^2/t^2} + \frac{y^2}{b^2/t^2} = 1$

$$e = \frac{\sqrt{(a^2/t^2) - (b^2/t^2)}}{a/t} = \frac{\sqrt{a^2 - b^2}}{a}$$

54. From the sketch we have:

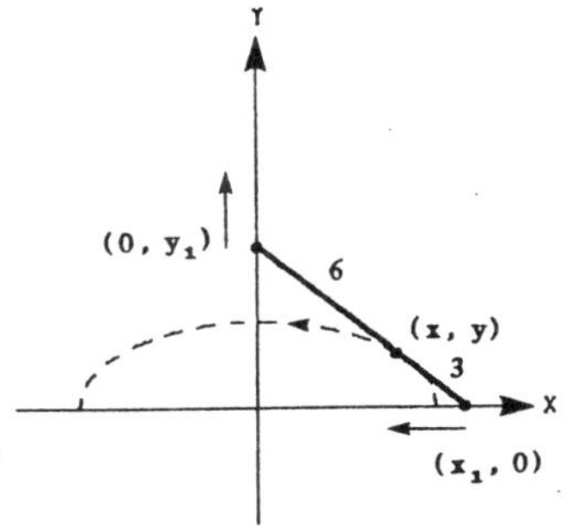

(1) $(x - 0)^2 + (y - y_1)^2 = 36$

(2) $(x - x_1)^2 + (y - 0)^2 = 9$

(3) $\dfrac{y_1 - y}{y} = \dfrac{x}{x_1 - x}$

From equations (1) and (3) we have

$x^2 + \dfrac{x^2y^2}{(x_1 - x)^2} = 36, \quad x^2 + \dfrac{x^2y^2}{9 - y^2} = 36$

[since $(x - x_1)^2 = 9 - y^2$ from equation 2]

$x^2(9 - y^2) + x^2y^2 = 36(9) - 36y^2, \quad 9x^2 + 36y^2 = 36(9), \quad \dfrac{x^2}{36} + \dfrac{y^2}{9} = 1$

55. $a = \dfrac{5}{2}, \quad b = 2, \quad c = \sqrt{(\dfrac{5}{2})^2 - (2)^2} = \dfrac{3}{2}$

The tacks should be placed 1.5' from the center.

56. $e = \dfrac{c}{a}, \quad 0.0167 = \dfrac{c}{14{,}957{,}000}, \quad c \approx 249{,}782$

Least distance: $a - c = 14{,}707{,}218$ kilometers

Greatest distance: $a + c = 15{,}206{,}782$ kilometers

57. $e = \dfrac{c}{a}$

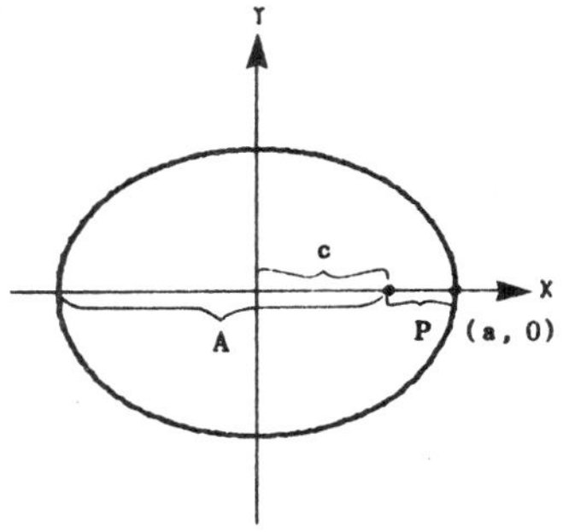

$A + P = 2a, \quad a = \dfrac{A + P}{2}$

$c = a - P = \dfrac{A + P}{2} - P = \dfrac{A - P}{2}$

$e = \dfrac{c}{a} = \dfrac{(A - P)/2}{(A + P)/2} = \dfrac{A - P}{A + P}$

58. $A = 583 + 4000 = 4583, \quad P = 132 + 4000 = 4132$

$e = \dfrac{4583 - 4132}{4583 + 4132} = \dfrac{451}{8715} = 0.052$

10.3 Hyperbolas

1. $\dfrac{x^2}{9} - \dfrac{y^2}{4} = 1,$ Center: $(0, 0)$, $a = 3$, $b = 2$

horizontal transverse axis, matches graph (e)

2. $\dfrac{y^2}{9} - \dfrac{x^2}{4} = 1,$ Center: $(0, 0)$, $a = 3$, $b = 2$

vertical transverse axis, matches graph (a)

3. $\dfrac{y^2}{1} - \dfrac{x^2}{16} = 1,$ Center: $(0, 0)$, $a = 1$, $b = 4$

vertical transverse axis, matches graph (f)

4. $\dfrac{y^2}{16} - \dfrac{x^2}{1} = 1,$ Center: $(0, 0)$, $a = 4$, $b = 1$

vertical transverse axis, matches graph (c)

5. $\dfrac{(x - 2)^2}{9} - \dfrac{y^2}{4} = 1,$ Center: $(2, 0)$, $a = 3$, $b = 2$

horizontal transverse axis, matches graph (d)

6. $\dfrac{(x + 1)^2}{16} - \dfrac{(y - 3)^2}{9} = 1,$ Center: $(-1, 3)$, $a = 4$, $b = 3$

horizontal transverse axis, matches graph (b)

7. $x^2 - y^2 = 1$

$a = 1$, $b = 1$, $c = \sqrt{2}$

Center: $(0, 0)$

Vertices: $(\pm 1, 0)$

Foci: $(\pm\sqrt{2}, 0)$

Asymptotes: $y = \pm x$

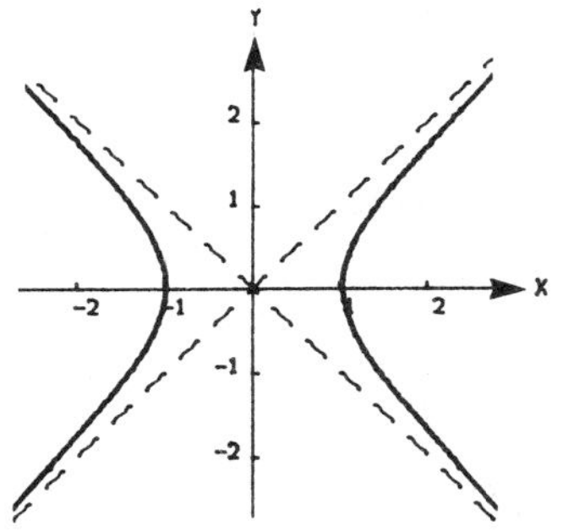

8. $\dfrac{x^2}{9} - \dfrac{y^2}{16} = 1$

$a = 3$, $b = 4$, $c = 5$

Center: $(0, 0)$

Vertices: $(\pm 3, 0)$

Foci: $(\pm 5, 0)$

Asymptotes: $y = \pm \dfrac{4}{3}x$

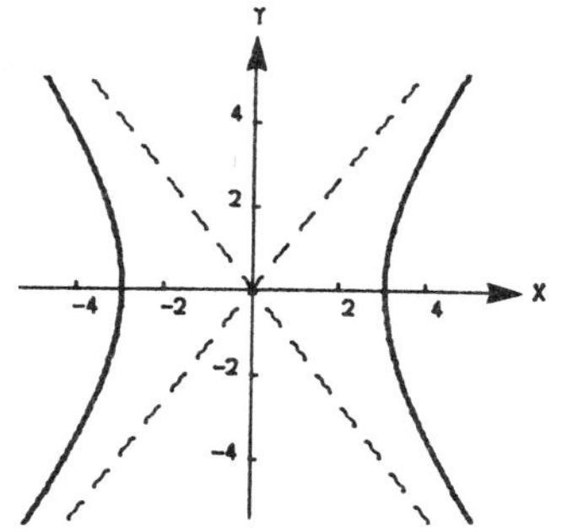

9. $\dfrac{y^2}{1} - \dfrac{x^2}{4} = 1$

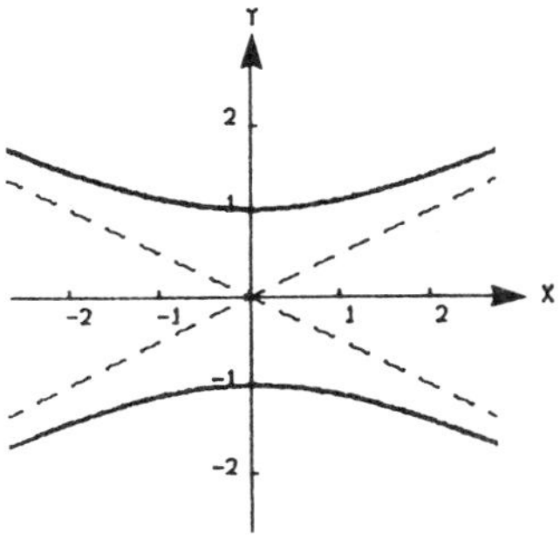

$a = 1, \quad b = 2, \quad c = \sqrt{5}$

Center: (0, 0)

Vertices: $(0, \pm 1)$

Foci: $(0, \pm\sqrt{5})$

Asymptotes: $y = \pm\dfrac{1}{2}x$

10. $\dfrac{y^2}{9} - \dfrac{x^2}{1} = 1$

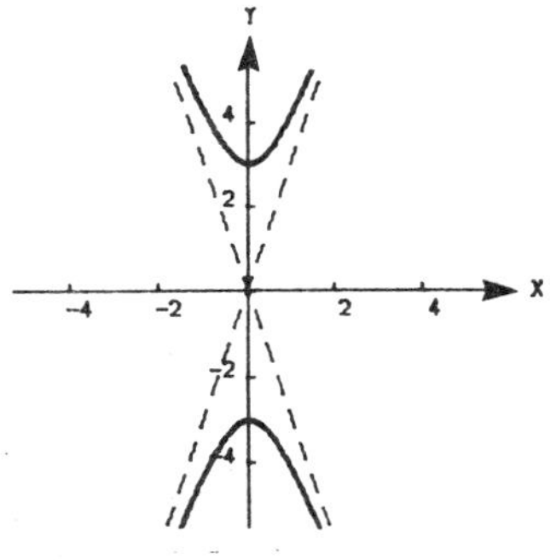

$a = 3, \quad b = 1, \quad c = \sqrt{10}$

Center: (0, 0)

Vertices: $(0, \pm 3)$

Foci: $(0, \pm\sqrt{10})$

Asymptotes: $y = \pm 3x$

11. $\dfrac{y^2}{25} - \dfrac{x^2}{144} = 1$

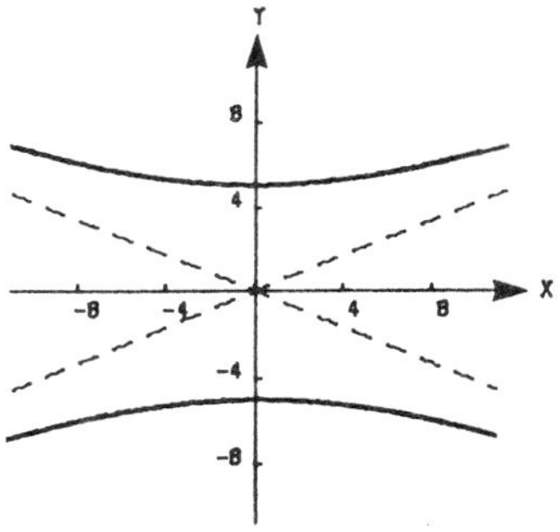

$a = 5, \quad b = 12, \quad c = 13$

Center: (0, 0)

Vertices: $(0, \pm 5)$

Foci: $(0, \pm 13)$

Asymptotes: $y = \pm\dfrac{5}{12}x$

12. $\dfrac{x^2}{36} - \dfrac{y^2}{4} = 1$

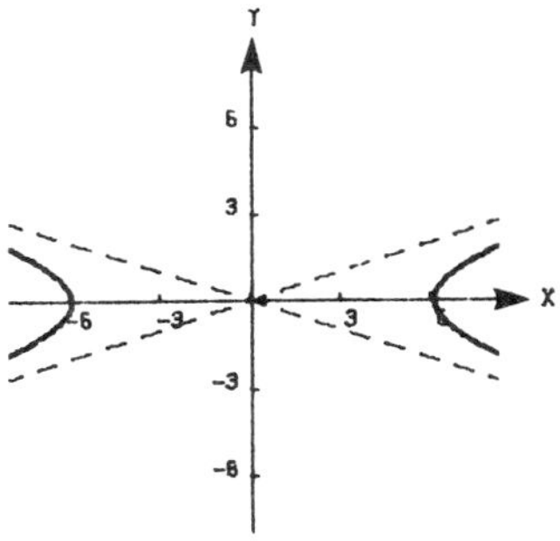

$a = 6, \quad b = 2, \quad c = 2\sqrt{10}$

Center: (0, 0)

Vertices: $(\pm 6, 0)$

Foci: $(\pm 2\sqrt{10}, 0)$

Asymptotes: $y = \pm\dfrac{1}{3}x$

13. $2x^2 - 3y^2 = 6, \qquad \dfrac{x^2}{3} - \dfrac{y^2}{2} = 1$

$a = \sqrt{3}, \quad b = \sqrt{2}, \quad c = \sqrt{5}$

Center: $(0, 0)$

Vertices: $(\pm\sqrt{3}, 0)$

Foci: $(\pm\sqrt{5}, 0)$

Asymptotes: $y = \pm\sqrt{2/3}\,x$

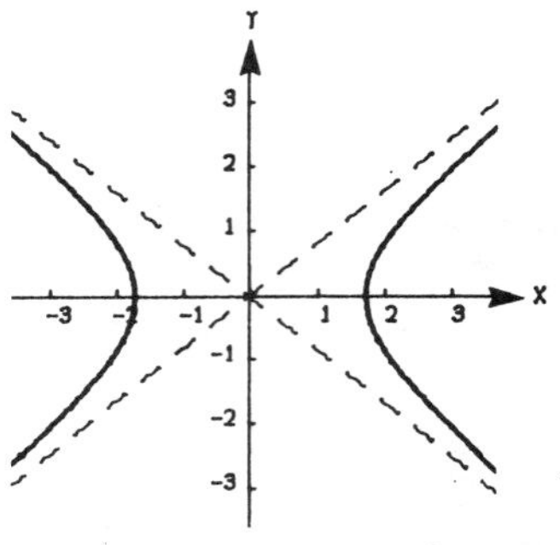

14. $3y^2 - 5x^2 = 15, \qquad \dfrac{y^2}{5} - \dfrac{x^2}{3} = 1$

$a = \sqrt{5}, \quad b = \sqrt{3}, \quad c = 2\sqrt{2}$

Center: $(0, 0)$

Vertices: $(0, \pm\sqrt{5})$

Foci: $(0, \pm 2\sqrt{2})$

Asymptotes: $y = \pm\dfrac{\sqrt{5}}{\sqrt{3}}x$

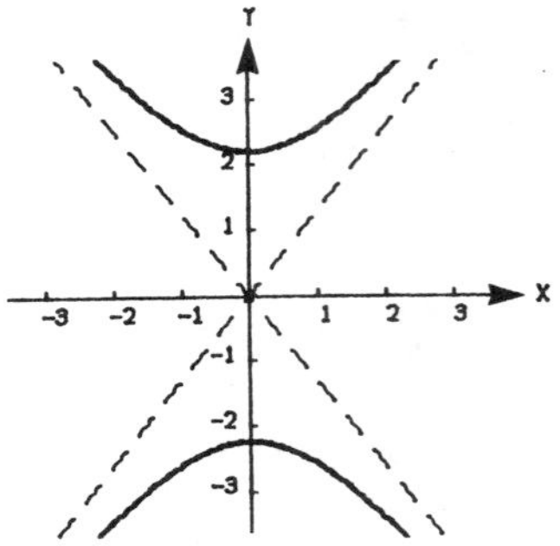

15. $5y^2 - 4x^2 = 20, \qquad \dfrac{y^2}{4} - \dfrac{x^2}{5} = 1$

$a = 2, \quad b = \sqrt{5}, \quad c = 3$

Center: $(0, 0)$

Vertices: $(0, \pm 2)$

Foci: $(0, \pm 3)$

Asymptotes: $y = \pm\dfrac{2}{\sqrt{5}}x$

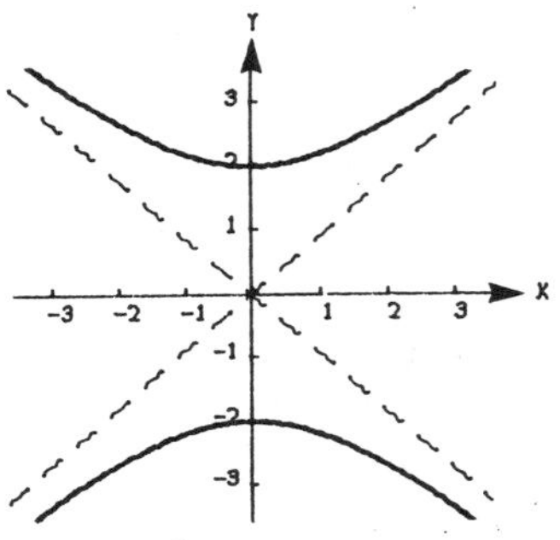

16. $7x^2 - 3y^2 = 21, \qquad \dfrac{x^2}{3} - \dfrac{y^2}{7} = 1$

$a = \sqrt{3}, \quad b = \sqrt{7}, \quad c = \sqrt{10}$

Center: $(0, 0)$

Vertices: $(\pm\sqrt{3}, 0)$

Foci: $(\pm\sqrt{10}, 0)$

Asymptotes: $y = \pm\dfrac{\sqrt{7}}{\sqrt{3}}x$

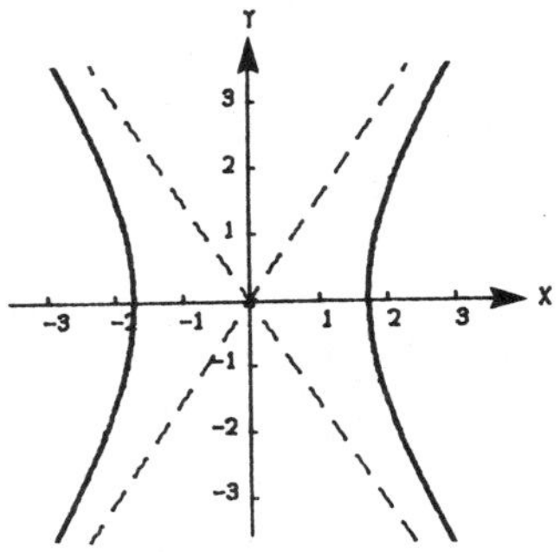

17. $\dfrac{(x-1)^2}{4} - \dfrac{(y+2)^2}{1} = 1$

$a = 2, \quad b = 1, \quad c = \sqrt{5}$

Center: $(1, -2)$

Vertices: $(-1, -2)$ $(3, -2)$

Foci: $(1 \pm \sqrt{5}, -2)$

Asymptotes: $y = -2 \pm \dfrac{1}{2}(x - 1)$

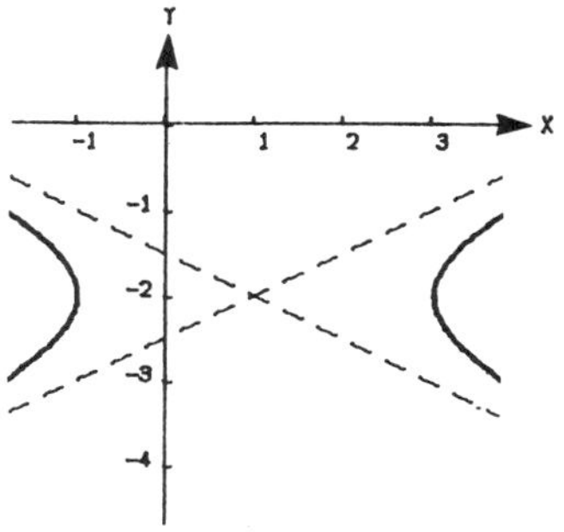

18. $\dfrac{(x+1)^2}{144} - \dfrac{(y-4)^2}{25} = 1$

$a = 12, \quad b = 5, \quad c = 13$

Center: $(-1, 4)$

Vertices: $(-13, 4)$, $(11, 4)$

Foci: $(-14, 4)$, $(12, 4)$

Asymptotes: $y = 4 \pm \dfrac{5}{12}(x + 1)$

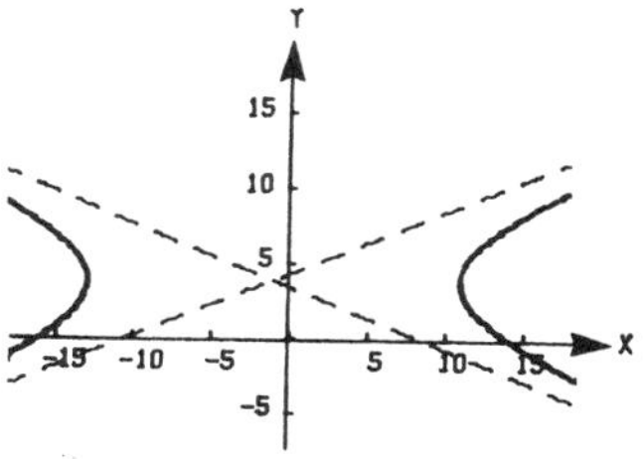

19. $(y+6)^2 - (x-2)^2 = 1$

$a = 1, \quad b = 1, \quad c = \sqrt{2}$

Center: $(2, -6)$

Vertices: $(2, -5)$, $(2, -7)$

Foci: $(2, -6 \pm \sqrt{2})$

Asymptotes: $y = -6 \pm (x - 2)$

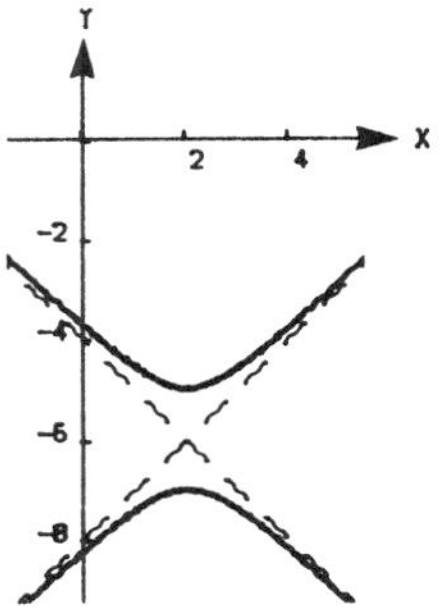

20. $\dfrac{(y-1)^2}{1/4} - \dfrac{(x+3)^2}{1/9} = 1$

$a = \dfrac{1}{2}, \quad b = \dfrac{1}{3}, \quad c = \dfrac{\sqrt{13}}{6}$

Center: $(-3, 1)$

Vertices: $(-3, \frac{1}{2})$, $(-3, \frac{3}{2})$

Foci: $(-3, 1 \pm \frac{1}{6}\sqrt{13})$

Asymptotes: $y = 1 \pm \dfrac{3}{2}(x + 3)$

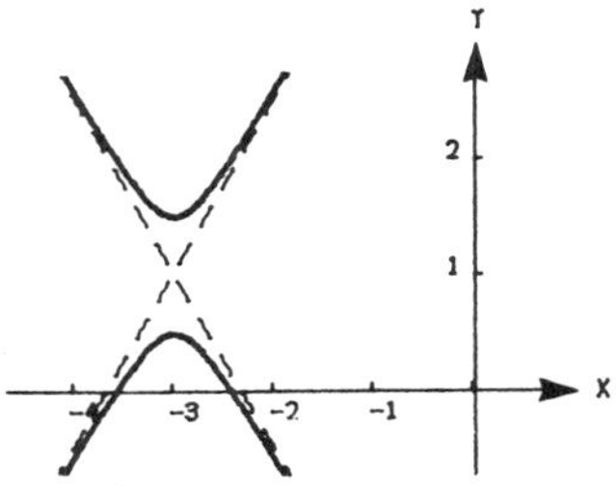

21. $9x^2 - y^2 - 36x - 6y + 18 = 0$

$9(x^2 - 4x + 4) - (y^2 + 6y + 9) = -18 + 36 - 9$

$\frac{(x - 2)^2}{1} - \frac{(y + 3)^2}{9} = 1$

$a = 1, \quad b = 3, \quad c = \sqrt{10}$

Center: $(2, -3)$

Vertices: $(1, -3), (3, -3)$

Foci: $(2 \pm \sqrt{10}, -3)$

Asymptotes: $y = -3 \pm 3(x - 2)$

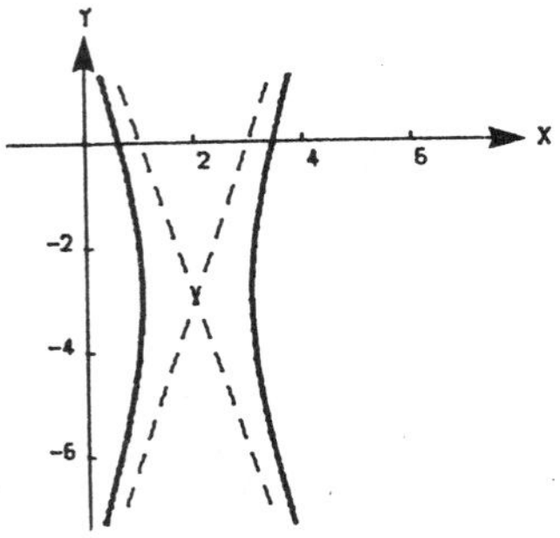

22. $x^2 - 9y^2 + 36y - 72 = 0$

$x^2 - 9(y^2 - 4y + 4) = 72 - 36$

$\frac{x^2}{36} - \frac{(y - 2)^2}{4} = 1$

$a = 6, \quad b = 2, \quad c = 2\sqrt{10}$

Center: $(0, 2)$

Vertices: $(-6, 2), (6, 2)$

Foci: $(\pm 2\sqrt{10}, 2)$

Asymptotes: $y = 2 \pm \frac{1}{3}x$

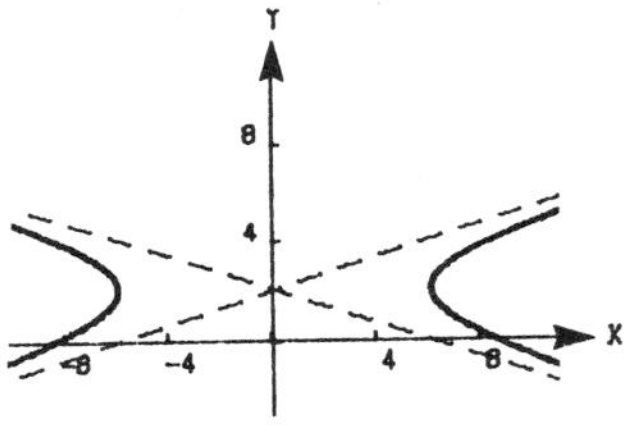

23. $9y^2 - x^2 + 2x + 54y + 62 = 0$

$9(y^2 + 6y + 9) - (x^2 - 2x + 1) = -62 - 1 + 81$

$\frac{(y + 3)^2}{2} - \frac{(x - 1)^2}{18} = 1$

$a = \sqrt{2}, \quad b = 3\sqrt{2}, \quad c = 2\sqrt{5}$

Center: $(1, -3)$

Vertices: $(1, -3 \pm \sqrt{2})$

Foci: $(1, -3 \pm 2\sqrt{5})$

Asymptotes: $y = -3 \pm \frac{1}{3}(x - 1)$

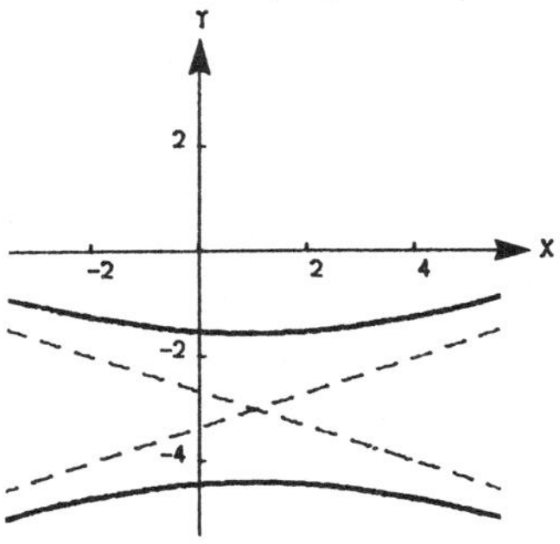

24. $16y^2 - x^2 + 2x + 64y + 63 = 0$

$16(y^2 + 4y + 4) - (x^2 - 2x + 1) = -63 + 64 - 1$

$16(y + 2)^2 - (x - 1)^2 = 0$

$y + 2 = \pm\frac{1}{4}(x - 1)$

Degenerate hyperbola
is two intersecting lines

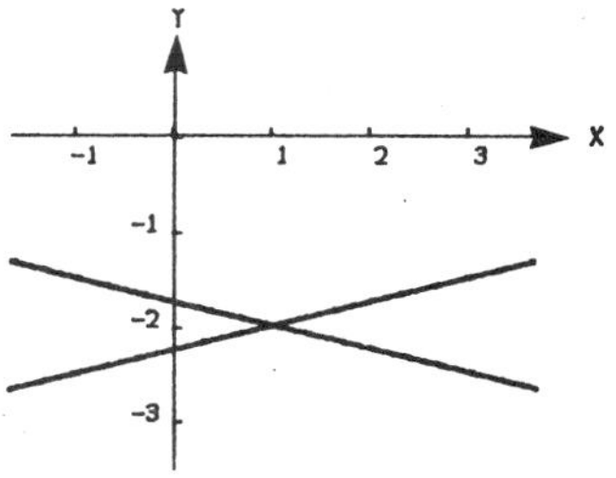

25. $x^2 - 9y^2 + 2x - 54y - 80 = 0$

$(x^2 + 2x + 1) - 9(y^2 + 6y + 9) = 80 + 1 - 81$

$(x + 1)^2 - 9(y + 3)^2 = 0$

$y + 3 = \pm\frac{1}{3}(x + 1)$

Degenerate hyperbola
is two intersecting lines

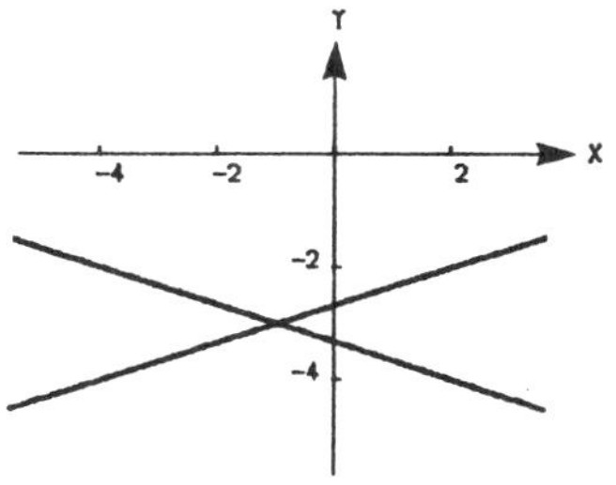

26. $9x^2 - y^2 + 54x + 10y + 55 = 0$

$9(x^2 + 6x + 9) - (y^2 - 10y + 25) = -55 + 81 - 25$

$$\frac{(x + 3)^2}{1/9} - \frac{(y - 5)^2}{1} = 1$$

$a = \frac{1}{3}$, $\quad b = 1$, $\quad c = \frac{\sqrt{10}}{3}$

Center: $(-3, 5)$

Vertices: $(-3 \pm \frac{1}{3}, 5)$

Foci: $(-3 \pm \frac{\sqrt{10}}{3}, 5)$

Asymptotes: $y = 5 \pm 3(x + 3)$

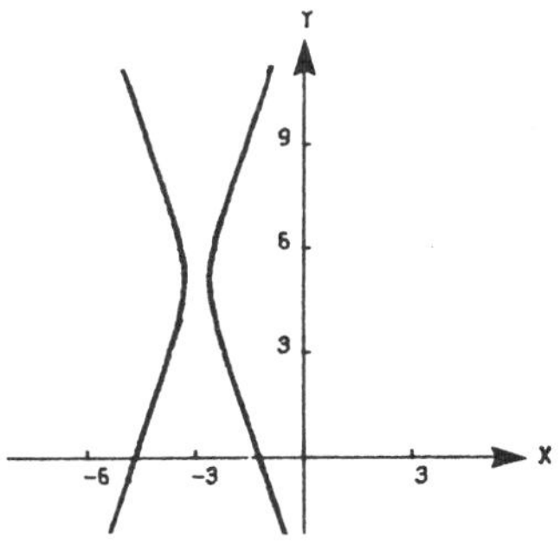

27. Transverse axis vertical, Center: $(0, 0)$, $a = 2$, $c = 4$

$b^2 = c^2 - a^2 = 12$. Therefore, the equation is: $\frac{y^2}{4} - \frac{x^2}{12} = 1$

28. Transverse axis horizontal, Center: $(0, 0)$, $a = 3$, $c = 5$

$b^2 = c^2 - a^2 = 16$. Therefore, the equation is: $\frac{x^2}{9} - \frac{y^2}{16} = 1$

29. Transverse axis horizontal, Center: (0, 0), $a = 1$, $b = 3$

Therefore, the equation is $\frac{x^2}{1} - \frac{y^2}{9} = 1$

30. Transverse axis vertical, Center: (0, 0), $a = 3$,

slopes of asymptotes are $\pm\frac{a}{b} = \pm 3$ and thus, $b = 1$

Therefore, the equation is $\frac{y^2}{9} - \frac{x^2}{1} = 1$

31. Transverse axis horizontal, Center: (3, 2), $a = 3$,

slopes of asymptotes are $\pm\frac{b}{a} = \pm\frac{2}{3}$ and thus, $b = 2$

Therefore, the equation is: $\frac{(x - 3)^2}{9} - \frac{(y - 2)^2}{4} = 1$

32. Transverse axis vertical, Center: (2, 0), $a = 3$, $c = 5$

$b^2 = c^2 - a^2 = 16$. Therefore, the equation is: $\frac{y^2}{9} - \frac{(x - 2)^2}{16} = 1$

33. Transverse axis vertical, Center: (2, 0), $a = 3$. Therefore, the equation is of the form $\frac{y^2}{9} - \frac{(x - 2)^2}{b^2} = 1$. Substituting the coordinates of the point (0, 5), we have $\frac{25}{9} - \frac{4}{b^2} = 1$ or $b^2 = \frac{9}{4}$

Therefore, the equation is: $\frac{y^2}{9} - \frac{(x - 2)^2}{9/4} = 1$

34. Transverse axis horizontal, Center: (0, 0) (since asymptotes intersect at the origin), $c = 10$, slopes of the asymptotes are $\pm\frac{b}{a} = \pm\frac{3}{4}$ or $b = \frac{3}{4}a$. Also $c^2 = a^2 + b^2 = 100$. Solving these two equations, we have $a^2 = 64$ and $b^2 = 36$. Therefore, the equation is: $\frac{x^2}{64} - \frac{y^2}{36} = 1$

35. The transverse axis is horizontal since (2, 2) and (10, 2) are the foci (see the definition of a hyperbola).

Also the center is (6, 2), $c = 4$, $2a = 6$ and $b^2 = c^2 - a^2 = 7$

Therefore, the equation is: $\frac{(x - 6)^2}{9} - \frac{(y - 2)^2}{7} = 1$

36. The transverse axis is vertical since $(-3, 0)$ and $(-3, 3)$ are the foci.

Also the center is $(-3, \frac{3}{2})$, $c = \frac{3}{2}$, $2a = 2$, and $b^2 = c^2 - a^2 = \frac{5}{4}$.

Therefore, the equation is: $\dfrac{[y - (3/2)]^2}{1} - \dfrac{(x + 3)^2}{5/4} = 1$

37. (a) $\dfrac{x^2}{9} - y^2 = 1, \quad \dfrac{2x}{9} - 2yy' = 0, \quad \dfrac{x}{9y} = y'$

At $x = 6$: $y = \pm\sqrt{3}$ and $y' = \dfrac{\pm 6}{9\sqrt{3}} = \dfrac{\pm 2\sqrt{3}}{9}$

At $(6, \sqrt{3})$: $y - \sqrt{3} = \dfrac{2\sqrt{3}}{9}(x - 6)$ or $2x - 3\sqrt{3}y - 3 = 0$

At $(6, -\sqrt{3})$: $y + \sqrt{3} = \dfrac{-2\sqrt{3}}{9}(x - 6)$ or $2x + 3\sqrt{3}y - 3 = 0$

(b) From part (a) we know that the slopes of the normal lines

must be $\mp \dfrac{9}{2\sqrt{3}}$

At $(6, \sqrt{3})$: $y - \sqrt{3} = -\dfrac{9}{2\sqrt{3}}(x - 6)$ or $9x + 2\sqrt{3}y - 60 = 0$

At $(6, -\sqrt{3})$: $y + \sqrt{3} = \dfrac{9}{2\sqrt{3}}(x - 6)$ or $9x - 2\sqrt{3}y - 60 = 0$

38. (a) $\dfrac{y^2}{4} - \dfrac{x^2}{2} = 1, \quad y^2 - 2x^2 = 4, \quad 2yy' - 4x = 0, \quad y' = \dfrac{4x}{2y} = \dfrac{2x}{y}$

At $x = 4$: $y = \pm 6$ and $y' = \dfrac{\pm 2(4)}{6} = \pm\dfrac{4}{3}$

At $(4, 6)$: $y - 6 = \dfrac{4}{3}(x - 4)$ or $4x - 3y + 2 = 0$

At $(4, -6)$: $y + 6 = -\dfrac{4}{3}(x - 4)$ or $4x + 3y + 2 = 0$

(b) From part (a) we know that the slopes of the normal lines

must be $\mp \dfrac{3}{4}$

At $(4, 6)$: $y - 6 = -\dfrac{3}{4}(x - 4)$ or $3x + 4y - 36 = 0$

At $(4, -6)$: $y + 6 = \dfrac{3}{4}(x - 4)$ or $3x - 4y - 36 = 0$

39. (a) $y = \sqrt{x^2 - 1}$

$$V = \pi \int_1^2 (\sqrt{x^2 - 1})^2 \, dx$$

$$= \pi\left(\frac{x^3}{3} - x\right)\Big]_1^2 = \frac{4}{3}\pi$$

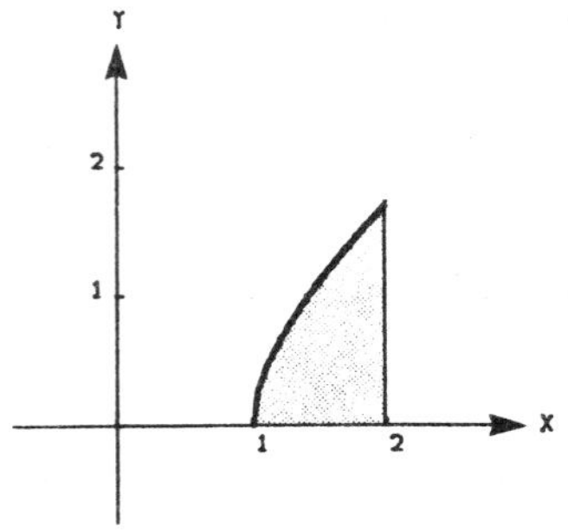

(b) $x^2 - y^2 = 1, \quad y' = \frac{x}{y}, \quad \sqrt{1 + (y')^2} = \frac{\sqrt{2x^2 - 1}}{y}$

$$S = 2\pi \int_1^2 y\left(\frac{\sqrt{2x^2 - 1}}{y}\right) dx = \sqrt{2}\pi \int_1^2 \sqrt{(\sqrt{2}x)^2 - 1}\,(\sqrt{2})\,dx$$

$$= \frac{\sqrt{2}\pi}{2}\Big[\sqrt{2}x\sqrt{2x^2 - 1} - \ln(\sqrt{2}x + \sqrt{2x^2 - 1})\Big]_1^2$$

$$= \pi(2\sqrt{7} - 1) + \frac{\sqrt{2}\pi}{2}\ln\left(\frac{\sqrt{2} + 1}{2\sqrt{2} + \sqrt{7}}\right) \approx 11.66$$

40. (a) $y = \frac{3}{4}\sqrt{x^2 - 16}$

$$V = \pi \int_4^5 \left[\frac{3}{4}\sqrt{x^2 - 16}\right]^2 dx$$

$$= \frac{9\pi}{16}\left(\frac{x^3}{3} - 16x\right)\Big]_4^5$$

$$= \frac{39\pi}{16}$$

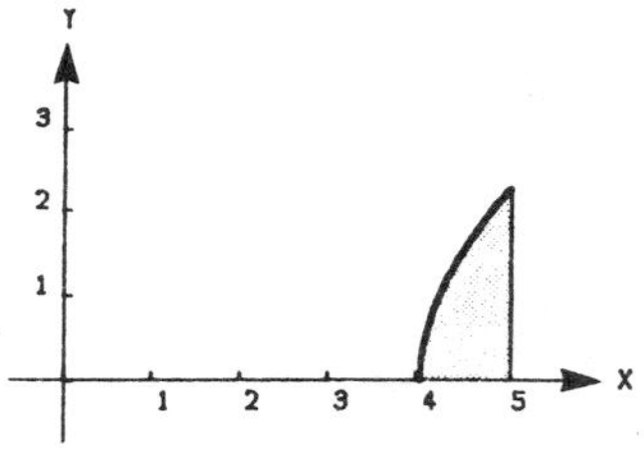

(b) $\frac{x^2}{16} - \frac{y^2}{9} = 1, \quad y' = \frac{9x}{16y}, \quad \sqrt{1 + (y')^2} = \frac{\sqrt{225x^2 - 2304}}{16y}$

$$S = 2\pi \int_4^5 y\left(\frac{\sqrt{225x^2 - 2304}}{16y}\right) dx = \frac{\pi}{8}\int_4^5 \sqrt{(15x)^2 - (48)^2}\,dx$$

$$= \frac{\pi}{240}\Big[15x\sqrt{(15x)^2 - (48)^2} - 48\ln(15x + \sqrt{(15x)^2 - (48)^2})\Big]_4^5$$

$$= \frac{\pi}{240}\Big[75\sqrt{3321} - 48\ln(75 + \sqrt{3321}) - 2160 + 48\ln 96\Big] \approx 28.099$$

41. Time for sound of bullet hitting target to reach (x, y):

$$\frac{2c}{v_m} + \frac{\sqrt{(x - c)^2 + y^2}}{v_s}$$

Time for sound of rifle to reach (x, y): $\dfrac{\sqrt{(x + c)^2 + y^2}}{v_s}$

Since the times are the same, we have

$$\frac{2c}{v_m} + \frac{\sqrt{(x - c)^2 + y^2}}{v_s} = \frac{\sqrt{(x + c)^2 + y^2}}{v_s}$$

$$\frac{4c^2}{v_m^2} + \frac{4c}{v_m v_s}\sqrt{(x - c)^2 + y^2} + \frac{(x - c)^2 + y^2}{v_s^2} = \frac{(x + c)^2 + y^2}{v_s^2}$$

$$\sqrt{(x - c)^2 + y^2} = \frac{v_m^2 x - v_s^2 c}{v_s v_{m2}}$$

$$\left(1 - \frac{v_m^2}{v_s^2}\right)x^2 + y^2 = \left(\frac{v_s^2}{v_m^2} - 1\right)c^2$$

$$\frac{x^2}{c^2 v_x^2/v_m^2} - \frac{y^2}{c^2(v_m^2 - v_s^2)/v_m^2} = 1$$

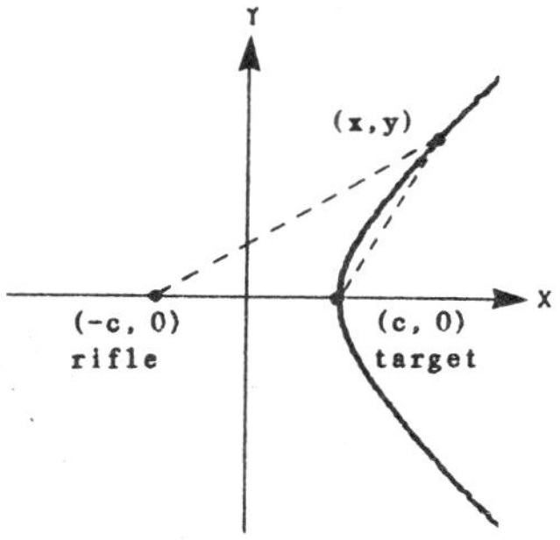

42. $\dfrac{x^2}{a^2} - \dfrac{y^2}{b^2} = 1,\quad \dfrac{2x}{a^2} - \dfrac{2yy'}{b^2} = 0$ or $y' = \dfrac{b^2x}{a^2y},\quad y - y_0 = \dfrac{b^2x_0}{a^2y_0}(x - x_0)$

$a^2y_0y - a^2y_0^2 = b^2x_0x - b^2x_0^2,\quad \dfrac{x_0x}{a^2} - \dfrac{y_0y}{b^2} = 1$

43. $\dfrac{x^2}{a^2} + \dfrac{2y^2}{b^2} = 1 \Rightarrow \dfrac{2y^2}{b^2} = 1 - \dfrac{x^2}{a^2},\quad c^2 = a^2 - b^2$

$$\frac{x^2}{a^2 - b^2} - \frac{2y^2}{b^2} = 1 \Rightarrow \frac{2y^2}{b^2} = \frac{x^2}{a^2 - b^2} - 1$$

$$1 - \frac{x^2}{a^2} = \frac{x^2}{a^2 - b^2} - 1 \Rightarrow 2 = x^2\left(\frac{1}{a^2} + \frac{1}{a^2 - b^2}\right)$$

$$x^2 = \frac{2a^2(a^2 - b^2)}{2a^2 - b^2} \Rightarrow x = \pm\frac{\sqrt{2}\, a\sqrt{a^2 - b^2}}{\sqrt{2a^2 - b^2}} = \frac{\sqrt{2}\, ac}{\sqrt{2a^2 - b^2}}$$

$$\frac{2y^2}{b^2} = 1 - \frac{1}{a^2}\left(\frac{2a^2c^2}{2a^2 - b^2}\right) \Rightarrow \frac{2y^2}{b^2} = \frac{b^2}{2a^2 - b^2}$$

$$y^2 = \frac{b^4}{2(2a^2 - b^2)} \Rightarrow y = \pm\frac{b^2}{\sqrt{2}\sqrt{2a^2 - b^2}}$$

43. (continued)

There are four points of intersection:

$$\left(\frac{\sqrt{2}\,ac}{\sqrt{2a^2 - b^2}}, \pm\frac{b^2}{\sqrt{2}\,\sqrt{2a^2 - b^2}}\right), \left(-\frac{\sqrt{2}\,ac}{\sqrt{2a^2 - b^2}}, \pm\frac{b^2}{\sqrt{2}\,\sqrt{2a^2 - b^2}}\right)$$

$$\frac{x^2}{a^2} + \frac{2y^2}{b^2} = 1 \Longrightarrow \frac{2x}{a^2} + \frac{4yy'}{b^2} = 0 \Longrightarrow y'_e = -\frac{b^2x}{2a^2y}$$

$$\frac{x^2}{a^2 - b^2} - \frac{2y^2}{b^2} = 1 \Longrightarrow \frac{2x}{c^2} - \frac{4yy'}{b^2} = 0 \Longrightarrow y'_h = \frac{b^2x}{2c^2y}$$

At $\left(\frac{\sqrt{2}\,ac}{\sqrt{2a^2 - b^2}}, \frac{b^2}{\sqrt{2}\,\sqrt{2a^2 - b^2}}\right)$ the slopes of the tangent lines are:

$$y'_e = \frac{-b^2\left(\frac{\sqrt{2}\,ac}{\sqrt{2a^2 - b^2}}\right)}{2a^2\left(\frac{b^2}{\sqrt{2}\,\sqrt{2a^2 - b^2}}\right)} = -\frac{c}{a}$$

$$y'_h = \frac{b^2\left(\frac{\sqrt{2}\,ac}{\sqrt{2a^2 - b^2}}\right)}{2c^2\left(\frac{b^2}{\sqrt{2}\,\sqrt{2a^2 - b^2}}\right)} = \frac{a}{c}$$

Since the slopes are negative reciprocals, the tangent lines are perpendicular. Similarly, the curves are perpendicular at the other three points of intersection.

44. circle

45. ellipse

46. hyperbola

47. parabola

48. ellipse

49. hyperbola

50. parabola

51. circle

52. parabola

53. circle

54. ellipse

55. hyperbola

56. ellipse

10.4 Rotation and the general second-degree equation

1. $xy + 1 = 0, \quad A = 0,\ B = 1,\ C = 0,\ D = 0,\ E = 0,\ F = 1$

$\cot 2\theta = \dfrac{A - C}{B} = 0, \quad 2\theta = \dfrac{\pi}{2}$ and $\theta = \dfrac{\pi}{4}$

$\sin(\dfrac{\pi}{4}) = \cos(\dfrac{\pi}{4}) = \dfrac{\sqrt{2}}{2}$

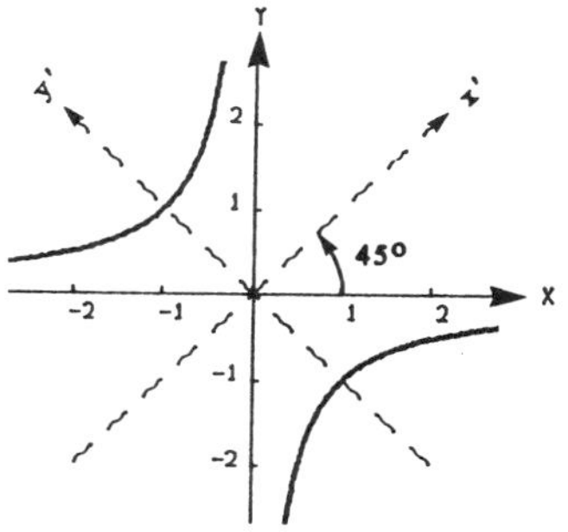

$A' = 0 + (\dfrac{\sqrt{2}}{2})(\dfrac{\sqrt{2}}{2}) + 0 = \dfrac{1}{2}$

$C' = 0 - (\dfrac{\sqrt{2}}{2})(\dfrac{\sqrt{2}}{2}) + 0 = -\dfrac{1}{2}$

$D' = E' = 0, \quad F' = 1$

$\dfrac{x'^2}{2} - \dfrac{y'^2}{2} + 1 = 0$ or $\dfrac{y'^2}{2} - \dfrac{x'^2}{2} = 1$

2. $xy - 4 = 0$

$A = 0,\ B = 1,\ C = 0,\ D = 0,\ E = 0,\ F = -4$

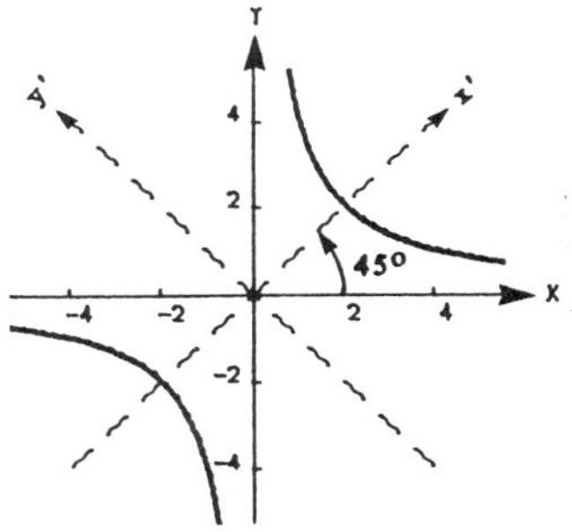

$\cot 2\theta = \dfrac{A - C}{B} = 0, \quad 2\theta = \dfrac{\pi}{2}$ and $\theta = \dfrac{\pi}{4}$

As in Exercise 1, $A' = \dfrac{1}{2},\ C' = -\dfrac{1}{2},\ F' = -4$

$\dfrac{x'^2}{2} - \dfrac{y'^2}{2} - 4 = 0, \quad \dfrac{x'^2}{8} - \dfrac{y'^2}{8} = 1$

3. $9x^2 + 24xy + 16y^2 + 90x - 130y = 0$

$A = 9,\ B = 24,\ C = 16,\ D = 90,\ E = -130,\ F = 0$

$\cot 2\theta = \dfrac{A - C}{B} = -\dfrac{7}{24}$ or $\theta = \dfrac{1}{2}\operatorname{arccot}(-\dfrac{7}{24}) \approx 53.13°$

since $\cot 2\theta = -\dfrac{7}{24}$, $\cos 2\theta = -\dfrac{7}{25}$

$\sin\theta = \dfrac{\sqrt{1 + (7/25)}}{\sqrt{2}} = \dfrac{4}{5}$

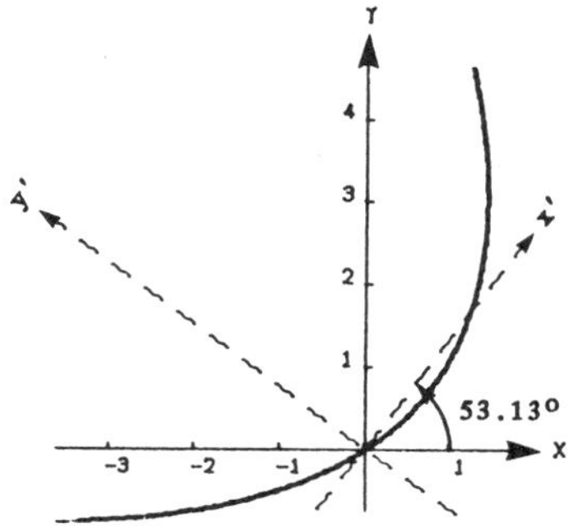

$\cos\theta = \dfrac{\sqrt{1 - (7/25)}}{\sqrt{2}} = \dfrac{3}{5}$

$A' = 9(\dfrac{9}{25}) + 24(\dfrac{4}{5})(\dfrac{3}{5}) + 16(\dfrac{16}{25}) = 25$

$C' = 9(\dfrac{16}{25}) - 24(\dfrac{4}{5})(\dfrac{3}{5}) + 16(\dfrac{9}{25}) = 0$

3. (continued)

$D' = 90(\frac{3}{5}) - 130(\frac{4}{5}) = -50$

$E' = -90(\frac{4}{5}) - 130(\frac{3}{5}) = -150$

$F' = 0$

$25x'^2 - 50x' - 150y' = 0$

$y' = \frac{x'^2}{6} - \frac{x'}{3}$

4. $9x^2 + 24xy + 16y^2 + 80x - 60y = 0$

$A = 9,\ B = 24,\ C = 16,\ D = 80,\ E = -60,\ F = 0$

From the solution to Exercise 3, we have

$\theta \approx 53.13°,\ \sin\theta = \frac{4}{5},\ \cos\theta = \frac{3}{5}$

$A' = 25$ and $C' = 0$

$D' = 80(\frac{3}{5}) + (-60)(\frac{4}{5}) = 0$

$E' = -80(\frac{4}{5}) - 60(\frac{3}{5}) = -110$

$25x'^2 - 100y' = 0$ or $x'^2 = 4y'$

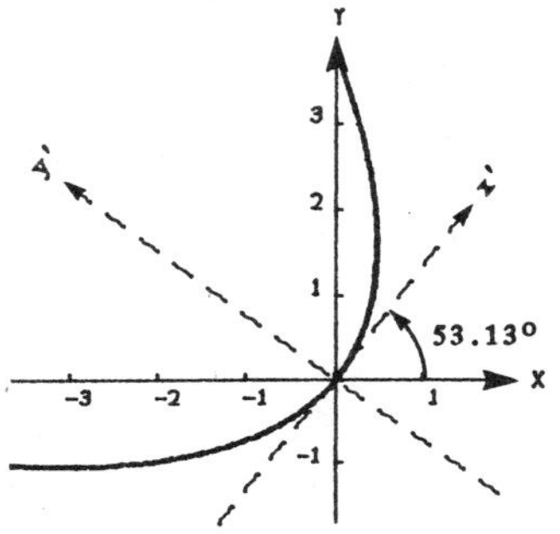

5. $x^2 - 10xy + y^2 + 1 = 0$

$A = 1,\ B = -10,\ C = 1,\ D = 0,\ E = 0,\ F = 1$

$\cot 2\theta = \frac{A - C}{B} = 0,\quad 2\theta = \frac{\pi}{2},\ \theta = \frac{\pi}{4}$

$\sin\theta = \cos\theta = \frac{\sqrt{2}}{2}$

$A' = (\frac{\sqrt{2}}{2})^2 - 10(\frac{\sqrt{2}}{2})(\frac{\sqrt{2}}{2}) + (\frac{\sqrt{2}}{2})^2 = -4$

$C' = (\frac{\sqrt{2}}{2})^2 + 10(\frac{\sqrt{2}}{2})(\frac{\sqrt{2}}{2}) + (\frac{\sqrt{2}}{2})^2 = 6$

$D' = E' = 0,\quad F' = 1$

$-4x'^2 + 6y'^2 + 1 = 0,\quad \frac{x'^2}{1/4} - \frac{y'^2}{1/6} = 1$

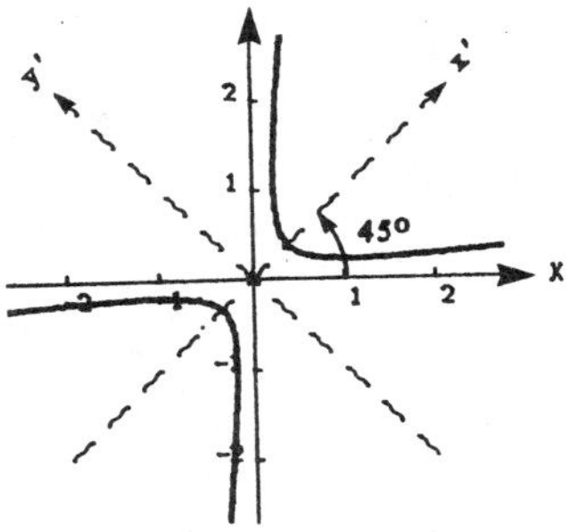

6. $xy + x - 2y + 3 = 0$

$A = 0,\ B = 1,\ C = 0,\ D = 1,\ E = -2,\ F = 3$

$\cot 2\theta = 0,\ 2\theta = \frac{\pi}{2},\ \theta = \frac{\pi}{4},\ \sin\theta = \cos\theta = \frac{\sqrt{2}}{2}$

$A' = 0 + \frac{1}{2} + 0 = \frac{1}{2}$

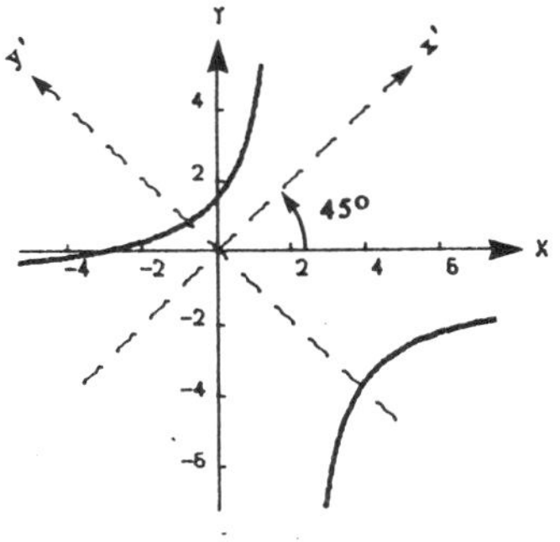

$C' = 0 - \frac{1}{2} + 0 = -\frac{1}{2}$

$D' = \frac{\sqrt{2}}{2} - 2\left(\frac{\sqrt{2}}{2}\right) = \frac{-\sqrt{2}}{2}$

$E' = \frac{-\sqrt{2}}{2} - 2\left(\frac{\sqrt{2}}{2}\right) = \frac{-3\sqrt{2}}{2}$

$F' = 3$

$\frac{1}{2}x'^2 - \frac{1}{2}y'^2 - \frac{\sqrt{2}}{2}x' - \frac{3\sqrt{2}}{2}y' + 3 = 0$

$$\frac{[y' + (3\sqrt{2}/2)]^2}{10} - \frac{[x' - (\sqrt{2}/2)]^2}{10} = 1$$

7. $xy - 2y - 4x = 0, \qquad A = 0,\ B = 1,\ C = 0,\ D = -4,\ E = -2,\ F = 0$

$\cot 2\theta = 0,\ 2\theta = \frac{\pi}{2},\ \theta = \frac{\pi}{4}$

$\sin\theta = \cos\theta = \frac{\sqrt{2}}{2}$

$A' = 0 + \frac{1}{2} + 0 = \frac{1}{2}$

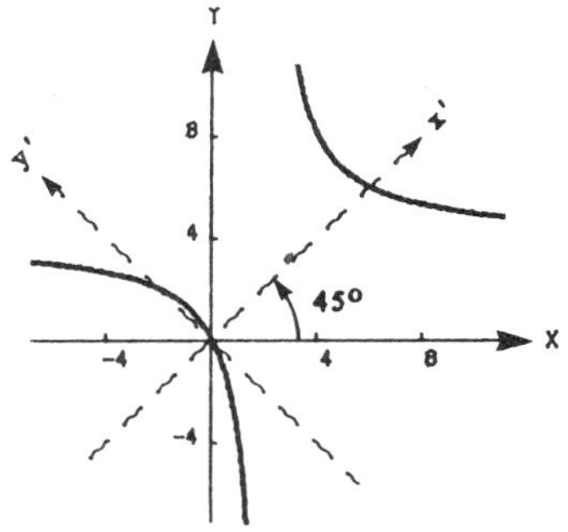

$C' = 0 - \frac{1}{2} + 0 = -\frac{1}{2}$

$D' = -4\left(\frac{\sqrt{2}}{2}\right) - 2\left(\frac{\sqrt{2}}{2}\right) = -3\sqrt{2}$

$E' = 4\left(\frac{\sqrt{2}}{2}\right) - 2\left(\frac{\sqrt{2}}{2}\right) = \sqrt{2}$

$F' = 0$

$\frac{1}{2}x'^2 - \frac{1}{2}y'^2 - 3\sqrt{2}\,x' + \sqrt{2}\,y' = 0$

$$\frac{(x' - 3\sqrt{2})^2}{16} - \frac{(y' - \sqrt{2})^2}{16} = 1$$

8. $2x^2 - 3xy - 2y^2 + 10 = 0$

$A = 2,\ B = -3,\ C = -2,\ D = 0,\ E = 0,\ F = 10$

$\cot 2\theta = -\frac{4}{3},\ \theta \approx 71.57^\circ,\ \cos 2\theta = -\frac{4}{5}$

$\sin\theta = \frac{\sqrt{1 + (4/5)}}{\sqrt{2}} = \frac{3}{\sqrt{10}}$

$\cos\theta = \frac{\sqrt{1 - (4/5)}}{\sqrt{2}} = \frac{1}{\sqrt{10}}$

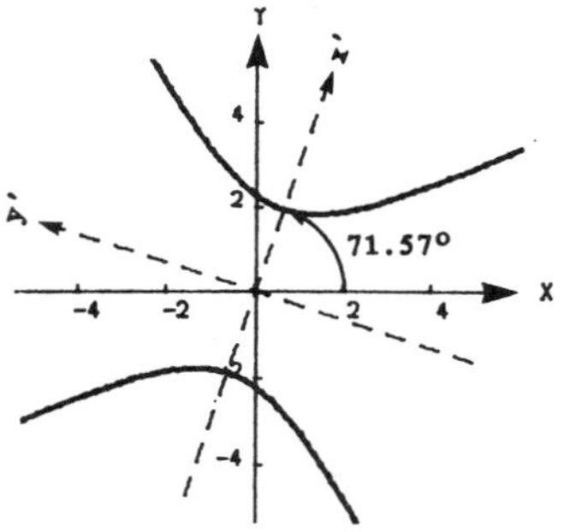

$A' = 2(\frac{1}{\sqrt{10}})^2 - 3(\frac{3}{\sqrt{10}})(\frac{1}{\sqrt{10}}) - 2(\frac{3}{\sqrt{10}})^2 = -\frac{5}{2}$

$C' = 2(\frac{3}{\sqrt{10}})^2 + 3(\frac{3}{\sqrt{10}})(\frac{1}{\sqrt{10}}) - 2(\frac{1}{\sqrt{10}})^2 = \frac{5}{2}$

$D' = E' = 0,\quad F' = 10$

$-\frac{5}{2}x'^2 + \frac{5}{2}y'^2 + 10 = 0$

$\frac{x'^2}{4} - \frac{y'^2}{4} = 1$

9. $5x^2 - 2xy + 5y^2 - 12 = 0$

$A = 5,\ B = -2,\ C = 5,\ D = 0,\ E = 0,\ F = -12$

$\cot 2\theta = 0,\ 2\theta = \frac{\pi}{2},\ \theta = \frac{\pi}{4}$

$\sin\theta = \cos\theta = \frac{\sqrt{2}}{2}$

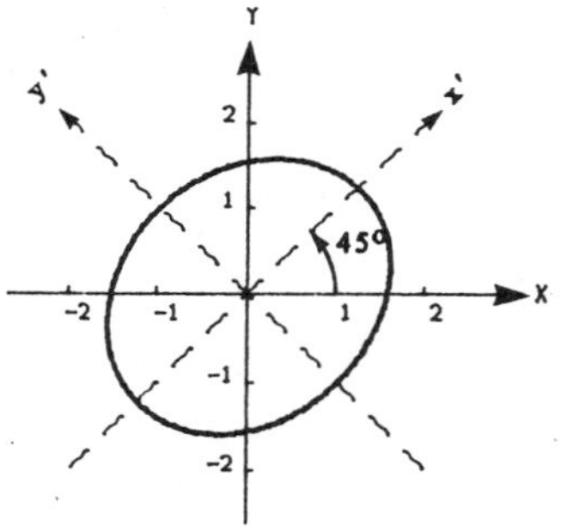

$A' = 5(\frac{\sqrt{2}}{2})^2 - 2(\frac{\sqrt{2}}{2})^2 + 5(\frac{\sqrt{2}}{2})^2 = 4$

$C' = 5(\frac{\sqrt{2}}{2})^2 + 2(\frac{\sqrt{2}}{2})^2 + 5(\frac{\sqrt{2}}{2})^2 = 6$

$D' = E' = 0,\quad F' = -12$

$4x'^2 + 6y'^2 - 12 = 0,\quad \frac{x'^2}{3} + \frac{y'^2}{2} = 1$

10. $13x^2 + 6\sqrt{3}xy + 7y^2 - 16 = 0$

$A = 13,\ B = 6\sqrt{3},\ C = 7,\ D = 0,\ E = 0,\ F = -16$

$\cot 2\theta = \dfrac{1}{\sqrt{3}},\ \theta = 30^\circ$

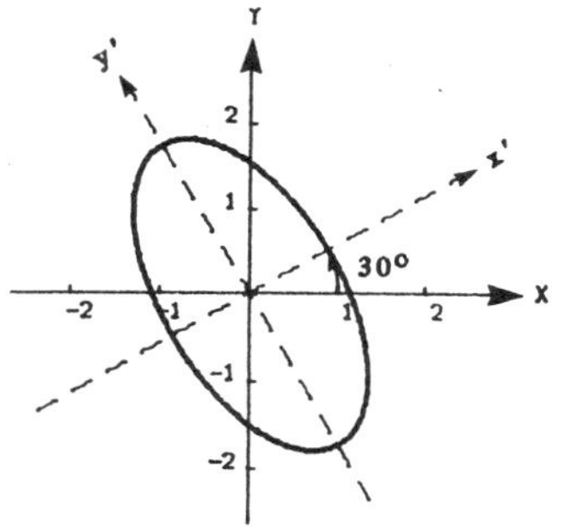

$\sin\theta = \dfrac{1}{2},\ \cos\theta = \dfrac{\sqrt{3}}{2}$

$A' = 13(\frac{3}{4}) + 6\sqrt{3}(\frac{1}{2})(\frac{\sqrt{3}}{2}) + 7(\frac{1}{4}) = 16$

$C' = 13(\frac{1}{4}) - 6\sqrt{3}(\frac{1}{2})(\frac{\sqrt{3}}{2}) + 7(\frac{3}{4}) = 4$

$D' = E' = 0,\quad F' = -16$

$16x'^2 + 4y'^2 - 16 = 0,\quad \dfrac{x'^2}{1} + \dfrac{y'^2}{4} = 1$

11. $3x^2 - 2\sqrt{3}xy + y^2 + 2x + 2\sqrt{3}y = 0$

$A = 3,\ B = -2\sqrt{3},\ C = 1,\ D = 2,\ E = 2\sqrt{3},\ F = 0$

$\cot 2\theta = -\dfrac{1}{\sqrt{3}},\ \theta = 60^\circ$

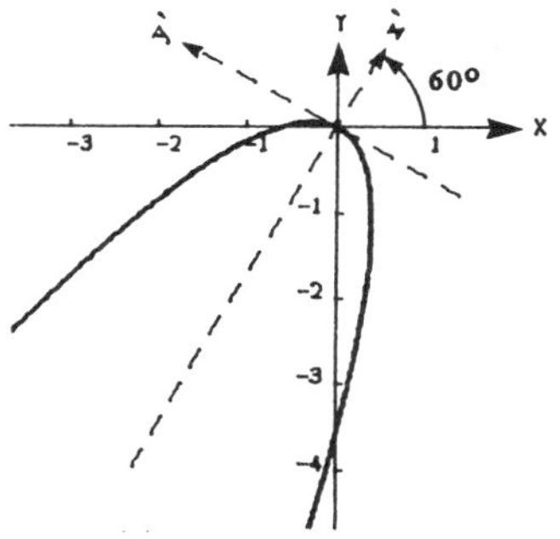

$\sin\theta = \dfrac{\sqrt{3}}{2},\ \cos\theta = \dfrac{1}{2}$

$A' = 3(\frac{1}{4}) - 2\sqrt{3}(\frac{\sqrt{3}}{2})(\frac{1}{2}) + (\frac{3}{4}) = 0$

$C' = 3(\frac{3}{4}) + 2\sqrt{3}(\frac{\sqrt{3}}{2})(\frac{1}{2}) + (\frac{1}{4}) = 4$

$D' = 2(\frac{1}{2}) + 2\sqrt{3}(\frac{\sqrt{3}}{2}) = 4$

$E' = -2(\frac{\sqrt{3}}{2}) + 2\sqrt{3}(\frac{1}{2}) = 0,\quad F' = 0$

$4y'^2 + 4x' = 0,\quad x' = -(y')^2$

12. $16x^2 - 24xy + 9y^2 - 60x - 80y + 100 = 0$

$A = 16,\ B = 24,\ C = 9,\ D = -60,\ E = -80,\ F = 100$

$\cot 2\theta = -\frac{7}{24}$, from Exercise 3, we have

$\theta \approx 53.13^\circ,\ \sin\theta = \frac{4}{5},\ \cos\theta = \frac{3}{5}$

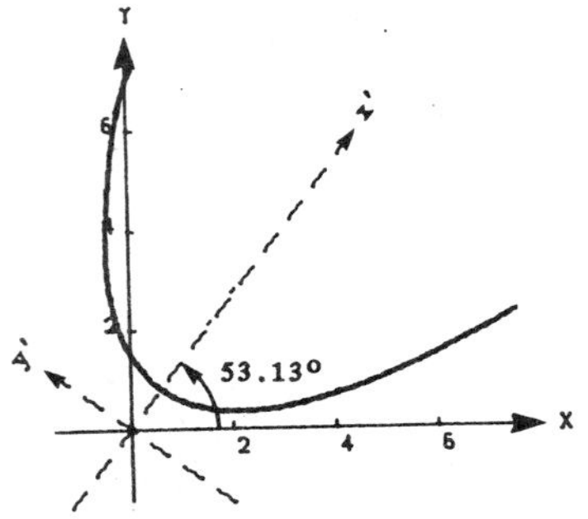

$A' = 16(\frac{9}{25}) - 24(\frac{4}{5})(\frac{3}{5}) + 9(\frac{16}{25}) = 0$

$C' = 16(\frac{16}{25}) + 24(\frac{4}{5})(\frac{3}{5}) + 9(\frac{9}{25}) = 25$

$D' = -60(\frac{3}{5}) - 80(\frac{4}{5}) = -100$

$E' = 60(\frac{4}{5}) - 80(\frac{3}{5}) = 0,\quad F' = 100$

$25y'^2 - 100x + 100 = 0,\quad y'^2 = 4(x' - 1)$

13. $17x^2 + 32xy - 7y^2 - 75 = 0$

$A = 17,\ B = 32,\ C = -7,\ D = 0,\ E = 0,\ F = -75$

$\cot 2\theta = \frac{3}{4},\ \cos 2\theta = \frac{3}{5},\ \theta \approx 26.57^\circ$

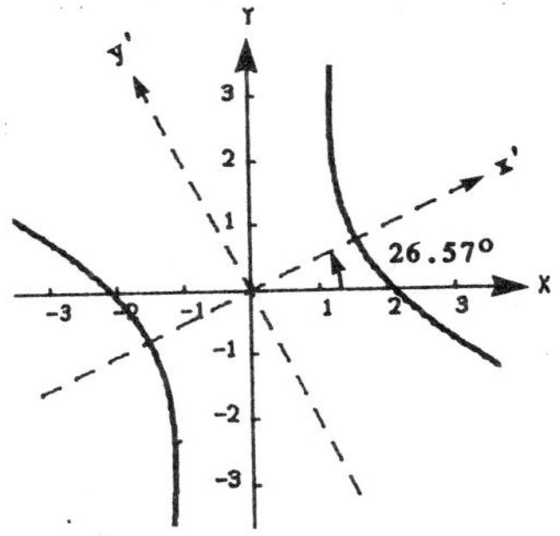

$\sin\theta = \dfrac{\sqrt{1 - (3/5)}}{\sqrt{2}} = \dfrac{\sqrt{2}}{\sqrt{10}}$

$\cos\theta = \dfrac{\sqrt{1 + (3/5)}}{\sqrt{2}} = 2\dfrac{\sqrt{2}}{\sqrt{10}}$

$A' = 17(\frac{8}{10}) + 32(\frac{2\sqrt{2}}{\sqrt{10}})(\frac{\sqrt{2}}{\sqrt{10}}) - 7(\frac{2}{10}) = 25$

$C' = 17(\frac{2}{10}) - 32(\frac{2\sqrt{2}}{\sqrt{10}})(\frac{\sqrt{2}}{\sqrt{10}}) - 7(\frac{8}{10}) = -15$

$D' = E' = 0,\quad F' = -75$

$25x'^2 - 15y'^2 - 75 = 0,\quad \dfrac{x'^2}{3} - \dfrac{y'^2}{5} = 1$

14. $40x^2 + 36xy + 25y^2 - 52 = 0$

$A = 40,\ B = 36,\ C = 25,\ D = 0,\ E = 0,\ F = -52$

$\cot 2\theta = \frac{5}{12},\ \cos 2\theta = \frac{5}{13},\quad \theta \approx 33.69^\circ$

$\sin\theta = \frac{\sqrt{1 - (5/13)}}{\sqrt{2}} = \frac{2}{\sqrt{13}}$

$\cos\theta = \frac{\sqrt{1 + (5/13)}}{\sqrt{2}} = \frac{3}{\sqrt{13}}$

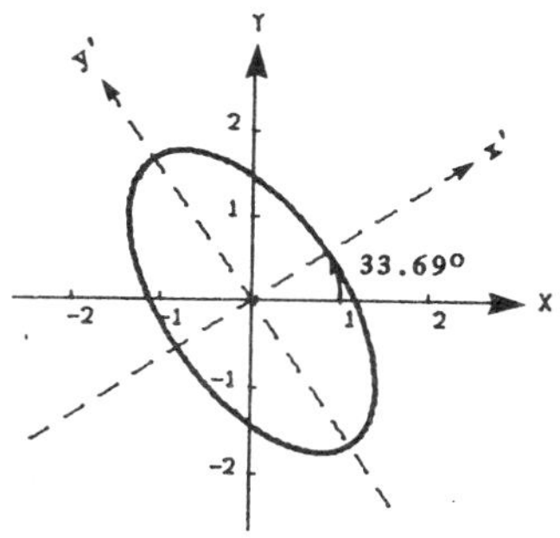

$A' = 40(\frac{9}{13}) + 36(\frac{2}{\sqrt{13}})(\frac{3}{\sqrt{13}}) + 25(\frac{4}{13}) = 52$

$C' = 40(\frac{4}{13}) - 36(\frac{2}{\sqrt{13}})(\frac{3}{\sqrt{13}}) + 25(\frac{9}{13}) = 13$

$D' = E' = 0,\quad F' = -52$

$52x'^2 + 13y'^2 - 52 = 0,\quad \frac{x'^2}{1} + \frac{y'^2}{4} = 1$

15. $32x^2 + 50xy + 7y^2 - 52 = 0$

$A = 32,\ B = 50,\ C = 7,\ D = 0,\ E = 0,\ F = -52$

$\cot 2\theta = \frac{1}{2},\ \cos 2\theta = \frac{1}{\sqrt{5}},\ \theta \approx 31.72^\circ$

$\sin\theta = \frac{\sqrt{1 - (1/\sqrt{5})}}{\sqrt{2}} = \sqrt{\frac{\sqrt{5} - 1}{2\sqrt{5}}}$

$\cos\theta = \frac{\sqrt{1 + (1/\sqrt{5})}}{\sqrt{2}} = \sqrt{\frac{\sqrt{5} + 1}{2\sqrt{5}}}$

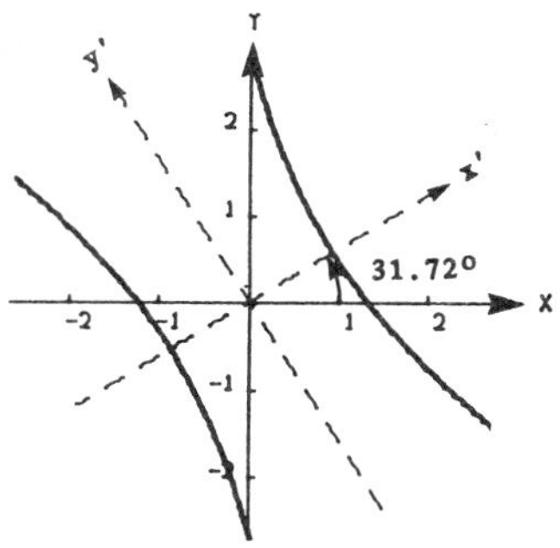

$A' = 32(\frac{\sqrt{5} + 1}{2\sqrt{5}}) + 50\sqrt{\frac{\sqrt{5} - 1}{2\sqrt{5}}}(\sqrt{\frac{\sqrt{5} + 1}{2\sqrt{5}}})$

$+ 7(\frac{\sqrt{5} - 1}{2\sqrt{5}}) = \frac{39 + 25\sqrt{5}}{2} \approx 47.451$

$C' = 32(\frac{\sqrt{5} - 1}{2\sqrt{5}}) - 50\sqrt{\frac{\sqrt{5} - 1}{2\sqrt{5}}}(\sqrt{\frac{\sqrt{5} + 1}{2\sqrt{5}}})$

$+ 7(\frac{\sqrt{5} + 1}{2\sqrt{5}}) = \frac{39 - 25\sqrt{5}}{2} \approx -8.451$

$D' = E' = 0,\quad F' = -52$

$47.451x'^2 - 8.451y'^2 - 52 = 0,\quad \frac{x'^2}{1.096} - \frac{y'^2}{6.153} = 1$

16. $4x^2 - 12xy + 9y^2 + (4\sqrt{3} - 12)x - (6\sqrt{13} + 8)y - 91 = 0$

$A = 4,\ B = -12,\ C = 9,\ D = 4\sqrt{13} - 12$

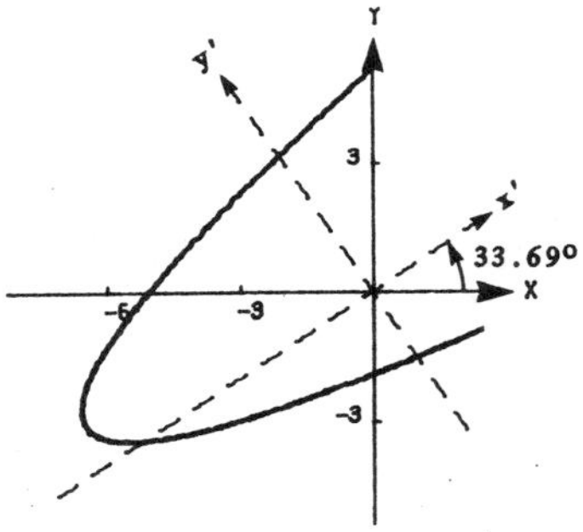

$E = -(6\sqrt{13} + 8),\ F = -91$

$\cot 2\theta = \dfrac{5}{12},\ \cos 2\theta = \dfrac{5}{13}$

$\theta \approx 33.69^\circ$

$\sin\theta = \dfrac{\sqrt{1 - (5/13)}}{\sqrt{2}} = \dfrac{2}{\sqrt{13}}$

$\cos\theta = \dfrac{\sqrt{1 + (5/13)}}{\sqrt{2}} = \dfrac{3}{\sqrt{13}}$

$A' = 4(\dfrac{9}{13}) - 12(\dfrac{2}{\sqrt{13}})(\dfrac{3}{\sqrt{13}}) + 9(\dfrac{4}{13}) = 0$

$C' = 4(\dfrac{4}{13}) + 12(\dfrac{2}{\sqrt{13}})(\dfrac{3}{\sqrt{13}}) + 9(\dfrac{9}{13}) = 13$

$D' = (4\sqrt{13} - 12)(\dfrac{3}{\sqrt{13}}) - (6\sqrt{13} + 8)(\dfrac{2}{\sqrt{13}}) = -4\sqrt{13}$

$E' = -(4\sqrt{13} - 12)(\dfrac{2}{\sqrt{13}}) - (6\sqrt{13} + 8)(\dfrac{3}{\sqrt{13}}) = -26$

$F' = -91$

$13y'^2 - 4\sqrt{13}x - 26y - 91 = 0,\quad (y' - 1)^2 = \dfrac{4}{\sqrt{13}}(x' + 2\sqrt{13})$

17. $B^2 - 4AC = (-24)^2 - 4(16)(9) = 0$ Parabola

18. $B^2 - 4AC = (-4)^2 - 4(1)(-2) = 24$ Hyperbola

19. $B^2 - 4AC = (-8)^2 - 4(13)(7) = -300$ Ellipse or Circle

20. $B^2 - 4AC = (4)^2 - 4(2)(5) = -24$ Ellipse or Circle

21. $B^2 - 4AC = (-6)^2 - 4(1)(-5) = 56$ Hyperbola

22. $B^2 - 4AC = (-60)^2 - 4(36)(25) = 0$ Parabola

23. $B^2 - 4AC = (4)^2 - 4(1)(4) = 0$ Parabola

24. $B^2 - 4AC = (1)^2 - 4(1)(4) = -15$ Ellipse or Circle

25. $(x')^2 + (y')^2 = [x\cos\theta + y\sin\theta]^2 + [y\cos\theta - x\sin\theta]^2$

$= x^2\cos^2\theta + 2xy\cos\theta\sin\theta + y^2\sin^2\theta + y^2\cos^2\theta$

$\quad - 2xy\cos\theta\sin\theta + x^2\sin^2\theta$

$= x^2(\cos^2\theta + \sin^2\theta) + y^2(\sin^2\theta + \cos^2\theta)$

$= x^2 + y^2 = r^2$

26. $Ax^2 + Bxy + Cy^2 + Dx + Ey + F = 0$

$A'(x')^2 + C'(y')^2 + D'x' + E'y' + F' = 0$

(a) $F = F'$

Substitute $x = x'\cos\theta - y'\sin\theta$ and $y = x'\sin\theta + y'\cos\theta$ into the original equation and collect like terms.

(b) $A' + C' = (A\cos^2\theta + B\cos\theta\sin\theta + C\sin^2\theta)$

$\quad + (A\sin^2\theta - B\cos\theta\sin\theta + C\cos^2\theta)$

$= A(\cos^2\theta + \sin^2\theta) + C(\sin^2\theta + \cos^2\theta) = A + C$

(c) $A' = A\cos^2\theta + B\cos\theta\sin\theta + C\sin^2\theta$

$= \dfrac{1}{2}[A + C - (C - A)\cos 2\theta + B\sin 2\theta]$

$B' = 2(C - A)\sin\theta\cos\theta + B(\cos^2\theta - \sin^2\theta)$

$= (C - A)\sin 2\theta + B\cos 2\theta$

$C' = A\sin^2\theta - B\sin\theta\cos\theta + C\cos^2\theta$

$= \dfrac{1}{2}[A + C + (C - A)\cos 2\theta - B\sin 2\theta]$

Thus in the $x'y'$-coordinate system the discriminant is:

$B'^2 - 4A'C' = [(C - A)\sin 2\theta + B\cos 2\theta]^2 - ((A + C)^2$

$\quad - [(C - A)\cos 2\theta - B\sin 2\theta]^2)$

$= (C - A)^2\sin^2 2\theta + 2B(C - A)\sin 2\theta\cos 2\theta + B^2\cos^2 2\theta - (A + C)^2$

$\quad + (C - A)^2\cos^2 2\theta - 2B(C - A)\sin 2\theta\cos 2\theta + B^2\sin^2 2\theta$

$= (C - A)^2 + B^2 - (A + C)^2$

$= C^2 - 2AC + A^2 + B^2 - A^2 - 2AC - C^2 = B^2 - 4AC$

Review Exercises for Chapter 10

1. $4x^2 + y^2 = 4$

$\frac{x^2}{1} + \frac{y^2}{4} = 1$

Matches graph (h)

2. $x^2 = 4y$

Matches graph (a)

3. $4x^2 - y^2 = 4$

$\frac{x^2}{1} - \frac{y^2}{4} = 1$

Matches graph (e)

4. $y^2 = -4x$

Matches graph (i)

5. $x^2 + 4y^2 = 4$

$\frac{x^2}{4} + \frac{y^2}{1} = 1$

Matches graph (f)

6. $y^2 - 4x^2 = 4$

$\frac{y^2}{4} - \frac{x^2}{1} = 1$

Matches graph (b)

7. $x^2 = -6y$

Matches graph (c)

8. $x^2 + 5y^2 = 10$

$\frac{x^2}{10} + \frac{y^2}{2} = 1$

Matches graph (j)

9. $x^2 - 5y^2 = -5$

$\frac{y^2}{1} - \frac{x^2}{5} = 1$

Matches graph (g)

10. $y^2 - 8x = 0$

$y^2 = 8x$

Matches graph (d)

11. $16x^2 + 16y^2 - 16x + 24y - 3 = 0$

$(x^2 - x + \frac{1}{4}) + (y^2 + \frac{3}{2}y + \frac{9}{16})$

$= \frac{3}{16} + \frac{1}{4} + \frac{9}{16}$

$(x - \frac{1}{2})^2 + (y + \frac{3}{4})^2 = 1$

Circle

Center: $(\frac{1}{2}, -\frac{3}{4})$

Radius: 1

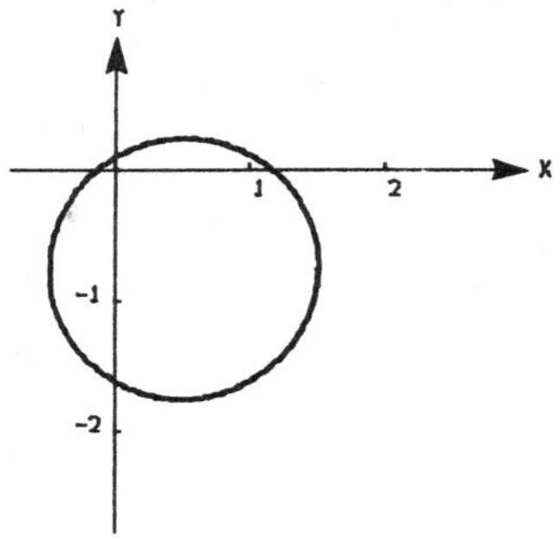

12. $y^2 - 12y - 8x + 20 = 0$

$y^2 - 12y + 36 = 8x - 20 + 36$

$(y - 6)^2 = 4(2)(x + 2)$

Parabola

Vertex: $(-2, 6)$

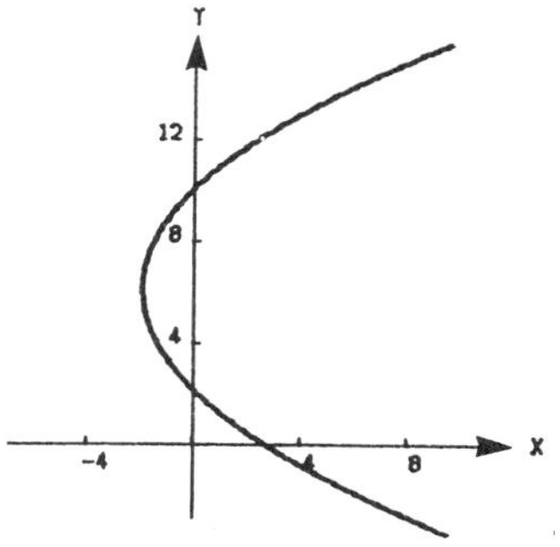

13. $3x^2 - 2y^2 + 24x + 12y + 24 = 0$

$3(x^2 + 8x + 16) - 2(y^2 - 6y + 9)$

$= -24 + 48 - 18$

$\frac{(x + 4)^2}{2} - \frac{(y - 3)^2}{3} = 1$

Hyperbola

Center: $(-4, 3)$

Vertices: $(-4 \pm \sqrt{2}, 3)$

Asymptotes: $y = 3 \pm \sqrt{3/2}(x + 4)$

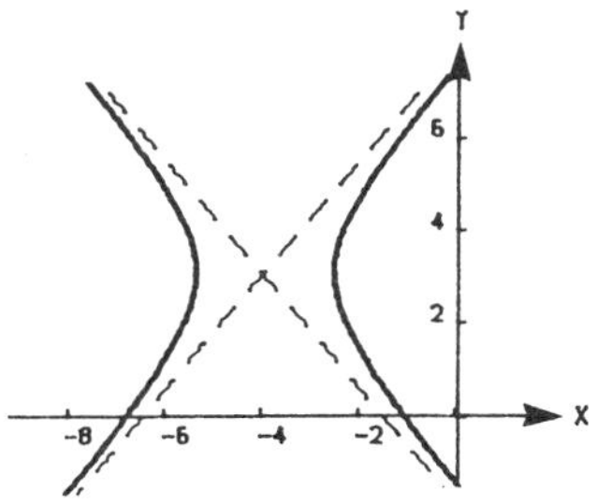

14. $4x^2 + y^2 - 16x + 15 = 0$

$4(x^2 - 4x + 4) + y^2 = -15 + 16$

$\frac{(x - 2)^2}{1/4} + \frac{y^2}{1} = 1$

Ellipse

Center: $(2, 0)$

Vertices: $(2, \pm 2)$

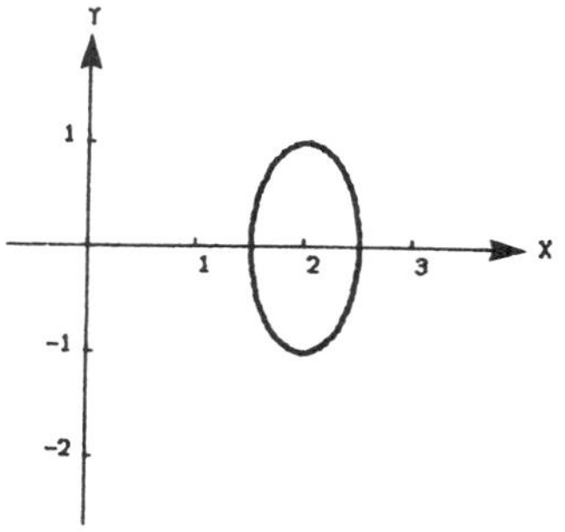

15. $3x^2 + 2y^2 - 12x + 12y + 29 = 0$

$3(x^2 - 4x + 4) + 2(y^2 + 6y + 9) = -29 + 12 + 18$

$\frac{(x - 2)^2}{1/3} + \frac{(y + 3)^2}{1/2} = 1$

Ellipse

Center: $(2, -3)$

Vertices: $(2, -3 \pm \frac{\sqrt{2}}{2})$

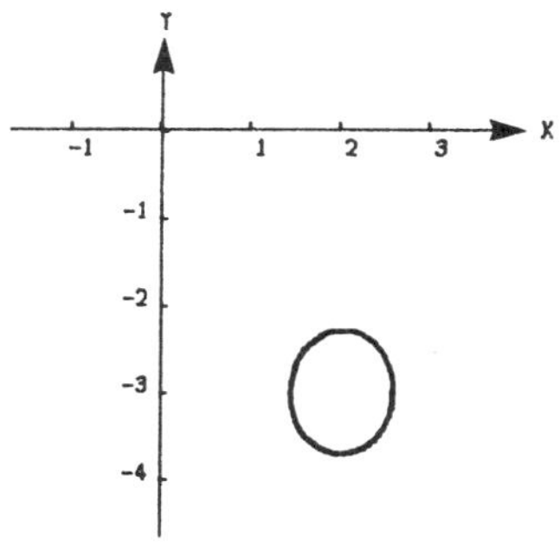

16. $4x^2 - 4y^2 - 4x + 8y - 11 = 0$

$$4\left(x^2 - x + \frac{1}{4}\right) - 4(y^2 - 2y + 1) = 11 + 1 - 4$$

$$\frac{[x - (1/2)]^2}{2} - \frac{(y - 1)^2}{2} = 1$$

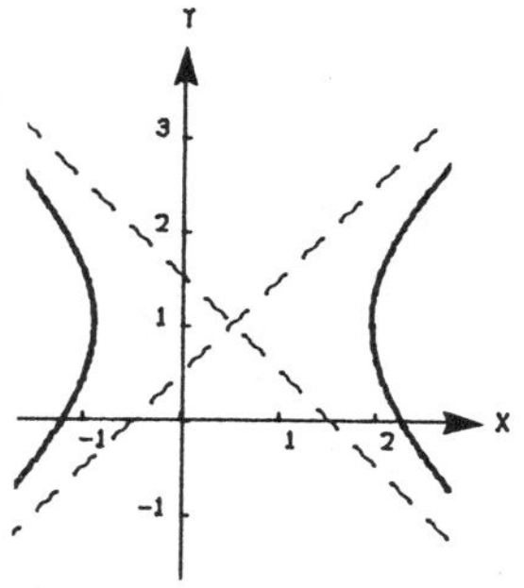

Hyperbola

Center: $\left(\frac{1}{2}, 1\right)$

Vertices: $\left(\frac{1}{2} \pm \sqrt{2}, 1\right)$

Asymptotes: $y = 1 \pm \left(x - \frac{1}{2}\right)$

17. $x^2 - 6x + 2y + 9 = 0$

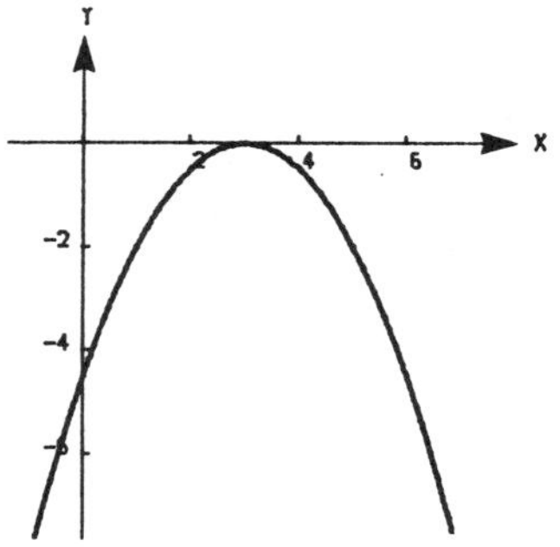

$$x^2 - 6x + 9 = -2y - 9 + 9$$

$$(x - 3)^2 = -4\left(\frac{1}{2}\right)y$$

Parabola

Vertex: $(3, 0)$

18. $x^2 + y^2 - 2x - 4y + 5 = 0$

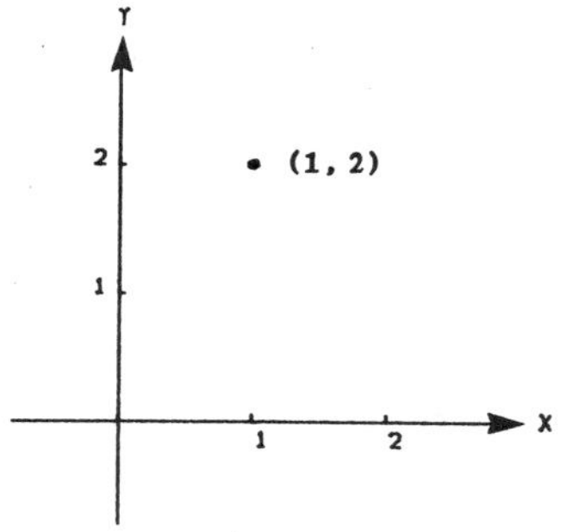

$$(x^2 - 2x + 1) + (y^2 - 4y + 4)$$

$$= -5 + 1 + 4$$

$$(x - 1)^2 + (y - 2)^2 = 0$$

Single point

$(1, 2)$

19. $x^2 + y^2 + 2xy + 2\sqrt{2}x - 2\sqrt{2}y + 2 = 0$

$A = 1,\ B = 2,\ C = 1,\ D = 2\sqrt{2},\ E = -2\sqrt{2},\ F = 2$

$$\cot 2\theta = 0, \quad 2\theta = \frac{\pi}{2}, \quad \theta = \frac{\pi}{4}$$

$$\sin\theta = \frac{\sqrt{2}}{2}, \quad \cos\theta = \frac{\sqrt{2}}{2}$$

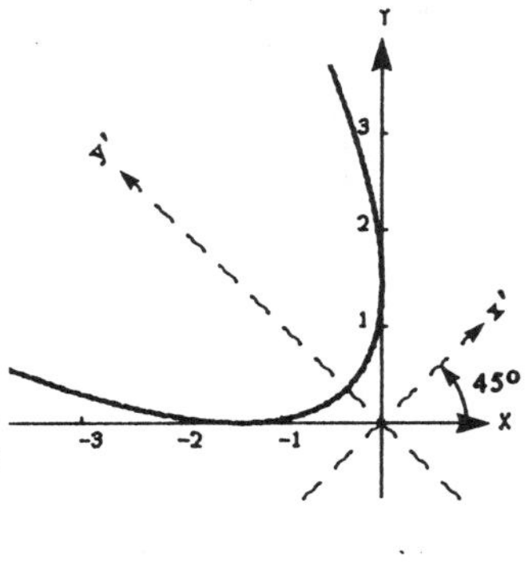

$$A' = \left(\frac{\sqrt{2}}{2}\right)^2 + 2\left(\frac{\sqrt{2}}{2}\right)^2 + \left(\frac{\sqrt{2}}{2}\right)^2 = 2$$

$$C' = \left(\frac{\sqrt{2}}{2}\right)^2 - 2\left(\frac{\sqrt{2}}{2}\right)^2 + \left(\frac{\sqrt{2}}{2}\right)^2 = 0$$

19. (continued)

$$D' = 2\sqrt{2}\left(\frac{\sqrt{2}}{2}\right) - 2\sqrt{2}\left(\frac{\sqrt{2}}{2}\right) = 0$$

$$E' = -2\sqrt{2}\left(\frac{\sqrt{2}}{2}\right) - 2\sqrt{2}\left(\frac{\sqrt{2}}{2}\right) = -4, \qquad F' = 2$$

$$2x'^2 - 4y' + 2 = 0, \quad x'^2 = 4\left(\frac{1}{2}\right)\left(y' - \frac{1}{2}\right)$$

Parabola

Vertex: $(x', y') = \left(0, \frac{1}{2}\right)$ or $(x, y) = \left(-\frac{1}{\sqrt{8}}, \frac{1}{\sqrt{8}}\right)$

20. $9x^2 + 6y^2 + 4xy - 20 = 0$

$A = 9,\ B = 4,\ C = 6,\ D = 0,\ E = 0,\ F = -20$

$\cot 2\theta = \frac{3}{4}, \quad \cos 2\theta = \frac{3}{5}, \quad \theta \approx 26.6^\circ$

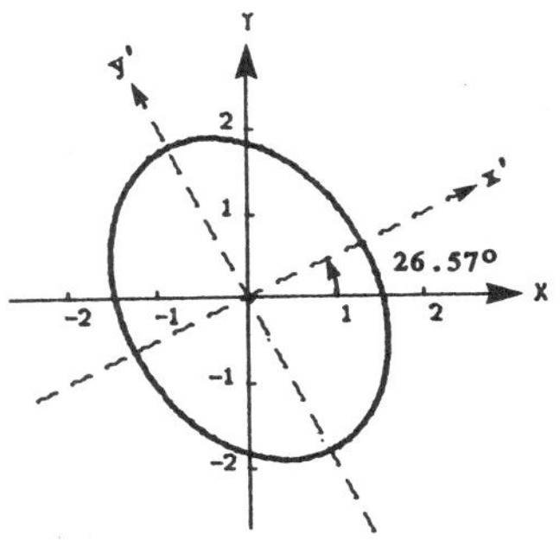

$$\sin\theta = \frac{\sqrt{1 - (3/5)}}{\sqrt{2}} = \frac{1}{\sqrt{5}}$$

$$\cos\theta = \frac{\sqrt{1 + (3/5)}}{\sqrt{2}} = \frac{2}{\sqrt{5}}$$

$$A' = 9\left(\frac{4}{5}\right) + 4\left(\frac{1}{\sqrt{5}}\right)\left(\frac{2}{\sqrt{5}}\right) + 6\left(\frac{1}{5}\right) = 10$$

$$C' = 9\left(\frac{1}{5}\right) - 4\left(\frac{1}{\sqrt{5}}\right)\left(\frac{2}{\sqrt{5}}\right) + 6\left(\frac{4}{5}\right) = 5$$

$D' = E' = 0, \quad F' = -20$

$$10x'^2 + 5y'^2 - 20 = 0, \qquad \frac{x'^2}{2} + \frac{y'^2}{4} = 1$$

Ellipse

Vertices: $(x', y') = (0, \pm 2)$

21. $4x^2 + 9y^2 - 8x + 9y + 4 = 0$

$$4(x^2 - 2x + 1) + 9\left(y^2 + y + \frac{1}{4}\right) = -4 + 4 + \frac{9}{4}$$

$$\frac{(x-1)^2}{9/16} + \frac{(y + (1/2))^2}{1/4} = 1$$

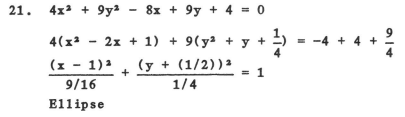

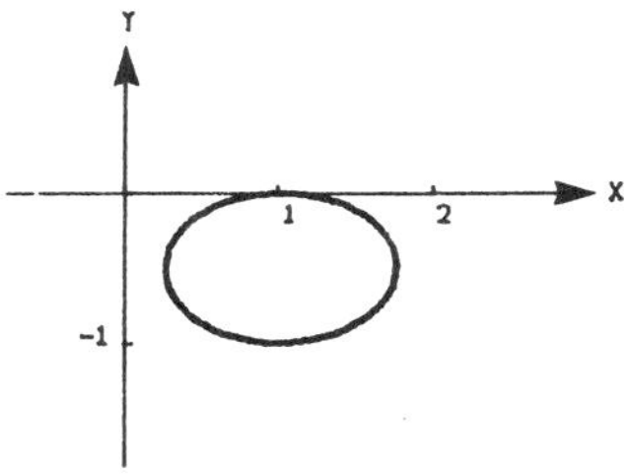

Ellipse

Center: $\left(1, -\frac{1}{2}\right)$

Vertices: $\left(1 \pm \frac{3}{4}, -\frac{1}{2}\right)$

22. $9x^2 + 4y^2 - 36x + 8y + 31 = 0$

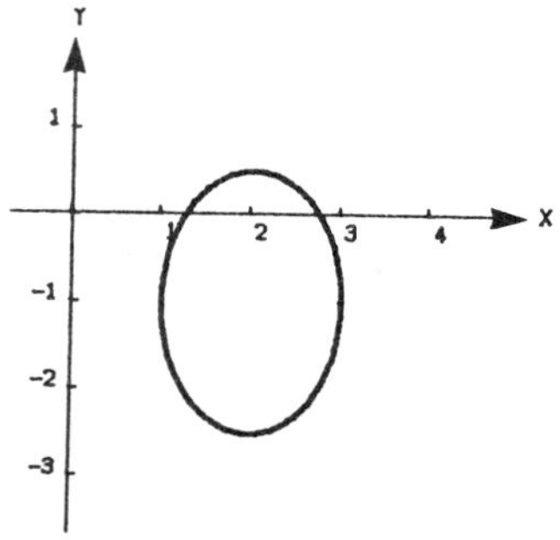

$9(x^2 - 4x + 4) + 4(y^2 + 2y + 1)$

$= -31 + 36 + 4$

$\dfrac{(x - 2)^2}{1} + \dfrac{(y + 1)^2}{9/4} = 1$

Ellipse

Center: $(2, -1)$

Vertices: $(2, -1 \pm \frac{3}{2})$

23. $\dfrac{y^2}{1} - \dfrac{x^2}{8} = 1$, since center = $(0, 0)$, $a = 1$, $c = 3$, $b^2 = 8$

transverse axis vertical

24. $\dfrac{x^2}{4} - \dfrac{(y - 2)^2}{12} = 1$, since center = $(0, 2)$, $a = 2$, $c = 4$, $b^2 = 12$

transverse axis horizontal

25. $\dfrac{(x - 2)^2}{25} + \dfrac{y^2}{21} = 1$, since center = $(2, 0)$, $c = 2$, $a = 5$, $b^2 = 21$

major axis horizontal

26. $p = 2$, $(x - 4)^2 = -4(2)(y - 2)$, parabola opens downward

27. A parabola with vertex at $(0, 0)$, $p = \sqrt{2}$ and opening toward the positive x' axis if the axis has been rotated 45°.

Therefore, $y'^2 = 4\sqrt{2}x'$. From the proof of Theorem 10.8,

$x' = x\cos\theta + y\sin\theta = \dfrac{x + y}{\sqrt{2}}$, $y' = -x\cos\theta + y\cos\theta = \dfrac{-x + y}{\sqrt{2}}$

Therefore, $(\dfrac{-x + y}{\sqrt{2}})^2 = 4\sqrt{2}(\dfrac{x + y}{\sqrt{2}})$ or $x^2 - 2xy + y^2 - 8x - 8y = 0$

28. $\dfrac{(x - 2)^2}{3} + \dfrac{(y - 2)^2}{4} = 1$, since center = $(2, 2)$, $c = 1$, $a = 2$, $b^2 = 3$

major axis vertical

29. Substituting the values of the coordinates of the given points into $\dfrac{x^2}{b^2} + \dfrac{y^2}{a^2} = 1$, we obtain the system $\dfrac{1}{b^2} + \dfrac{4}{a^2} = 1$, $\dfrac{4}{b^2} = 1$

Solving the system, we have $a^2 = \dfrac{16}{3}$ and $b^2 = 4$, $\dfrac{x^2}{4} + \dfrac{y^2}{16/3} = 1$

30. The transverse axis is horizontal with $c = 4$. Also from the slopes of the asymptotes $\frac{\pm b}{a} = \pm 2$ or $\pm b = \pm 2a$. Now $c^2 = a^2 + b^2 = a^2 + (2a)^2 = 16$

Therefore, $a^2 = \frac{16}{5}$, $b^2 = \frac{64}{5}$ and $\frac{x^2}{16/5} - \frac{y^2}{64/5} = 1$, $\frac{5x^2}{16} - \frac{5y^2}{64} = 1$

31. The transverse axis is horizontal with $c = 4$ and $2a = 4$. Therefore,

$a^2 = 4$, $b^2 = c^2 - a^2 = 12$, $\frac{x^2}{4} - \frac{y^2}{12} = 1$

32. The parabola opens downward with vertex $(0, 2)$. Therefore,

$x^2 = -4p(y - 2)$. Substituting the coordinates of the point $(-1, 0)$, we have: $1 = (-4p)(-2)$ or $p = \frac{1}{8}$. Therefore, $x^2 = -\frac{1}{2}(y - 2)$

33. $y = x^2 - 2x + 2$ and the slope of the line perpendicular to $y = x - 2$ is -1. Therefore, $y' = 2x - 2 = -1$ then $x = \frac{1}{2}$. At $x = \frac{1}{2}$, $y = \frac{5}{4}$

Therefore, $y - \frac{5}{4} = -1(x - \frac{1}{2})$ or $4x + 4y - 7 = 0$

34. The slope of $x - 2y - 12 = 0$ is $\frac{1}{2}$. $x^2 + 4y^2 - 4x - 8y - 24 = 0$,

$2x + 8yy' - 4 - 8y' = 0$, $y' = \frac{2 - x^2}{4(y - 1)} = \frac{1}{2}$ implies $x = 4 - 2y$

Substituting this into the equation of the ellipse, we have

$(4 - 2y)^2 + 4y^2 - 4(4 - 2y) - 24 = 0$, $(y - 3)(y + 1) = 0$

If $y = -1$, $x = 6$ and we have $y + 1 = \frac{1}{2}(x - 6)$ or $x - 2y - 8 = 0$.

If $y = 3$, $x = -2$ and we have $y - 3 = \frac{1}{2}(x + 2)$ or $x - 2y + 8 = 0$

35. $\frac{x^2}{a^2} - \frac{y^2}{4} = 1$, $\frac{2x}{a^2} - \frac{2y'}{4} = 0$, $y' = \frac{4x}{a^2y} = 2$, the slope of $2x - y - 4 = 0$

Therefore, $x = \frac{1}{2}a^2y$. Substituting this into the equation of the line, we have $a^2y - y - 4 = 0$ or $y = \frac{4}{a^2 - 1}$. Now, substituting for x and y in the hyperbola: $\frac{a^4[4/(a^2 - 1)]^2}{4a^2} - \frac{[4/(a^2 - 1)]^2}{4} = 1$,

$\frac{4}{(a^2 - 1)^2}(a^2 - 1) = 1$, $4 = a^2 - 1$, $a = \sqrt{5}$

36. $y = \frac{1}{200}x^2$

(a) $x^2 = 200y$, $x^2 = 4(50)y$, focus: $(0, 50)$

(b) $x = 10\sqrt{2y}$, $x' = \frac{5\sqrt{2}}{\sqrt{y}}$, $\sqrt{1 + (x')^2} = \sqrt{1 + \frac{50}{y}}$

$$S = 2\pi \int_0^{100} 10\sqrt{2}\sqrt{y}\sqrt{1 + \frac{50}{y}}\, dy = 20\sqrt{2}\pi \int_0^{100} \sqrt{y + 50}\, dy$$

$$= \frac{40\sqrt{2}\pi}{3}(y + 50)^{3/2}\Big]_0^{100} = \frac{20,000\pi}{3}(3^{3/2} - 1)$$

37. $A = 4\int_0^a \frac{b}{a}\sqrt{a^2 - x^2}\, dx = \frac{4b}{a}(\frac{1}{2})\left[x\sqrt{a^2 - x^2} + a^2 \arcsin(\frac{x}{a})\right]_0^a = \pi ab$

38. (a) <u>Disc</u>

$$V = 2\pi \int_0^b \frac{a^2}{b^2}(b^2 - y^2)\, dy = \frac{2\pi a^2}{b^2}\int_0^b (b^2 - y^2)\, dy$$

$$= \frac{2\pi a^2}{b^2}\left[b^2 y - \frac{1}{3}y^3\right]_0^b = \frac{4}{3}\pi a^2 b$$

(b) $S = 4\pi \int_0^b \frac{a}{b}\sqrt{b^2 - y^2}\left(\frac{\sqrt{b^4 + (a^2 - b^2)y^2}}{b\sqrt{b^2 - y^2}}\right) dy = \frac{4\pi a}{b^2}\int_0^b \sqrt{b^4 + c^2y^2}\, dy$

$$= \frac{2\pi a}{b^2 c}\left[cy\sqrt{b^4 + c^2y^2} + b^4 \ln|cy + \sqrt{b^4 + c^2y^2}|\right]_0^b$$

$$= \frac{2\pi a}{b^2 c}\left[b^2 c\sqrt{b^2 + c^2} + b^4 \ln|cb + b\sqrt{b^2 + c^2}| - b^4 \ln(b^2)\right]$$

$$= 2\pi a^2 + \frac{\pi ab^2}{c}\ln(\frac{c + a}{e})^2 = 2\pi a^2 + (\frac{\pi b^2}{e})\ln(\frac{1 + e}{1 - e})$$

39. (a) <u>Disc</u>

$$V = 2\pi \int_0^a \frac{b^2}{a^2}(a^2 - x^2)\, dx = \frac{2\pi b^2}{a^2}\int_0^a (a^2 - x^2)\, dx$$

$$= \frac{2\pi b^2}{a^2}\left[a^2 x - \frac{1}{3}x^3\right]_0^a = \frac{4}{3}\pi ab^2$$

39. (b) $S = 2(2\pi)\int_0^a \frac{b}{a}\sqrt{a^2 - x^2}\left(\frac{\sqrt{a^4 - (a^2 - b^2)x^2}}{a\sqrt{a^2 - x^2}}\right) dx$

$= \frac{4\pi b}{a^2}\int_0^a \sqrt{a^4 - c^2x^2}\, dx$

$= \frac{2\pi b}{a^2 c}\left[cx\sqrt{a^4 - c^2x^2} + a^4 \arcsin\left(\frac{cx}{a^2}\right)\right]_0^a$

$= \frac{2\pi b}{a^2 c}\left[a^2 c\sqrt{a^2 - c^2} + a^4 \arcsin\left(\frac{c}{a}\right)\right]$

$= 2\pi b^2 + 2\pi\left(\frac{ab}{e}\right) \arcsin(e)$

40. $\frac{x^2}{9} + \frac{y^2}{4} = 1, \quad a = 3, \quad b = 2, \quad c = \sqrt{5}, \quad e = \frac{\sqrt{5}}{3}$

<u>Prolate</u> <u>spheroid</u>:

$V = \frac{4\pi}{3}(3)(4) = 16\pi$

$S = 2\pi(4) + 2\pi\left(\frac{(3)(2)}{\sqrt{5}/3}\right) \arcsin\left(\frac{\sqrt{5}}{3}\right)$

$= 8\pi + \left(\frac{36\pi}{\sqrt{5}}\right) \arcsin\left(\frac{\sqrt{5}}{3}\right) \approx 67.673$

<u>Oblate</u> <u>spheroid</u>:

$V = \frac{4\pi}{3}(9)2 = 24\pi$

$S = 18\pi + \pi\left(\frac{4}{\sqrt{5}/3}\right) \ln\left(\frac{1 + (\sqrt{5}/3)}{1 - (\sqrt{5}/3)}\right)$

$= 18\pi + \frac{12\pi}{\sqrt{5}} \ln\left(\frac{3 + \sqrt{5}}{3 - \sqrt{5}}\right) \approx 89.001$

41. $\frac{x^2}{9} + \frac{y^2}{4} = 1, \quad a = 3, \quad b = 2, \quad c = \sqrt{5}, \quad e = \frac{\sqrt{5}}{3}$

By example 5 of Section 10.2: $C = 12\int_0^{\pi/2} \sqrt{1 - (5/9)\sin^2\theta}\, d\theta$

$C \approx \frac{12(\pi/2)}{12}\Big[\sqrt{1 - (5/9)\sin^2(0)} + 4\sqrt{1 - (5/9)\sin^2(\pi/8)}$

$+ 2\sqrt{1 - (5/9)\sin^2(\pi/4)} + 4\sqrt{1 - (5/9)\sin^2(3\pi/8)}$

$+ \sqrt{1 - (5/9)\sin^2(\pi/2)}\Big] \approx 15.87$ (Simpson's Rule, $n = 4$)

42. $V = (\pi ab)(\text{length}) = 12\pi(16)$ [from Exercise 37] $= 192\pi\ \text{ft}^3$

43. Want $\frac{3}{4}$ of the total area, (12π from Exercise 37) covered

Find h so that

$$2\int_0^h \frac{4}{3}\sqrt{9 - y^2}\, dy = 3\pi$$

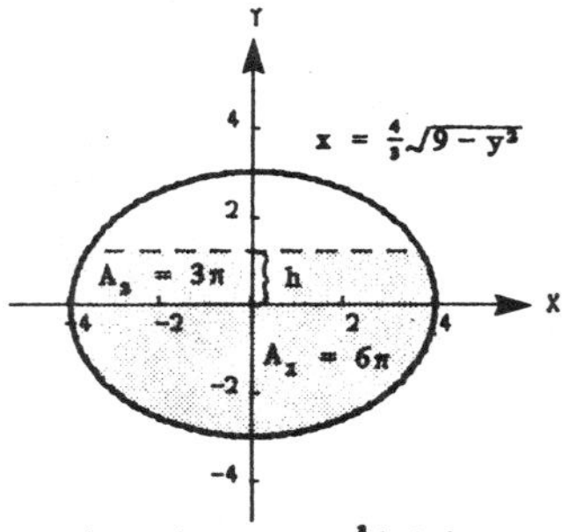

$A_1 + A_2 = 9\pi = \frac{3}{4}(12\pi)$

$$\int_0^h \sqrt{9 - y^2}\, dy = \frac{9\pi}{8}$$

$$\frac{1}{2}\left[y\sqrt{9 - y^2} + 9\arcsin\left(\frac{y}{3}\right)\right]_0^h = \frac{9\pi}{8}$$

$$h\sqrt{9 - h^2} + 9\arcsin\left(\frac{h}{3}\right) = \frac{9\pi}{4}$$

By Newton's Method $h \approx 1.212$

Therefore, the total height of the milk is $1.212 + 3 = 4.212$ ft.

44. $F = 2(62.4)\int_{-3}^{3}(3 - y)\frac{4}{3}\sqrt{9 - y^2}\, dy$

$$= \frac{8}{3}(62.4)\left[3\int_{-3}^{3}\sqrt{9 - y^2}\, dy - \int_{-3}^{3} y\sqrt{9 - y^2}\, dy\right]$$

$$= \frac{8}{3}(62.4)\left[\frac{3}{2}\left(y\sqrt{9 - y^2} + 9\arcsin\frac{y}{3}\right) + \frac{1}{3}(9 - y^2)^{3/2}\right]_{-3}^{3}$$

$$= \frac{8}{3}(62.4)\left[\frac{3}{2}\left(\frac{9\pi}{2}\right) - \frac{3}{2}\left(-\frac{9\pi}{2}\right)\right] = \frac{8}{3}(62.4)\left(\frac{27\pi}{2}\right)$$

$$\approx 7057.274$$

45. Area of ends $= 2(12\pi)$ [from Exercise 37] $= 24\pi$

Area of sides = (perimeter)(length)

$$= 16\int_0^{\pi/2}\left(\sqrt{1 - (7/16)\sin^2\theta}\, d\theta\right)(16)$$

[from example 5 of Section 10.2]

$$\approx 256\left(\frac{\pi/2}{12}\right)\left[\sqrt{1 - (7/16)\sin^2(0)} + 4\sqrt{1 - (7/16)\sin^2(\pi/8)} + 2\sqrt{1 - (7/16)\sin^2(\pi/4)} + 4\sqrt{1 - (7/16)\sin^2(3\pi/8)} + \sqrt{1 - (7/16)\sin^2(\pi/2)}\right] \approx 353.28$$

Total area $= 24\pi + 353.28 = 428.68\ \text{ft}^2$

11 Plane curves, parametric equations, and polar coordinates

11.1 Plane curves and parametric equations

1. $x = 3t - 1, \quad y = 2t + 1$

$y = 2\left(\frac{x + 1}{3}\right) + 1$

$2x - 3y + 5 = 0$

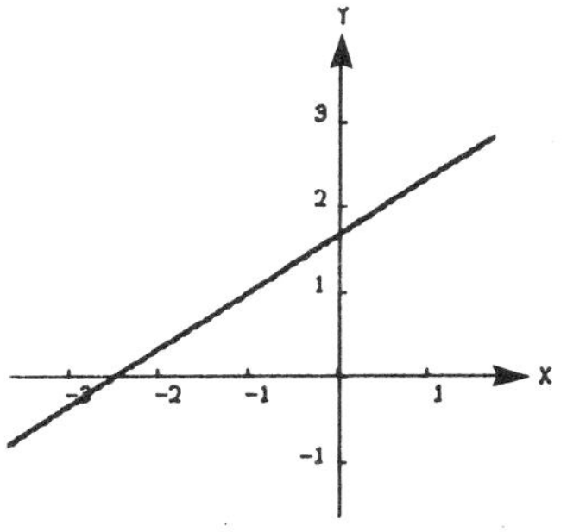

2. $x = 3 - 2t, \quad y = 2 + 3t$

$y = 2 + 3\left(\frac{3 - x}{2}\right)$

$3x + 2y = 13$

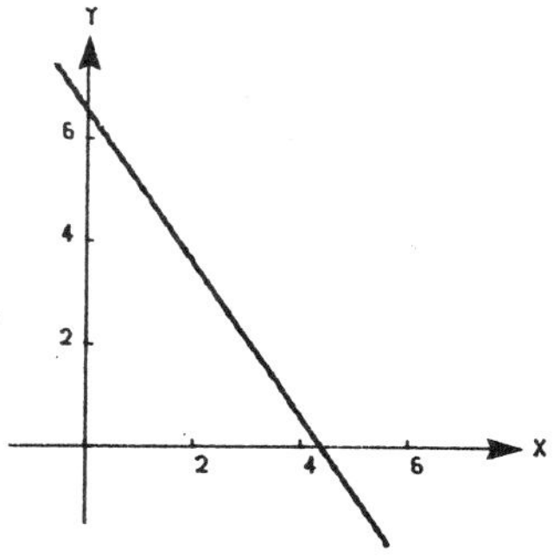

3. $x = \sqrt{t}, \quad y = 1 - t$

$y = 1 - x^2, \ x \geq 0$

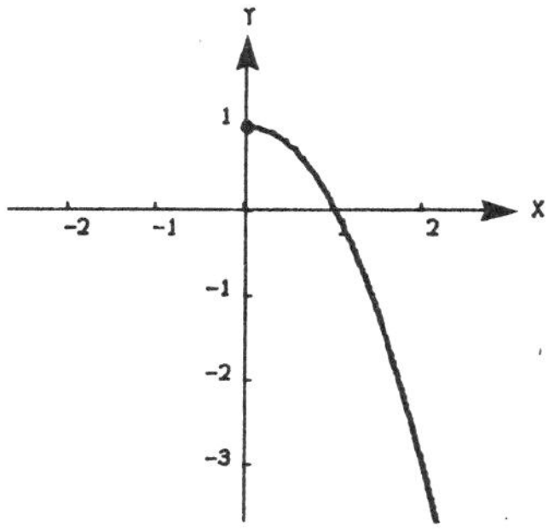

4. $x = \sqrt[3]{t}, \quad y = 1 - t$

$y = 1 - x^3$

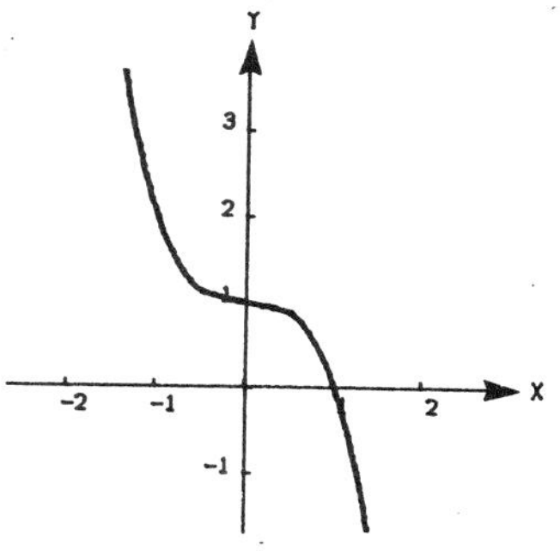

5. $x = t + 1, \quad y = t^2$
$y = (x - 1)^2$

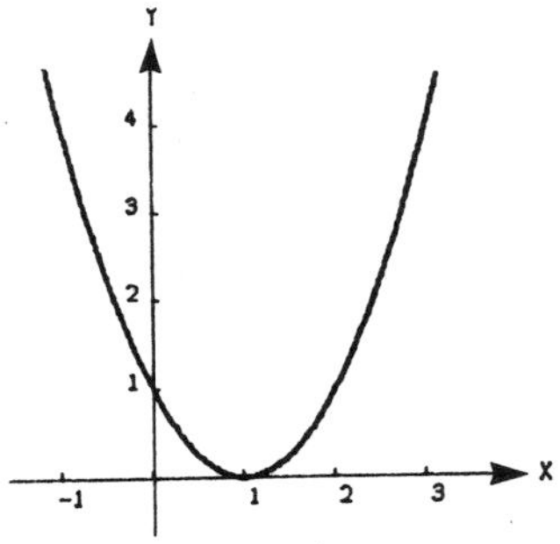

6. $x = t + 1, \quad y = t^3$
$y = (x - 1)^3$

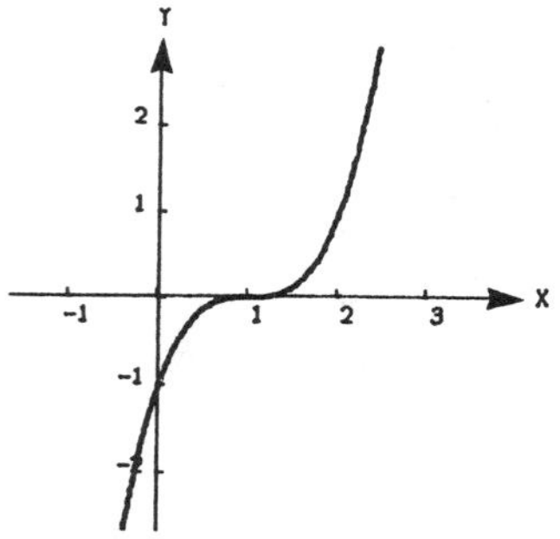

7. $x = t^3, \quad y = \frac{1}{2}t^2$

$x = t^3$ implies $t = x^{1/3}$

$y = \frac{1}{2}x^{2/3}$

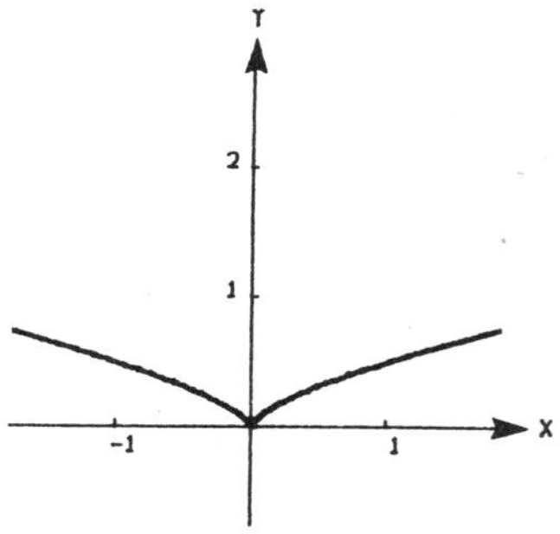

8. $x = 1 + \frac{1}{t}, \quad y = t - 1$

$x = 1 + \frac{1}{t}$ implies $t = \frac{1}{x - 1}$

$y = \frac{1}{x - 1} - 1, \quad y = \frac{2 - x}{x - 1}$

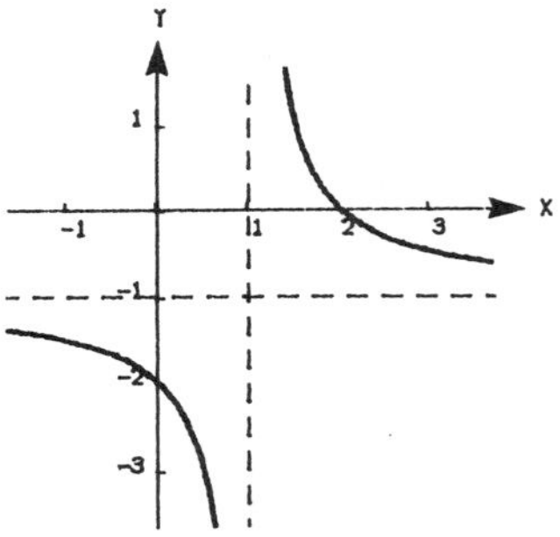

9. $x = t - 1, \quad y = \frac{t}{t - 1}$

$y = \frac{x + 1}{x}$

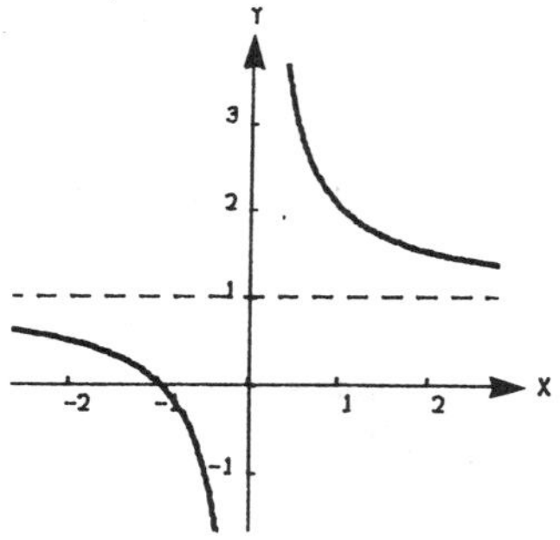

10. $x = t^2 + t, \quad y = t^2 - t$

Subtracting the second equation from the first, we have

$x - y = 2t \quad \text{or} \quad t = \dfrac{x - y}{2}$

$y = \dfrac{(x - y)^2}{4} - \dfrac{x - y}{2}$

$0 = x^2 - 2xy + y^2 - 2x - 2y$

t	0	1	−1	2	−2
x	0	2	0	6	2
y	0	0	2	2	6

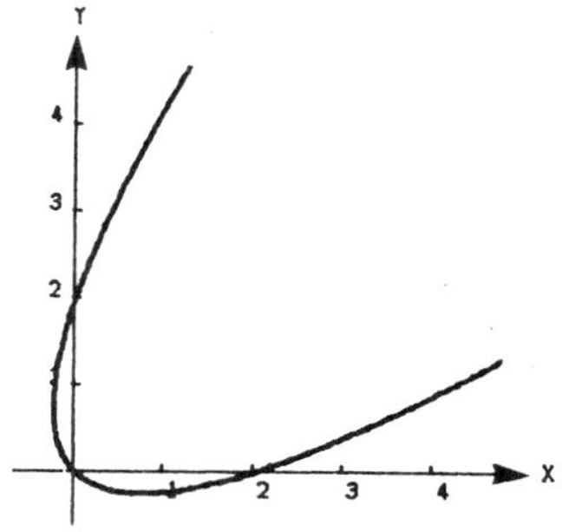

11. $x = 2t, \quad y = |t - 2|$

$y = \left|\dfrac{x}{2} - 2\right| = \dfrac{|x - 4|}{2}$

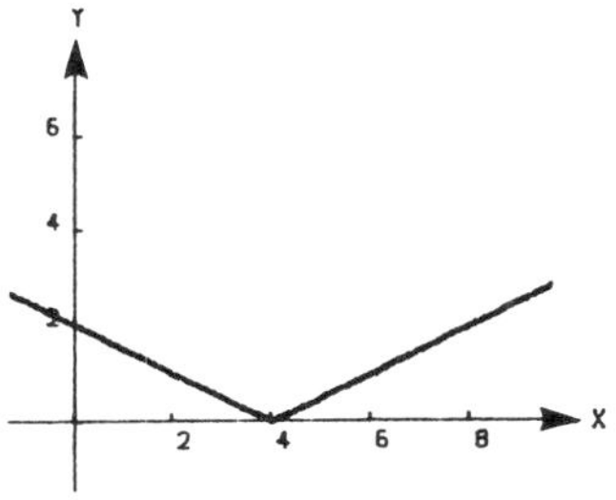

12. $x = |t - 1|, \quad y = t + 2$

$x = |(y - 2) - 1| = |y - 3|$

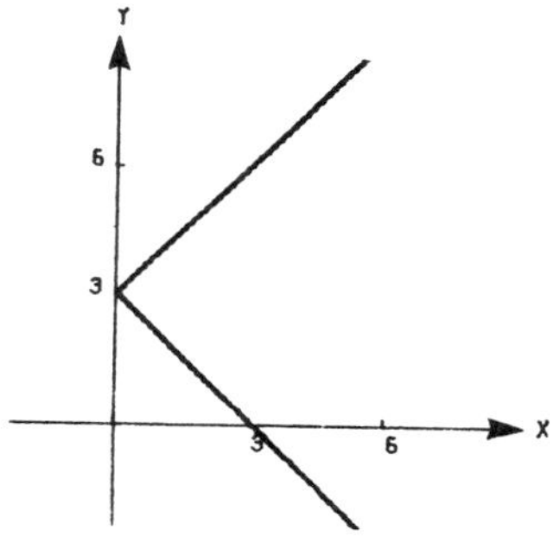

13. $x = \sec\theta, \quad y = \cos\theta$

$xy = 1, \quad y = \dfrac{1}{x}, \quad -1 \le y \le 1$

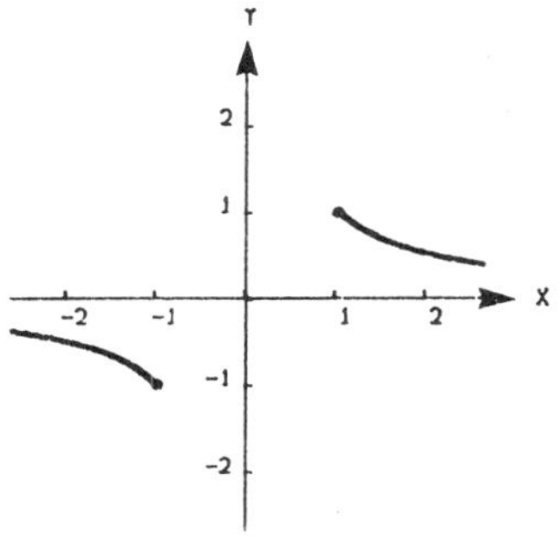

14. $x = \tan^2\theta, \quad y = \sec^2\theta$

$y = 1 + x, \quad x \ge 0, \quad y \ge 1$

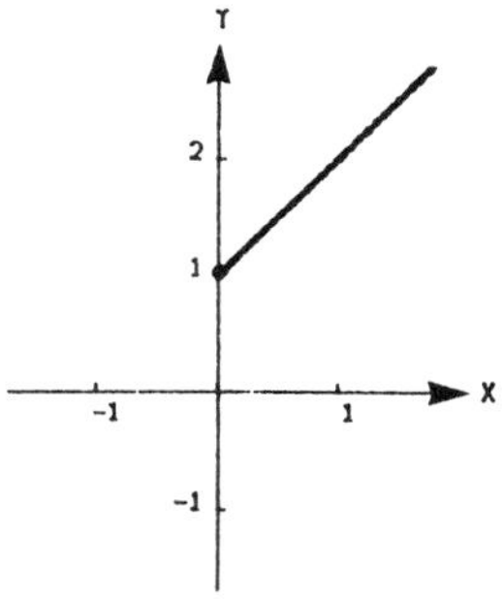

15. $x = 3\cos\theta, \quad y = 3\sin\theta$

Squaring both equations and adding, we have: $x^2 + y^2 = 9$

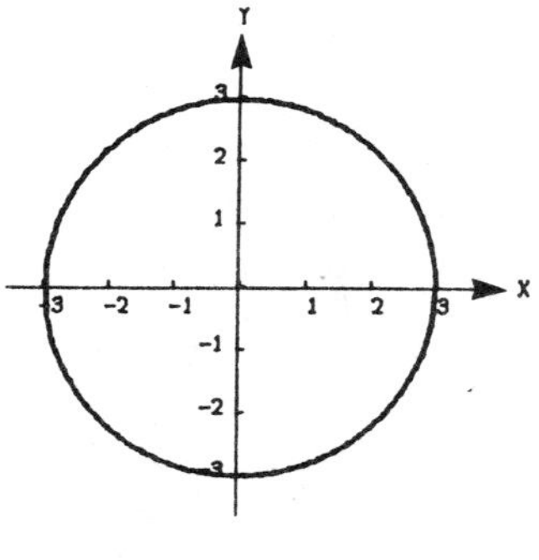

16. $x = \cos\theta, \quad y = 3\sin\theta$

$x^2 = \cos^2\theta, \quad \dfrac{y^2}{9} = \sin^2\theta$

$\dfrac{x^2}{1} + \dfrac{y^2}{9} = 1$

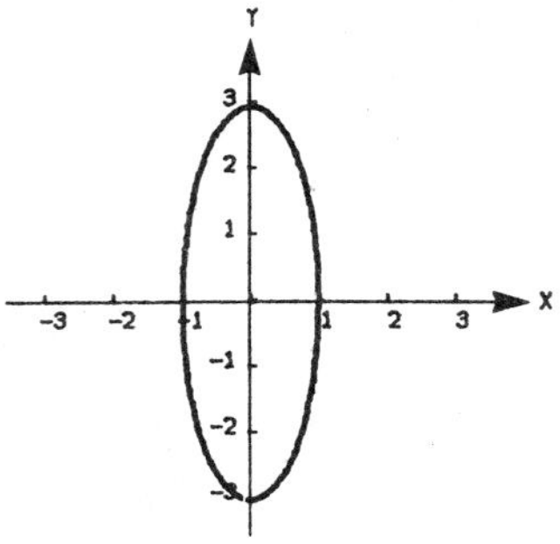

17. $x = 4\sin 2\theta, \quad y = 2\cos 2\theta$

$\dfrac{x^2}{16} = \sin^2 2\theta, \quad \dfrac{y^2}{4} = \cos^2 2\theta$

$\dfrac{x^2}{16} + \dfrac{y^2}{4} = 1$

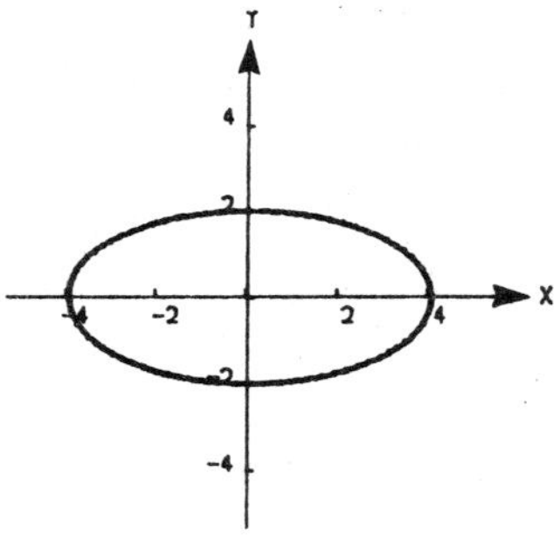

18. $x = \cos\theta, \quad y = 2\sin 2\theta$

$\theta = \arccos x, \quad y = 4\sin\theta\cos\theta$

$y = 4x\sin(\arccos x)$

$y = \pm 4x\sqrt{1 - x^2}, \quad y^2 = 16x^2(1 - x^2)$

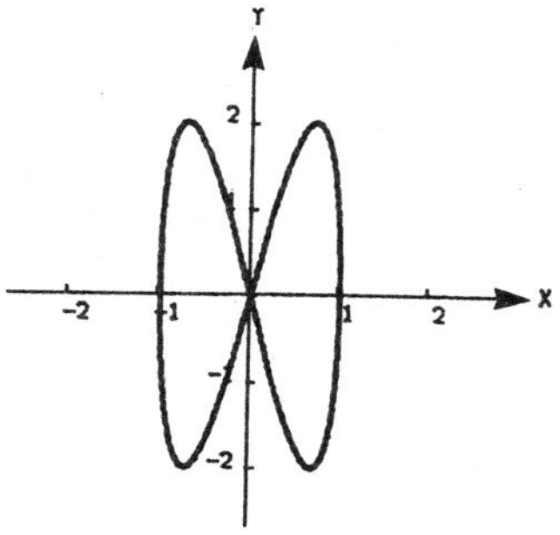

19. $x = \cos\theta, \quad y = 2\sin^2\theta$

$$x^2 + \frac{y}{2} = 1$$

$$y = 2 - 2x^2, \quad -1 \leq x \leq 1$$

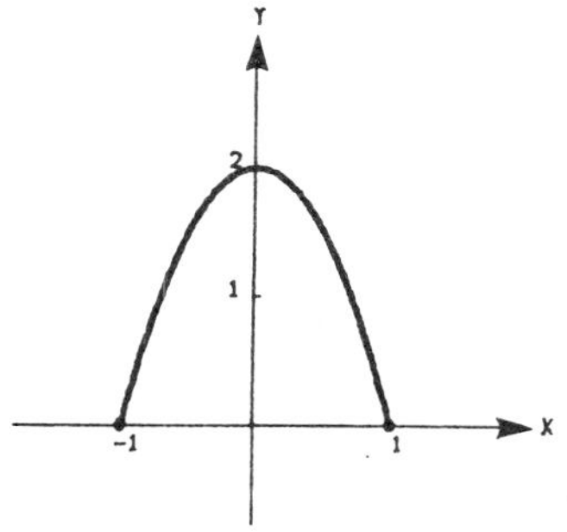

20. $x = 4\cos^2\theta, \quad y = 2\sin\theta$

$$\frac{x}{4} = \cos^2\theta, \quad \frac{y^2}{4} = \sin^2\theta$$

$$\frac{x}{4} + \frac{y^2}{4} = 1$$

$$x = 4 - y^2, \quad x \geq 0$$

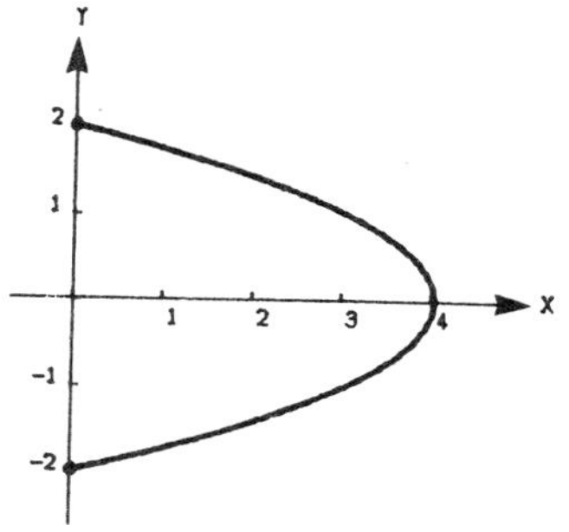

21. $x = 4 + 2\cos\theta, \quad y = -1 + \sin\theta$

$$\frac{(x-4)^2}{4} = \cos^2\theta$$

$$\frac{(y+1)^2}{1} = \sin^2\theta$$

$$\frac{(x-4)^2}{4} + \frac{(y+1)^2}{1} = 1$$

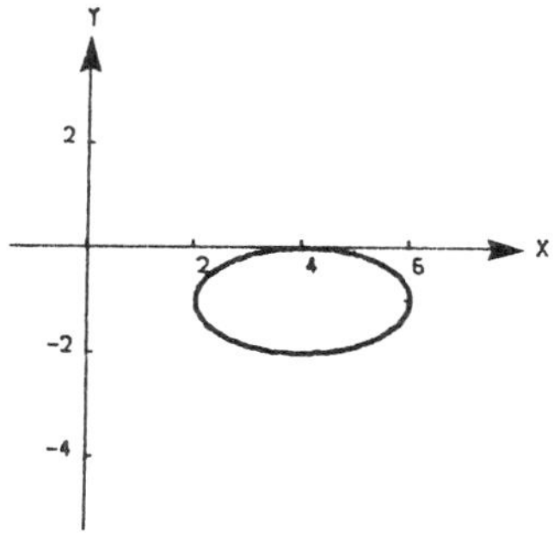

22. $x = 4 + 2\cos\theta, \; y = -1 + 2\sin\theta$

$$(x-4)^2 = 4\cos^2\theta$$

$$(y+1)^2 = 4\sin^2\theta$$

$$(x-4)^2 + (y+1)^2 = 4$$

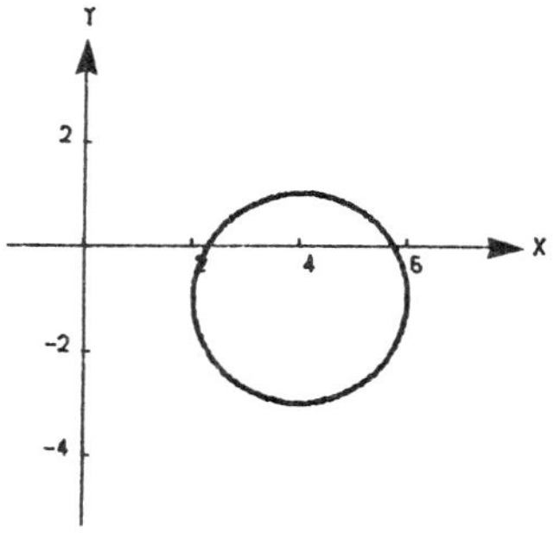

23. $x = 4 + 2\cos\theta,\ y = -1 + 4\sin\theta$

$$\frac{(x-4)^2}{4} = \cos^2\theta$$

$$\frac{(y+1)^2}{16} = \sin^2\theta$$

$$\frac{(x-4)^2}{4} + \frac{(y+1)^2}{16} = 1$$

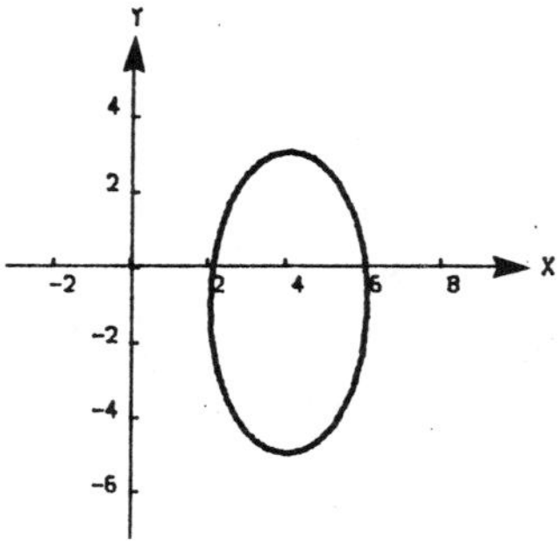

24. $x = \sec\theta,\ y = \tan\theta$

$$x^2 = \sec^2\theta,\ y^2 = \tan^2\theta$$

$$x^2 - y^2 = 1$$

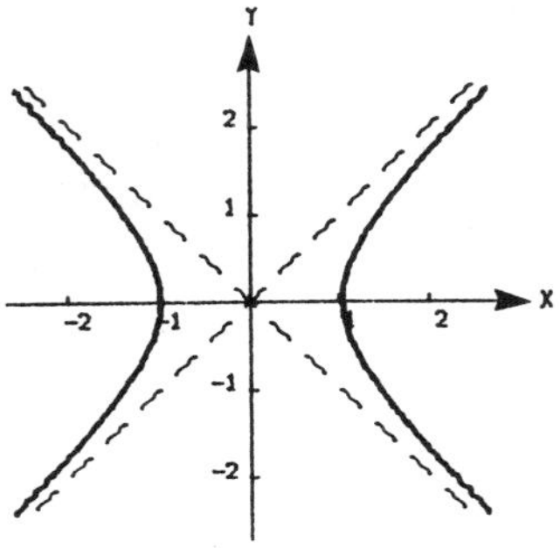

25. $x = 4\sec\theta,\quad y = 3\tan\theta$

$$\frac{x^2}{16} = \sec^2\theta,\ \frac{y^2}{9} = \tan^2\theta$$

$$\frac{x^2}{16} - \frac{y^2}{9} = 1$$

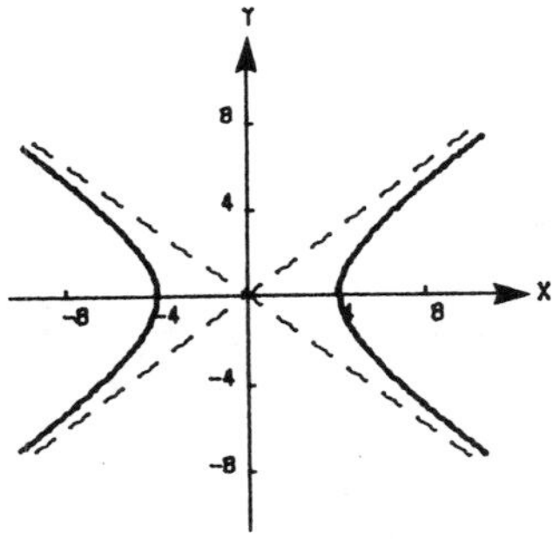

26. $x = \cos^3\theta,\ y = \sin^3\theta$

$$x^{2/3} = \cos^2\theta,\ y^{2/3} = \sin^2\theta$$

$$x^{2/3} + y^{2/3} = 1$$

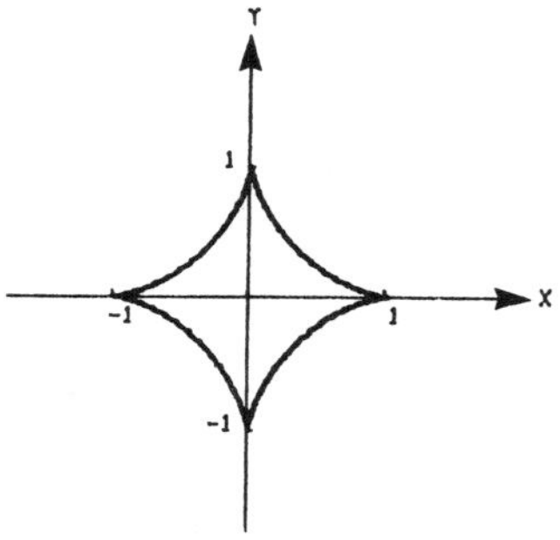

27. $x = t^3, \; y = 3 \ln t$

$y = 3 \ln \sqrt[3]{x} = \ln x$

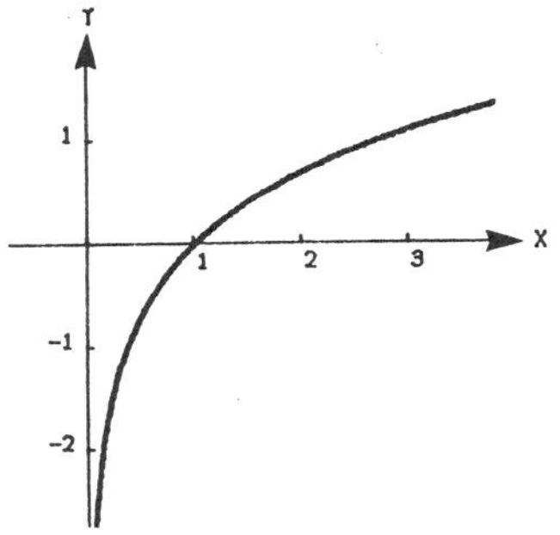

28. $x = e^{2t}, \quad y = e^t$

$y^2 = x, \quad y > 0$

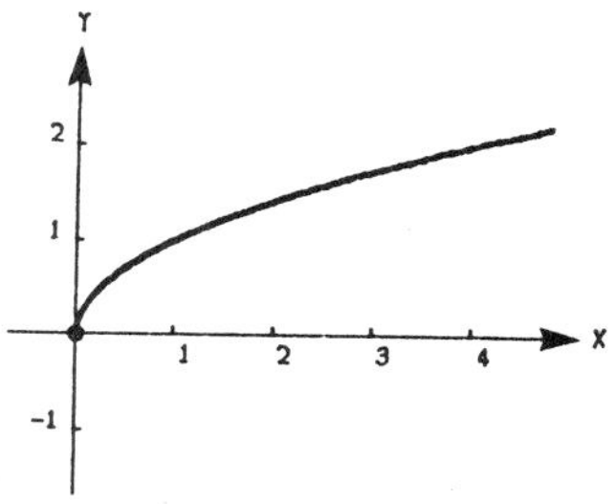

29. $x = e^{-t}, \quad y = e^{3t}$

$e^t = \dfrac{1}{x}, \quad e^t = \sqrt[3]{y}$

$\sqrt[3]{y} = \dfrac{1}{x}, \quad y = \dfrac{1}{x^3}, \quad x > 0, \quad y > 0$

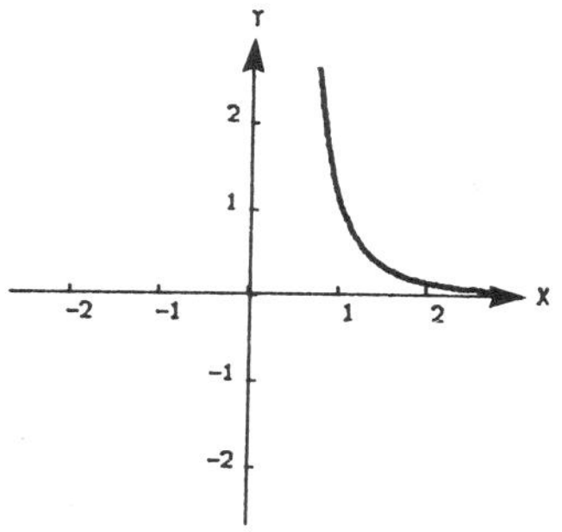

30. $x = \ln 2t, \quad y = t^2$

$t = \dfrac{e^x}{2}, \quad y = \dfrac{e^{2x}}{4} = \dfrac{1}{4} e^{2x}$

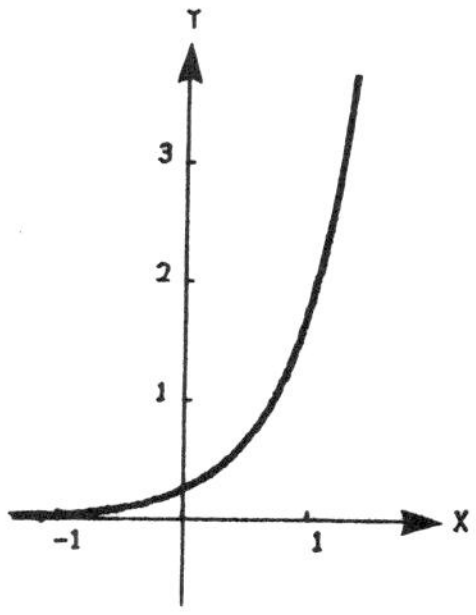

31. By eliminating the parameters in (a)-(d), we get $y = 2x + 1$. They differ from each other in orientation and in restricted domains.

(a) $x = t, \quad y = 2t + 1$

(b) $x = \cos\theta, \quad y = 2\cos\theta + 1$

$-1 \leq x \leq 1 \qquad -1 \leq y \leq 3$

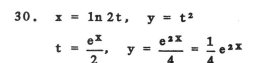

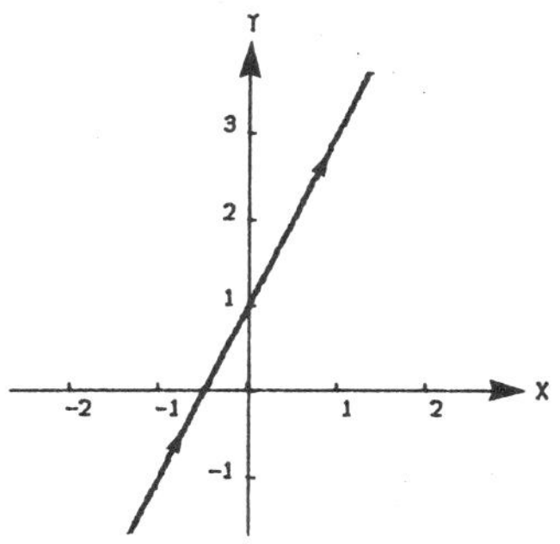

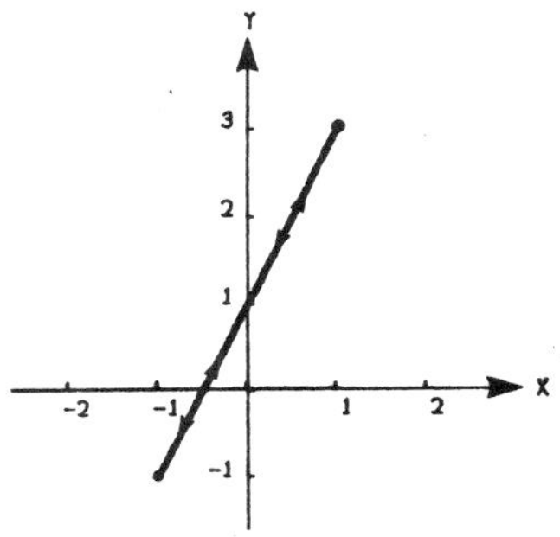

31. (c) $x = e^{-t}$, $y = 2e^{-t} + 1$

$x > 0$, $y > 1$

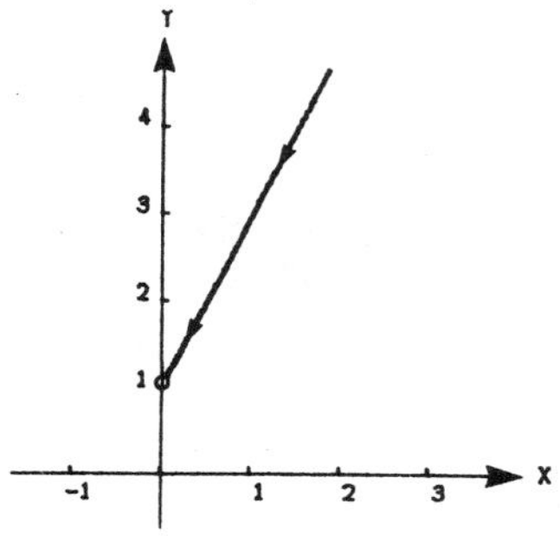

(d) $x = e^{t}$, $y = 2e^{t} + 1$

$x > 0$, $y > 1$

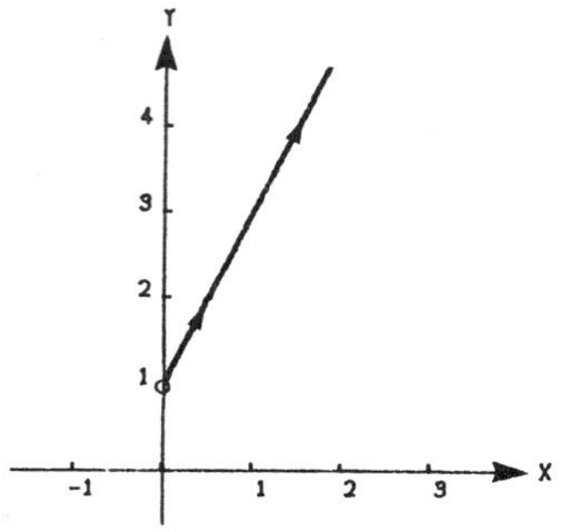

32. By eliminating the parameters in (a)-(d), we get $x^2 + y^2 = 4$. They differ from each other in orientation and in restricted domains.

(a) $x = 2\cos\theta$, $y = 2\sin\theta$

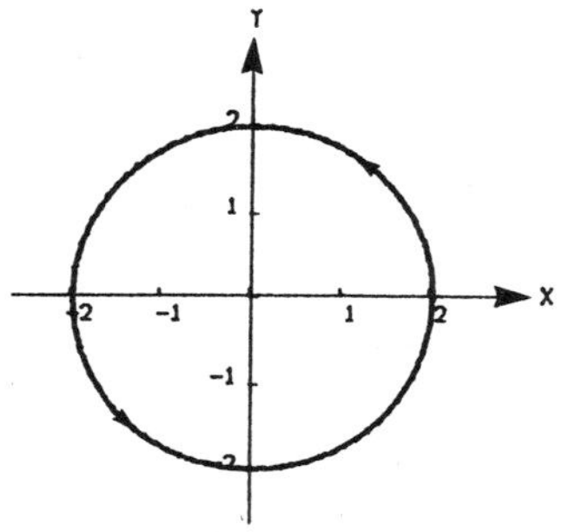

(b) $x = \dfrac{\sqrt{4t^2 - 1}}{|t|}$, $y = \dfrac{1}{t}$

$x \geq 0$, $y \neq 0$

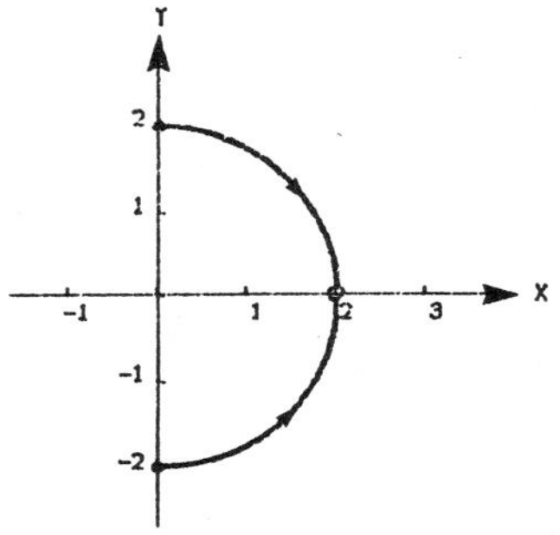

(c) $x = \sqrt{t}$, $y = \sqrt{4 - t}$

$x \geq 0$, $y \geq 0$

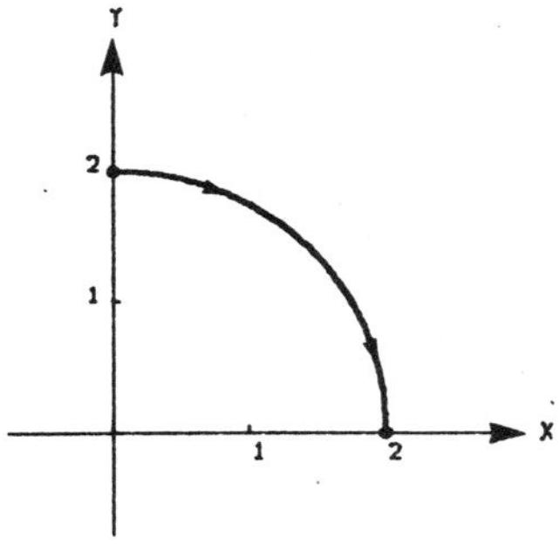

(d) $x = -\sqrt{4 - e^{2t}}$, $y = e^{t}$

$-2 < x \leq 0$, $y > 0$

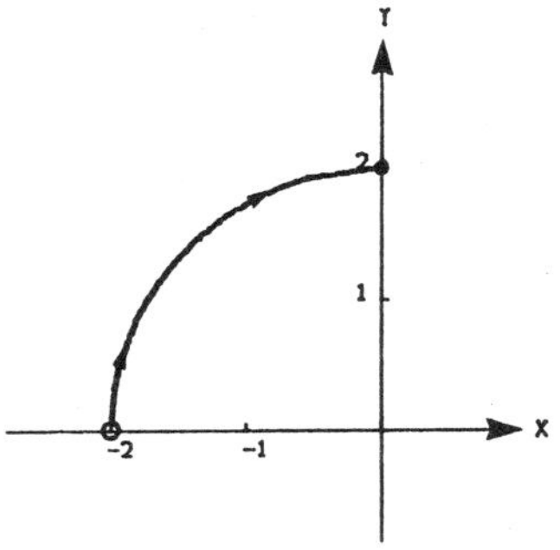

33. $x = 2(\theta - \sin\theta)$, $y = 2(1 - \cos\theta)$

θ	0	$\pi/6$	$\pi/4$	$\pi/3$	$\pi/2$	$3\pi/4$	π	$5\pi/4$	$3\pi/2$	$7\pi/4$	2π
x	0	0.047	0.157	0.362	1.142	3.298	6.283	9.268	11.423	12.410	12.566
y	0	0.268	0.586	1	2	3.414	4	3.414	2	0.586	0

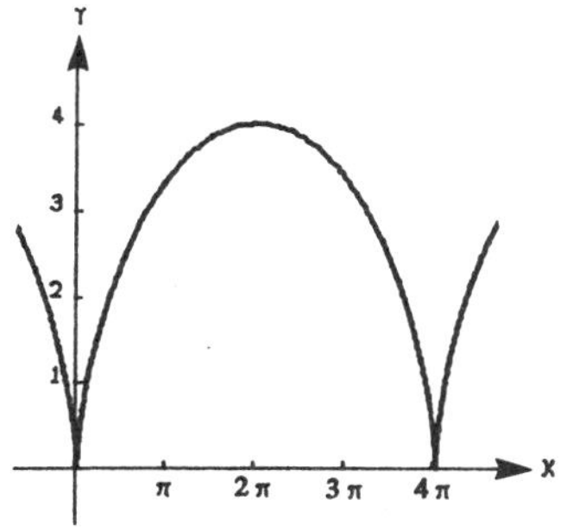

34. $x = \theta + \sin\theta$, $y = 1 - \cos\theta$

θ	0	$\pi/6$	$\pi/4$	$\pi/3$	$\pi/2$	...	π	$3\pi/2$	2π
x	0	1.02	1.49	1.91	2.52		3.14	3.71	6.28
y	0	0.13	0.29	0.5	1		2	1	0

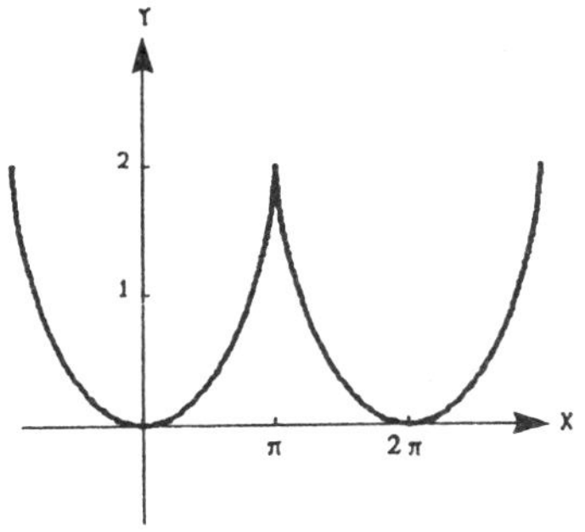

35. $x = 2\cot\theta$, $y = 2\sin^2\theta$

θ	$\pi/18$	$\pi/6$	$\pi/4$	$\pi/3$	$\pi/2$
x	11.34	3.46	2	1.15	0
y	0.06	0.05	1	1.5	2

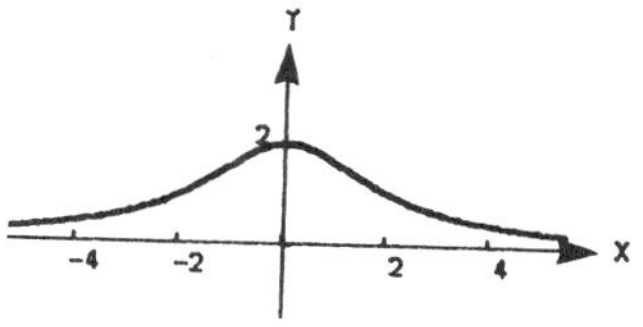

36. $x = 2\theta - \sin\theta, \quad y = 2 - \cos\theta$

θ	0	$\pi/6$	$\pi/4$	$\pi/3$	$\pi/2$	$3\pi/4$	π	$3\pi/2$	2π	$5\pi/4$	$7\pi/4$
x	0	0.55	0.86	1.23	2.14	4.01	6.28	10.42	12.57	8.56	11.70
y	1	1.13	1.29	1.50	2	2.71	3	2	1	2.71	1.29

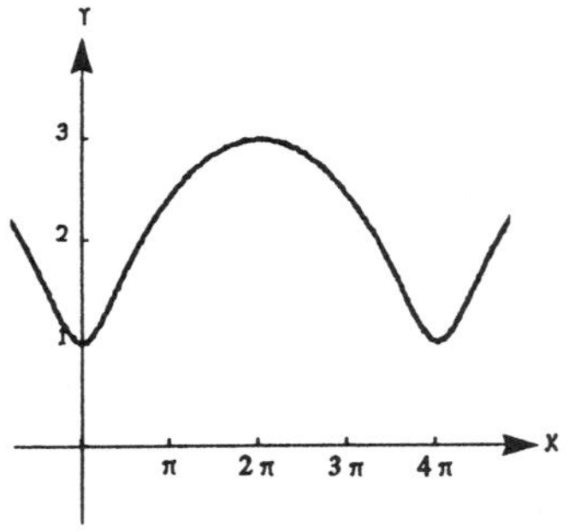

37. $x = \theta - \frac{3}{2}\sin\theta, \quad y = 1 - \frac{3}{2}\cos\theta$

θ	0	$\pi/6$	$\pi/4$	$\pi/3$	$\pi/2$	$3\pi/4$	π	$5\pi/4$	$3\pi/2$	$7\pi/4$	2π
x	0	−0.23	−0.28	−0.25	0.07	1.30	3.14	4.99	6.21	6.56	6.28
y	−0.5	−0.30	−0.06	0.25	1.00	2.06	2.50	2.06	1.00	−0.06	−0.50

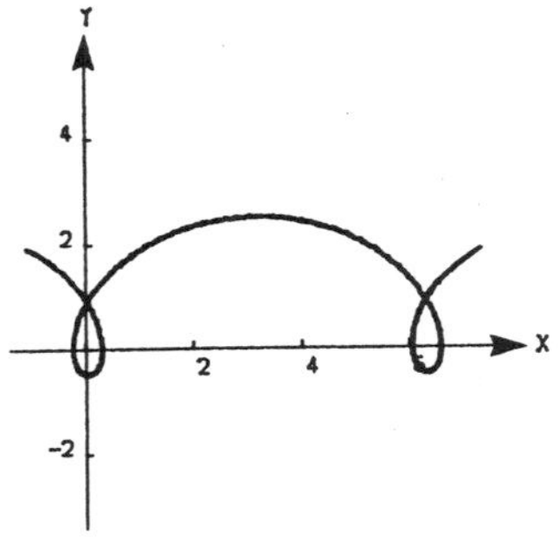

38. $x = \dfrac{3t}{1 + t^3}$, $y = \dfrac{3t^2}{1 + t^3}$ undefined when $t = -1$

t	-10	-5	-2	-1.4	-1.2	-1	-0.8	-0.4
x	0.03	0.12	0.86	2.41	4.95	-	-4.90	-1.28
y	-0.30	-0.60	-1.71	-3.37	-5.93	-	3.33	0.51

t	0	0.4	1	2	3	10
x	0	1.13	1.5	0.67	0.32	0.03
y	0	0.45	1.5	1.33	0.96	0.3

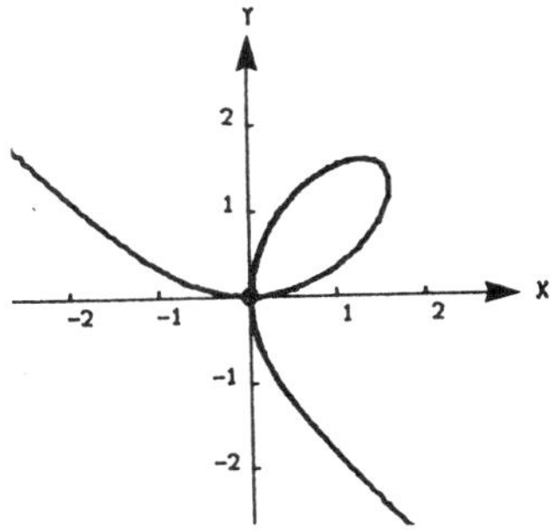

39. $x = x_1 + t(x_2 - x_1)$, $y = y_1 + t(y_2 - y_1)$, $\dfrac{x - x_1}{x_2 - x_1} = t$

$y = y_1 + \left(\dfrac{x - x_1}{x_2 - x_1}\right)(y_2 - y_1)$, $y - y_1 = \dfrac{y_2 - y_1}{x_2 - x_1}(x - x_1)$

$y - y_1 = m(x - x_1)$

40. $x = r\cos\theta$, $y = r\sin\theta$, $\cos\theta = \dfrac{x}{r}$, $\sin\theta = \dfrac{y}{r}$

$\cos^2\theta + \sin^2\theta = \dfrac{x^2}{r^2} + \dfrac{y^2}{r^2} = 1$, $x^2 + y^2 = r^2$

41. $x = h + r\cos\theta$, $y = k + r\sin\theta$, $\cos\theta = \dfrac{x - h}{r}$, $\sin\theta = \dfrac{y - k}{r}$

$\cos^2\theta + \sin^2\theta = \dfrac{(x - h)^2}{r^2} + \dfrac{(y - k)^2}{r^2} = 1$, $(x - h)^2 + (y - k)^2 = r^2$

42. $x = h + a\cos\theta$, $y = k + b\sin\theta$, $\dfrac{x - h}{a} = \cos\theta$,

$\dfrac{y - k}{b} = \sin\theta$, $\dfrac{(x - h)^2}{a^2} + \dfrac{(y - k)^2}{b^2} = 1$

43. $x = h + a\sec\theta$, $y = k + b\tan\theta$, $\dfrac{x - h}{a} = \sec\theta$

$\dfrac{y - k}{b} = \tan\theta$, $\dfrac{(x - h)^2}{a^2} - \dfrac{(y - k)^2}{b^2} = 1$

44. $x = h + a\sqrt{t + 1}$, $y = k + b\sqrt{t}$, $\dfrac{x - h}{a} = \sqrt{t + 1}$, $\dfrac{y - k}{b} = \sqrt{t}$

$\dfrac{(x - h)^2}{a^2} - 1 = t$, $\dfrac{(y - k)^2}{b^2} = t$, $\dfrac{(x - h)^2}{a^2} - \dfrac{(y - k)^2}{b^2} = 1$

45. From Exercise 39 we have:

$x = 5t$
$y = -2t$
Solution not unique

46. From Exercise 39 we have:

$x = 1 + 4t$
$y = 4 - 6t$
Solution not unique

47. From Exercise 41 we have:

$x = 2 + 4\cos\theta$
$y = 1 + 4\sin\theta$
Solution not unique

48. From Exercise 41 we have:

$x = -3 + 3\cos\theta$
$y = 1 + 3\sin\theta$
Solution not unique

49. From Exercise 42 we have:

$x = 5\cos\theta$
$y = 3\sin\theta$
$b = 3$
Center: (0, 0)
Solution not unique

50. From Exercise 42 we have:

$x = 4 + 5\cos\theta$
$y = 2 + 4\sin\theta$
$b = 4$
Center: (4, 2)
Solution not unique

51. From Exercise 43 we have:

$x = 4\sec\theta$
$y = 3\tan\theta$
$b = 3$
Center: (0, 0)
Solution not unique

52. From Exercise 43 we have:

$x = \sqrt{3}\tan\theta$
$y = \sec\theta$
$b = \sqrt{3}$
Center: (0, 0)
Solution not unique
The transverse axis is vertical, therefore, x and y are interchanged.

53. $y = x^3$

<u>Examples</u>

$x = t, \quad y = t^3$
$x = \sqrt[3]{t}, \quad y = t$
$x = \tan t, \quad y = \tan^3 t$

54. $y = x^2$

<u>Examples</u>

$x = t, \quad y = t^2$
$x = t^2, \quad y = t^4$
$x = \sin t, \quad y = \sin^2 t$

55. When the circle has rolled θ radians, we know that the center is at $(a\theta, a)$.

$$\sin\theta = \sin(180^\circ - \theta) = \frac{|AC|}{b} = \frac{|BD|}{b}$$

or $|BD| = b\sin\theta$

$$\cos\theta = -\cos(180^\circ - \theta) = \frac{|AP|}{-b}$$

or $|AP| = -b\cos\theta$

Therefore, $x = a\theta - b\sin\theta$ and $y = a - b\cos\theta$

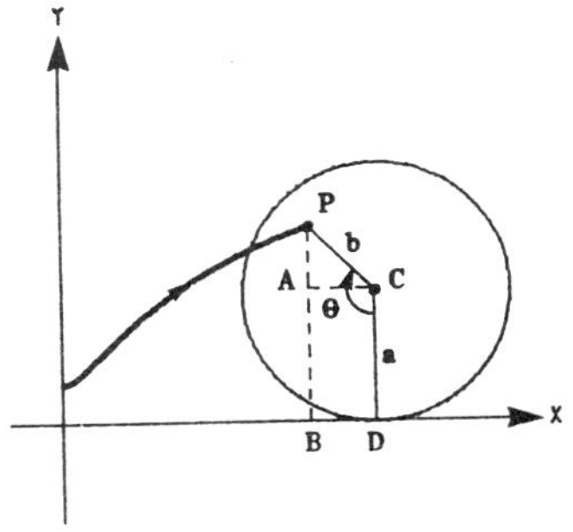

56. $x = 3\cos\theta - \cos 3\theta$

$y = 3\sin\theta - \sin 3\theta$

11.2 Parametric equations and calculus

1. $x = 2t, \quad y = 3t - 1$

$$\frac{dy}{dx} = \frac{3}{2}$$

$$\frac{d^2y}{dx^2} = 0$$

2. $x = \sqrt{t}, \quad y = 3t - 1$

$$\frac{dy}{dx} = \frac{3}{1/(2\sqrt{t})}$$

$$= 6\sqrt{t} = 6 \text{ when } t = 1$$

$$\frac{d^2y}{dx^2} = \frac{3/\sqrt{t}}{1/(2\sqrt{t})} = 6$$

3. $x = t + 1, \quad y = t^2 + 3t$

$$\frac{dy}{dx} = \frac{2t + 3}{1} = 1 \text{ when } t = -1$$

$$\frac{d^2y}{dx^2} = 2$$

4. $x = t^2 + 3t, \quad y = t + 1$

$$\frac{dy}{dx} = \frac{1}{2t + 3} = \frac{1}{3} \text{ when } t = 0$$

$$\frac{d^2y}{dx^2} = \frac{-2/(2t + 3)^2}{2t + 3} = \frac{-2}{(2t + 3)^3}$$

$$= -\frac{2}{27} \text{ when } t = 0$$

5. $x = 2\cos\theta, \quad y = 2\sin\theta, \quad \dfrac{dy}{dx} = \dfrac{2\cos\theta}{-2\sin\theta} = -\cot\theta = -1 \text{ when } \theta = \dfrac{\pi}{4}$

$$\frac{d^2y}{dx^2} = \frac{\csc^2\theta}{-2\sin\theta} = \frac{-\csc^3\theta}{2} = -\sqrt{2} \text{ when } \theta = \frac{\pi}{4}$$

6. $x = \cos\theta, \quad y = 3\sin\theta, \quad \dfrac{dy}{dx} = \dfrac{3\cos\theta}{-\sin\theta} = -3\cot\theta$, undefined when $\theta = 0$

$$\frac{d^2y}{dx^2} = \frac{3\csc^2\theta}{-\sin\theta} = \frac{-3}{\sin^3\theta}, \text{ undefined when } \theta = 0$$

7. $x = 2 + \sec\theta, \quad y = 1 + 2\tan\theta$

$$\frac{dy}{dx} = \frac{2\sec^2\theta}{\sec\theta\tan\theta} = \frac{2\sec\theta}{\tan\theta} = 2\csc\theta = 4 \text{ when } \theta = \frac{\pi}{6}$$

$$\frac{d^2y}{dx^2} = \frac{-2\csc\theta\cot\theta}{\sec\theta\tan\theta} = -2\cot^3\theta = -6\sqrt{3} \text{ when } \theta = \frac{\pi}{6}$$

8. $x = \sqrt{t}, \quad y = \sqrt{t-1}, \quad \dfrac{dy}{dx} = \dfrac{1/(2\sqrt{t-1})}{1/(2\sqrt{t})} = \dfrac{\sqrt{t}}{\sqrt{t-1}} = \sqrt{2}$ at $t = 2$

$$\frac{d^2y}{dx^2} = \frac{[\sqrt{t-1}/(2\sqrt{t}) - \sqrt{t}(2/\sqrt{t-1})]/(t-1)}{1/(2\sqrt{t})} = \frac{-1}{(t-1)^{3/2}} = -1 \text{ when } t = 2$$

9. $x = \cos^3\theta, \quad y = \sin^3\theta, \quad \dfrac{dy}{dx} = \dfrac{3\sin^2\theta\cos\theta}{-3\cos^2\theta\sin\theta} = -\tan\theta = -1$ when $\theta = \dfrac{\pi}{4}$

$$\frac{d^2y}{dx^2} = \frac{-\sec^2\theta}{-3\cos^2\theta\sin\theta} = \frac{1}{3\cos^4\theta\sin\theta} = \frac{4\sqrt{2}}{3} \text{ when } \theta = \frac{\pi}{4}$$

10. $x = \theta - \sin\theta, \quad y = 1 - \cos\theta, \quad \dfrac{dy}{dx} = \dfrac{\sin\theta}{1 - \cos\theta} = 0$ when $\theta = \pi$

$$\frac{d^2y}{dx^2} = \frac{[(1-\cos\theta)\cos\theta - \sin^2\theta]/(1-\cos\theta)^2}{(1-\cos\theta)}$$

$$= \frac{-1}{(1-\cos\theta)^2} = -\frac{1}{4} \text{ when } \theta = \pi$$

11. $x = 2t, \quad y = t^2 - 1, \quad \dfrac{dy}{dx} = \dfrac{2t}{2} = t$

when $t = 2$ at $(4, 3)$, $\dfrac{dy}{dx} = 2$

$y - 3 = 2(x - 4), \quad y = 2x - 5$

12. $x = t - 1, \quad y = \dfrac{1}{t} + 1, \quad \dfrac{dy}{dx} = -\dfrac{1}{t^2} = -1$

when $t = 1$ at $(0, 2)$, $y = -x + 2$

13. $x = t^2 - t + 2, \quad y = t^3 - 3t, \quad \dfrac{dy}{dx} = \dfrac{3t^2 - 3}{2t - 1} = \dfrac{3(t+1)(t-1)}{2t - 1}$

when $t = -1$ at $(4, 2)$, $\dfrac{dy}{dx} = 0, \quad y = 2$

14. $x = 4\cos\theta, \quad y = 3\sin\theta, \quad \left(-\frac{4}{\sqrt{2}}, \frac{3}{\sqrt{2}}\right)$

$$\frac{dy}{dx} = \frac{3\cos\theta}{-4\sin\theta} = -\frac{3}{4}\cot\theta = \frac{3}{4} \text{ when } \theta = \frac{3\pi}{4}$$

$$y - \frac{3}{\sqrt{2}} = \frac{3}{4}\left(x + \frac{4}{\sqrt{2}}\right), \quad 3x - 4y + 12\sqrt{2} = 0$$

15. $x = 2\cot\theta, \quad y = 2\sin^2\theta$ when $\theta = \frac{\pi}{4}$ at $(2, 1)$

$$\frac{dy}{dx} = \frac{4\sin\theta\cos\theta}{-2\csc^2\theta} = -\sin 2\theta\sin^2\theta = -\frac{1}{2} \text{ at } \theta - \frac{\pi}{4}$$

$$y - 1 = -\frac{1}{2}(x - 2), \quad x + 2y - 4 = 0$$

16. $x = 2 - 3\cos\theta, \quad y = 3 + 2\sin\theta, \quad \theta = \frac{5\pi}{3}$ at $\left(\frac{1}{2}, 3 - \sqrt{3}\right)$

$$\frac{dy}{dx} = \frac{2\cos\theta}{3\sin\theta} = \frac{-2\sqrt{3}}{9} \text{ at } \theta = \frac{5\pi}{3}, \quad y - 3 + \sqrt{3} = \frac{-2\sqrt{3}}{9}\left(x - \frac{1}{2}\right)$$

$$2x - 3\sqrt{3}y - 10 + 9\sqrt{3} = 0$$

17. $x = 1 - t, \quad y = t^2, \quad \frac{dy}{dt} = 2t = 0$ when $t = 0$ at $(1, 0)$

18. $x = t + 1, \quad y = t^2 + 3t, \quad \frac{dy}{dt} = 2t + 3 = 0$ when $t = -\frac{3}{2}$ at $\left(\frac{-1}{2}, \frac{-9}{4}\right)$

19. $x = 1 - t, \quad y = t^3 - 3t, \quad \frac{dy}{dt} = (3t^2 - 3) = 3(t - 1)(t + 1) = 0$

when $t = \pm 1, \quad t = 1$ at $(0, -2), \quad t = -1$ at $(2, 2)$

20. $x = t^2 - t + 2, \quad y = t^3 - 3t, \quad \frac{dy}{dt} = 3t^2 - 3 = 3(t + 1)(t - 1) = 0$

when $t = \pm 1, \quad t = 1$ at $(2, -2), \quad t = -1$ at $(4, 2)$

21. $x = \cot\theta, \quad y = 2\sin\theta\cos\theta = \sin 2\theta, \quad \frac{dy}{d\theta} = 2\cos 2\theta = 0$

when $\theta = \frac{\pi}{4}, \frac{3\pi}{4}, \quad \theta = \frac{\pi}{4}$ at $(1, 1), \quad \theta = \frac{3\pi}{4}$ at $(-1, -1)$

22. $x = \dfrac{3t}{1 + t^3}$, $y = \dfrac{3t^2}{1 + t^3}$, $\dfrac{dy}{dt} = \dfrac{3t(2 - t^3)}{(1 + t^3)^2} = 0$

when $t = 0$, $t = \sqrt[3]{2}$, $t = 0$ at $(0, 0)$, $t = \sqrt[3]{2}$ at $(2^{1/3}, 2^{2/3})$

23. $x = 2\theta$, $y = 2(1 - \cos\theta)$, $\dfrac{dy}{d\theta} = 2\sin\theta = 0$

when $\theta = 0, \pm\pi, \pm 2\pi$ at $(4n\pi, 0)$, $(2(2n - 1)\pi, 4)$, where n is an integer

24. $x = \cos\theta + \theta\sin\theta$, $y = \sin\theta - \theta\cos\theta$, $\dfrac{dy}{d\theta} = \theta\sin\theta = 0$

when $\theta = 0, \pi, 2\pi, 3\pi, \ldots$, at $(-1, (2n - 1)\pi)$, $(1, 2n\pi)$

where n is an integer

25. $x = e^{-t}\cos t$, $y = e^{-t}\sin t$, $0 \le t \le \dfrac{\pi}{2}$

$\dfrac{dx}{dt} = -e^{-t}(\sin t + \cos t)$, $\dfrac{dy}{dt} = e^{-t}(\cos t - \sin t)$

$$s = \int_0^{\pi/2} \sqrt{\left(\frac{dx}{dt}\right)^2 + \left(\frac{dy}{dt}\right)^2}\, dt = \int_0^{\pi/2} \sqrt{2e^{-2t}}\, dt = -\sqrt{2}\int_0^{\pi/2} e^{-t}(-1)\, dt$$

$$= -\sqrt{2}\, e^{-t}\Big]_0^{\pi/2} = \sqrt{2}(1 - e^{-\pi/2}) \approx 1.12$$

26. $x = t^2$, $y = 4t^3 - 1$, $-1 \le t \le 1$, $\dfrac{dx}{dt} = 2t$, $\dfrac{dy}{dt} = 12t^2$

$$s = \int_{-1}^{1} \sqrt{4t^2 + 144t^4}\, dt = 2\int_0^1 2t\sqrt{1 + 36t^2}\, dt$$

$$= \frac{1}{18}\int_0^1 (1 + 36t^2)^{1/2}(72t)\, dt = \frac{1}{27}(1 + 36t^2)^{3/2}\Big]_0^1 \approx 8.30$$

27. $x = t^2$, $y = 2t$, $0 \le t \le 2$

$\dfrac{dx}{dt} = 2t$, $\dfrac{dy}{dt} = 2$, $\left(\dfrac{dx}{dt}\right)^2 + \left(\dfrac{dy}{dt}\right)^2 = 4t^2 + 4 = 4(t^2 + 1)$

$$s = 2\int_0^2 \sqrt{t^2 + 1}\, dt = \Big[t\sqrt{t^2 + 1} + \ln\left|t + \sqrt{t^2 + 1}\right|\Big]_0^2$$

$$= 2\sqrt{5} + \ln(2 + \sqrt{5}) \approx 5.916$$

28. $x = \arcsin t, \quad y = \ln\sqrt{1 - t^2}, \quad 0 \le t \le \frac{1}{2}$

$$\frac{dx}{dt} = \frac{1}{\sqrt{1 - t^2}}, \quad \frac{dy}{dt} = \frac{1}{2}\left(\frac{-2t}{1 - t^2}\right) = -\frac{t}{1 - t^2}$$

$$s = \int_0^{1/2} \sqrt{\left(\frac{dx}{dt}\right)^2 + \left(\frac{dy}{dt}\right)^2}\, dt = \int_0^{1/2} \sqrt{\frac{1}{(1 - t^2)^2}}\, dt$$

$$= \int_0^{1/2} \frac{1}{1 - t^2}\, dt = -\frac{1}{2}\ln\left|\frac{t - 1}{t + 1}\right|\Bigg]_0^{1/2} = -\frac{1}{2}\ln\left(\frac{1}{3}\right) = \frac{1}{2}\ln(3) \approx 0.549$$

29. $x = \sqrt{t}, \quad y = 3t - 1, \quad \frac{dx}{dt} = \frac{1}{2\sqrt{t}}, \quad \frac{dy}{dt} = 3$

$$s = \int_0^1 \sqrt{\frac{1}{4t} + 9}\, dt = \frac{1}{2}\int_0^1 \frac{\sqrt{1 + 36t}}{\sqrt{t}}\, dt = \frac{1}{6}\int_0^6 \sqrt{1 + u^2}\, du$$

$$= \frac{1}{12}\left[\ln\left(\sqrt{1 + u^2} + u\right) + u\sqrt{1 + u^2}\right]_0^6 = \frac{1}{12}\left[\ln\left(\sqrt{37} + 6\right) + 6\sqrt{37}\right]$$

$$\approx 3.249$$

$$u = 6\sqrt{t}, \qquad u' = \frac{3}{\sqrt{t}}$$

30. $x = t, \quad y = \frac{t^5}{10} + \frac{1}{6t^3}, \quad \frac{dx}{dt} = 1, \quad \frac{dy}{dt} = \frac{t^4}{2} - \frac{1}{2t^4}$

$$s = \int_1^2 \sqrt{1 + \left(\frac{t^4}{2} - \frac{1}{2t^4}\right)^2}\, dt = \int_1^2 \sqrt{\left(\frac{t^4}{2} + \frac{1}{2t^4}\right)^2}\, dt$$

$$= \int_1^2 \left(\frac{t^4}{2} + \frac{1}{2t^4}\right) dt = \frac{t^5}{10} - \frac{1}{6t^3}\Bigg]_1^2 = \frac{779}{240}$$

31. $x = a\cos^3\theta, \quad y = a\sin^3\theta, \quad \frac{dx}{d\theta} = -3a\cos^2\theta\sin\theta, \quad \frac{dy}{d\theta} = 3a\sin^2\theta\cos\theta$

$$s = 4\int_0^{\pi/2} \sqrt{9a^2\cos^4\theta\sin^2\theta + 9a^2\sin^4\theta\cos^2\theta}\, d\theta$$

$$= 12a\int_0^{\pi/2} \sin\theta\cos\theta\sqrt{\cos^2\theta + \sin^2\theta}\, d\theta$$

$$= 6a\int_0^{\pi/2} \sin 2\theta\, d\theta = -3a\cos 2\theta\Bigg]_0^{\pi/2} = 6a$$

32. $x = a\cos\theta,\quad y = a\sin\theta,\quad \dfrac{dx}{d\theta} = -a\sin\theta,\quad \dfrac{dy}{d\theta} = a\cos\theta$

$$s = 4\int_0^{\pi/2}\sqrt{a^2\sin^2\theta + a^2\cos^2\theta}\,d\theta = 4a\int_0^{\pi/2}d\theta = 4a\theta\Big]_0^{\pi/2} = 2\pi a$$

33. $x = a(\theta - \sin\theta),\quad y = a(1 - \cos\theta),\quad \dfrac{dx}{d\theta} = a(1 - \cos\theta),\quad \dfrac{dy}{d\theta} = a\sin\theta$

$$s = 2\int_0^{\pi}\sqrt{a^2(1 - \cos\theta)^2 + a^2\sin^2\theta}\,d\theta = 2\sqrt{2}a\int_0^{\pi}\sqrt{1 - \cos\theta}\,d\theta$$

$$= 2\sqrt{2}a\int_0^{\pi}\frac{\sin\theta}{\sqrt{1 + \cos\theta}}\,d\theta = \left[-4\sqrt{2}a\sqrt{1 + \cos\theta}\right]_0^{\pi} = 8a$$

34. $x = 3\cos\theta,\quad y = 4\sin\theta,\quad \dfrac{dx}{d\theta} = -3\sin\theta,\quad \dfrac{dy}{d\theta} = 4\cos\theta$

$$s = 4\int_0^{\pi/2}\sqrt{(-3\sin\theta)^2 + (4\cos\theta)^2}\,d\theta = 4\int_0^{\pi/2}\sqrt{9\sin^2\theta + 16\cos^2\theta}\,d\theta$$

$$\approx \frac{\pi}{9}[4 + 4(3.94095) + 2(3.77492) + 4(3.53553) + 2(3.27872)$$

$$+ 4(3.07716) + 3] \approx 22.103$$

35. $x = t,\quad y = 2t,\quad \dfrac{dx}{dt} = 1,\quad \dfrac{dy}{dt} = 2$

(a) $S = 2\pi\displaystyle\int_0^4 2t\sqrt{1 + 4}\,dt = 4\sqrt{5}\pi\int_0^4 t\,dt = 2\sqrt{5}t^2\Big]_0^4 = 32\pi\sqrt{5}$

(b) $S = 2\pi\displaystyle\int_0^4 t\sqrt{1 + 4}\,dt = 2\sqrt{5}\pi\int_0^4 t\,dt = \sqrt{5}\pi t^2\Big]_0^4 = 16\pi\sqrt{5}$

36. $x = t,\quad y = 4 - 2t,\quad \dfrac{dx}{dt} = 1,\quad \dfrac{dy}{dt} = -2$

(a) $S = 2\pi\displaystyle\int_0^2 (4 - 2t)\sqrt{1 + 4}\,dt = 2\sqrt{5}\pi(4t - t^2)\Big]_0^2 = 8\pi\sqrt{5}$

(b) $S = 2\pi\displaystyle\int_0^2 t\sqrt{1 + 4}\,dt = \sqrt{5}\pi t^2\Big]_0^2 = 4\pi\sqrt{5}$

37. $x = 4\cos\theta, \quad y = 4\sin\theta, \quad \dfrac{dx}{d\theta} = -4\sin\theta, \quad \dfrac{dy}{d\theta} = 4\cos\theta$

$$S = 2\pi\int_0^{\pi/2} 4\cos\theta\sqrt{(-4\sin\theta)^2 + (4\cos\theta)^2}\,d\theta = 32\pi\int_0^{\pi/2}\cos\theta\,d\theta$$

$$= 32\pi\sin\theta\Big]_0^{\pi/2} = 32\pi$$

38. $x = t^3, \quad y = t + 2, \quad \dfrac{dx}{dt} = 3t^2, \quad \dfrac{dy}{dt} = 1$

$$S = 2\pi\int_1^2 t^3\sqrt{9t^4 + 1}\,dt = \frac{\pi}{27}(9t^4 + 1)^{3/2}\Big]_1^2$$

$$= \frac{\pi}{27}(145\sqrt{145} - 10\sqrt{10}) \approx 199.48$$

39. $x = a\cos^3\theta, \quad y = a\sin^3\theta, \quad \dfrac{dx}{d\theta} = -3a\cos^2\theta\sin\theta, \quad \dfrac{dy}{d\theta} = 3a\sin^2\theta\cos\theta$

$$S = 4\pi\int_0^{\pi/2} a\sin^3\theta\sqrt{9a^2\cos^4\theta\sin^2\theta + 9a^2\sin^4\theta\cos^2\theta}\,d\theta$$

$$= 12a^2\pi\int_0^{\pi/2}\sin^4\theta\cos\theta\,d\theta = \frac{12\pi a^2}{5}\Big[\sin^5\theta\Big]_0^{\pi/2} = \frac{12}{5}\pi a^2$$

40. $x = a\cos\theta, \quad y = b\sin\theta, \quad \dfrac{dx}{d\theta} = -a\sin\theta, \quad \dfrac{dy}{d\theta} = b\cos\theta$

(a) $$S = 4\pi\int_0^{\pi/2} b\sin\theta\sqrt{a^2\sin^2\theta + b^2\cos^2\theta}\,d\theta$$

$$= 4\pi\int_0^{\pi/2} ab\sin\theta\sqrt{1 - (\frac{a^2 - b^2}{a})\cos^2\theta}\,d\theta$$

$$= \frac{-4ab\pi}{e}\int_0^{\pi/2}(-e\sin\theta)\sqrt{1 - e^2\cos^2\theta}\,d\theta$$

$$= \frac{-2ab\pi}{e}\Big[e\cos\theta\sqrt{1 - e^2\cos^2\theta} + \arcsin(e\cos\theta)\Big]_0^{\pi/2}$$

$$= \frac{2ab\pi}{e}[e\sqrt{1 - e^2} + \arcsin(e)]$$

$$= 2\pi b^2 + (\frac{2\pi a^2 b}{\sqrt{a^2 - b^2}})\arcsin(\frac{\sqrt{a^2 - b^2}}{a})$$

$$= 2\pi b^2 + 2\pi(\frac{ab}{e})\arcsin(e)$$

(e: eccentricity)

40. (continued)

(b) $S = 4\pi \int_0^{\pi/2} a\cos\theta\sqrt{a^2\sin^2\theta + b^2\cos^2\theta}\,d\theta$

$= 4\pi \int_0^{\pi/2} a\cos\theta\sqrt{b^2 + c^2\sin^2\theta}\,d\theta$

$= \frac{4a\pi}{c}\int_0^{\pi/2} c\cos\theta\sqrt{b^2 + c^2\sin^2\theta}\,d\theta$

$= \frac{2a\pi}{c}\left[c\sin\theta\sqrt{b^2 + c^2\sin^2\theta} + b^2\ln\left|c\sin\theta + \sqrt{b^2 + c^2\sin^2\theta}\right|\right]_0^{\pi/2}$

$= \frac{2\pi a}{c}\left[c\sqrt{b^2 + c^2} + b^2\ln\left|c + \sqrt{b^2 + c^2}\right| - b^2\ln b\right]$

$= 2\pi a^2 + \frac{2\pi ab^2}{\sqrt{a^2 - b^2}}\ln\left|\frac{a + \sqrt{a^2 - b^2}}{b}\right| = 2\pi a^2 + \left(\frac{\pi b^2}{e}\right)\ln\left|\frac{1 + e}{1 - e}\right|$

(e: eccentricity)

41. $x = \cos^3\theta,\quad y = \sin^3\theta,\quad \frac{dx}{d\theta} = -3\cos^2\theta\sin\theta$

$A = 4\int_{\pi/2}^{0} \sin^3\theta(-3\cos^2\theta\sin\theta)\,d\theta = -12\int_{\pi/2}^{0}\sin^4\theta\cos^2\theta\,d\theta$

$= \left[2\sin^3\theta\cos^3\theta + \frac{3}{16}\sin 4\theta - \frac{3}{4}\theta\right]_{\pi/2}^{0} = \frac{3\pi}{8}$

42. $x = a\cos\theta,\quad y = b\sin\theta,\quad \frac{dx}{d\theta} = -a\sin\theta$

$A = 4\int_{\pi/2}^{0}(b\sin\theta)(-a\sin\theta)\,d\theta = -4ab\int_{\pi/2}^{0}\sin^2\theta$

$= -4ab\int_{\pi/2}^{0}\frac{1 - \cos 2\theta}{2}\,d\theta = -2ab\left(\theta - \frac{\sin 2\theta}{2}\right)\Big]_{\pi/2}^{0} = ab\pi$

43. $x = 2\sin^2\theta$

$y = 2\sin^2\theta\tan\theta$

$\dfrac{dx}{d\theta} = 4\sin\theta\cos\theta$

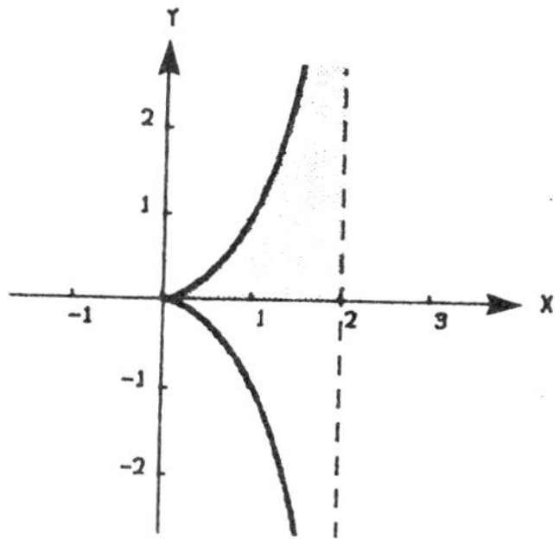

θ	0	$\pi/6$	$\pi/4$	$\pi/3$	$\pi/2$
x	0	0.50	1.00	1.50	2.00
y	0	0.29	1.00	2.60	∞

$$A = \int_0^{\pi/2} 2\sin^2\theta\tan\theta(4\sin\theta\cos\theta)\,d\theta = 8\int_0^{\pi/2}\sin^4\theta\,d\theta$$

$$= 8\left[\frac{-\sin^3\theta\cos\theta}{4} - \frac{3}{8}\sin\theta\cos\theta + \frac{3}{8}\theta\right]_0^{\pi/2} = \frac{3\pi}{2}$$

44. $x = 2\cot\theta,\quad y = 2\sin^2\theta,\quad \dfrac{dx}{d\theta} = -2\csc^2\theta$

$$A = 2\int_{\pi/2}^{0}(2\sin^2\theta)(-2\csc^2\theta)\,d\theta = 8\int_0^{\pi/2}d\theta = 8\theta\Big]_0^{\pi/2} = 4\pi$$

45. $x = \sqrt{t},\quad y = 4 - t,\quad 0 \le t \le 4$

$$A = \int_0^4 (4 - t)\frac{1}{2\sqrt{t}}\,dt = \frac{1}{2}\int_0^4 (4t^{-1/2} - t^{1/2})\,dt$$

$$= \frac{1}{2}(8\sqrt{t} - \frac{2}{3}t\sqrt{t})\Big]_0^4 = \frac{16}{3}$$

$$\bar{x} = \frac{3}{16}\int_0^4 (4 - t)\sqrt{t}(\frac{1}{2\sqrt{t}})\,dt = \frac{3}{32}\int_0^4 (4 - t)\,dt$$

$$= \frac{3}{32}(4t - \frac{t^2}{2})\Big]_0^4 = \frac{3}{4}$$

$$\bar{y} = \frac{3}{32}\int_0^4 (4 - t)^2\frac{1}{2\sqrt{t}}\,dt = \frac{3}{64}\int_0^4 [16t^{-1/2} - 8t^{1/2} + t^{3/2}]\,dt$$

$$= \frac{3}{64}\left[32\sqrt{t} - \frac{16}{3}t\sqrt{t} + \frac{2}{5}t^2\sqrt{t}\right]_0^4 = \frac{8}{5}$$

$$(\bar{x}, \bar{y}) = (\frac{3}{4}, \frac{8}{5})$$

46. $x = \sqrt{4-t}$, $y = \sqrt{t}$, $\dfrac{dx}{dt} = -\dfrac{1}{2\sqrt{4-t}}$, $0 \le t \le 4$

$$A = \int_4^0 \sqrt{t}\left(-\frac{1}{2\sqrt{4-t}}\right) dt = \int_0^2 \sqrt{4-u^2}\, du$$

$$= \frac{1}{2}\left[u\sqrt{4-u^2} + 4\arcsin\frac{u}{2}\right]_0^2 = \pi$$

Let $u = \sqrt{4-t}$, then $du = -\dfrac{1}{2\sqrt{4-t}}\,dt$ and $\sqrt{t} = \sqrt{4-u^2}$

$$\overline{x} = \frac{1}{\pi}\int_4^0 \sqrt{4-t}\,\sqrt{t}\left(-\frac{1}{2\sqrt{4-t}}\right) dt = -\frac{1}{2\pi}\int_4^0 \sqrt{t}\,dt$$

$$= -\frac{1}{2\pi}\,\frac{2}{3}t^{3/2}\Big]_4^0 = \frac{8}{3\pi}$$

$$\overline{y} = \frac{1}{2\pi}\int_4^0 (\sqrt{t})^2\left(-\frac{1}{2\sqrt{4-t}}\right) dt = -\frac{1}{4\pi}\int_4^0 \frac{t}{\sqrt{4-t}}\,dt$$

$$= -\frac{1}{4\pi}\left[\frac{-2(8+t)}{3}\sqrt{4-t}\right]_4^0 = \frac{8}{3\pi}$$

$$(\overline{x}, \overline{y}) = \left(\frac{8}{3\pi}, \frac{8}{3\pi}\right)$$

47. $x = 3\cos\theta$, $y = 3\sin\theta$, $\dfrac{dx}{d\theta} = -3\sin\theta$

$$V = 2\pi\int_0^{\pi/2}(3\sin\theta)^2(-3\sin\theta)\,d\theta = -54\pi\int_0^{\pi/2}\sin^3\theta\,d\theta$$

$$= -54\pi\int_0^{\pi/2}(1-\cos^2\theta)\sin\theta\,d\theta$$

$$= -54\pi\left[-\cos\theta + \frac{\cos^3\theta}{3}\right]_0^{\pi/2} = 36\pi$$

48. $x = \cos\theta$, $y = 3\sin\theta$, $\dfrac{dx}{d\theta} = -\sin\theta$

$$V = 2\pi\int_0^{\pi/2}(3\sin\theta)^2(-\sin\theta)\,d\theta = -18\pi\int_0^{\pi/2}\sin^3\theta\,d\theta$$

$$= -18\pi\left[-\cos\theta + \frac{\cos^3\theta}{3}\right]_0^{\pi/2} = 12\pi$$

49. $x = a\cos\phi, \quad y = a\sin\phi$

$$S = 2\pi \int_0^{\theta} a\sin\phi\sqrt{a^2\sin^2\phi + a^2\cos^2\phi}\, d\phi$$

$$= 2\pi a^2 \int_0^{\theta} \sin\phi\, d\phi$$

$$= -2\pi a^2 \cos\phi \Big]_0^{\theta} = 2\pi a^2(1 - \cos\theta)$$

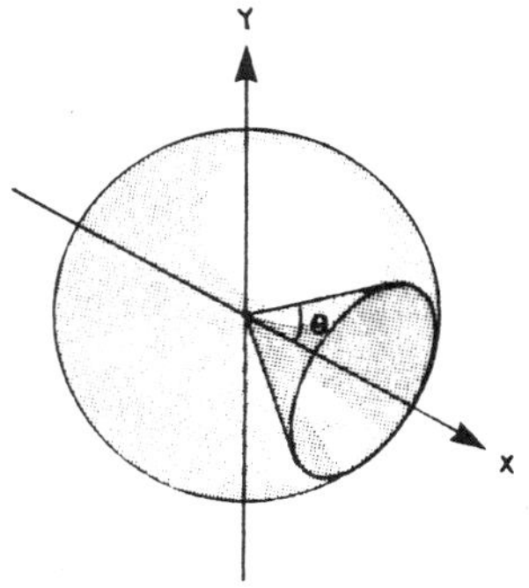

11.3 Polar coordinates and polar graphs

1. $(4, \frac{3\pi}{6})$, $x = 4\cos(\frac{3\pi}{6}) = 0$

$y = 4\sin(\frac{3\pi}{6}) = 4$, $(0, 4)$

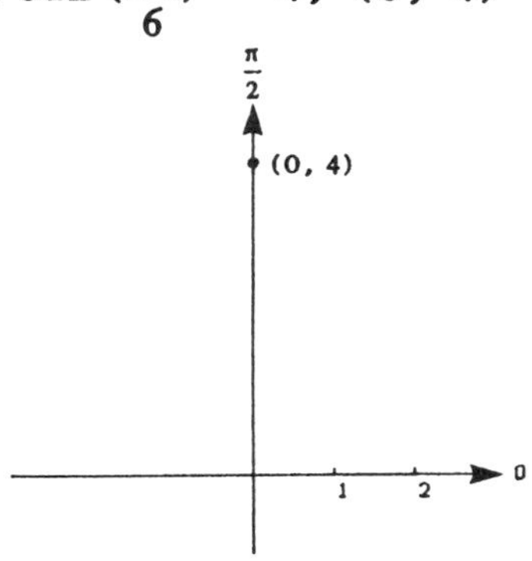

2. $(4, \frac{3\pi}{2})$, $x = 4\cos(\frac{3\pi}{2}) = 0$

$y = 4\sin(\frac{3\pi}{2}) = -4$, $(0, -4)$

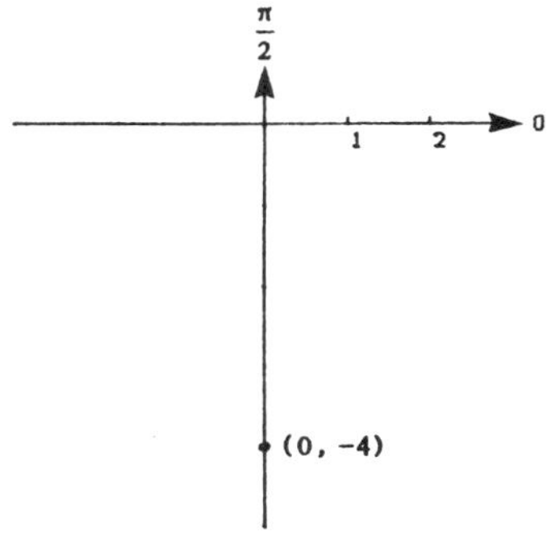

3. $(-1, \frac{5\pi}{4})$, $x = -1\cos(\frac{5\pi}{4}) = \frac{\sqrt{2}}{2}$

$y = -1\sin(\frac{5\pi}{4}) = \frac{\sqrt{2}}{2}$, $(\frac{\sqrt{2}}{2}, \frac{\sqrt{2}}{2})$

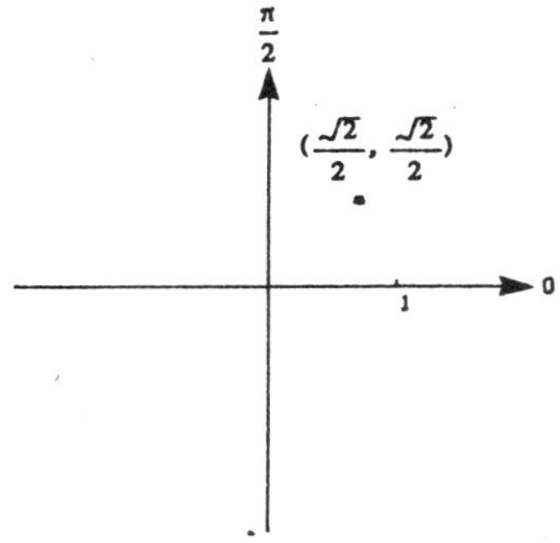

4. $(0, -\pi)$, $x = 0\cos(-\pi) = 0$

$y = 0\sin(-\pi) = 0$, $(0, 0)$

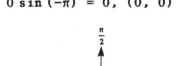

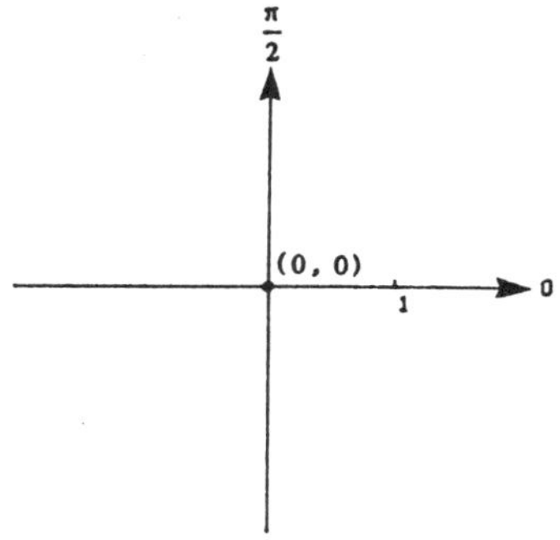

5. $(4, -\frac{\pi}{3})$, $x = 4\cos(-\frac{\pi}{3}) = 2$

$y = 4\sin(-\frac{\pi}{3}) = -2\sqrt{3}$, $(2, -2\sqrt{3})$

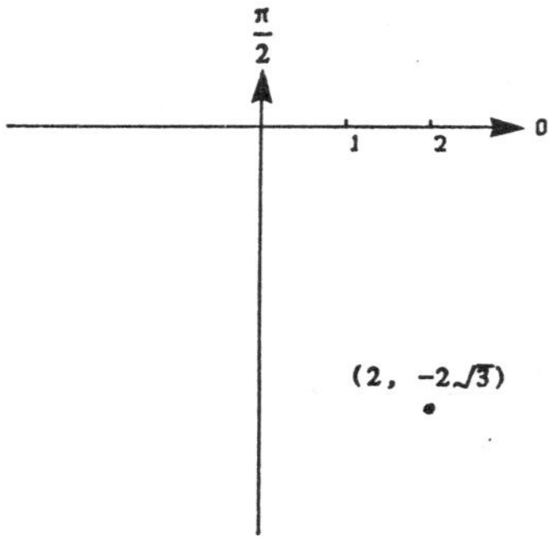

6. $(-1, \frac{-3\pi}{4})$, $x = -1\cos(\frac{-3\pi}{4}) = \frac{\sqrt{2}}{2}$

$y = -1\sin(\frac{-3\pi}{4}) = \frac{\sqrt{2}}{2}$, $(\frac{\sqrt{2}}{2}, \frac{\sqrt{2}}{2})$

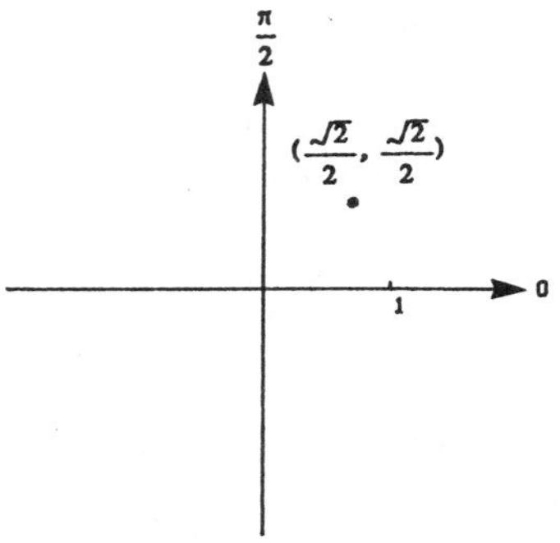

7. $(\sqrt{2}, 2.36)$

$x = \sqrt{2}\cos(2.36) = -1.004$

$y = \sqrt{2}\sin(2.36) = 0.996$

$(-1.004, 0.996)$

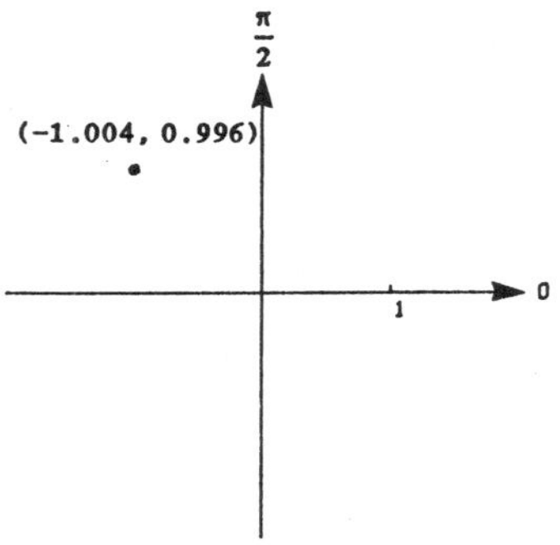

8. $(-3, -1.57)$

$x = -3\cos(-1.57) = -0.0024$

$y = -3\sin(-1.57) = 3$

$(-0.0024, 3)$

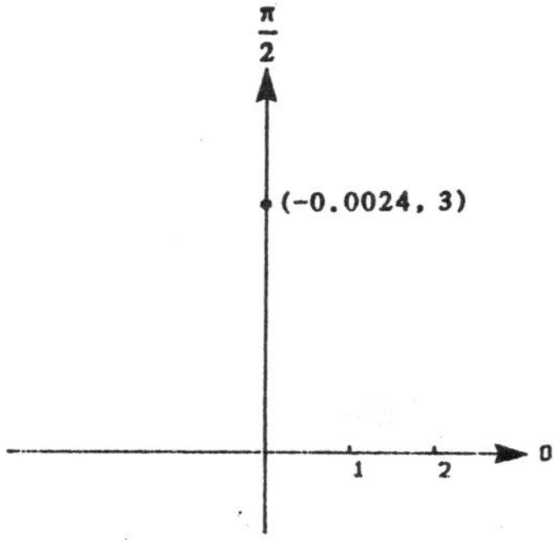

9. $(1, 1)$, $r = \sqrt{2}$, $\tan\theta = 1$, $\theta = \frac{\pi}{4}$, $(\sqrt{2}, \frac{\pi}{4})$, $(-\sqrt{2}, \frac{5\pi}{4})$

10. $(0, -5)$, $r = 5$, $\tan\theta$ undefined, $\theta = \frac{\pi}{2}$, $(5, \frac{3\pi}{2})$, $(-5, \frac{\pi}{2})$

11. $(-3, 4)$, $r = \sqrt{9 + 16} = 5$, $\tan\theta = -\frac{4}{3}$, $\theta \approx 2.214$

$(5, 2.214)$, $(-5, 5.356)$

12. $(3, -1)$, $r = \sqrt{9 + 1} = \sqrt{10}$, $\tan\theta = -\frac{1}{3}$, $\theta \approx -0.322$

$(-\sqrt{10}, 2.820)$, $(\sqrt{10}, 5.961)$

13. $(-\sqrt{3}, -\sqrt{3})$, $r = \sqrt{3 + 3} = \sqrt{6}$, $\tan\theta = 1$, $\theta = \dfrac{\pi}{4}$

$(\sqrt{6}, \dfrac{5\pi}{4})$, $(-\sqrt{6}, \dfrac{\pi}{4})$

14. $(-2, 0)$, $r = 2$, $\tan\theta = 0$, $\theta = 0$, $(2, \pi)$, $(-2, 0)$

15. $(4, 6)$, $r = \sqrt{16 + 36} = 2\sqrt{13}$, $\tan\theta = \dfrac{3}{2}$, $\theta \approx 0.983$

$(2\sqrt{13}, 0.983)$, $(-2\sqrt{13}, 4.124)$

16. $(5, 12)$, $r = \sqrt{25 + 144} = 13$, $\tan\theta = \dfrac{12}{5}$, $\theta \approx 1.176$

$(13, 1.176)$, $(-13, 4.318)$

17. $x^2 + y^2 = 9$, $r = 3$

18. $x^2 + y^2 = a^2$, $r = a$

19. $x^2 + y^2 - 2ax = 0$
$r^2 - 2ar\cos\theta = 0$
$r(r - 2a\cos\theta) = 0$
$r = 2a\cos\theta$

20. $x^2 + y^2 - 2ay = 0$
$r^2 - 2ar\sin\theta = 0$
$r(r - 2a\sin\theta) = 0$
$r = 2a\sin\theta$

21. $y = 4$
$r\sin\theta = 4$
$r = 4\csc\theta$

22. $y = b$
$r\sin\theta = b$
$r = b\csc\theta$

23. $x = 10$
$r\cos\theta = 10$
$r = 10\sec\theta$

24. $x = a$
$r\cos\theta = a$
$r = a\sec\theta$

25. $3x - y + 2 = 0$, $3r\cos\theta - r\sin\theta + 2 = 0$

$r(3\cos\theta - \sin\theta) = -2$, $r = \dfrac{-2}{3\cos\theta - \sin\theta}$

26. $4x + 7y - 2 = 0$, $4r\cos\theta + 7r\sin\theta - 2 = 0$

$r(4\cos\theta + 7\sin\theta) = 2$, $r = \dfrac{2}{4\cos\theta + 7\sin\theta}$

27. $xy = 4$, $(r\cos\theta)(r\sin\theta) = 4$, $r^2 = 4\sec\theta\csc\theta = 8\csc 2\theta$

28. $y = x$, $r\cos\theta = r\sin\theta$, $1 = \tan\theta$, $\theta = \dfrac{\pi}{4}$

29. $y^2 = 9x$, $r^2\sin^2\theta = 9r\cos\theta$, $r = \dfrac{9\cos\theta}{\sin^2\theta}$, $r = 9\csc^2\theta\cos\theta$

30. $y^2 - 8x - 16 = 0, \quad r^2 \sin^2\theta - 8r\cos\theta = 16$

$r^2 - r^2\cos^2\theta - 8r\cos\theta - 16 = 0, \quad r^2\cos^2\theta + 8r\cos\theta + 16 = r^2$

$(r\cos\theta + 4)^2 = r^2, \quad r = \pm(r\cos\theta + 4)$

$r = \dfrac{4}{1 - \cos\theta}$ or $r = \dfrac{-4}{1 + \cos\theta}$

31. $x^2 - 4ay - 4a^2 = 0, \quad r^2\cos^2\theta - 4ar\sin\theta - 4a^2 = 0,$

$r^2(1 - \sin^2\theta) - 4ar\sin\theta - 4a^2 = 0, \quad r^2 = r^2\sin^2\theta + 4ar\sin\theta + 4a^2,$

$r^2 = (r\sin\theta + 2a)^2, \quad r = \pm(r\sin\theta + 2a)$

$r = \dfrac{2a}{1 - \sin\theta}$ or $r = \dfrac{-2a}{1 + \sin\theta}$

32. $(x^2 + y^2)^2 - 9(x^2 - y^2) = 0, \quad (r^2)^2 - 9(r^2\cos^2\theta - r^2\sin^2\theta) = 0$

$r^2[r^2 - 9(\cos 2\theta)] = 0, \quad r^2 = 9\cos 2\theta$

33. $r = 4\sin\theta, \quad r^2 = 4r\sin\theta, \quad x^2 + y^2 = 4y, \quad x^2 + y^2 - 4y = 0$

34. $r = 4\cos\theta, \quad \sqrt{x^2 + y^2} = 4(\frac{x}{r}), \quad x^2 + y^2 = 4x, \quad x^2 + y^2 - 4x = 0$

35. $\theta = \frac{\pi}{6}, \quad \tan\theta = \frac{\sqrt{3}}{3}, \quad \frac{y}{x} = \frac{\sqrt{3}}{3}, \quad y = \frac{\sqrt{3}}{3}x, \quad \sqrt{3}x - 3y = 0$

36. $r = 4, \quad r^2 = 16, \quad x^2 + y^2 = 16$

37. $r = 2\csc\theta, \quad r\sin\theta = 2, \quad y = 2$

38. $r^2 = \sin 2\theta = 2\sin\theta\cos\theta, \quad r^2 = 2(\frac{y}{r})(\frac{x}{r}) = \frac{2xy}{r^2}$

$r^4 = 2xy, \quad (x^2 + y^2)^2 = 2xy$

39. $r = 1 - 2\sin\theta, \quad r^2 = r - 2r\sin\theta, \quad x^2 + y^2 = \sqrt{x^2 + y^2} - 2y$

$(x^2 + y^2 + 2y)^2 = x^2 + y^2$

40. $r = \dfrac{1}{1 - \cos\theta}, \quad r - r\cos\theta = 1, \quad \sqrt{x^2 + y^2} - x = 1$

$x^2 + y^2 = 1 + 2x + x^2, \quad y^2 = 2x + 1$

41. $r = \dfrac{6}{2 - 3\sin\theta} = \dfrac{6}{2 - 3(y/r)} = \dfrac{6r}{2r - 3y}$ Therefore, $2r^2 - 3ry = 6r$

$2r = 6 + 3y, \quad 4r^2 = 36 + 36y + 9y^2, \quad 4x^2 + 4y^2 = 36 + 36y + 9y^2$

$4x^2 - 5y^2 - 36y - 36 = 0$

42. First, $\sin 3\theta = \sin(\theta + 2\theta) = \sin\theta\cos 2\theta + \cos\theta\sin 2\theta$

$= \sin\theta(2\cos^2\theta - 1) + \cos\theta(2\sin\theta\cos\theta) = \sin\theta(4\cos^2\theta - 1)$

Therefore, if we multiply the given equation by r, we have

$r^2 = 2r\sin 3\theta = 2r\sin\theta(4\cos^2\theta - 1)$

$x^2 + y^2 = 2y\left[\dfrac{4x}{x^2 + y^2} - 1\right] = 2y\left(\dfrac{3x^2 - y^2}{x^2 + y^2}\right)$ or $(x^2 + y^2)^2 = 2y(3x^2 - y^2)$

43. $r^2 = 4\sin\theta, \quad r^3 = 4r\sin\theta, \quad (x^2 + y^2)^{3/2} = 4y, \quad (x^2 + y^2)^3 = 16y^2$

44. $r = \dfrac{6}{2\cos\theta - 3\sin\theta} = \dfrac{6}{2(x/r) - 3(y/r)} = \dfrac{6r}{2x - 3y}$

or $1 = \dfrac{6}{2x - 3y}, \quad 2x - 3y = 6$

45. $r = 5$

Circle of radius 5

Centered at the pole

Symmetric to polar axis

$\theta = \dfrac{\pi}{2}$ and pole

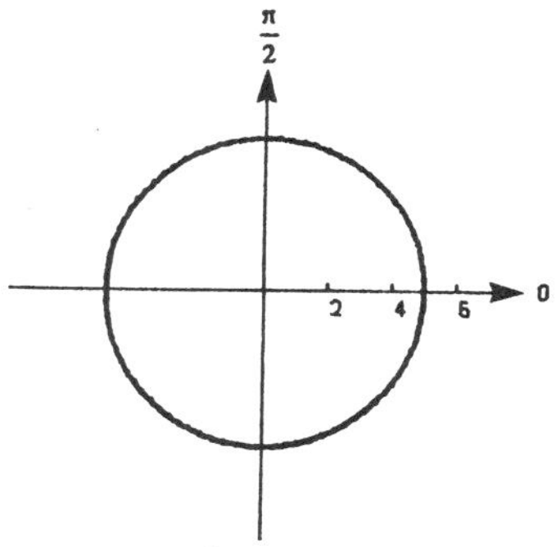

46. $r = -2$

Circle of radius 2

Centered at the pole

Symmetric to polar axis

$\theta = \dfrac{\pi}{2}$ and pole

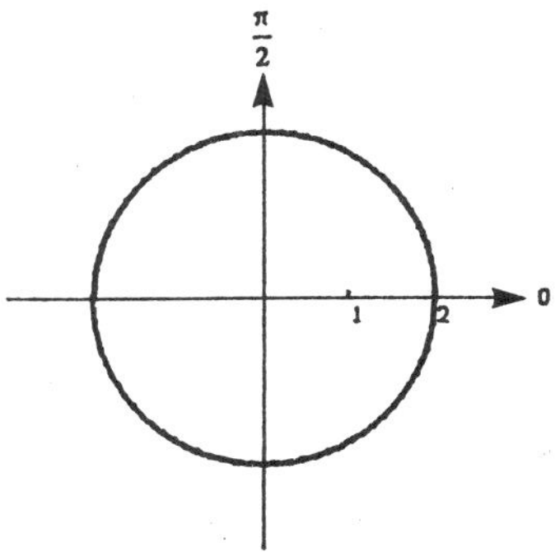

47. $r = \theta$

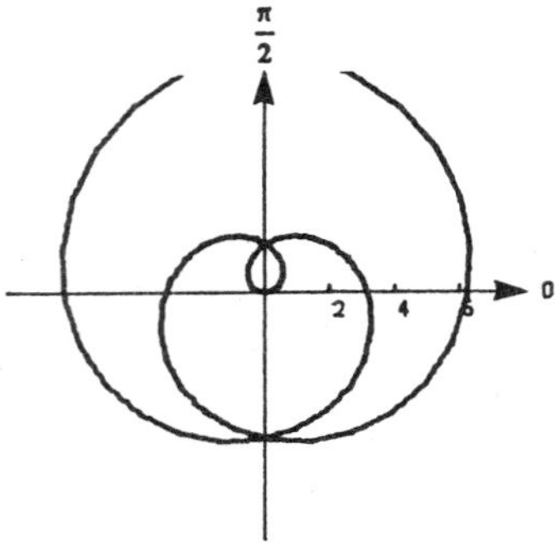

Spiral

Symmetric to the vertical axis

θ	0	$\pi/6$	$\pi/2$	$3\pi/4$	...
r	0	$\pi/6$	$\pi/2$	$3\pi/4$	...

48. $r = \dfrac{\theta}{\pi}$

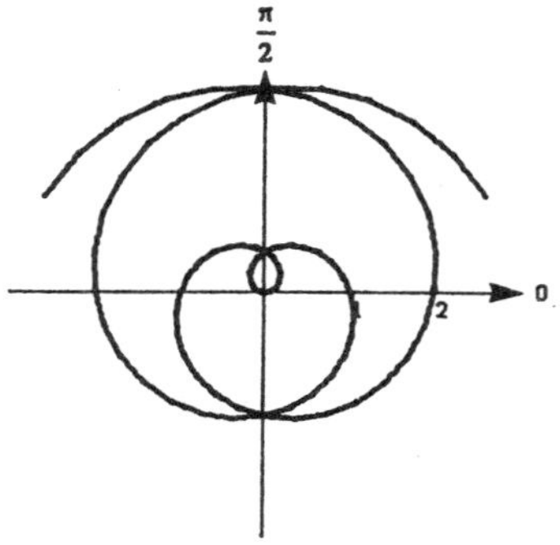

Spiral

Symmetric to the vertical axis

θ	0	$\pi/6$	$\pi/4$	$\pi/3$	$\pi/2$	π
r	0	1/6	1/4	1/3	1/2	1

49. $r = \sin\theta$

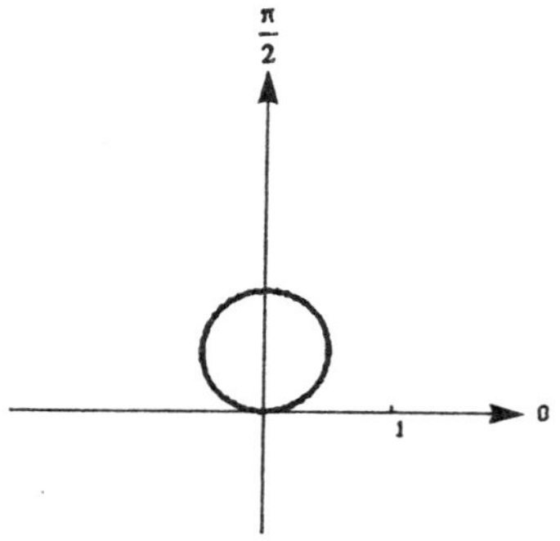

Symmetric to the vertical axis

Circle

$r = \dfrac{1}{2}$

Center: $(0, \dfrac{1}{2})$

50. $r = 3\cos\theta$

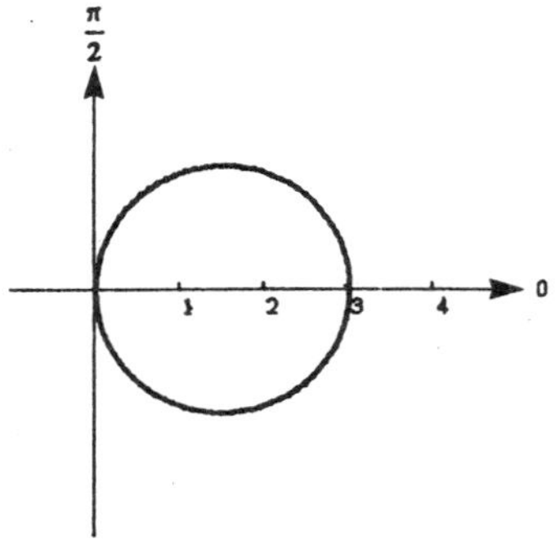

Symmetric to the polar axis

Circle

$r = \dfrac{3}{2}$

Center: $(\dfrac{3}{2}, 0)$

51. $r = 2 \sec \theta$

$r \cos \theta = 2$

$x = 2$

Symmetric to the polar axis

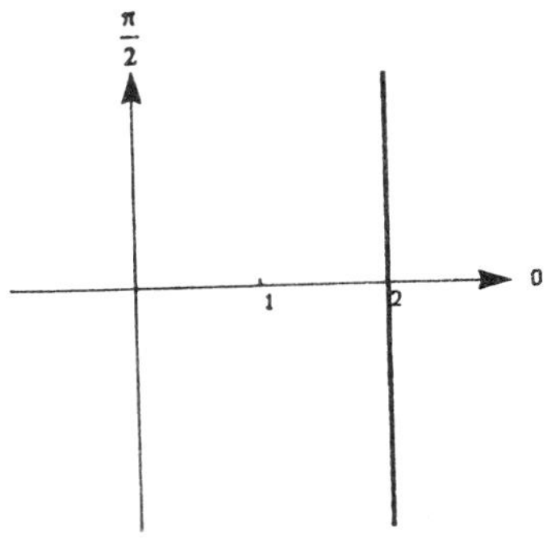

52. $r = 3 \csc \theta$

$r \sin \theta = 3$

$y = 3$

Symmetric to the vertical axis

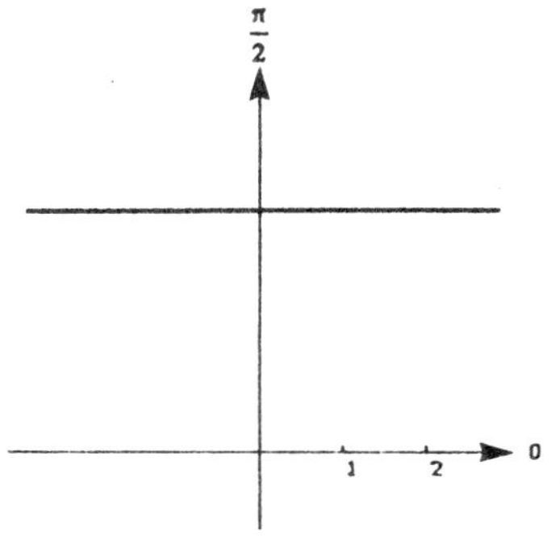

53. $r = 4(2 + \sin \theta)$

Symmetric to the vertical axis

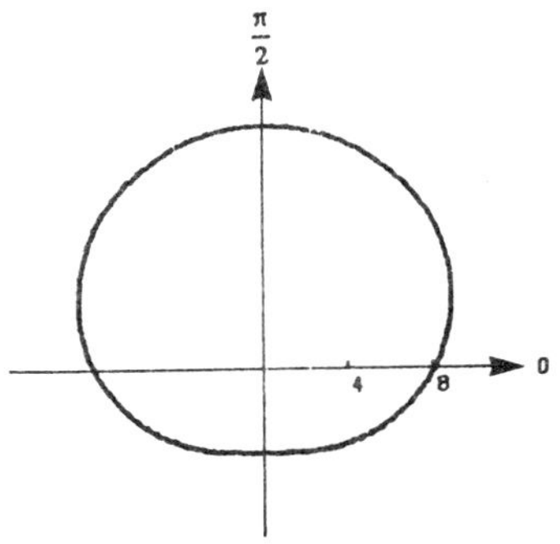

54. $r = 2 + \cos \theta$

Symmetric to the polar axis

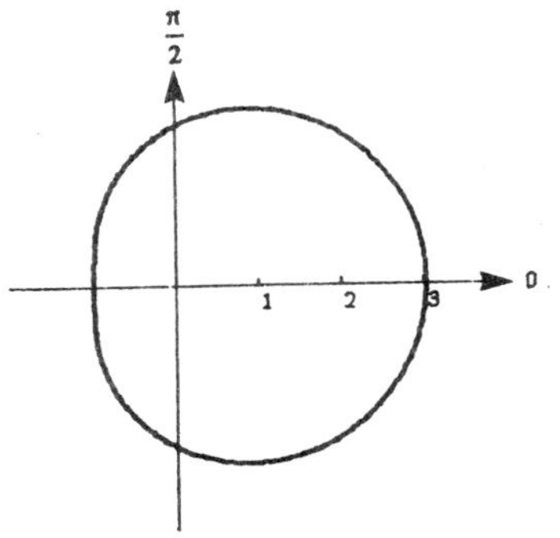

11.4 Tangent lines and curve sketching in polar coordinates

1. $r = 3(1 - \cos\theta)$

$$\frac{dy}{dx} = \frac{3\sin\theta(\sin\theta) + 3\cos\theta(1 - \cos\theta)}{3\sin\theta(\cos\theta) - 3\sin\theta(1 - \cos\theta)} = \frac{(1 + 2\cos\theta)(1 - \cos\theta)}{s\text{ in}\,\theta(2\cos\theta - 1)}$$

at $\theta = \frac{\pi}{2}$, $\frac{dy}{dx} = -1$

2. $r = 2(1 - \sin\theta)$, $\frac{dy}{dx} = \frac{-2\cos\theta\sin\theta + 2\cos\theta(1 - \sin\theta)}{-2\cos\theta\cos\theta - 2\sin\theta(1 - \sin\theta)}$

at $\theta = \frac{\pi}{6}$, $\frac{dy}{dx} = 0$

3. $r = 2 + 3\sin\theta$, $\frac{dy}{dx} = \frac{3\cos\theta\sin\theta + \cos\theta(2 + 3\sin\theta)}{3\cos\theta\cos\theta - \sin\theta(2 + 3\sin\theta)}$

at $\theta = \frac{\pi}{2}$, $\frac{dy}{dx} = 0$

4. $r = 3 - 2\cos\theta$, $\frac{dy}{dx} = \frac{2\sin\theta\sin\theta + \cos\theta(3 - 2\cos\theta)}{2\sin\theta\cos\theta - \sin\theta(3 - 2\cos\theta)}$

at $\theta = 0$, $\frac{dy}{dx}$ is undefined

5. $r = \theta$, $\frac{dy}{dx} = \frac{\sin\theta + \theta\cos\theta}{\cos\theta - \theta\sin\theta}$, at $\theta = \pi$, $\frac{dy}{dx} = \pi$

6. $r = 3\cos\theta$, $\frac{dy}{dx} = \frac{-3\sin\theta\sin\theta + 3\cos\theta\cos\theta}{-3\sin\theta\cos\theta - 3\cos\theta\sin\theta}$

at $\theta = \frac{\pi}{4}$, $\frac{dy}{dx} = 0$

7. $r = 3\sin\theta$, $\frac{dy}{dx} = \frac{3\sin\theta\cos\theta + 3\cos\theta\sin\theta}{-3\sin^2\theta + 3\cos^2\theta} = \frac{2\sin\theta\cos\theta}{1 - 2\sin^2\theta}$

at $\theta = \frac{\pi}{3}$, $\frac{dy}{dx} = -\sqrt{3}$

8. $r = 4$, $\frac{dy}{dx} = \frac{4\cos\theta}{-4\sin\theta} = -\cot\theta$, at $\theta = \frac{\pi}{4}$, $\frac{dy}{dx} = -1$

9. $r = 2\sec\theta$, $r\cos\theta = 2$, $x = 2$, vertical line, $\frac{dy}{dx}$ is undefined

10. $r = \dfrac{6}{2\sin\theta - 3\cos\theta}$, $2r\sin\theta - 3r\cos\theta = 6$, $2y - 3x = 6$

$y = \dfrac{3}{2}x + 3$, $\dfrac{dy}{dx} = \dfrac{3}{2}$, at $\theta = \pi$, $\dfrac{dy}{dx} = \dfrac{3}{2}$

11. $r = 1 + \sin\theta$

$\dfrac{dy}{d\theta} = (1 + \sin\theta)\cos\theta + \cos\theta\sin\theta = \cos\theta(1 + 2\sin\theta) = 0$

$\cos\theta = 0$, $\sin\theta = -\dfrac{1}{2}$, $\theta = \dfrac{\pi}{2}, \dfrac{3\pi}{2}$, $\theta = \dfrac{7\pi}{6}, \dfrac{11\pi}{6}$

Horizontal: $(2, \dfrac{\pi}{2})$, $(\dfrac{1}{2}, \dfrac{7\pi}{6})$, $(\dfrac{1}{2}, \dfrac{11\pi}{6})$

$\dfrac{dx}{d\theta} = -(1 + \sin\theta)\sin\theta + \cos^2\theta = (1 + \sin\theta)(1 - 2\sin\theta) = 0$

$\sin\theta = -1$, $\sin\theta = \dfrac{1}{2}$, $\theta = \dfrac{3\pi}{2}$, $\theta = \dfrac{\pi}{6}, \dfrac{5\pi}{6}$

Vertical: $(\dfrac{3}{2}, \dfrac{\pi}{6})$, $(\dfrac{3}{2}, \dfrac{5\pi}{6})$

12. $r = a\sin\theta$, $\dfrac{dy}{d\theta} = a\sin\theta\cos\theta + a\cos\theta\sin\theta = 2a\sin\theta\cos\theta = 0$

$\theta = 0, \dfrac{\pi}{2}, \pi, \dfrac{3\pi}{2}$

$\dfrac{dx}{d\theta} = -a\sin^2\theta + a\cos^2\theta = a(1 - 2\sin^2\theta) = 0$

$\sin\theta = \pm\dfrac{1}{\sqrt{2}}$, $\theta = \dfrac{\pi}{4}, \dfrac{3\pi}{4}, \dfrac{5\pi}{4}, \dfrac{7\pi}{4}$

Horizontal: $(0, 0)$, $(a, \dfrac{\pi}{2})$

Vertical: $(\dfrac{a\sqrt{2}}{2}, \dfrac{\pi}{4})$, $(\dfrac{a\sqrt{2}}{2}, \dfrac{3\pi}{4})$

13. $r = 2\csc\theta + 3$

$\dfrac{dy}{d\theta} = (2\csc\theta + 3)\cos\theta + (-2\csc\theta\cot\theta)\sin\theta = 3\cos\theta = 0$

$\theta = \dfrac{\pi}{2}, \dfrac{3\pi}{2}$

Horizontal: $(5, \dfrac{\pi}{2})$, $(1, \dfrac{3\pi}{2})$

14. $r = a \sin\theta \cos^2\theta$

$$\frac{dy}{d\theta} = a \sin\theta \cos^3\theta + [-2a \sin^2\theta \cos\theta + a \cos^3\theta] \sin\theta$$

$$= 2a[\sin\theta \cos^3\theta - \sin^3\theta \cos\theta] = 0$$

$\tan^2\theta = 0, \quad \theta = 0, \pi$

Horizontal: (0, 0)

15. $r = 4$

Circle of radius 4

Centered at the pole

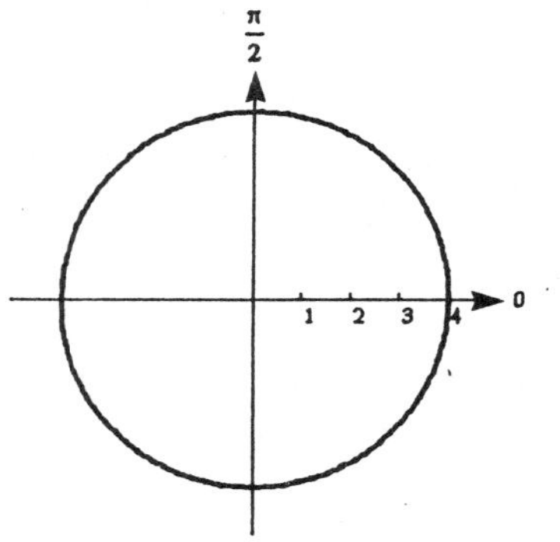

16. $r = -2$

Circle of radius 2

Centered at the pole

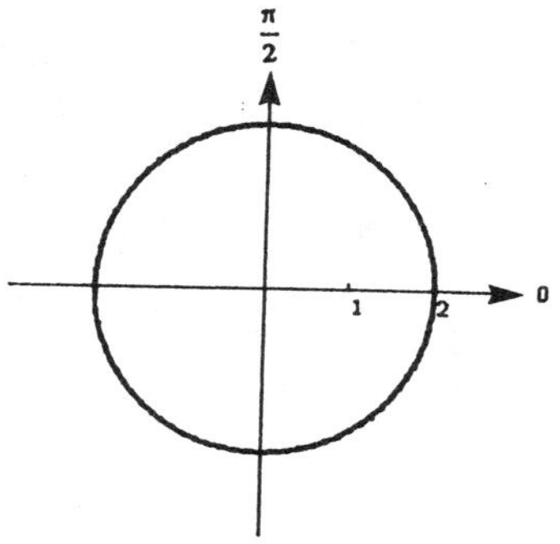

17. $r = 3 \sin\theta$

$r^2 = 3r \sin\theta$

$x^2 + y^2 = 3y$

$x^2 + (y - \frac{3}{2})^2 = \frac{9}{4}$

Circle

$r = \frac{3}{2}$, Center: $(0, \frac{3}{2})$

Tangent at the pole: $\theta = 0$

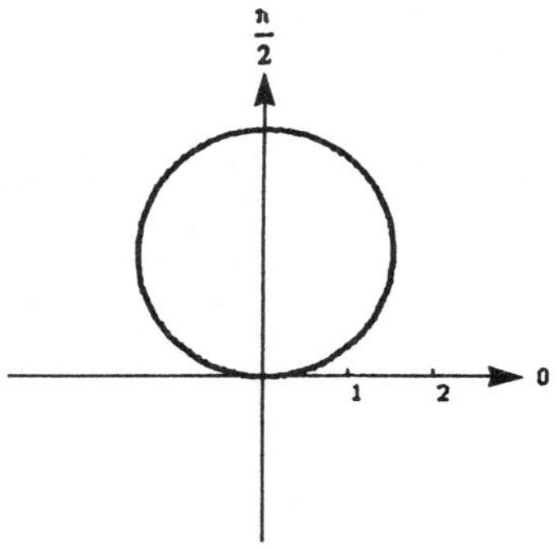

18. $r = 3\cos\theta$, $\quad r^2 = 3r\cos\theta$

$x^2 + y^2 = 3x$, $\quad (x - \frac{3}{2})^2 + y^2 = \frac{9}{4}$

Circle

$r = \frac{3}{2}$, Center: $(\frac{3}{2}, 0)$

Tangent at the pole: $\theta = \frac{\pi}{2}$

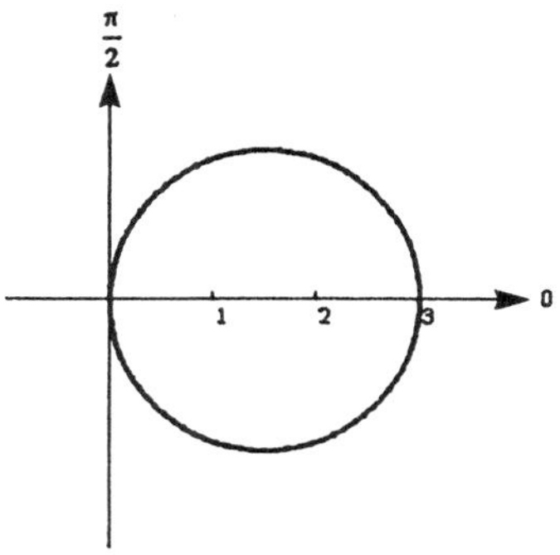

19. $r = 3(1 - \cos\theta)$, Cardioid

Symmetric to polar axis since r is a function of $\cos\theta$

θ	0	$\pi/3$	$\pi/2$	$2\pi/3$	π
r	0	3/2	3	9/2	6

Tangent at the pole: $\theta = 0$

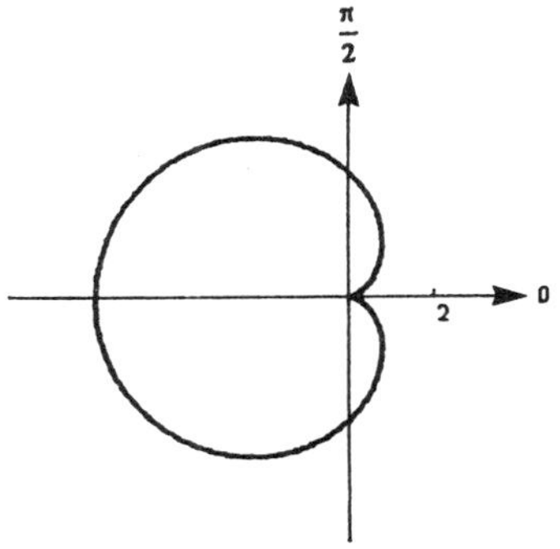

20. $r = 2(1 - \sin\theta)$, Cardioid

Symmetric to vertical axis since r is a function of $\sin\theta$

θ	$-\pi/2$	$-\pi/6$	0	$\pi/6$	$\pi/2$
r	4	3	2	1	0

Tangent at the pole: $\theta = \frac{\pi}{2}$

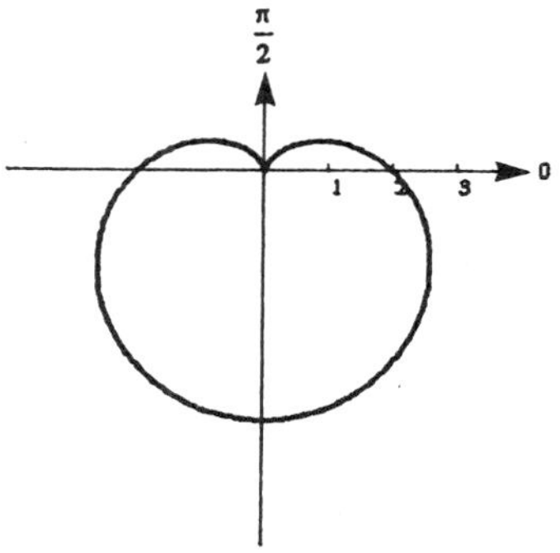

21. $r = 4(1 + \sin\theta)$, Cardioid

Symmetric to vertical axis

θ	0	$\pi/6$	$\pi/2$	$5\pi/6$	π	$3\pi/2$
r	4	6	8	6	4	0

Tangent at the pole: $\theta = \frac{3\pi}{2}$

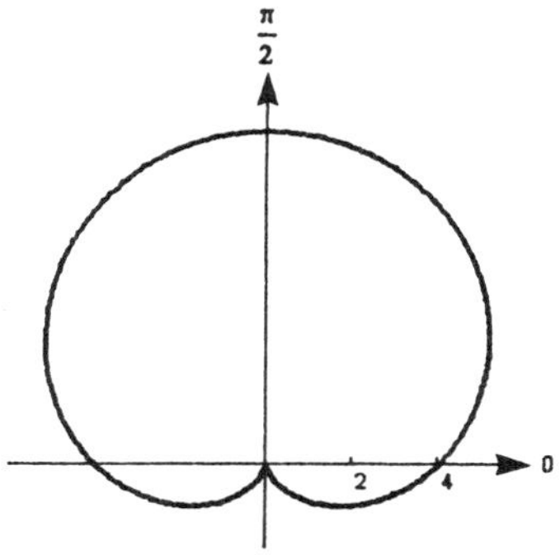

22. $r = 1 + \cos\theta$

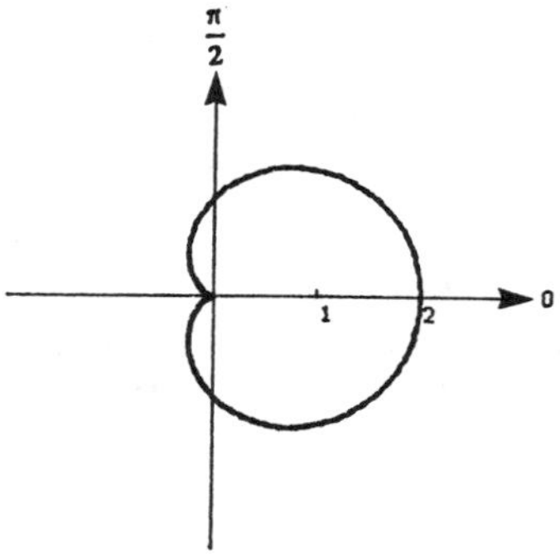

Cardioid

Symmetric to polar axis

θ	0	$\pi/3$	$\pi/2$	$2\pi/3$	π
r	2	3/2	1	1/2	0

23. $r = 2 + 3\sin\theta$

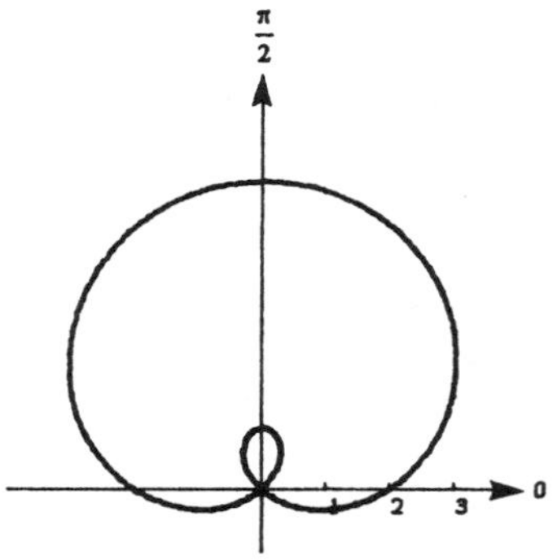

Limacon with two loops since form $r = a + b\sin\theta$ with $b > a$

Symmetric to the vertical axis since r is a function of $\sin\theta$

Relative extrema: $(-1, \frac{3\pi}{2})$, $(5, \frac{\pi}{2})$

θ	$-\pi$	$-\pi/6$	0	$\pi/6$	$\pi/2$	π	2.412	$3\pi/2$
r	2	1/2	2	7/2	5	2	0	-1

Tangents at the pole: $\theta = \arcsin(-\frac{2}{3})$ and $\theta = \pi + \arcsin(\frac{2}{3})$

24. $r = 4 + 5\cos\theta$

Limacon with two loops since form $r = a + b\cos\theta$ with $b > a$

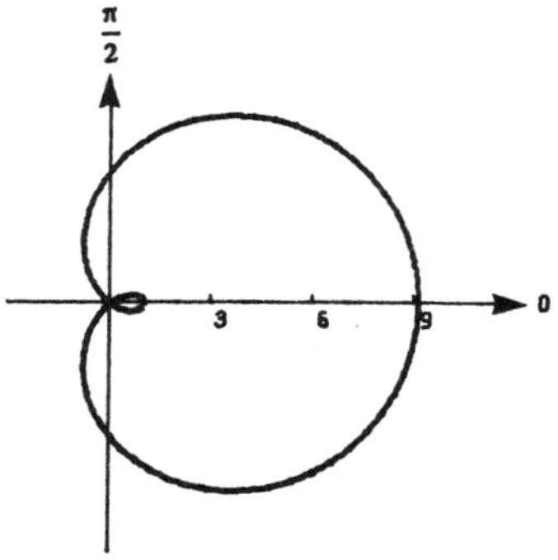

Symmetric to the polar axis since r is a function of $\cos\theta$

Relative extrema: $(9, 0)$, $(-1, \pi)$

θ	0	$\pi/3$	$\pi/2$	$2\pi/3$	2.498	π
r	9	13/2	4	3/2	0	-1

25. $r = 3 - 4\cos\theta$

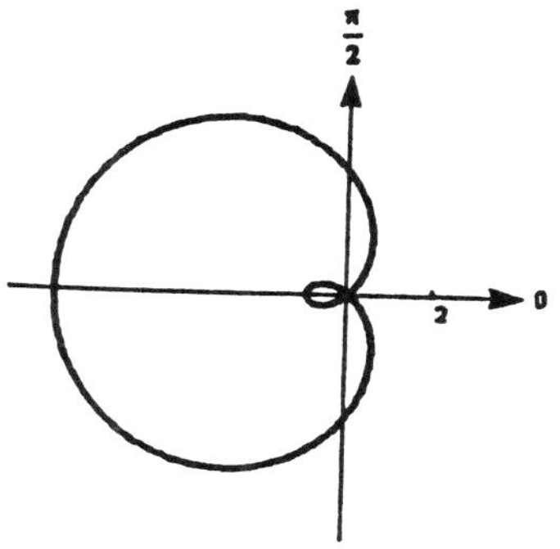

Limacon with two loops

Symmetric to polar axis

θ	0	0.723	$\pi/3$	$\pi/2$	$2\pi/3$	π
r	−1	0	1	3	5	7

Tangents at the pole: $\theta = \arccos\frac{3}{4}$, $\theta = 2\pi - \arccos\frac{3}{4}$

26. $r = 2(1 - 2\sin\theta)$

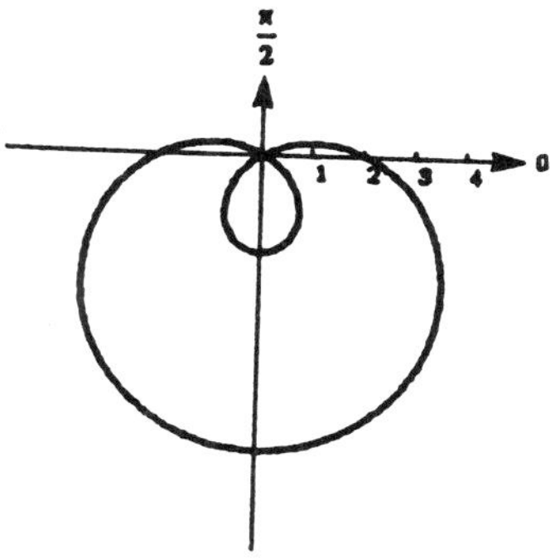

Limacon with two loops

Symmetric to the vertical axis

θ	0	$\pi/6$	$\pi/2$	$5\pi/6$	π	$3\pi/2$
r	2	0	−2	0	2	6

Tangents at the pole: $\theta = \frac{\pi}{6}$, $\theta = \frac{5\pi}{6}$

27. $r = 3 - 2\cos\theta$

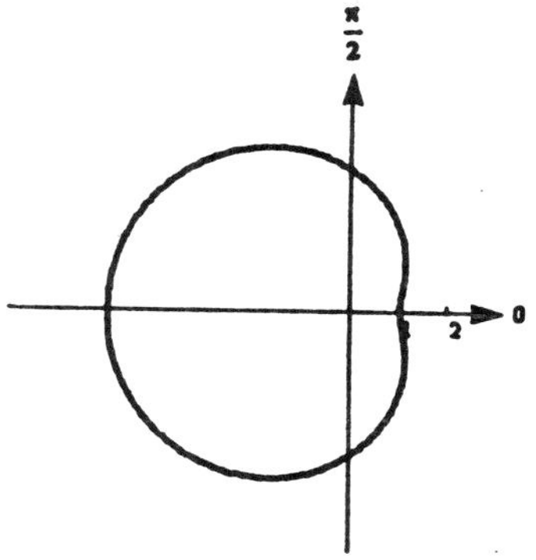

Limacon

Symmetric to polar axis

θ	0	$\pi/3$	$\pi/2$	$2\pi/3$	π
r	1	2	3	4	5

28. $r = 5 - 4\sin\theta$

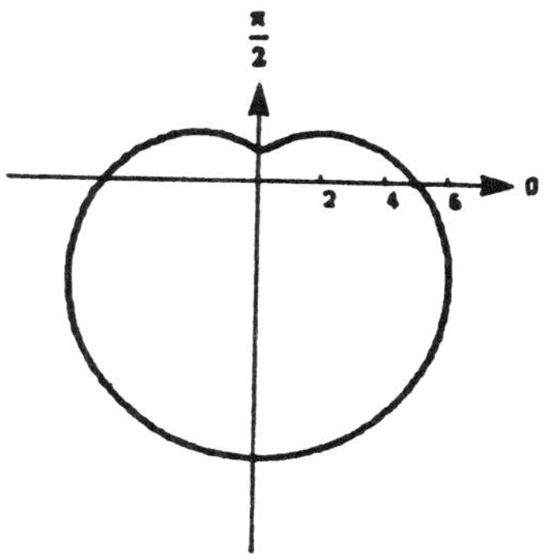

Limacon

Symmetric to the vertical axis

θ	$-\pi/2$	$-\pi/6$	0	$\pi/6$	$\pi/2$
r	9	7	5	3	1

29. $r = 2 + \sin\theta$

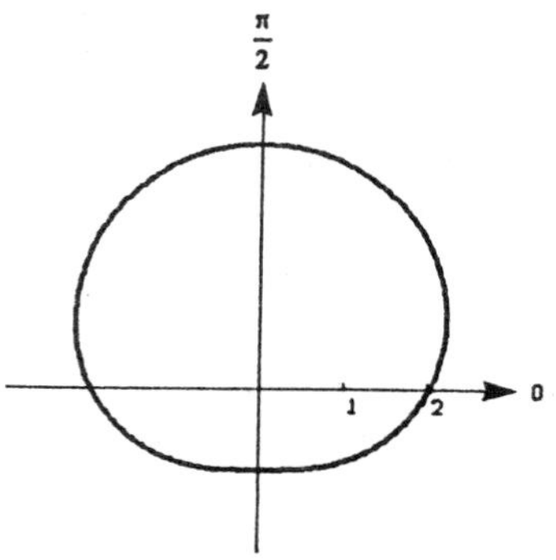

Limacon

Symmetric to the vertical axis

θ	$-\pi/2$	$-\pi/6$	0	$\pi/6$	$\pi/2$
r	1	3/2	2	5/2	3

30. $r = 4 + 3\cos\theta$

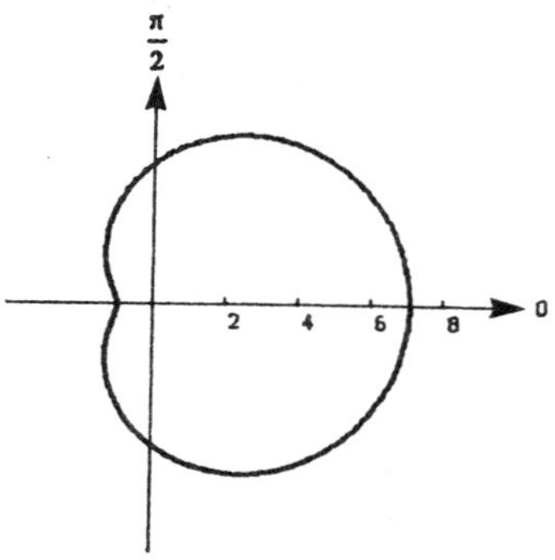

Limacon

Symmetric to polar axis

θ	0	$\pi/3$	$\pi/2$	$2\pi/3$	π
r	7	11/2	4	5/2	1

31. $r = 2\cos(3\theta)$

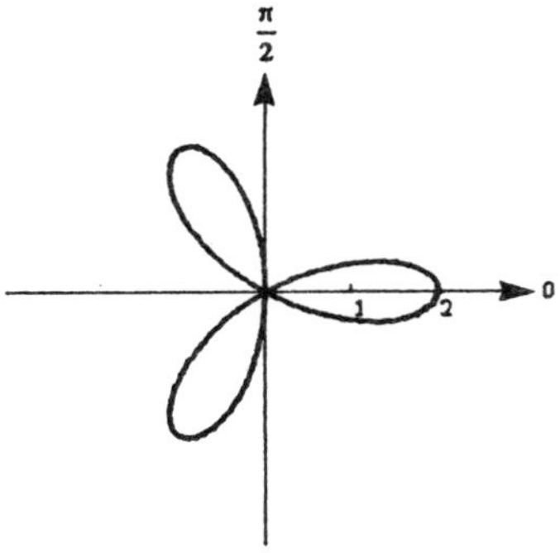

Rose curve with 3 petals

Symmetric to the polar axis

Relative extrema: $(2, 0)$, $(-2, \frac{\pi}{3})$, $(2, \frac{2\pi}{3})$

θ	0	$\pi/2$	$\pi/6$	$\pi/4$	$\pi/3$	$5\pi/12$	$\pi/2$
r	2	$\sqrt{2}$	0	$-\sqrt{2}$	-2	$-\sqrt{2}$	0

Tangents at the pole: $\theta = \frac{\pi}{6}, \frac{\pi}{2}, \frac{5\pi}{6}$

32. $r = -\sin(5\theta)$

Rose curve with 5 petals

Symmetric to the vertical axis

Relative extrema occur when

$\frac{dr}{d\theta} = -5\cos(5\theta) = 0$ at

$\theta = \frac{\pi}{10}, \frac{3\pi}{10}, \frac{5\pi}{10}, \frac{7\pi}{10}, \frac{9\pi}{10}$

Tangents at the pole: $\theta = 0, \frac{\pi}{5}, \frac{2\pi}{5}, \frac{3\pi}{5}, \frac{4\pi}{5}$

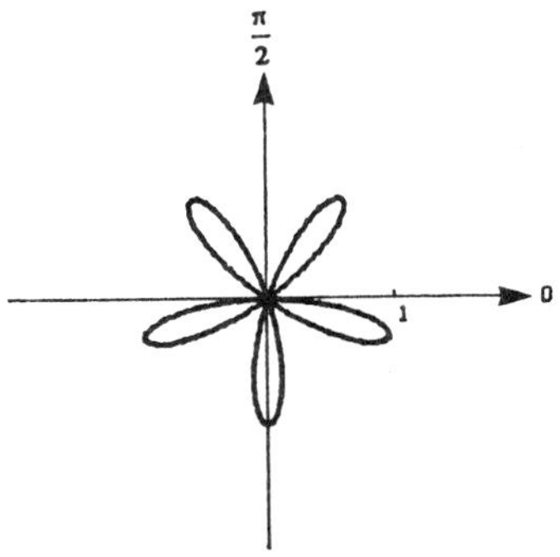

33. $r = 3\sin 2\theta$

Rose curve with 4 petals

Symmetric to the polar axis, vertical axis, and pole

Relative extrema: $(\pm 3, \frac{\pi}{4})$, $(\pm 3, \frac{5\pi}{4})$

Tangents at the pole: $\theta = 0, \frac{\pi}{2}$

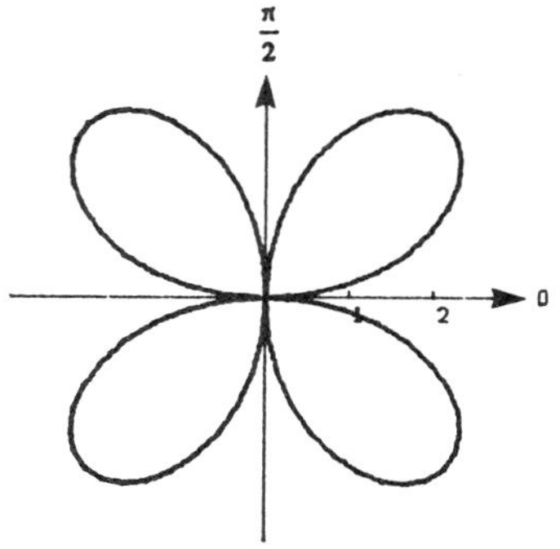

34. $r = 3\cos 2\theta$

Rose curve with 4 petals

Symmetric to the polar axis, vertical axis, and pole

Relative extrema: $(3, 0)$, $(-3, \frac{\pi}{2})$,

$(3, \pi)$, $(-3, \frac{3\pi}{2})$

Tangents at the pole: $\theta = \frac{\pi}{4}, \frac{3\pi}{4}$

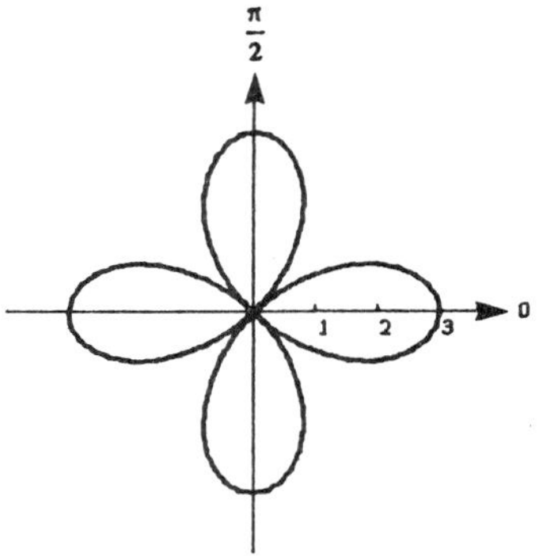

35. $r = 2\sec\theta$

$r\cos\theta = 2$

$x = 2$

Vertical line

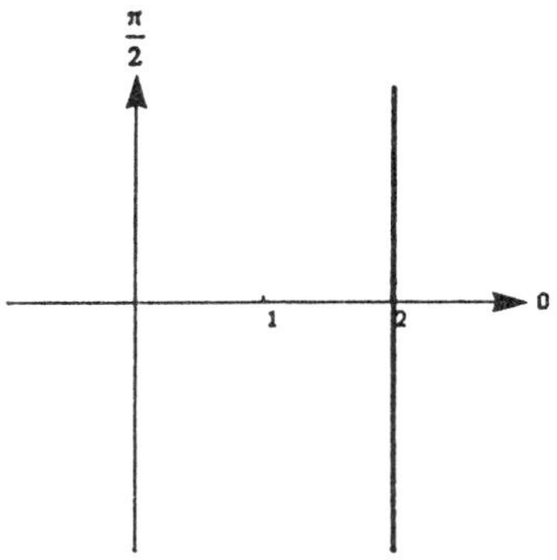

36. $r = 3\csc\theta = \dfrac{3}{\sin\theta}$

$r\sin\theta = 3$

$y = 3$

Horizontal line

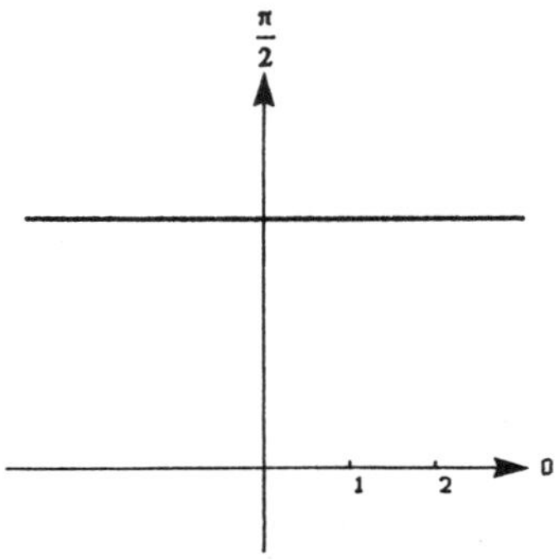

37. $r = \dfrac{3}{\sin\theta - 2\cos\theta}$

$r\sin\theta - 2r\cos\theta = 3$

$y - 2x = 3$

Line

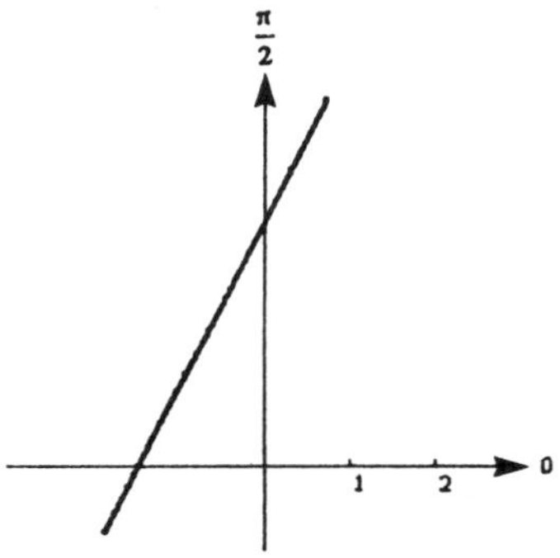

38. $r = \dfrac{6}{2\sin\theta - 3\cos\theta}$

$2r\sin\theta - 3r\cos\theta = 6$

$2y - 3x = 6$

Line

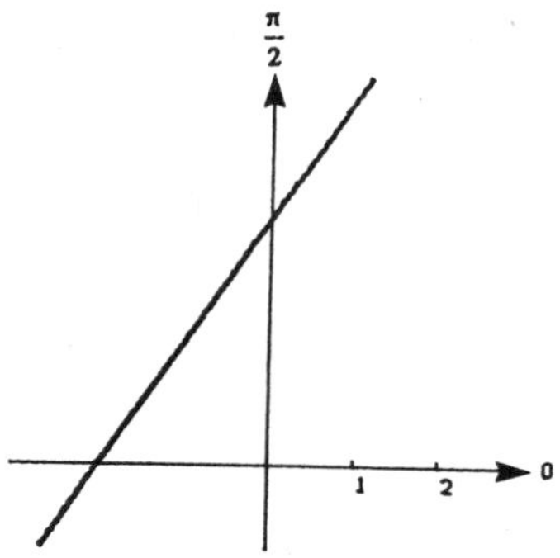

39. $r^2 = 4\cos(2\theta)$

Lemniscate

Symmetric to the polar axis, vertical axis, and pole

Relative maxima: $(\pm 2, 0)$

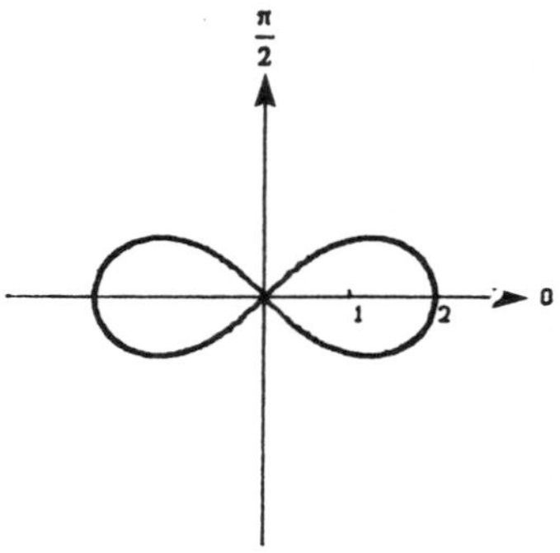

θ	0	$\pi/6$	$\pi/4$
r	± 2	$\pm\sqrt{2}$	0

Tangents at the pole: $\theta = \dfrac{\pi}{4}, \dfrac{3\pi}{4}$

40. $r^2 = 4\sin\theta$

Lemniscate

Symmetric to the polar axis, vertical axis, and pole

Relative extrema: $(\pm 2, \frac{\pi}{2})$

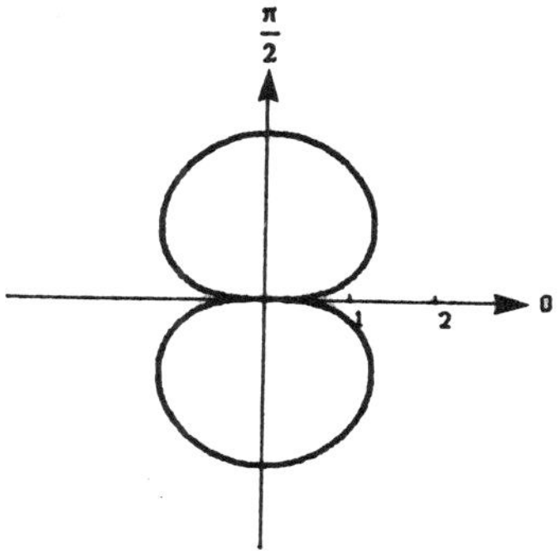

θ	0	π/6	π/2	5π/6	π
r	0	$\pm\sqrt{2}$	±2	$\pm\sqrt{2}$	0

Tangent at the pole: $\theta = 0$

41. $r^2 = 4\sin 2\theta$

Lemniscate

Symmetric to the pole

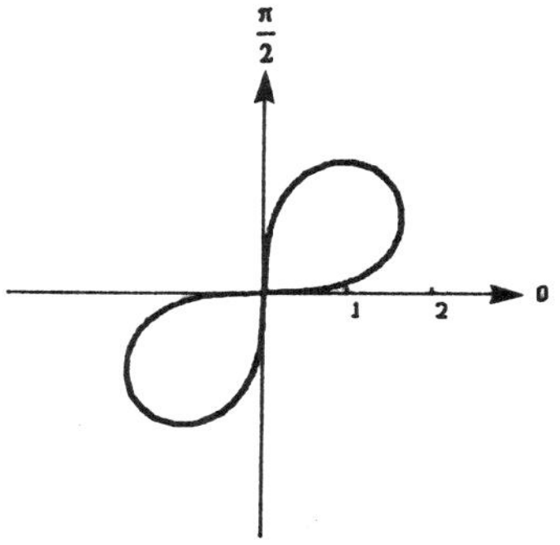

θ	0	π/4	π/2	5π/4	π
r	0	±2	0	±2	0

Tangents at the pole: $\theta = 0, \frac{\pi}{2}$

42. $r^2 = \cos(3\theta)$

Rose curve with 6 petals

Symmetric to the polar axis, vertical axis, and pole

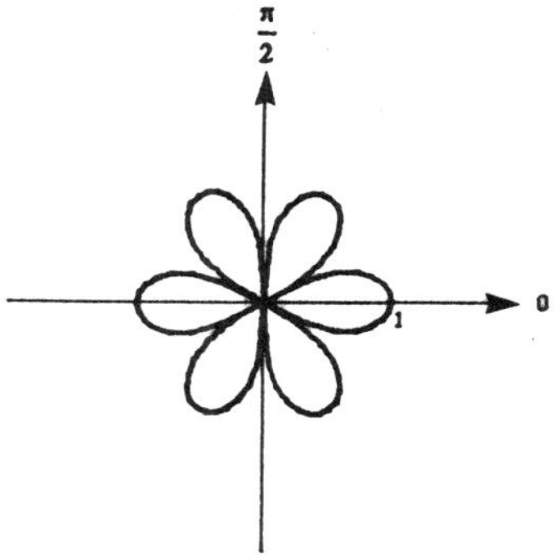

Relative extrema occur when

$\cos(3\theta) > 0$ and $\frac{dr}{d\theta} = \frac{-3\sin(3\theta)}{2r} = 0$

at $\theta = 0, \frac{2\pi}{3}, \frac{4\pi}{3}$

Relative extrema: $(\pm 1, 0)$, $(\pm 1, \frac{2\pi}{3})$, $(\pm 1, \frac{4\pi}{3})$

θ	0	π/9	π/6
r	±1	$\pm\sqrt{2}/2$	0

Tangents at the pole: $\theta = \frac{\pi}{6}, \frac{\pi}{2}, \frac{5\pi}{6}$

43. $r = \sin\theta\cos^2\theta$

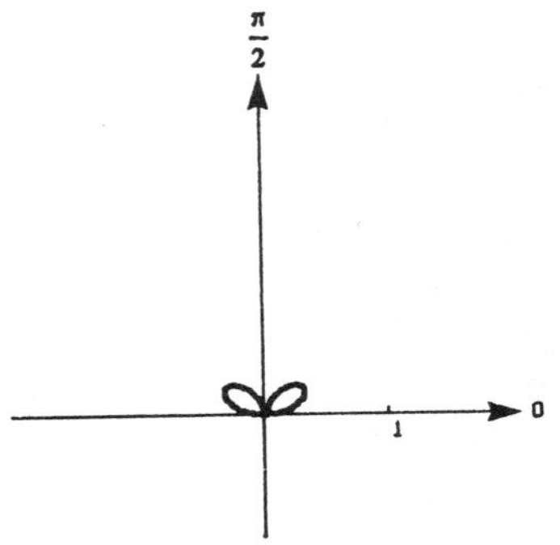

Bifolium

Symmetric to the vertical axis

θ	0	$\pi/4$	$\pi/2$	$3\pi/4$	π
r	0	$\dfrac{1}{2\sqrt{2}}$	0	$\dfrac{1}{2\sqrt{2}}$	0

Tangents at the pole: $\theta = 0, \dfrac{\pi}{2}$

44. $r^2 = \dfrac{1}{\theta}$

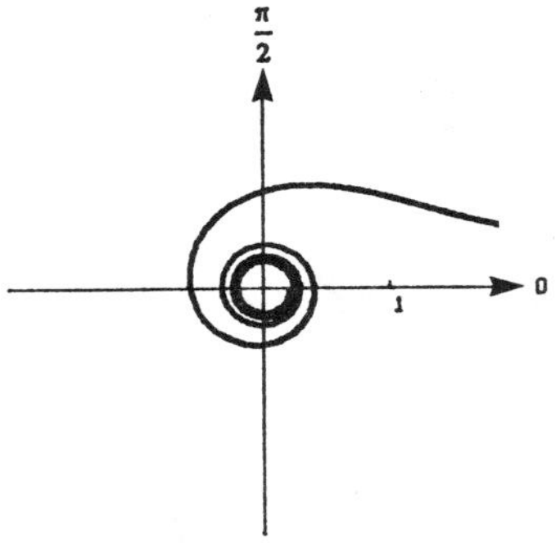

Lituus

θ	$\pi/4$	$\pi/2$	π	$3\pi/4$	$2\pi/3$
r	$\pm\dfrac{2}{\sqrt{\pi}}$	$\pm\sqrt{2/\pi}$	$\pm\dfrac{1}{\sqrt{\pi}}$	$\pm\dfrac{2}{\sqrt{3\pi}}$	$\pm\sqrt{3/2\pi}$

45. $r = 2\theta$

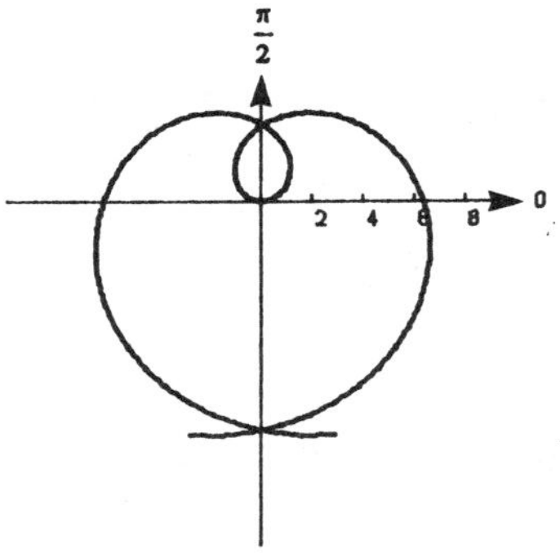

Spiral of Archimedes

Symmetric to the vertical axis

θ	0	$\pi/4$	$\pi/2$	$3\pi/4$	π	$5\pi/4$	$3\pi/2$
r	0	$\pi/2$	π	$3\pi/2$	2π	$5\pi/2$	3π

46. $r = \dfrac{1}{\theta}$

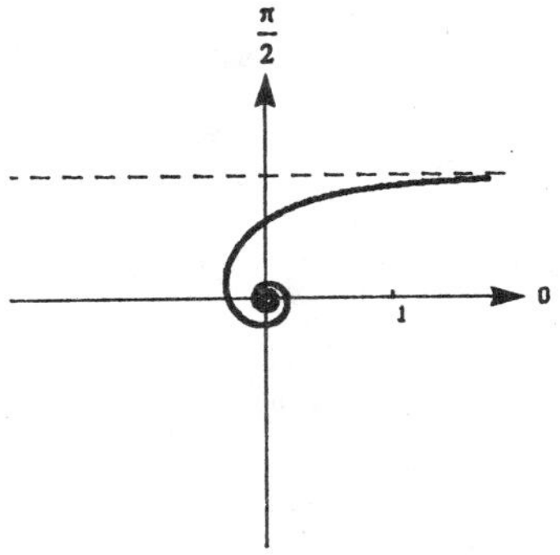

Hyperbolic spiral

θ	$\pi/4$	$\pi/2$	$3\pi/4$	π	$5\pi/4$	$3\pi/2$
r	$4/\pi$	$2/\pi$	$4/3\pi$	$1/\pi$	$4/5\pi$	$2/3\pi$

47. $r = 2\cos(\frac{3\theta}{2})$

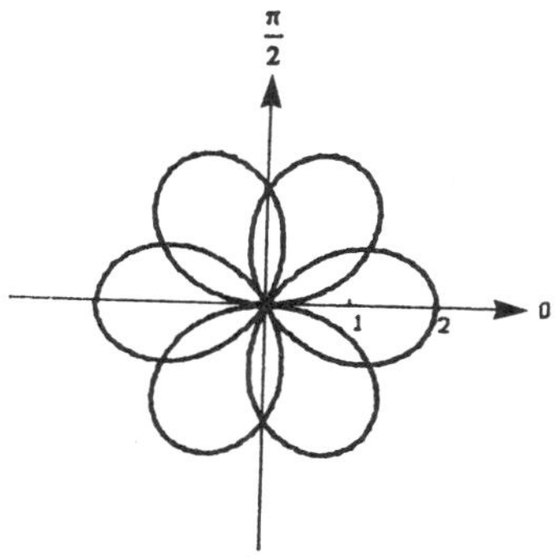

Symmetric to the polar axis

Relative extrema occur when

$\frac{dr}{d\theta} = -3\sin(\frac{3\theta}{2}) = 0$ at

$\theta = 0, \frac{2\pi}{3}, \frac{4\pi}{3}, 2\pi, \frac{8\pi}{3}, \frac{10\pi}{3}$

Relative extrema: $(2, 0), (-2, \frac{2\pi}{3}),$

$(2, \frac{4\pi}{3}), (-2, 2\pi), (2, \frac{8\pi}{3}), (-2, \frac{10\pi}{3})$

Tangents at the pole: $\theta = \frac{\pi}{3}, \pi, \frac{5\pi}{3}$

48. $r = 3\sin(\frac{5\theta}{2})$

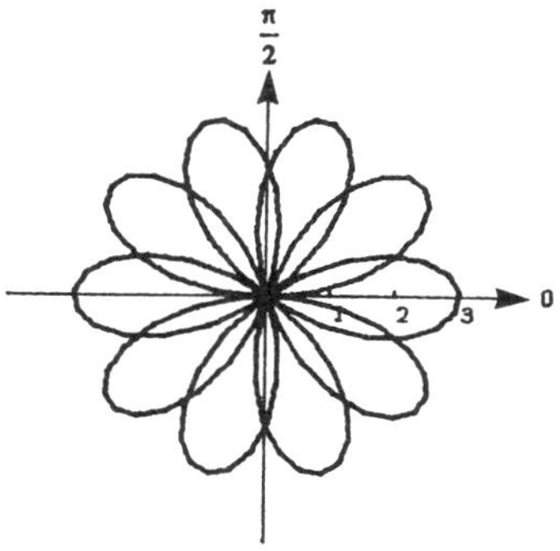

Symmetric to the vertical axis

Relative extrema occur when

$\frac{dr}{d\theta} = \frac{15}{2}\cos(\frac{5\theta}{2}) = 0$ at

$\theta = \frac{\pi}{5}, \frac{3\pi}{5}, \pi, \frac{7\pi}{5}, \frac{9\pi}{5}, \frac{11\pi}{5},$

$\frac{13\pi}{5}, \frac{15\pi}{5}, \frac{17\pi}{5}, \frac{19\pi}{5}$

Tangents at the pole: $\theta = 0, \frac{2\pi}{5}, \frac{4\pi}{5}, \frac{6\pi}{5}, \frac{8\pi}{5}$

49. Since $r = 2 - \sec\theta = 2 - \frac{1}{\cos\theta}$,

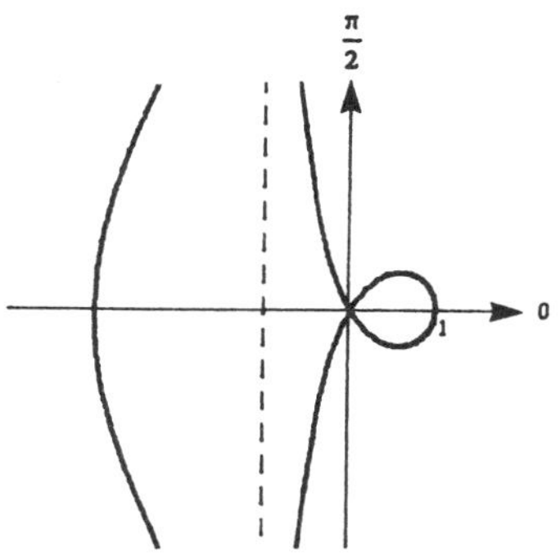

the graph has polar axis symmetry and the tangents at the pole are

$\theta = \frac{\pi}{3}, \frac{-\pi}{3}$

Furthermore, $r \longrightarrow -\infty$ as $\theta \longrightarrow \pi/2^-$
$r \longrightarrow \infty$ as $\theta \longrightarrow -\pi/2^+$

Also, $r = 2 - \frac{1}{\cos\theta} = 2 - \frac{r}{r\cos\theta} = 2 - \frac{r}{x}$

$rx = 2x - r, \quad r = \frac{2x}{1 + x}$ Thus, $r \longrightarrow \pm\infty$ as $x \longrightarrow -1$

50. Since $r = 2 + \csc\theta = 2 + \dfrac{1}{\sin\theta}$, the graph has symmetry with respect to the vertical axis. Furthermore,

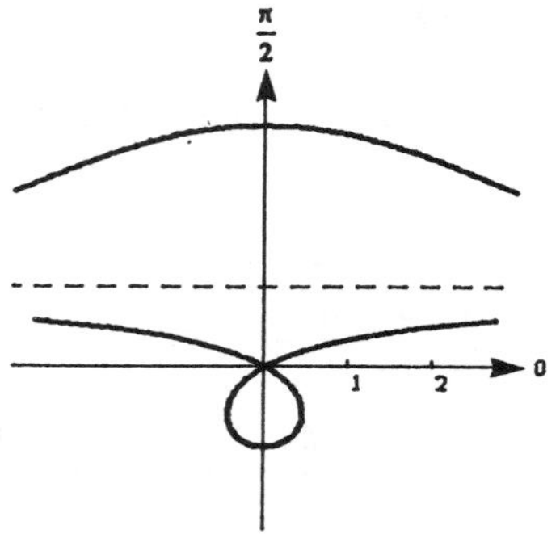

$r \longrightarrow \infty$ as $\theta \longrightarrow 0^+$

$r \longrightarrow \infty$ as $\theta \longrightarrow \pi^-$

Also, $r = 2 + \dfrac{1}{\sin\theta} = 2 + \dfrac{r}{y}$

$ry = 2y + r$, $r = \dfrac{2y}{y-1}$ Thus, $r \longrightarrow \pm\infty$ as $y \longrightarrow 1$

51. $r = \dfrac{2}{\theta}$

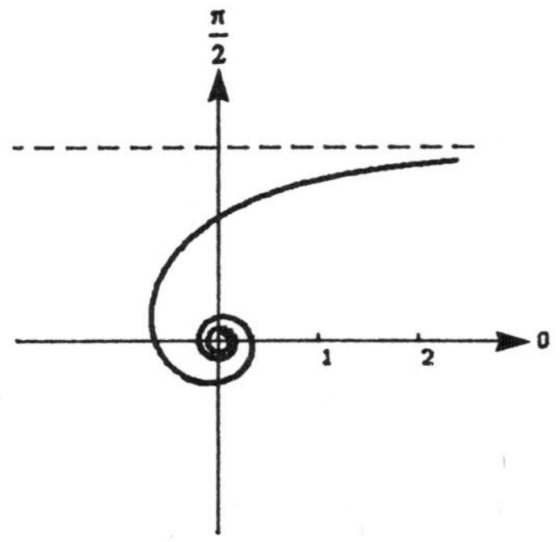

Hyperbolic spiral

$r \longrightarrow \infty$ as $\theta \longrightarrow 0$

$r = \dfrac{2}{\theta} \Rightarrow \theta = \dfrac{2}{r} = \dfrac{2\sin\theta}{r\sin\theta} = \dfrac{2\sin\theta}{y}$

$y = \dfrac{2\sin\theta}{\theta}$

$$\lim_{\theta \to 0} \frac{2\sin\theta}{\theta} = \lim_{\theta \to 0} \frac{2\cos\theta}{1} = 2$$

52. $r = 2\cos 2\theta \sec\theta$

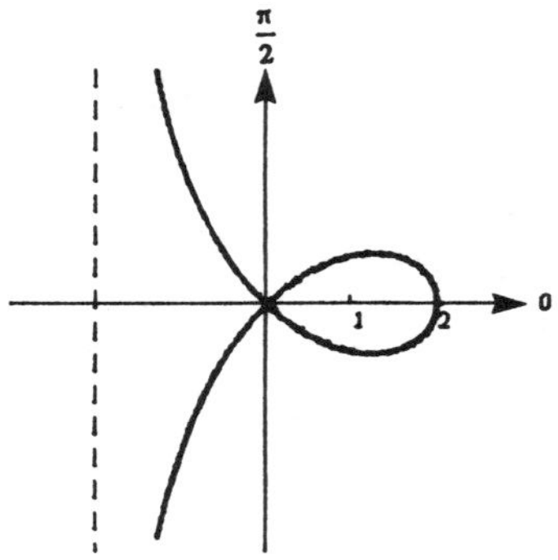

Strophoid

$r \longrightarrow -\infty$ as $\theta \longrightarrow \dfrac{\pi^-}{2}$

$r \longrightarrow \infty$ as $\theta \longrightarrow \dfrac{-\pi^+}{2}$

$r = 2\cos 2\theta \sec\theta$

$= 2(2\cos^2\theta - 1)\sec\theta$

$r\cos\theta = 4\cos^2\theta - 2$

$x = 4\cos^2\theta - 2$

$$\lim_{\theta \to \pm\pi/2} (4\cos^2\theta - 2) = -2$$

54. (a) $\sin(\theta - \frac{\pi}{2}) = \sin\theta\cos(\frac{\pi}{2}) - \cos\theta\sin(\frac{\pi}{2}) = -\cos\theta$

$r = f[\sin(\theta - \frac{\pi}{2})] = f(-\cos\theta)$

(b) $\sin(\theta - \pi) = \sin\theta\cos\pi - \cos\theta\sin\pi = -\sin\theta$

$r = f[\sin(\theta - \pi)] = f(-\sin\theta)$

(c) $\sin(\theta - \frac{3\pi}{2}) = \sin\theta\cos(\frac{3\pi}{2}) - \cos\theta\sin(\frac{3\pi}{2}) = \cos\theta$

$r = f[\sin(\theta - \frac{3\pi}{2})] = f(\cos\theta)$

55. $r = 2 - \sin\theta$

(a) $r = 2 - \sin(\theta - \frac{\pi}{4}) = 2 - \frac{\sqrt{2}}{2}(\sin\theta + \cos\theta)$

(b) $r = 2 - (-\cos\theta) = 2 + \cos\theta$

(c) $r = 2 - (-\sin\theta) = 2 + \sin\theta$

(d) $r = 2 - \cos\theta$

56. $r = 2\sin 2\theta = 4\sin\theta\cos\theta$

(a) $r = 4\sin(\theta - \frac{\pi}{6})\cos(\theta - \frac{\pi}{6})$

(b) $r = 4\sin(\theta - \frac{\pi}{2})\cos(\theta - \frac{\pi}{2}) = -4\sin\theta\cos\theta$

(c) $r = 4\sin(\theta - \frac{3\pi}{2})\cos(\theta - \frac{3\pi}{2}) = -4\sin\theta\cos\theta$

(d) $r = 4\sin(\theta - \pi)\cos(\theta - \pi) = 4\sin\theta\cos\theta$

57. (a) $r = 1 - \sin\theta$

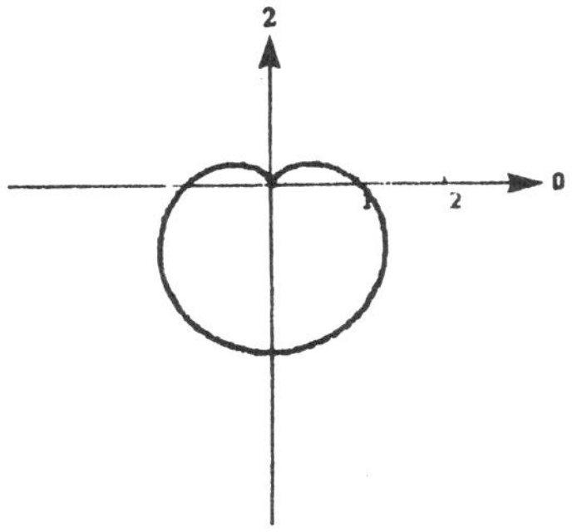

(b) $r = 1 - \sin(\theta - \frac{\pi}{4})$

rotate the graph of $r = 1 - \sin\theta$

through the angle $\frac{\pi}{4}$

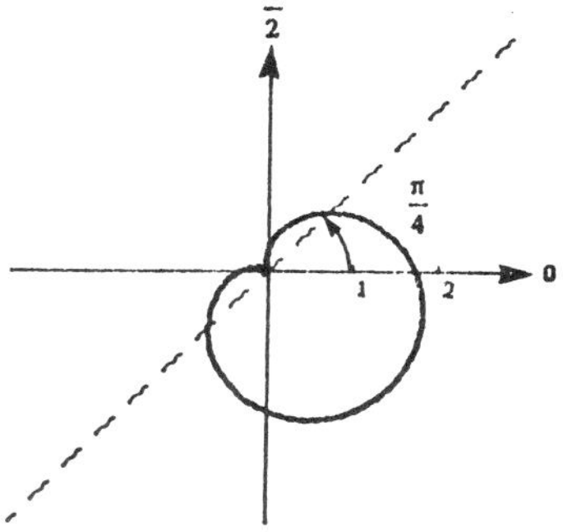

58. (a) $r = 3 \sec \theta$, $r \cos \theta = 3$

$x = 3$

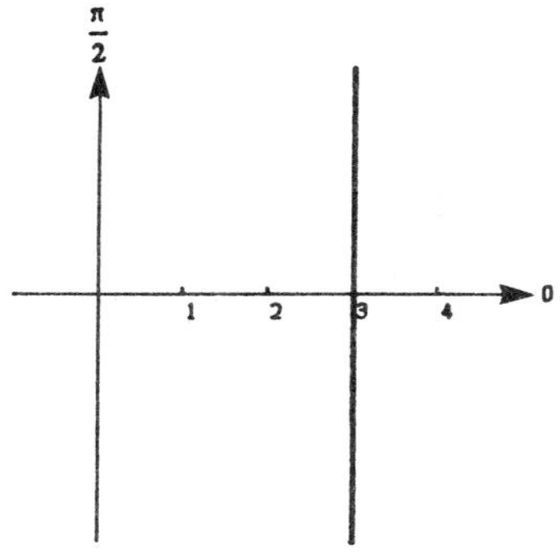

(b) $r = 3 \sec (\theta - \frac{\pi}{4})$

rotate the graph of $r = 3 \sec \theta$

through the angle $\frac{\pi}{4}$

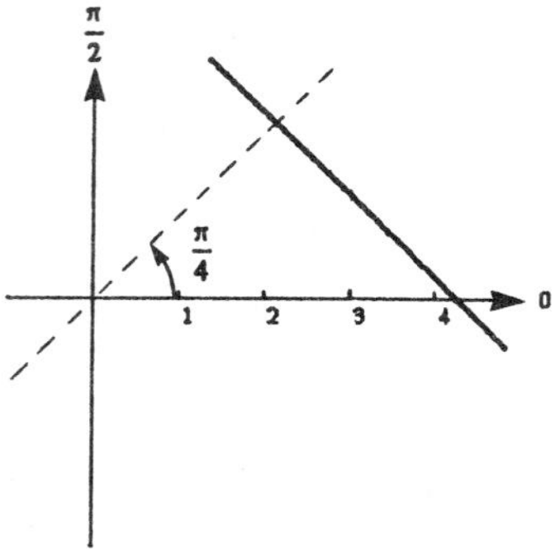

(c) $r = 3 \sec (\theta + \frac{\pi}{3})$

rotate the graph of $r = 3 \sec \theta$

through the angle $-\frac{\pi}{3}$

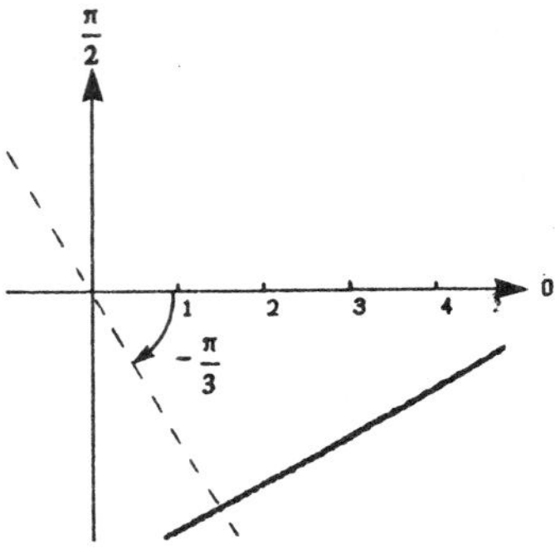

11.5 Polar equations for conics

1. $r = \dfrac{2}{1 - \cos \theta}$

Parabola since $e = 1$

Vertex: $(1, \pi)$

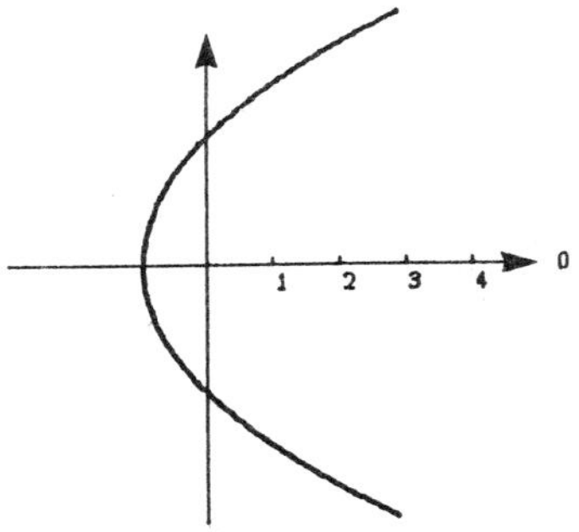

2. $r = \dfrac{4}{1 + \sin \theta}$

Parabola since $e = 1$

Vertex: $(2, \frac{\pi}{2})$

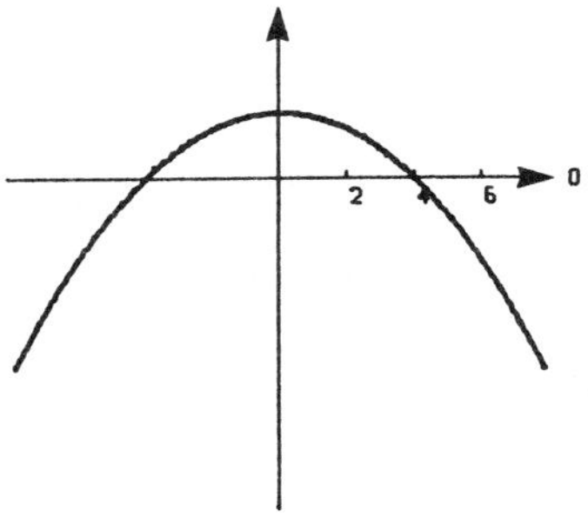

3. $r = \dfrac{5}{1 + \sin\theta}$

Parabola since $e = 1$

Vertex: $\left(\dfrac{5}{2}, \dfrac{\pi}{2}\right)$

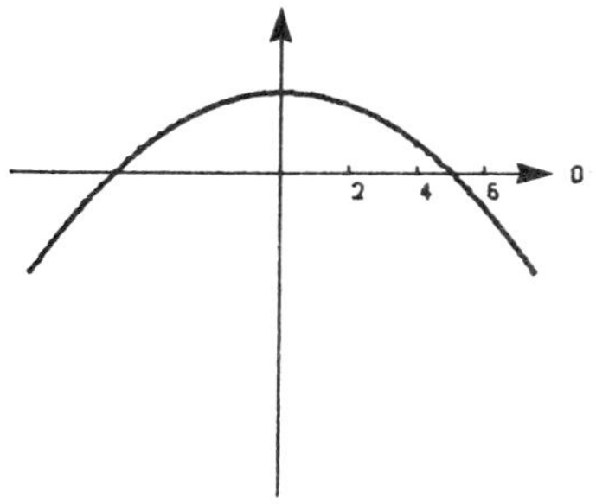

4. $r = \dfrac{6}{1 + \cos\theta}$

Parabola since $e = 1$

Vertex: $(3, 0)$

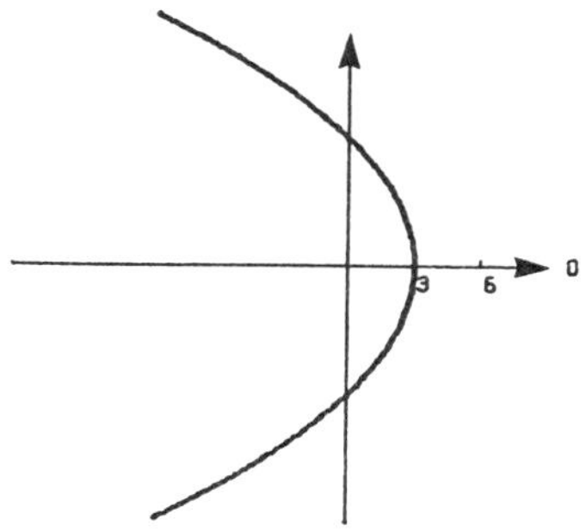

5. $r = \dfrac{2}{2 - \cos\theta} = \dfrac{1}{1 - \frac{1}{2}\cos\theta}$

Ellipse since $e = \dfrac{1}{2} < 1$

Vertices: $(2, 0)$, $\left(\dfrac{2}{3}, \pi\right)$

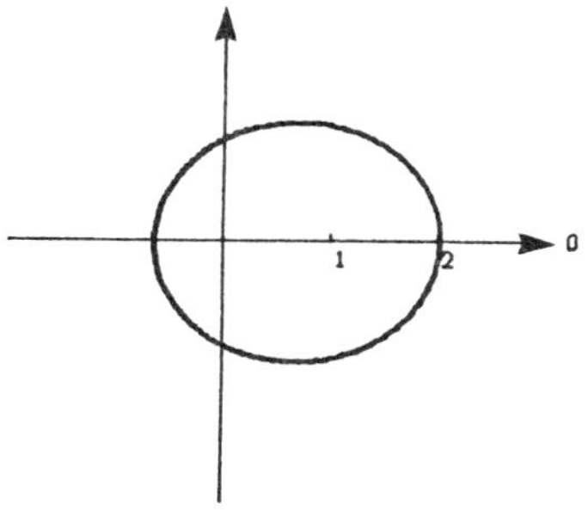

6. $r = \dfrac{3}{3 + 2\sin\theta} = \dfrac{1}{1 + \frac{2}{3}\sin\theta}$

Ellipse since $e = \dfrac{2}{3} < 1$

Vertices: $\left(\dfrac{3}{5}, \dfrac{\pi}{2}\right)$, $\left(3, \dfrac{3\pi}{2}\right)$

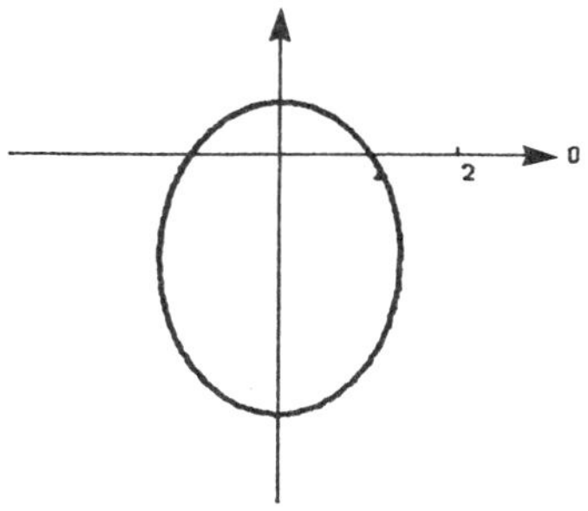

7. $r(2 + \sin\theta) = 4$

$r = \dfrac{4}{2 + \sin\theta} = \dfrac{2}{1 + \frac{1}{2}\sin\theta}$

Ellipse since $e = \dfrac{1}{2} < 1$

Vertices: $\left(\dfrac{4}{3}, \dfrac{\pi}{2}\right)$, $\left(4, \dfrac{3\pi}{2}\right)$

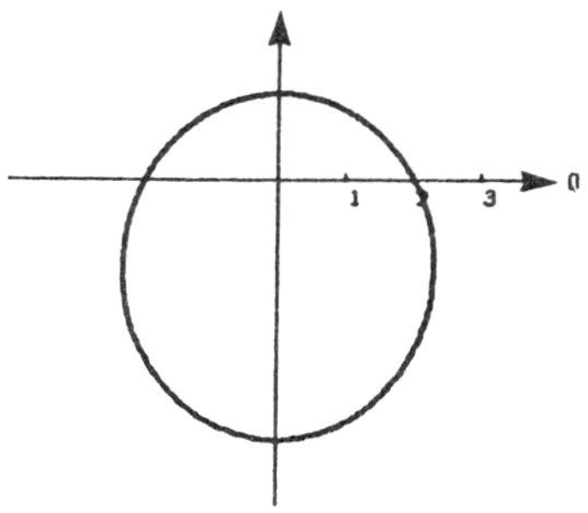

8. $r(3 - 2\cos\theta) = 6$

$$r = \frac{6}{3 - 2\cos\theta} = \frac{2}{1 - \frac{2}{3}\cos\theta}$$

Ellipse since $e = \frac{2}{3} < 1$

Vertices: $(6, 0)$, $(\frac{6}{5}, \pi)$

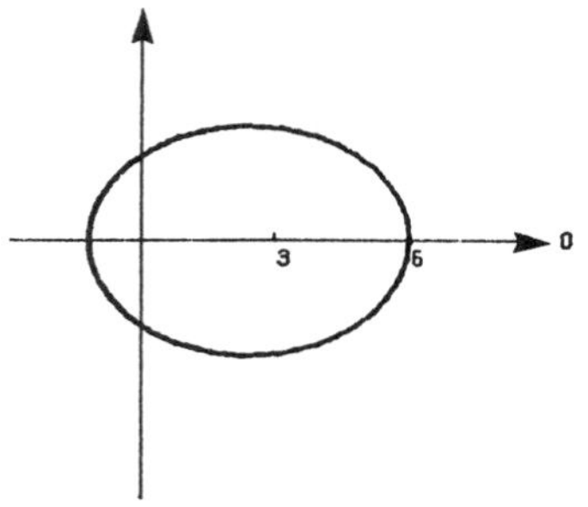

9. $r = \dfrac{-1}{1 - \sin\theta}$

Parabola since $e = 1$

Vertex: $(-\frac{1}{2}, \frac{3\pi}{2})$

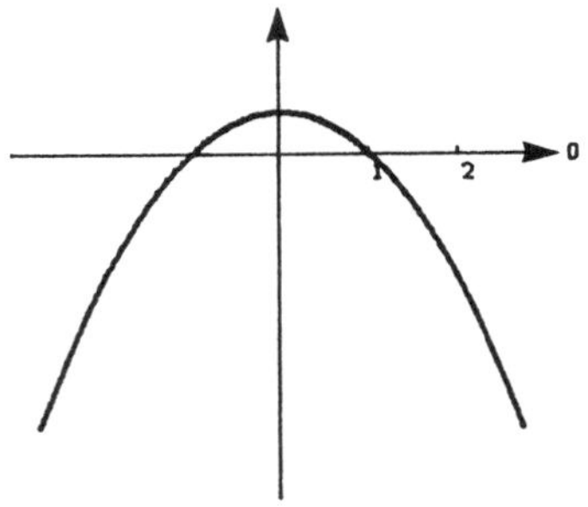

10. $r = \dfrac{-3}{2 + 4\sin\theta} = \dfrac{-3/2}{1 + 2\sin\theta}$

Hyperbola since $e = 2 > 1$

Vertices: $(-\frac{1}{2}, \frac{\pi}{2})$, $(\frac{3}{2}, \frac{3\pi}{2})$

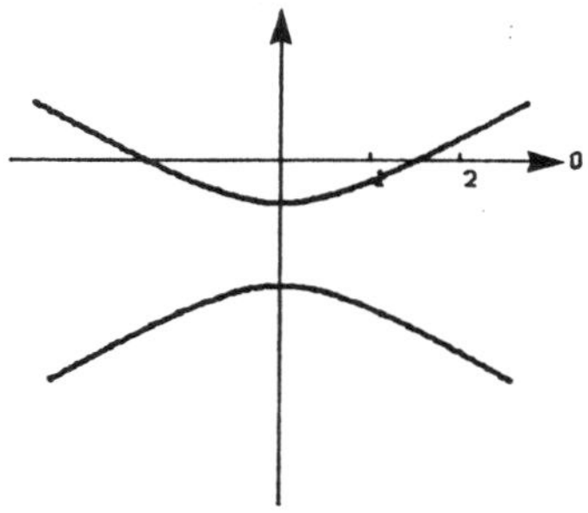

11. $r = \dfrac{5}{-1 + 2\cos\theta} = \dfrac{-5}{1 - 2\cos\theta}$

Hyperbola since $e = 2 > 1$

Vertices: $(5, 0)$, $(-\frac{5}{3}, \pi)$

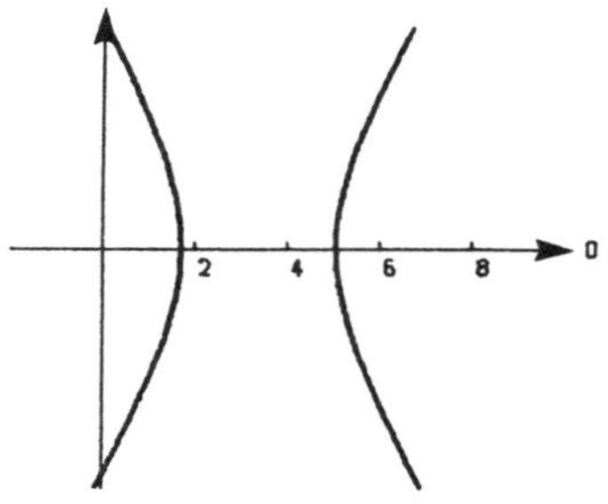

12. $r = \dfrac{3}{-4 + 2\cos\theta} = \dfrac{-3/4}{1 - \frac{1}{2}\cos\theta}$

Ellipse since $e = \dfrac{1}{2} < 1$

Vertices: $\left(-\dfrac{3}{2}, 0\right)$, $\left(-\dfrac{1}{2}, \pi\right)$

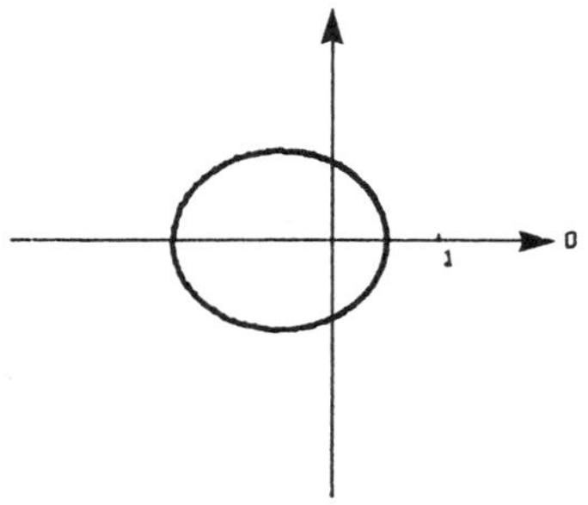

13. $r = \dfrac{3}{2 - 6\cos\theta} = \dfrac{3/2}{1 - 3\cos\theta}$

Hyperbola since $e = 3 > 1$

Vertices: $\left(-\dfrac{3}{4}, 0\right)$, $\left(\dfrac{3}{8}, \pi\right)$

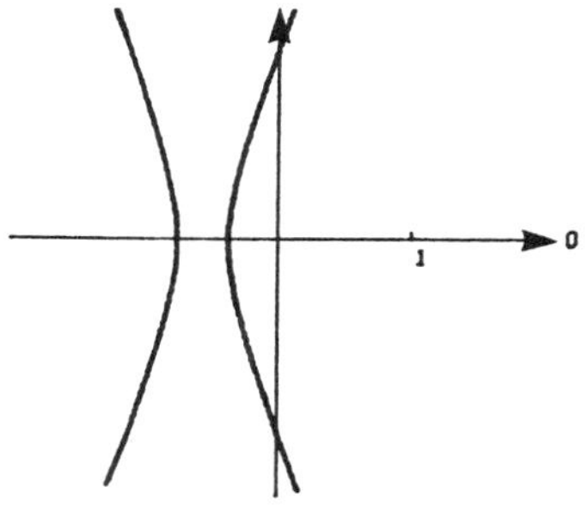

14. $r = \dfrac{4}{1 - 2\cos\theta}$

Hyperbola since $e = 2 > 1$

Vertices: $(-4, 0)$, $\left(\dfrac{4}{3}, \pi\right)$

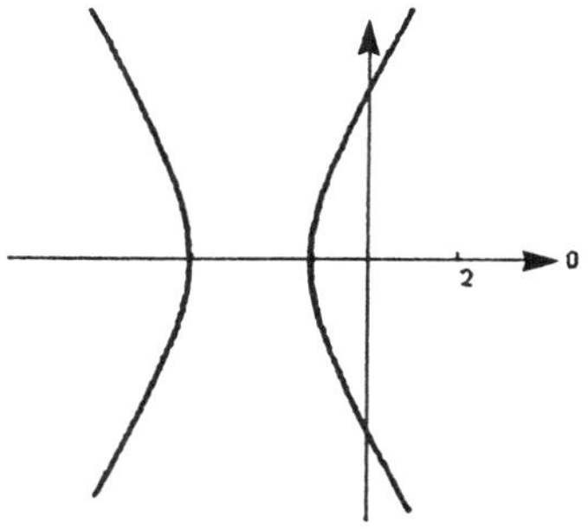

15. $r = \dfrac{3}{2 + 6\sin\theta} = \dfrac{3/2}{1 + 3\sin\theta}$

Hyperbola since $e = 3 > 1$

Vertices: $\left(\dfrac{3}{8}, \dfrac{\pi}{2}\right)$, $\left(-\dfrac{3}{4}, \dfrac{3\pi}{2}\right)$

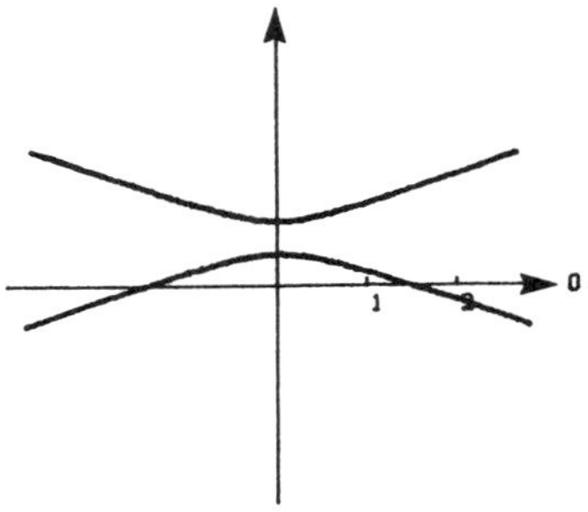

16. $r = \dfrac{4}{1 + 2\cos\theta}$

Hyperbola since $e = 2 > 1$

Vertices: $(\frac{4}{3}, 0)$, $(-4, \pi)$

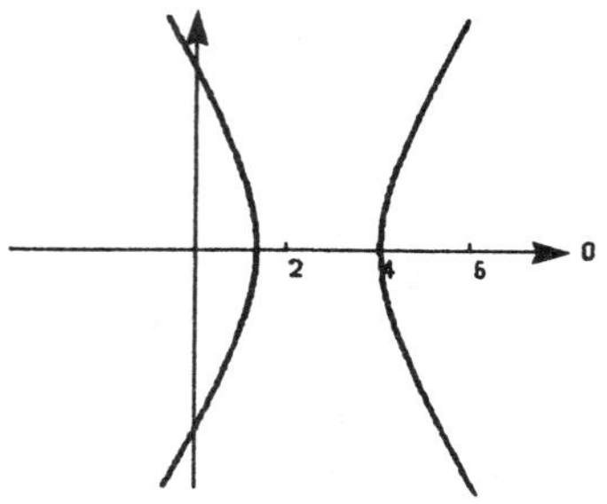

17. Ellipse, $e = \frac{1}{2}$, $y = 1$, $p = 1$

$$r = \frac{ep}{1 + e\sin\theta} = \frac{1/2}{1 + \frac{1}{2}\sin\theta} = \frac{1}{2 + \sin\theta}$$

18. Ellipse, $e = \frac{3}{4}$, $y = -2$, $p = -2$

$$r = \frac{ep}{1 - e\sin\theta} = \frac{-2(3/4)}{1 - \frac{3}{4}\sin\theta} = \frac{-6}{4 - 3\sin\theta}$$

19. Parabola

$e = 1$, $x = -1$, $p = -1$

$$r = \frac{ep}{1 - e\cos\theta} = \frac{-1}{1 - \cos\theta}$$

20. Parabola

$e = 1$, $y = 1$, $p = 1$

$$r = \frac{ep}{1 + e\sin\theta} = \frac{1}{1 + \sin\theta}$$

21. Hyperbola

$e = 2$, $x = 1$, $p = 1$

$$r = \frac{ep}{1 + e\cos\theta} = \frac{2}{1 + 2\cos\theta}$$

22. Hyperbola

$e = \frac{3}{2}$, $x = -1$, $p = -1$

$$r = \frac{ep}{1 - e\cos\theta}$$

$$= \frac{-3/2}{1 - \frac{3}{2}\cos\theta} = \frac{-3}{2 - 3\cos\theta}$$

23. Parabola

Vertex at $(1, -\frac{\pi}{2})$

$e = 1$, $p = 2$, $r = \dfrac{2}{1 - \sin\theta}$

24. Parabola

Directrix is $x = 8$

$e = 1$, $p = 8$, $r = \dfrac{8}{1 + \cos\theta}$

25. Ellipse

Vertices at $(6, \frac{\pi}{2})$ and $(2, \frac{3\pi}{2})$, $e = \frac{1}{2}$, $p = -6$

$$r = \frac{ep}{1 - e\sin\theta} = \frac{-6/2}{1 - \frac{1}{2}\sin\theta} = \frac{-6}{2 - \sin\theta}$$

26. Ellipse

Vertex at $(2, 0)$

Directrix is $x = 6$, $p = 6$, $e = \frac{1}{2}$

$$r = \frac{ep}{1 + e\cos\theta} = \frac{6/2}{1 + \frac{1}{2}\cos\theta} = \frac{6}{2 + \cos\theta}$$

27. Hyperbola

Vertex at $(2, \frac{3\pi}{2})$

Directrix is $y = -3$, $e = 2$, $p = -3$

$$r = \frac{ep}{1 - e\sin\theta} = \frac{-6}{1 - 2\sin\theta}$$

28. Hyperbola

Vertices at $(2, 0)$ and $(6, 0)$, $e = 2$, $p = 3$

$$r = \frac{ep}{1 + e\cos\theta} = \frac{6}{1 + 2\cos\theta}$$

29. $\frac{x^2}{a^2} + \frac{y^2}{b^2} = 1$, $x^2b^2 + y^2a^2 = a^2b^2$, $b^2r^2\cos^2\theta + a^2r^2\sin^2\theta = a^2b^2$

$r^2[b^2\cos^2\theta + a^2(1 - \cos^2\theta)] = a^2b^2$, $r^2[a^2 + \cos^2\theta(b^2 - a^2)] = a^2b^2$

$$r^2 = \frac{a^2b^2}{a^2 + (b^2 - a^2)\cos^2\theta} = \frac{a^2b^2}{a^2 - c^2\cos^2\theta} = \frac{b^2}{1 - (c/a)^2\cos^2\theta}$$

$$= \frac{b^2}{1 - e^2\cos^2\theta}$$

30. $\frac{x^2}{a^2} - \frac{y^2}{b^2} = 1$, $x^2b^2 - y^2a^2 = a^2b^2$, $b^2r^2\cos^2\theta - a^2r^2\sin^2\theta = a^2b^2$

$r^2[b^2\cos^2\theta - a^2(1 - \cos^2\theta)] = a^2b^2$, $r^2[-a^2 + \cos^2\theta(a^2 + b^2)] = a^2b^2$

$$r^2 = \frac{a^2b^2}{-a^2 + c^2\cos^2\theta} = \frac{b^2}{-1 + (c^2/a^2)\cos^2\theta} = \frac{-b^2}{1 - e^2\cos^2\theta}$$

31. $a = 5, \quad c = 4, \quad e = \frac{4}{5}, \quad b = 3$

$$r^2 = \frac{9}{1 - (16/25)\cos^2\theta}$$

32. $a = 4, \quad c = 5, \quad b = 3, \quad e = \frac{5}{4}$

$$r^2 = \frac{-9}{1 - (25/16)\cos^2\theta}$$

33. $a = 3, \quad b = 4, \quad c = 5, \quad e = \frac{5}{3}$

$$r^2 = \frac{-16}{1 - (25/9)\cos^2\theta}$$

34. $a = 2, \quad b = 1, \quad c = \sqrt{3}, \quad e = \frac{\sqrt{3}}{2}$

$$r^2 = \frac{1}{1 - (3/4)\cos^2\theta}$$

35. $r = \dfrac{2}{1 - \cos(\theta - (\pi/4))}$

Rotate the graph of

$$r = \frac{2}{1 - \cos\theta}$$

through the angle $\dfrac{\pi}{4}$

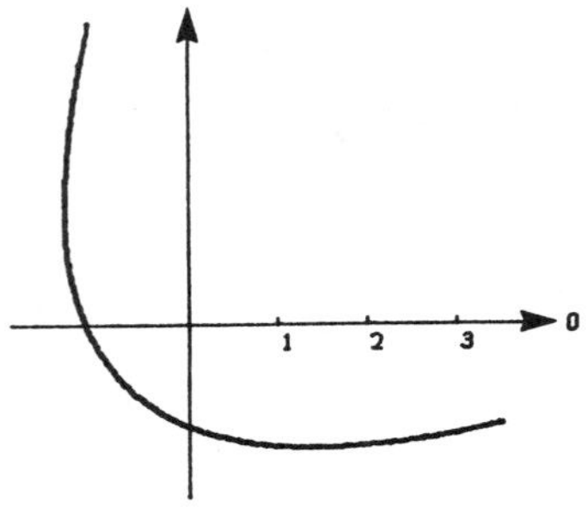

36. $r = \dfrac{4}{1 + \sin(\theta - (\pi/3))}$

Rotate the graph of

$$r = \frac{4}{1 + \sin\theta}$$

through the angle $\dfrac{\pi}{3}$

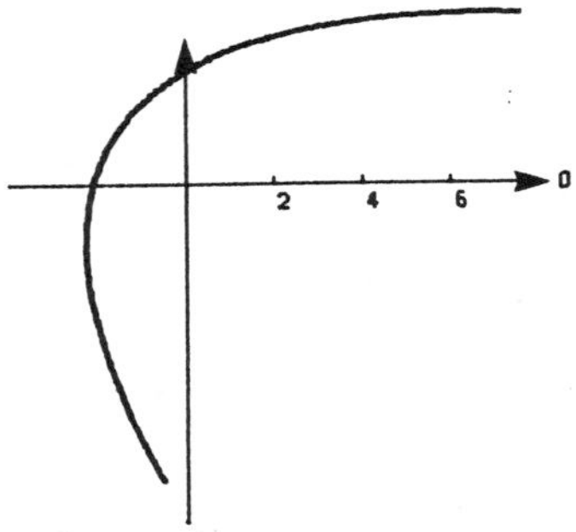

37. $r = \dfrac{4}{2 + \sin(\theta + (\pi/6))}$

Rotate the graph of

$$r = \frac{4}{2 + \sin\theta}$$

through the angle $-\dfrac{\pi}{6}$

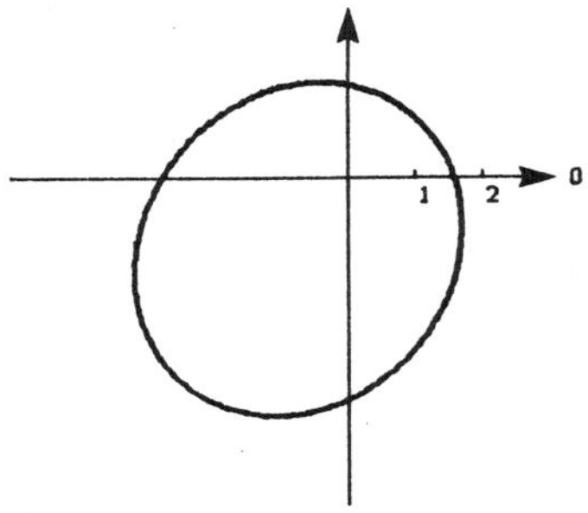

38. $r = \dfrac{4}{1 + 2\cos(\theta + (2\pi/3))}$

Rotate the graph of

$r = \dfrac{4}{1 + 2\cos\theta}$

through the angle $-\dfrac{2\pi}{3}$

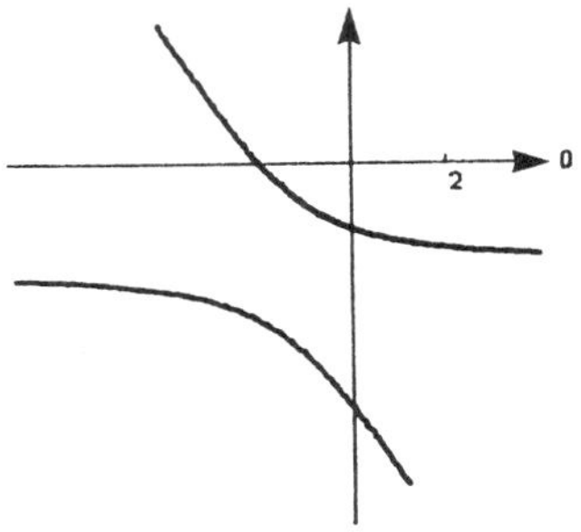

39. $r = 2(1 - \cos\theta)$, $\theta = \pi$, $\dfrac{dr}{d\theta} = 2\sin\theta$

$\tan\psi = \dfrac{r}{dr/d\theta} = \dfrac{2(1 - \cos\pi)}{2\sin\pi} = \dfrac{2}{0}$ undefined, $\psi = \dfrac{\pi}{2}$

40. $r = 3(1 - \cos\theta)$, $\theta = \dfrac{3\pi}{4}$, $\dfrac{dr}{d\theta} = 3\sin\theta$

$\tan\psi = \dfrac{r}{dr/d\theta} = \dfrac{3(1 - \cos(3\pi/4))}{3\sin(3\pi/4)} = \dfrac{1 + (\sqrt{2}/2)}{\sqrt{2}/2} = \dfrac{2 + \sqrt{2}}{\sqrt{2}}$

$\psi = \arctan\left(\dfrac{2 + \sqrt{2}}{\sqrt{2}}\right) \approx 1.1781$ rad

41. $r = 2\cos 3\theta$, $\theta = \dfrac{\pi}{6}$, $\dfrac{dr}{d\theta} = -6\sin 3\theta$

$\tan\psi = \dfrac{r}{dr/d\theta} = \dfrac{2\cos(\pi/2)}{-6\sin(\pi/2)} = 0$, $\psi = \arctan 0 = 0$

42. $r = 4\sin 2\theta$, $\theta = \dfrac{\pi}{6}$, $\dfrac{dr}{d\theta} = 8\cos 2\theta$

$\tan\psi = \dfrac{r}{dr/d\theta} = \dfrac{4\sin(\pi/3)}{8\cos(\pi/3)} = \dfrac{\sqrt{3}/2}{2(1/2)} = \dfrac{\sqrt{3}}{2}$

$\psi = \arctan\dfrac{\sqrt{3}}{2} \approx 0.7137$ rad

43. $r = \dfrac{6}{1 - \cos\theta} = 6(1 - \cos\theta)^{-1}$, $\theta = \dfrac{2\pi}{3}$

$\dfrac{dr}{d\theta} = -6(1 - \cos\theta)^{-2}\sin\theta = \dfrac{-6\sin\theta}{(1 - \cos\theta)^2}$

$\tan\psi = \dfrac{r}{dr/d\theta} = \dfrac{4}{(-4\sqrt{3}/3)} = -\sqrt{3}$

$\psi = \dfrac{\pi}{3}$ (since ψ is defined to be $0 \le \psi \le \dfrac{\pi}{2}$)

44. $r = 5,\quad \theta = \frac{\pi}{6},\quad \frac{dr}{d\theta} = 0$

$\tan \psi = \frac{r}{dr/d\theta} = \frac{5}{0}$ undefined, $\psi = \frac{\pi}{2}$

11.6
Area and arc length in polar coordinates

1. $r = 1 + \cos\theta$, $r = 1 - \cos\theta$. Solving simultaneously, $1 + \cos\theta = 1 - \cos\theta$, $2\cos\theta = 0$, $\theta = \frac{\pi}{2}, \frac{3\pi}{2}$. Replacing r by -r and θ by $\theta + \pi$ in the first equation and solving, $-1 + \cos\theta = 1 - \cos\theta$, $\cos\theta = 1$, $\theta = 0$. Both curves pass through the pole, $(0, \pi)$ and $(0, 0)$, respectively. Points of intersection: $(1, \frac{\pi}{2})$, $(1, \frac{3\pi}{2})$, $(0, 0)$

2. $r = 3(1 + \sin\theta)$, $r = 3(1 - \sin\theta)$. Solving simultaneously, $3(1 + \sin\theta) = 3(1 - \sin\theta)$, $2\sin\theta = 0$, $\theta = 0, \pi$. Replacing r by -r and θ by $\theta + \pi$ in the first equation and solving, $-3(1 - \sin\theta) = 3(1 - \sin\theta)$, $\sin\theta = 1$, $\theta = \frac{\pi}{2}$. Both curves pass through the pole, $(0, \frac{3\pi}{2})$ and $(0, \frac{\pi}{2})$ respectively. Points of intersection: $(3, 0)$, $(3, \pi)$, $(0, 0)$

3. $r = 1 + \cos\theta$, $r = 1 - \sin\theta$. Solving simultaneously, $1 + \cos\theta = 1 - \sin\theta$, $\cos\theta = -\sin\theta$, $\tan\theta = -1$, $\theta = \frac{3\pi}{4}, \frac{7\pi}{4}$. Replacing r by -r and θ by $\theta + \pi$ in the first equation and solving, $-1 + \cos\theta = 1 - \sin\theta$, $\sin\theta + \cos\theta = 2$, which has no solution. Both curves pass through the pole, $(0, \pi)$ and $(0, \frac{\pi}{2})$, respectively. Points of intersection: $(\frac{2 - \sqrt{2}}{2}, \frac{3\pi}{4})$, $(\frac{2 + \sqrt{2}}{2}, \frac{7\pi}{4})$, $(0, 0)$

4. $r = 2 - 3\cos\theta$, $r = \cos\theta$. Solving simultaneously, $2 - 3\cos\theta = \cos\theta$, $\cos\theta = \frac{1}{2}$, $\theta = \frac{\pi}{3}, \frac{-\pi}{3}$. Both curves pass through the pole, $(0, \arccos\frac{2}{3})$ and $(0, \frac{\pi}{2})$, respectively. Points of intersection: $(\frac{1}{2}, \frac{\pi}{3})$, $(\frac{1}{2}, \frac{-\pi}{3})$, $(0, 0)$

5. $r = 4 - 5\sin\theta$, $r = 3\sin\theta$. Solving simultaneously, $4 - 5\sin\theta = 3\sin\theta$, $\sin\theta = \frac{1}{2}$, $\theta = \frac{\pi}{6}, \frac{5\pi}{6}$. Both curves pass through the pole, $(0, \arcsin\frac{4}{5})$ and $(0, 0)$, respectively. Points of intersection: $(\frac{3}{2}, \frac{\pi}{6})$, $(\frac{3}{2}, \frac{5\pi}{6})$, $(0, 0)$

6. $r = 1 + \cos\theta$, $r = 3\cos\theta$. Solving simultaneously, $1 + \cos\theta = 3\cos\theta$, $\cos\theta = \frac{1}{2}$, $\theta = \frac{\pi}{3}, \frac{5\pi}{3}$. Both curves pass through the pole: $(0, \pi)$ and $(0, \frac{\pi}{2})$, respectively. Points of intersection: $(\frac{3}{2}, \frac{\pi}{3})$, $(\frac{3}{2}, \frac{5\pi}{3})$, $(0, 0)$

7. $r = \frac{\theta}{2}$, $r = 2$. Solving simultaneously, we have $\frac{\theta}{2} = 2$, $\theta = 4$. Points of intersection: $(2, 4)$, $(-2, -4)$

8. $\theta = \frac{\pi}{4}$, $r = 2$. Line of slope 1 passing through the pole and a circle of radius 2 centered at the pole. Points of intersection: $(2, \frac{\pi}{4})$, $(-2, \frac{\pi}{4})$

9. $r = 4\sin 2\theta$, $r = 2$. From Section 11.4, we know $r = 4\sin 2\theta$ is the equation of a rose curve with 4 petals and is symmetric to the polar axis, the vertical axis and the pole. Also, $r = 2$ is the equation of a circle of radius 2 centered at the pole. Solving simultaneously, $4\sin 2\theta = 2$, $2\theta = \frac{\pi}{6}, \frac{5\pi}{6}$, $\theta = \frac{\pi}{12}, \frac{5\pi}{12}$. Therefore, the points of intersection for one petal are $(2, \frac{\pi}{12})$, $(2, \frac{5\pi}{12})$. By symmetry, the other points of intersection are $(2, \frac{7\pi}{12})$, $(2, \frac{11\pi}{12})$, $(2, \frac{13\pi}{12})$, $(2, \frac{17\pi}{12})$, $(2, \frac{19\pi}{12})$, $(2, \frac{23\pi}{12})$

10. $r = 3 + \sin\theta$, $r = 2\csc\theta$. The graph of $r = 3 + \sin\theta$ is a limacon symmetrical to the vertical axis and the graph of $r = 2\csc\theta$ is the horizontal line $y = 3$. Therefore, there are two points of intersection. Solving simultaneously, $3 + \sin\theta = 2\csc\theta$, $\sin^2\theta + 3\sin\theta - 2 = 0$,

$\sin\theta = \dfrac{-3 \pm \sqrt{17}}{2}$, $\theta = \arcsin\left(\dfrac{\sqrt{17} - 3}{2}\right) \approx 0.596$.

Points of intersection: $\left(\dfrac{\sqrt{17} + 3}{2}, \arcsin\left(\dfrac{\sqrt{17} - 3}{2}\right)\right)$,

$\left(\dfrac{\sqrt{17} + 3}{2}, \pi - \arcsin\left(\dfrac{\sqrt{17} - 3}{2}\right)\right)$, $(3.56, 0.596)$, $(3.56, 2.545)$

11. $r = 2 + 3\cos\theta$, $r = \dfrac{\sec\theta}{2}$. The graph of $r = 2 + 3\cos\theta$ is a limacon with a double loop $(b > a)$ and is symmetrical to the polar axis. The graph of $r = \dfrac{\sec\theta}{2}$ is the vertical line $x = \dfrac{1}{2}$. Therefore, there are four points of intersection. Solving simultaneously,

$2 + 3\cos\theta = \dfrac{\sec\theta}{2}$, $6\cos^2\theta + 4\cos\theta - 1 = 0$, $\cos\theta = \dfrac{-2 \pm \sqrt{10}}{6}$,

$\theta = \arccos\left(\dfrac{-2 + \sqrt{10}}{6}\right) \approx 1.376$. Replacing r by $-r$ and θ by $\theta + \pi$ and solving $-2 + 3\cos\theta = \dfrac{\sec\theta}{2}$, $6\cos^2\theta - 4\cos\theta - 1 = 0$, $\cos\theta = \dfrac{2 \pm \sqrt{10}}{6}$,

$\theta = \arccos\left(\dfrac{2 + \sqrt{10}}{6}\right) \approx 0.535$. Points of intersection: $(0.581, \pm 0.535)$, $(2.581, \pm 1.376)$

12. $r = 3(1 - \cos\theta)$, $r = \dfrac{6}{1 - \cos\theta}$. The graph of $3(1 - \cos\theta)$ is a cardioid with polar axis symmetry. The graph of $r = \dfrac{6}{1 - \cos\theta}$ is a parabola with focus at the pole, vertex $(3, \pi)$, and polar axis symmetry. Therefore, there are two points of intersection. Solving simultaneously,

$3(1 - \cos\theta) = \dfrac{6}{1 - \cos\theta}$, $(1 - \cos\theta)^2 = 2$, $\cos\theta = 1 \pm \sqrt{2}$,

$\theta = \arccos(1 - \sqrt{2})$. Points of intersection: $(3\sqrt{2}, \arccos(1 - \sqrt{2}))$, $(3\sqrt{2}, 2\pi - \arccos(1 - \sqrt{2}))$, $(4.243, 1.998)$, $(4.243, 4.285)$

13. $A = 2\left[\frac{1}{2}\int_0^{\pi/6}(2\cos 3\theta)^2\,d\theta\right] = 2\left[\theta + \frac{1}{6}\sin 6\theta\right]_0^{\pi/6} = \frac{\pi}{3}$

14. $A = 2\left[\frac{1}{2}\int_0^{\pi/4}(4\sin 2\theta)^2\,d\theta\right] = 8\left[\theta - \frac{1}{4}\sin 4\theta\right]_0^{\pi/4} = 2\pi$

15. $A = 2\left[\frac{1}{2}\int_0^{\pi/3}(\cos(\frac{3\theta}{2}))^2\,d\theta\right] = \frac{1}{2}\left[\theta + \frac{1}{3}\sin 3\theta\right]_0^{\pi/3} = \frac{\pi}{6}$

16. $A = 2\left[\frac{1}{2}\int_0^{\pi/10}(\cos 5\theta)^2\,d\theta\right] = \frac{1}{2}\left[\theta + \frac{1}{10}\sin 10\theta\right]_0^{\pi/10} = \frac{\pi}{20}$

17. $A = 2\left[\frac{1}{2}\int_{-\pi/2}^{\pi/2}(1 - \sin\theta)^2\,d\theta\right] = \left[\frac{3}{2}\theta + 2\cos\theta - \frac{1}{4}\sin 2\theta\right]_{-\pi/2}^{\pi/2} = \frac{3\pi}{2}$

18. $A = 2\left[\frac{1}{2}\int_0^{\pi/2}(1 - \sin\theta)^2\,d\theta\right] = \left[\frac{3}{2}\theta + 2\cos\theta - \frac{1}{4}\sin 2\theta\right]_0^{\pi/2} = \frac{3\pi - 8}{4}$

19. $r = \frac{3}{2 - \cos\theta} = \frac{3/2}{1 - (1/2)\cos\theta}$. We have $e = \frac{1}{2}$, $a = 2$, $b = \sqrt{3}$, $c = 1$

$A = \pi ab = 2\pi\sqrt{3}$

20. $r = \frac{2}{3 - 2\sin\theta} = \frac{2/3}{1 - (2/3)\sin\theta}$. We have $e = \frac{2}{3}$, $a = \frac{6}{5}$, $b = \frac{2\sqrt{5}}{5}$, $c = \frac{4}{5}$

$A = \pi ab = \frac{12\sqrt{5}\pi}{25}$

21. $A = 2\left[\frac{1}{2}\int_{2\pi/3}^{\pi}(1 + 2\cos\theta)^2\,d\theta\right] = \left[3\theta + 4\sin\theta + \sin 2\theta\right]_{2\pi/3}^{\pi} = \frac{2\pi - 3\sqrt{3}}{2}$

22. $A = 2\left[\frac{1}{2}\int_{\arcsin(-3/4)}^{3\pi/2}(3 + 4\sin\theta)^2\,d\theta\right]$

$= \left[17\theta - 24\cos\theta - 4\sin(2\theta)\right]_{\arcsin(-3/4)}^{3\pi/2} \approx 0.3806$

23. The area inside the outer loop is

$$2\left[\frac{1}{2}\int_0^{2\pi/3}(1 + 2\cos\theta)^2\, d\theta\right] = \left[3\theta + 4\sin\theta + \sin 2\theta\right]_0^{2\pi/3} = \frac{4\pi + 3\sqrt{3}}{2}$$

From the result of Exercise 21, the area between the loops is

$$A = \left(\frac{4\pi + 3\sqrt{3}}{2}\right) - \left(\frac{2\pi - 3\sqrt{3}}{2}\right) = \pi + 3\sqrt{3}$$

24. Four times the area in Exercise 23, $A = 4(\pi + 3\sqrt{3})$

25. From Exercise 9, the points of intersection of the circle with one petal are $(2, \frac{\pi}{12})$, $(2, \frac{5\pi}{12})$. The area within one petal is

$$A = \frac{1}{2}\int_0^{\pi/12}(4\sin 2\theta)^2\, d\theta + \frac{1}{2}\int_{\pi/12}^{5\pi/12}(2)^2\, d\theta + \frac{1}{2}\int_{5\pi/12}^{\pi/2}(4\sin 2\theta)^2\, d\theta$$

$$= 16\int_0^{\pi/12}\sin^2(2\theta)\, d\theta + 2\int_{\pi/12}^{5\pi/12} d\theta \text{ (by symmetry of the petal)}$$

$$= 8\left[\theta - \frac{1}{4}\sin 4\theta\right]_0^{\pi/12} + \left[2\theta\right]_{\pi/12}^{5\pi/12} = \frac{4\pi}{3} - \sqrt{3}$$

$$\text{Total area} = 4\left(\frac{4\pi}{3} - \sqrt{3}\right) = \frac{16\pi}{3} - 4\sqrt{3} = \frac{4}{3}(4\pi - 3\sqrt{3})$$

26. $$A = 4\left[\frac{1}{2}\int_0^{\pi/2}9(1 - \sin\theta)^2\, d\theta\right] = 18\int_0^{\pi/2}(1 - \sin\theta)^2\, d\theta = \frac{9}{2}(3\pi - 8)$$

(from Exercise 18)

27. $$A = 4\left[\frac{1}{2}\int_0^{\pi/2}(3 - 2\sin\theta)^2\, d\theta\right] = 2\left[11\theta + 12\cos\theta - \sin(2\theta)\right]_0^{\pi/2}$$

$$= 11\pi - 24$$

28. $$A = 2\left[\frac{1}{2}\int_{\pi/4}^{5\pi/4}(3 - 2\sin\theta)^2\, d\theta\right] = \left[11\theta + 12\cos\theta - \sin(2\theta)\right]_{\pi/4}^{5\pi/4}$$

$$= 11\pi - 12\sqrt{2}$$

29. $A = 2\left[\frac{1}{2}\int_0^{\pi/6}(4\sin\theta)^2\,d\theta + \frac{1}{2}\int_{\pi/6}^{\pi/2}(2)^2\,d\theta\right]$

$= 16\left[\frac{1}{2}\theta - \frac{1}{4}\sin(2\theta)\right]_0^{\pi/6} + \left[4\theta\right]_{\pi/6}^{\pi/2} = \frac{8\pi}{3} - 2\sqrt{3}$

$= \frac{2}{3}(4\pi - 3\sqrt{3})$

30. $A = 2\left[\frac{1}{2}\int_{\pi/6}^{\pi/2}(3\sin\theta)^2\,d\theta - \frac{1}{2}\int_{\pi/6}^{\pi/2}(2-\sin\theta)^2\,d\theta\right]$

$= \int_{\pi/6}^{\pi/2}(-4\cos 2\theta + 4\sin\theta)\,d\theta = \left[-2\sin(2\theta) - 4\cos\theta\right]_{\pi/6}^{\pi/2} = 3\sqrt{3}$

31. $A = 2\left[\frac{1}{2}\int_0^{\pi}[a(1+\cos\theta)]^2\,d\theta\right] - \frac{a^2\pi}{4}$

$= a^2\left[\frac{3}{2}\theta + 2\sin\theta + \frac{\sin 2\theta}{2}\right]_0^{\pi} - \frac{a^2\pi}{4}$

$= \frac{3a^2\pi}{2} - \frac{a^2\pi}{4} = \frac{5a^2\pi}{4}$

32. Area = Area of $r = 2a\cos\theta$ − Area of sector − Area between

$r = 2a\cos\theta$ and the lines $\theta = \frac{\pi}{3}$, $\theta = -\frac{\pi}{3}$

$A = \pi a^2 - (\frac{\pi}{3})a^2 - 2\left[\frac{1}{2}\int_{\pi/3}^{\pi/2}(2a\cos\theta)^2\,d\theta\right]$

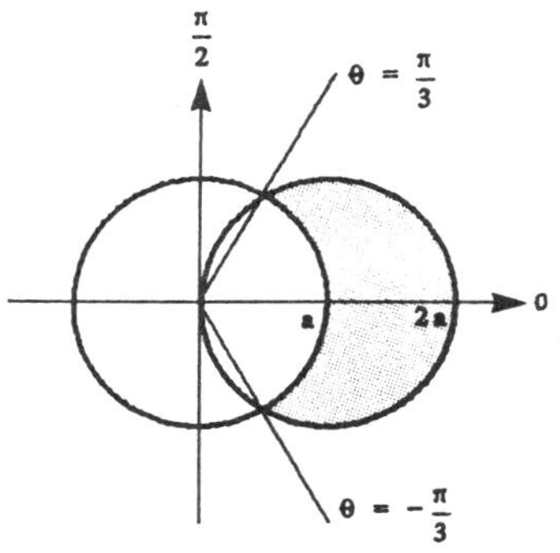

$= \frac{2\pi a^2}{3} - 2a^2\int_{\pi/3}^{\pi/2}(1 + \cos 2\theta)\,d\theta$

$= \frac{2\pi a^2}{3} - 2a^2\left[\theta + \frac{\sin 2\theta}{2}\right]_{\pi/3}^{\pi/2}$

$= \frac{2\pi a^2}{3} - 2a^2\left[\frac{\pi}{2} - \frac{\pi}{3} - \frac{\sqrt{3}}{4}\right]$

$= \frac{2\pi a^2 + 3\sqrt{3}}{6}$

33. $A = \frac{\pi a^2}{8} + \frac{1}{2}\int_{\pi/2}^{\pi} [a(1 + \cos\theta)]^2\, d\theta$

$= \frac{\pi a^2}{8} + \frac{a^2}{2}\int_{\pi/2}^{\pi} (\frac{3}{2} + 2\cos\theta + \frac{\cos 2\theta}{2})\, d\theta$

$= \frac{\pi a^2}{8} + \frac{a^2}{2}\left[\frac{3}{2}\theta + 2\sin\theta + \frac{\sin 2\theta}{4}\right]_{\pi/2}^{\pi}$

$= \frac{\pi a^2}{8} + \frac{a^2}{2}\left[\frac{3\pi}{2} - \frac{3\pi}{4} - 2\right]$

$= \frac{a^2}{2}[\pi - 2]$

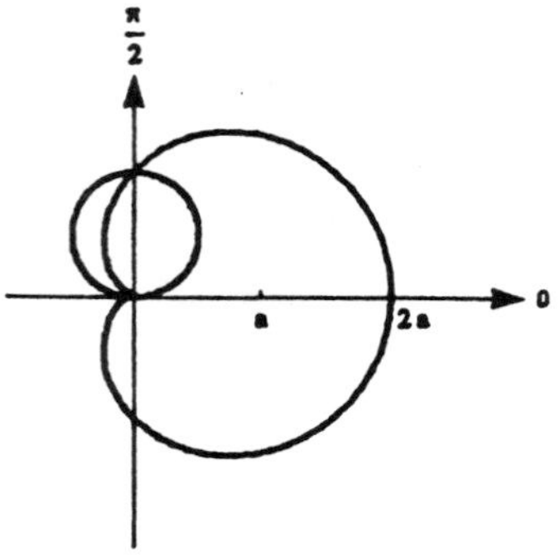

34. $r = \frac{ab}{a\sin\theta + b\cos\theta}$ ⇒ $ay + bx = ab$

$\frac{y}{b} + \frac{x}{a} = 1$

Assume $a > 0$ and $b > 0$

$A = \frac{1}{2}ab$

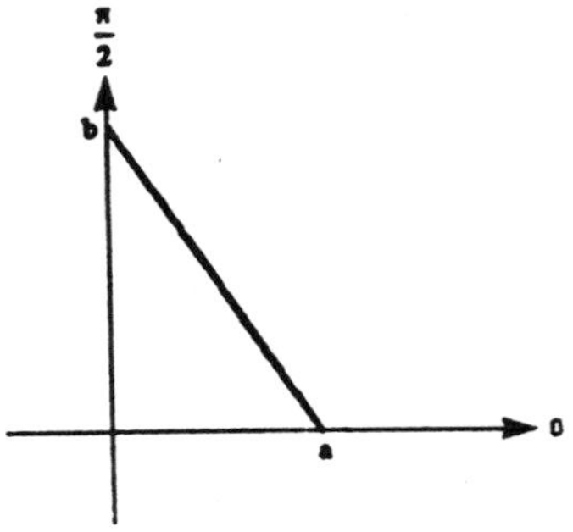

35. $r = a,\quad r' = 0,\quad s = \int_0^{2\pi} \sqrt{a^2 + 0^2}\, d\theta = a\theta\Big]_0^{2\pi} = 2\pi a$

36. $r = a\cos\theta,\quad r' = -a\sin\theta$

$s = \int_0^{2\pi} \sqrt{(a\cos\theta)^2 + (-a\sin\theta)^2}\, d\theta = a\theta\Big]_0^{2\pi} = 2\pi a$

37. $r = 1 + \sin\theta,\quad r' = \cos\theta$

$s = 2\int_{\pi/2}^{3\pi/2} \sqrt{(1 + \sin\theta)^2 + (\cos\theta)^2}\, d\theta = 2\sqrt{2}\int_{\pi/2}^{3\pi/2} \sqrt{1 + \sin\theta}\, d\theta$

$= 2\sqrt{2}\int_{\pi/2}^{3\pi/2} \frac{-\cos\theta}{\sqrt{1 - \sin\theta}}\, d\theta = 4\sqrt{2}\sqrt{1 - \sin\theta}\Big]_{\pi/2}^{3\pi/2}$

$= 4\sqrt{2}(0 - \sqrt{2}) = 8$

38. $r = 5(1 + \cos\theta),\quad r' = -5\sin\theta$

$s = 2\int_0^{\pi} \sqrt{[5(1 + \cos\theta)]^2 + (-5\sin\theta)^2}\, d\theta = 10\sqrt{2}\int_0^{\pi} \sqrt{1 + \cos\theta}\, d\theta$

$= 10\sqrt{2}\int_0^{\pi} \frac{\sin\theta}{\sqrt{1 - \cos\theta}}\, d\theta = 20\sqrt{2}\sqrt{1 - \cos\theta}\Big]_0^{\pi} = 40$

39. $r = 2\theta, \quad r' = 2,$

$$s = \int_0^{\pi/2} \sqrt{4\theta^2 + 4}\, d\theta = 2 \int_0^{\pi/2} \sqrt{\theta^2 + 1}\, d\theta$$

$$= \left[\theta\sqrt{\theta^2 + 1} + \ln|\theta + \sqrt{\theta^2 + 1}|\right]_0^{\pi/2} = \frac{\pi}{4}\sqrt{\pi^2 + 4} + \ln\left(\frac{\pi + \sqrt{\pi^2 + 4}}{2}\right)$$

40. $r = \sec\theta, \quad r' = \sec\theta\tan\theta$

$$s = \int_0^{\pi/3} \sqrt{\sec^2\theta + \sec^2\theta\tan^2\theta}\, d\theta = \int_0^{\pi/3} \sec^2\theta = \tan\theta\Big]_0^{\pi/3} = \sqrt{3}$$

41. $r = \dfrac{1}{\theta}, \quad r' = -\dfrac{1}{\theta^2}$

$$s = \int_\pi^{2\pi} \sqrt{\frac{1}{\theta^2} + \frac{1}{\theta^4}}\, d\theta = \int_\pi^{2\pi} \frac{\sqrt{\theta^2 + 1}}{\theta^2}\, d\theta$$

$$= \left[-\frac{\sqrt{\theta^2 + 1}}{\theta} + \ln|\theta + \sqrt{\theta^2 + 1}|\right]_\pi^{2\pi}$$

$$= \left(-\frac{\sqrt{4\pi^2 + 1}}{2\pi} + \ln|2\pi + \sqrt{4\pi^2 + 1}|\right) - \left(-\frac{\sqrt{\pi^2 + 1}}{\pi} + \ln|\pi + \sqrt{\pi^2 + 1}|\right)$$

$$= \ln\left[\frac{2\pi + \sqrt{4\pi^2 + 1}}{\pi + \sqrt{\pi^2 + 1}}\right] + \frac{2\sqrt{\pi^2 + 1} - \sqrt{4\pi^2 + 1}}{2\pi}$$

42. $r = e^\theta, \quad r' = e^\theta$

$$s = \int_0^\pi \sqrt{(e^\theta)^2 + (e^\theta)^2}\, d\theta = \sqrt{2}\int_0^\pi e^\theta\, d\theta = \sqrt{2}e^\theta\Big]_0^\pi = \sqrt{2}(e^\pi - 1)$$

43. $r = 2\cos\theta, \quad r' = -2\sin\theta$

$$S = 2\pi\int_0^{\pi/2} 2\cos\theta\sin\theta\sqrt{4\cos^2\theta + 4\sin^2\theta}\, d\theta$$

$$= 8\pi\int_0^{\pi/2} \sin\theta\cos\theta\, d\theta = 4\pi\sin^2\theta\Big]_0^{\pi/2} = 4\pi$$

44. $r = a\cos\theta, \quad r' = -a\sin\theta$

$$S = 2\pi\int_0^{\pi/2} a\cos\theta(\cos\theta)\sqrt{a^2\cos^2\theta + a^2\sin^2\theta}\, d\theta$$

$$= 2\pi a^2\int_0^{\pi/2} \cos^2\theta\, d\theta = \pi a^2\int_0^{\pi/2} (1 + \cos 2\theta)\, d\theta$$

$$= \pi a^2\left(\theta + \frac{\sin 2\theta}{2}\right)\Big]_0^{\pi/2} = \frac{\pi^2 a^2}{2}$$

45. $r = e^{a\theta}, \qquad r' = ae^{a\theta}$

$$S = 2\pi \int_0^{\pi/2} e^{a\theta} \cos\theta \sqrt{(e^{a\theta})^2 + (ae^{a\theta})^2}\, d\theta$$

$$= 2\pi\sqrt{1 + a^2} \int_0^{\pi/2} e^{2a\theta} \cos\theta\, d\theta$$

$$= 2\pi\sqrt{1 + a^2}\ \frac{e^{2a\theta}}{4a^2 + 1}(2a\cos\theta + \sin\theta)\Big]_0^{\pi/2}$$

$$= \frac{2\pi\sqrt{1 + a^2}}{4a^2 + 1}(e^{\pi a} - 2a)$$

46. $r = a(1 + \cos\theta), \qquad r' = -a\sin\theta$

$$S = 2\pi \int_0^{\pi} a(1 + \cos\theta)\sin\theta \sqrt{a^2(1 + \cos\theta)^2 + a^2\sin^2\theta}\, d\theta$$

$$= 2\pi a^2 \int_0^{\pi} \sin\theta(1 + \cos\theta)\sqrt{2 + 2\cos\theta}\, d\theta$$

$$= -2\sqrt{2}\,\pi a^2 \int_0^{\pi} (1 + \cos\theta)^{3/2}(-\sin\theta)\, d\theta$$

$$= -\frac{4\sqrt{2}\,\pi a^2}{5}\Big[(1 + \cos\theta)^{5/2}\Big]_0^{\pi}$$

$$= \frac{32\pi a^2}{5}$$

47. $r = 4\cos 2\theta, \qquad r' = -8\sin 2\theta$

$$S = 2\pi \int_0^{\pi/4} 4\cos 2\theta \sin\theta\sqrt{16\cos^2 2\theta + 64\sin^2 2\theta}\, d\theta$$

$$= 32\pi \int_0^{\pi/4} \cos 2\theta \sin\theta\sqrt{\cos^2 2\theta + 4\sin^2 2\theta}\, d\theta$$

$$\approx \frac{2\pi^2}{3}[0 + 4(0.2162) + 2(0.4278) + 4(0.4012) + 0] \quad \text{(Simpson's Rule, } n = 4)$$

$$\approx 21.88$$

48. $r = \theta, \quad r' = 1$

$$S = 2\pi \int_0^{\pi} \theta\sin\theta\sqrt{\theta^2 + 1}\, d\theta$$

$$\approx \frac{\pi^2}{6}[0 + 4(0.7062) + 2(2.9250) + 4(4.2645) + 0] \quad \text{(Simpson's Rule, } n = 4)$$

$$\approx 42.33$$

50. Ellipse

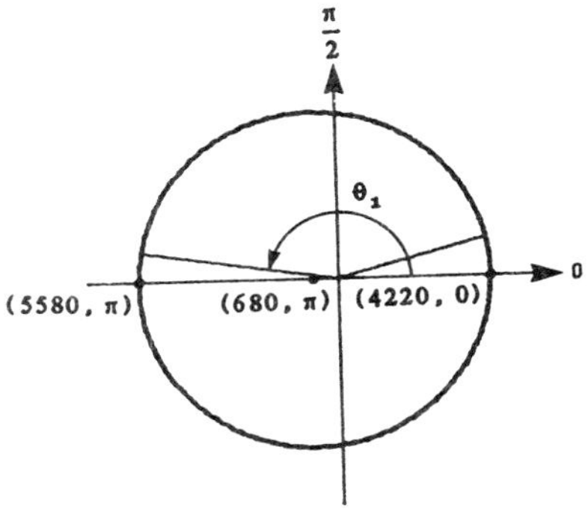

Focus at pole

Vertices: (4220, 0), (5580, π)

Center: (680, π)

a = 4900, c = 680

b ≈ 4852.59, e ≈ 0.1388

(a)

$$A = \frac{23{,}547{,}600}{2}\left[\frac{-680\sin\theta}{4900 + 680\cos\theta} + \frac{9800}{4852.59}\arctan\left(\frac{4852.59\tan(\theta/2)}{5580}\right)\right]_0^{\pi/12}$$

$\approx 2{,}337{,}606\ \text{mi}^2$

(b) The area of one half the ellipse is $\frac{1}{2}\pi ab \approx 37{,}349{,}886\ \text{mi}^2$.

Find θ_1 so that the area swept out as θ ranges from θ_1 to π is equivalent to finding θ_1 so that the area swept out as θ ranges from 0 to θ_1 is $37{,}349{,}886 - 2{,}337{,}606 = 35{,}012{,}280$, $\theta \approx 2.99$

(c) We approximate the distance traveled by

$s = r\theta = 4220(\frac{\pi}{12}) \approx 1105$ mi.

$s = r\theta = 5580(\pi - 2.99) \approx 846$ mi.

Review Exercises for Chapter 11

1. $x = 1 + 4t$, $y = 2 - 3t$, $t = \frac{x-1}{4}$

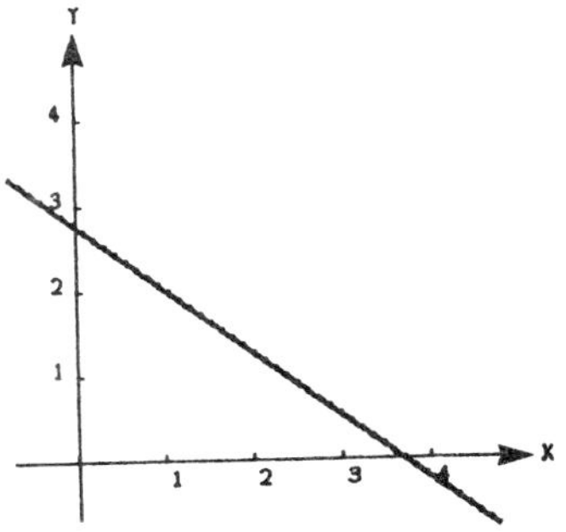

$y = 2 - \frac{3}{4}(x - 1) = -\frac{3}{4}x + \frac{11}{4}$

$\frac{dy}{dx} = -\frac{3}{4}$

No horizontal tangents

2. $x = e^t, \quad y = e^{-t}, \quad t = \ln x$

$y = e^{-\ln x} = e^{\ln(1/x)} = \frac{1}{x}$

$x > 0, \quad \frac{dy}{dx} = -\frac{1}{x^2}$

No horizontal tangents

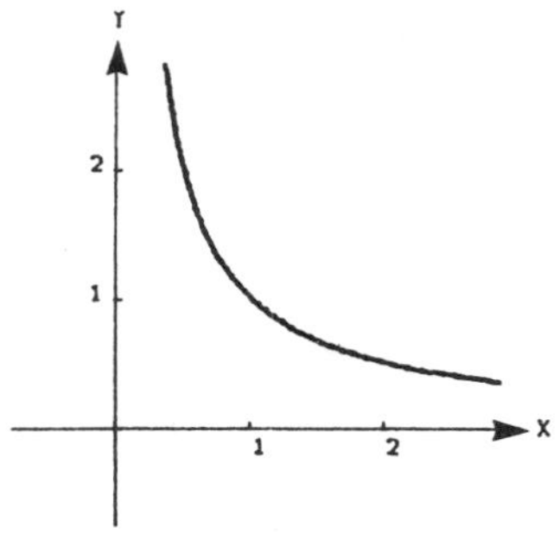

3. $x = 3 + 2\cos\theta, \quad y = 2 + 5\sin\theta$

$$\frac{(x-3)^2}{4} + \frac{(y-2)^2}{25} = 1$$

$$\frac{dy}{dx} = \frac{5\cos\theta}{-2\sin\theta} = -2.5\cot\theta = 0$$

when $\theta = \frac{\pi}{2}, \frac{3\pi}{2}$

Points of horizontal tangency:

$(3, 7), (3, -3)$

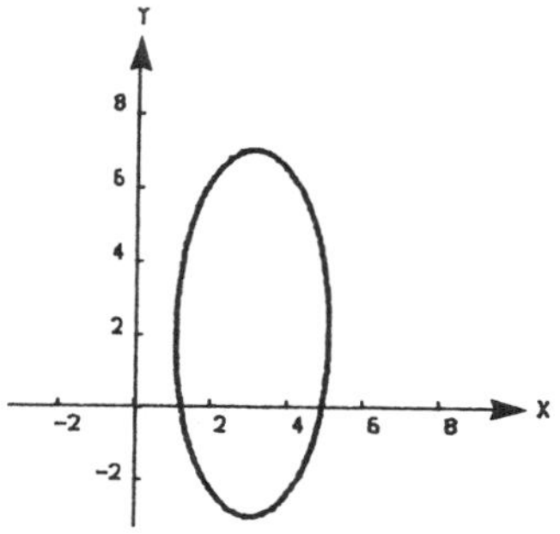

4. $x = t^2 - 3t + 2, \quad y = t^3 - 3t^2 + 2$

$$\frac{dy}{dx} = \frac{3t^2 - 6t}{2t - 3} = \frac{3t(t-2)}{2t-3} = 0$$

when $t = 0, \quad t = 2$

Points of horizontal tangency:

$(2, 2), (0, -2)$

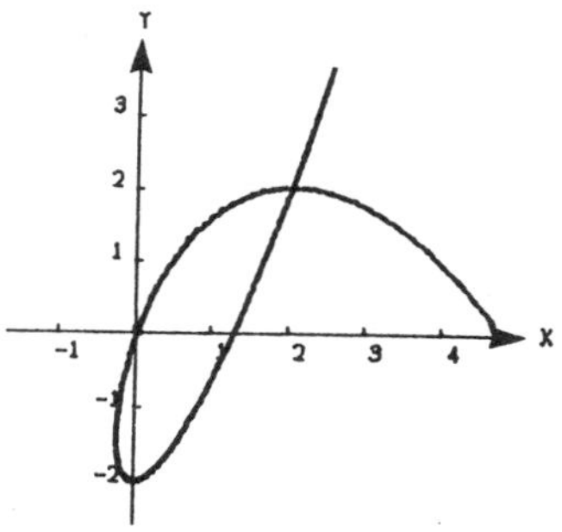

5. $x = \frac{1}{t}, \quad y = 2t + 3, \quad t = \frac{1}{x}$

$y = \frac{2}{x} + 3, \quad \frac{dy}{dx} = -\frac{2}{x^2} = -2t^2$

No horizontal tangents

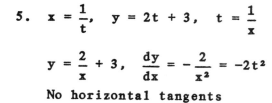

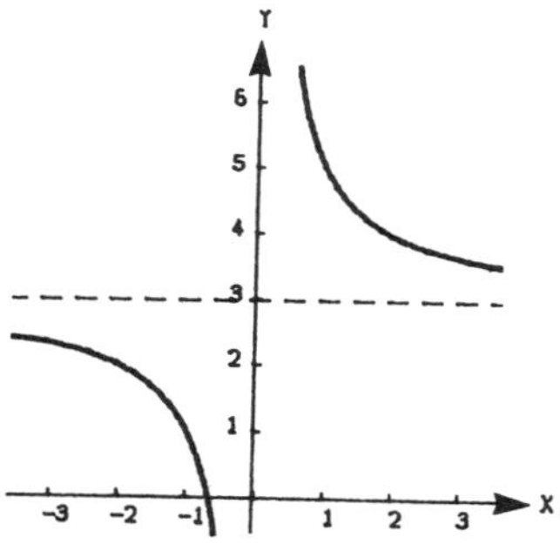

6. $x = 2t - 1, \quad y = \dfrac{1}{t^2 - 2t}, \quad t = \dfrac{x + 1}{2}$

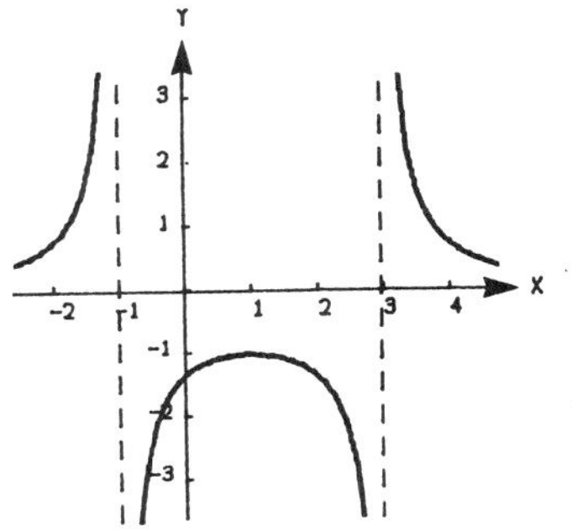

$$y = \frac{1}{[(x + 1)/2]^2 - 2[(x + 1)/2]}$$

$$y = \frac{4}{(x - 3)(x + 1)}$$

$$\frac{dy}{dx} = \frac{-(t^2 - 2t)^{-2}\,(2t - 2)}{2}$$

$$= \frac{1 - t}{t^2(t - 2)^2} = 0 \text{ when } t = 1$$

Point of horizontal tangency: $(1, -1)$

7. $x = \dfrac{1}{2t + 1}, \quad y = \dfrac{2t(t + 1)}{2t + 1} = \dfrac{1 - x^2}{2x}$

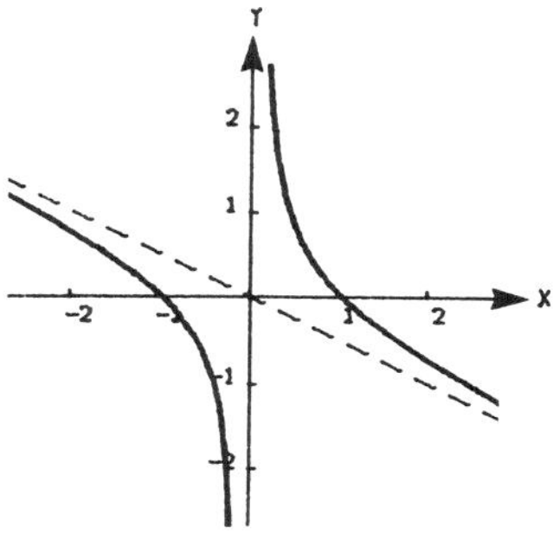

$$\frac{dy}{dx} = \frac{[(2t+1)(4t+2) - (2t^2+2t)2]/(2t+1)^2}{-2/(2t + 1)^2}$$

$$\frac{dy}{dx} = -(2t^2 + 2t + 1) < 0 \text{ all } x$$

No horizontal tangents

8. $x = \cot\theta, \quad y = \sin 2\theta$

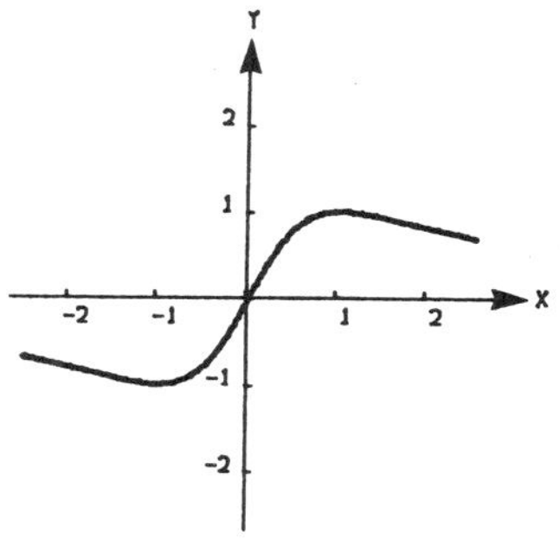

$$y = \sin 2\theta = 2\sin\theta\cos\theta$$

$$y = 2\sin(\operatorname{arccot} x)\cos(\operatorname{arccot} x)$$

$$= 2\left(\frac{1}{\sqrt{x^2 + 1}}\right)\left(\frac{x}{\sqrt{x^2 + 1}}\right) = \frac{2x}{x^2 + 1}$$

$$\frac{dy}{dx} = \frac{2\cos 2\theta}{-\csc^2\theta} = 0 \text{ when } 2\theta = \frac{\pm\pi}{2} \text{ or } \theta = \frac{\pm\pi}{4}$$

Points of horizontal tangency: $(1, 1), (-1, -1)$

9. $x = \cos^3\theta, \quad y = 4\sin^3\theta$

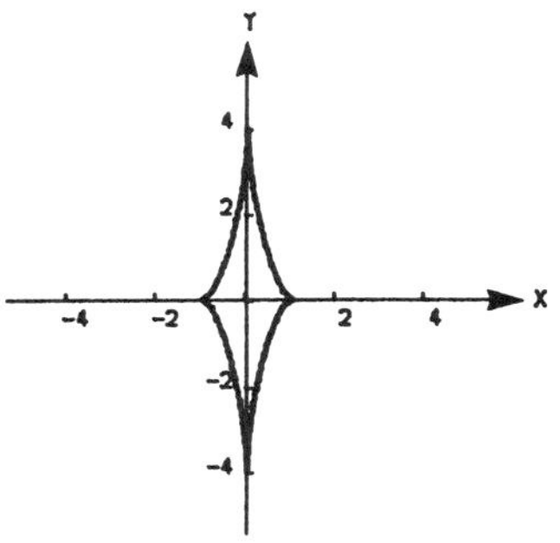

$$\frac{dy}{dx} = \frac{12\sin^2\theta\cos\theta}{3\cos^2\theta(-\sin\theta)}$$

$$= \frac{-4\sin\theta}{\cos\theta} = -4\tan\theta = 0 \text{ when } \theta = 0, \pi$$

Points of horizontal tangency: $(\pm 1, 0)$

$$x^{2/3} + \left(\frac{y}{4}\right)^{2/3} = 1$$

10. $x = 2\theta - \sin\theta, \quad y = 2 - \cos\theta$

$\dfrac{dy}{dx} = \dfrac{\sin\theta}{2 - \cos\theta} = 0$ when $\theta = n\pi$

Points of horizontal tangency:

$(4n\pi, 1), (2(2n + 1)\pi, 3)$

n is an integer

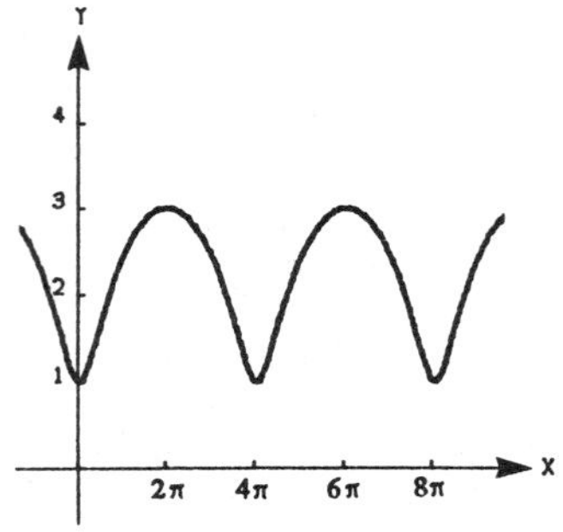

11. $\dfrac{(x + 3)^2}{16} + \dfrac{(y - 4)^2}{9} = 1$, let $\dfrac{(x + 3)^2}{16} = \cos^2\theta$ and $\dfrac{(y - 4)^2}{9} = \sin^2\theta$

$x = -3 + 4\cos\theta, \quad y = 4 + 3\sin\theta$

12. $\dfrac{y^2}{16} - \dfrac{x^2}{9} = 1$, let $\dfrac{y^2}{16} = \sec^2\theta$ and $\dfrac{x^2}{9} = \tan^2\theta$

then $x = 3\tan\theta$ and $y = 4\sec\theta$

13. $x = a(\theta - \sin\theta), \quad y = 1(1 - \cos\theta)$

$\cos\theta = \dfrac{a - y}{a}, \quad \theta = \arccos\left(\dfrac{a - y}{a}\right)$

$x = a\arccos\left(\dfrac{a - y}{a}\right) - a\sin\left[\arccos\left(\dfrac{a - y}{a}\right)\right]$

$x = a\arccos\left(\dfrac{a - y}{a}\right) \pm \sqrt{2ay - y^2}$

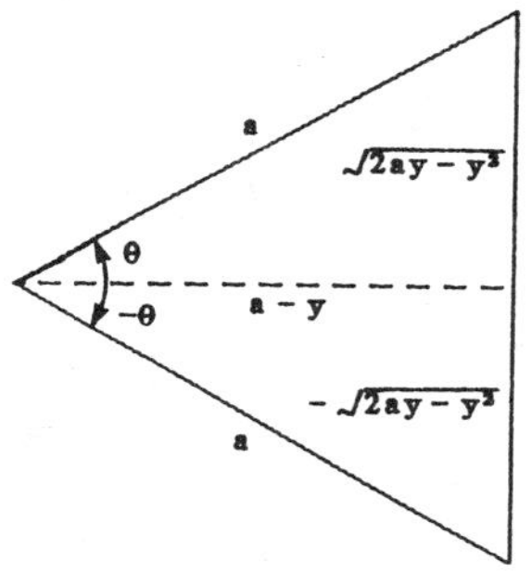

14. $x = t + u = r\cos\theta + r\theta\sin\theta$

$= r(\cos\theta + \theta\sin\theta)$

$y = v - w = r\sin\theta - r\theta\cos\theta$

$= r(\sin\theta - \theta\cos\theta)$

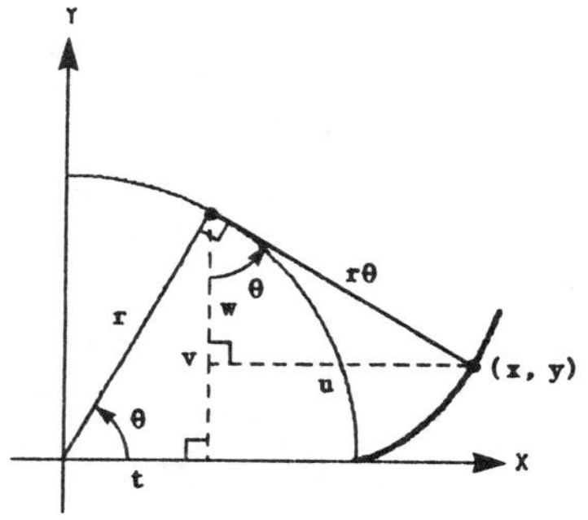

15. $x = r(\cos\theta + \theta\sin\theta), \quad y = r(\sin\theta - \theta\cos\theta)$

$$\frac{dx}{d\theta} = r\theta\cos\theta, \quad \frac{dy}{d\theta} = r\theta\sin\theta,$$

$$s = r\int_0^{\pi} \sqrt{\theta^2\cos^2\theta + \theta^2\sin^2\theta}\, d\theta = r\int_0^{\pi} \theta\, d\theta = \frac{r}{2}\Big[\theta^2\Big]_0^{\pi} = \frac{1}{2}\pi^2 r$$

16. $x = 6\cos\theta, \quad y = 6\sin\theta, \quad \frac{dx}{d\theta} = -6\sin\theta, \quad \frac{dy}{d\theta} = 6\cos\theta$

$$s = \int_0^{\pi} \sqrt{36\sin^2\theta + 36\cos^2\theta}\, d\theta = 6\theta\Big]_0^{\pi} = 6\pi$$

17. $r = -2(1 + \cos\theta)$

Cardioid

Symmetric to polar axis

θ	0	$\pi/3$	$\pi/2$	$2\pi/3$	π
r	−4	−3	−2	−1	0

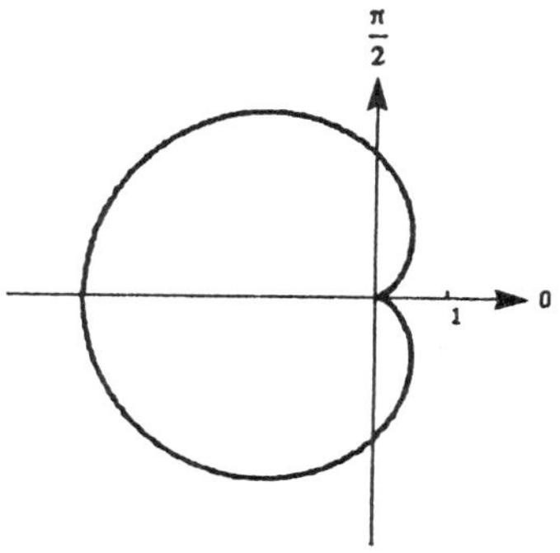

18. $r = 3 - 4\cos\theta$

Limacon

Symmetric to polar axis

θ	0	$\pi/3$	$\pi/2$	$2\pi/3$	π
r	−1	1	3	5	7

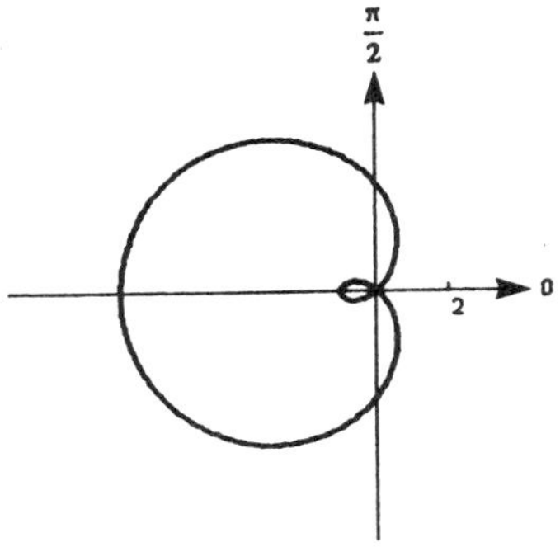

19. $r = 4 - 3\cos\theta$

Limacon

Symmetric to polar axis

θ	0	$\pi/3$	$\pi/2$	$2\pi/3$	π
r	1	5/2	4	11/2	7

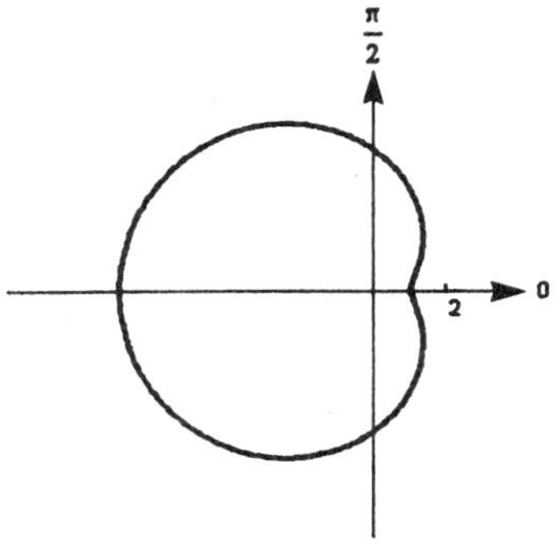

20. $r = \cos(5\theta)$

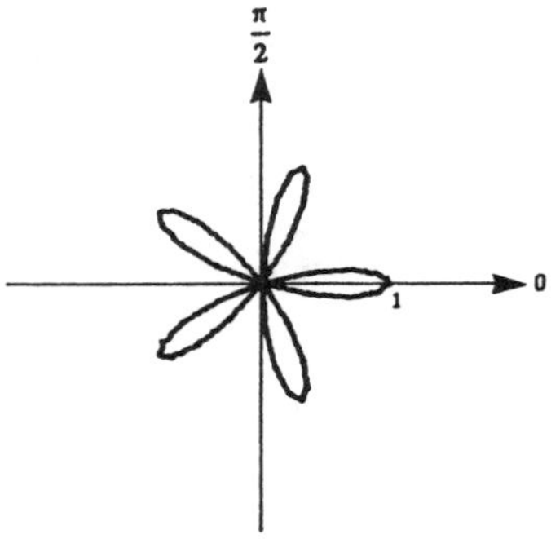

Rose curve with 5 petals

Symmetric to polar axis

Relative extrema: $(1, 0)$, $(-1, \frac{\pi}{5})$, $(1, \frac{2\pi}{5})$, $(-1, \frac{3\pi}{5})$, $(1, \frac{4\pi}{5})$

Tangents at the pole: $\theta = \frac{\pi}{10}, \frac{3\pi}{10}, \frac{\pi}{2}, \frac{7\pi}{10}, \frac{9\pi}{10}$

21. $r = -3\cos(2\theta)$

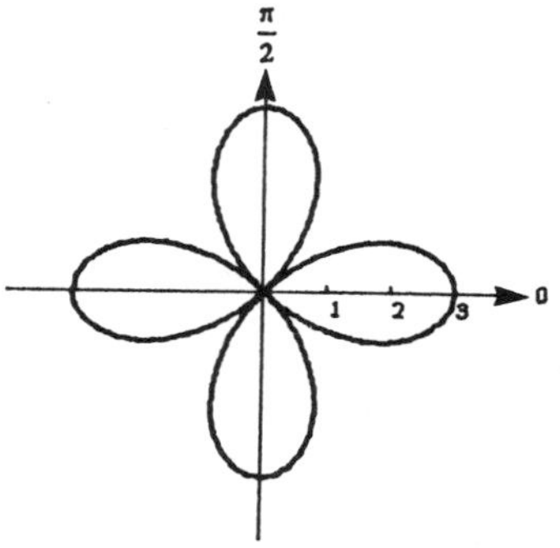

Rose curve with 4 petals

Symmetric to polar axis, vertical axis, and pole

Relative extrema: $(-3, 0)$, $(3, \frac{\pi}{2})$, $(-3, \pi)$, $(3, \frac{3\pi}{2})$

Tangents at the pole: $\theta = \frac{\pi}{4}, \frac{3\pi}{4}$

22. $r^2 = \cos(2\theta)$

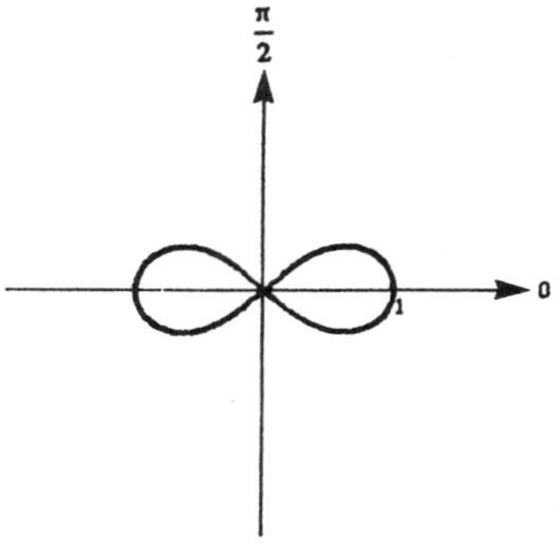

Lemniscate

Symmetric to the polar axis

Relative extrema: $(\pm 1, 0)$

θ	0	$\pi/6$	$\pi/4$
r	± 1	$\pm\sqrt{2}/2$	0

Tangents at the pole: $\theta = \frac{\pi}{4}, \frac{-\pi}{4}$

23. $r = 4$

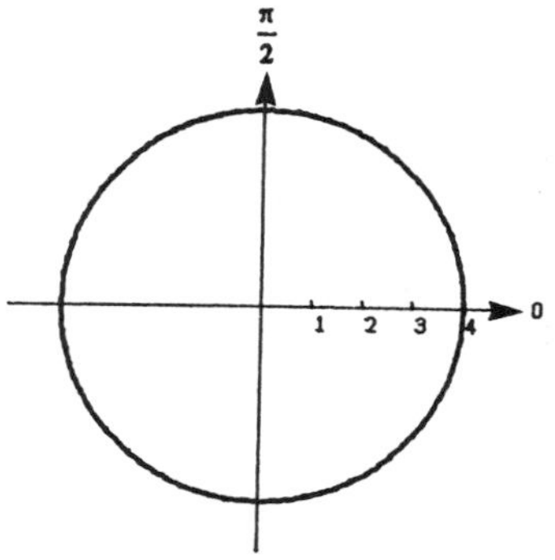

Circle of radius 4

Centered at the pole

24. $r = 2\theta$

Spiral

Symmetric to the vertical axis

θ	0	$\pi/4$	$\pi/2$	$3\pi/4$	π	$5\pi/4$	$3\pi/2$
r	0	$\pi/2$	π	$3\pi/2$	2π	$5\pi/2$	3π

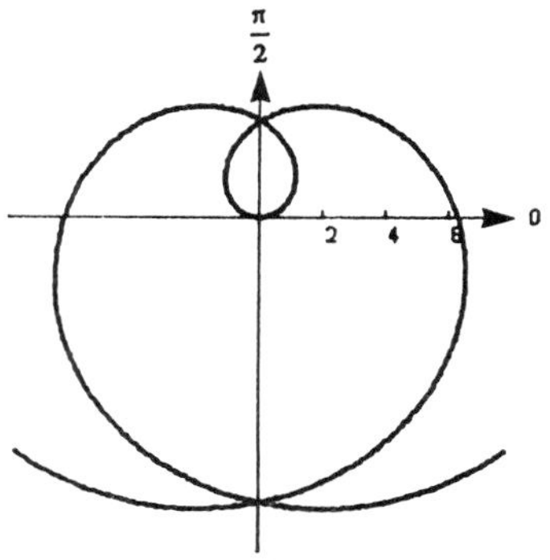

25. $r = -\sec\theta = \dfrac{-1}{\cos\theta}$

$r\cos\theta = -1$

$x = -1$

Vertical line

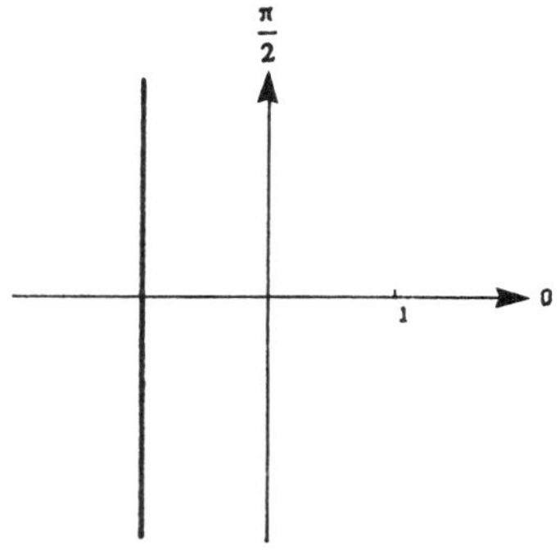

26. $r = 3\csc\theta$

$r\sin\theta = 3$

$y = 3$

Horizontal line

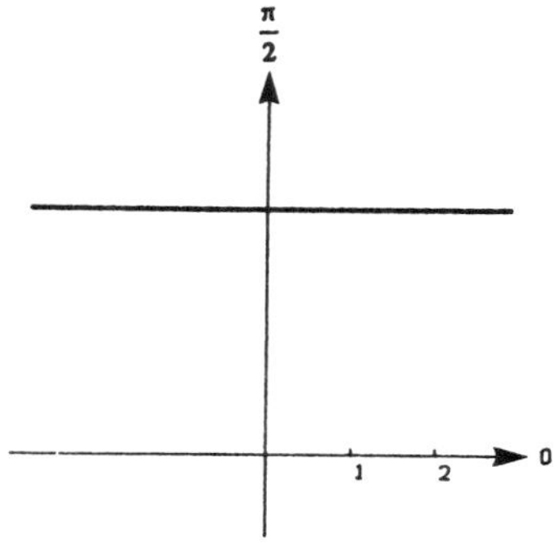

27. $r^2 = 4\sin^2(2\theta)$

$r = \pm 2\sin(2\theta)$

Rose curve with 4 petals

Symmetric to the polar axis,
vertical axis, and pole

Relative extrema: $\left(\pm 2, \dfrac{\pi}{4}\right)$, $\left(\pm 2, \dfrac{5\pi}{4}\right)$

Tangents at the pole: $\theta = 0, \dfrac{\pi}{2}$

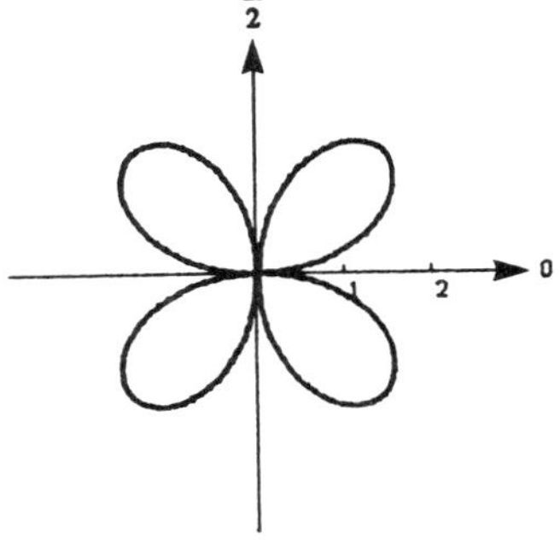

28. $r = 2 \sin\theta \cos^2\theta$

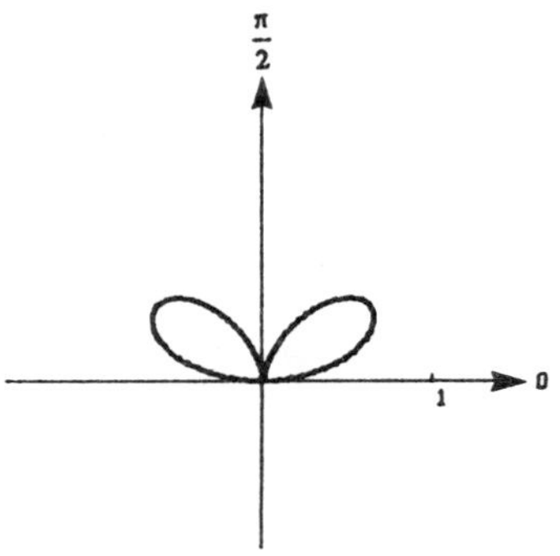

Bifolium

Symmetric to the vertical axis

$$\frac{dr}{d\theta} = 2\cos\theta(1 - 3\sin^2\theta) = 0$$

when $\theta = \frac{\pi}{2}$, $\sin\theta = \frac{1}{\sqrt{3}}$ or $\theta \approx 0.6155$

Relative extrema: (0.77, 0.62), (0.77, 2.53)

θ	0	$\pi/6$	$\pi/3$	$\pi/2$
r	0	0.75	0.43	0

Tangents at the pole: $\theta = 0, \frac{\pi}{2}$

29. $r = \dfrac{2}{1 - \sin\theta}$

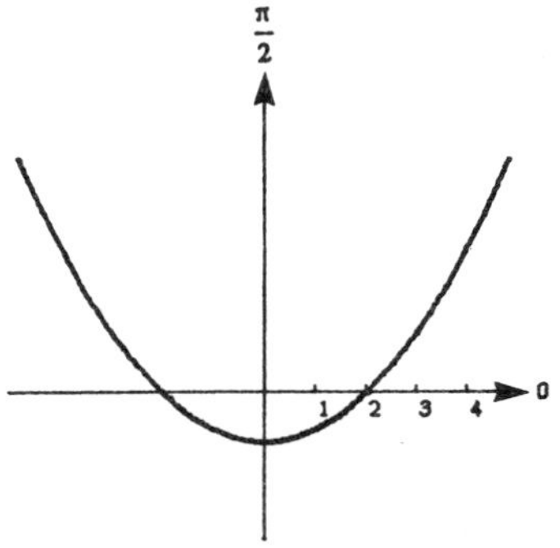

Parabola

Focus at the pole

Vertex: $(1, \frac{3\pi}{2})$

30. $r = \dfrac{4}{5 - 3\cos\theta}$

$= \dfrac{4/5}{1 - (3/5)\cos\theta}$

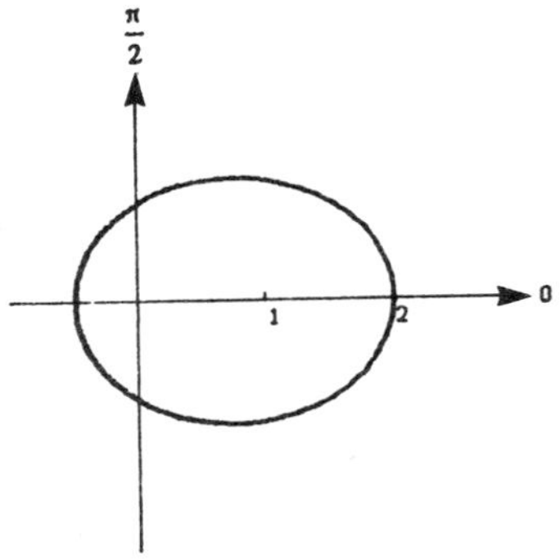

Ellipse

Focus at the pole

$e = \frac{3}{5}$

Vertices: $(2, 0)$, $(\frac{1}{2}, \pi)$

31. $r = 4\cos 2\theta \sec\theta$

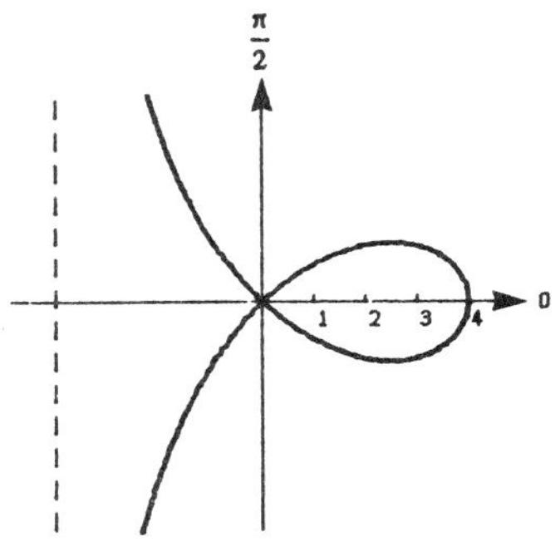

Strophoid

Symmetric to the polar axis

$r \longrightarrow -\infty$ as $\theta \longrightarrow \frac{\pi^-}{2}$

$r \longrightarrow -\infty$ as $\theta \longrightarrow \frac{-\pi^+}{2}$

(See Exercise 37)

32. $r = 4(\sec\theta - \cos\theta)$

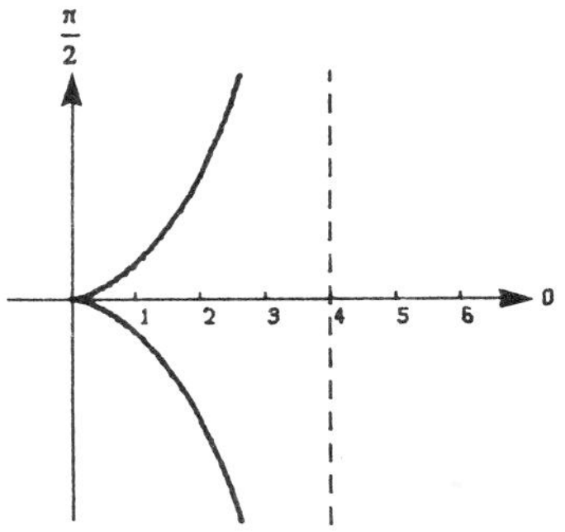

Semicubical parabola

Symmetric to the polar axis

$r \longrightarrow \infty$ as $\theta \longrightarrow \frac{\pi^-}{2}$

$r \longrightarrow -\infty$ as $\theta \longrightarrow \frac{3\pi^-}{2}$

θ	0	$\pi/6$	$\pi/3$	$\pi/2$	$2\pi/3$	$5\pi/6$	$3\pi/2$
r	0	1.155	6	$+\infty$	−6	−1.155	$-\infty$

33. $r = 3\cos\theta,\quad r^2 = 3r\cos\theta,\quad x^2 + y^2 = 3x,\quad x^2 + y^2 - 3x = 0$

34. $r = 4\sec\left(\theta - \frac{\pi}{3}\right) = \dfrac{4}{\cos[\theta - (\pi/3)]} = \dfrac{4}{(1/2)\cos\theta + (\sqrt{3}/2)\sin\theta}$

$r(\cos\theta + \sqrt{3}\sin\theta) = 8,\quad x + \sqrt{3}y = 8$

35. $r = -2(1 + \cos\theta),\quad x^2 + y^2 = -2\sqrt{x^2 + y^2} - 2x$

$(x^2 + y^2 + 2x)^2 = 4(x^2 + y^2)$

36. $r = 1 + \tan\theta,\quad \sqrt{x^2 + y^2} = 1 + \frac{y}{x},\quad y = x(\sqrt{x^2 + y^2} - 1)$

$(y + x)^2 = x^2(x^2 + y^2)$

37. $r = 4\cos 2\theta \sec\theta = 4(2\cos^2\theta - 1)\left(\frac{1}{\cos\theta}\right)$, $r\cos\theta = 8\cos^2\theta - 4$,

$x = 8\left(\frac{x^2}{x^2 + y^2}\right) - 4$, $x^3 + xy^2 = 4x^2 - 4y^2$, $y^2 = x^2\left(\frac{4 - x}{4 + x}\right)$

38. $\theta = \frac{3\pi}{4}$, $\tan\theta = -1$, $\frac{y}{x} = -1$, $y = -x$

39. $(x^2 + y^2)^2 = ax^2y$, $r^4 = a(r^2\cos^2\theta)(r\sin\theta)$, $r = a\cos^2\theta\sin\theta$

40. $x^2 + y^2 - 4x = 0$, $r^2 - 4r\cos\theta = 0$, $r = 4\cos\theta$

41. $x^2 + y^2 = a^2(\arctan\frac{y}{x})^2$, $r^2 = a^2\theta^2$

42. $(x^2 + y^2)(\arctan\frac{y}{x})^2 = a^2$, $r^2\theta^2 = a^2$

43. $r = \frac{A}{1 - \cos\theta}$, $r = 2$ when $\theta = \pi$, $2 = \frac{A}{2}$ or $A = 4$, $r = \frac{4}{1 - \cos\theta}$

44. Ellipse

Focus at pole

Center: (2, 0)

$a = 3$, $c = 2$, $e = \frac{2}{3}$

$$r = \frac{A}{1 - (2/3)\cos\theta}$$

where $r = 5$ when $\theta = 0$

$$5 = \frac{A}{1 - (2/3)} = \frac{A}{1/3} \text{ or } A = \frac{5}{3}$$

$$r = \frac{5/3}{1 - (2/3)\cos\theta} = \frac{5}{3 - 2\cos\theta}$$

45. Hyperbola

Focus at the pole

Center: (4, 0)

$a = 3$, $c = 4$, $e = \frac{4}{3}$

$$r = \frac{A}{1 + (4/3)\cos\theta}$$

where $r = 1$ when $\theta = 0$

$$1 = \frac{A}{1 + (4/3)} = \frac{A}{7/3} \text{ or } A = \frac{7}{3}$$

$$r = \frac{7/3}{1 + (4/3)\cos\theta} = \frac{7}{3 + 4\cos\theta}$$

46. Circle

Center: (0, 5) and passing through the origin

$x^2 + (y - 5)^2 = 25$

$x^2 + y^2 - 10y = 0$

$r^2 - 10r\sin\theta = 0$

$r = 10\sin\theta$

47. In rectangular form we have $\frac{x}{3} + \frac{y}{4} = 1$ or $4x + 3y = 12$

Therefore, in polar coordinates, $4r\cos\theta + 3r\sin\theta = 12$ or

$$r = \frac{12}{4\cos\theta + 3\sin\theta}$$

48. $y = \sqrt{3}x$, $r\sin\theta = \sqrt{3}r\cos\theta$, $\tan\theta = \sqrt{3}$, $\theta = \frac{\pi}{3}$

49. The graph has polar symmetry and the tangents at the pole are $\theta = \frac{\pi}{3}, -\frac{\pi}{3}$.

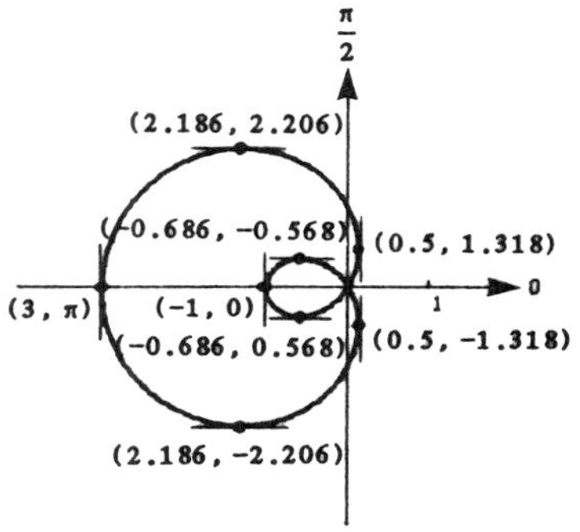

$$\frac{dy}{dx} = \frac{2\sin^2\theta + (1 - 2\cos\theta)\cos\theta}{2\sin\theta\cos\theta - (1 - 2\cos\theta)\sin\theta}$$

Horizontal tangents:

$$-4\cos^2\theta + \cos\theta + 2 = 0$$

$$\cos\theta = \frac{-1 \pm \sqrt{1 + 32}}{-8} = \frac{1 \pm \sqrt{33}}{8}$$

When $\cos\theta = \frac{1 \pm \sqrt{33}}{8}$, $r = 1 - 2\left(\frac{1 \pm \sqrt{33}}{4}\right) = \frac{3 \mp \sqrt{33}}{4}$

$$\left(\frac{3 - \sqrt{33}}{4}, \arccos\left(\frac{1 + \sqrt{33}}{8}\right)\right) \approx (-0.686, 0.568)$$

$$\left(\frac{3 - \sqrt{33}}{4}, -\arccos\left(\frac{1 + \sqrt{33}}{8}\right)\right) \approx (-0.686, -0.568)$$

$$\left(\frac{3 + \sqrt{33}}{4}, \arccos\left(\frac{1 - \sqrt{33}}{8}\right)\right) \approx (2.186, 2.206)$$

$$\left(\frac{3 + \sqrt{33}}{4}, -\arccos\left(\frac{1 - \sqrt{33}}{8}\right)\right) \approx (2.186, -2.206)$$

Vertical tangents: $\sin\theta(4\cos\theta - 1) = 0$, $\sin\theta = 0$ or $\cos\theta = \frac{1}{4}$,

$\theta = 0, \pi$ or $\theta = \pm\arccos\left(\frac{1}{4}\right)$, $(-1, 0)$, $(3, \pi)$,

$\left(\frac{1}{2}, \pm\arccos\frac{1}{4}\right) \approx (0.5, \pm 1.318)$

50. $r^2 = 4\sin(2\theta)$, $2r\left(\dfrac{dr}{d\theta}\right) = 8\cos(2\theta)$

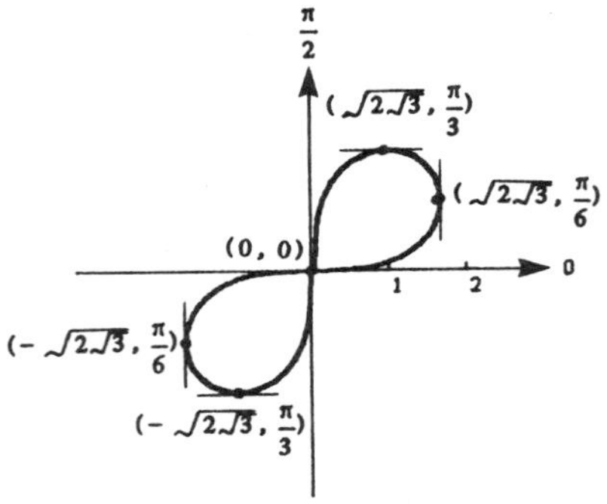

or $\dfrac{dr}{d\theta} = \dfrac{4\cos(2\theta)}{r}$

Tangents at the pole: $\theta = 0, \dfrac{\pi}{2}$

$$\frac{dy}{dx} = \frac{r\cos\theta + (4\cos 2\theta\sin\theta)/r}{-r\sin\theta + (4\cos 2\theta\cos\theta)/r}$$

$$= \frac{\cos(2\theta)\sin\theta + \sin(2\theta)\cos\theta}{\cos(2\theta)\cos\theta - \sin(2\theta)\sin\theta}$$

Horizontal tangents: $\dfrac{dy}{dx} = 0$ when $\cos(2\theta)\sin\theta + \sin(2\theta)\cos\theta = 0$,

$\tan\theta = -\tan(2\theta)$, $\theta = 0, \dfrac{\pi}{3}$, $(0, 0)$, $\left(\pm\sqrt{2\sqrt{3}}, \dfrac{\pi}{3}\right)$

Vertical tangents when $\cos 2\theta\cos\theta - \sin 2\theta\sin\theta = 0$,

$\tan 2\theta\tan\theta = 1$, $\theta = 0, \dfrac{\pi}{6}$, $(0, 0)$, $\left(\pm\sqrt{2\sqrt{3}}, \dfrac{\pi}{6}\right)$

51. The points $\left(1, \dfrac{\pi}{2}\right)$ and $\left(1, \dfrac{3\pi}{2}\right)$ are the two points of intersection (other than the pole). The slope of the graph of $r = 1 + \cos\theta$ is

$$m_1 = \frac{dy}{dx} = \frac{r'\sin\theta + r\cos\theta}{r'\cos\theta - r\sin\theta} = \frac{-\sin^2\theta + \cos\theta(1 + \cos\theta)}{-\sin\theta\cos\theta - \sin\theta(1 + \cos\theta)}$$

At $\left(1, \dfrac{\pi}{2}\right)$, $m_1 = \dfrac{-1}{-1} = 1$ and at $\left(1, \dfrac{3\pi}{2}\right)$, $m_1 = \dfrac{-1}{1} = -1$

The slope of the graph of $r = 1 - \cos\theta$ is

$$m_2 = \frac{dy}{dx} = \frac{\sin^2\theta + \cos\theta(1 - \cos\theta)}{\sin\theta\cos\theta - \sin\theta(1 - \cos\theta)}$$

At $\left(1, \dfrac{\pi}{2}\right)$, $m_2 = \dfrac{1}{-1} = -1$ and at $\left(1, \dfrac{3\pi}{2}\right)$, $m_2 = \dfrac{1}{1} = 1$

In both cases $m_1 = \dfrac{-1}{m_2}$ and we conclude that the graphs are orthogonal at $\left(1, \dfrac{\pi}{2}\right)$ and $\left(1, \dfrac{3\pi}{2}\right)$.

52. The points of intersection are $(\frac{a}{\sqrt{2}}, \frac{\pi}{4})$ and $(-\frac{a}{\sqrt{2}}, \frac{5\pi}{4})$.

For $r = a\sin\theta$, $m_1 = \frac{dy}{dx} = \frac{a\cos\theta\sin\theta + a\sin\theta\cos\theta}{a\cos^2\theta - a\sin^2\theta} = \frac{2\sin\theta\cos\theta}{\cos 2\theta}$

At $(\frac{a}{\sqrt{2}}, \frac{\pi}{4})$, m_1 is undefined and at $(-\frac{a}{\sqrt{2}}, \frac{5\pi}{4})$, m_1 is undefined.

The tangent lines are vertical.

For $r = a\cos\theta$, $m_2 = \frac{dy}{dx} = \frac{-a\sin^2\theta + a\cos^2\theta}{-a\sin\theta\cos\theta - a\cos\theta\sin\theta} = \frac{\cos 2\theta}{-2\sin\theta\cos\theta}$

At $(\frac{a}{\sqrt{2}}, \frac{\pi}{4})$, $m_2 = 0$ and at $(-\frac{a}{\sqrt{2}}, \frac{5\pi}{4})$, $m_2 = 0$.

The tangent lines are horizontal.

Therefore, the graphs are orthogonal at $(\frac{a}{\sqrt{2}}, \frac{\pi}{4})$ and $(-\frac{a}{\sqrt{2}}, \frac{5\pi}{4})$.

53. $A = 2\left[\frac{1}{2}\int_0^{\pi}(2 + \cos\theta)^2\,d\theta\right] = \left[\frac{9}{2}\theta + 4\sin\theta + \frac{1}{4}\sin 2\theta\right]_0^{\pi} = \frac{9\pi}{2}$

54. $A = 2\left[\frac{1}{2}\int_{\pi/2}^{3\pi/2}[5(1 - \sin\theta]^2\,d\theta\right] = 25\left[\frac{3}{2}\theta + 2\cos\theta - \frac{\sin 2\theta}{2}\right]_{\pi/2}^{3\pi/2} = \frac{75\pi}{2}$

55. $A = 2\left[\frac{1}{2}\int_0^{\pi/2}(\sin\theta\cos^2\theta)^2\,d\theta\right]$

$= \left[\frac{1}{8}(\frac{1}{2}\theta - \frac{1}{8}\sin 4\theta) + \frac{1}{16}(\frac{\sin^3 2\theta}{3})\right]_0^{\pi/2} = \frac{\pi}{32}$

56. $A = 3\left[\frac{1}{2}\int_0^{\pi/3}(4\sin 3\theta)^2\,d\theta\right] = 12\left[\theta - \frac{1}{6}\sin 6\theta\right]_0^{\pi/3} = 4\pi$

57. $A = 2\left[\frac{1}{2}\int_0^{\pi/2}a^2\sin 2\theta\,d\theta\right] = \frac{-a^2}{2}\cos 2\theta\Big]_0^{\pi/2} = \frac{a^2}{2} + \frac{a^2}{2} = a^2$

58. $A = 2\left[\frac{1}{2}\int_0^{\pi/12}2a^2\sin 2\theta\,d\theta + \frac{1}{2}\int_{\pi/12}^{5\pi/12}a^2\,d\theta + \frac{1}{2}\int_{5\pi/12}^{\pi/2}2a^2\sin 2\theta\,d\theta\right]$

$= 2\int_0^{\pi/12}2a^2\sin 2\theta\,d\theta + a^2\int_{\pi/12}^{5\pi/12}d\theta = \left[-2a^2\cos 2\theta\right]_0^{\pi/12} + \left[a^2\theta\right]_0^{5\pi/12}$

$= \frac{a^2}{3}(6 - 3\sqrt{3} + \pi)$

59. $A = 2\left[\frac{1}{2}\int_0^{\pi/3} 4\, d\theta + \frac{1}{2}\int_{\pi/3}^{\pi/2} (4\cos\theta)^2\, d\theta\right] = \left[4\theta\right]_0^{\pi/3} + 8\left[\theta + \frac{1}{2}\sin 2\theta\right]_{\pi/3}^{\pi/2}$

$= \frac{8\pi - 6\sqrt{3}}{3}$

60. $A = \frac{1}{2}\int_0^{\pi} e^{2\theta}\, d\theta = \frac{1}{4}e^{2\theta}\Big]_0^{\pi} = \frac{1}{4}(e^{2\pi} - 1)$

61. $s = 2\int_0^{\pi} \sqrt{a^2(1 - \cos\theta)^2 + a^2\sin^2\theta}\, d\theta = 2\sqrt{2}a\int_0^{\pi} \sqrt{1 - \cos\theta}\, d\theta$

$= 2\sqrt{2}a\int_0^{\pi} \frac{\sin\theta}{\sqrt{1 + \cos\theta}}\, d\theta = -4\sqrt{2}a(1 + \cos\theta)^{1/2}\Big]_0^{\pi} = 8a$

62. $r = a\cos 2\theta,\quad \frac{dr}{d\theta} = -2a\sin 2\theta$

$s = 8\int_0^{\pi/4} \sqrt{a^2\cos^2 2\theta + 4a^2\sin^2 2\theta}\, d\theta$

$= 8a\int_0^{\pi/4} \sqrt{1 + 3\sin^2 2\theta}\, d\theta$ (Simpson's Rule: $n = 4$)

$\approx \frac{\pi a}{6}[1 + 4(1.1997) + 2(1.5811) + 4(1.8870) + 2] \approx 9.69a$

63. Circle: $r = 3\sin\theta$

$\frac{dy}{dx} = \frac{3\cos\theta\sin\theta + 3\sin\theta\cos\theta}{3\cos\theta\cos\theta - 3\sin\theta\sin\theta}$

$= \frac{\sin 2\theta}{\cos^2\theta - \sin^2\theta} = \tan 2\theta$ at $\theta = \frac{\pi}{6}$, $\frac{dy}{dx} = \sqrt{3}$

Limacon: $r = 4 - 5\sin\theta$

$\frac{dy}{dx} = \frac{-5\cos\theta\sin\theta + (4 - 5\sin\theta)\cos\theta}{-5\cos\theta\cos\theta - (4 - 5\sin\theta)\sin\theta}$ at $\theta = \frac{\pi}{6}$, $\frac{dy}{dx} = \frac{\sqrt{3}}{9}$

Let α be the angle between the curves, $\tan\alpha = \frac{\sqrt{3} - (\sqrt{3}/9)}{1 + (1/3)} = \frac{2\sqrt{3}}{3}$,

therefore, $\alpha = \arctan\left(\frac{2\sqrt{3}}{3}\right)$.

12 Vectors and the geometry of space

12.1 Vectors in the plane

1.(a) $\mathbf{v} = \langle 5 - 1, 3 - 1\rangle = \langle 4, 2\rangle$
(b)

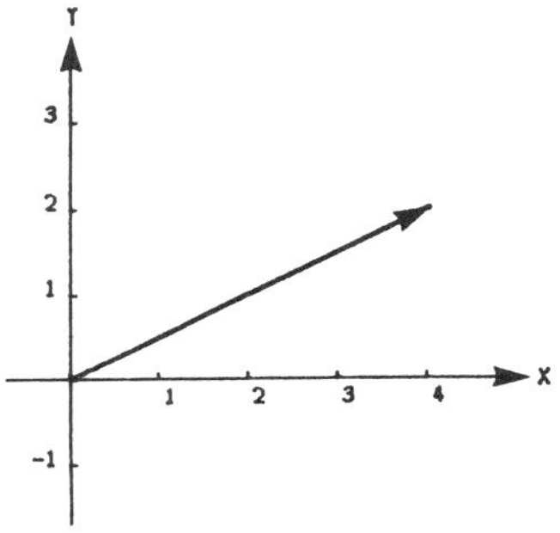

2.(a) $\mathbf{v} = \langle 3 - 3, -2 - 4\rangle = \langle 0, -6\rangle$
(b)

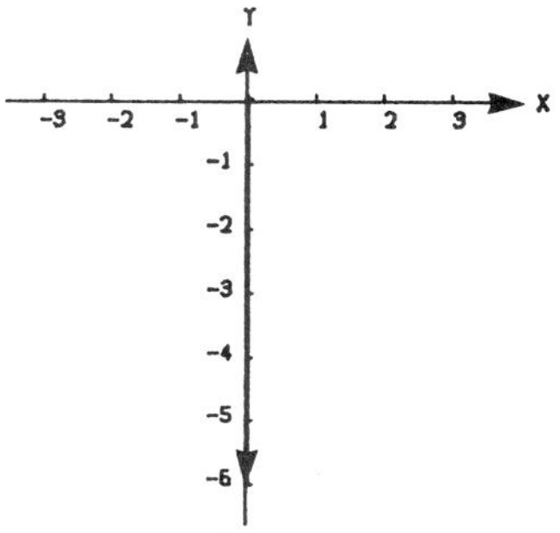

3.(a) $\mathbf{v} = \langle -4 - 3, -2 - (-2)\rangle$
$= \langle -7, 0\rangle$
(b)

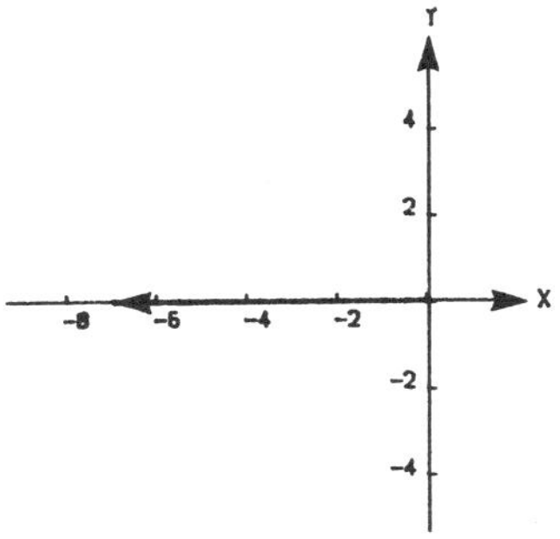

4.(a) $\mathbf{v} = \langle -1 - 2, 3 - 1\rangle$
$= \langle -3, 2\rangle$
(b)

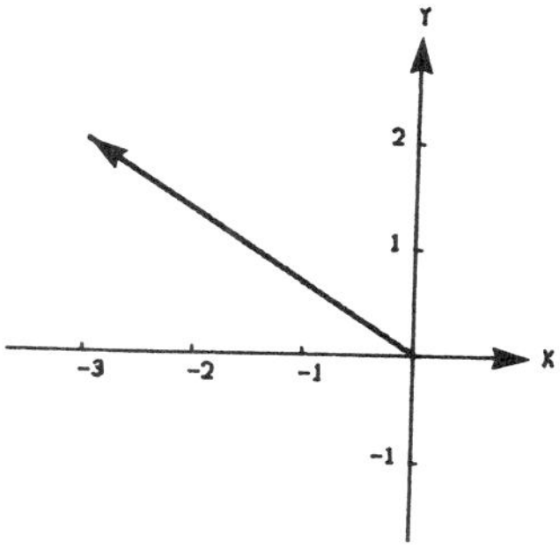

5.(b) $\mathbf{v} = \langle 5 - 1, 5 - 2\rangle = \langle 4, 3\rangle$
(a) and (c)

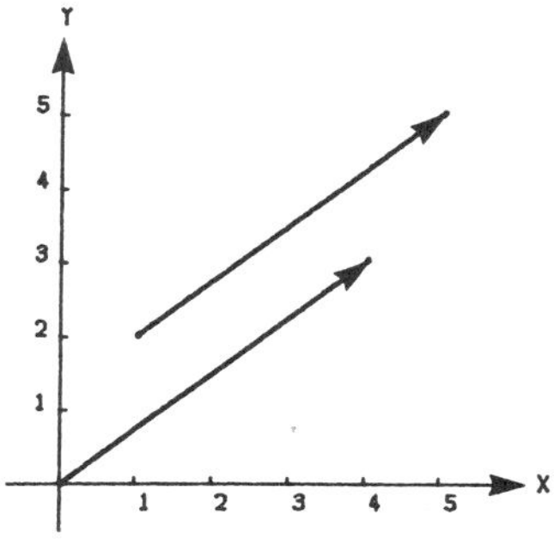

6.(b) $\mathbf{v} = \langle 4 - 3, 7 - (-5)\rangle = \langle 1, 12\rangle$
(a) and (c)

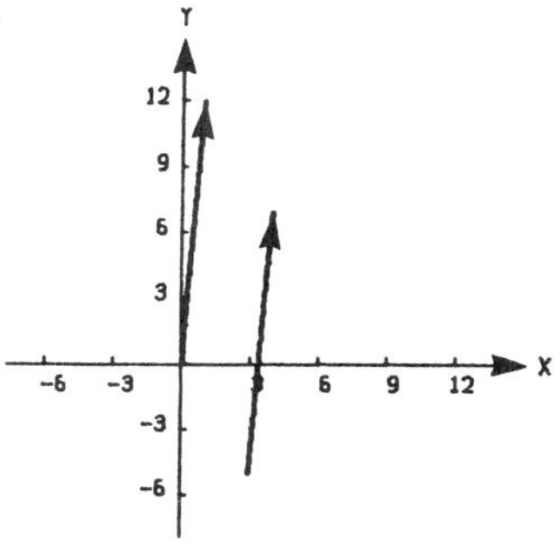

7.(b) $\mathbf{v} = \langle 6 - 10, -1 - 2\rangle$
$= \langle -4, -3\rangle$

(a) and (c)

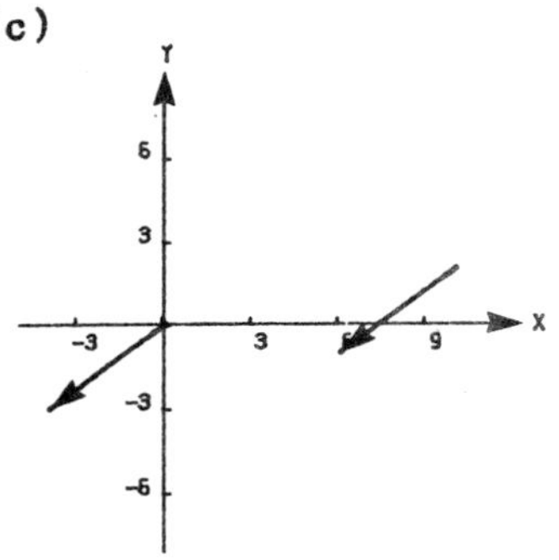

8.(b) $\mathbf{v} = \langle -5 - 0, -1 - (-4)\rangle$
$= \langle -5, 3\rangle$

(a) and (c)

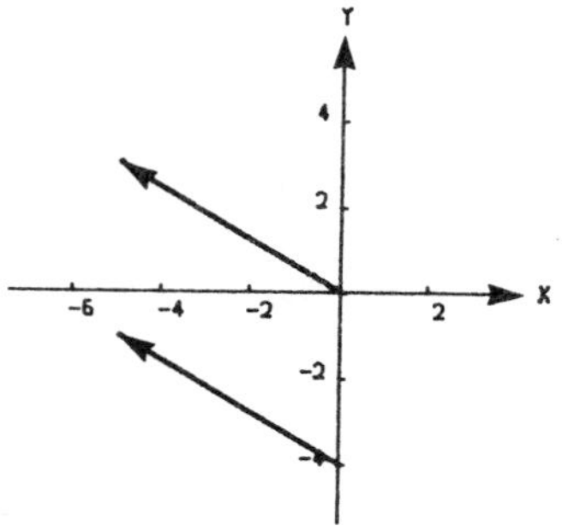

9.(b) $\mathbf{v} = \langle 6 - 6, 6 - 2\rangle = \langle 0, 4\rangle$

(a) and (c)

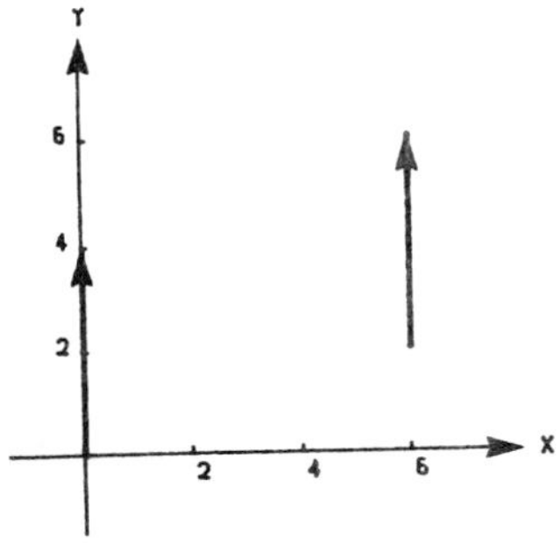

10.(b) $\mathbf{v} = \langle -3 - 7, -1 - (-1)\rangle$
$= \langle -10, 0\rangle$

(a) and (c)

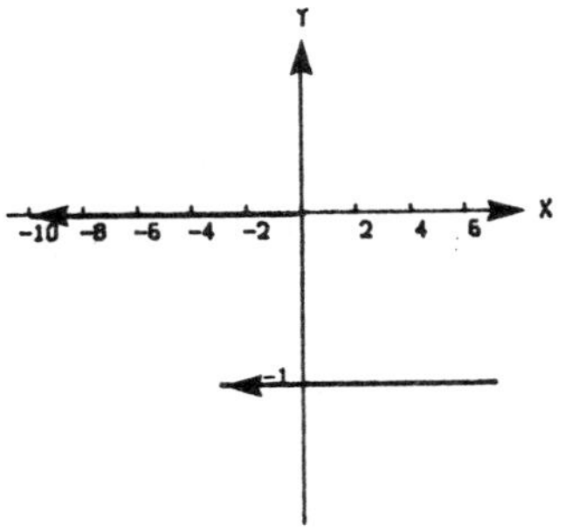

11.(b) $\mathbf{v} = \langle \frac{1}{2} - \frac{3}{2}, 3 - \frac{4}{3}\rangle$

$= \langle -1, \frac{5}{3}\rangle$

(a) and (c)

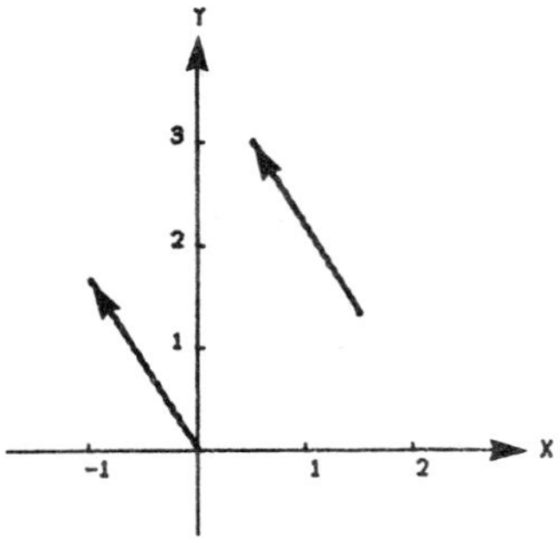

12.(b) $\mathbf{v} = \langle 0.84 - 0.12, 1.25 - 0.60\rangle$
$= \langle 0.72, 0.65\rangle$

(a) and (c)

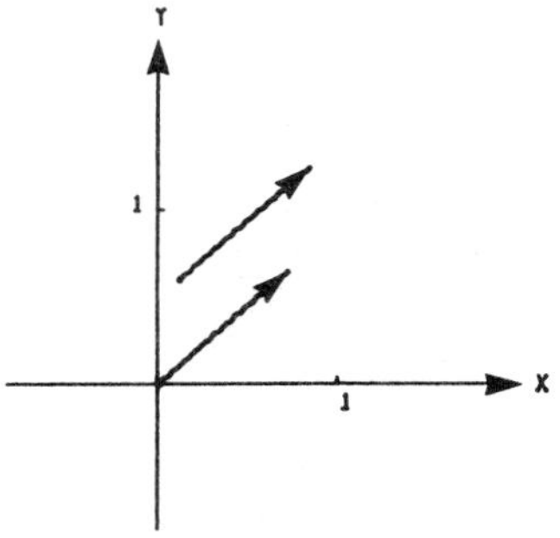

13.(a) $2\mathbf{v} = \langle 4, 6\rangle$

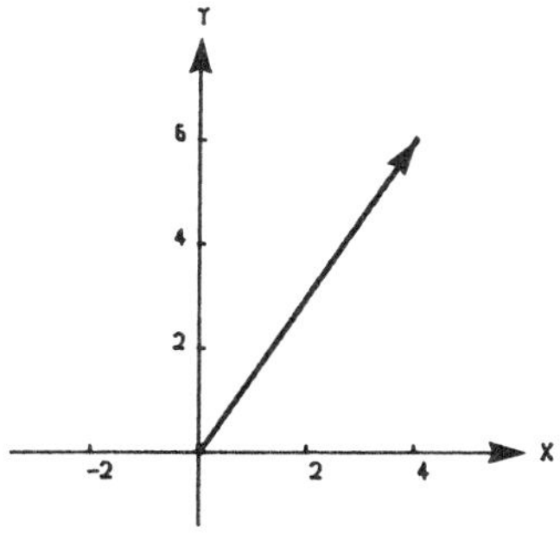

(b) $-3\mathbf{v} = \langle -6, -9\rangle$

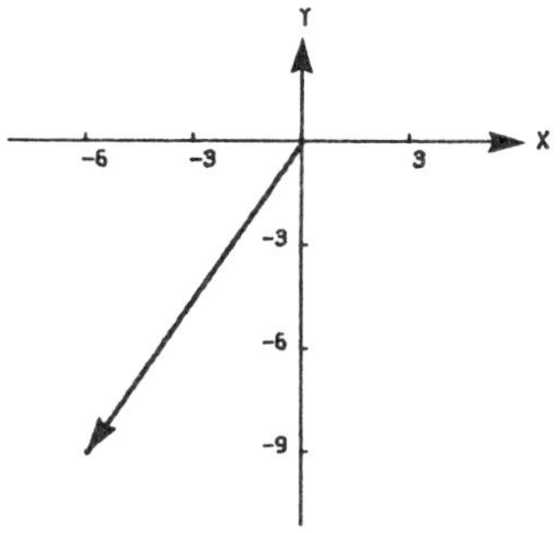

(c) $\frac{7}{2}\mathbf{v} = \langle 7, \frac{21}{2}\rangle$

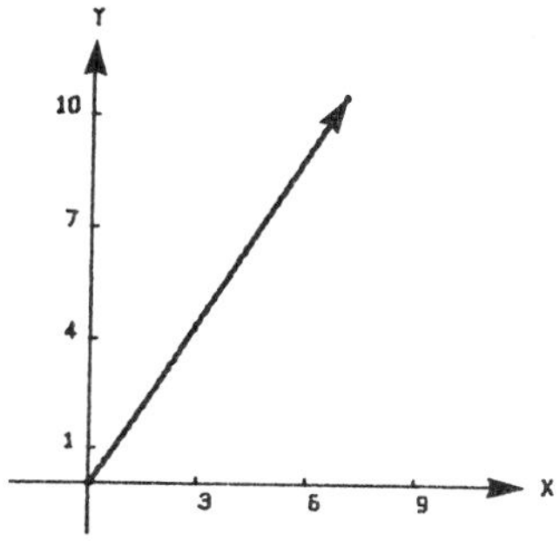

(d) $\frac{2}{3}\mathbf{v} = \langle \frac{4}{3}, 2\rangle$

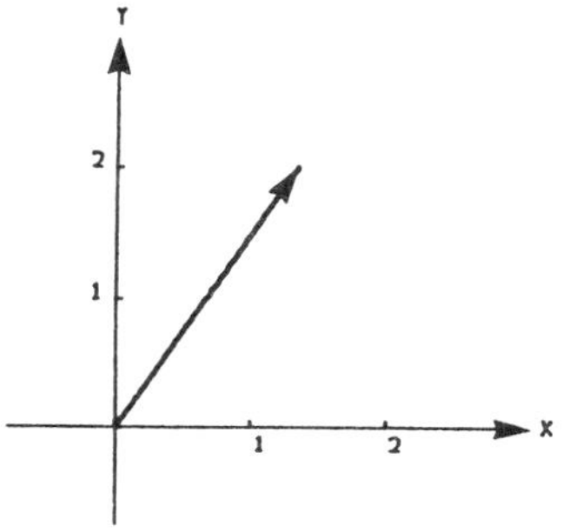

14.(a) $4\mathbf{v} = \langle -4, 20\rangle$

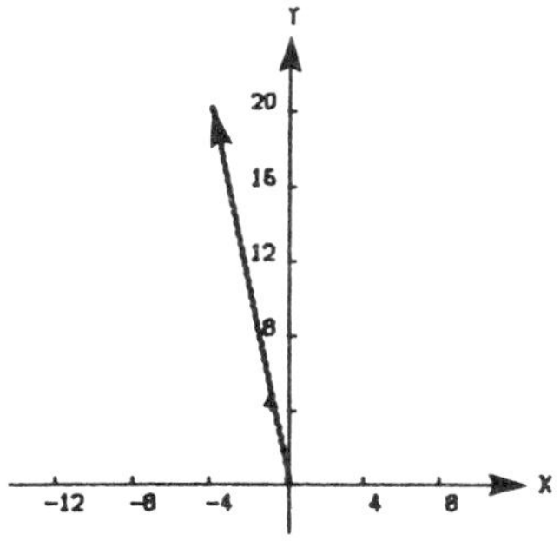

(b) $-\frac{1}{2}\mathbf{v} = \langle \frac{1}{2}, -\frac{5}{2}\rangle$

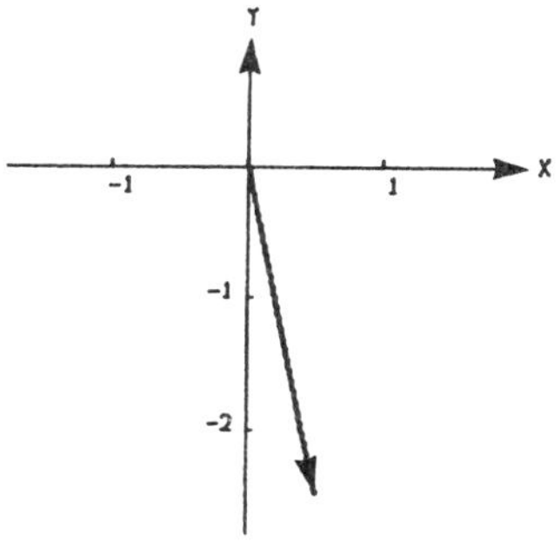

(c) $0\mathbf{v} = \langle 0, 0\rangle$

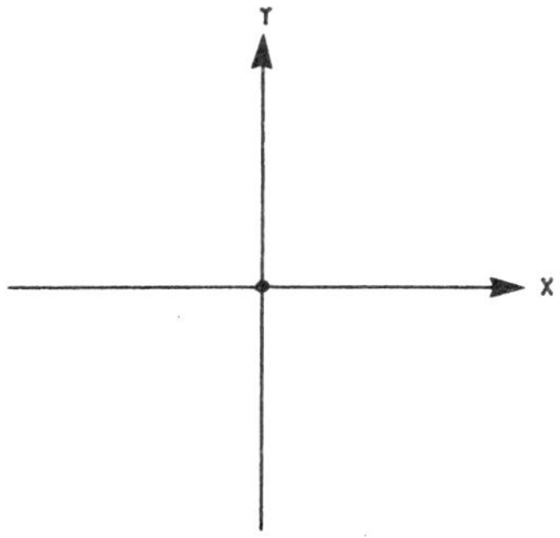

(d) $-6\mathbf{v} = \langle 6, -30\rangle$

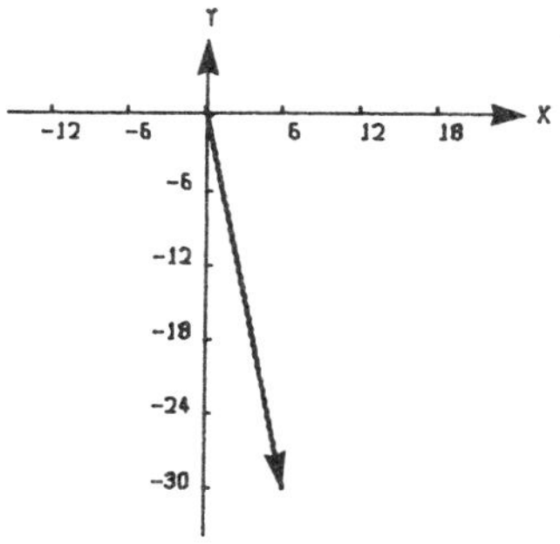

15. $\mathbf{v} = \frac{3}{2}(2\mathbf{i} - \mathbf{j})$

$= 3\mathbf{i} - \frac{3}{2}\mathbf{j} = \langle 3, \frac{-3}{2} \rangle$

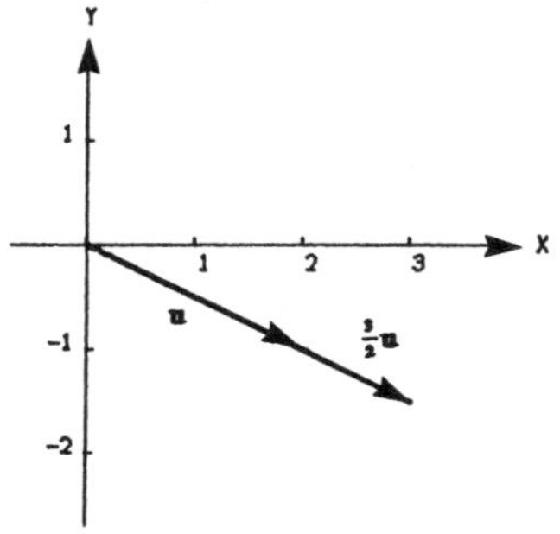

16. $\mathbf{v} = (2\mathbf{i} - \mathbf{j}) + (\mathbf{i} + 2\mathbf{j})$
$= 3\mathbf{i} + \mathbf{j} = \langle 3, 1 \rangle$

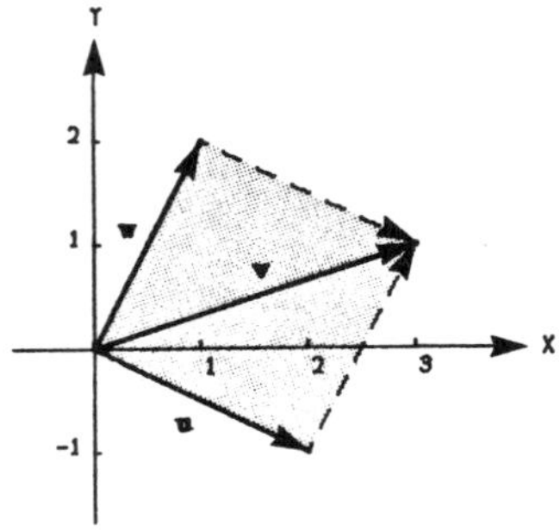

17. $\mathbf{v} = (2\mathbf{i} - \mathbf{j}) + 2(\mathbf{i} + 2\mathbf{j})$
$= 4\mathbf{i} + 3\mathbf{j} = \langle 4, 3 \rangle$

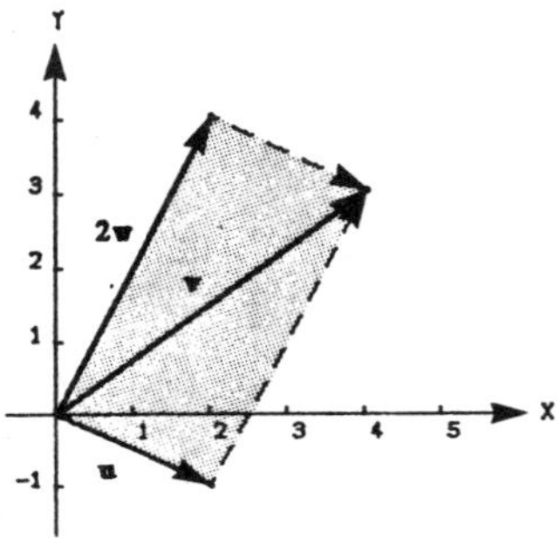

18. $\mathbf{v} = -(2\mathbf{i} - \mathbf{j}) + (\mathbf{i} + 2\mathbf{j})$
$= -\mathbf{i} + 3\mathbf{j} = \langle -1, 3 \rangle$

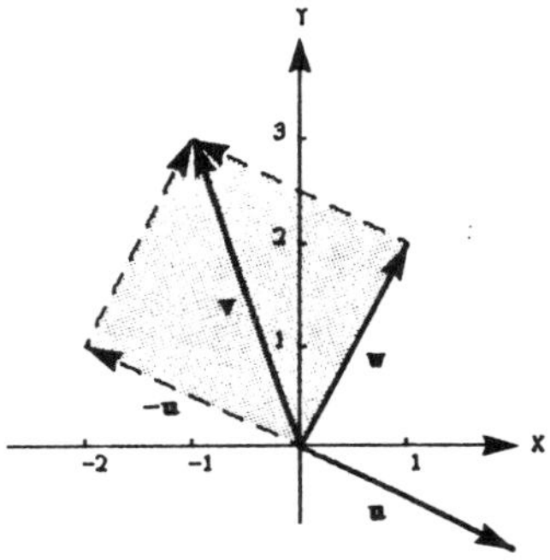

19. $\mathbf{v} = \frac{1}{2}[3(2\mathbf{i} - \mathbf{j}) + (\mathbf{i} + 2\mathbf{j})]$

$= \frac{1}{2}(7\mathbf{i} - \mathbf{j}) = \frac{7\mathbf{i} - \mathbf{j}}{2} = \langle \frac{7}{2}, -\frac{1}{2} \rangle$

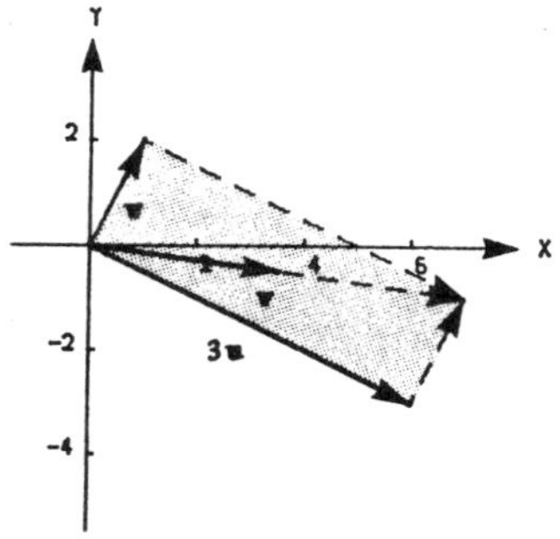

20. $\mathbf{v} = (2\mathbf{i} - \mathbf{j}) - 2(\mathbf{i} + 2\mathbf{j})$
$= -5\mathbf{j} = \langle 0, -5 \rangle$

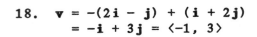

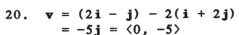

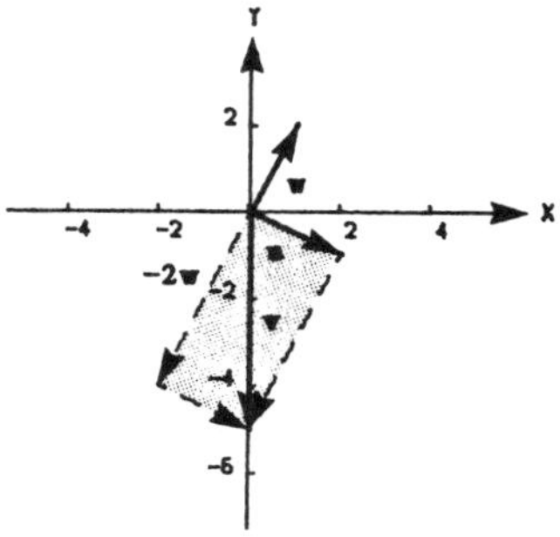

For exercises 21 - 26: $a\mathbf{u} + b\mathbf{w} = a(\mathbf{i} + 2\mathbf{j}) + b(\mathbf{i} - \mathbf{j}) = (a + b)\mathbf{i} + (2a - b)\mathbf{j}$

21. $\mathbf{v} = 2\mathbf{i} + \mathbf{j}$, therefore $a + b = 2$, $2a - b = 1$
Solving simultaneously we have $a = 1$, $b = 1$

22. $\mathbf{v} = 3\mathbf{j}$, therefore $a + b = 0$, $2a - b = 3$
Solving simultaneously we have $a = 1$, $b = -1$

23. $\mathbf{v} = 3\mathbf{i}$, therefore $a + b = 3$, $2a - b = 0$
Solving simultaneously we have $a = 1$, $b = 2$

24. $\mathbf{v} = 3\mathbf{i} + 3\mathbf{j}$, therefore $a + b = 3$, $2a - b = 3$
Solving simultaneously we have $a = 2$, $b = 1$

25. $\mathbf{v} = \mathbf{i} + \mathbf{j}$, therefore $a + b = 1$, $2a - b = 1$

Solving simultaneously we have $a = \frac{2}{3}$, $b = \frac{1}{3}$

26. $\mathbf{v} = -\mathbf{i} + 7\mathbf{j}$, therefore $a + b = -1$, $2a - b = 7$
Solving simultaneously we have $a = 2$, $b = -3$

27. $u_1 - 4 = -1$, $u_2 - 2 = 3$, $u_1 = 3$, $u_2 = 5$, $Q = (3, 5)$

28. $u_1 - 3 = 4$, $u_2 - 2 = -9$, $u_1 = 7$, $u_2 = -7$, $Q = (7, -7)$

29. $\|\mathbf{v}\| = \sqrt{16 + 9} = 5$

30. $\|\mathbf{v}\| = \sqrt{144 + 25} = 13$

31. $\|\mathbf{v}\| = \sqrt{36 + 25} = \sqrt{61}$

32. $\|\mathbf{v}\| = \sqrt{100 + 9} = \sqrt{109}$

33. $\|\mathbf{v}\| = \sqrt{0 + 16} = 4$

34. $\|\mathbf{v}\| = \sqrt{1 + 1} = \sqrt{2}$

35.(a) $\|\mathbf{v}\| = \sqrt{1 + 1/4} = \dfrac{\sqrt{5}}{2}$

(b) $\|\mathbf{v}\| = \sqrt{4 + 9} = \sqrt{13}$

(c) $\mathbf{u} + \mathbf{v} = \langle 3, \frac{7}{2}\rangle$

$\|\mathbf{u} + \mathbf{v}\| = \sqrt{9 + 49/4} = \dfrac{\sqrt{85}}{2}$

(d) $\dfrac{\mathbf{u}}{\|\mathbf{u}\|} = \dfrac{2}{\sqrt{5}}\langle 1, \frac{1}{2}\rangle$

$\left\|\dfrac{\mathbf{u}}{\|\mathbf{u}\|}\right\| = 1$

(e) $\dfrac{\mathbf{v}}{\|\mathbf{v}\|} = \dfrac{1}{\sqrt{13}}\langle 2, 3\rangle$

$\left\|\dfrac{\mathbf{v}}{\|\mathbf{v}\|}\right\| = 1$

(f) $\dfrac{\mathbf{u} + \mathbf{v}}{\|\mathbf{u} + \mathbf{v}\|} = \dfrac{2}{\sqrt{85}}\langle 3, \frac{7}{2}\rangle$

$\left\|\dfrac{\mathbf{u} + \mathbf{v}}{\|\mathbf{u} + \mathbf{v}\|}\right\| = 1$

36.(a) $\|\mathbf{u}\| = \sqrt{4 + 16} = 2\sqrt{5}$

(b) $\|\mathbf{v}\| = \sqrt{25 + 25} = 5\sqrt{2}$

(c) $\mathbf{u} + \mathbf{v} = \langle 7, 1\rangle$, $\quad \|\mathbf{u} + \mathbf{v}\| = \sqrt{49 + 1} = 5\sqrt{2}$

(d) $\dfrac{\mathbf{u}}{\|\mathbf{u}\|} = \dfrac{1}{2\sqrt{5}}\langle 2, -4\rangle$, $\quad \left\|\dfrac{\mathbf{u}}{\|\mathbf{u}\|}\right\| = 1$

(e) $\dfrac{\mathbf{v}}{\|\mathbf{v}\|} = \dfrac{1}{5\sqrt{2}}\langle 5, 5\rangle$, $\quad \left\|\dfrac{\mathbf{v}}{\|\mathbf{v}\|}\right\| = 1$

(f) $\dfrac{\mathbf{u} + \mathbf{v}}{\|\mathbf{u} + \mathbf{v}\|} = \dfrac{1}{5\sqrt{2}}\langle 7, 1\rangle$, $\left\|\dfrac{\mathbf{u} + \mathbf{v}}{\|\mathbf{u} + \mathbf{v}\|}\right\| = 1$

37. $\mathbf{u} = \langle 2, 1\rangle$, $\|\mathbf{u}\| = \sqrt{5} \approx 2.236$, $\quad \mathbf{v} = \langle 5, 4\rangle$, $\|\mathbf{v}\| = \sqrt{41} \approx 6.403$

$\mathbf{u} + \mathbf{v} = \langle 7, 5\rangle$, $\quad \|\mathbf{u} + \mathbf{v}\| = \sqrt{74} \approx 8.602$

$\|\mathbf{u} + \mathbf{v}\| \le \|\mathbf{u}\| + \|\mathbf{v}\|$

38. $\mathbf{u} = \langle -3, 2\rangle$, $\|\mathbf{u}\| = \sqrt{13} \approx 3.606$, $\quad \mathbf{v} = \langle 1, -2\rangle$, $\|\mathbf{v}\| = \sqrt{5} \approx 2.236$

$\mathbf{u} + \mathbf{v} = \langle -2, 0\rangle$, $\quad \|\mathbf{u} + \mathbf{v}\| = 4$

$\|\mathbf{u} + \mathbf{v}\| \le \|\mathbf{u}\| + \|\mathbf{v}\|$

39. $\dfrac{\mathbf{u}}{\|\mathbf{u}\|} = \dfrac{1}{\sqrt{2}}\langle 1, 1\rangle$, $\quad 4\left(\dfrac{\mathbf{u}}{\|\mathbf{u}\|}\right) = 2\sqrt{2}\langle 1, 1\rangle$, $\quad \mathbf{v} = \langle 2\sqrt{2}, 2\sqrt{2}\rangle$

40. $\dfrac{\mathbf{u}}{\|\mathbf{u}\|} = \dfrac{1}{\sqrt{2}}\langle -1, 1\rangle$, $\quad 4\left(\dfrac{\mathbf{u}}{\|\mathbf{u}\|}\right) = 2\sqrt{2}\langle -1, 1\rangle$, $\quad \mathbf{v} = \langle -2\sqrt{2}, 2\sqrt{2}\rangle$

41. $\dfrac{\mathbf{u}}{\|\mathbf{u}\|} = \dfrac{1}{2\sqrt{3}}\langle \sqrt{3}, 3\rangle$, $\quad 2\left(\dfrac{\mathbf{u}}{\|\mathbf{u}\|}\right) = \dfrac{1}{\sqrt{3}}\langle \sqrt{3}, 3\rangle$, $\quad \mathbf{v} = \langle 1, \sqrt{3}\rangle$

42. $\dfrac{\mathbf{u}}{\|\mathbf{u}\|} = \dfrac{1}{3}\langle 0, 3\rangle$, $\quad 3\left(\dfrac{\mathbf{u}}{\|\mathbf{u}\|}\right) = \langle 0, 3\rangle$, $\quad \mathbf{v} = \langle 0, 3\rangle$

43. $y = x^3$, $\quad y' = 3x^2 = 3$ at $x = 1$

(a) $m = 3$, $\quad$ Let $\mathbf{w} = \langle 1, 3\rangle$, $\quad \dfrac{\mathbf{w}}{\|\mathbf{w}\|} = \dfrac{1}{\sqrt{10}}\langle 1, 3\rangle$

(b) $m = -\dfrac{1}{3}$, $\quad$ Let $\mathbf{w} = \langle 3, -1\rangle$, $\quad \dfrac{\mathbf{w}}{\|\mathbf{w}\|} = \dfrac{1}{\sqrt{10}}\langle 3, -1\rangle$

44. $y = x^3, \quad y' = 3x^2 = 12$ at $x = -2$

(a) $m = 12, \quad$ Let $\mathbf{w} = \langle 1, 12\rangle, \quad \dfrac{\mathbf{w}}{\|\mathbf{w}\|} = \dfrac{1}{\sqrt{145}}\langle 1, 12\rangle$

(b) $m = -\dfrac{1}{12}, \quad$ Let $\mathbf{w} = \langle 12, -1\rangle, \quad \dfrac{\mathbf{w}}{\|\mathbf{w}\|} = \dfrac{1}{\sqrt{145}}\langle 12, -1\rangle$

45. $f(x) = \sqrt{25 - x^2}, \quad f'(x) = \dfrac{-x}{\sqrt{25 - x^2}} = \dfrac{-3}{4}$ at $x = 3$

(a) $m = -\dfrac{3}{4}, \quad$ Let $\mathbf{w} = \langle -4, 3\rangle, \quad \dfrac{\mathbf{w}}{\|\mathbf{w}\|} = \dfrac{1}{5}\langle -4, 3\rangle$

(b) $m = \dfrac{4}{3}, \quad$ Let $\mathbf{w} = \langle 3, 4\rangle, \quad \dfrac{\mathbf{w}}{\|\mathbf{w}\|} = \dfrac{1}{5}\langle 3, 4\rangle$

46. $f(x) = \tan x, \quad f'(x) = \sec^2 x = 2$ at $x = \dfrac{\pi}{4}$

(a) $m = 2, \quad$ Let $\mathbf{w} = \langle 1, 2\rangle, \quad \dfrac{\mathbf{w}}{\|\mathbf{w}\|} = \dfrac{1}{\sqrt{5}}\langle 1, 2\rangle$

(b) $m = -\dfrac{1}{2}, \quad$ Let $\mathbf{w} = \langle -2, 1\rangle, \quad \dfrac{\mathbf{w}}{\|\mathbf{w}\|} = \dfrac{1}{\sqrt{5}}\langle -2, 1\rangle$

47. $\mathbf{v} = 3[(\cos 0^\circ)\mathbf{i} + (\sin 0^\circ)\mathbf{j}] = 3\mathbf{i} = \langle 3, 0\rangle$

48. $\mathbf{v} = (\cos 45^\circ)\mathbf{i} + (\sin 45^\circ)\mathbf{j} = \dfrac{\sqrt{2}}{2}\mathbf{i} + \dfrac{\sqrt{2}}{2}\mathbf{j} = \left\langle \dfrac{\sqrt{2}}{2}, \dfrac{\sqrt{2}}{2}\right\rangle$

49. $\mathbf{v} = 2[(\cos 150^\circ)\mathbf{i} + (\sin 150^\circ)\mathbf{j}] = -\sqrt{3}\,\mathbf{i} + \mathbf{j} = \langle -\sqrt{3}, 1\rangle$

50. $\mathbf{v} = (\cos 3.5)\mathbf{i} + (\sin 3.5)\mathbf{j} = -0.9364\mathbf{i} - 0.3508\mathbf{j} = \langle -0.9364, -0.3508\rangle$

51. $\mathbf{u} = \mathbf{i}, \quad \mathbf{v} = \dfrac{3\sqrt{2}}{2}\mathbf{i} + \dfrac{3\sqrt{2}}{2}\mathbf{j}, \quad \mathbf{u} + \mathbf{v} = \left(\dfrac{2 + 3\sqrt{2}}{2}\right)\mathbf{i} + \dfrac{3\sqrt{2}}{2}\mathbf{j}$

52. $\mathbf{u} = 4\mathbf{i}, \quad \mathbf{v} = \mathbf{i} + \sqrt{3}\,\mathbf{j}, \quad \mathbf{u} + \mathbf{v} = 5\mathbf{i} + \sqrt{3}\,\mathbf{j}$

53. $\mathbf{u} = 2(\cos 4)\mathbf{i} + 2(\sin 4)\mathbf{j}, \quad \mathbf{v} = (\cos 2)\mathbf{i} + (\sin 2)\mathbf{j}$

$\mathbf{u} + \mathbf{v} = (2\cos 4 + \cos 2)\mathbf{i} + (2\sin 4 + \sin 2)\mathbf{j}$

54. $\mathbf{u} = 5[\cos(-0.5)]\mathbf{i} + 5[\sin(-0.5)]\mathbf{j} = 5[\cos(0.5)]\mathbf{i} - 5[\sin 0.5]\mathbf{j}$

$\mathbf{v} = 5[\cos(0.5)]\mathbf{i} + 5[\sin(0.5)]\mathbf{j}, \quad \mathbf{u} + \mathbf{v} = 10[\cos(0.5)]\mathbf{i}$

55. $\mathbf{u} = \frac{\sqrt{2}}{2}\mathbf{i} + \frac{\sqrt{2}}{2}\mathbf{j}$, $\mathbf{u} + \mathbf{v} = \sqrt{2}\,\mathbf{j}$, $\mathbf{v} = -\frac{\sqrt{2}}{2}\mathbf{i} + \frac{\sqrt{2}}{2}\mathbf{j}$

56. $\mathbf{u} = 2\sqrt{3}\,\mathbf{i} + 2\mathbf{j}$, $\mathbf{u} + \mathbf{v} = -3\mathbf{i} + 3\sqrt{3}\,\mathbf{j}$, $\mathbf{v} = (-3 - 2\sqrt{3})\mathbf{i} + (3\sqrt{3} - 2)\mathbf{j}$

57. $\|\mathbf{u}\| = 150$

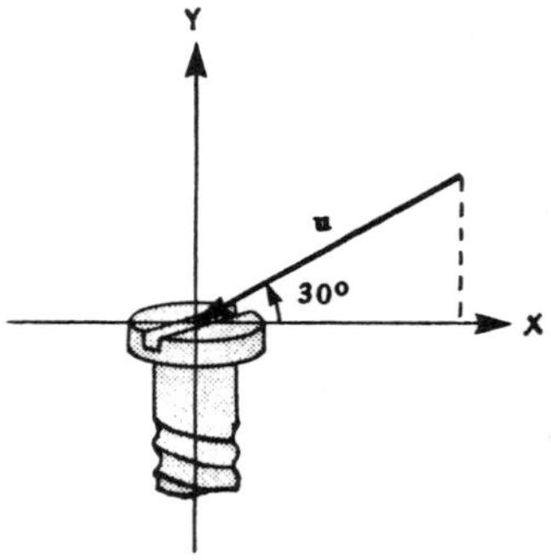

$\mathbf{u} = -150[(\cos 30°)\mathbf{i} + (\sin 30°)\mathbf{j}]$

Horizontal component $= -150 \cos 30°$
$= -75\sqrt{3}$ lb

Vertical component $= -150 \sin 30°$
$= -75$ lb

58. To lift the weight vertically, the sum of the vertical components of **u** and **v** must be 100 and the sum of the horizontal components must be 0.

Thus $\|\mathbf{u}\| \sin 120° + \|\mathbf{v}\| \sin 70° = 100$ or $\|\mathbf{u}\|(\frac{\sqrt{3}}{2}) + \|\mathbf{v}\| \sin 70° = 100$

and $\|\mathbf{u}\| \cos 120° + \|\mathbf{v}\| \cos 70° = 0$ or $\|\mathbf{u}\|(\frac{-1}{2}) + \|\mathbf{v}\| \cos 70° = 0$

If we multiply the second equation by $\sqrt{3}$ and then add the two equations, we have

$\|\mathbf{v}\|[\sqrt{3} \cos 70° + \sin 70°] = 100$, $\|\mathbf{v}\| \approx 65.27$ lb

$65.27(\cos 70°) = \frac{\|\mathbf{u}\|}{2}$, $\|\mathbf{u}\| \approx 44.65$ lb

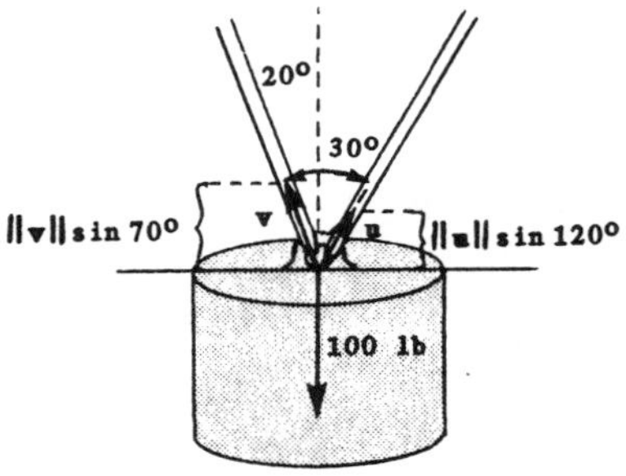

Tension in each rope:

$\|\mathbf{u}\| = 44.65$ lb, $\|\mathbf{v}\| = 65.27$ lb

Vertical component of each man's force:

$\|\mathbf{u}\| \sin 120° \approx 38.67$ lb

$\|\mathbf{v}\| \sin 70° \approx 61.33$ lb

59. $\mathbf{u} = 580[(-\cos 32^\circ)\mathbf{i} + (\sin 32^\circ)\mathbf{j}]$, $\mathbf{v} = 60[(\cos 45^\circ)\mathbf{i} + (\sin 45^\circ)\mathbf{j}]$

$\mathbf{u} + \mathbf{v} = (60 \cos 45^\circ - 580 \cos 32^\circ)\mathbf{i} + (60 \sin 45^\circ + 580 \sin 32^\circ)\mathbf{j}$

$= -449.44\mathbf{i} + 349.78\mathbf{j}$

Direction north of West

$\cos\theta = \dfrac{(\mathbf{u} + \mathbf{v})\cdot(-\mathbf{i})}{\|\mathbf{u} + \mathbf{v}\|}$

$= \dfrac{449.44}{569.51} = 0.7892$

$\theta = \arccos(0.7892) = 37.89^\circ \approx 38^\circ$ N of W

speed: $\|\mathbf{u} + \mathbf{v}\| = 569.51$ mph

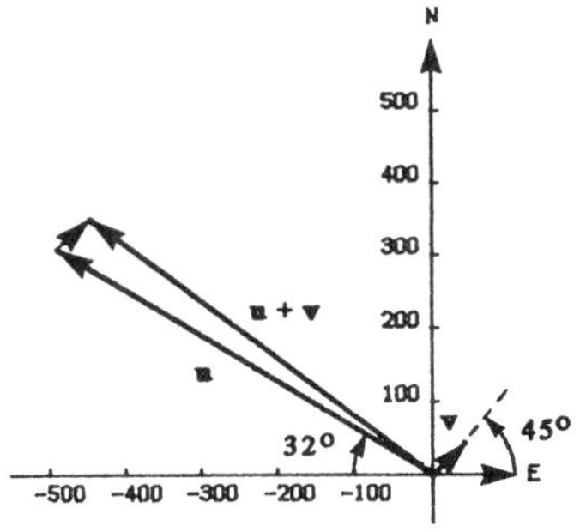

60. Horizontal component $= \|\mathbf{v}\| \cos\theta = 80 \cos 50^\circ = 51.42$ ft/s

Vertical component $= \|\mathbf{v}\| \sin\theta = 80 \sin 50^\circ = 61.28$ ft/s

61. (-4, -1) (6, 5) (10, 3)

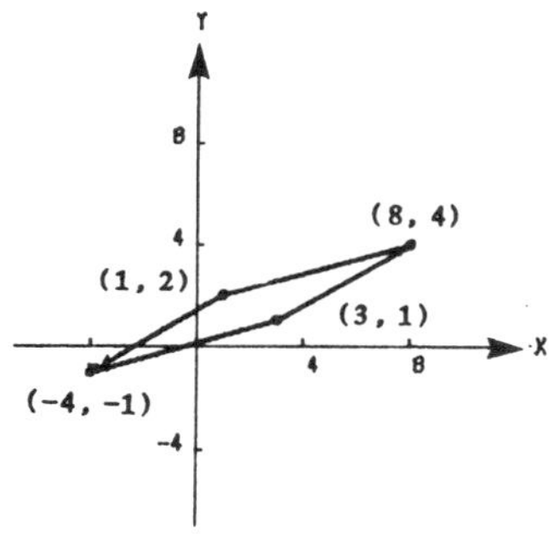

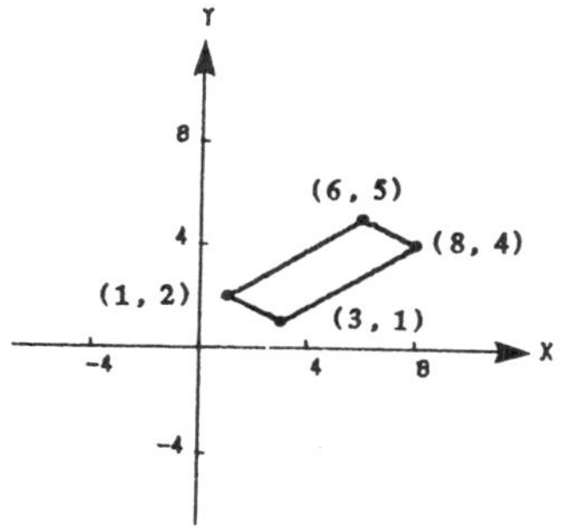

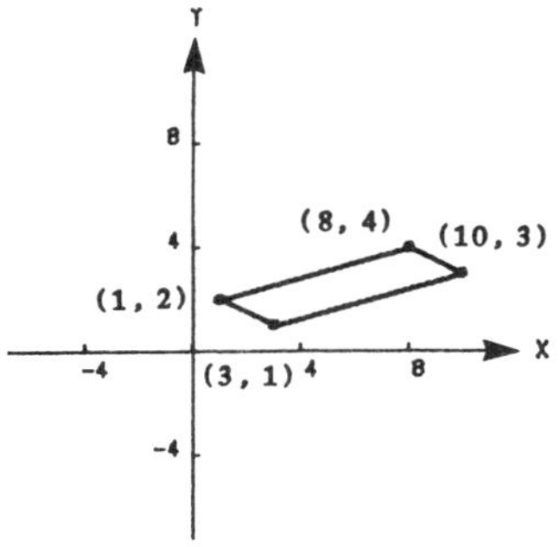

62. The two diagonals are: $\mathbf{u} + \mathbf{v}$ and $\mathbf{v} - \mathbf{u}$

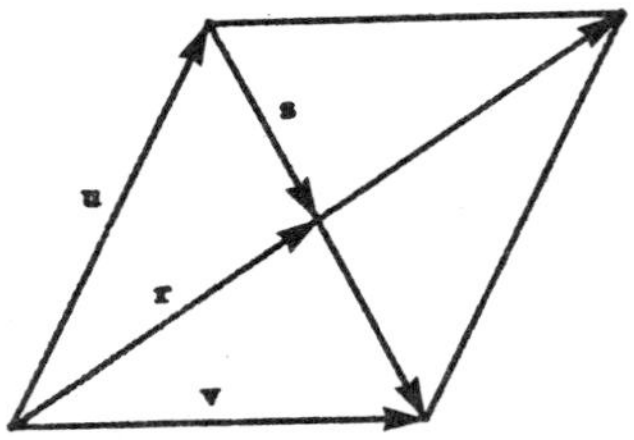

Therefore $\mathbf{r} = x(\mathbf{u} + \mathbf{v})$ and $\mathbf{s} = y(\mathbf{v} - \mathbf{u})$

But $\mathbf{u} = \mathbf{r} - \mathbf{s} = x(\mathbf{u} + \mathbf{v}) - y(\mathbf{v} - \mathbf{u})$
$= (x + y)\mathbf{u} + (x - y)\mathbf{v}$

Therefore $x + y = 1$ and $x - y = 0$

Solving we have $x = y = \frac{1}{2}$

63. $\mathbf{u} = \langle 6, 3\rangle$, $\quad \frac{1}{3}\mathbf{u} = \langle 2, 1\rangle$

$P_1 = (1, 2) + (2, 1) = (3, 3)$, $\quad P_2 = (1, 2) + 2(2, 1) = (5, 4)$

64. $\|\mathbf{u}\| = \sqrt{\cos^2\theta + \sin^2\theta} = 1$, $\quad \|\mathbf{v}\| = \sqrt{\sin^2\theta + \cos^2\theta} = 1$

65. $\mathbf{v} = \left(\frac{a + b}{2} - \frac{a}{2}\right)\mathbf{i} + \frac{c}{2}\mathbf{j}$

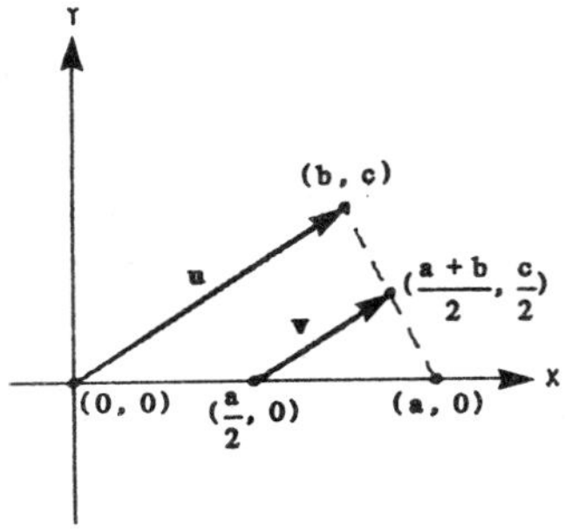

$= \frac{b}{2}\mathbf{i} + \frac{c}{2}\mathbf{j}$

$= \frac{1}{2}(b\mathbf{i} + c\mathbf{j})$

$= \frac{1}{2}\mathbf{u}$

12.2 Space coordinates and vectors in space

1.

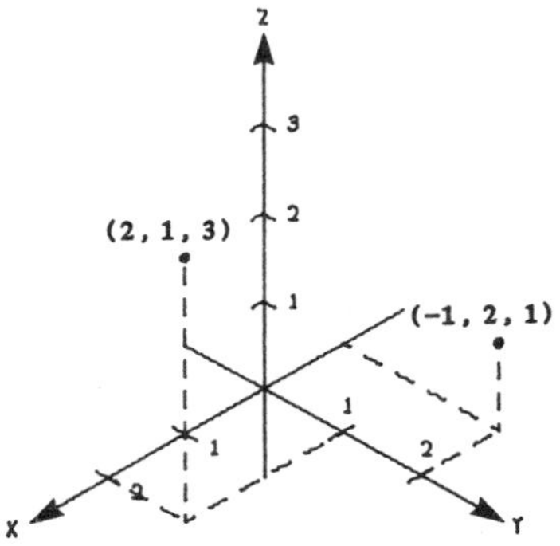

2.

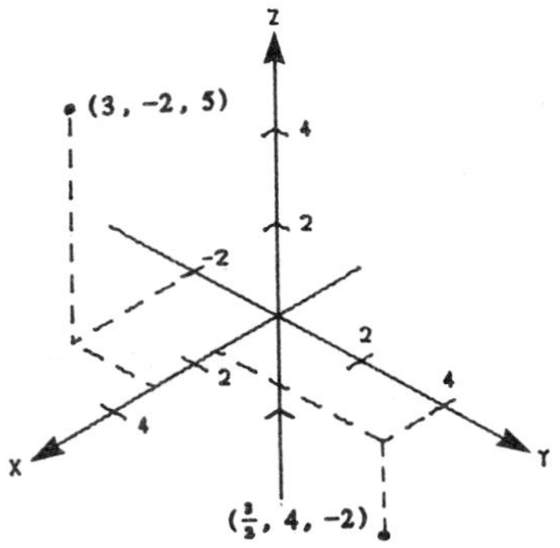

3.

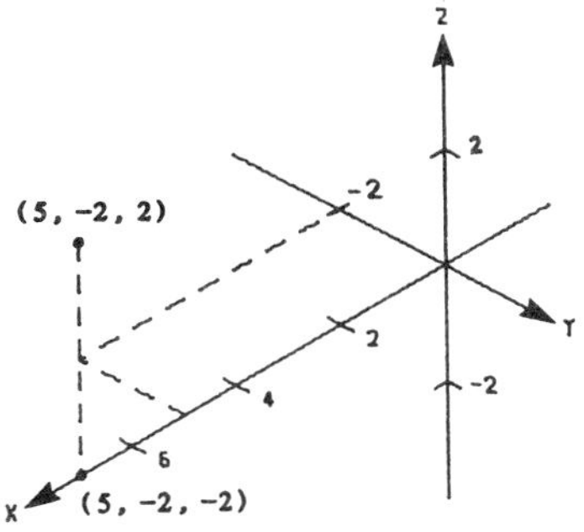

4.

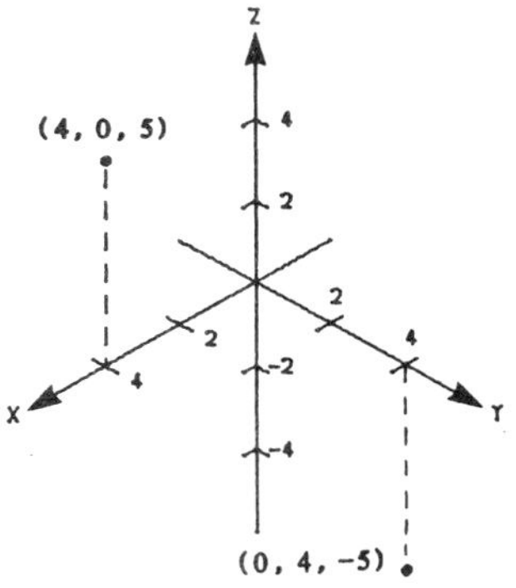

5. A(0, 0, 0) B(2, 2, 1) C(2, -4, 4), $|AB| = \sqrt{4 + 4 + 1} = 3$

$|AC| = \sqrt{4 + 16 + 16} = 6$, $|BC| = \sqrt{0 + 36 + 9} = 3\sqrt{5}$

$|BC|^2 = |AB|^2 + |AC|^2$, right triangle

6. A(5, 3, 4) B(7, 1, 3) C(3, 5, 3), $|AB| = \sqrt{4 + 4 + 1} = 3$

$|AC| = \sqrt{4 + 4 + 1} = 3$, $|BC| = \sqrt{16 + 16 + 0} = 4\sqrt{2}$

Since $|AB| = |AC|$ the triangle is isosceles.

7. A(1, -3, -2) B(5, -1, 2) C(-1, 1, 2), $|AB| = \sqrt{16 + 4 + 16} = 6$

$|AC| = \sqrt{4 + 16 + 16} = 6$, $|BC| = \sqrt{36 + 4 + 0} = 2\sqrt{10}$

Since $|AB| = |AC|$ the triangle is isosceles.

8. A(5, 0, 0) B(0, 2, 0) C(0, 0, -3), $|AB| = \sqrt{25 + 4 + 0} = \sqrt{29}$

$|AC| = \sqrt{25 + 0 + 9} = \sqrt{34}$, $|BC| = \sqrt{0 + 4 + 9} = \sqrt{13}$, neither

9. $\left(\frac{5 + (-2)}{2}, \frac{-9 + 3}{2}, \frac{7 + 3}{2}\right) = \left(\frac{3}{2}, -3, 5\right)$

10. $\left(\frac{4 + 8}{2}, \frac{0 + 8}{2}, \frac{-6 + 20}{2}\right) = (6, 4, 7)$

11. Center: (0, 2, 5), radius: 2

$(x - 0)^2 + (y - 2)^2 + (z - 5)^2 = 4$, $x^2 + y^2 + z^2 - 4y - 10z + 25 = 0$

12. Center: (4, -1, 1), radius: 5

$(x - 4)^2 + (y + 1)^2 + (z - 1)^2 = 25$, $x^2 + y^2 + z^2 - 8x + 2y - 2z - 7 = 0$

13. Center: $(1, 3, 0)$, radius: $\sqrt{10}$

$(x - 1)^2 + (y - 3)^2 + (z - 0)^2 = 10, \quad x^2 + y^2 + z^2 - 2x - 6y = 0$

14. Center: $(-2, 1, 1)$, radius: 1

$(x + 2)^2 + (y - 1)^2 + (z - 1)^2 = 1, \quad x^2 + y^2 + z^2 + 4x - 2y - 2z + 5 = 0$

15. $x^2 + y^2 + z^2 - 2x + 6y + 8z + 1 = 0$

$(x^2 - 2x + 1) + (y^2 + 6y + 9) + (z^2 + 8z + 16) = -1 + 1 + 9 + 16$

$(x - 1)^2 + (y + 3)^2 + (z + 4)^2 = 25$

Center: $(1, -3, -4)$, radius: 5

16. $x^2 + y^2 + z^2 + 9x - 2y + 10z + 19 = 0$

$(x^2 + 9x + \frac{81}{4}) + (y^2 - 2y + 1) + (z^2 + 10z + 25) = -19 + \frac{81}{4} + 1 + 25$

$(x + \frac{9}{2})^2 + (y - 1)^2 + (z + 5)^2 = \frac{109}{4}$

Center: $(-\frac{9}{2}, 1, -5)$, radius: $\frac{\sqrt{109}}{2}$

17. $9x^2 + 9y^2 + 9z^2 - 6x + 18y + 1 = 0$

$x^2 + y^2 + z^2 - \frac{2}{3}x + 2y + \frac{1}{9} = 0$

$(x^2 - \frac{2}{3}x + \frac{1}{9}) + (y^2 + 2y + 1) + z^2 = -\frac{1}{9} + \frac{1}{9} + 1$

$(x - \frac{1}{3})^2 + (y + 1)^2 + (z - 0)^2 = 1$

Center: $(\frac{1}{3}, -1, 0)$, radius: 1

18. $4x^2 + 4y^2 + 4z^2 - 4x - 32y + 8z + 33 = 0$

$x^2 + y^2 + z^2 - x - 8y + 2z + \frac{33}{4} = 0$

$(x^2 - x + \frac{1}{4}) + (y^2 - 8y + 16) + (z^2 + 2z + 1) = -\frac{33}{4} + \frac{1}{4} + 16 + 1$

$(x - \frac{1}{2})^2 + (y - 4)^2 + (z + 1)^2 = 9$

Center: $(\frac{1}{2}, 4, -1)$, radius: 3

19.(a) $\mathbf{v} = (2 - 4)\mathbf{i} + (4 - 2)\mathbf{j} + (3 - 1)\mathbf{k} = -2\mathbf{i} + 2\mathbf{j} + 2\mathbf{k} = \langle -2, 2, 2 \rangle$

(b)

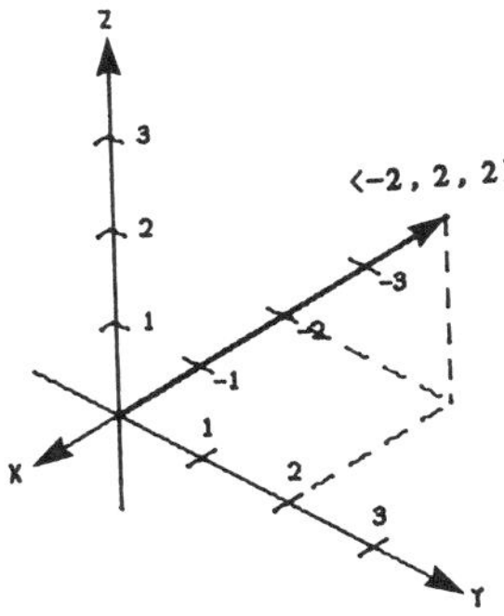

20.(a) $\mathbf{v} = (4 - 0)\mathbf{i} + (0 - 5)\mathbf{j} + (3 - 1)\mathbf{k} = 4\mathbf{i} - 5\mathbf{j} + 2\mathbf{k} = \langle 4, -5, 2 \rangle$

(b)

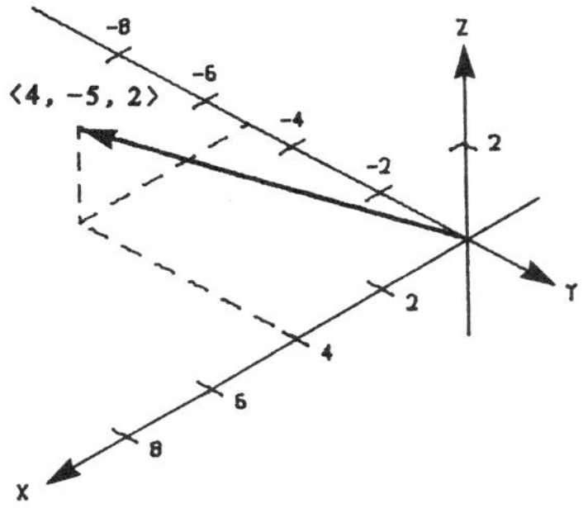

21.(a) $\mathbf{v} = (0 - 3)\mathbf{i} + (3 - 3)\mathbf{j} + (3 - 0)\mathbf{k} = -3\mathbf{i} + 3\mathbf{k} = \langle -3, 0, 3 \rangle$

(b)

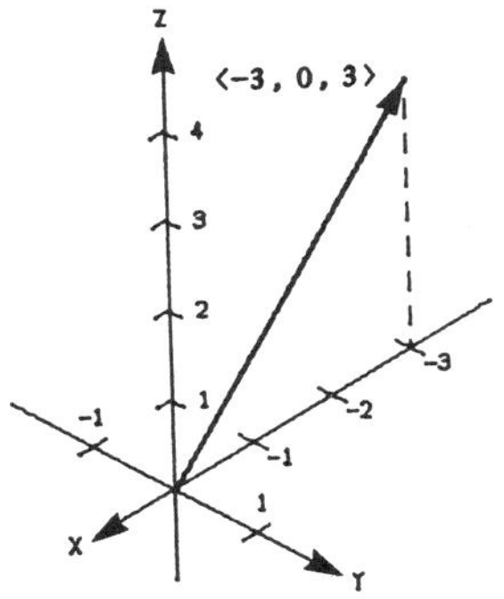

22.(a) $\mathbf{v} = (2 - 2)\mathbf{i} + (3 - 3)\mathbf{j} + (4 - 0)\mathbf{k} = 4\mathbf{k} = \langle 0, 0, 4 \rangle$

(b)

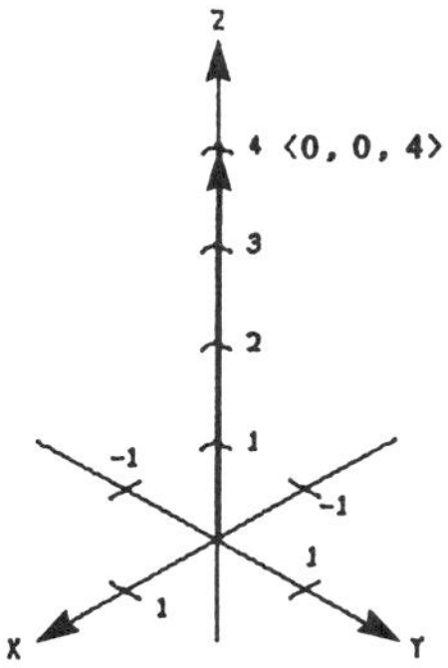

23.(b) $\mathbf{v} = (3 + 1)\mathbf{i} + (3 - 2)\mathbf{j} + (4 - 3)\mathbf{k} = 4\mathbf{i} + \mathbf{j} + \mathbf{k} = \langle 4, 1, 1\rangle$

(a) and (c)

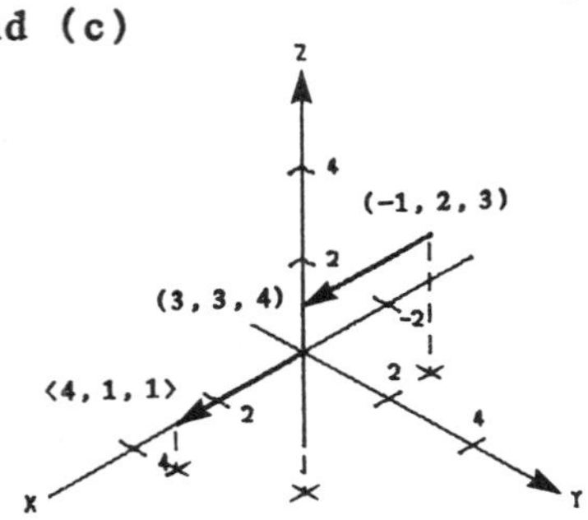

24.(b) $\mathbf{v} = (-4 - 2)\mathbf{i} + (3 + 1)\mathbf{j} + (7 + 2)\mathbf{k} = -6\mathbf{i} + 4\mathbf{j} + 9\mathbf{k} = \langle -6, 4, 9\rangle$

(a) and (c)

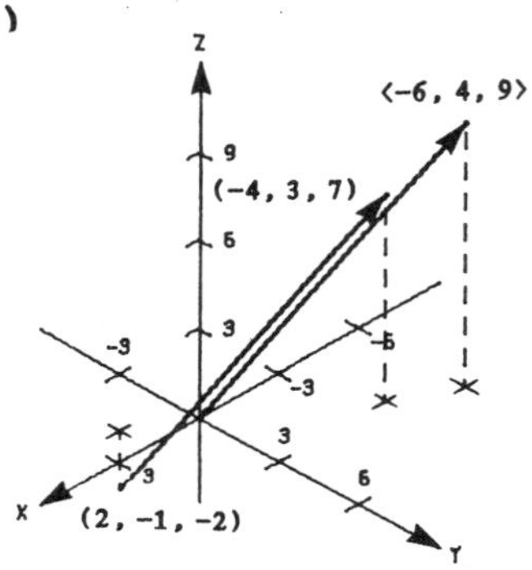

25.(a) $2\mathbf{v} = \langle 2, 4, 4\rangle$

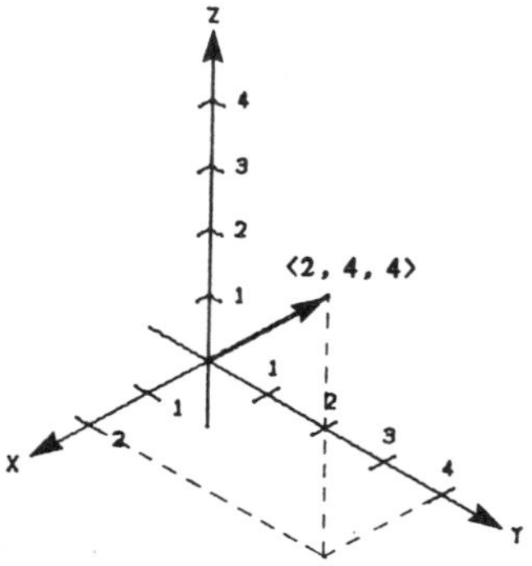

(b) $-\mathbf{v} = \langle -1, -2, -2\rangle$

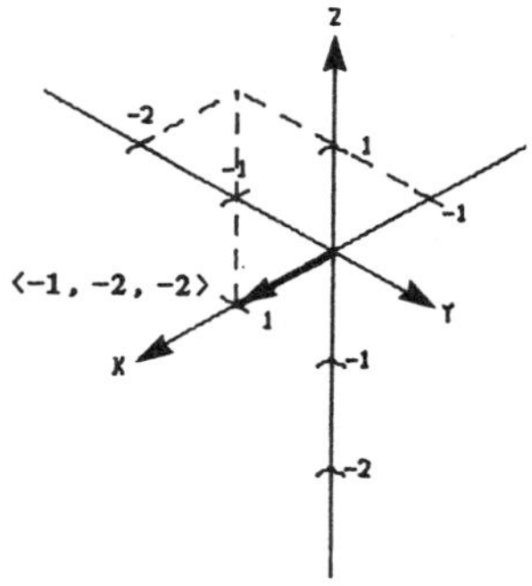

(c) $\frac{3}{2}\mathbf{v} = \langle \frac{3}{2}, 3, 3\rangle$

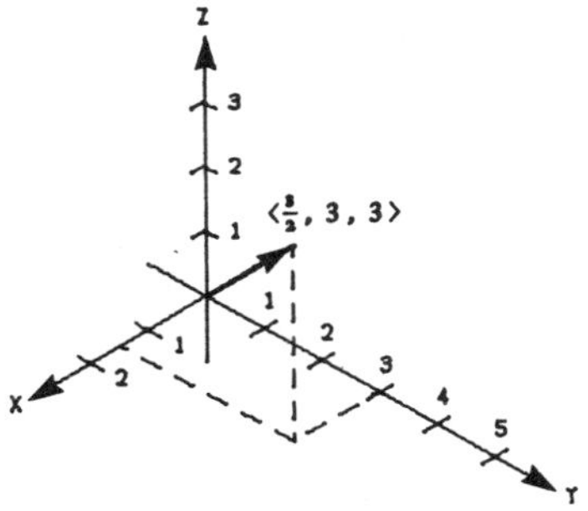

(d) $0\mathbf{v} = \langle 0, 0, 0\rangle$

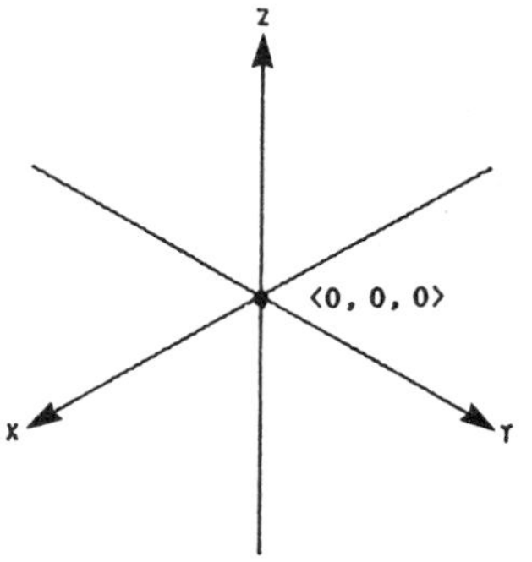

26.(a) $-\mathbf{v} = \langle -2, 2, -1 \rangle$

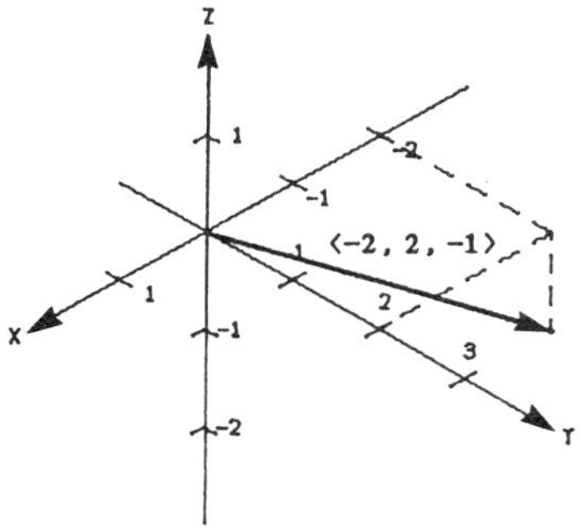

(b) $2\mathbf{v} = \langle 4, -4, 2 \rangle$

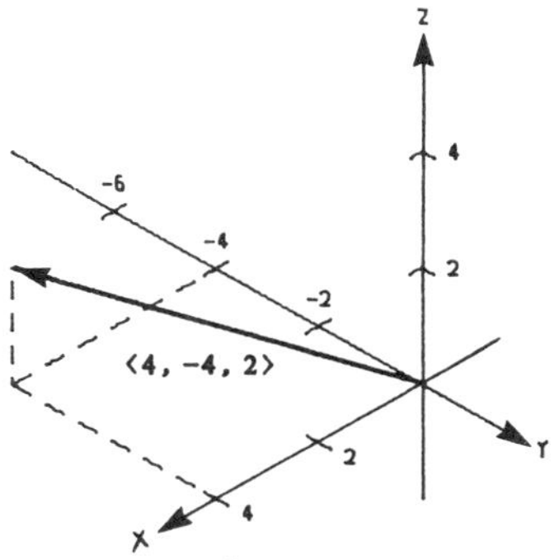

(c) $\frac{1}{2}\mathbf{v} = \langle 1, -1, \frac{1}{2} \rangle$

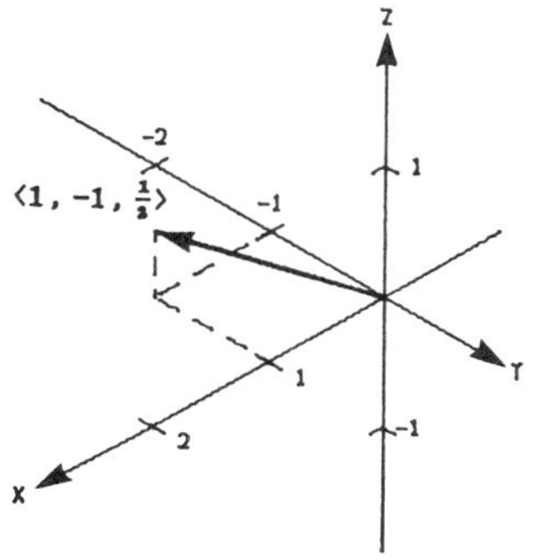

(d) $\frac{5}{2}\mathbf{v} = \langle 5, -5, \frac{5}{2} \rangle$

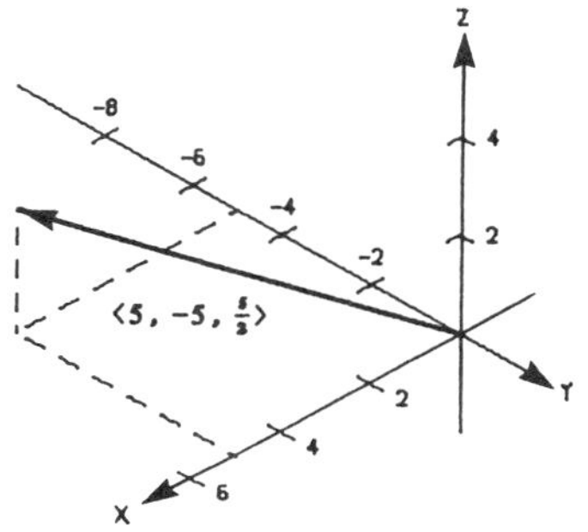

27. $\mathbf{u} - \mathbf{v} = \langle 1, 2, 3 \rangle - \langle 2, 2, -1 \rangle = \langle -1, 0, 4 \rangle$

28. $\mathbf{u} - \mathbf{v} + 2\mathbf{w} = \langle 1, 2, 3 \rangle - \langle 2, 2, -1 \rangle + \langle 8, 0, -8 \rangle = \langle 7, 0, -4 \rangle$

29. $2\mathbf{u} + 4\mathbf{v} - \mathbf{w} = \langle 2, 4, 6 \rangle + \langle 8, 8, -4 \rangle - \langle 4, 0, -4 \rangle = \langle 6, 12, 6 \rangle$

30. $5\mathbf{u} - 3\mathbf{v} - \frac{1}{2}\mathbf{w} = \langle 5, 10, 15 \rangle - \langle 6, 6, -3 \rangle - \langle 2, 0, -2 \rangle = \langle -3, 4, 20 \rangle$

31. $2\mathbf{z} - 3\mathbf{u} = 2\langle z_1, z_2, z_3 \rangle - 3\langle 1, 2, 3 \rangle = \langle 4, 0, -4 \rangle$

$2z_1 - 3 = 4 \Rightarrow z_1 = \frac{7}{2}$

$2z_2 - 6 = 0 \Rightarrow z_2 = 3$

$2z_3 - 9 = -4 \Rightarrow z_3 = \frac{5}{2}$

$\mathbf{z} = \langle \frac{7}{2}, 3, \frac{5}{2} \rangle$

32. $2\mathbf{u} + \mathbf{v} - \mathbf{w} + 3\mathbf{z} = 2\langle 1, 2, 3\rangle + \langle 2, 2, -1\rangle - \langle 4, 0, -4\rangle + 3\langle z_1, z_2, z_3\rangle$

$= \langle 0, 0, 0\rangle$

$\langle 0, 6, 9\rangle + \langle 3z_1, 3z_2, 3z_3\rangle = \langle 0, 0, 0\rangle$

$0 + 3z_1 = 0 \Rightarrow z_1 = 0,\qquad 6 + 3z_2 = 0 \Rightarrow z_2 = -2$

$9 + 3z_3 = 0 \Rightarrow z_3 = -3,\qquad \mathbf{z} = \langle 0, -2, -3\rangle$

33. (a) and (b) are parallel since $\langle -6, -4, 10\rangle = -2\langle 3, 2, -5\rangle$

$\langle 2, \frac{4}{3}, -\frac{10}{3}\rangle = \frac{2}{3}\langle 3, 2, -5\rangle$

34. (b) and (d) are parallel since $-\mathbf{i} + \frac{4}{3}\mathbf{j} - \frac{3}{2}\mathbf{k} = -2(\frac{1}{2}\mathbf{i} - \frac{2}{3}\mathbf{j} + \frac{3}{4}\mathbf{k})$

$\frac{3}{4}\mathbf{i} - \mathbf{j} + \frac{9}{8}\mathbf{k} = \frac{3}{2}(\frac{1}{2}\mathbf{i} - \frac{3}{2}\mathbf{j} + \frac{3}{4}\mathbf{k})$

35. $\mathbf{z} = -3\mathbf{i} + 4\mathbf{j} + 2\mathbf{k}$, (a) is parallel since $-6\mathbf{i} + 8\mathbf{j} + 4\mathbf{k} = 2\mathbf{z}$

36. $\mathbf{z} = -4\mathbf{i} - 5\mathbf{j} + 6\mathbf{k} = \langle -4, -5, 6\rangle$, (b) is parallel since $\langle 8, 10, -12\rangle = -2\mathbf{z}$

37. $P(0, -2, -5)$, $Q(3, 4, 4)$, $R(2, 2, 1)$

$\overrightarrow{PQ} = \langle 3, 6, 9\rangle$, $\overrightarrow{PR} = \langle 2, 4, 6\rangle$, $\langle 3, 6, 9\rangle = \frac{3}{2}\langle 2, 4, 6\rangle$

Therefore $\overrightarrow{PQ}$ and $\overrightarrow{PR}$ are parallel and the points are collinear.

38. $P(1, -1, 5)$, $Q(0, -1, 6)$, $R(3, -1, 3)$

$\overrightarrow{PQ} = \langle -1, 0, 1\rangle$, $\overrightarrow{PR} = \langle 2, 0, -2\rangle$, $\langle -1, 0, 1\rangle = -\frac{1}{2}\langle 2, 0, -2\rangle$

Therefore $\overrightarrow{PQ}$ and $\overrightarrow{PR}$ are parallel and the points are collinear.

39. $P(1, 2, 4)$, $Q(2, 5, 0)$, $R(0, 1, 5)$

$\overrightarrow{PQ} = \langle 1, 3, -4\rangle$, $\overrightarrow{PR} = \langle -1, -1, 1\rangle$

Since $\overrightarrow{PQ}$ and $\overrightarrow{PR}$ are not parallel, the points are not collinear.

40. $P(0, 0, 0)$, $Q(1, 3, -2)$, $R(2, -6, 4)$

$\overrightarrow{PQ} = \langle 1, 3, -2\rangle$, $\overrightarrow{PR} = \langle 2, -6, 4\rangle$

Since $\overrightarrow{PQ}$ and $\overrightarrow{PR}$ are not parallel, the points are not collinear.

41. $A(2, 9, 1)$, $B(3, 11, 4)$, $C(0, 10, 2)$, $D(1, 12, 5)$

$|AB| = \sqrt{1 + 4 + 9} = \sqrt{14}$, $|AC| = \sqrt{4 + 1 + 1} = \sqrt{6}$

$|BD| = \sqrt{4 + 1 + 1} = \sqrt{6}$, $|CD| = \sqrt{1 + 4 + 9} = \sqrt{14}$

Since $|AB| = |CD|$ and $|AC| = |BD|$ the given points form the vertices of a parallelogram.

42. $A(1, 1, -3)$, $B(9, -1, 2)$, $C(11, 2, 1)$, $D(3, 4, -4)$

$|AB| = \sqrt{64 + 4 + 25} = \sqrt{93}$, $|AD| = \sqrt{4 + 9 + 1} = \sqrt{14}$

$|BC| = \sqrt{4 + 9 + 1} = \sqrt{14}$, $|CD| = \sqrt{64 + 4 + 25} = \sqrt{93}$

Since $|AB| = |CD|$ and $|AD| = |BC|$ the given points form the vertices of a parallelogram.

43. $(q_1, q_2, q_3) - (0, 6, 2) = (3, -5, 6)$, $Q = (3, 1, 8)$

44. $(q_1, q_2, q_3) - (3, 0, -\frac{2}{3}) = (0, \frac{1}{2}, -\frac{1}{3})$, $Q = (3, \frac{1}{2}, -1)$

45. $\|\mathbf{v}\| = 0$

46. $\|\mathbf{v}\| = \sqrt{1 + 0 + 9} = \sqrt{10}$

47. $\mathbf{v} = \langle 1, 3, -5\rangle$

$\|\mathbf{v}\| = \sqrt{0 + 9 + 25} = \sqrt{34}$

48. $\mathbf{v} = \langle 1, 3, -2\rangle$

$\|\mathbf{v}\| = \sqrt{1 + 9 + 4} = \sqrt{14}$

49. $\mathbf{v} = \langle 1, -2, -3\rangle$

$\|\mathbf{v}\| = \sqrt{1 + 4 + 9} = \sqrt{14}$

50. $\mathbf{v} = \langle -4, 3, 7\rangle$

$\|\mathbf{v}\| = \sqrt{16 + 9 + 49} = \sqrt{74}$

51. $\mathbf{u} = \langle 2, -1, 2\rangle$, $\|\mathbf{u}\| = \sqrt{4 + 1 + 4} = 3$

$\frac{\mathbf{u}}{\|\mathbf{u}\|} = \langle \frac{2}{3}, -\frac{1}{3}, \frac{2}{3}\rangle$, $-\frac{\mathbf{u}}{\|\mathbf{u}\|} = \langle -\frac{2}{3}, \frac{1}{3}, -\frac{2}{3}\rangle$

52. $\mathbf{u} = \langle 6, 0, 8\rangle$, $\|\mathbf{u}\| = \sqrt{36 + 0 + 64} = 10$

$\dfrac{\mathbf{u}}{\|\mathbf{u}\|} = \langle \dfrac{3}{5}, 0, \dfrac{4}{5}\rangle$, $-\dfrac{\mathbf{u}}{\|\mathbf{u}\|} = \langle -\dfrac{3}{5}, 0, -\dfrac{4}{5}\rangle$

53. $\mathbf{u} = \langle 3, 2, -5\rangle$, $\|\mathbf{u}\| = \sqrt{9 + 4 + 25} = \sqrt{38}$

$\dfrac{\mathbf{u}}{\|\mathbf{u}\|} = \langle \dfrac{3}{\sqrt{38}}, \dfrac{2}{\sqrt{38}}, -\dfrac{5}{\sqrt{38}}\rangle$, $-\dfrac{\mathbf{u}}{\|\mathbf{u}\|} = \langle -\dfrac{3}{\sqrt{38}}, -\dfrac{2}{\sqrt{38}}, \dfrac{5}{\sqrt{38}}\rangle$

54. $\mathbf{u} = \langle 8, 0, 0\rangle$, $\|\mathbf{u}\| = 8$

$\dfrac{\mathbf{u}}{\|\mathbf{u}\|} = \langle 1, 0, 0\rangle$, $-\dfrac{\mathbf{u}}{\|\mathbf{u}\|} = \langle -1, 0, 0\rangle$

55. $c\mathbf{u} = \langle c, 2c, 3c\rangle$, $\|c\mathbf{u}\| = \sqrt{c^2 + 4c^2 + 9c^2} = 1$

$14c^2 = 1$, $c = \pm\dfrac{1}{\sqrt{14}} = \pm\dfrac{\sqrt{14}}{14}$

56. $c\mathbf{v} = \langle 2c, 2c, -c\rangle$, $\|c\mathbf{v}\| = \sqrt{4c^2 + 4c^2 + c^2} = 1$

$9c^2 = 1$, $c = \pm\dfrac{1}{3}$

57. $c\mathbf{v} = \langle 2c, 2c, -c\rangle$, $\|c\mathbf{v}\| = \sqrt{4c^2 + 4c^2 + c^2} = 5$

$9c^2 = 25$, $c = \pm\dfrac{5}{3}$

58. $c\mathbf{u} = \langle c, 2c, 3c\rangle$, $\|c\mathbf{u}\| = \sqrt{c^2 + 4c^2 + 9c^2} = 3$

$14c^2 = 9$, $c = \pm\dfrac{3\sqrt{14}}{14}$

59. $\mathbf{v} = \langle -3, -6, 3\rangle$, $\dfrac{2}{3}\mathbf{v} = \langle -2, -4, 2\rangle$

$(4, 3, 0) + (-2, -4, 2) = (2, -1, 2)$

60. $\mathbf{v} = \langle 8, 0, -2\rangle$, $\dfrac{2}{3}\mathbf{v} = \langle \dfrac{16}{3}, 0, -\dfrac{4}{3}\rangle$

$(-2, 1, 6) + (\dfrac{16}{3}, 0, -\dfrac{4}{3}) = (\dfrac{10}{3}, 1, \dfrac{14}{3})$

61. $\mathbf{v} = \langle 5, 6, -3\rangle$, $\dfrac{2}{3}\mathbf{v} = \langle \dfrac{10}{3}, 4, -2\rangle$

$(1, 2, 5) + (\dfrac{10}{3}, 4, -2) = (\dfrac{13}{3}, 6, 3)$

62. $\mathbf{v} = \langle 21, 11, -6\rangle$, $\quad \frac{2}{3}\mathbf{v} = \langle 14, \frac{22}{3}, -4\rangle$

$(-9, -8, 5) + (14, \frac{22}{3}, -4) = (5, -\frac{2}{3}, 1)$

63. $\mathbf{v} = 2[\cos(\pm 30^\circ)\mathbf{j} + \sin(\pm 30^\circ)\mathbf{k}]$

$= \sqrt{3}\mathbf{j} \pm \mathbf{k}$

$= \langle 0, \sqrt{3}, \pm 1\rangle$

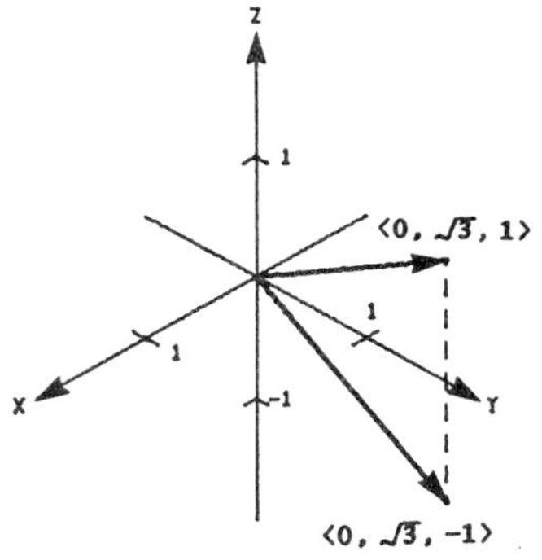

64. $\mathbf{v} = 5(\cos 45^\circ \mathbf{i} + \sin 45^\circ \mathbf{k})$

$= \frac{5\sqrt{2}}{2}(\mathbf{i} + \mathbf{k})$

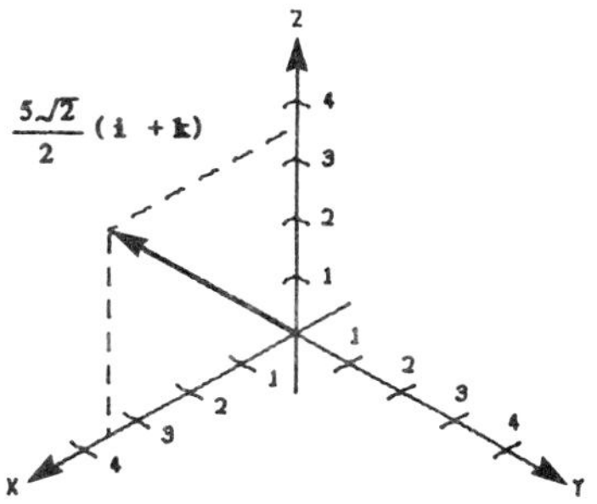

65. Let α be the angle between $\mathbf{v}$ and the coordinate axes.

$\mathbf{v} = (\cos\alpha)\mathbf{i} + (\cos\alpha)\mathbf{j} + (\cos\alpha)\mathbf{k}$

$\|\mathbf{v}\| = \sqrt{3}\cos\alpha = 1$

$\cos\alpha = \frac{1}{\sqrt{3}} = \frac{\sqrt{3}}{3}$

$\mathbf{v} = \frac{\sqrt{3}}{3}(\mathbf{i} + \mathbf{j} + \mathbf{k})$

$= \frac{\sqrt{3}}{3}\langle 1, 1, 1\rangle$

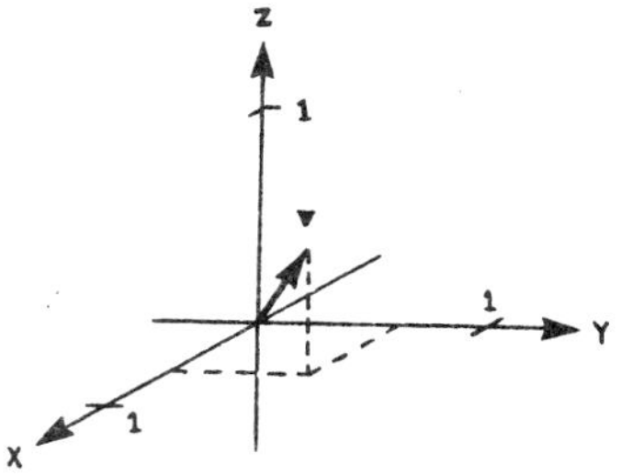

66. Multiplying the vector in Exercise 65 by the scalar 5 we have

$\mathbf{v} = \frac{5\sqrt{3}}{3}(\mathbf{i} + \mathbf{j} + \mathbf{k})$

$= \frac{5\sqrt{3}}{3}\langle 1, 1, 1\rangle$

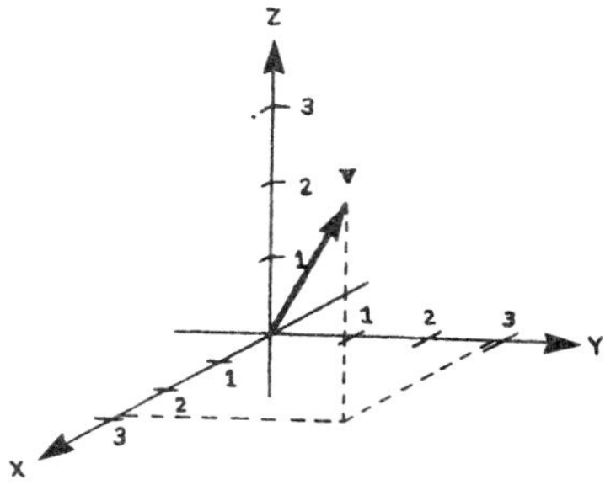

67.(a)

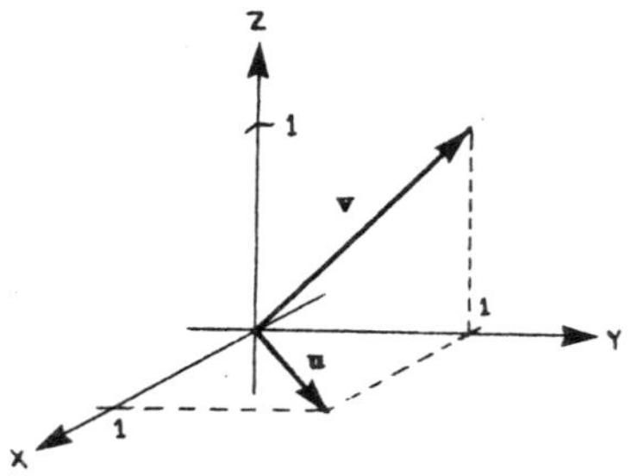

(b) $\mathbf{w} = a\mathbf{u} + b\mathbf{v}$

$= a\mathbf{i} + (a + b)\mathbf{j} + b\mathbf{k} = 0$

$a = 0, \quad a + b = 0, \quad b = 0$

Thus a and b are both zero

(c) $a\mathbf{i} + (a + b)\mathbf{j} + b\mathbf{k} = \mathbf{i} + 2\mathbf{j} + \mathbf{k}$

$a = 1, \quad b = 1$

$\mathbf{w} = \mathbf{u} + \mathbf{v}$

(d) $a\mathbf{i} + (a + b)\mathbf{j} + b\mathbf{k} = \mathbf{i} + 2\mathbf{j} + 3\mathbf{k}$

$a = 1, \quad a + b = 2, \quad b = 3$

Not possible

68. $\sqrt{(x - x_1)^2 + (y - y_1)^2 + (z - z_1)^2} = 4$

Sphere: $(x - x_1)^2 + (y - y_1)^2 + (z - z_1)^2 = 16$

69. $550 = \|c(75\mathbf{i} - 50\mathbf{j} - 100\mathbf{k})\|, \qquad 302500 = 18125c^2$

$c^2 = 16.689655, \qquad c \approx 4.085$

$\mathbf{F} \approx 4.085(75\mathbf{i} - 50\mathbf{j} - 100\mathbf{k}) \approx 306\mathbf{i} - 204\mathbf{j} - 409\mathbf{k}$

70. $\overrightarrow{PQ} = \langle 0, -18, 6\sqrt{55}\rangle$

$\mathbf{F} = c\langle 0, -18, 6\sqrt{55}\rangle$

$c6\sqrt{55} = \dfrac{25}{3}$

$c = \dfrac{5\sqrt{55}}{198}$

$\mathbf{F} = \langle 0, -\dfrac{5\sqrt{55}}{11}, \dfrac{25}{3}\rangle$

$\|\mathbf{F}\| \approx 8.99$ lb

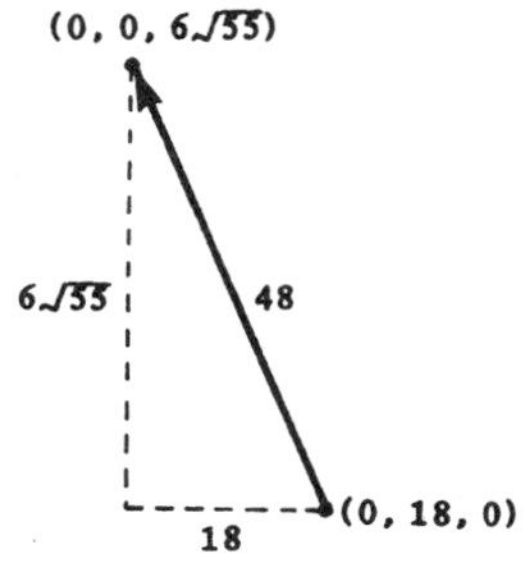

12.3 The dot product of two vectors

1. $\mathbf{u} = \langle 3, 4\rangle$, $\mathbf{v} = \langle 2, -3\rangle$

(a) $\mathbf{u}\cdot\mathbf{v} = 3(2) + 4(-3) = -6$
(b) $\mathbf{u}\cdot\mathbf{u} = 3(3) + 4(4) = 25$
(c) $\|\mathbf{u}\|^2 = 25$
(d) $(\mathbf{u}\cdot\mathbf{v})\mathbf{v} = -6\langle 2, -3\rangle = \langle -12, 18\rangle$
(e) $\mathbf{u}\cdot(2\mathbf{v}) = 2(\mathbf{u}\cdot\mathbf{v}) = 2(-6) = -12$

2. $\mathbf{u} = \langle 5, 12\rangle$, $\mathbf{v} = \langle -3, 2\rangle$

(a) $\mathbf{u}\cdot\mathbf{v} = 5(-3) + 12(2) = 9$
(b) $\mathbf{u}\cdot\mathbf{u} = 5(5) + 12(12) = 169$
(c) $\|\mathbf{u}\|^2 = 169$
(d) $(\mathbf{u}\cdot\mathbf{v})\mathbf{v} = 9\langle -3, 2\rangle = \langle -27, 18\rangle$
(e) $\mathbf{u}\cdot(2\mathbf{v}) = 2(\mathbf{u}\cdot\mathbf{v}) = 2(9) = 18$

3. $\mathbf{u} = \langle 2, -3, 4\rangle$, $\mathbf{v} = \langle 0, 6, 5\rangle$

(a) $\mathbf{u}\cdot\mathbf{v} = 2(0) + (-3)(6) + (4)(5) = 2$
(b) $\mathbf{u}\cdot\mathbf{u} = 2(2) + (-3)(-3) + 4(4) = 29$
(c) $\|\mathbf{u}\|^2 = 29$
(d) $(\mathbf{u}\cdot\mathbf{v})\mathbf{v} = 2\langle 0, 6, 5\rangle = \langle 0, 12, 10\rangle$
(e) $\mathbf{u}\cdot(2\mathbf{v}) = 2(\mathbf{u}\cdot\mathbf{v}) = 2(2) = 4$

4. $\mathbf{u} = \mathbf{i}$, $\mathbf{v} = \mathbf{i}$

(a) $\mathbf{u}\cdot\mathbf{v} = 1$
(b) $\mathbf{u}\cdot\mathbf{u} = 1$
(c) $\|\mathbf{u}\|^2 = 1$
(d) $(\mathbf{u}\cdot\mathbf{v})\mathbf{v} = \mathbf{i}$
(e) $\mathbf{u}\cdot(2\mathbf{v}) = 2(\mathbf{u}\cdot\mathbf{v}) = 2$

5. $\mathbf{u} = 2\mathbf{i} - \mathbf{j} + \mathbf{k}$, $\mathbf{v} = \mathbf{i} - \mathbf{k}$

(a) $\mathbf{u}\cdot\mathbf{v} = 2(1) + (-1)(0) + 1(-1) = 1$
(b) $\mathbf{u}\cdot\mathbf{u} = 2(2) + (-1)(-1) + (1)(1) = 6$
(c) $\|\mathbf{u}\|^2 = 6$
(d) $(\mathbf{u}\cdot\mathbf{v})\mathbf{v} = \mathbf{v} = \mathbf{i} - \mathbf{k}$
(e) $\mathbf{u}\cdot(2\mathbf{v}) = 2(\mathbf{u}\cdot\mathbf{v}) = 2$

6. $\mathbf{u} = 2\mathbf{i} + \mathbf{j} - 2\mathbf{k}$, $\mathbf{v} = \mathbf{i} - 3\mathbf{j} + 2\mathbf{k}$

(a) $\mathbf{u}\cdot\mathbf{v} = 2(1) + 1(-3) + (-2)(2) = -5$
(b) $\mathbf{u}\cdot\mathbf{u} = 2(2) + 1(1) + (-2)(-2) = 9$
(c) $\|\mathbf{u}\|^2 = 9$
(d) $(\mathbf{u}\cdot\mathbf{v})\mathbf{v} = -5(\mathbf{i} - 3\mathbf{j} + 2\mathbf{k}) = -5\mathbf{i} + 15\mathbf{j} - 10\mathbf{k}$
(e) $\mathbf{u}\cdot(2\mathbf{v}) = 2(\mathbf{u}\cdot\mathbf{v}) = 2(-5) = -10$

7. $\mathbf{u} = \langle 1, 1\rangle$, $\mathbf{v} = \langle 2, -2\rangle$

$$\cos\theta = \frac{\mathbf{u}\cdot\mathbf{v}}{\|\mathbf{u}\|\ \|\mathbf{v}\|} = \frac{0}{\sqrt{2}\sqrt{8}} = 0$$

$$\theta = \frac{\pi}{2}$$

8. $\mathbf{u} = \langle 3, 1\rangle$, $\mathbf{v} = \langle 2, -1\rangle$

$$\cos\theta = \frac{\mathbf{u}\cdot\mathbf{v}}{\|\mathbf{u}\|\ \|\mathbf{v}\|} = \frac{5}{\sqrt{10}\sqrt{5}} = \frac{1}{\sqrt{2}}$$

$$\theta = \frac{\pi}{4}$$

9. $\mathbf{u} = 3\mathbf{i} + \mathbf{j}, \quad \mathbf{v} = -2\mathbf{i} + 4\mathbf{j}$

$$\cos\theta = \frac{\mathbf{u}\cdot\mathbf{v}}{\|\mathbf{u}\|\ \|\mathbf{v}\|} = \frac{-2}{\sqrt{10}\sqrt{20}} = \frac{-1}{5\sqrt{2}}$$

$$\theta = \arccos\left(-\frac{1}{5\sqrt{2}}\right) \approx 98.13^\circ$$

10. $\mathbf{u} = \cos\left(\frac{\pi}{6}\right)\mathbf{i} + \sin\left(\frac{\pi}{6}\right)\mathbf{j} = \frac{\sqrt{3}}{2}\mathbf{i} + \frac{1}{2}\mathbf{j}$

$$\mathbf{v} = \cos\left(\frac{3\pi}{4}\right)\mathbf{i} + \sin\left(\frac{3\pi}{4}\right)\mathbf{j} = -\frac{\sqrt{2}}{2}\mathbf{i} + \frac{\sqrt{2}}{2}\mathbf{j}$$

$$\cos\theta = \frac{\mathbf{u}\cdot\mathbf{v}}{\|\mathbf{u}\|\ \|\mathbf{v}\|} = \frac{\sqrt{3}}{2}\left(-\frac{\sqrt{2}}{2}\right) + \frac{1}{2}\left(\frac{\sqrt{2}}{2}\right) = \frac{\sqrt{2}}{4}(1 - \sqrt{3})$$

$$\theta = \arccos\left[\frac{\sqrt{2}}{4}(1 - \sqrt{3})\right] = 105^\circ$$

11. $\mathbf{u} = \langle 1, 1, 1\rangle, \quad \mathbf{v} = \langle 2, 1, -1\rangle$

$$\cos\theta = \frac{\mathbf{u}\cdot\mathbf{v}}{\|\mathbf{u}\|\ \|\mathbf{v}\|} = \frac{2}{\sqrt{3}\sqrt{6}} = \frac{\sqrt{2}}{3}$$

$$\theta = \arccos\frac{\sqrt{2}}{3} \approx 61.87^\circ$$

12. $\mathbf{u} = 2\mathbf{i} + 3\mathbf{j} + \mathbf{k}, \quad \mathbf{v} = -3\mathbf{i} + 2\mathbf{j}$

$$\cos\theta = \frac{\mathbf{u}\cdot\mathbf{v}}{\|\mathbf{u}\|\ \|\mathbf{v}\|} = 0$$

$$\theta = \frac{\pi}{2}$$

13. $\mathbf{u} = 3\mathbf{i} + 4\mathbf{j}, \quad \mathbf{v} = -2\mathbf{j} + 3\mathbf{k}$

$$\cos\theta = \frac{\mathbf{u}\cdot\mathbf{v}}{\|\mathbf{u}\|\ \|\mathbf{v}\|} = \frac{-8}{5\sqrt{13}} = \frac{-8\sqrt{13}}{65}$$

$$\theta = \arccos\left(-\frac{8\sqrt{13}}{65}\right) \approx 116.34^\circ$$

14. $\mathbf{u} = 2\mathbf{i} - 3\mathbf{j} + \mathbf{k}, \quad \mathbf{v} = \mathbf{i} - 2\mathbf{j} + \mathbf{k}$

$$\cos\theta = \frac{\mathbf{u}\cdot\mathbf{v}}{\|\mathbf{u}\|\ \|\mathbf{v}\|} = \frac{9}{\sqrt{14}\sqrt{6}} = \frac{9}{2\sqrt{21}} = \frac{9\sqrt{21}}{42}$$

$$\theta = \arccos\left(\frac{9\sqrt{21}}{42}\right) \approx 10.89^\circ$$

15. $\mathbf{u} = \langle 4, 0\rangle, \quad \mathbf{v} = \langle 1, 1\rangle$

$\mathbf{u} \neq c\mathbf{v}$ ⇨ not parallel

$\mathbf{u}\cdot\mathbf{v} = 4 \neq 0$ ⇨ not orthogonal

neither

16. $\mathbf{u} = \langle 2, 18\rangle, \quad \mathbf{v} = \langle \frac{3}{2}, -\frac{1}{6}\rangle$

$\mathbf{u} \neq c\mathbf{v}$ ⇨ not parallel

$\mathbf{u}\cdot\mathbf{v} = 0$ ⇨ orthogonal

17. $\mathbf{u} = \langle 4, 3\rangle, \quad \mathbf{v} = \langle \frac{1}{2}, -\frac{2}{3}\rangle$

$\mathbf{u} \neq c\mathbf{v}$ ⇨ not parallel

$\mathbf{u}\cdot\mathbf{v} = 0$ ⇨ orthogonal

18. $\mathbf{u} = -\frac{1}{3}(\mathbf{i} - 2\mathbf{j}), \quad \mathbf{v} = 2\mathbf{i} - 4\mathbf{j}$

$\mathbf{u} = -\frac{1}{6}\mathbf{v}$ ⇨ parallel

19. $\mathbf{u} = \mathbf{j} + 6\mathbf{k}$, $\mathbf{v} = \mathbf{i} - 2\mathbf{j} - \mathbf{k}$

$\mathbf{u} \neq c\mathbf{v}$ ⇨ not parallel

$\mathbf{u}\cdot\mathbf{v} = -8 \neq 0$ ⇨ not orthogonal

neither

20. $\mathbf{u} = -2\mathbf{i} + 3\mathbf{j} - \mathbf{k}$, $\mathbf{v} = 2\mathbf{i} + \mathbf{j} - \mathbf{k}$

$\mathbf{u} \neq c\mathbf{v}$ ⇨ not parallel

$\mathbf{u}\cdot\mathbf{v} = 0$ ⇨ orthogonal

21. $\mathbf{u} = \langle 2, -3, 1\rangle$,

$\mathbf{v} = \langle -1, -1, -1\rangle$

$\mathbf{u} \neq c\mathbf{v}$ ⇨ not parallel

$\mathbf{u}\cdot\mathbf{v} = 0$ ⇨ orthogonal

22. $\mathbf{u} = \langle \cos\theta, \sin\theta, -1\rangle$

$\mathbf{v} = \langle \sin\theta, -\cos\theta, 0\rangle$

$\mathbf{u} \neq c\mathbf{v}$ ⇨ not parallel

$\mathbf{u}\cdot\mathbf{v} = 0$ ⇨ orthogonal

23. $\mathbf{u} = \mathbf{i} + 2\mathbf{j} + 2\mathbf{k}$, $\|\mathbf{u}\| = 3$

$\cos\alpha = \frac{1}{3}$, $\cos\beta = \frac{2}{3}$, $\cos\gamma = \frac{2}{3}$

$\cos^2\alpha + \cos^2\beta + \cos^2\gamma = \frac{1}{9} + \frac{4}{9} + \frac{4}{9} = 1$

24. $\mathbf{u} = 3\mathbf{i} - \mathbf{j} + 5\mathbf{k}$, $\|\mathbf{u}\| = \sqrt{35}$

$\cos\alpha = \frac{3}{\sqrt{35}}$, $\cos\beta = -\frac{1}{\sqrt{35}}$, $\cos\gamma = \frac{5}{\sqrt{35}}$

$\cos^2\alpha + \cos^2\beta + \cos^2\gamma = \frac{9}{35} + \frac{1}{35} + \frac{25}{35} = 1$

25. $\mathbf{u} = \langle 0, 6, -4\rangle$, $\|\mathbf{u}\| = \sqrt{52} = 2\sqrt{13}$

$\cos\alpha = 0$, $\cos\beta = \frac{3}{\sqrt{13}}$, $\cos\gamma = -\frac{2}{\sqrt{13}}$

$\cos^2\alpha + \cos^2\beta + \cos^2\gamma = 0 + \frac{9}{13} + \frac{4}{13} = 1$

26. $\mathbf{u} = \langle a, b, c\rangle$, $\|\mathbf{u}\| = \sqrt{a^2 + b^2 + c^2}$, $\cos\alpha = \frac{a}{\sqrt{a^2 + b^2 + c^2}}$

$\cos\beta = \frac{b}{\sqrt{a^2 + b^2 + c^2}}$, $\cos\gamma = \frac{c}{\sqrt{a^2 + b^2 + c^2}}$

$$\cos^2\alpha + \cos^2\beta + \cos^2\gamma = \frac{a^2}{a^2 + b^2 + c^2} + \frac{b^2}{a^2 + b^2 + c^2} + \frac{c^2}{a^2 + b^2 + c^2}$$
$$= 1$$

27. $\mathbf{u} = \langle 2, 3\rangle$, $\mathbf{v} = \langle 5, 1\rangle$

(a) $\mathbf{w}_1 = \left(\frac{\mathbf{u}\cdot\mathbf{v}}{\|\mathbf{v}\|^2}\right)\mathbf{v} = \frac{13}{26}\langle 5, 1\rangle$

$= \langle \frac{5}{2}, \frac{1}{2}\rangle$

(b) $\mathbf{w}_2 = \mathbf{u} - \mathbf{w}_1 = \langle -\frac{1}{2}, \frac{5}{2}\rangle$

28. $\mathbf{u} = \langle 2, -3\rangle$, $\mathbf{v} = \langle 3, 2\rangle$

(a) $\mathbf{w}_1 = \left(\frac{\mathbf{u}\cdot\mathbf{v}}{\|\mathbf{v}\|^2}\right)\mathbf{v} = 0\mathbf{v} = \langle 0, 0\rangle$

(b) $\mathbf{w}_2 = \mathbf{u} - \mathbf{w}_1 = \langle 2, -3\rangle$

29. $\mathbf{u} = \langle 2, 1, 2\rangle$, $\mathbf{v} = \langle 0, 3, 4\rangle$

(a) $\mathbf{w}_1 = \left(\frac{\mathbf{u}\cdot\mathbf{v}}{\|\mathbf{v}\|^2}\right)\mathbf{v} = \frac{11}{25}\langle 0, 3, 4\rangle$

$= \langle 0, \frac{33}{25}, \frac{44}{25}\rangle$

(b) $\mathbf{w}_2 = \mathbf{u} - \mathbf{w}_1 = \langle 2, -\frac{8}{25}, \frac{6}{25}\rangle$

30. $\mathbf{u} = \langle 0, 4, 1\rangle$, $\mathbf{v} = \langle 0, 2, 3\rangle$

(a) $\mathbf{w}_1 = \left(\frac{\mathbf{u}\cdot\mathbf{v}}{\|\mathbf{v}\|^2}\right)\mathbf{v} = \frac{11}{13}\langle 0, 2, 3\rangle$

$= \langle 0, \frac{22}{13}, \frac{33}{13}\rangle$

(b) $\mathbf{w}_2 = \mathbf{u} - \mathbf{w}_1 = \langle 0, \frac{30}{13}, -\frac{20}{13}\rangle$

31. $\mathbf{u} = \langle 1, 1, 1\rangle$, $\mathbf{v} = \langle -2, -1, 1\rangle$

(a) $\mathbf{w}_1 = \left(\frac{\mathbf{u}\cdot\mathbf{v}}{\|\mathbf{v}\|^2}\right)\mathbf{v} = \frac{-2}{6}\langle -2, -1, 1\rangle$

$= \langle \frac{2}{3}, \frac{1}{3}, -\frac{1}{3}\rangle$

(b) $\mathbf{w}_2 = \mathbf{u} - \mathbf{w}_1 = \langle \frac{1}{3}, \frac{2}{3}, \frac{4}{3}\rangle$

32. $\mathbf{u} = \langle -2, -1, 1\rangle$, $\mathbf{v} = \langle 1, 1, 1\rangle$

(a) $\mathbf{w}_1 = \left(\frac{\mathbf{u}\cdot\mathbf{v}}{\|\mathbf{v}\|^2}\right)\mathbf{v} = -\frac{2}{3}\langle 1, 1, 1\rangle$

$= \langle -\frac{2}{3}, -\frac{2}{3}, -\frac{2}{3}\rangle$

(b) $\mathbf{w}_2 = \mathbf{u} - \mathbf{w}_1 = \langle -\frac{4}{3}, -\frac{1}{3}, \frac{5}{3}\rangle$

33. $\mathbf{u} = \langle 5, -4, 3\rangle$, $\mathbf{v} = \langle 1, 0, 0\rangle$

(a) $\mathbf{w}_1 = \left(\frac{\mathbf{u}\cdot\mathbf{v}}{\|\mathbf{v}\|^2}\right)\mathbf{v}$

$= 5\langle 1, 0, 0\rangle = \langle 5, 0, 0\rangle$

(b) $\mathbf{w}_2 = \mathbf{u} - \mathbf{w}_1 = \langle 0, -4, 3\rangle$

34. $\mathbf{u} = \langle 5, -4, 3\rangle$, $\mathbf{v} = \langle 0, 1, 0\rangle$

(a) $\mathbf{w}_1 = \left(\frac{\mathbf{u}\cdot\mathbf{v}}{\|\mathbf{v}\|^2}\right)\mathbf{v} = -4\ \langle 0, 1, 0\rangle$

$= \langle 0, -4, 0\rangle$

(b) $\mathbf{w}_2 = \mathbf{u} - \mathbf{w}_1 = \langle 5, 0, 3\rangle$

35.(a) Gravitational Force $\mathbf{F} = -32{,}000\mathbf{j}$

$\mathbf{v} = \cos 15^\circ \mathbf{i} + \sin 15^\circ \mathbf{j}$

$$\mathbf{w}_1 = \frac{\mathbf{F}\cdot\mathbf{v}}{\|\mathbf{v}\|^2}\mathbf{v} = (\mathbf{F}\cdot\mathbf{v})\mathbf{v} = (-32{,}000)(\sin 15^\circ)\mathbf{v}$$

$$\approx -8282.2(\cos 15^\circ \mathbf{i} + \sin 15^\circ \mathbf{j})$$

$\|\mathbf{w}_1\| = 8282.2$ lb

(b) $\mathbf{w}_2 = \mathbf{F} - \mathbf{w}_1 = -32{,}000\mathbf{j} + 8282.2(\cos 15^\circ \mathbf{i} + \sin 15^\circ \mathbf{j})$

$\|\mathbf{w}_2\| = 30{,}909.6$ lb

36. $\mathbf{v} = -75\mathbf{i} + 50\mathbf{j} + 100\mathbf{k}$, $\|\mathbf{v}\| = \sqrt{18125} = 25\sqrt{29}$

$$\cos\alpha = -\frac{75}{25\sqrt{29}} = -\frac{3}{\sqrt{29}}, \quad \alpha \approx 123.85^\circ$$

$$\cos\beta = \frac{50}{25\sqrt{29}} = \frac{2}{\sqrt{29}}, \quad \beta \approx 68.20^\circ$$

$$\cos\gamma = \frac{100}{25\sqrt{29}} = \frac{4}{\sqrt{29}}, \quad \gamma \approx 42.03^\circ$$

37. $\mathbf{F} = 85(\frac{1}{2}\mathbf{i} + \frac{\sqrt{3}}{2}\mathbf{j})$, $\mathbf{v} = 10\mathbf{i}$, $W = \mathbf{F}\cdot\mathbf{v} = 425$ ft·lbs

38. $\mathbf{F} = 15(\frac{\sqrt{3}}{2}\mathbf{i} + \frac{1}{2}\mathbf{j})$, $\mathbf{v} = 50\mathbf{i}$, $W = \mathbf{F}\cdot\mathbf{v} = 375\sqrt{3}$ ft·lbs

39. $\overrightarrow{PQ} = \langle 4, 7, 5\rangle$, $\mathbf{v} = \langle 1, 4, 8\rangle$, $W = \overrightarrow{PQ}\cdot\mathbf{v} = 72$

40. $\overrightarrow{PQ} = \langle -4, 2, 10\rangle$, $\mathbf{v} = \langle -2, 3, 6\rangle$, $W = \overrightarrow{PQ}\cdot\mathbf{v} = 74$

41. $\mathbf{u} = \langle 3240, 1450, 2235\rangle$, $\mathbf{v} = \langle 2.22, 1.85, 3.25\rangle$, $\mathbf{u}\cdot\mathbf{v} = \$17{,}139.05$

This gives the amount that the farmer earned on his crops for that year.

42. Let s = length of a side

$\mathbf{v} = \langle s, s, s\rangle$, $\|\mathbf{v}\| = s\sqrt{3}$

$$\cos\alpha = \cos\beta = \cos\gamma = \frac{s}{s\sqrt{3}} = \frac{1}{\sqrt{3}}$$

$$\alpha = \beta = \gamma = \arccos\left(\frac{1}{\sqrt{3}}\right) \approx 54.7^\circ$$

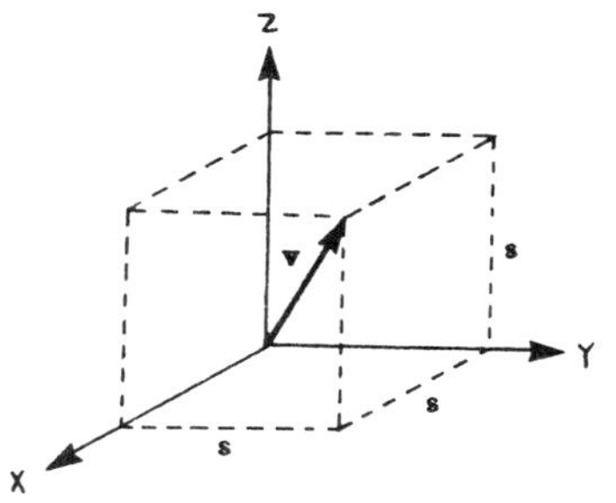

43.(a) $\mathbf{u}\cdot\mathbf{v} = 0 \Longrightarrow$ $\mathbf{u}$ and $\mathbf{v}$ are orthogonal and $\theta = \frac{\pi}{2}$

(b) $\mathbf{u}\cdot\mathbf{v} > 0 \Longrightarrow \cos\theta > 0 \Longrightarrow 0 < \theta < \frac{\pi}{2}$

(c) $\mathbf{u}\cdot\mathbf{v} < 0 \Longrightarrow \cos\theta < 0 \Longrightarrow \frac{\pi}{2} < \theta < \pi$

44.(a) $\left(\frac{\mathbf{u}\cdot\mathbf{v}}{\|\mathbf{v}\|^2}\right)\mathbf{v} = \mathbf{u} \Longrightarrow \mathbf{u} = c\mathbf{v} \Longrightarrow$ $\mathbf{u}$ and $\mathbf{v}$ are parallel

(b) $\left(\frac{\mathbf{u}\cdot\mathbf{v}}{\|\mathbf{v}\|^2}\right)\mathbf{v} = 0 \Longrightarrow \mathbf{u}\cdot\mathbf{v} = 0 \Longrightarrow$ $\mathbf{u}$ and $\mathbf{v}$ are orthogonal

45. In a rhombus $\|\mathbf{u}\| = \|\mathbf{v}\|$

The diagonals are $\mathbf{u} + \mathbf{v}$ and $\mathbf{u} - \mathbf{v}$

$(\mathbf{u} + \mathbf{v})\cdot(\mathbf{u} - \mathbf{v}) = (\mathbf{u} + \mathbf{v})\cdot\mathbf{u} - (\mathbf{u} + \mathbf{v})\cdot\mathbf{v}$

$= \mathbf{u}\cdot\mathbf{u} + \mathbf{v}\cdot\mathbf{u} - \mathbf{u}\cdot\mathbf{v} - \mathbf{v}\cdot\mathbf{v}$

$= \|\mathbf{u}\|^2 - \|\mathbf{v}\|^2 = 0$

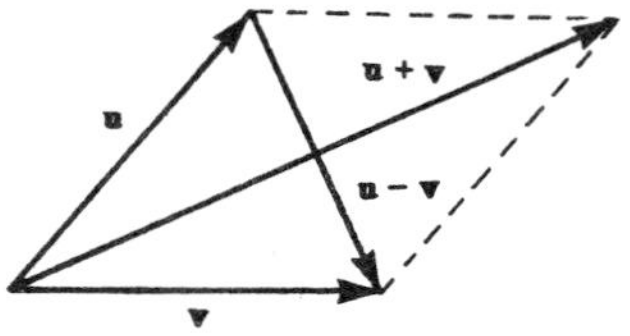

Therefore the diagonals are orthogonal

46. $\|\mathbf{u} - \mathbf{v}\|^2 = (\mathbf{u} - \mathbf{v})\cdot(\mathbf{u} - \mathbf{v}) = (\mathbf{u} - \mathbf{v})\cdot\mathbf{u} - (\mathbf{u} - \mathbf{v})\cdot\mathbf{v}$

$= \mathbf{u}\cdot\mathbf{u} - \mathbf{v}\cdot\mathbf{u} - \mathbf{u}\cdot\mathbf{v} + \mathbf{v}\cdot\mathbf{v} = \|\mathbf{u}\|^2 - \mathbf{u}\cdot\mathbf{v} - \mathbf{u}\cdot\mathbf{v} + \|\mathbf{v}\|^2$

$= \|\mathbf{u}\|^2 + \|\mathbf{v}\|^2 - 2\mathbf{u}\cdot\mathbf{v}$

47. Property 2: $\mathbf{u}\cdot(\mathbf{v} + \mathbf{w}) = \mathbf{u}\cdot\mathbf{v} + \mathbf{u}\cdot\mathbf{w}$

Let $\mathbf{u} = \langle u_1, u_2, u_3\rangle$, $\mathbf{v} = \langle v_1, v_2, v_3\rangle$, $\mathbf{w} = \langle w_1, w_2, w_3\rangle$

$\mathbf{u}\cdot(\mathbf{v} + \mathbf{w}) = \langle u_1, u_2, u_3\rangle\cdot\langle v_1 + w_1, v_2 + w_2, v_3 + w_3\rangle$

$= u_1(v_1 + w_1) + u_2(v_2 + w_2) + u_3(v_3 + w_3)$

$= u_1v_1 + u_1w_1 + u_2v_2 + u_2w_2 + u_3v_3 + u_3w_3$

$= (u_1v_1 + u_2v_2 + u_3v_3) + (u_1w_1 + u_2w_2 + u_3w_3)$

$= \mathbf{u}\cdot\mathbf{v} + \mathbf{u}\cdot\mathbf{w}$

47. (continued)

Property 3: $c(\mathbf{u}\cdot\mathbf{v}) = (c\mathbf{u})\cdot\mathbf{v} = \mathbf{u}\cdot(c\mathbf{v})$

Let $\mathbf{u} = \langle u_1, u_2, u_3\rangle$, $\mathbf{v} = \langle v_1, v_2, v_3\rangle$, c a scalar

$$c(\mathbf{u}\cdot\mathbf{v}) = c(u_1v_1 + u_2v_2 + u_3v_3)$$
$$= cu_1v_1 + cu_2v_2 + cu_3v_3$$
$$= (cu_1)v_1 + (cu_2)v_2 + (cu_3)v_3 = (c\mathbf{u})\cdot\mathbf{v}$$
$$= u_1(cv_1) + u_2(cv_2) + u_3(cv_3) = \mathbf{u}(c\mathbf{v})$$

Property 4: $\mathbf{0}\cdot\mathbf{v} = 0$

Let $\mathbf{v} = \langle v_1, v_2, v_3\rangle$, $\mathbf{0} = \langle 0, 0, 0\rangle$

$$\mathbf{0}\cdot\mathbf{v} = 0(v_1) + 0(v_2) + 0(v_3) = 0$$

48. Since $\mathbf{u}$ is orthogonal to both $\mathbf{v}$ and $\mathbf{w}$ we have

$\mathbf{u}\cdot\mathbf{v} = 0$ and $\mathbf{u}\cdot\mathbf{w} = 0$

$$\mathbf{u}\cdot(c\mathbf{v} + d\mathbf{w}) = \mathbf{u}\cdot(c\mathbf{v}) + \mathbf{u}\cdot(d\mathbf{w}) = c(\mathbf{u}\cdot\mathbf{v}) + d(\mathbf{u}\cdot\mathbf{w}) = c(0) + d(0) = 0$$

Therefore $\mathbf{u}$ is orthogonal to $c\mathbf{v} + d\mathbf{w}$

12.4
The cross product of two vectors in space

1. $\mathbf{j}\times\mathbf{i} = \begin{vmatrix} \mathbf{i} & \mathbf{j} & \mathbf{k} \\ 0 & 1 & 0 \\ 1 & 0 & 0 \end{vmatrix} = -\mathbf{k}$

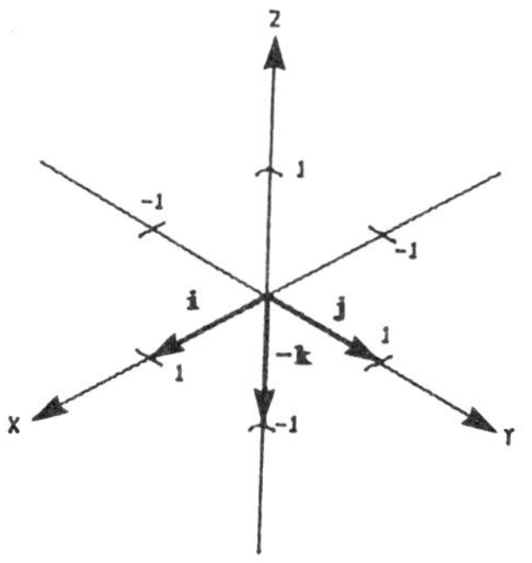

2. $\mathbf{i}\times\mathbf{j} = \begin{vmatrix} \mathbf{i} & \mathbf{j} & \mathbf{k} \\ 1 & 0 & 0 \\ 0 & 1 & 0 \end{vmatrix} = \mathbf{k}$

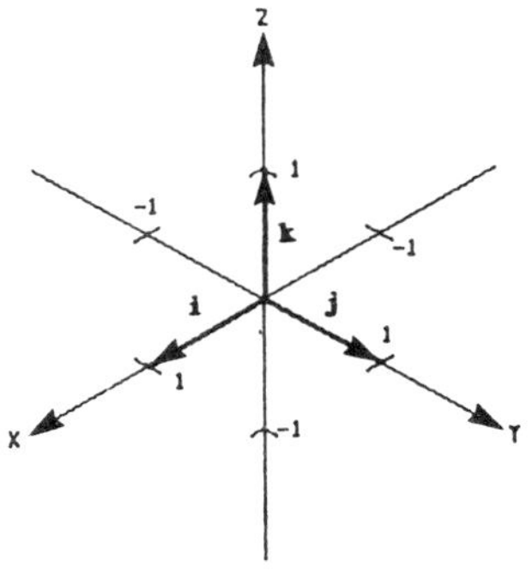

3. $\mathbf{j} \times \mathbf{k} = \begin{vmatrix} \mathbf{i} & \mathbf{j} & \mathbf{k} \\ 0 & 1 & 0 \\ 0 & 0 & 1 \end{vmatrix} = \mathbf{i}$

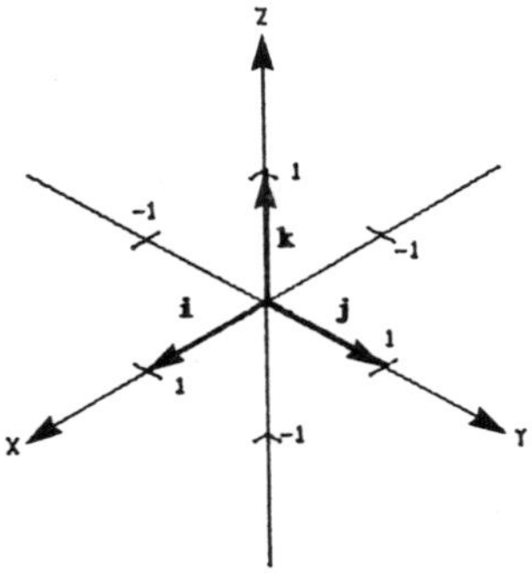

4. $\mathbf{k} \times \mathbf{j} = \begin{vmatrix} \mathbf{i} & \mathbf{j} & \mathbf{k} \\ 0 & 0 & 1 \\ 0 & 1 & 0 \end{vmatrix} = -\mathbf{i}$

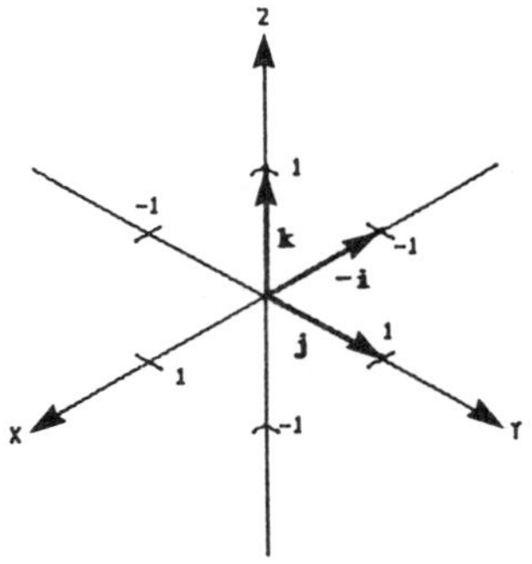

5. $\mathbf{i} \times \mathbf{k} = \begin{vmatrix} \mathbf{i} & \mathbf{j} & \mathbf{k} \\ 1 & 0 & 0 \\ 0 & 0 & 1 \end{vmatrix} = -\mathbf{j}$

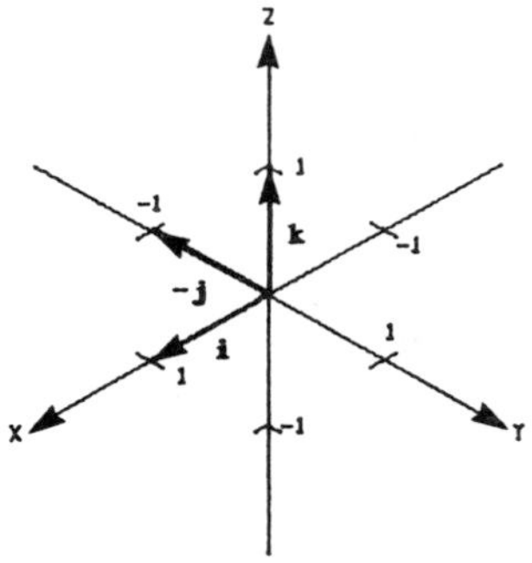

6. $\mathbf{k} \times \mathbf{i} = \begin{vmatrix} \mathbf{i} & \mathbf{j} & \mathbf{k} \\ 0 & 0 & 1 \\ 1 & 0 & 0 \end{vmatrix} = \mathbf{j}$

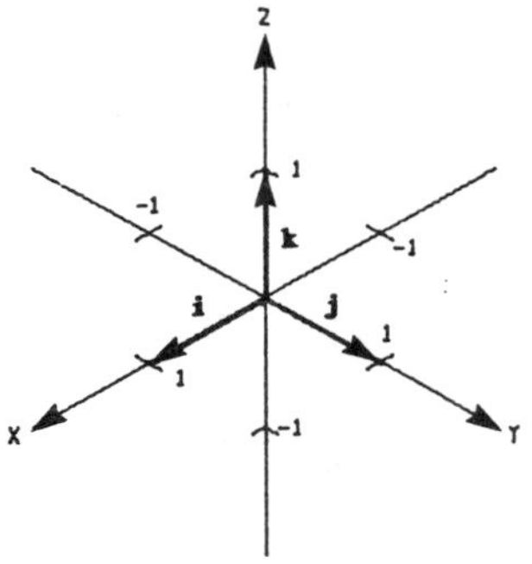

7. $\mathbf{u} = \langle 2, -3, 1 \rangle, \quad \mathbf{v} = \langle 1, -2, 1 \rangle$

$$\mathbf{u} \times \mathbf{v} = \begin{vmatrix} \mathbf{i} & \mathbf{j} & \mathbf{k} \\ 2 & -3 & 1 \\ 1 & -2 & 1 \end{vmatrix} = -\mathbf{i} - \mathbf{j} - \mathbf{k} = \langle -1, -1, -1 \rangle$$

$\mathbf{u} \cdot (\mathbf{u} \times \mathbf{v}) = 2(-1) + (-3)(-1) + (1)(-1) = 0 \implies \mathbf{u} \perp \mathbf{u} \times \mathbf{v}$

$\mathbf{v} \cdot (\mathbf{u} \times \mathbf{v}) = 1(-1) + (-2)(-1) + (1)(-1) = 0 \implies \mathbf{v} \perp \mathbf{u} \times \mathbf{v}$

8. $\mathbf{u} = \langle -1, 1, 2 \rangle, \quad \mathbf{v} = \langle 0, 1, 0 \rangle$

$$\mathbf{u} \times \mathbf{v} = \begin{vmatrix} \mathbf{i} & \mathbf{j} & \mathbf{k} \\ -1 & 1 & 2 \\ 0 & 1 & 0 \end{vmatrix} = -2\mathbf{i} - \mathbf{k} = \langle -2, 0, -1 \rangle$$

$\mathbf{u} \cdot (\mathbf{u} \times \mathbf{v}) = (-1)(-2) + (1)(0) + (2)(-1) = 0 \implies \mathbf{u} \perp \mathbf{u} \times \mathbf{v}$

$\mathbf{v} \cdot (\mathbf{u} \times \mathbf{v}) = (0)(-2) + (1)(0) + (0)(-1) = 0 \implies \mathbf{v} \perp \mathbf{u} \times \mathbf{v}$

9. $\mathbf{u} = \langle 12, -3, 0\rangle, \quad \mathbf{v} = \langle -2, 5, 0\rangle$

$$\mathbf{u} \times \mathbf{v} = \begin{vmatrix} \mathbf{i} & \mathbf{j} & \mathbf{k} \\ 12 & -3 & 0 \\ -2 & 5 & 0 \end{vmatrix} = 54\mathbf{k} = \langle 0, 0, 54\rangle$$

$$\mathbf{u}\cdot(\mathbf{u} \times \mathbf{v}) = 12(0) + (-3)(0) + 0(54) = 0 \Rightarrow \mathbf{u} \perp \mathbf{u} \times \mathbf{v}$$

$$\mathbf{v}\cdot(\mathbf{u} \times \mathbf{v}) = -2(0) + 5(0) + 0(54) = 0 \Rightarrow \mathbf{v} \perp \mathbf{u} \times \mathbf{v}$$

10. $\mathbf{u} = \langle -10, 0, 6\rangle, \quad \mathbf{v} = \langle 7, 0, 0\rangle$

$$\mathbf{u} \times \mathbf{v} = \begin{vmatrix} \mathbf{i} & \mathbf{j} & \mathbf{k} \\ -10 & 0 & 6 \\ 7 & 0 & 0 \end{vmatrix} = 42\mathbf{j} = \langle 0, 42, 0\rangle$$

$$\mathbf{u}\cdot(\mathbf{u} \times \mathbf{v}) = (-10)(0) + (0)(42) + 6(0) = 0 \Rightarrow \mathbf{u} \perp \mathbf{u} \times \mathbf{v}$$

$$\mathbf{v}\cdot(\mathbf{u} \times \mathbf{v}) = 7(0) + (0)(42) + (0)(0) = 0 \Rightarrow \mathbf{v} \perp \mathbf{u} \times \mathbf{v}$$

11. $\mathbf{u} = \mathbf{i} + \mathbf{j} + \mathbf{k}, \quad \mathbf{v} = 2\mathbf{i} + \mathbf{j} - \mathbf{k}$

$$\mathbf{u} \times \mathbf{v} = \begin{vmatrix} \mathbf{i} & \mathbf{j} & \mathbf{k} \\ 1 & 1 & 1 \\ 2 & 1 & -1 \end{vmatrix} = -2\mathbf{i} + 3\mathbf{j} - \mathbf{k} = \langle -2, 3, -1\rangle$$

$$\mathbf{u}\cdot(\mathbf{u} \times \mathbf{v}) = 1(-2) + 1(3) + 1(-1) = 0 \Rightarrow \mathbf{u} \perp \mathbf{u} \times \mathbf{v}$$

$$\mathbf{v}\cdot(\mathbf{u} \times \mathbf{v}) = 2(-2) + 1(3) + (-1)(-1) = 0 \Rightarrow \mathbf{v} \perp \mathbf{u} \times \mathbf{v}$$

12. $\mathbf{u} = \mathbf{j} + 6\mathbf{k}, \quad \mathbf{v} = \mathbf{i} - 2\mathbf{j} + \mathbf{k}$

$$\mathbf{u} \times \mathbf{v} = \begin{vmatrix} \mathbf{i} & \mathbf{j} & \mathbf{k} \\ 0 & 1 & 6 \\ 1 & -2 & 1 \end{vmatrix} = 13\mathbf{i} + 6\mathbf{j} - \mathbf{k} = \langle 13, 6, -1\rangle$$

$$\mathbf{u}\cdot(\mathbf{u} \times \mathbf{v}) = 0(13) + 1(6) + 6(-1) = 0 \Rightarrow \mathbf{u} \perp \mathbf{u} \times \mathbf{v}$$

$$\mathbf{v}\cdot(\mathbf{u} \times \mathbf{v}) = 1(13) + (-2)(6) + 1(-1) = 0 \Rightarrow \mathbf{v} \perp \mathbf{u} \times \mathbf{v}$$

13. $\mathbf{u} = -3\mathbf{i} + 2\mathbf{j} - 5\mathbf{k}, \quad \mathbf{v} = \frac{1}{2}\mathbf{i} - \frac{3}{4}\mathbf{j} + \frac{1}{10}\mathbf{k}$

$$\mathbf{u} \times \mathbf{v} = \begin{vmatrix} \mathbf{i} & \mathbf{j} & \mathbf{k} \\ -3 & 2 & -5 \\ \frac{1}{2} & -\frac{3}{4} & \frac{1}{10} \end{vmatrix} = -\frac{71}{20}\mathbf{i} - \frac{11}{5}\mathbf{j} + \frac{5}{4}\mathbf{k} = \langle -\frac{71}{20}, -\frac{11}{5}, \frac{5}{4} \rangle$$

$$\mathbf{u}\cdot(\mathbf{u} \times \mathbf{v}) = (-3)(-\frac{71}{20}) + 2(-\frac{11}{5}) - 5(\frac{5}{4}) = 0 \Rightarrow \mathbf{u} \perp \mathbf{u} \times \mathbf{v}$$

$$\mathbf{v}\cdot(\mathbf{u} \times \mathbf{v}) = (\frac{1}{2})(-\frac{71}{20}) + (-\frac{3}{4})(-\frac{11}{5}) + (\frac{1}{10})(\frac{5}{4}) = 0 \Rightarrow \mathbf{v} \perp \mathbf{u} \times \mathbf{v}$$

14. $\mathbf{u} = \frac{2}{3}\mathbf{k}, \quad \mathbf{v} = \frac{1}{2}\mathbf{i} + 6\mathbf{k},$

$$\mathbf{u} \times \mathbf{v} = \begin{vmatrix} \mathbf{i} & \mathbf{j} & \mathbf{k} \\ 0 & 0 & \frac{2}{3} \\ \frac{1}{2} & 0 & 6 \end{vmatrix} = \frac{1}{3}\mathbf{j} = \langle 0, \frac{1}{3}, 0 \rangle$$

$$\mathbf{u}\cdot(\mathbf{u} \times \mathbf{v}) = (0)(0) + (0)(\frac{1}{3}) + (\frac{2}{3})(0) = 0 \Rightarrow \mathbf{u} \perp \mathbf{u} \times \mathbf{v}$$

$$\mathbf{v}\cdot(\mathbf{u} \times \mathbf{v}) = (\frac{1}{2})(0) + (0)(\frac{1}{3}) + (6)(0) = 0 \Rightarrow \mathbf{v} \perp \mathbf{u} \times \mathbf{v}$$

15. $\mathbf{u} = \mathbf{j}, \quad \mathbf{v} = \mathbf{j} + \mathbf{k}, \quad \mathbf{u} \times \mathbf{v} = \begin{vmatrix} \mathbf{i} & \mathbf{j} & \mathbf{k} \\ 0 & 1 & 0 \\ 0 & 1 & 1 \end{vmatrix} = \mathbf{i}, \quad A = \|\mathbf{u} \times \mathbf{v}\| = \|\mathbf{i}\| = 1$

16. $\mathbf{u} = \mathbf{i} + \mathbf{j} + \mathbf{k}, \quad \mathbf{v} = \mathbf{j} + \mathbf{k}, \quad \mathbf{u} \times \mathbf{v} = \begin{vmatrix} \mathbf{i} & \mathbf{j} & \mathbf{k} \\ 1 & 1 & 1 \\ 0 & 1 & 1 \end{vmatrix} = -\mathbf{j} + \mathbf{k}$

$$A = \|\mathbf{u} \times \mathbf{v}\| = \|-\mathbf{j} + \mathbf{k}\| = \sqrt{2}$$

17. $\mathbf{u} = \langle 3, 2, -1 \rangle, \quad \mathbf{v} = \langle 1, 2, 3 \rangle, \quad \mathbf{u} \times \mathbf{v} = \begin{vmatrix} \mathbf{i} & \mathbf{j} & \mathbf{k} \\ 3 & 2 & -1 \\ 1 & 2 & 3 \end{vmatrix} = \langle 8, -10, 4 \rangle$

$$A = \|\mathbf{u} \times \mathbf{v}\| = \|\langle 8, -10, 4 \rangle\| = \sqrt{180} = 6\sqrt{5}$$

18. $\mathbf{u} = \langle 2, -1, 0 \rangle, \quad \mathbf{v} = \langle -1, 2, 0 \rangle, \quad \mathbf{u} \times \mathbf{v} = \begin{vmatrix} \mathbf{i} & \mathbf{j} & \mathbf{k} \\ 2 & -1 & 0 \\ -1 & 2 & 0 \end{vmatrix} = \langle 0, 0, 5 \rangle$

$$A = \|\mathbf{u} \times \mathbf{v}\| = \|\langle 0, 0, 5 \rangle\| = 5$$

19. $A(1, 1, 1)$, $B(2, 3, 4)$, $C(6, 5, 2)$, $D(7, 7, 5)$

$\overrightarrow{AB} = \langle 1, 2, 3 \rangle$, $\overrightarrow{AC} = \langle 5, 4, 1 \rangle$, $\overrightarrow{AB}$ and $\overrightarrow{AC}$ are adjacent sides

and $\overrightarrow{AB} \times \overrightarrow{AC} = \begin{vmatrix} \mathbf{i} & \mathbf{j} & \mathbf{k} \\ 1 & 2 & 3 \\ 5 & 4 & 1 \end{vmatrix} = -10\mathbf{k} + 14\mathbf{j} - 6\mathbf{k}$

$A = \|\overrightarrow{AB} \times \overrightarrow{AC}\| = \sqrt{332} = 2\sqrt{83}$

20. $A(2, -1, 1)$, $B(5, 1, 4)$, $C(0, 1, 1)$, $D(3, 3, 4)$

$\overrightarrow{AB} = \langle 3, 2, 3 \rangle$, $\overrightarrow{AC} = \langle -2, 2, 0 \rangle$, $\overrightarrow{CD} = \langle 3, 2, 3 \rangle$, $\overrightarrow{BD} = \langle -2, 2, 0 \rangle$

$\overrightarrow{AB}$ and $\overrightarrow{AC}$ are adjacent sides and $\overrightarrow{AB} \times \overrightarrow{AC} = \begin{vmatrix} \mathbf{i} & \mathbf{j} & \mathbf{k} \\ 3 & 2 & 3 \\ -2 & 2 & 0 \end{vmatrix} = -6\mathbf{i} - 6\mathbf{j} + 10\mathbf{k}$

$A = \|\overrightarrow{AB} \times \overrightarrow{AC}\| = \sqrt{172} = 2\sqrt{43}$

21. $A(0, 0, 0)$, $B(1, 2, 3)$, $C(-3, 0, 0)$

$\overrightarrow{AB} = \langle 1, 2, 3 \rangle$, $\overrightarrow{AC} = \langle -3, 0, 0 \rangle$, $\overrightarrow{AB} \times \overrightarrow{AC} = \begin{vmatrix} \mathbf{i} & \mathbf{j} & \mathbf{k} \\ 1 & 2 & 3 \\ -3 & 0 & 0 \end{vmatrix} = -9\mathbf{j} + 6\mathbf{k}$

$A = \frac{1}{2}\|\overrightarrow{AB} \times \overrightarrow{AC}\| = \frac{1}{2}\sqrt{117} = \frac{3}{2}\sqrt{13}$

22. $A(2, -3, 4)$, $B(0, 1, 2)$, $C(-1, 2, 0)$

$\overrightarrow{AB} = \langle -2, 4, -2 \rangle$, $\overrightarrow{AC} = \langle -3, 5, -4 \rangle$

$\overrightarrow{AB} \times \overrightarrow{AC} = \begin{vmatrix} \mathbf{i} & \mathbf{j} & \mathbf{k} \\ -2 & 4 & -2 \\ -3 & 5 & -4 \end{vmatrix} = -6\mathbf{i} - 2\mathbf{j} + 2\mathbf{k}$, $A = \frac{1}{2}\|\overrightarrow{AB} \times \overrightarrow{AC}\| = \frac{1}{2}\sqrt{44} = \sqrt{11}$

23. $A(1, 3, 5)$, $B(3, 3, 0)$, $C(-2, 0, 5)$

$\overrightarrow{AB} = \langle 2, 0, -5 \rangle$, $\overrightarrow{AC} = \langle -3, -3, 0 \rangle$

$\overrightarrow{AB} \times \overrightarrow{AC} = \begin{vmatrix} \mathbf{i} & \mathbf{j} & \mathbf{k} \\ 2 & 0 & -5 \\ -3 & -3 & 0 \end{vmatrix} = -15\mathbf{i} + 15\mathbf{j} - 6\mathbf{k}$

$A = \frac{1}{2}\|\overrightarrow{AB} \times \overrightarrow{AC}\| = \frac{1}{2}\sqrt{486} = \frac{9}{2}\sqrt{6}$

24. $A(1, 2, 0)$, $B(-2, 1, 0)$, $C(0, 0, 0)$

$\overrightarrow{AB} = \langle -3, -1, 0\rangle$, $\overrightarrow{AC} = \langle -1, -2, 0\rangle$, $\overrightarrow{AB} \times \overrightarrow{AC} = \begin{vmatrix} \mathbf{i} & \mathbf{j} & \mathbf{k} \\ -3 & -1 & 0 \\ -1 & -2 & 0 \end{vmatrix} = 5\mathbf{i}$

$A = \frac{1}{2}\|\overrightarrow{AB} \times \overrightarrow{AC}\| = \frac{5}{2}$

25. $\mathbf{u}\cdot(\mathbf{v} \times \mathbf{w}) = \begin{vmatrix} 1 & 0 & 0 \\ 0 & 1 & 0 \\ 0 & 0 & 1 \end{vmatrix} = 1$

26. $\mathbf{u}\cdot(\mathbf{v} \times \mathbf{w}) = \begin{vmatrix} 1 & 1 & 1 \\ 2 & 1 & 0 \\ 0 & 0 & 1 \end{vmatrix} = -1$

27. $\mathbf{u}\cdot(\mathbf{v} \times \mathbf{w}) = \begin{vmatrix} 2 & 0 & 1 \\ 0 & 3 & 0 \\ 0 & 0 & 1 \end{vmatrix} = 6$

28. $\mathbf{u}\cdot(\mathbf{v} \times \mathbf{w}) = \begin{vmatrix} 2 & 0 & 0 \\ 1 & 1 & 1 \\ 0 & 2 & 2 \end{vmatrix} = 0$

29. $\mathbf{u}\cdot(\mathbf{v} \times \mathbf{w}) = \begin{vmatrix} 1 & 1 & 0 \\ 0 & 1 & 1 \\ 1 & 0 & 1 \end{vmatrix} = 2$

$V = |\mathbf{u}\cdot(\mathbf{v} \times \mathbf{w})| = 2$

30. $\mathbf{u}\cdot(\mathbf{v} \times \mathbf{w}) = \begin{vmatrix} 1 & 3 & 1 \\ 0 & 5 & 5 \\ 4 & 0 & 4 \end{vmatrix} = 60$

$V = |\mathbf{u}\cdot(\mathbf{v} \times \mathbf{w})| = 60$

31. $\mathbf{u} = \langle 3, 0, 0\rangle$, $\mathbf{v} = \langle 0, 5, 1\rangle$, $\mathbf{w} = \langle 2, 0, 5\rangle$

$\mathbf{u}\cdot(\mathbf{v} \times \mathbf{w}) = \begin{vmatrix} 3 & 0 & 0 \\ 0 & 5 & 1 \\ 2 & 0 & 5 \end{vmatrix} = 75$, $V = |\mathbf{u}\cdot(\mathbf{v} \times \mathbf{w})| = 75$

32. $\mathbf{u} = \langle 1, 1, 0\rangle$, $\mathbf{v} = \langle 1, 0, 2\rangle$, $\mathbf{w} = \langle 0, 1, 1\rangle$

$\mathbf{u}\cdot(\mathbf{v} \times \mathbf{w}) = \begin{vmatrix} 1 & 1 & 0 \\ 1 & 0 & 2 \\ 0 & 1 & 1 \end{vmatrix} = -3$, $V = |\mathbf{u}\cdot(\mathbf{v} \times \mathbf{w})| = 3$

33. Let $\mathbf{u} = \langle u_1, u_2, u_3\rangle$, $\mathbf{v} = \langle v_1, v_2, v_3\rangle$

$$\mathbf{u} \times \mathbf{v} = \begin{vmatrix} \mathbf{i} & \mathbf{j} & \mathbf{k} \\ u_1 & u_2 & u_3 \\ v_1 & v_2 & v_3 \end{vmatrix} = (u_2v_3 - u_3v_2)\mathbf{i} - (u_1v_3 - u_3v_1)\mathbf{j} + (u_1v_2 - u_2v_1)\mathbf{k}$$

$$\mathbf{v} \times \mathbf{u} = \begin{vmatrix} \mathbf{i} & \mathbf{j} & \mathbf{k} \\ v_1 & v_2 & v_3 \\ u_1 & u_2 & u_3 \end{vmatrix} = (u_3v_2 - u_2v_3)\mathbf{i} - (u_3v_1 - u_1v_3)\mathbf{j} + (u_2v_1 - u_1v_2)\mathbf{k}$$

$\mathbf{u} \times \mathbf{v} = -(\mathbf{v} \times \mathbf{u})$

34. Let $\mathbf{u} = \langle u_1, u_2, u_3\rangle$, $\mathbf{v} = \langle v_1, v_2, v_3\rangle$, c is a scalar

$$(c\mathbf{u}) \times \mathbf{v} = \begin{vmatrix} \mathbf{i} & \mathbf{j} & \mathbf{k} \\ cu_1 & cu_2 & cu_3 \\ v_1 & v_2 & v_3 \end{vmatrix}$$

$$= (cu_2v_3 - cu_3v_2)\mathbf{i} - (cu_1v_3 - cu_3v_1)\mathbf{j} + (cu_1v_2 - cu_2v_1)\mathbf{k}$$

$$= c[(u_2v_3 - u_3v_2)\mathbf{i} - (u_1v_3 - u_3v_1)\mathbf{j} + (u_1v_2 - u_2v_1)\mathbf{k}] = c(\mathbf{u} \times \mathbf{v})$$

35. Let $\mathbf{u} = \langle a_1, b_1, c_1\rangle$, $\mathbf{v} = \langle a_2, b_2, c_2\rangle$, $\mathbf{w} = \langle a_3, b_3, c_3\rangle$

$$\mathbf{v} \times \mathbf{w} = \begin{vmatrix} \mathbf{i} & \mathbf{j} & \mathbf{k} \\ a_2 & b_2 & c_2 \\ a_3 & b_3 & c_3 \end{vmatrix} = (b_2c_3 - b_3c_2)\mathbf{i} - (a_2c_3 - a_3c_2)\mathbf{j} + (a_2b_3 - a_3b_2)\mathbf{k}$$

$$\mathbf{u} \times (\mathbf{v} \times \mathbf{w}) = \begin{vmatrix} \mathbf{i} & \mathbf{j} & \mathbf{k} \\ a_1 & b_1 & c_1 \\ (b_2c_3 - b_3c_2) & (a_3c_2 - a_2c_3) & (a_2b_3 - a_3b_2) \end{vmatrix}$$

$$\mathbf{u} \times (\mathbf{v} \times \mathbf{w}) = [b_1(a_2b_3 - a_3b_2) - c_1(a_3c_2 - a_2c_3)]\mathbf{i}$$

$$- [a_1(a_2b_3 - a_3b_2) - c_1(b_2c_3 - b_3c_2)]\mathbf{j}$$

$$+ [a_1(a_3c_2 - a_2c_3) - b_1(b_2c_3 - b_3c_2)]\mathbf{k}$$

$$= [a_2(a_1a_3 + b_1b_3 + c_1c_3) - a_3(a_1a_2 + b_1b_2 + c_1c_2)]\mathbf{i}$$

$$+ [b_2(a_1a_3 + b_1b_3 + c_1c_3) - b_3(a_1a_2 + b_1b_2 + c_1c_2)]\mathbf{j}$$

$$+ [c_2(a_1a_3 + b_1b_3 + c_1c_3) - c_3(a_1a_2 + b_1b_2 + c_1c_2)]\mathbf{k}$$

$$= (\mathbf{u}\cdot\mathbf{w})\mathbf{v} - (\mathbf{u}\cdot\mathbf{v})\mathbf{w}$$

36. $\mathbf{u}\cdot(\mathbf{v} \times \mathbf{w}) = \begin{vmatrix} u_1 & u_2 & u_3 \\ v_1 & v_2 & v_3 \\ w_1 & w_2 & w_3 \end{vmatrix}$, $(\mathbf{u} \times \mathbf{v})\cdot\mathbf{w} = \mathbf{w}\cdot(\mathbf{u} \times \mathbf{v}) = \begin{vmatrix} w_1 & w_2 & w_3 \\ u_1 & u_2 & u_3 \\ v_1 & v_2 & v_3 \end{vmatrix}$

Since two rows are interchanged, the value of the determinant is unchanged and $\mathbf{u}\cdot(\mathbf{v} \times \mathbf{w}) = (\mathbf{u} \times \mathbf{v})\cdot\mathbf{w}$

37. $\mathbf{u} \times \mathbf{v} = (u_2v_3 - u_3v_2)\mathbf{i} - (u_1v_3 - u_3v_1)\mathbf{j} + (u_1v_2 - u_2v_1)\mathbf{k}$

$$(\mathbf{u} \times \mathbf{v})\cdot\mathbf{u} = (u_2v_3 - u_3v_2)u_1 + (u_3v_1 - u_1v_3)u_2 + (u_1v_2 - u_2v_1)u_3 = 0$$

$$(\mathbf{u} \times \mathbf{v})\cdot\mathbf{v} = (u_2v_3 - u_3v_2)v_1 + (u_3v_1 - u_1v_3)v_2 + (u_1v_2 - u_2v_1)v_3 = 0$$

Thus $\mathbf{u} \times \mathbf{v} \perp \mathbf{u}$ and $\mathbf{u} \times \mathbf{v} \perp \mathbf{v}$

38. If $\mathbf{u}$ and $\mathbf{v}$ are scalar multiples of each other $\mathbf{u} = c\mathbf{v}$ for some scalar c.

$$\mathbf{u} \times \mathbf{v} = (c\mathbf{v}) \times \mathbf{v} = c(\mathbf{v} \times \mathbf{v}) = c(\mathbf{0}) = \mathbf{0}$$

If $\mathbf{u} \times \mathbf{v} = \mathbf{0}$ then $\|\mathbf{u}\|\ \|\mathbf{v}\| \sin\theta = 0$

(Assume $\mathbf{u} \neq \mathbf{0}$, $\mathbf{v} \neq \mathbf{0}$) Thus $\sin\theta = 0$, $\theta = 0$ and $\mathbf{u}$ and $\mathbf{v}$ are parallel. Therefore $\mathbf{u} = c\mathbf{v}$ for some scalar c.

39. $\|\mathbf{u} \times \mathbf{v}\| = \|\mathbf{u}\|\ \|\mathbf{v}\| \sin\theta$

If $\mathbf{u}$ and $\mathbf{v}$ are orthogonal, $\theta = \frac{\pi}{2}$ and $\sin\theta = 1$.

Therefore $\|\mathbf{u} \times \mathbf{v}\| = \|\mathbf{u}\|\ \|\mathbf{v}\|$

12.5
Lines and planes in space

1. Point: $(0, 0, 0)$ Direction vector: $\mathbf{v} = \langle 1, 2, 3\rangle$

Direction numbers: 1, 2, 3, Parametric: $x = t$, $y = 2t$, $z = 3t$

Symmetric: $x = \frac{y}{2} = \frac{z}{3}$

2. Point: $(0, 0, 0)$ Direction vector: $\mathbf{v} = \langle -2, \frac{5}{2}, 1\rangle$

Direction numbers: -4, 5, 2, Parametric: $x = -4t$, $y = 5t$, $z = 2t$

Symmetric: $\frac{x}{-4} = \frac{y}{5} = \frac{z}{2}$

3. Point: $(-2, 0, 3)$ Direction vector: $\mathbf{v} = \langle 2, 4, -2\rangle$

Direction numbers: 2, 4, -2, Parametric: $x = -2 + 2t$, $y = 4t$, $z = 3 - 2t$, Symmetric: $\frac{x + 2}{2} = \frac{y}{4} = \frac{z - 3}{-2}$

4. Point: $(-2, 0, 3)$ Direction vector: $\mathbf{v} = \langle 6, 3, 0\rangle$

Direction numbers: 2, 1, 0, Parametric: $x = -2 + 2t$, $y = t$, $z = 3$

Symmetric: $\frac{x + 2}{2} = y$, $z = 3$

5. $(5, -3, -2)$ $\left(\frac{-2}{3}, \frac{2}{3}, 1\right)$ Direction vector: $\mathbf{v} = \frac{17}{3}\mathbf{i} - \frac{11}{3}\mathbf{j} - 3\mathbf{k}$

Direction numbers: 17, −11, −9, Parametric: $x = 5 + 17t$, $y = -3 - 11t$

$z = -2 - 9t$, Symmetric: $\frac{x - 5}{17} = \frac{y + 3}{-11} = \frac{z + 2}{-9}$

6. $(1, 0, 1)$, $(1, 3, -2)$ Direction vector: $\mathbf{v} = 3\mathbf{j} - 3\mathbf{k}$

Direction numbers: 0, 1, −1, Parametric: $x = 1$, $y = t$, $z = 1 - t$

Symmetric: $y = 1 - z$, $x = 1$

7. $(1, 0, 1)$ Direction vector: $\mathbf{v} = 3\mathbf{i} - 2\mathbf{j} + \mathbf{k}$

Direction numbers: 3, −2, 1, Parametric: $x = 1 + 3t$, $y = -2t$,

$z = 1 + t$, Symmetric: $\frac{x - 1}{3} = \frac{y}{-2} = \frac{z - 1}{1}$

8. $(-3, 5, 4)$, Direction numbers: 3, −2, 1

Parametric: $x = -3 + 3t$, $y = 5 - 2t$, $z = 4 + t$

Symmetric: $\frac{x + 3}{3} = \frac{y - 5}{-2} = z - 4$

9. $(2, 3, 4)$, Direction vector: $\mathbf{v} = \mathbf{k}$, Direction numbers: 0, 0, 1

Parametric: $x = 2$, $y = 3$, $z = 4 + t$, No symmetric form

10. $(2, 3, 4)$, Direction vector: $\mathbf{v} = 3\mathbf{i} + 2\mathbf{j} - \mathbf{k}$

Direction numbers: 3, 2, −1, Parametric: $x = 2 + 3t$, $y = 3 + 2t$,

$z = 4 - t$, Symmetric: $\frac{x - 2}{3} = \frac{y - 3}{2} = \frac{z - 4}{-1}$

11. $(-2, 3, 1)$, Direction vector: $\mathbf{v} = 4\mathbf{i} - \mathbf{k}$

Direction numbers: 4, 0, −1, Parametric: $x = -2 + 4t$, $y = 3$,

$z = 1 - t$, Symmetric: $\frac{x + 2}{4} = \frac{z - 1}{-1}$, $y = 3$

(a) on line (b) on line

(c) not on line $(y \neq 3)$ (d) on line

(e) not on line $\left(\frac{6 + 2}{4} \neq \frac{-2 - 1}{-1}\right)$

12. (2, 0, −3) (4, 2, −2), Direction vector: $\mathbf{v} = 2\mathbf{i} + 2\mathbf{j} + \mathbf{k}$

Direction numbers: 2, 2, 1, Parametric: $x = 2 + 2t$, $y = 2t$,

$z = -3 + t$, Symmetric: $\frac{x - 2}{2} = \frac{y}{2} = \frac{z + 3}{1}$

(a) not on line $(1 \neq \frac{1}{2} \neq 1)$ (b) on line

(c) on line (d) not on line $(\frac{-3}{2} = \frac{-3}{2} \neq -1)$

(e) on line

13. At the point of intersection, the coordinates for one line equal the corresponding coordinates for the other line. Thus

(i) $4t + 2 = 2s + 2$, (ii) $3 = 2s + 3$, (iii) $-t + 1 = s + 1$

From (ii), we find that $s = 0$ and consequently, from (iii), $t = 0$. Letting $s = t = 0$, we see that equation (i) is satisfied and therefore the two lines intersect. Substituting zero for s or for t, we obtain the point (2, 3, 1).

$\mathbf{u} = 4\mathbf{i} - \mathbf{k}$, $\mathbf{v} = 2\mathbf{i} + 2\mathbf{j} + \mathbf{k}$, $\cos\theta = \frac{|\mathbf{u}\cdot\mathbf{v}|}{\|\mathbf{u}\|\,\|\mathbf{v}\|} = \frac{8 - 1}{\sqrt{17}\sqrt{9}} = \frac{7}{3\sqrt{17}} = \frac{7\sqrt{17}}{51}$

14. By equating like variables we have

(i) $-3t + 1 = 3s + 1$, (ii) $4t + 1 = 2s + 4$, (iii) $2t + 4 = -s + 1$

From (i) we have $s = -t$, and consequently from (ii), $t = \frac{1}{2}$ and

from (iii) $t = -3$. The lines do not intersect.

15. Writing the equations of the lines in parametric form we have:

$x = 3t$, $y = 2 - t$, $z = -1 + t$, $x = 1 + 4t$, $y = -2 + t$, $z = -3 - 3t$

For the x coordinates to be equal $3t = 1 + 4t$ or $t = -1$

Substituting $t = -1$ into the equations for y we have

L_1: $y = 2 - (-1) = 3$, L_2: $y = -2 - 1 = -3$ The lines do not intersect.

16. Writing the equations of the lines in parametric form we have:

$x = 2 - 3t, \quad y = 2 + 6t, \quad z = 3 + t$

$x = 3 + 2s, \quad y = -5 + s, \quad z = -2 + 4s$

By equating like variables we have:

$2 - 3t = 3 + 2s, \quad 2 + 6t = -5 + s, \quad 3 + t = -2 + 4s$

Thus $t = -1$, $s = 1$ and the point of intersection is $(5, -4, 2)$.

$\mathbf{u} = \langle -3, 6, 1\rangle, \quad \mathbf{v} = \langle 2, 1, 4\rangle$

$$\cos\theta = \frac{|\mathbf{u}\cdot\mathbf{v}|}{\|\mathbf{u}\|\ \|\mathbf{v}\|} = \frac{4}{\sqrt{46}\ \sqrt{21}} = \frac{4}{\sqrt{966}} = \frac{2\sqrt{966}}{483}$$

17. Point: $(2, 1, 2)$, $\mathbf{n} = \mathbf{i} = \langle 1, 0, 0\rangle$

$1(x - 2) + 0(y - 1) + 0(z - 2) = 0, \quad x - 2 = 0$

18. Point: $(1, 0, -3)$, $\mathbf{n} = \mathbf{k} = \langle 0, 0, 1\rangle$

$0(x - 1) + 0(y - 0) + 1(z - (-3)) = 1, \quad z + 3 = 0$

19. $(3, 2, 2)$, Normal vector: $\mathbf{v} = 2\mathbf{i} + 3\mathbf{j} - \mathbf{k}$

$2(x - 3) + 3(y - 2) - 1(z - 2) = 0, \quad 2x + 3y - z = 10$

20. $(3, 2, 2)$, Normal vector: $\mathbf{v} = 4\mathbf{i} + \mathbf{j} - 3\mathbf{k}$

$4(x - 3) + (y - 2) - 3(z - 2) = 0, \quad 4x + y - 3z = 8$

21. Let $\mathbf{u}$ be the vector from $(0, 0, 0)$ to $(1, 2, 3)$: $\mathbf{u} = \mathbf{i} + 2\mathbf{j} + 3\mathbf{k}$

Let $\mathbf{v}$ be the vector from $(0, 0, 0)$ to $(-2, 3, 3)$: $\mathbf{v} = -2\mathbf{i} + 3\mathbf{j} + 3\mathbf{k}$

$$\text{Normal vector: } \mathbf{u}\times\mathbf{v} = \begin{vmatrix} \mathbf{i} & \mathbf{j} & \mathbf{k} \\ 1 & 2 & 3 \\ -2 & 3 & 3 \end{vmatrix} = -3\mathbf{i} + (-9)\mathbf{j} + 7\mathbf{k}, \quad 3x + 9y - 7z = 0$$

22. Let $\mathbf{u}$ be the vector from (1, 2, −3) to (2, 3, 1): $\mathbf{u} = \mathbf{i} + \mathbf{j} + 4\mathbf{k}$

Let $\mathbf{v}$ be the vector from (1, 2, −3) to (0, −2, −1): $\mathbf{v} = -\mathbf{i} - 4\mathbf{j} + 2\mathbf{k}$

Normal vector: $\mathbf{u} \times \mathbf{v} = \begin{vmatrix} \mathbf{i} & \mathbf{j} & \mathbf{k} \\ 1 & 1 & 4 \\ -1 & -4 & 2 \end{vmatrix} = 3(6\mathbf{i} - 2\mathbf{j} - \mathbf{k})$

$6(x - 1) - 2(y - 2) - (z + 3) = 0, \quad 6x - 2y - z = 5$

23. Let $\mathbf{u}$ be the vector from (1, 2, 3) to (3, 2, 1): $\mathbf{u} = 2\mathbf{i} - 2\mathbf{k}$

Let $\mathbf{v}$ be the vector from (1, 2, 3) to (−1, −2, 2): $\mathbf{v} = -2\mathbf{i} - 4\mathbf{j} - \mathbf{k}$

Normal vector: $(\frac{1}{2}\mathbf{u}) \times (-\mathbf{v}) = \begin{vmatrix} \mathbf{i} & \mathbf{j} & \mathbf{k} \\ 1 & 0 & -1 \\ 2 & 4 & 1 \end{vmatrix} = 4\mathbf{i} - 3\mathbf{j} + 4\mathbf{k}$

$4(x - 1) - 3(y - 2) + 4(z - 3) = 0, \quad 4x - 3y + 4z = 10$

24. (1, 2, 3), Normal vector: $\mathbf{v} = \mathbf{i}$, $1(x - 1) = 0$, $x = 1$

25. (1, 2, 3), Normal vector: $\mathbf{v} = \mathbf{k}$, $1(z - 3) = 0$, $z = 3$

26. The plane passes through the three points (0, 0, 0) (0, 1, 0) ($\sqrt{3}$, 0, 1). The vector from (0, 0, 0) to (0, 1, 0): $\mathbf{u} = \mathbf{j}$

The vector from (0, 0, 0) to ($\sqrt{3}$, 0, 1): $\mathbf{v} = \sqrt{3}\,\mathbf{i} + \mathbf{k}$

Normal vector: $\mathbf{u} \times \mathbf{v} = \begin{vmatrix} \mathbf{i} & \mathbf{j} & \mathbf{k} \\ 0 & 1 & 0 \\ \sqrt{3} & 0 & 1 \end{vmatrix} = \mathbf{i} - \sqrt{3}\,\mathbf{k}, \quad x - \sqrt{3}\,z = 0$

27. The direction vectors for the lines are:

$\mathbf{u} = -2\mathbf{i} + \mathbf{j} + \mathbf{k}, \quad \mathbf{v} = -3\mathbf{i} + 4\mathbf{j} - \mathbf{k}$

Normal vector: $\mathbf{u} \times \mathbf{v} = \begin{vmatrix} \mathbf{i} & \mathbf{j} & \mathbf{k} \\ -2 & 1 & 1 \\ -3 & 4 & -1 \end{vmatrix} = -5(\mathbf{i} + \mathbf{j} + \mathbf{k})$

Point of intersection of the lines: (−1, 5, 1)

$(x + 1) + (y - 5) + (z - 1) = 0, \quad x + y + z = 5$

28. The direction vector of the line is: $\mathbf{u} = 2\mathbf{i} - \mathbf{j} + \mathbf{k}$

Choose any point on the line, say (0, 4, 0) and let $\mathbf{v}$ be the vector from (0, 4, 0) to the given point (2, 2, 1): $\mathbf{v} = 2\mathbf{i} - 2\mathbf{j} + \mathbf{k}$

Normal vector: $\mathbf{u} \times \mathbf{v} = \begin{vmatrix} \mathbf{i} & \mathbf{j} & \mathbf{k} \\ 2 & -1 & 1 \\ 2 & -2 & 1 \end{vmatrix} = \mathbf{i} - 2\mathbf{k}$

$(x - 2) - 2(z - 1) = 0, \quad x - 2z = 0$

29. Let $\mathbf{v}$ be the vector from (−1, 1, −1) to (2, 2, 1): $\mathbf{v} = 3\mathbf{i} + \mathbf{j} + 2\mathbf{k}$

Let $\mathbf{n}$ be a vector normal to the plane $2x - 3y + z = 3$: $\mathbf{n} = 2\mathbf{i} - 3\mathbf{j} + \mathbf{k}$

$\mathbf{v} \times \mathbf{n} = \begin{vmatrix} \mathbf{i} & \mathbf{j} & \mathbf{k} \\ 3 & 1 & 2 \\ 2 & -3 & 1 \end{vmatrix} = 7\mathbf{i} + \mathbf{j} - 11\mathbf{k}$

$7(x - 2) + 1(y - 2) - 11(z - 1) = 0, \quad 7x + y - 11z - 5 = 0$

30. Let $\mathbf{v}$ be the vector from (3, 2, 1) to (3, 1, −5): $\mathbf{v} = -\mathbf{j} - 6\mathbf{k}$

Let $\mathbf{n}$ be the normal to the given plane: $\mathbf{n} = 6\mathbf{i} + 7\mathbf{j} + 2\mathbf{k}$

Since $\mathbf{v}$ and $\mathbf{n}$ both lie in plane P, the normal vector to P is:

$\mathbf{v} \times \mathbf{n} = \begin{vmatrix} \mathbf{i} & \mathbf{j} & \mathbf{k} \\ 0 & -1 & -6 \\ 6 & 7 & 2 \end{vmatrix} = 40\mathbf{i} - 36\mathbf{j} + 6\mathbf{k} = 2(20\mathbf{i} - 18\mathbf{j} + 3\mathbf{k})$

$20(x - 3) - 18(y - 2) + 3(z - 1) = 0, \quad 20x - 18y + 3z = 27$

31. Let $\mathbf{u} = \mathbf{i}$ and let $\mathbf{v}$ be the vector from (1, −2, −1) to (2, 5, 6): $\mathbf{v} = \mathbf{i} + 7\mathbf{j} + 7\mathbf{k}$. Since $\mathbf{u}$ and $\mathbf{v}$ both lie in the plane P, the normal vector to P is:

$\mathbf{u} \times \mathbf{v} = \begin{vmatrix} \mathbf{i} & \mathbf{j} & \mathbf{k} \\ 1 & 0 & 0 \\ 1 & 7 & 7 \end{vmatrix} = -7\mathbf{j} + 7\mathbf{k} = -7(\mathbf{j} - \mathbf{k})$

$[y - (-2)] - [z - (-1)] = 0, \quad y - z = -1$

32. Let $\mathbf{u} = \mathbf{k}$ and let $\mathbf{v}$ be the vector from (4, 2, 1) to (−3, 5, 7):

$\mathbf{v} = -7\mathbf{i} + 3\mathbf{j} + 6\mathbf{k}$. Since $\mathbf{u}$ and $\mathbf{v}$ both lie in the plane P, the normal vector to P is:

$$\mathbf{u} \times \mathbf{v} = \begin{vmatrix} \mathbf{i} & \mathbf{j} & \mathbf{k} \\ 0 & 0 & 1 \\ -7 & 3 & 6 \end{vmatrix} = -3\mathbf{i} - 7\mathbf{j} = -(3\mathbf{i} + 7\mathbf{j})$$

$3(x - 4) + 7(y - 2) = 0, \qquad 3x + 7y = 26$

33. The normal vectors to the planes are: $\mathbf{n}_1 = \langle 5, -3, 1\rangle$, $\mathbf{n}_2 = \langle 1, 4, 7\rangle$

$$\cos\theta = \frac{|\mathbf{n}_1 \cdot \mathbf{n}_2|}{\|\mathbf{n}_1\| \, \|\mathbf{n}_2\|} = 0 \qquad \text{Thus } \theta = \frac{\pi}{2}$$

and the planes are orthogonal.

34. The normal vectors to the planes are: $\mathbf{n}_1 = \langle 3, 1, -4\rangle$,

$\mathbf{n}_2 = \langle -9, -3, 12\rangle$. Since $\mathbf{n}_2 = -3\mathbf{n}_1$ the planes are parallel.

35. The normal vectors to the planes are:

$\mathbf{n}_1 = \mathbf{i} - 3\mathbf{j} + 6\mathbf{k}, \quad \mathbf{n}_2 = 5\mathbf{i} + \mathbf{j} - \mathbf{k}$

$$\cos\theta = \frac{|\mathbf{n}_1 \cdot \mathbf{n}_2|}{\|\mathbf{n}_1\| \, \|\mathbf{n}_2\|} = \frac{|5 - 3 - 6|}{\sqrt{46}\,\sqrt{27}} = \frac{4\sqrt{138}}{414}$$

Therefore $\theta = \arccos\left(\frac{4\sqrt{138}}{414}\right) \approx 83.5°$

36. The normal vectors to the planes are:

$\mathbf{n}_1 = 3\mathbf{i} + 2\mathbf{j} - \mathbf{k}, \quad \mathbf{n}_2 = \mathbf{i} - 4\mathbf{j} + 2\mathbf{k}$

$$\cos\theta = \frac{|\mathbf{n}_1 \cdot \mathbf{n}_2|}{\|\mathbf{n}_1\| \, \|\mathbf{n}_2\|} = \frac{|3 - 8 - 2|}{\sqrt{14}\,\sqrt{21}} = \frac{1}{\sqrt{6}}$$

Therefore $\theta = \arccos\left(\frac{1}{\sqrt{6}}\right) \approx 65.9°$

37. The normal vectors to the planes are: $\mathbf{n}_1 = \langle 1, -5, -1\rangle$,

$\mathbf{n}_2 = \langle 5, -25, -5\rangle$. Since $\mathbf{n}_2 = 5\mathbf{n}_1$, the planes are parallel.

38. The normal vectors to the planes are: $\mathbf{n}_1 = \langle 2, 0, -1\rangle$, $\mathbf{n}_2 = \langle 4, 1, 8\rangle$. $\cos\theta = \dfrac{|\mathbf{n}_1\cdot\mathbf{n}_2|}{\|\mathbf{n}_1\|\,\|\mathbf{n}_2\|} = 0$

Thus $\theta = \dfrac{\pi}{2}$ and the planes are orthogonal.

39. The normal vectors to the planes are: $\mathbf{n}_1 = \langle 1, 3, 1\rangle$, $\mathbf{n}_2 = \langle 1, 0, -5\rangle$. $\cos\theta = \dfrac{|\mathbf{n}_1\cdot\mathbf{n}_2|}{\|\mathbf{n}_1\|\,\|\mathbf{n}_2\|} = \dfrac{4}{\sqrt{11}\,\sqrt{26}} = \dfrac{2\sqrt{286}}{143}$

Therefore $\theta = \arccos\left(\dfrac{2\sqrt{286}}{143}\right) \approx 76.3^\circ$

40. The normal vectors to the planes are: $\mathbf{n}_1 = \langle 2, 1, 0\rangle$, $\mathbf{n}_2 = \langle 1, 0, -5\rangle$. $\cos\theta = \dfrac{|\mathbf{n}_1\cdot\mathbf{n}_2|}{\|\mathbf{n}_1\|\,\|\mathbf{n}_2\|} = \dfrac{2}{\sqrt{5}\,\sqrt{26}} = \dfrac{\sqrt{130}}{65}$

Therefore $\theta = \arccos\left(\dfrac{\sqrt{130}}{65}\right) \approx 79.9^\circ$

41. $4x + 2y + 6z = 12$

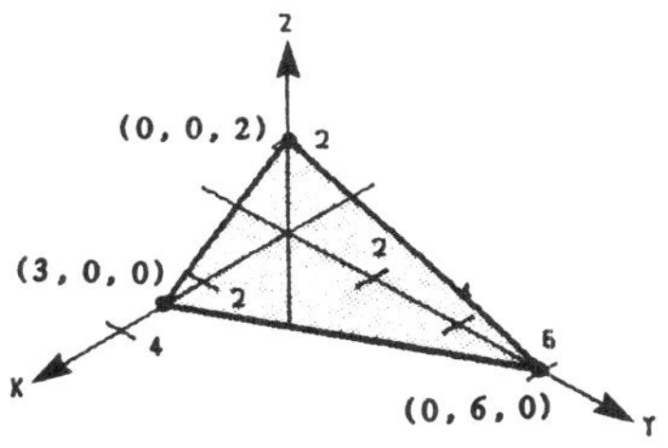

42. $3x + 6y + 2z = 6$

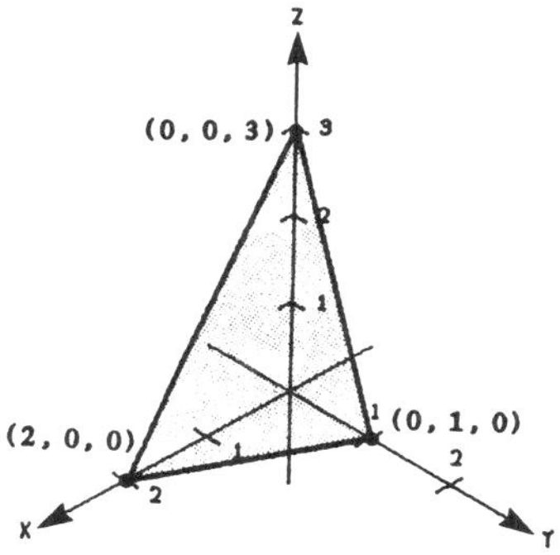

43. $2x - y + 3z = 4$

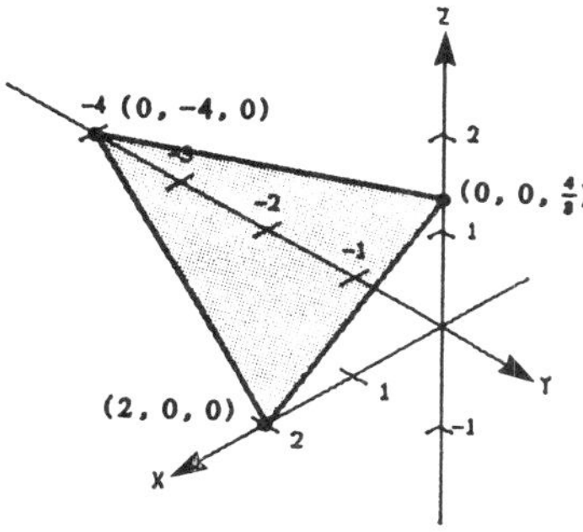

44. $2x - y + z = 4$

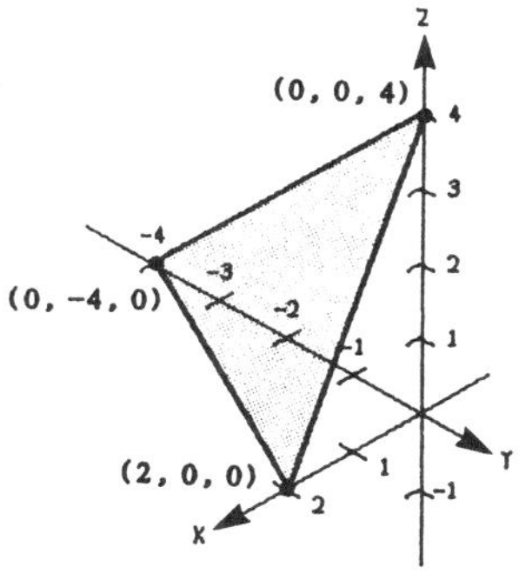

45. $y + z = 5$

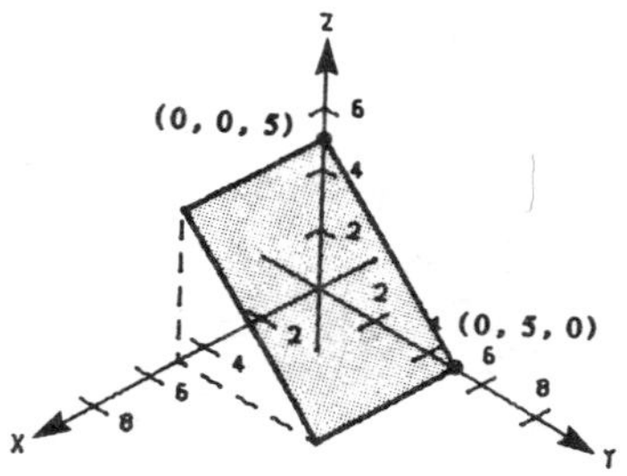

46. $x + 2y = 4$

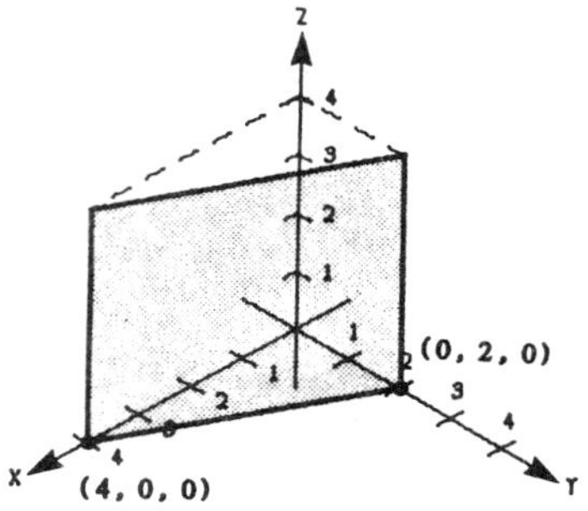

47. $2x + y - z = 6$

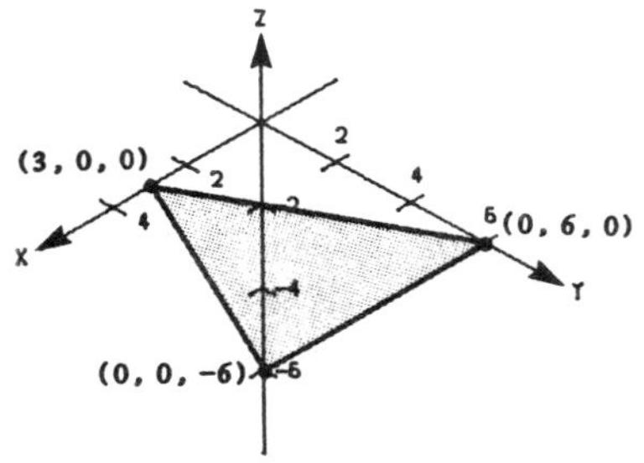

48. $x - 3z = 3$

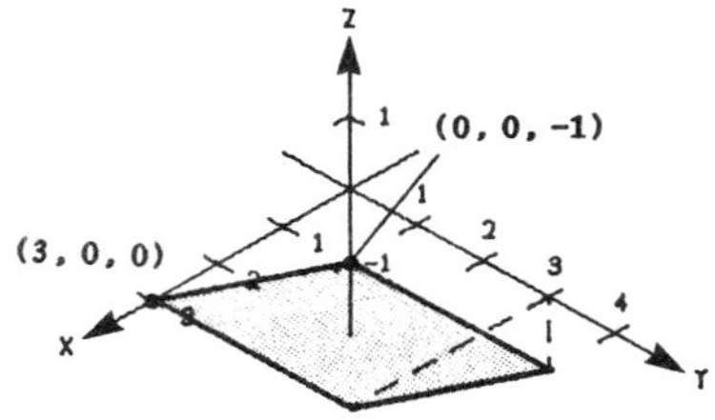

49. The normals to the planes are: $\mathbf{n}_1 = 3\mathbf{i} + 2\mathbf{j} - \mathbf{k}$, $\mathbf{n}_2 = \mathbf{i} - 4\mathbf{j} + 2\mathbf{k}$

The direction vector for the line is:

$$\mathbf{n}_2 \times \mathbf{n}_1 = \begin{vmatrix} \mathbf{i} & \mathbf{j} & \mathbf{k} \\ 1 & -4 & 2 \\ 3 & 2 & -1 \end{vmatrix} = 7(\mathbf{j} + 2\mathbf{k})$$

Now find a point of intersection of the planes:

$$\begin{array}{r} 6x + 4y - 2z = 14 \\ \underline{x - 4y + 2z = 0} \\ 7x \qquad\qquad = 14 \\ x = 2 \end{array}$$

Substituting 2 for x in the second equation we have: $-4y + 2z = -2$ or $z = 2y - 1$. Letting $y = 1$, a point of intersection is $(2, 1, 1)$.

$$x = 2, \quad y = 1 + t, \quad z = 1 + 2t$$

50. The normals to the planes are: $\mathbf{n}_1 = \mathbf{i} - 3\mathbf{j} + 6\mathbf{k}$, $\mathbf{n}_2 = 5\mathbf{i} + \mathbf{j} - \mathbf{k}$

The direction vector for the line is:

$$\mathbf{n}_1 \times \mathbf{n}_2 = \begin{vmatrix} \mathbf{i} & \mathbf{j} & \mathbf{k} \\ 1 & -3 & 6 \\ 5 & 1 & -1 \end{vmatrix} = -3\mathbf{i} + 31\mathbf{j} + 16\mathbf{k}$$

Now find a point of intersection of the planes:

$$\begin{array}{r} x - 3y + 6z = 4 \\ \underline{15x + 3y - 3z = 12} \\ 16x + \quad\quad 3z = 16 \end{array}$$

Let $z = 0$ then $x = 1$ and $y = -5x + 4 + z = -1$.
A point of intersection is $(1, -1, 0)$.

$$x = 1 - 3t, \ y = -1 + 31t, \ z = 16t$$

51. Writing the equation of the line in parametric form and substituting into the equation of the plane we have:

$$x = \frac{1}{2} + t, \quad y = \frac{-3}{2} - t, \quad z = -1 + 2t$$

$$2\left(\frac{1}{2} + t\right) - 2\left(\frac{-3}{2} - t\right) + (-1 + 2t) = 12, \qquad t = \frac{3}{2}$$

Substituting $t = 3/2$ into the parametric equations for the line we have the point of intersection $(2, -3, 2)$.

52. Writing the equation of the line in parametric form and substituting into the equation of the plane we have:

$$x = 1 + 4t, \quad y = 2t, \quad z = 3 + 6t$$

$$2(1 + 4t) + 3(2t) = -5, \qquad t = \frac{-1}{2}$$

Substituting $t = -1/2$ into the parametric equations for the line we have the point of intersection $(-1, -1, 0)$.

53. Writing the equation of the line in parametric form and then substituting into the equation of the plane we have:

$$x = 1 + 3t, \quad y = -1 - 2t, \quad z = 3 + t$$

$$2(1 + 3t) + 3(-1 - 2t) = 10$$

$-1 = 10$, contradiction, therefore the lines do not intersect.

54. Writing the equation of the line in parametric form and substituting into the equation of the plane we have:

$$x = 4 + 2t, \quad y = -1 - 3t, \quad z = -2 + 5t$$

$$5(4 + 2t) + 3(-1 - 3t) = 17, \qquad t = 0$$

Substituting $t = 0$ into the parametric equations for the line we have the point of intersection $(4, -1, -2)$.

55. $\mathbf{u} = \langle 4, 0, -1\rangle$ is the direction vector for the line. To find a point P on the line, let $t = 0$. $P = (-2, 3, 1)$.

Then $\overrightarrow{PQ} = \langle 12, 0, -3\rangle$ and $\overrightarrow{PQ} \times \mathbf{u} = \begin{vmatrix} \mathbf{i} & \mathbf{j} & \mathbf{k} \\ 12 & 0 & -3 \\ 4 & 0 & -1 \end{vmatrix} = \mathbf{0}$

The point lies on the line, therefore $d = 0$

56. $\mathbf{u} = \langle 2, 2, 1\rangle$ is the direction vector for the line. To find a point P on the line, let $t = 0$. $P = (2, 0, -3)$.

Then $\overrightarrow{PQ} = \langle 2, 1, 1\rangle$ and $\overrightarrow{PQ} \times \mathbf{u} = \begin{vmatrix} \mathbf{i} & \mathbf{j} & \mathbf{k} \\ 2 & 1 & 1 \\ 2 & 2 & 1 \end{vmatrix} = \langle -1, 0, 2\rangle$

$$d = \frac{\|\overrightarrow{PQ} \times \mathbf{u}\|}{\|\mathbf{u}\|} = \frac{\sqrt{5}}{\sqrt{9}} = \frac{\sqrt{5}}{3}$$

57. Point: $(0, 0, 0)$, Plane: $2x + 3y + z - 12 = 0$.

Normal to the plane: $\mathbf{n} = 2\mathbf{i} + 3\mathbf{j} + \mathbf{k}$, From Example 5, we have:

$$d = \frac{|-12|}{\sqrt{4 + 9 + 1}} = \frac{12}{\sqrt{14}} = \frac{6\sqrt{14}}{7}$$

58. Point: $(1, 2, 3)$, Plane: $2x - y + z - 4 = 0$

Normal to the plane: $\mathbf{n} = 2\mathbf{i} - \mathbf{j} + \mathbf{k}$, From Example 5, we have:

$$d = \frac{|2(1) - 2 + 3 - 4|}{\sqrt{4 + 1 + 1}} = \frac{\sqrt{6}}{6}$$

59. The normal vectors to the planes are $\mathbf{n}_1 = \langle 1, -3, 4\rangle$, $\mathbf{n}_2 = \langle 1, -3, 4\rangle$.

Since $\mathbf{n}_1 = \mathbf{n}_2$ the planes are parallel. Choose a point in each plane.

$P = (10, 0, 0)$ is a point in $x - 3y + 4z = 10$

$Q = (6, 0, 0)$ is a point in $x - 3y + 4z = 6$

$\overrightarrow{PQ} = \langle -4, 0, 0\rangle, \quad d = \dfrac{|\overrightarrow{PQ} \cdot \mathbf{n}_1|}{\|\mathbf{n}_1\|} = \dfrac{4}{\sqrt{26}} = \dfrac{2\sqrt{26}}{13}$

60. The normal vectors to the planes are $\mathbf{n}_1 = \langle 2, 0, -4\rangle$, $\mathbf{n}_2 = \langle 2, 0, -4\rangle$.

Since $\mathbf{n}_1 = \mathbf{n}_2$ the planes are parallel. Chose a point in each plane.

$P = (2, 0, 0)$ is a point in $2x - 4z = 4$

$Q = (5, 0, 0)$ is a point in $2x - 4z = 10$

$\overrightarrow{PQ} = \langle 3, 0, 0\rangle, \quad d = \dfrac{|\overrightarrow{PQ} \cdot \mathbf{n}_1|}{\|\mathbf{n}_1\|} = \dfrac{6}{\sqrt{20}} = \dfrac{3\sqrt{5}}{5}$

61. Let $\mathbf{n}_1 = \langle 1, 2, 3\rangle$ and $\mathbf{n}_2 = \langle -1, 1, 1\rangle$ be vectors along the given lines, respectively. Let $\mathbf{v}$ be the vector from the arbitrary points $(0, 0, 0)$ and $(0, 5, 0)$: $\mathbf{v} = \langle 0, 5, 0\rangle$. The component of $\mathbf{v}$ in the direction of $\mathbf{n}_1 \times \mathbf{n}_2$ will be the actual distance between the lines.

$$\mathbf{n}_1 \times \mathbf{n}_2 = \begin{vmatrix} \mathbf{i} & \mathbf{j} & \mathbf{k} \\ 1 & 2 & 3 \\ -1 & 1 & 1 \end{vmatrix} = \langle -1, -4, 3\rangle$$

$$d = \frac{|\mathbf{v}\cdot(\mathbf{n}_1 \times \mathbf{n}_2)|}{\|\mathbf{n}_1 \times \mathbf{n}_2\|} = \frac{|-20|}{\sqrt{1 + 16 + 9}} = \frac{20}{\sqrt{26}} \approx 3.92$$

62. Let $\mathbf{n}_1 = \langle 3, -1, 1\rangle$ and $\mathbf{n}_1 = \langle 4, 1, -3\rangle$ be vectors along the given lines, respectively. Let $\mathbf{v}$ be the vector from the arbitrary points $(0, 2, -1)$ and $(1, -2, -3)$: $\mathbf{v} = \langle 1, -4, -2\rangle$. The component of $\mathbf{v}$ in the direction of $\mathbf{n}_1 \times \mathbf{n}_2$ will be the actual distance between the lines.

$$\mathbf{n}_2 \times \mathbf{n}_2 = \begin{vmatrix} \mathbf{i} & \mathbf{j} & \mathbf{k} \\ 3 & -1 & 1 \\ 4 & 1 & -3 \end{vmatrix} = \langle 2, 13, 7\rangle$$

$$d = \frac{|\mathbf{v}\cdot(\mathbf{n}_1 \times \mathbf{n}_2)|}{\|\mathbf{n}_1 \times \mathbf{n}_2\|} = \frac{64}{\sqrt{222}} = \frac{32\sqrt{222}}{111} \approx 4.30$$

12.6 Surfaces in space

1. **Ellipsoid**
   Matches graph c

2. **Hyperboloid of two sheets**
   Matches graph e

3. **Hyperboloid of one sheet**
   Matches graph f

4. **Elliptic cone**
   Matches graph g

5. **Elliptic paraboloid**
   Matches graph d

6. **Hyperbolic paraboloid**
   Matches graph b

7. **Hyperbolic paraboloid**
   Matches graph a

8. **Sphere**
   Matches graph h

9. $z = 3$
   Plane parallel to the xy-coordinate plane

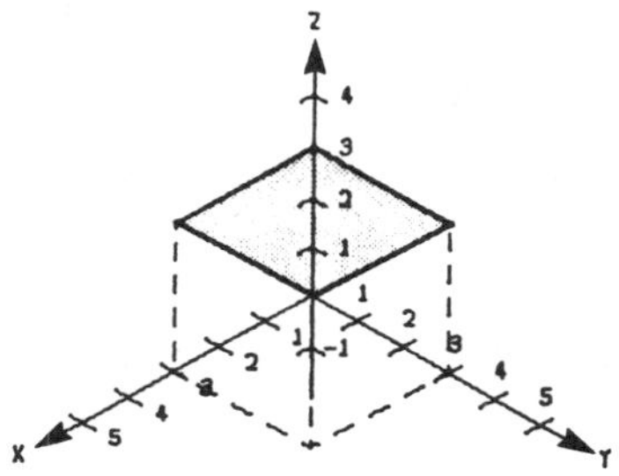

10. $x = 4$
    Plane parallel to the yz-coordinate plane

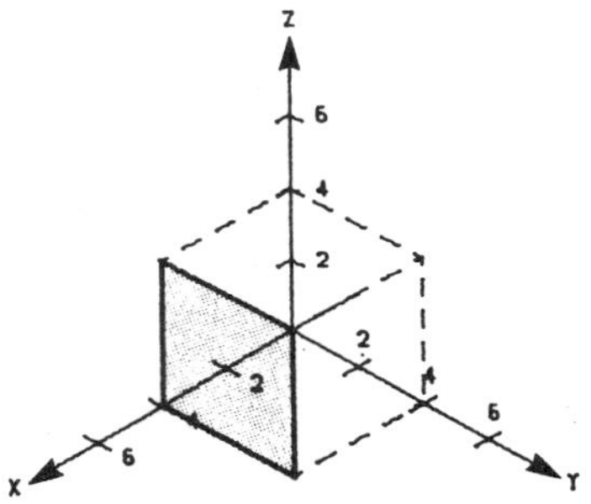

11. $y^2 + z^2 = 9$

Since the x-coordinate is missing we have a cylindrical surface with rulings parallel to the x-axis. The generating curve is a circle.

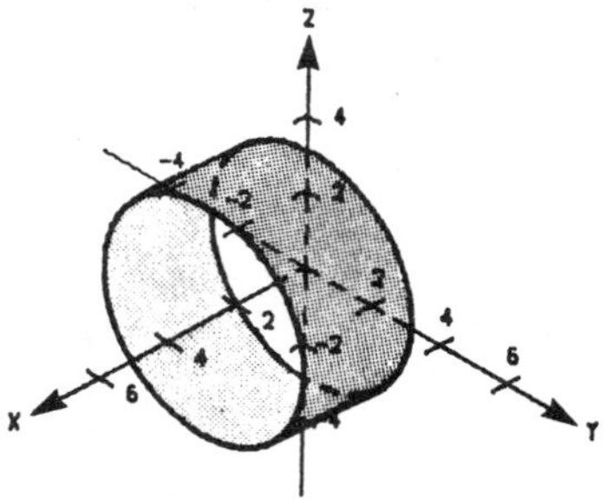

12. $x^2 + z^2 = 16$

Since the y-coordinate is missing we have a cylindrical surface with rulings parallel to the y-axis. The generating curve is a circle.

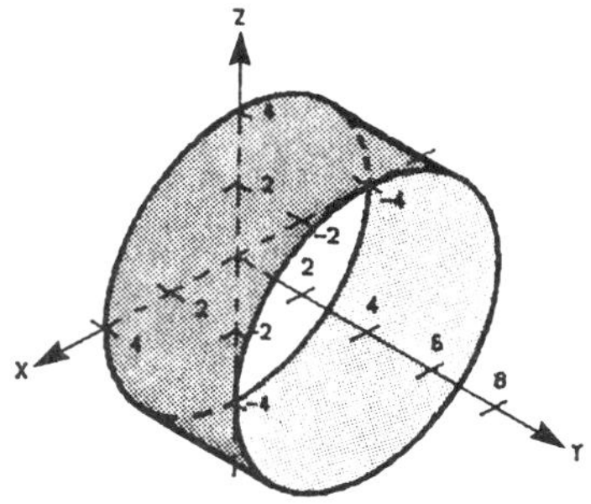

13. $y = x^2$

Since the z-coordinate is missing we have a cylindrical surface with rulings parallel to the z-axis. The generating curve is a parabola.

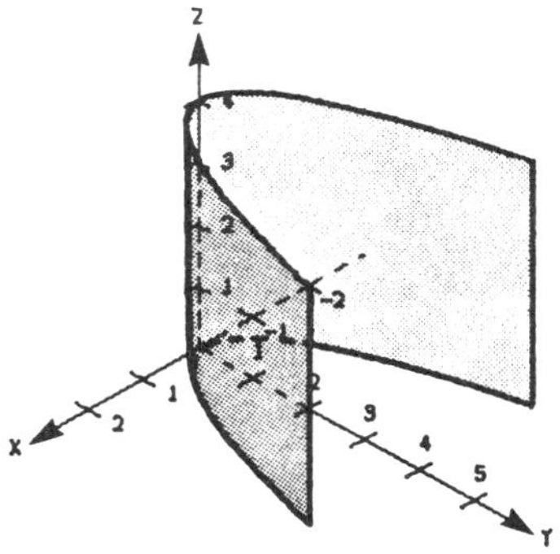

14. $z = 4 - y^2$

Since the x-coordinate is missing we have a cylindrical surface with rulings parallel to the x-axis. The generating curve is a parabola.

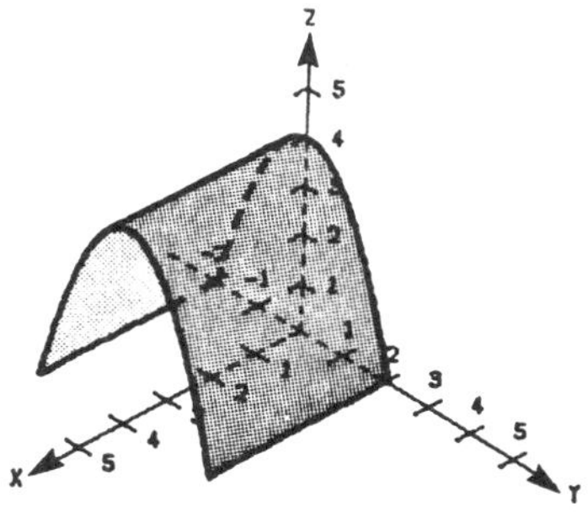

15. $4x^2 + y^2 = 4, \qquad \frac{x^2}{1} + \frac{y^2}{4} = 1$

Since the z-coordinate is missing we have a cylindrical surface with rulings parallel to the z-axis. The generating curve is an ellipse.

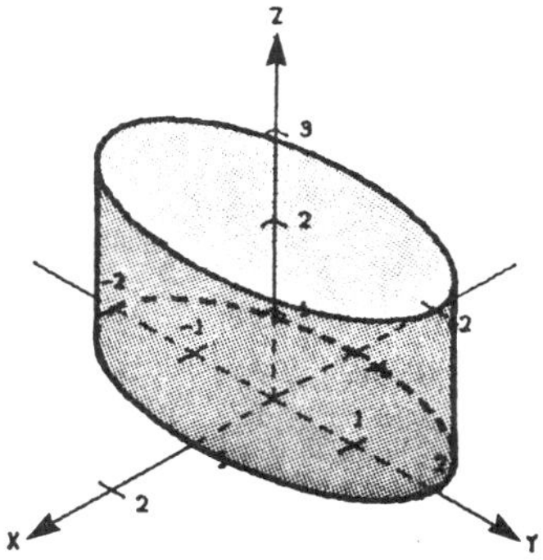

16. $z = \sin y$

Since the x-coordinate is missing we have a cylindrical surface with rulings parallel to the x-axis. The generating curve is the sine curve.

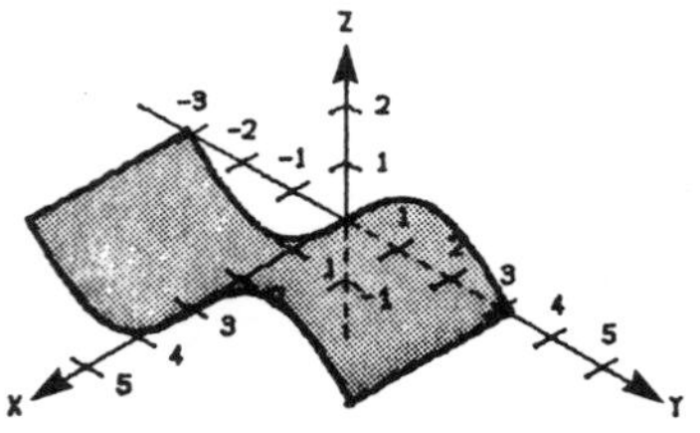

17. $y^2 - z^2 = 4, \qquad \frac{y^2}{4} - \frac{z^2}{4} = 1$

Since the x-coordinate is missing we have a cylindrical surface with rulings parallel to the x-axis. The generating curve is a hyperbola.

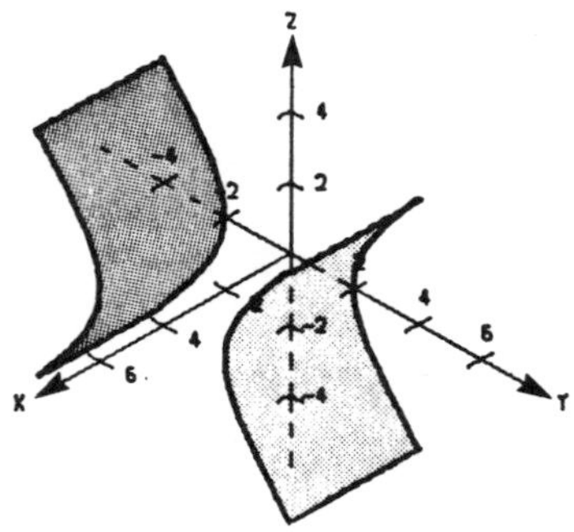

18. $z = e^y$

Since the x-coordinate is missing we have a cylindrical surface with rulings parallel to the x-axis. The generating curve is the exponential curve.

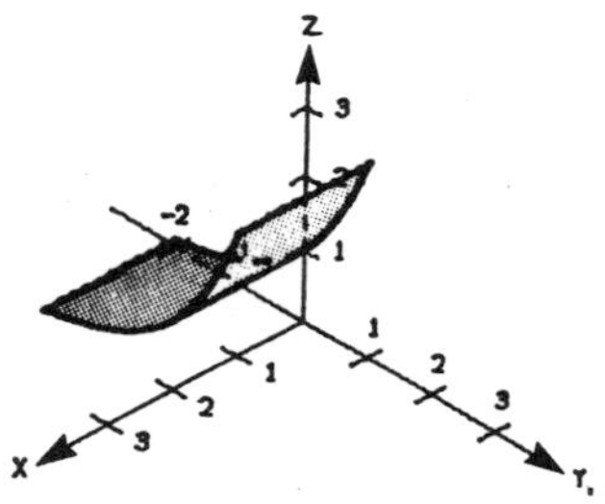

19. $\frac{x^2}{1} + \frac{y^2}{4} + \frac{z^2}{1} = 1$

Ellipsoid

xy-trace: $\frac{x^2}{1} + \frac{y^2}{4} = 1$

xz-trace: $x^2 + z^2 = 1$

yz-trace: $\frac{y^2}{4} + \frac{z^2}{1} = 1$

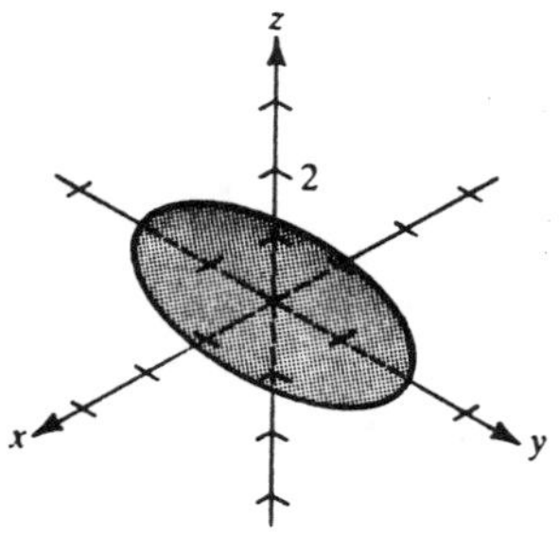

20. $\frac{x^2}{9} + \frac{y^2}{16} + \frac{z^2}{16} = 1$

Ellipsoid

xy-trace: $\frac{x^2}{9} + \frac{y^2}{16} = 1$

xz-trace: $\frac{x^2}{9} + \frac{z^2}{16} = 1$

yz-trace: $y^2 + z^2 = 16$

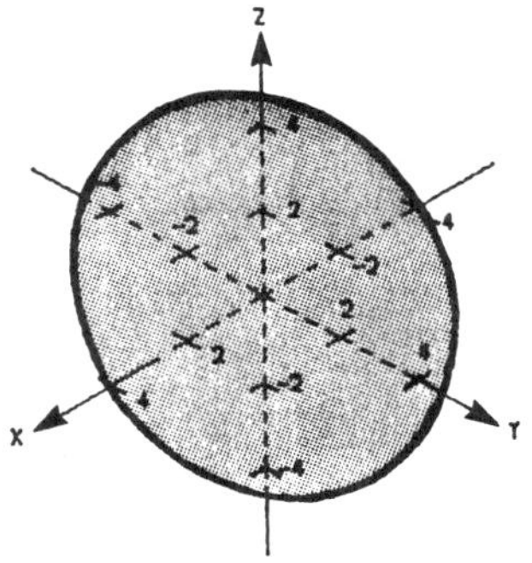

21. $16x^2 - y^2 + 16^2 = 4$

$4x^2 - \frac{y^2}{4} + 4z^2 = 1$

Hyperboloid of one sheet

xy-trace: $4x^2 - \frac{y^2}{4} = 1$

xz-trace: $4(x^2 + z^2) = 1$

yz-trace: $\frac{-y^2}{4} + 4z^2 = 1$

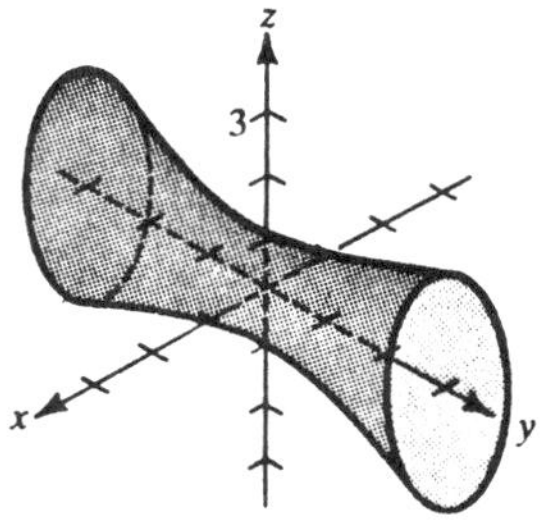

22. $9x^2 + 4y^2 - 8z^2 = 72$

$\frac{x^2}{8} + \frac{y^2}{18} - \frac{z^2}{9} = 1$

Hyperboloid of one sheet

xy-trace: $\frac{x^2}{8} + \frac{y^2}{18} = 1$

xz-trace: $\frac{x^2}{8} - \frac{z^2}{9} = 1$

yz-trace: $\frac{y^2}{18} - \frac{z^2}{9} = 1$

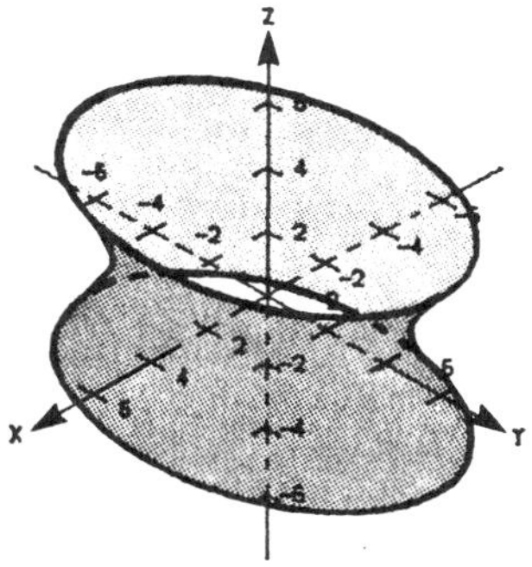

23. $x^2 - y + z^2 = 0$

Elliptical paraboloid

xy-trace: $y = x^2$
xz-trace: $x^2 + z^2 = 0$, point(0, 0, 0)
yz-trace: $y = z^2$

$y = 1$: $x^2 + z^2 = 1$

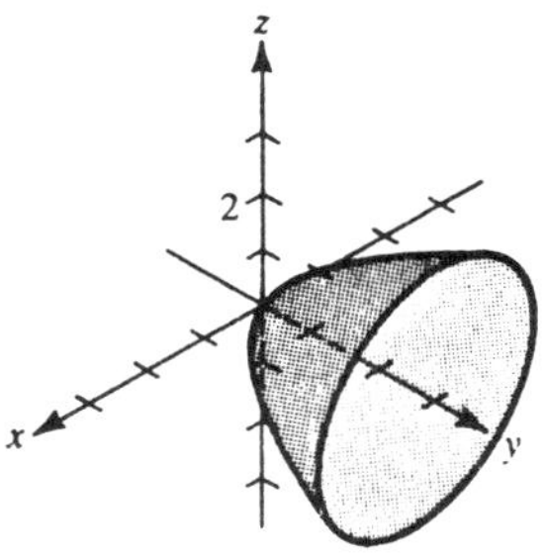

24. $z = 4x^2 + y^2$

Elliptic paraboloid

xy-trace: point(0, 0, 0)
xz-trace: $z = 4x^2$
yz-trace: $z = y^2$

$z = 4$: $x^2 + \dfrac{y^2}{4} = 1$

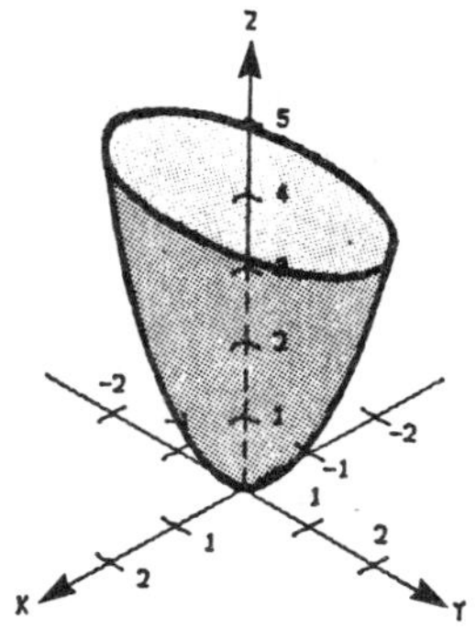

25. $x^2 - y^2 + z = 0$

Hyperboloic paraboloid

xy-trace: $y = \pm x$
xz-trace: $z = -x^2$
yz-trace: $z = y^2$

$y = \pm 1$: $z = 1 - x^2$

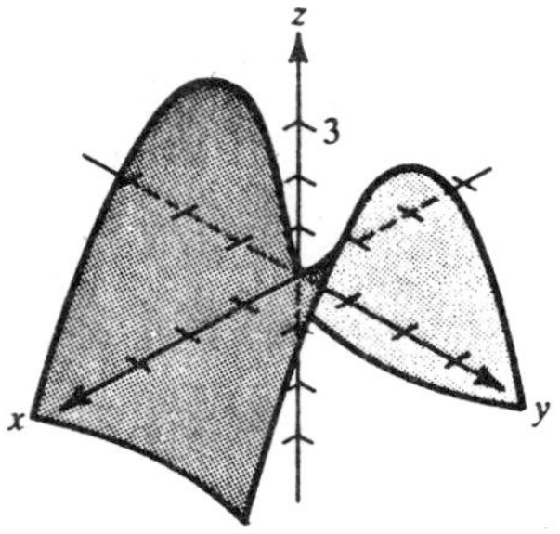

26. $z^2 - x^2 - \dfrac{y^2}{4} = 1$

Hyperboloid of two sheets

xy-trace: none

xz-trace: $z^2 - x^2 = 1$

yz-trace: $z^2 - \dfrac{y^2}{4} = 1$

$z = \pm\sqrt{10}$: $\dfrac{x^2}{9} + \dfrac{y^2}{36} = 1$

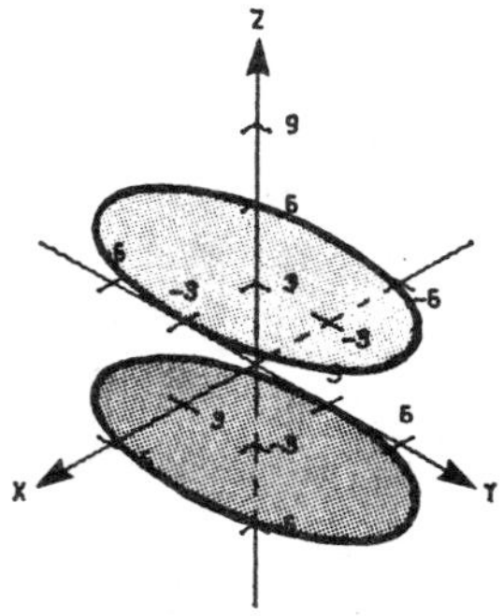

27. $4x^2 - y^2 + 4z^2 = -16$

Hyperboloid of two sheets

xy-trace: $\dfrac{y^2}{16} - \dfrac{x^2}{4} = 1$

xz-trace: none

yz-trace: $\dfrac{y^2}{16} - \dfrac{z^2}{16} = 1$

$y = 16$: $x^2 + z^2 = 64$

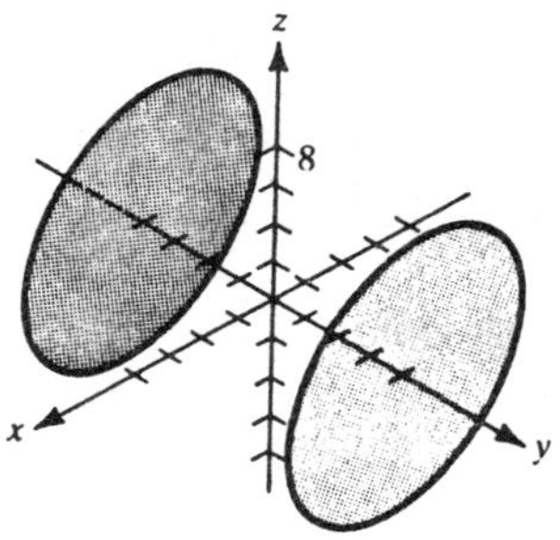

28. $z^2 = x^2 + \frac{y^2}{4}$, Cone

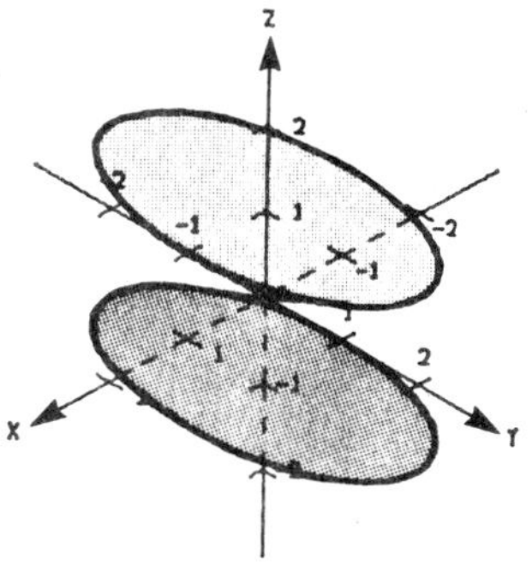

xy-trace: point(0, 0, 0)

xz-trace: $z = \pm x$

yz-trace: $z = \frac{\pm 1}{2} y$

$z = \pm 1$: $x^2 + \frac{y^2}{4} = 1$

29. $z^2 = x^2 + 4y^2$, Cone

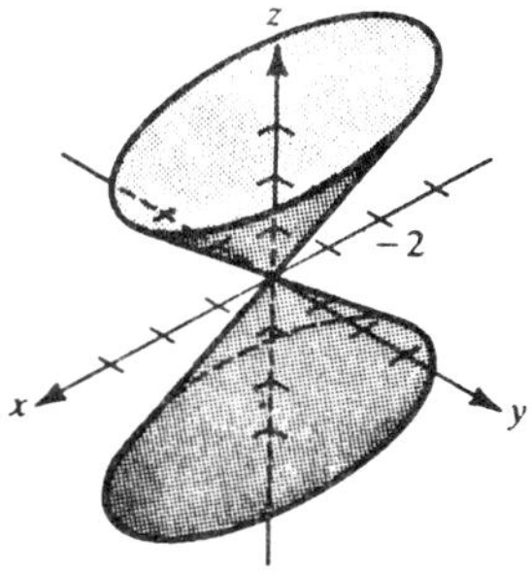

$x = 0$: $z = \pm 2y$
$y = 0$: $z = \pm x$

$z = 4$: $1 = \frac{x^2}{16} + \frac{y^2}{4}$

xy-trace: point(0, 0, 0)
yz-trace: $z = \pm 2y$
xz-trace: $z = \pm x$

$z = 1$: $x^2 + 4y^2 = 1$

30. $4y = x^2 + z^2$

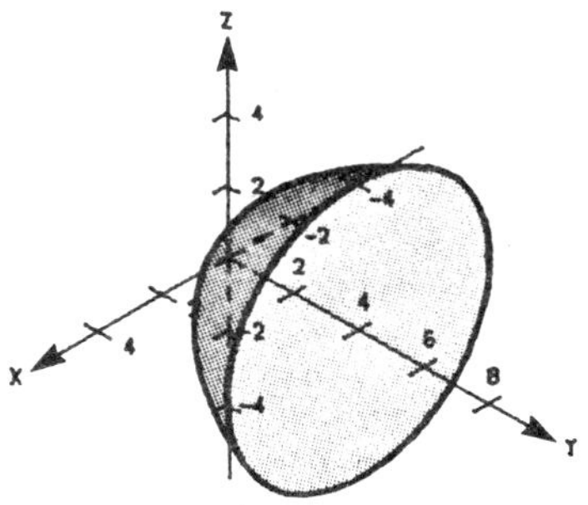

Paraboloid of revolution

xy-trace: $y = \frac{1}{4}x^2$

yz-trace: $y = \frac{1}{4}z^2$

xz-trace: point(0, 0, 0)

$y = 4$: $x^2 + y^2 = 16$

31. $3z = -y^2 + x^2$

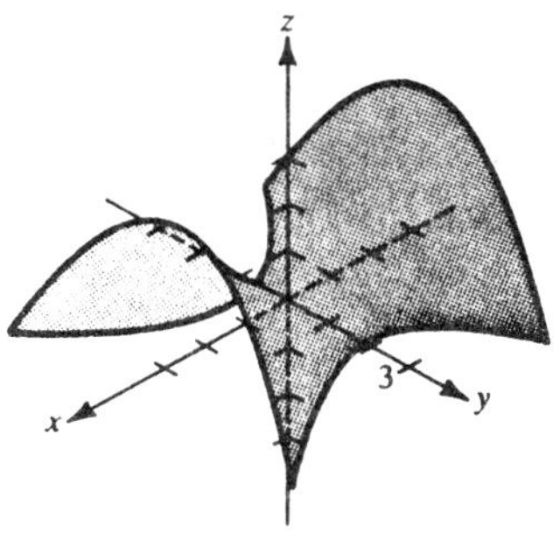

Hyperbolic paraboloid

xy-trace: $y = \pm x$

xz-trace: $z = \frac{1}{3}x^2$

yz-trace: $z = -\frac{1}{3}y^2$

32. $z^2 = 2x^2 + 2y^2$

Cone

xy-trace: point(0, 0, 0)
xz-trace: $z = \pm \sqrt{2}x$
yz-trace: $z = \pm \sqrt{2}y$

$z = 4$: $x^2 + y^2 = 8$

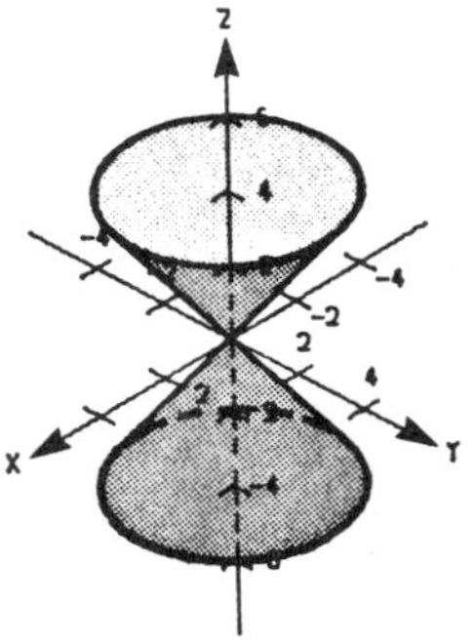

33. $16x^2 + 9y^2 + 16z^2 - 32x - 36y + 36 = 0$

$$16(x^2 - 2x + 1) + 9(y^2 - 4y + 4) + 16z^2 = -36 + 16 + 36$$

$$16(x - 1)^2 + 9(y - 2)^2 + 16z^2 = 16$$

$$\frac{(x - 1)^2}{1} + \frac{(y - 2)^2}{(16/9)} + \frac{z^2}{1} = 1$$

Ellipsoid with center (1, 2, 0)

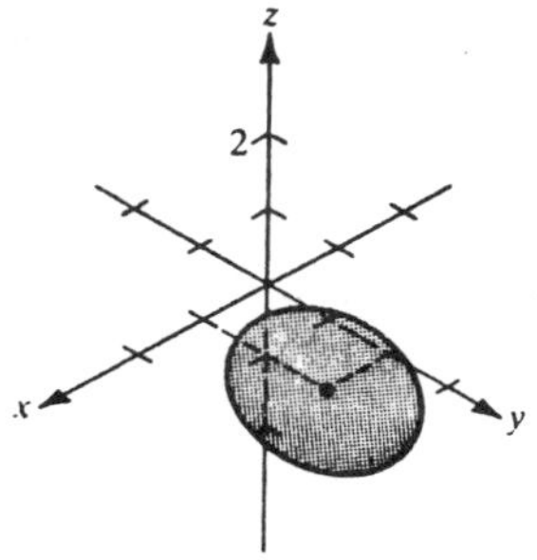

34. $4x^2 + y^2 - 4z^2 - 16x - 6y - 16z + 9 = 0$

$$4(x^2 - 4x + 4) + (y^2 - 6y + 9) - 4(z^2 + 4z + 4) = -9 + 16 + 9 - 16$$

$$4(x - 2)^2 + (y - 3)^2 - 4(z + 2)^2 = 0$$

$$\frac{(x - 2)^2}{(1/4)} + \frac{(y - 3)^2}{1} - \frac{(z + 2)^2}{(1/4)} = 0$$

Elliptic cone with center (2, 3, −2)

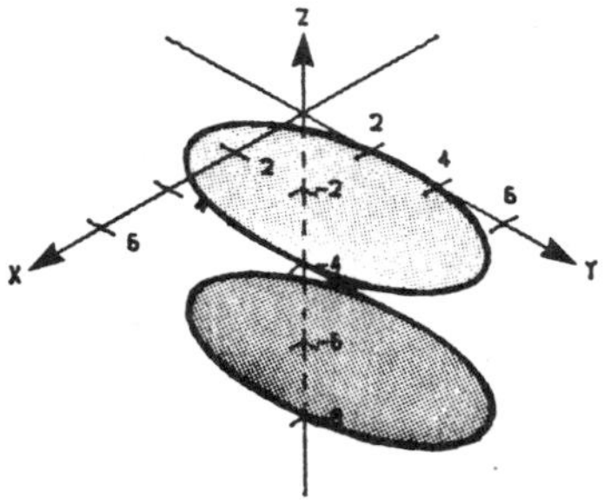

35. $z = 2\sqrt{x^2 + y^2}$
$z = 2$
$2\sqrt{x^2 + y^2} = 2$
$x^2 + y^2 = 1$

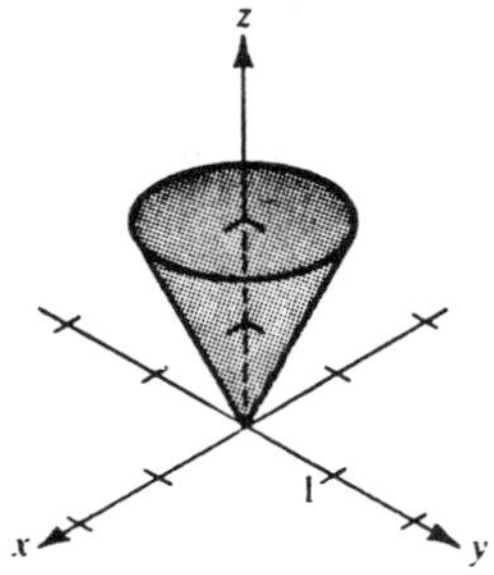

36. $z = \sqrt{4 - x^2}$
$y = \sqrt{4 - x^2}$.
$x = 0, \quad y = 0, \quad z = 0$

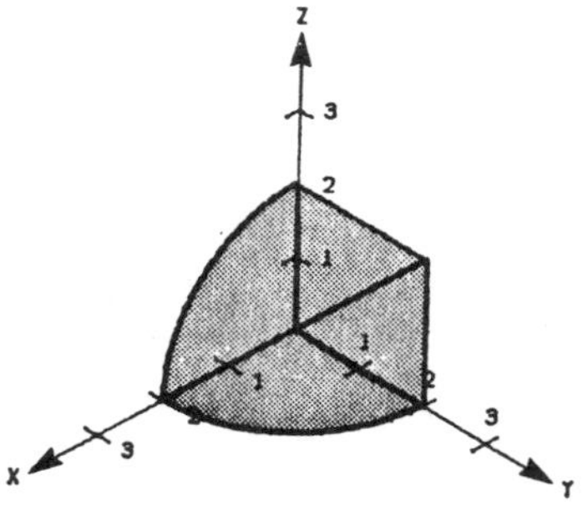

37. $x^2 + y^2 = 1$
$x + y = 2$
$z = 0$

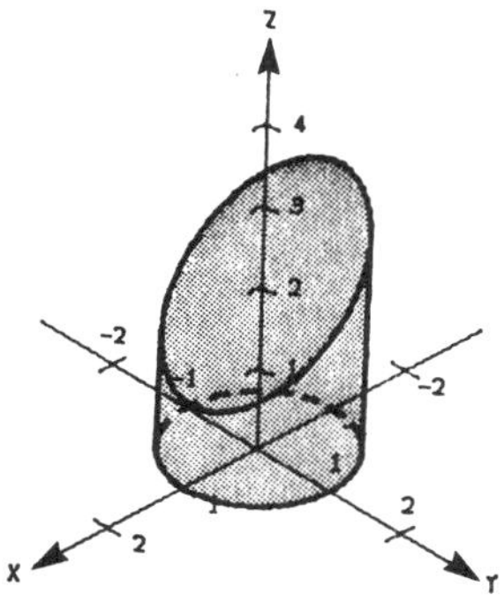

38. $x^2 + y^2 + z^2 = 4$
$z = \sqrt{x^2 + y^2}$
$z = 0$

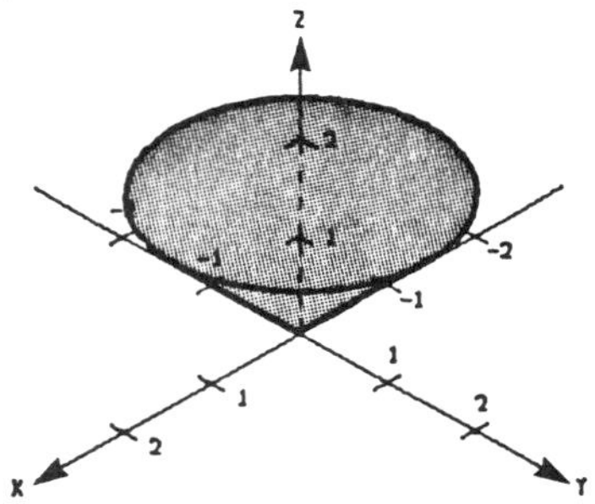

39. $z = \sqrt{4 - x^2 - y^2}$
$y = 2z$
$z = 0$

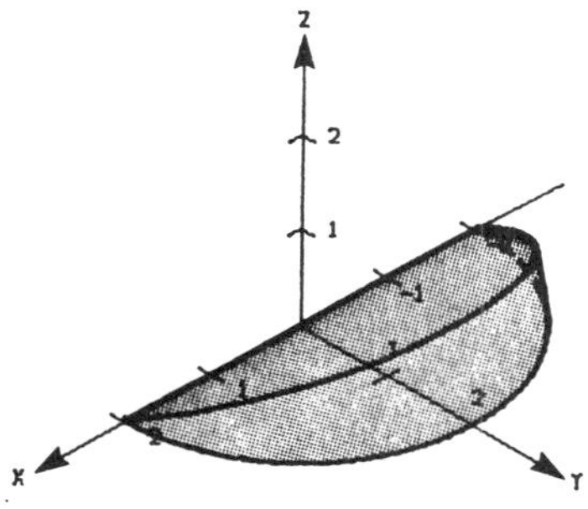

40. $z = \sqrt{x^2 + y^2}$
$z = 4 - x^2 - y^2$

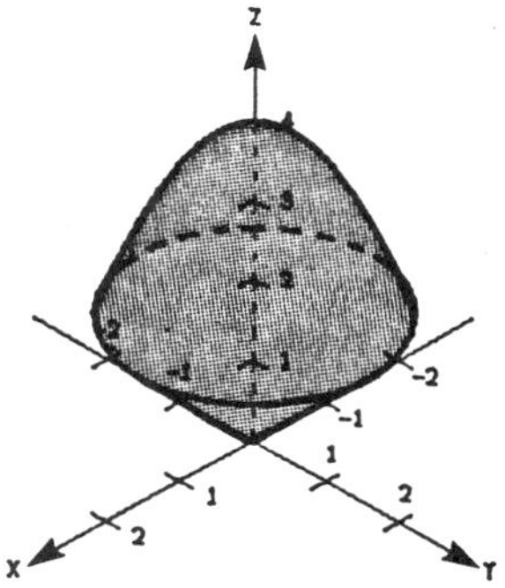

41. $x^2 + z^2 = [r(y)]^2$ But $z = r(y) = \pm 2\sqrt{y}$, therefore $x^2 + z^2 = 4y$

42. $x^2 + z^2 = [r(y)]^2$ But $z = r(y) = 2y$, therefore $x^2 + z^2 = 4y^2$

43. $x^2 + y^2 = [r(z)]^2$

But $y = r(z) = \dfrac{z}{2}$, therefore $x^2 + y^2 = \dfrac{z^2}{4}$, $4x^2 + 4y^2 = z^2$

44. $y^2 + z^2 = [r(x)]^2$ But $z = r(x) = \dfrac{1}{2}\sqrt{4 - x^2}$,

therefore $y^2 + z^2 = \dfrac{1}{4}(4 - x^2)$, $x^2 + 4y^2 + 4z^2 = 4$

45. $y^2 + z^2 = [r(x)]^2$

But $y = r(x) = \frac{2}{x}$, therefore $y^2 + z^2 = (\frac{2}{x})^2$, $y^2 + z^2 = \frac{4}{x^2}$

46. $x^2 + y^2 = [r(z)]^2$ But $y = r(z) = e^z$, therefore $x^2 + y^2 = e^{2z}$

47. $x^2 + y^2 - 2z = 0$, $x^2 + y^2 = (\sqrt{2z})^2$

Equation of generating curve: $y = \sqrt{2z}$ or $x = \sqrt{2z}$

48. $x^2 + z^2 = \sin^2 y$

Equation of generating curve: $x = \sin y$ or $z = \sin y$

49. $z = \frac{x^2}{2} + \frac{y^2}{4}$

(a) When $z = 2$ we have: $2 = \frac{x^2}{2} + \frac{y^2}{4}$ Major axis: $2\sqrt{8} = 4\sqrt{2}$

or: $1 = \frac{x^2}{4} + \frac{y^2}{8}$ Minor axis: $2\sqrt{4} = 4$

$c^2 = a^2 - b^2$, $c^2 = 4$, $c = 2$, Foci: $(0, \pm 2, 2)$

(b) When $z = 8$ we have: $8 = \frac{x^2}{2} + \frac{y^2}{4}$ Major axis: $2\sqrt{32} = 8\sqrt{2}$

or: $1 = \frac{x^2}{16} + \frac{y^2}{32}$ Minor axis: $2\sqrt{16} = 8$

$c^2 = 32 - 16 = 16$, $c = 4$, Foci: $(0, \pm 4, 8)$

50. $z = \frac{x^2}{2} + \frac{y^2}{4}$

(a) When $y = 4$ we have:

$z = \frac{x^2}{2} + 4$, $4(\frac{1}{2})(z - 4) = x^2$

Focus: $(0, 4, \frac{9}{2})$

(b) When $x = 2$ we have:

$z = 2 + \frac{y^2}{4}$, $4(z - 2) = y^2$

Focus: $(2, 0, 3)$

12.7 Cylindrical and spherical coordinates

1. (0, 5, 1) rectangular, $r = \sqrt{(0)^2 + (5)^2} = 5$

   $\theta = \arctan\frac{5}{0} = \frac{\pi}{2}$, $z = 1$, $(5, \frac{\pi}{2}, 1)$ cylindrical

2. $(2\sqrt{2}, -2\sqrt{2}, 4)$ rectangular, $r = \sqrt{(2\sqrt{2})^2 + (-2\sqrt{2})^2} = 4$

   $\theta = \arctan(-1) = -\frac{\pi}{4}$, $z = 4$, $(4, -\frac{\pi}{4}, 4)$ cylindrical

3. $(1, \sqrt{3}, 4)$ rectangular, $r = \sqrt{1^2 + (\sqrt{3})^2} = 2$

   $\theta = \arctan\sqrt{3} = \frac{\pi}{3}$, $z = 4$, $(2, \frac{\pi}{3}, 4)$ cylindrical

4. $(\sqrt{3}, -1, 2)$ rectangular, $r = \sqrt{(\sqrt{3})^2 + (-1)^2} = 2$

   $\theta = \arctan(\frac{-1}{\sqrt{3}}) = \frac{-\pi}{6}$, $z = 2$, $(2, \frac{-\pi}{6}, 2)$ cylindrical

5. (2, −2, −4) rectangular, $r = \sqrt{2^2 + (-2)^2} = 2\sqrt{2}$

   $\theta = \arctan(-1) = \frac{-\pi}{4}$, $z = -4$, $(2\sqrt{2}, \frac{-\pi}{4}, -4)$ cylindrical

6. (−3, 2, −1) rectangular, $r = \sqrt{(-3)^2 + 2^2} = \sqrt{13}$

   $\theta = \arctan(\frac{-2}{3}) = -\arctan\frac{2}{3}$, $z = -1$, $(\sqrt{13}, -\arctan\frac{2}{3}, -1)$ cylindrical

7. (5, 0, 2) cylindrical, $x = 5\cos 0 = 5$, $y = 5\sin 0 = 0$, $z = 2$

   (5, 0, 2) rectangular

8. $(4, \frac{\pi}{2}, -2)$ cylindrical, $x = 4\cos\frac{\pi}{2} = 0$, $y = 4\sin\frac{\pi}{2} = 4$, $z = -2$

   (0, 4, −2) rectangular

9. $(2, \frac{\pi}{3}, 2)$ cylindrical, $x = 2\cos\frac{\pi}{3} = 1$, $y = 2\sin\frac{\pi}{3} = \sqrt{3}$, $z = 2$

   $(1, \sqrt{3}, 2)$ rectangular

10. $(3, \frac{-\pi}{4}, 1)$ cylindrical, $x = 3\cos(\frac{-\pi}{4}) = \frac{3}{\sqrt{2}}$, $y = 3\sin(\frac{-\pi}{4}) = \frac{-3}{\sqrt{2}}$

$z = 1$, $(\frac{3}{\sqrt{2}}, \frac{-3}{\sqrt{2}}, 1)$ rectangular

11. $(4, \frac{7\pi}{6}, 3)$ cylindrical, $x = 4\cos\frac{7\pi}{6} = -2\sqrt{3}$, $y = 4\sin\frac{7\pi}{6} = -2$

$z = 3$, $(-2\sqrt{3}, -2, 3)$ rectangular

12. $(1, \frac{3\pi}{2}, 1)$ cylindrical, $x = \cos\frac{3\pi}{2} = 0$, $y = \sin\frac{3\pi}{2} = -1$, $z = 1$

$(0, -1, 1)$ rectangular

13. $(4, 0, 0)$ rectangular, $\rho = \sqrt{4^2 + 0^2 + 0^2} = 4$, $\theta = \arctan 0 = 0$,

$\phi = \arccos 0 = \frac{\pi}{2}$, $(4, 0, \frac{\pi}{2})$ spherical

14. $(1, 1, 1)$ rectangular, $\rho = \sqrt{1^2 + 1^2 + 1^2} = \sqrt{3}$, $\theta = \arctan 1 = \frac{\pi}{4}$

$\phi = \arccos\frac{1}{\sqrt{3}}$, $(\sqrt{3}, \frac{\pi}{4}, \arccos\frac{1}{\sqrt{3}})$ spherical

15. $(-2, 2\sqrt{3}, 4)$ rectangular, $\rho = \sqrt{(-2)^2 + (2\sqrt{3})^2 + 4^2} = 4\sqrt{2}$

$\theta = \arctan(-\sqrt{3}) = \frac{2\pi}{3}$, $\phi = \arccos\frac{1}{\sqrt{2}} = \frac{\pi}{4}$, $(4\sqrt{2}, \frac{2\pi}{3}, \frac{\pi}{4})$ spherical

16. $(2, 2, 4\sqrt{2})$ rectangular, $\rho = \sqrt{2^2 + 2^2 + (4\sqrt{2})^2} = 2\sqrt{10}$

$\theta = \arctan 1 = \frac{\pi}{4}$, $\phi = \arccos\frac{2}{\sqrt{5}}$, $(2\sqrt{10}, \frac{\pi}{4}, \arccos\frac{2}{\sqrt{5}})$ spherical

17. $(\sqrt{3}, 1, 2\sqrt{3})$ rectangular, $\rho = \sqrt{3 + 1 + 12} = 4$, $\theta = \arctan\frac{1}{\sqrt{3}} = \frac{\pi}{6}$

$\phi = \arccos\frac{\sqrt{3}}{2} = \frac{\pi}{6}$, $(4, \frac{\pi}{6}, \frac{\pi}{6})$ spherical

18. $(-4, 0, 0)$ rectangular, $\rho = \sqrt{(-4)^2 + 0^2 + 0^2} = 4$, $\theta = \arctan 0 = 0$,

$\phi = \arccos 0 = \frac{\pi}{2}$, $(4, 0, \frac{\pi}{2})$ spherical

19. $(4, \frac{\pi}{6}, \frac{\pi}{4})$ spherical, $x = 4\sin\frac{\pi}{4}\cos\frac{\pi}{6} = \sqrt{6}$

$y = 4\sin\frac{\pi}{4}\sin\frac{\pi}{6} = \sqrt{2}$, $z = 4\cos\frac{\pi}{4} = 2\sqrt{2}$, $(\sqrt{6}, \sqrt{2}, 2\sqrt{2})$ rectangular

20. $(12, \frac{3\pi}{4}, \frac{\pi}{9})$ spherical, $x = 12 \sin\frac{\pi}{9} \cos\frac{3\pi}{4} \approx -2.902$

$y = 12 \sin\frac{\pi}{9} \sin\frac{3\pi}{4} \approx 2.902$, $z = 12\cos\frac{\pi}{9} \approx 11.276$

$(-2.902, 2.902, 11.276)$ rectangular

21. $(12, \frac{-\pi}{4}, 0)$ spherical, $x = 12 \sin 0 \cos(\frac{-\pi}{4}) = 0$,

$y = 12 \sin 0 \sin(\frac{-\pi}{4}) = 0$, $z = 12\cos 0 = 12$, $(0, 0, 12)$ rectangular

22. $(9, \frac{\pi}{4}, \pi)$ spherical, $x = 9 \sin\pi \cos\frac{\pi}{4} = 0$, $y = 9\sin\pi \sin\frac{\pi}{4} = 0$,

$z = 9\cos\pi = -9$, $(0, 0, -9)$ rectangular

23. $(5, \frac{\pi}{4}, \frac{3\pi}{4})$ spherical, $x = 5\sin\frac{3\pi}{4}\cos\frac{\pi}{4} = \frac{5}{2}$, $y = 5\sin\frac{3\pi}{4}\sin\frac{\pi}{4} = \frac{5}{2}$,

$z = 5\cos\frac{3\pi}{4} = -\frac{5\sqrt{2}}{2}$, $(\frac{5}{2}, \frac{5}{2}, -\frac{5\sqrt{2}}{2})$ rectangular

24. $(6, \pi, \frac{\pi}{2})$ spherical, $x = 6\sin\frac{\pi}{2}\cos\pi = -6$, $y = 6\sin\frac{\pi}{2}\sin\pi = 0$,

$z = 6\cos\frac{\pi}{2} = 0$, $(-6, 0, 0)$ rectangular

25. $(4, \frac{\pi}{4}, 0)$ cylindrical, $\rho = \sqrt{4^2 + 0^2} = 4$, $\theta = \frac{\pi}{4}$, $\phi = \arccos 0 = \frac{\pi}{2}$,

$(4, \frac{\pi}{4}, \frac{\pi}{2})$ spherical

26. $(2, \frac{2\pi}{3}, -2)$ cylindrical, $\rho = \sqrt{2^2 + (-2)^2} = 2\sqrt{2}$, $\theta = \frac{2\pi}{3}$

$\phi = \arccos(\frac{-1}{\sqrt{2}}) = \frac{3\pi}{4}$, $(2\sqrt{2}, \frac{2\pi}{3}, \frac{3\pi}{4})$ spherical

27. $(4, \frac{-\pi}{6}, 6)$ cylindrical, $\rho = \sqrt{4^2 + 6^2} = 2\sqrt{13}$, $\theta = \frac{-\pi}{6}$

$\phi = \arccos\frac{3}{\sqrt{13}}$, $(2\sqrt{13}, \frac{-\pi}{6}, \arccos\frac{3}{\sqrt{13}})$ spherical

28. $(-4, \frac{\pi}{3}, 4)$ cylindrical, $\rho = \sqrt{(-4)^2 + 4^2} = 4\sqrt{2}$, $\theta = \frac{\pi}{3}$,

$\phi = \arccos\frac{1}{\sqrt{2}} = \frac{\pi}{4}$, $(4\sqrt{2}, \frac{\pi}{3}, \frac{\pi}{4})$ spherical

29. $(12, \pi, 5)$ cylindrical, $\rho = \sqrt{12^2 + 5^2} = 13$, $\theta = \pi$, $\phi = \arccos\dfrac{5}{13}$

$(13, \pi, \arccos\dfrac{5}{13})$ spherical

30. $(4, \dfrac{\pi}{2}, 3)$ cylindrical, $\rho = \sqrt{4^2 + 3^2} = 5$, $\theta = \dfrac{\pi}{2}$, $\phi = \arccos\dfrac{3}{5}$

$(5, \dfrac{\pi}{2}, \arccos\dfrac{3}{5})$ spherical

31. $(10, \dfrac{\pi}{6}, \dfrac{\pi}{2})$ spherical, $r = 10\sin\dfrac{\pi}{2} = 10$, $\theta = \dfrac{\pi}{6}$, $z = 10\cos\dfrac{\pi}{2} = 0$

$(10, \dfrac{\pi}{6}, 0)$ cylindrical

32. $(4, \dfrac{\pi}{18}, \dfrac{\pi}{2})$ spherical, $r = 4\sin\dfrac{\pi}{2} = 4$, $\theta = \dfrac{\pi}{18}$, $z = 4\cos\dfrac{\pi}{2} = 0$

$(4, \dfrac{\pi}{18}, 0)$ cylindrical

33. $(6, -\dfrac{\pi}{6}, 6)$ spherical, $r = 6\sin 6$, $\theta = -\dfrac{\pi}{6}$, $z = 6\cos 6$,

$(6\sin 6, -\dfrac{\pi}{6}, 6\cos 6)$ cylindrical

34. $(5, -\dfrac{5\pi}{6}, \pi)$ spherical, $r = 5\sin\pi = 0$, $\theta = -\dfrac{5\pi}{6}$, $z = 5\cos\pi = -5$

$(0, -\dfrac{5\pi}{6}, -5)$ cylindrical

35. $(8, \dfrac{7\pi}{6}, \dfrac{\pi}{6})$ spherical, $r = 8\sin\dfrac{\pi}{6} = 4$, $\theta = \dfrac{7\pi}{6}$, $z = 8\cos\dfrac{\pi}{6} = \dfrac{8\sqrt{3}}{2}$

$(4, \dfrac{7\pi}{6}, 4\sqrt{3})$ cylindrical

36. $(7, \dfrac{\pi}{4}, \dfrac{3\pi}{4})$ spherical, $r = 7\sin\dfrac{3\pi}{4} = \dfrac{7\sqrt{2}}{2}$, $\theta = \dfrac{\pi}{4}$

$z = 7\cos\dfrac{3\pi}{4} = -\dfrac{7\sqrt{2}}{2}$, $(\dfrac{7\sqrt{2}}{2}, \dfrac{\pi}{4}, -\dfrac{7\sqrt{2}}{2})$ cylindrical

37. $r = 5$, cylinder

matches graph d

38. $\theta = \frac{\pi}{4}$, plane

matches graph e

39. $\rho = 5$, sphere

matches graph c

40. $\phi = \frac{\pi}{4}$, cone

matches graph a

41. $r^2 = z$, $x^2 + y^2 = z$

paraboloid

matches graph f

42. $\rho = 4\sec\phi$, $z = 4$, plane

matches graph b

43. $r = 2$, $\sqrt{x^2 + y^2} = 2$,

$x^2 + y^2 = 4$

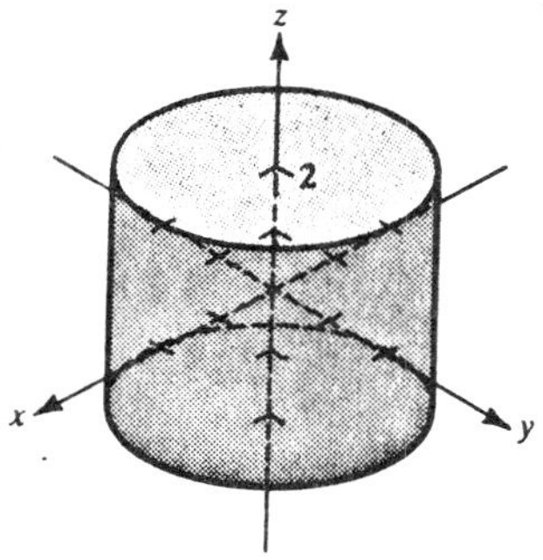

44. $z = 2$, same

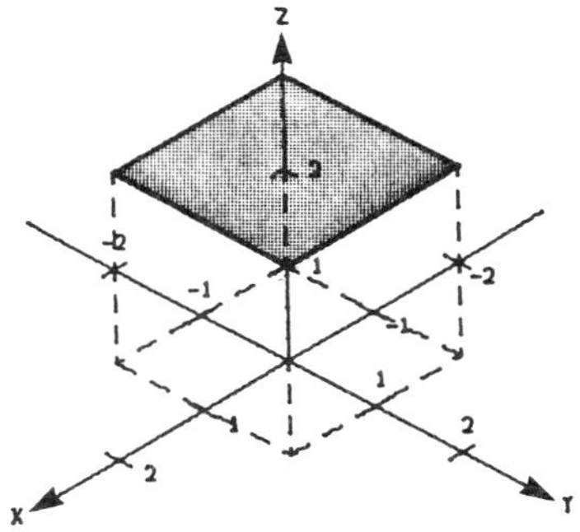

45. $\theta = \frac{\pi}{6}$, $\tan\frac{\pi}{6} = \frac{y}{x}$,

$\frac{1}{\sqrt{3}} = \frac{y}{x}$,

$x = \sqrt{3}y$, $x - \sqrt{3}y = 0$

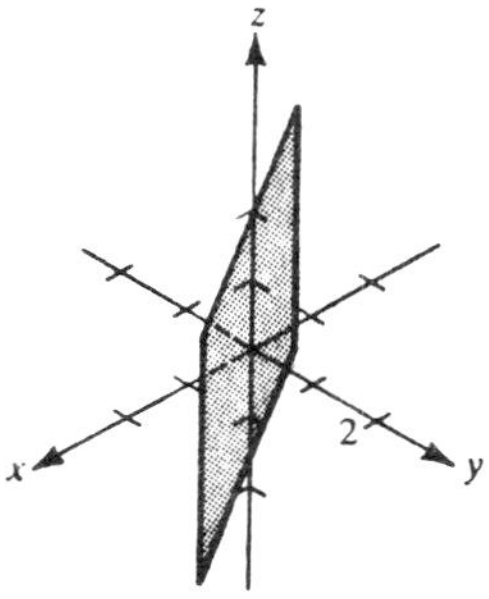

46. $r = \frac{z}{2}$, $\sqrt{x^2 + y^2} = \frac{z}{2}$,

$x^2 + y^2 - \frac{z^2}{4} = 0$

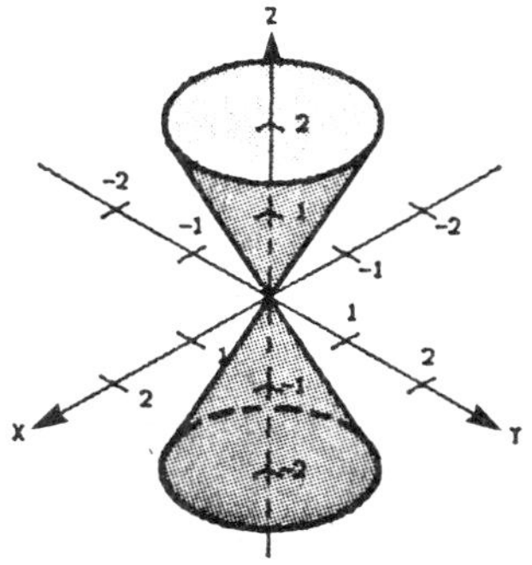

47. $r = 2\sin\theta$, $r^2 = 2r\sin\theta$,

$x^2 + y^2 = 2y$, $x^2 + y^2 - 2y = 0$

$x^2 + (y - 1)^2 = 1$

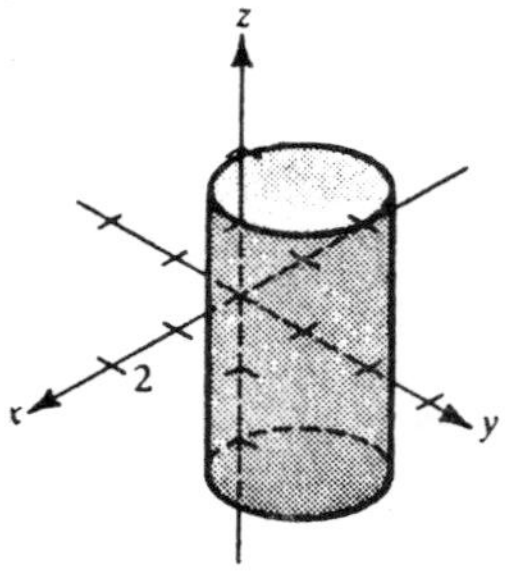

48. $r = 2\cos\theta$, $r^2 = 2r\cos\theta$,

$x^2 + y^2 = 2x$, $x^2 + y^2 - 2x = 0$

$(x - 1)^2 + y^2 = 1$

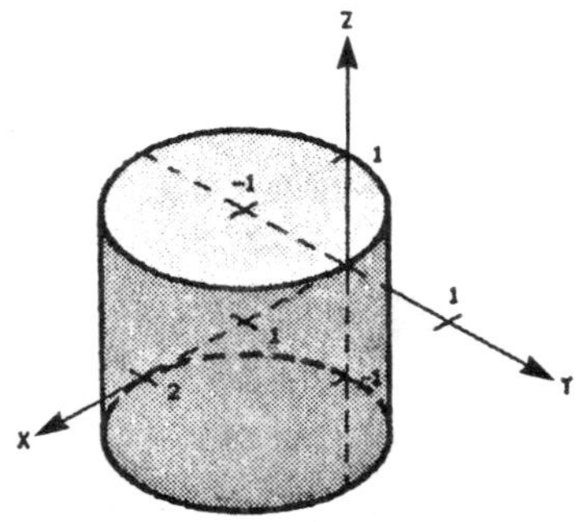

49. $r^2 + z^2 = 4$, $x^2 + y^2 + z^2 = 4$

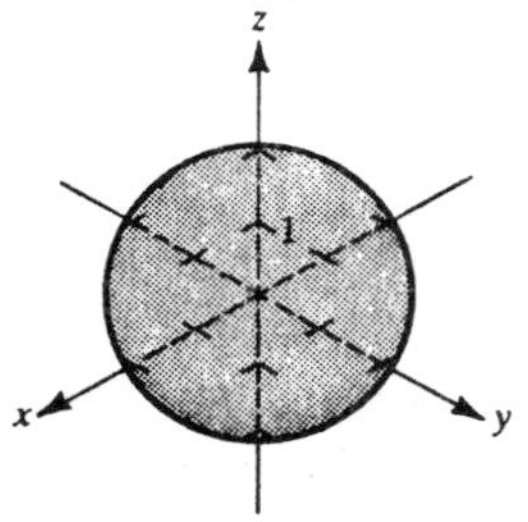

50. $z = r^2\sin^2\theta$, $z = y^2$

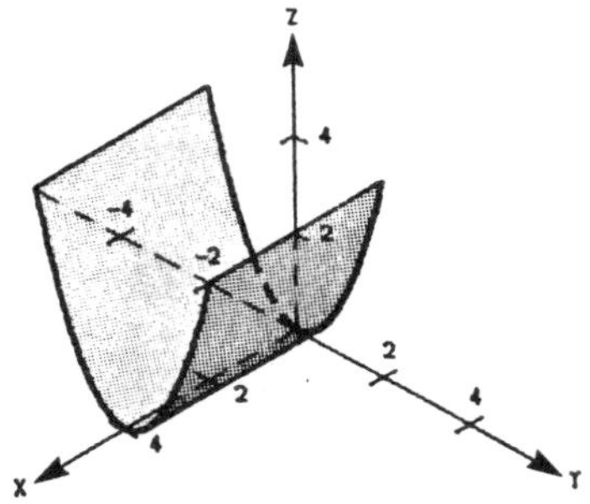

51. $\rho = 2$, $x^2 + y^2 + z^2 = 4$

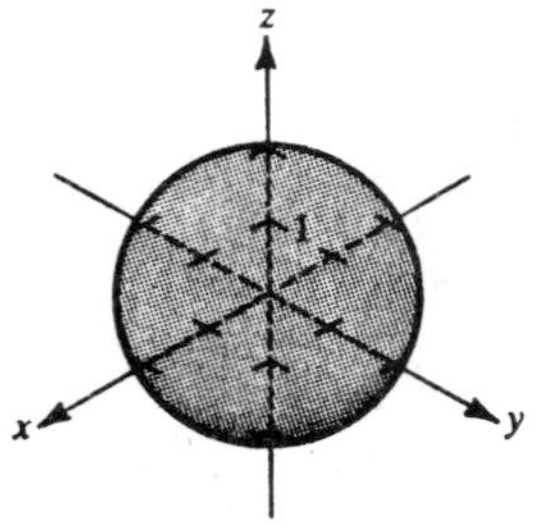

52. $\theta = \dfrac{3\pi}{4}$, $\tan\theta = \dfrac{y}{x}$,

$1 = \dfrac{y}{x}$, $x - y = 0$

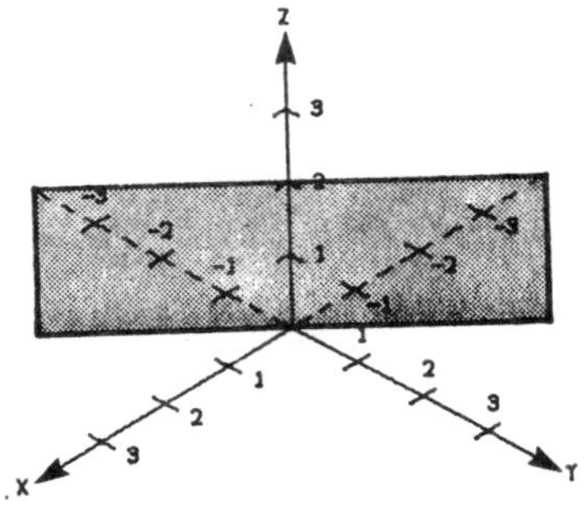

53. $\phi = \frac{\pi}{6}$, $\cos\phi = \frac{z}{\sqrt{x^2 + y^2 + z^2}}$

$\frac{\sqrt{3}}{2} = \frac{z}{\sqrt{x^2 + y^2 + z^2}}$

$\frac{3}{4} = \frac{z^2}{x^2 + y^2 + z^2}$

$3x^2 + 3y^2 - z^2 = 0$

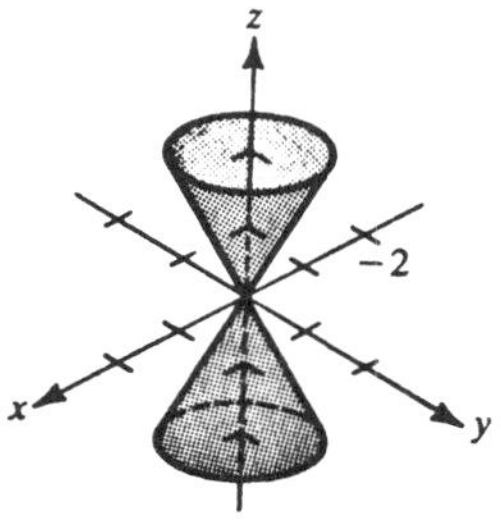

54. $\phi = \frac{\pi}{2}$, $\cos\phi = \frac{z}{\sqrt{x^2 + y^2 + z^2}}$

$0 = \frac{z}{\sqrt{x^2 + y^2 + z^2}}$

$z = 0$, xy plane

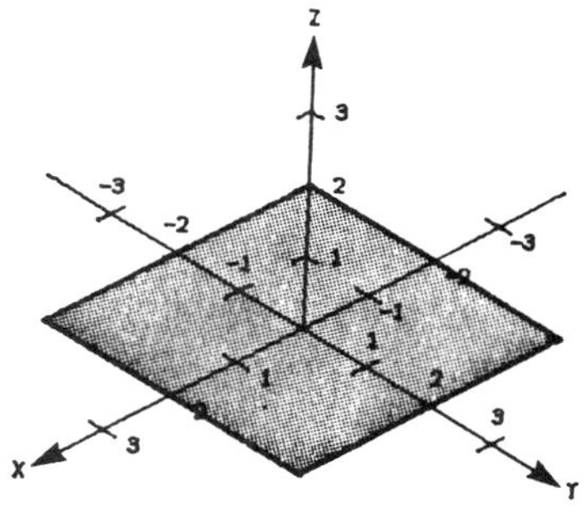

55. $\rho = 4\cos\phi$, $\sqrt{x^2 + y^2 + z^2}$

$= \frac{4z}{\sqrt{x^2 + y^2 + z^2}}$

$x^2 + y^2 + z^2 - 4z = 0$

$x^2 + y^2 + (z - 2)^2 = 4$

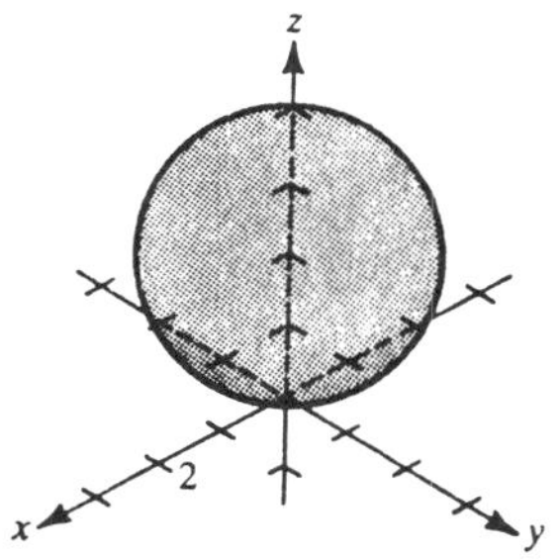

56. $\rho = 2\sec\phi$, $\rho\cos\phi = 2$

$z = 2$

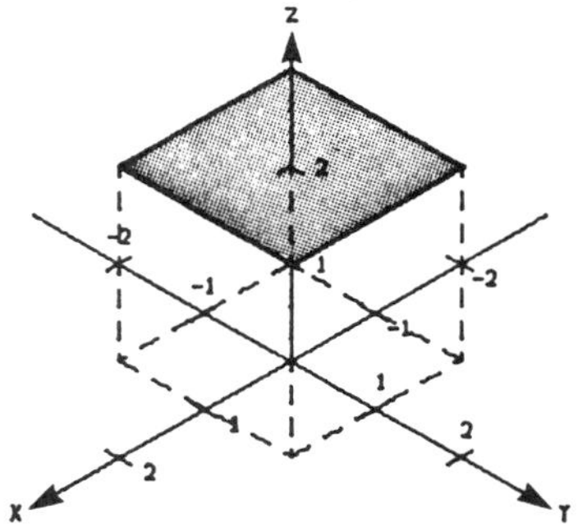

57. $\rho = \csc\phi$, $\rho\sin\phi = 1$,

$\sqrt{x^2 + y^2} = 1$, $x^2 + y^2 = 1$

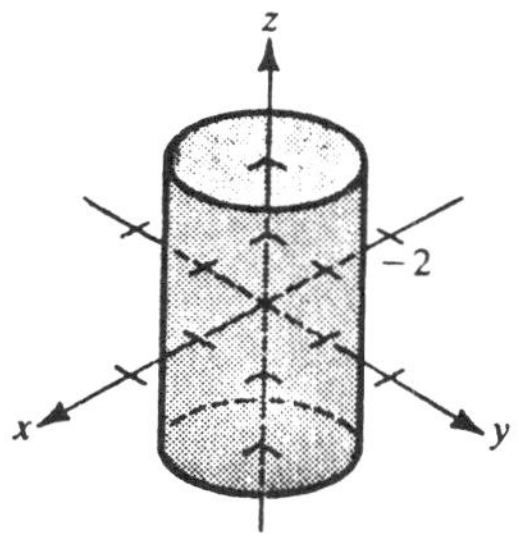

58. $\rho = 4\csc\phi\sec\theta = \frac{4}{\sin\phi\cos\theta}$

$\rho\sin\phi\cos\theta = 4$, $x = 4$

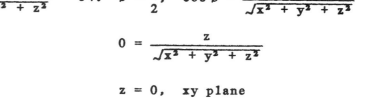

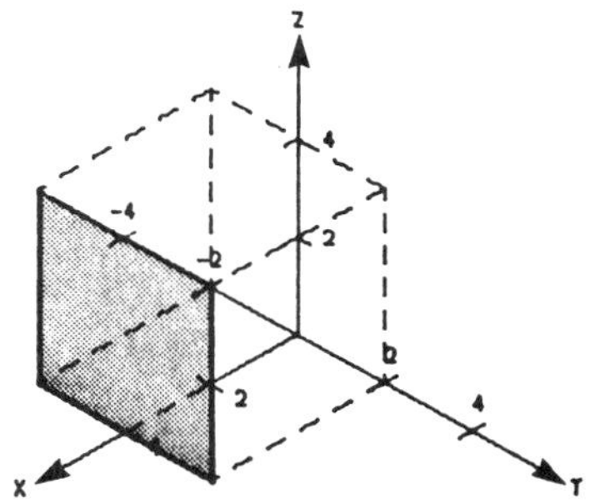

59. $x^2 + y^2 + z^2 = 16$

(a) $r^2 + z^2 = 16$

(b) $\rho^2 = 16, \quad \rho = 4$

60. $4(x^2 + y^2) = z^2$

(a) $4r^2 = z^2, \quad 2r = z$

(b) $4(\rho^2 \sin^2 \phi \cos^2 \theta + \rho^2 \sin^2 \phi \sin^2 \theta) = \rho^2 \cos^2 \phi$

$4 \sin^2 \phi = \cos^2 \phi, \quad \tan^2 \phi = \frac{1}{4}, \quad \tan \phi = \frac{1}{2}, \quad \phi = \arctan \frac{1}{2}$

61. $x^2 + y^2 + z^2 - 2z = 0$

(a) $r^2 + z^2 - 2z = 0, \quad r^2 + (z - 1)^2 = 1$

(b) $\rho^2 - 2\rho \cos \phi = 0, \quad \rho(\rho - 2 \cos \phi) = 0, \quad \rho = 2 \cos \phi$

62. $x^2 + y^2 = z$

(a) $r^2 = z$

(b) $\rho^2 \sin^2 \phi = \rho \cos \phi, \quad \rho \sin^2 \phi = \cos \phi, \quad \rho = \dfrac{\cos \phi}{\sin^2 \phi}$

$\rho = \csc \phi \cot \phi$

63. $x^2 + y^2 = 4y$

(a) $r^2 = 4r \sin \theta, \quad r = 4 \sin \theta$

(b) $\rho^2 \sin^2 \phi = 4\rho \sin \phi \sin \theta, \quad \rho \sin \phi (\rho \sin \phi - 4 \sin \theta) = 0,$

$\rho = \dfrac{4 \sin \theta}{\sin \phi}, \quad \rho = 4 \sin \theta \csc \phi$

64. $x^2 + y^2 = 16$

(a) $r^2 = 16, \quad r = 4$

(b) $\rho^2 \sin^2 \phi = 16, \quad \rho^2 \sin^2 \phi - 16 = 0, \quad (\rho \sin \phi - 4)(\rho \sin \phi + 4) = 0,$

$\rho = 4 \csc \phi$

65. $x^2 - y^2 = 9$

(a) $r^2 \cos^2\theta - r^2 \sin^2\theta = 9,\qquad r^2 = \dfrac{9}{\cos^2\theta - \sin^2\theta}$

(b) $\rho^2 \sin^2\phi \cos^2\theta - \rho^2 \sin^2\phi \sin^2\theta = 9,\qquad \rho^2 \sin^2\phi = \dfrac{9}{\cos^2\theta - \sin^2\theta}$

$\rho^2 = \dfrac{9\csc^2\phi}{\cos^2\theta - \sin^2\theta}$

66. $y = 4,$

(a) $r\sin\theta = 4,\qquad r = 4\csc\theta$

(b) $\rho\sin\phi\sin\theta = 4,\qquad \rho = 4\csc\phi\csc\theta$

67. $0 \le \theta \le 2\pi,\qquad 0 \le r \le a$

$r \le z \le a$

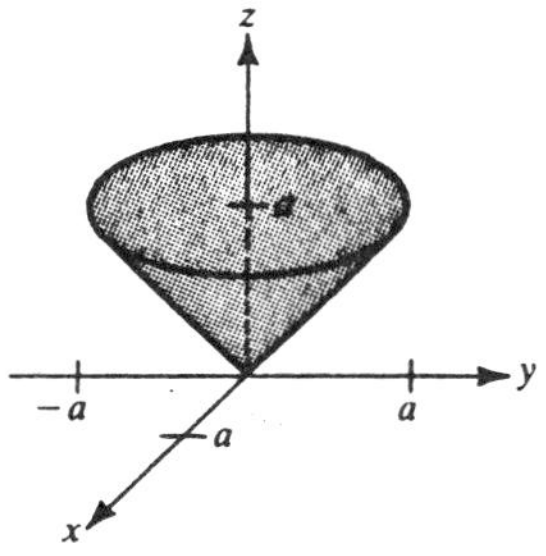

68. $0 \le \theta \le 2\pi,\qquad 2 \le r \le 4,$

$z^2 \le -r^2 + 6r - 8$

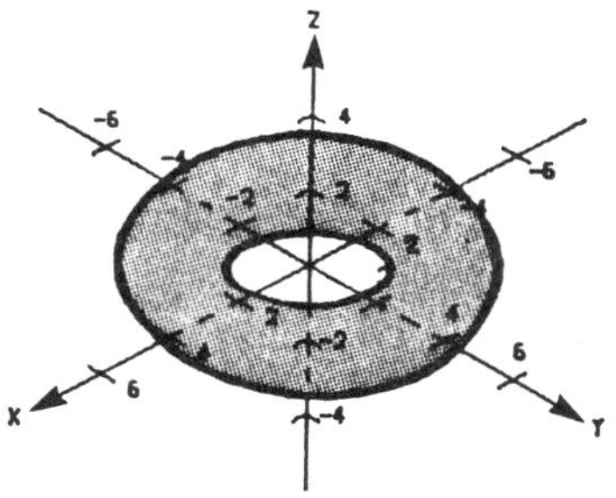

69. $0 \le \theta \le 2\pi,\qquad 0 \le \phi \le \dfrac{\pi}{6}$

$0 \le \rho \le a\sec\phi$

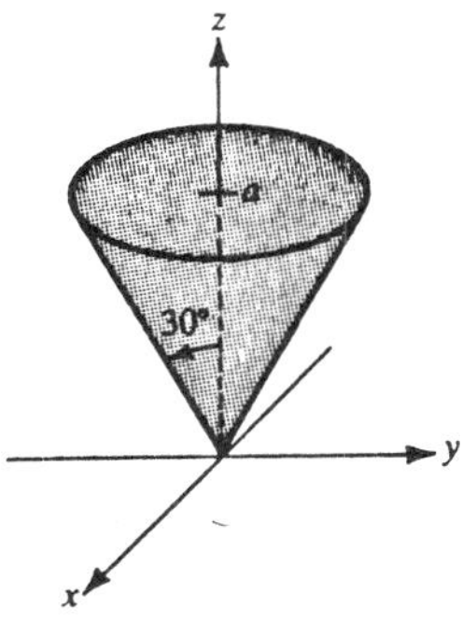

70. $0 \le \theta \le 2\pi,\qquad \dfrac{\pi}{4} \le \phi \le \dfrac{\pi}{2}$

$0 \le \rho \le 1$

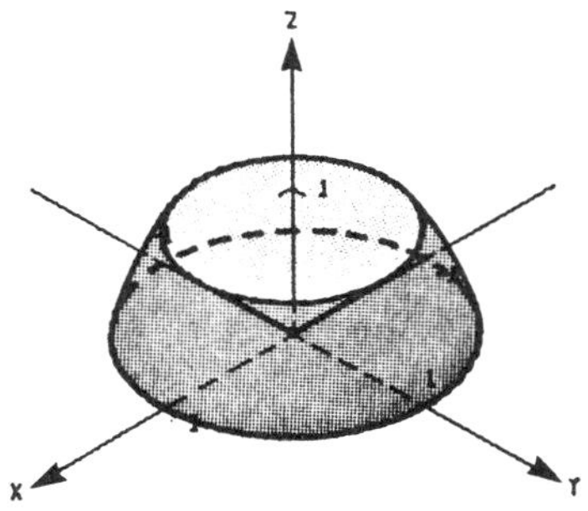

Review Exercises for Chapter 12

1. $P = (1, 2)$, $Q = (4, 1)$, $R = (5, 4)$

(a) $\mathbf{u} = \overrightarrow{PQ} = \langle 3, -1\rangle = 3\mathbf{i} - \mathbf{j}$

$\mathbf{v} = \overrightarrow{PR} = \langle 4, 2\rangle = 4\mathbf{i} + 2\mathbf{j}$

(b) $\|\mathbf{v}\| = \sqrt{4^2 + 2^2} = 2\sqrt{5}$

(c) $\mathbf{u}\cdot\mathbf{v} = 3(4) + (-1)(2) = 10$

(d) $2\mathbf{u} + \mathbf{v} = \langle 6, -2\rangle + \langle 4, 2\rangle$
$= \langle 10, 0\rangle = 10\mathbf{i}$

(e) $\mathbf{w}_1 = \text{proj}_{\mathbf{v}}(\mathbf{u}) = \left(\dfrac{\mathbf{u}\cdot\mathbf{v}}{\|\mathbf{v}\|^2}\right)\mathbf{v} = \dfrac{10}{(2\sqrt{5})^2}(4\mathbf{i} + 2\mathbf{j}) = 2\mathbf{i} + \mathbf{j}$

(f) $\mathbf{w}_2 = \mathbf{u} - \mathbf{w}_1 = (3\mathbf{i} - \mathbf{j}) - (2\mathbf{i} + \mathbf{j}) = \mathbf{i} - 2\mathbf{j}$

2. $P = (-2, -1)$, $Q = (5, -1)$, $R = (2, 4)$

(a) $\mathbf{u} = \overrightarrow{PQ} = \langle 7, 0\rangle = 7\mathbf{i}$

$\mathbf{v} = \overrightarrow{PR} = \langle 4, 5\rangle = 4\mathbf{i} + 5\mathbf{j}$

(b) $\|\mathbf{v}\| = \sqrt{4^2 + 5^2} = \sqrt{41}$

(c) $\mathbf{u}\cdot\mathbf{v} = 7(4) + 0(5) = 28$

(d) $2\mathbf{u} + \mathbf{v} = 14\mathbf{i} + (4\mathbf{i} + 5\mathbf{j})$
$= 18\mathbf{i} + 5\mathbf{j}$

(e) $\mathbf{w}_1 = \text{proj}_{\mathbf{v}}(\mathbf{u}) = \left(\dfrac{\mathbf{u}\cdot\mathbf{v}}{\|\mathbf{v}\|^2}\right)\mathbf{v} = \dfrac{28}{41}(4\mathbf{i} + 5\mathbf{j}) = \dfrac{112}{41}\mathbf{i} + \dfrac{140}{41}\mathbf{j}$

(f) $\mathbf{w}_2 = \mathbf{u} - \mathbf{w}_1 = 7\mathbf{i} - \left(\dfrac{112}{41}\mathbf{i} + \dfrac{140}{41}\mathbf{j}\right) = \dfrac{175}{41}\mathbf{i} - \dfrac{140}{41}\mathbf{j}$

3. $P = (5, 0, 0)$, $Q = (4, 4, 0)$, $R = (2, 0, 6)$

(a) $\mathbf{u} = \overrightarrow{PQ} = \langle -1, 4, 0\rangle = -\mathbf{i} + 4\mathbf{j}$

$\mathbf{v} = \overrightarrow{PR} = \langle -3, 0, 6\rangle = -3\mathbf{i} + 6\mathbf{k}$

(b) $\mathbf{u}\cdot\mathbf{v} = (-1)(-3) + 4(0) + 0(6) = 3$

(c) $\mathbf{u} \times \mathbf{v} = \begin{vmatrix} \mathbf{i} & \mathbf{j} & \mathbf{k} \\ -1 & 4 & 0 \\ -3 & 0 & 6 \end{vmatrix} = 24\mathbf{i} + 6\mathbf{j} + 12\mathbf{k}$

(d) $\mathbf{n} = 24\mathbf{i} + 6\mathbf{j} + 12\mathbf{k}$

$24(x - 5) + 6y + 12z = 0, \quad 4x + y + 2z - 20 = 0$

(e) $x = 5 - t, \quad y = 4t, \quad z = 0$ OR $x = 4 - t, \quad y = 4 + 4t, \quad z = 0$

4. $P = (2, -1, 3)$, $Q = (0, 5, 1)$, $R = (5, 5, 0)$

(a) $\mathbf{u} = \overrightarrow{PQ} = \langle -2, 6, -2 \rangle = -2\mathbf{i} + 6\mathbf{j} - 2\mathbf{k}$

$\mathbf{v} = \overrightarrow{PR} = \langle 3, 6, -3 \rangle = 3\mathbf{i} + 6\mathbf{j} - 3\mathbf{k}$

(b) $\mathbf{u}\cdot\mathbf{v} = (-2)(3) + (6)(6) + (-2)(-3) = 36$

(c) $$\mathbf{u} \times \mathbf{v} = \begin{vmatrix} \mathbf{i} & \mathbf{j} & \mathbf{k} \\ -2 & 6 & -2 \\ 3 & 6 & -3 \end{vmatrix} = -6\mathbf{i} - 12\mathbf{j} - 30\mathbf{k} = -6(\mathbf{i} + 2\mathbf{j} + 5\mathbf{k})$$

(d) $\mathbf{n} = \mathbf{i} + 2\mathbf{j} + 5\mathbf{k}$

$1(x - 2) + 2(y + 1) + 5(z - 3) = 0, \quad x + 2y + 5z = 15$

(e) $x = 2 - 2t, \quad y = -1 + 6t, \quad z = 3 - 2t$ OR

$x = -2t, \quad y = 5 + 6t, \quad z = 1 - 2t$

5. $$\mathbf{u} = 5\left(\cos\frac{3\pi}{4}\mathbf{i} + \sin\frac{3\pi}{4}\mathbf{j}\right) = \frac{5\sqrt{2}}{2}[-\mathbf{i} + \mathbf{j}]$$

$$\mathbf{v} = 2\left[\cos\frac{2\pi}{3}\mathbf{i} + \sin\frac{2\pi}{3}\mathbf{j}\right] = -\mathbf{i} + \sqrt{3}\,\mathbf{j}$$

$$\mathbf{u}\cdot\mathbf{v} = \frac{5\sqrt{2}}{2}(1 + \sqrt{3}), \quad \|\mathbf{u}\| = 5, \quad \|\mathbf{v}\| = 2$$

$$\cos\theta = \frac{|\mathbf{u}\cdot\mathbf{v}|}{\|\mathbf{u}\|\ \|\mathbf{v}\|} = \frac{(5\sqrt{2}/2)(1 + \sqrt{3})}{5(2)} = \frac{\sqrt{2} + \sqrt{6}}{4}$$

$$\theta = \arccos\frac{\sqrt{2} + \sqrt{6}}{4} = 15^\circ$$

6. $\mathbf{u} = \langle 4, -1, 5 \rangle$, $\mathbf{v} = \langle 3, 2, -2 \rangle$

$\mathbf{u}\cdot\mathbf{v} = 0 \implies \mathbf{u}$ is orthogonal to $\mathbf{v}$, $\theta = \dfrac{\pi}{2}$

7. $\mathbf{u} = \langle 10, -5, 15 \rangle$, $\mathbf{v} = \langle -2, 1, -3 \rangle$

$\mathbf{u} = -5\mathbf{v} \implies \mathbf{u}$ is parallel to $\mathbf{v}$ and in the opposite direction

$\theta = \pi$

8. $\mathbf{u} = \langle 1, 0, -3\rangle, \quad \mathbf{v} = \langle 2, -2, 1\rangle$

$\mathbf{u}\cdot\mathbf{v} = -1, \quad \|\mathbf{u}\| = \sqrt{10}, \quad \|\mathbf{v}\| = 3$

$\cos\theta = \dfrac{|\mathbf{u}\cdot\mathbf{v}|}{\|\mathbf{u}\|\ \|\mathbf{v}\|} = \dfrac{1}{3\sqrt{10}}, \quad \theta \approx 83.9^{\circ}$

9. $\mathbf{u} = 4[(\cos 135^{\circ})\mathbf{i} + (\sin 135^{\circ})\mathbf{j}] = 4[-\dfrac{\sqrt{2}}{2}\mathbf{i} + \dfrac{\sqrt{2}}{2}\mathbf{j}]$

$= -2\sqrt{2}\mathbf{i} + 2\sqrt{2}\mathbf{j}$

10. $\mathbf{u} = 8[(\cos 180^{\circ})\mathbf{i} + (\sin 180^{\circ})\mathbf{j}] = -8\mathbf{i}$

11. $\mathbf{v} = \mathbf{i} - 3\mathbf{j} + 4\mathbf{k}$

$\mathbf{u} = 3(\dfrac{\mathbf{v}}{\|\mathbf{v}\|}) = 3(\dfrac{\mathbf{i} - 3\mathbf{j} + 4\mathbf{k}}{\sqrt{26}}) = \dfrac{3}{\sqrt{26}}\mathbf{i} - \dfrac{9}{\sqrt{26}}\mathbf{j} + \dfrac{12}{\sqrt{26}}\mathbf{k}$

12. $\mathbf{u} = -\mathbf{i} + 2\mathbf{j} + 5\mathbf{k}, \quad \mathbf{v} = 7\mathbf{i} + \mathbf{j} + 2\mathbf{k}$

$$\mathbf{u}\times\mathbf{v} = \begin{vmatrix} \mathbf{i} & \mathbf{j} & \mathbf{k} \\ -1 & 2 & 5 \\ 7 & 1 & 2 \end{vmatrix} = -\mathbf{i} + 37\mathbf{j} - 15\mathbf{k}$$

$$\mathbf{w} = \frac{\mathbf{u}\times\mathbf{v}}{\|\mathbf{u}\times\mathbf{v}\|} = \frac{1}{\sqrt{1595}}(-\mathbf{i} + 37\mathbf{j} - 15\mathbf{k})$$

In Exercises 13–20, $\mathbf{u} = \langle 3, -2, 1\rangle, \quad \mathbf{v} = \langle 2, -4, -3\rangle, \quad \mathbf{w} = \langle -1, 2, 2\rangle$

13. $\|\mathbf{u}\| = \sqrt{3^2 + (-2)^2 + 1^2} = \sqrt{14}$

14. $\cos\theta = \dfrac{|\mathbf{u}\cdot\mathbf{v}|}{\|\mathbf{u}\|\ \|\mathbf{v}\|} = \dfrac{11}{\sqrt{14}\sqrt{29}}, \quad \theta = \arccos(\dfrac{11}{\sqrt{14}\sqrt{29}}) \approx 56.9^{\circ}$

15. $\mathbf{u}\cdot\mathbf{u} = 3(3) + (-2)(-2) + (1)(1) = 14 = (\sqrt{14})^2 = \|\mathbf{u}\|^2$

16. $$\mathbf{n} = \mathbf{v}\times\mathbf{w} = \begin{vmatrix} \mathbf{i} & \mathbf{j} & \mathbf{k} \\ 2 & -4 & -3 \\ -1 & 2 & 2 \end{vmatrix} = -2\mathbf{i} - \mathbf{j}, \quad \|\mathbf{n}\| = \sqrt{5}$$

$$\frac{\mathbf{n}}{\|\mathbf{n}\|} = \frac{1}{\sqrt{5}}(-2\mathbf{i} - \mathbf{j})$$

17. $\mathbf{u}\cdot(\mathbf{v} + \mathbf{w}) = \langle 3, -2, 1\rangle\cdot\langle 1, -2, -1\rangle = 6$

$\mathbf{u}\cdot\mathbf{v} + \mathbf{u}\cdot\mathbf{w} = \langle 3, -1, 1\rangle\cdot\langle 2, -4, -3\rangle + \langle 3, -2, 1\rangle\cdot\langle -1, 2, 2\rangle$
$= 11 + (-5) = 6$

Thus $\mathbf{u}\cdot(\mathbf{v} + \mathbf{w}) = \mathbf{u}\cdot\mathbf{v} + \mathbf{u}\cdot\mathbf{w}$

18. $\mathbf{u} \times (\mathbf{v} + \mathbf{w}) = \langle 3, -2, 1\rangle \times \langle 1, -2, -1\rangle = \begin{vmatrix} \mathbf{i} & \mathbf{j} & \mathbf{k} \\ 3 & -2 & 1 \\ 1 & -2 & -1 \end{vmatrix} = 4\mathbf{i} + 4\mathbf{j} - 4\mathbf{k}$

$\mathbf{u} \times \mathbf{v} = \begin{vmatrix} \mathbf{i} & \mathbf{j} & \mathbf{k} \\ 3 & -2 & 1 \\ 2 & -4 & -3 \end{vmatrix} = 10\mathbf{i} + 11\mathbf{j} - 8\mathbf{k}$

$\mathbf{u} \times \mathbf{w} = \begin{vmatrix} \mathbf{i} & \mathbf{j} & \mathbf{k} \\ 3 & -2 & 1 \\ -1 & 2 & 2 \end{vmatrix} = -6\mathbf{i} - 7\mathbf{j} + 4\mathbf{k}$

$(\mathbf{u} \times \mathbf{v}) + (\mathbf{u} \times \mathbf{w}) = 4\mathbf{i} + 4\mathbf{j} - 4\mathbf{k} = \mathbf{u} \times (\mathbf{v} + \mathbf{w})$

19. $V = |\mathbf{u}\cdot(\mathbf{v} \times \mathbf{w})| = |\langle 3, -2, 1\rangle\cdot\langle -2, -1, 0\rangle| = |-4| = 4$

20. $\text{Work} = |\mathbf{u}\cdot\mathbf{w}| = |-3 - 4 + 2| = 5$

21. $\mathbf{v} = \mathbf{j}$

(a) $x = 1, \quad y = 2 + t, \quad z = 3$ (b) None

22. Direction Numbers: 1, 1, 1

(a) $x = 1 + t, \quad y = 2 + t$
$z = 3 + t$ (b) $x - 1 = y - 2 = z - 3$

23. $3x - 3y - 7z = -4, \; x - y + 2z = 3$

Solving simultaneously we have $z = 1$. Substituting $z = 1$ into the second equation we have $y = x - 1$. Substituting for x in this equation we obtain two points on the line of intersection:

$(0, -1, 1) \quad (1, 0, 1)$

The direction vector of the line of intersection: $\mathbf{v} = \mathbf{i} + \mathbf{j}$

(a) $x = t, \quad y = -1 + t, \quad z = 1$ (b) $x = y + 1, \quad z = 1$

24. $\mathbf{u} \times \mathbf{v} = \begin{vmatrix} \mathbf{i} & \mathbf{j} & \mathbf{k} \\ 2 & -5 & 1 \\ -3 & 1 & 4 \end{vmatrix} = -21\mathbf{i} - 11\mathbf{j} - 13\mathbf{k}$

Direction Numbers: 21, 11, 13

(a) $x = 21t, \quad y = 1 + 11t$
$z = 4 + 13t$ (b) $\dfrac{x}{21} = \dfrac{y - 1}{11} = \dfrac{z - 4}{13}$

25. $\mathbf{n} = \mathbf{i} + \mathbf{j} + \mathbf{k}$, $(x - 1) + (y - 2) + (z - 3) = 0$, $x + y + z = 6$

26. $\mathbf{n} = \mathbf{i}$, $1(x - 4) + 0(y - 2) + 0(z - 1) = 0$, $x = 4$

27. The two lines are parallel as they have the same direction numbers: $-2, 1, 1$

Therefore a vector parallel to the plane is: $\mathbf{v} = -2\mathbf{i} + \mathbf{j} + \mathbf{k}$

A point on the first line is $(1, 0, -1)$ and a point on the second line is $(-1, 1, 2)$. The vector $\mathbf{u} = 2\mathbf{i} - \mathbf{j} - 3\mathbf{k}$ connecting these two points is also parallel to the plane. Therefore a normal to the plane is:

$$\mathbf{v} \times \mathbf{u} = \begin{vmatrix} \mathbf{i} & \mathbf{j} & \mathbf{k} \\ -2 & 1 & 1 \\ 2 & -1 & -3 \end{vmatrix} = -2\mathbf{i} - 4\mathbf{j} = -2(\mathbf{i} + 2\mathbf{j})$$

Equation of the plane: $(x - 1) + 2y = 0$
$x + 2y = 1$

28. $P = (-3, -4, 2)$, $Q = (-3, 4, 1)$, $R = (1, 1, -2)$

$\overrightarrow{PQ} = \langle 0, 8, -1\rangle$, $\overrightarrow{PR} = \langle 4, 5, -4\rangle$

$$\mathbf{n} = \overrightarrow{PQ} \times \overrightarrow{PR} = \begin{vmatrix} \mathbf{i} & \mathbf{j} & \mathbf{k} \\ 0 & 8 & -1 \\ 4 & 5 & -4 \end{vmatrix} = -27\mathbf{i} - 4\mathbf{j} - 32\mathbf{k}$$

$-27(x + 3) - 4(y + 4) - 32(z - 2) = 0$, $27x + 4y + 32z = -33$

29. $Q = (1, 0, 2)$, $2x - 3y + 6z = 6$

A point P on the plane is $(3, 0, 0)$

$\overrightarrow{PQ} = \langle -2, 0, 2\rangle$, $\mathbf{n} = \langle 2, -3, 6\rangle$, $D = \dfrac{|\overrightarrow{PQ} \cdot \mathbf{n}|}{\|\mathbf{n}\|} = \dfrac{8}{7}$

30. $Q = (0, 0, 0)$, $4x - 7y + z = 2$

A point P on the plane is $(0, 0, 2)$

$\overrightarrow{PQ} = \langle 0, 0, -2\rangle$, $\mathbf{n} = \langle 4, -7, 1\rangle$, $d = \dfrac{|\overrightarrow{PQ} \cdot \mathbf{n}|}{\|\mathbf{n}\|} = \dfrac{2}{\sqrt{66}} = \dfrac{\sqrt{66}}{33}$

31. $5x - 3y + z = 2, \quad 5x - 3y + z = -3$

A normal vector to the parallel planes is $\mathbf{n} = \langle 5, -3, 1\rangle$. $(0, 0, 2)$ is a point on the first plane. $(0, 0, -3)$ is a point on the second plane. The vector connecting these points is $\mathbf{v} = \langle 0, 0, 5\rangle$.

$$d = \frac{|\mathbf{v}\cdot\mathbf{n}|}{\|\mathbf{n}\|} = \frac{5}{\sqrt{35}} = \frac{\sqrt{35}}{7}$$

32. $3x + 2y - z = -1, \quad 6x + 4y - 2z = 1$

A normal vector to the parallel planes is $\mathbf{n} = \langle 3, 2, -1\rangle$. $(0, 0, 1)$ is a point on the first plane. $(0, 0, -1/2)$ is a point on the second plane. The vector connecting these points is $\mathbf{v} = \langle 0, 0, 3/2\rangle$.

$$d = \frac{|\mathbf{v}\cdot\mathbf{n}|}{\|\mathbf{n}\|} = \frac{3/2}{\sqrt{14}} = \frac{3\sqrt{14}}{28}$$

33. Direction vectors for the two lines: $\mathbf{u} = \langle 0, -2, 1\rangle$, $\mathbf{v} = \langle -1, 1, 0\rangle$

A vector perpendicular to the two lines:

$$\mathbf{u} \times \mathbf{v} = \begin{vmatrix} \mathbf{i} & \mathbf{j} & \mathbf{k} \\ 0 & -2 & 1 \\ -1 & 1 & 0 \end{vmatrix} = \langle -1, -1, -2\rangle$$

Point on Line 1: $(4, 0, 1)$. Point on Line 2: $(-1, 2, 3)$

The vector connecting the two points: $\mathbf{w} = \langle -5, 2, 2\rangle$

$$d = \frac{|\mathbf{w}\cdot(\mathbf{u} \times \mathbf{v})|}{\|\mathbf{u} \times \mathbf{v}\|} = \frac{1}{\sqrt{6}} = \frac{\sqrt{6}}{6}$$

34. Direction vectors for the two lines: $\mathbf{u} = \langle 1, -1, 3\rangle$, $\mathbf{v} = \langle -5, 2, -6\rangle$

A vector perpendicular to the two lines:

$$\mathbf{u} \times \mathbf{v} = \begin{vmatrix} \mathbf{i} & \mathbf{j} & \mathbf{k} \\ 1 & -1 & 3 \\ -5 & 2 & -6 \end{vmatrix} = \langle 0, -9, -3\rangle$$

Point on Line 1: $(4, 3, 7)$ Point on Line 2: $(-3, 7, -5)$

The vector connecting the two points:

$$\mathbf{w} = \langle -7, 4, -12\rangle, \quad d = \frac{|\mathbf{w}\cdot(\mathbf{u} \times \mathbf{v})|}{\|\mathbf{u} \times \mathbf{v}\|} = 0$$

The lines intersect at $(2, 5, 1)$

35. Direction vectors for the two lines: $\mathbf{u} = \langle 1, 2, 3\rangle$, $\mathbf{v} = \langle -1, 3, 2\rangle$.

A vector perpendicular to the two lines:

$$\mathbf{u} \times \mathbf{v} = \begin{vmatrix} \mathbf{i} & \mathbf{j} & \mathbf{k} \\ 1 & 2 & 3 \\ -1 & 3 & 2 \end{vmatrix} = \langle -5, -5, 5\rangle$$

Point on Line 1: $(1, 2, 3)$. Point on Line 2: $(0, -3, -4)$

The vector connecting the two points: $\mathbf{w} = \langle -1, -5, -7\rangle$

$$d = \frac{|\mathbf{w} \cdot (\mathbf{u} \times \mathbf{v})|}{\|\mathbf{u} \times \mathbf{v}\|} = \frac{5}{5\sqrt{3}} = \frac{\sqrt{3}}{3}$$

36. Direction vectors for the two lines: $\mathbf{u} = \langle 1, -2, 4\rangle$, $\mathbf{v} = \langle 3, 1, 1\rangle$.

A vector perpendicular to the two lines:

$$\mathbf{u} \times \mathbf{v} = \begin{vmatrix} \mathbf{i} & \mathbf{j} & \mathbf{k} \\ 1 & -2 & 4 \\ 3 & 1 & 1 \end{vmatrix} = \langle -6, 11, 7\rangle$$

Point on Line 1: $(2, -3, 1)$. Point on Line 2: $(0, 2, 3)$

The vector connecting the two points: $\mathbf{w} = \langle -2, 5, 2\rangle$

$$d = \frac{|\mathbf{w} \cdot (\mathbf{u} \times \mathbf{v})|}{\|\mathbf{u} \times \mathbf{v}\|} = \frac{81}{\sqrt{206}} = \frac{81\sqrt{206}}{206}$$

37. $x + 2y + 3z = 6$

Plane

Intercepts: $(6, 0, 0)$, $(0, 3, 0)$, $(0, 0, 2)$

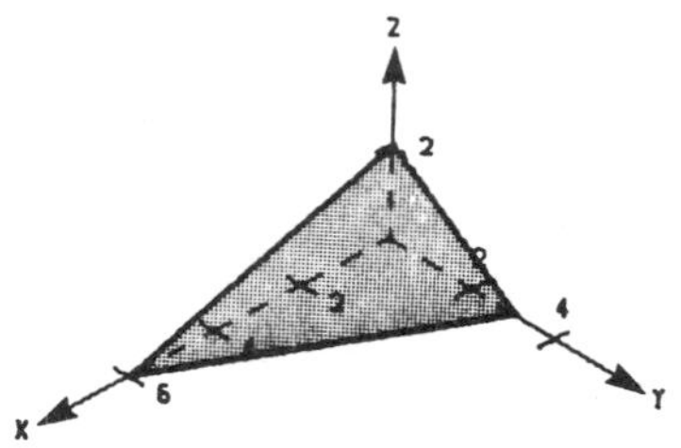

38. $y = z^2$

Since the x-coordinate is missing we have a cylindrical surface with rulings parallel to the x-axis. The generating curve is a parabola in the yz-coordinate plane

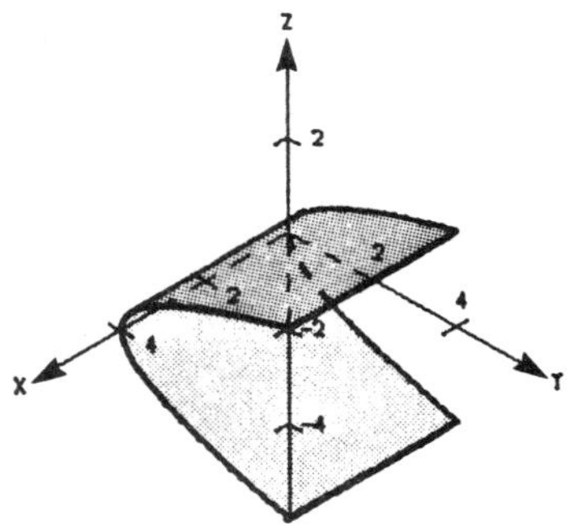

39. $x^2 + z^2 = 4$

Since the y-coordinate is missing we have a cylindrical surface with rulings parallel to the y-axis. The generating curve is the circle $x^2 + z^2 = 4$ in the xz-coordinate plane.

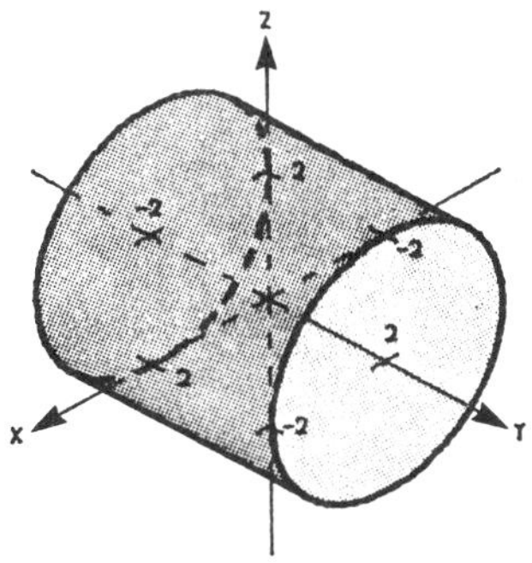

40. $x^2 + y^2 + z^2 - 2x + 4y - 6z + 5 = 0$

$(x - 1)^2 + (y + 2)^2 + (z - 3)^2 = 9$

Circle

Center: $(1, -2, 3)$

Radius: 3

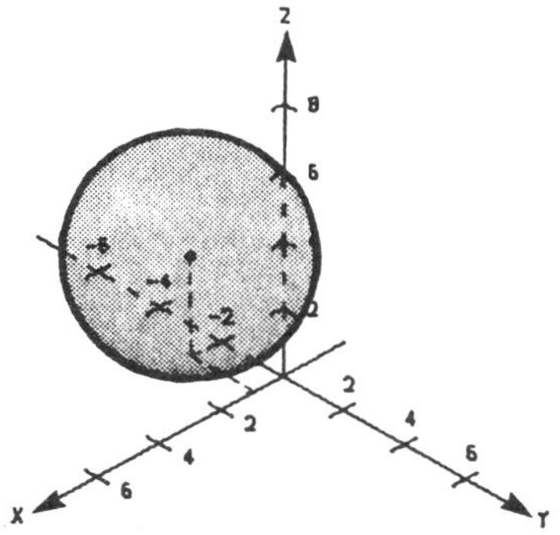

41. $16x^2 + 16y^2 - 9z^2 = 0$

Cone

xy-trace: point $(0, 0, 0)$

xz-trace: $z = \pm \dfrac{4x}{3}$

yz-trace: $z = \pm \dfrac{4y}{3}$

$z = 4, \quad x^2 + y^2 = 9$

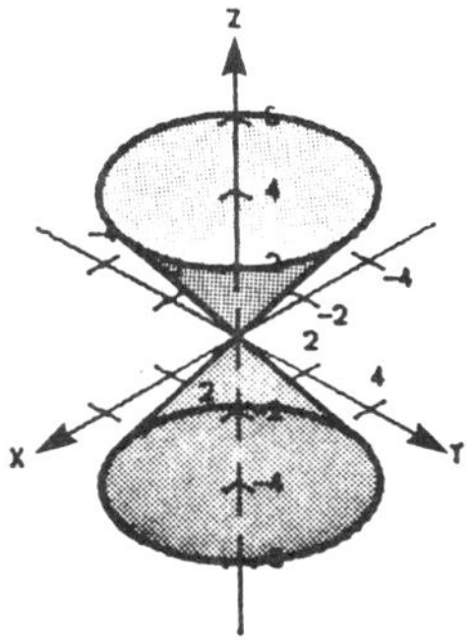

42. $\dfrac{x^2}{16} + \dfrac{y^2}{9} + z^2 = 1$

Ellipsoid

xy-trace: $\dfrac{x^2}{16} + \dfrac{y^2}{9} = 1$

xz-trace: $\dfrac{x^2}{16} + z^2 = 1$

yz-trace: $\dfrac{x^2}{9} + z^2 = 1$

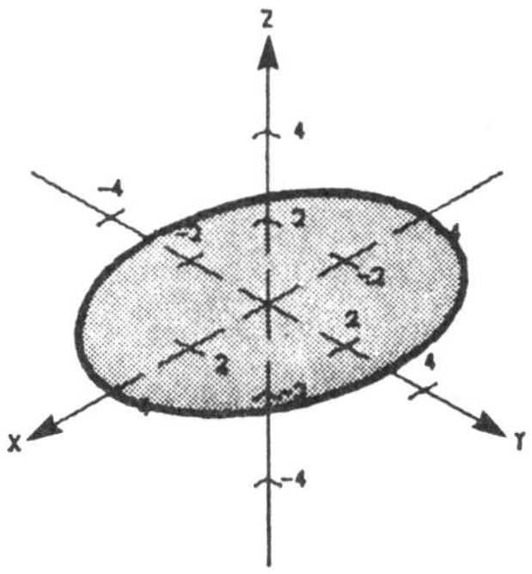

43. $y = \frac{1}{2}z$

Plane parallel to the x-axis

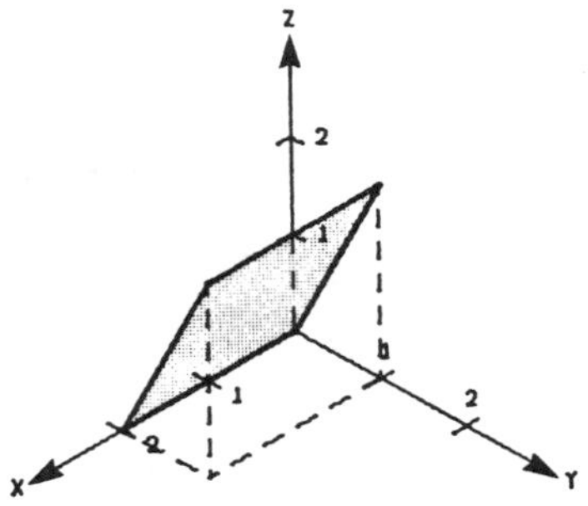

44. $\frac{x^2}{16} + \frac{y^2}{9} - z^2 = 1$

Hyperboloid of one sheet

xy-trace: $\frac{x^2}{16} + \frac{y^2}{9} = 1$

xz-trace: $\frac{x^2}{16} - z^2 = 1$

yz-trace: $\frac{y^2}{9} - z^2 = 1$

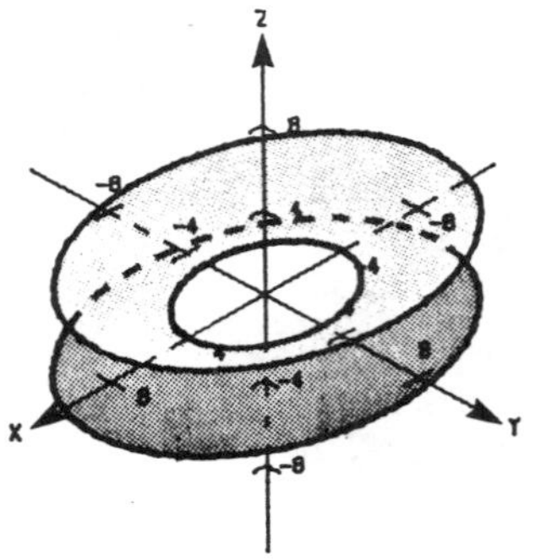

45. $\frac{x^2}{16} - \frac{y^2}{9} + z^2 = -1$

Hyperboloid of two sheets

xy-trace: $\frac{y^2}{9} - \frac{x^2}{16} = 1$

xz-trace: none

yz-trace: $\frac{y^2}{9} - z^2 = 1$

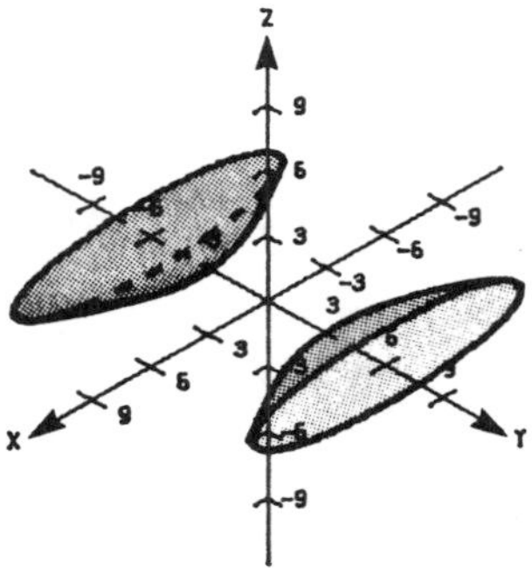

46. $\frac{x^2}{25} + \frac{y^2}{4} - \frac{z^2}{100} = 1$

Hyperboloid of one sheet

xy-trace: $\frac{x^2}{25} + \frac{y^2}{4} = 1$

xz-trace: $\frac{x^2}{25} - \frac{z^2}{100} = 1$

yz-trace: $\frac{y^2}{4} - \frac{z^2}{100} = 1$

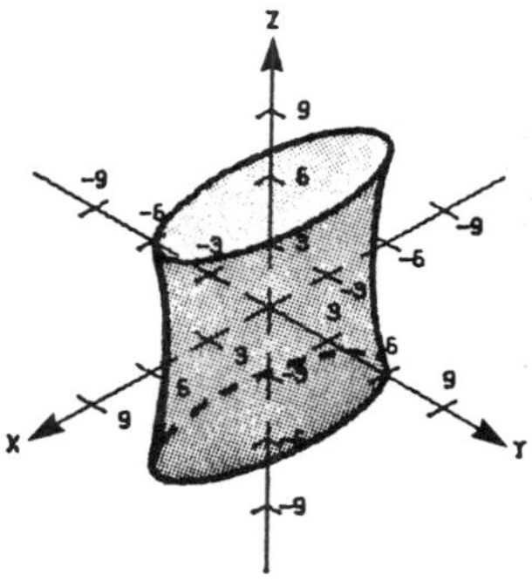

47. $y^2 - 4x^2 = z$

Hyperboloid paraboloid

xy-trace: $y = \pm 2x$
xz-trace: $z = -4x^2$
yz-trace: $z = y^2$

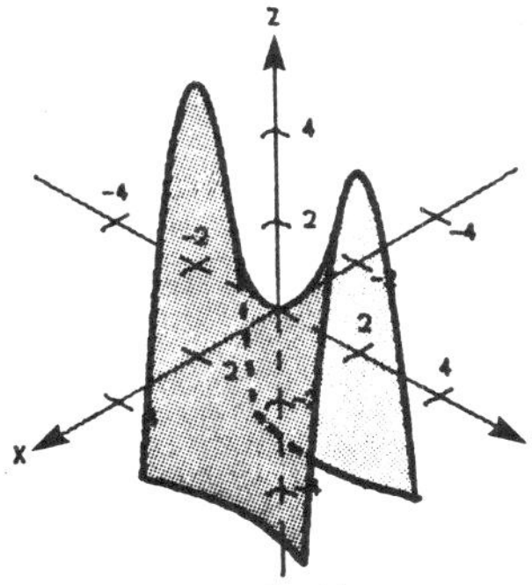

48. $y = \cos z$

Since the x coordinate is missing we have a cylindrical surface with rulings parallel to the x-axis. The generating curve is $y = \cos z$.

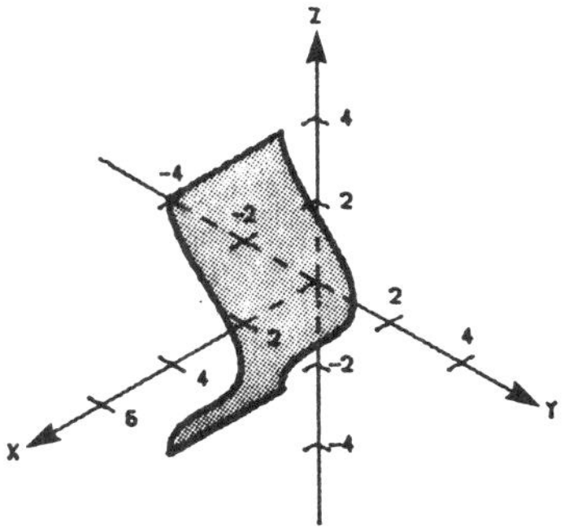

49. $(-2\sqrt{2}, 2\sqrt{2}, 2)$ rectangular

(a) $r = \sqrt{(-2\sqrt{2})^2 + (2\sqrt{2})^2} = 4$, $\theta = \arctan(-1) = \dfrac{3\pi}{4}$, $z = 2$

$(4, \dfrac{3\pi}{4}, 2)$ cylindrical

(b) $\rho = \sqrt{(-2\sqrt{2})^2 + (2\sqrt{2})^2 + (2)^2} = 2\sqrt{5}$, $\theta = \dfrac{3\pi}{4}$,

$\phi = \arccos \dfrac{2}{2\sqrt{5}} = \arccos \dfrac{1}{\sqrt{5}}$, $(2\sqrt{5}, \dfrac{3\pi}{4}, \arccos \dfrac{\sqrt{5}}{5})$ spherical

50. $(\dfrac{3/\sqrt{3}}{4}, \dfrac{3}{4}, \dfrac{3\sqrt{3}}{2})$ rectangular

(a) $r = \sqrt{(\dfrac{3/\sqrt{3}}{4})^2 + (\dfrac{3}{4})^2} = \dfrac{\sqrt{3}}{2}$, $\theta = \arctan \sqrt{3} = \dfrac{\pi}{3}$,

$z = \dfrac{3\sqrt{3}}{2}$, $(\dfrac{\sqrt{3}}{2}, \dfrac{\pi}{3}, \dfrac{3\sqrt{3}}{2})$ cylindrical

(b) $\rho = \sqrt{(\dfrac{3/\sqrt{3}}{4})^2 + (\dfrac{3}{4})^2 + (\dfrac{3\sqrt{3}}{2})^2} = \dfrac{\sqrt{30}}{2}$, $\theta = \dfrac{\pi}{3}$,

$\phi = \arccos \dfrac{3}{\sqrt{10}}$, $(\dfrac{\sqrt{30}}{2}, \dfrac{\pi}{3}, \arccos \dfrac{3}{\sqrt{10}})$ spherical

51. $x^2 - y^2 = 2z$

(a) cylindrical: $r^2 \cos^2\theta - r^2 \sin^2\theta = 2z$, $r^2 \cos 2\theta = 2z$

(b) spherical: $\rho^2 \sin^2\phi \cos^2\theta - \rho^2 \sin^2\phi \sin^2\theta = 2\rho \cos\phi$,

$\rho \sin^2\phi \cos 2\theta - 2\cos\phi = 0$, $\rho = 2 \sec 2\theta \cos\phi \csc^2\phi$

52. $x^2 + y^2 + z^2 = 16$

(a) cylindrical: $r^2 + z^2 = 16$

(b) spherical: $\rho = 4$

53. $r^2(\cos^2\theta - \sin^2\theta) + z^2 = 1$, $r^2\cos^2\theta - r^2\sin^2\theta + z^2 = 1$

$x^2 - y^2 + z^2 = 1$

54. $r^2 = 16z$, $x^2 + y^2 = 16z$, $x^2 + y^2 - 16z = 0$

55. $\rho = \csc\phi$, $\rho\sin\phi = 1$, $\sqrt{x^2 + y^2} = 1$, $x^2 + y^2 = 1$

56. $\rho = 5$, $\sqrt{x^2 + y^2 + z^2} = 5$, $x^2 + y^2 + z^2 = 25$

13 Vector-valued functions

13.1 Vector-valued functions

1. $x = 3t$

   $y = t - 1$

   $y = \frac{x}{3} - 1$

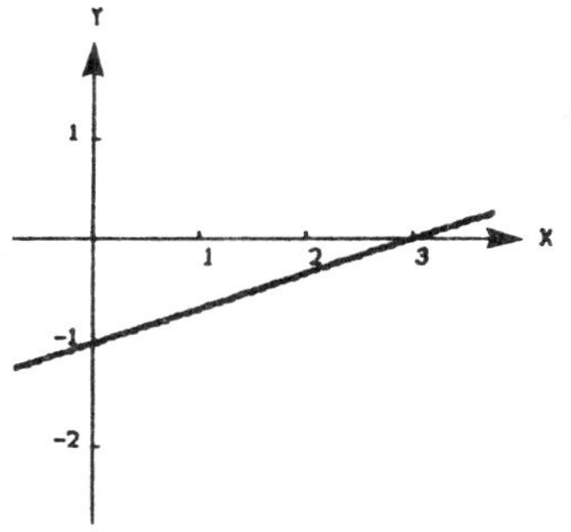

2. $x = 2\cos t$

   $y = 2\sin t$

   $x^2 + y^2 = 4$

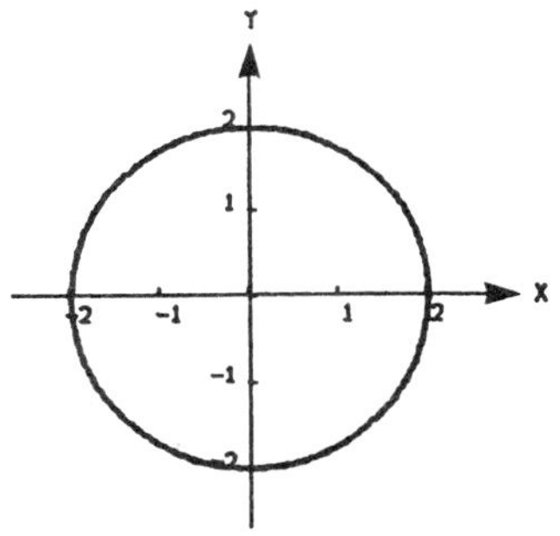

3. $x = -t + 1$
   $y = 4t + 2$
   $z = 2t + 3$

   line passing through the

   points: (0, 6, 5) (1, 2, 3)

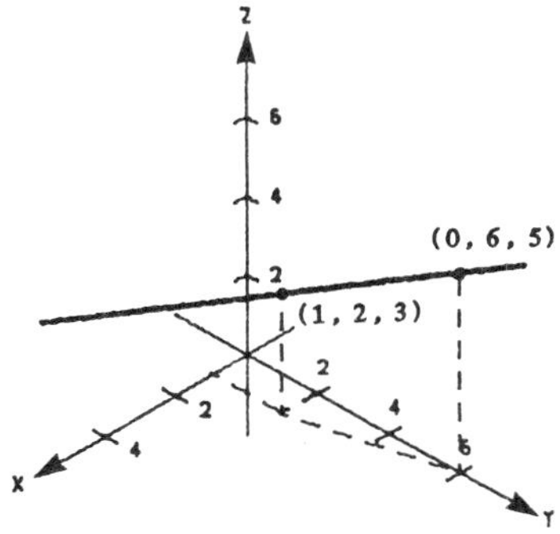

4. $x = t$
   $y = 2t - 5$
   $z = 3t$

   line passing through the

   points: (0, −5, 0) $(\frac{5}{2}, 0, \frac{15}{2})$

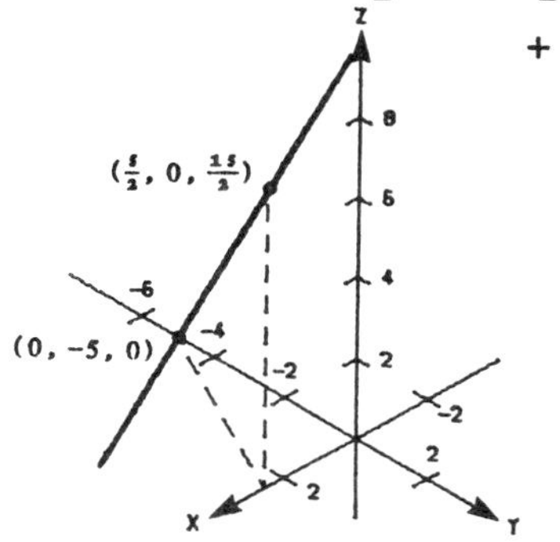

5. $x = 2\cos t,\quad y = 2\sin t,\quad z = t$

$$\frac{x^2}{4} + \frac{y^2}{4} = 1$$

$$z = t$$

Circular Helix

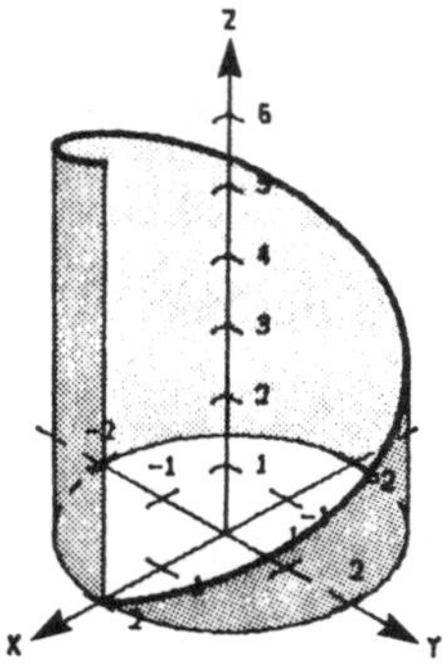

6. $x = 3\cos t,\quad y = 4\sin t,\quad z = \frac{t}{2}$

$$\frac{x^2}{9} + \frac{y^2}{16} = 1$$

$$z = \frac{t}{2}$$

Elliptic Helix

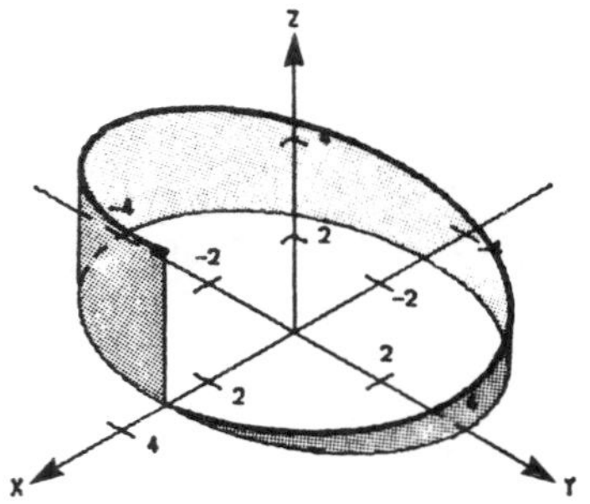

7. $x = 2\cos t,\quad y = 2\sin t,\quad z = e^{-t}$

$x^2 + y^2 = 4,\quad z = e^{-t}$

t	$-\pi/2$	$-\pi/4$	0	$\pi/4$	$\pi/2$	π
x	0	1.41	2	1.41	0	−2
y	−2	−1.41	0	1.41	2	0
z	4.81	2.19	1	0.46	0.21	0.04

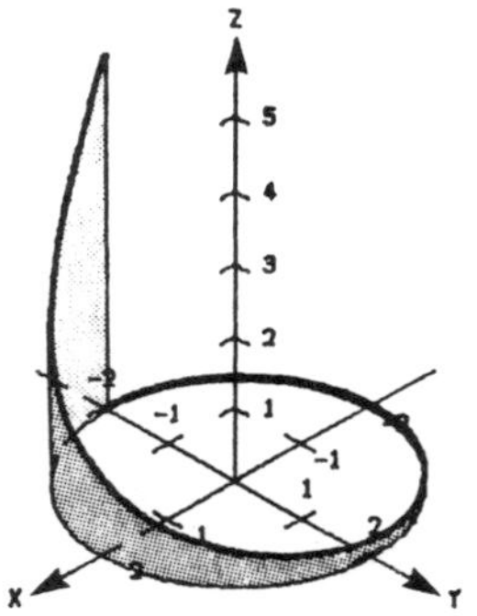

8. $x = t,\quad y = t^2,\quad z = \frac{3}{2}t$

$y = x^2,\quad z = \frac{3}{2}x$

t	−2	−1	0	1	2
x	−2	−1	0	1	2
y	4	1	0	1	4
z	−3	−3/2	0	3/2	3

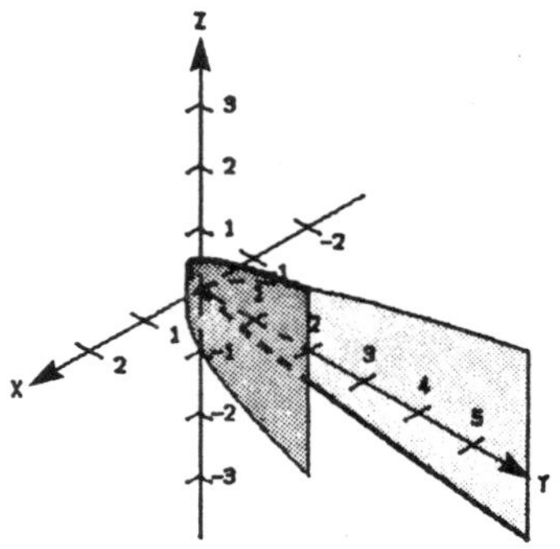

9. $x = t, \quad y = t^2, \quad z = \dfrac{2}{3}t^3$

$y = x^2, \quad z = \dfrac{2}{3}x^3$

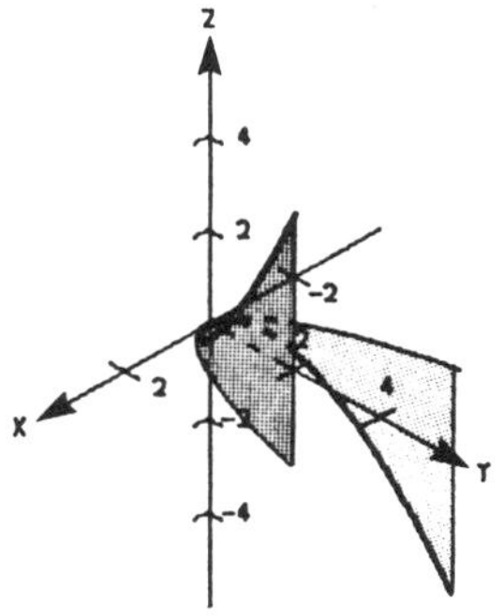

t	−2	−1	0	1	2
x	−2	−1	0	1	2
y	4	1	0	1	4
z	−16/3	−2/3	0	2/3	16/3

10. $x = \cos t + t \sin t$

$y = \sin t - t \cos t$

$z = t$

$x^2 + y^2 = 1 + t^2 = 1 + z^2$ OR

$x^2 + y^2 - z^2 = 1, \quad z = t$

Helix along a hyperboloid
of one sheet

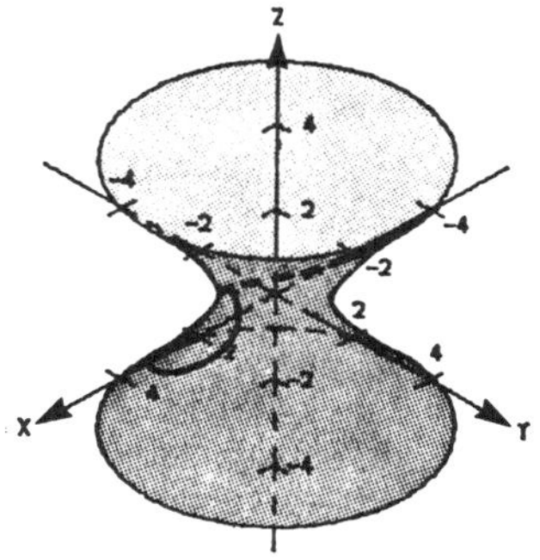

11. $z = x^2 + y^2, \quad x + y = 0$

Let $x = t$, then

$y = -x = -t$ and

$z = x^2 + y^2 = 2t^2$

Therefore $x = t, \quad y = -t, \quad z = 2t^2$

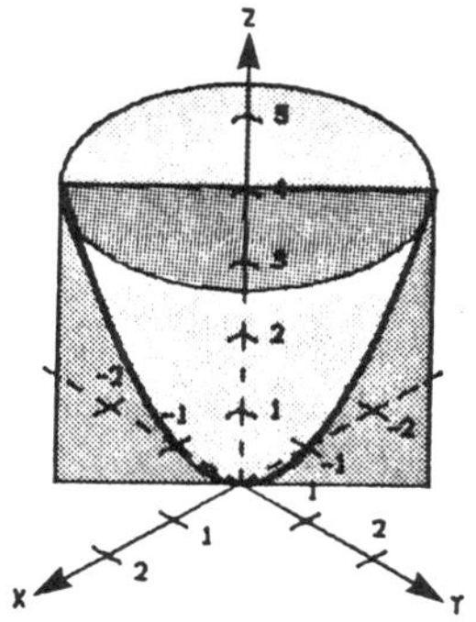

12. $z = x^2 + y^2, \quad z = 4$

Therefore $x^2 + y^2 = 4$ OR

$x = 2\cos t$

$y = 2\sin t$

$z = 4$

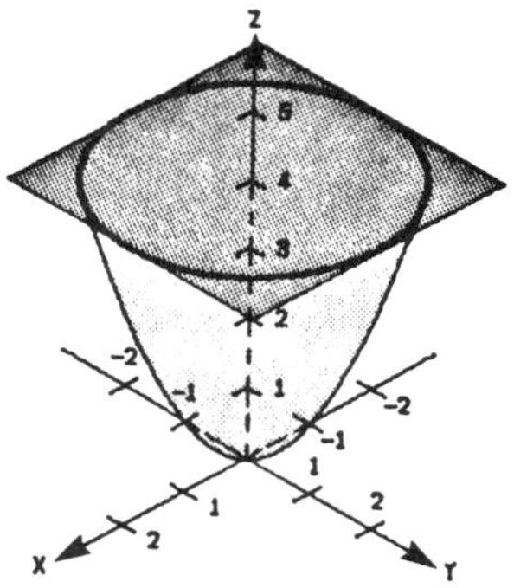

13. $x^2 + y^2 = 4$, $z = x^2$

$x = 2\sin t$, $y = 2\cos t$

$z = x^2 = 4\sin^2 t$

t	0	$\pi/6$	$\pi/4$	$\pi/2$	$3\pi/4$	π
x	0	1	$\sqrt{2}$	2	$\sqrt{2}$	0
y	2	$\sqrt{3}$	$\sqrt{2}$	0	$-\sqrt{3}$	-2
z	0	1	2	4	2	0

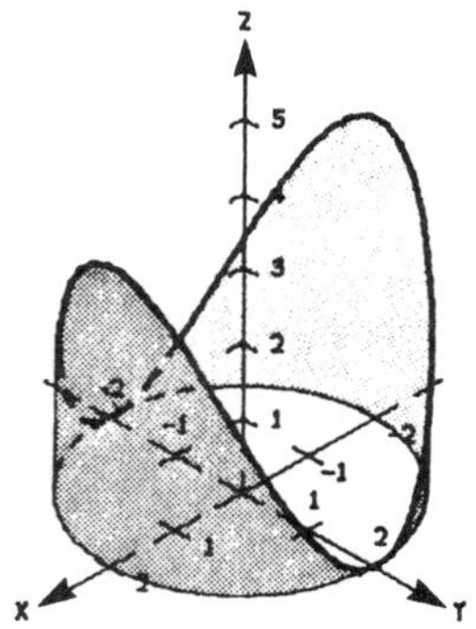

14. $4x^2 + y^2 + 4z^2 = 16$, $x = y^2$

If $y = t$, then $x = t^2$ and $z = \frac{1}{2}\sqrt{16 - 4t^4 - t^2}$

t	-1.3	-1.2	-1	0	1	1.2
x	1.69	1.44	1	0	1	1.44
y	-1.3	-1.2	-1	0	1	1.2
z	0.85	1.25	1.66	2	1.66	1.25

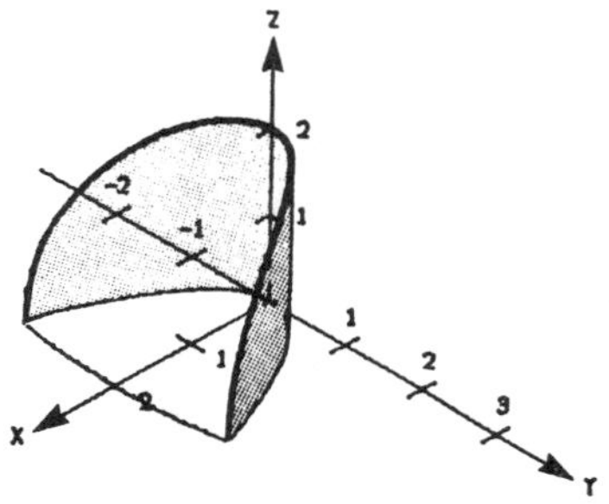

15. $x^2 + y^2 + z^2 = 4$, $x + z = 2$

Let $x = 1 + \sin t$, then $z = 2 - x = 1 - \sin t$ and $x^2 + y^2 + z^2 = 4$

$(1 + \sin t)^2 + y^2 + (1 - \sin t)^2 = 2 + 2\sin^2 t + y^2 = 4$

$y^2 = 2\cos^2 t$, $y = \pm\sqrt{2}\cos t$

$x = 1 + \sin t$, $y = \pm\sqrt{2}\cos t$

$z = 1 - \sin t$

t	$-\pi/2$	$-\pi/6$	0	$\pi/6$	$\pi/2$
x	0	1/2	1	3/2	2
y	0	$\pm\sqrt{6}/2$	$\pm\sqrt{2}$	$\pm\sqrt{6}/2$	0
z	2	3/2	1	1/2	0

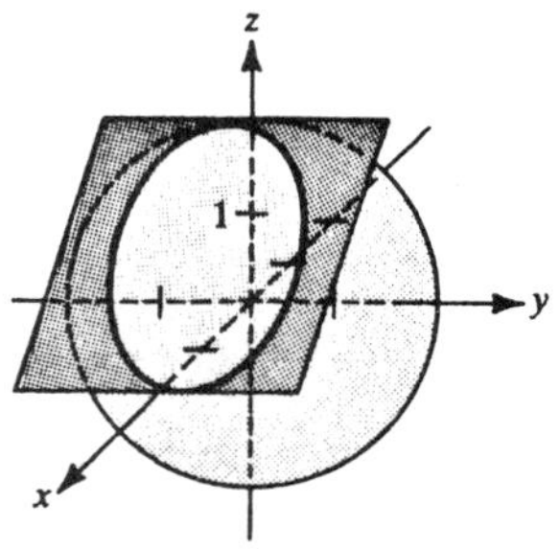

16. $x^2 + y^2 + z^2 = 10$, $x + y = 4$

Let $x = 2 + \sin t$, then $y = 2 - \sin t$ and $z = \sqrt{2(1 - \sin^2 t)} = \sqrt{2}\cos t$

t	$-\pi/2$	$-\pi/6$	0	$\pi/6$	$\pi/2$	π
x	1	3/2	2	5/2	3	2
y	3	5/2	2	3/2	1	2
z	0	$\frac{4-\sqrt{3}}{2}$	$\sqrt{2}$	$\frac{4+\sqrt{3}}{2}$	0	$-\sqrt{2}$

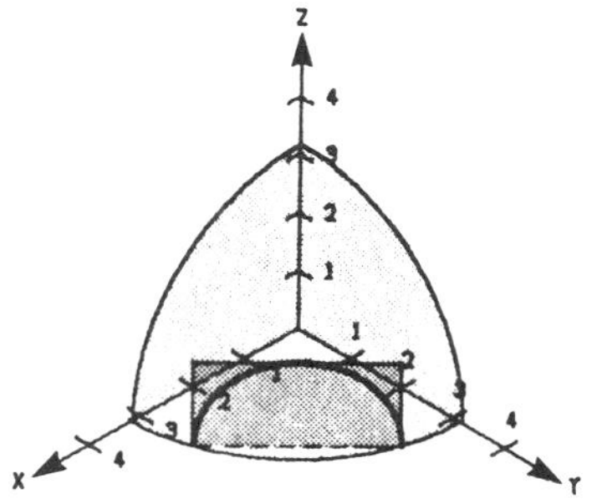

17. $x^2 + z^2 = 4$, $y^2 + z^2 = 4$

Subtracting we have

$x^2 - y^2 = 0$ or $y = \pm x$

Therefore in the first octant

if we let $x = t$ then

$x = t$, $y = t$, $z = \sqrt{4 - t^2}$

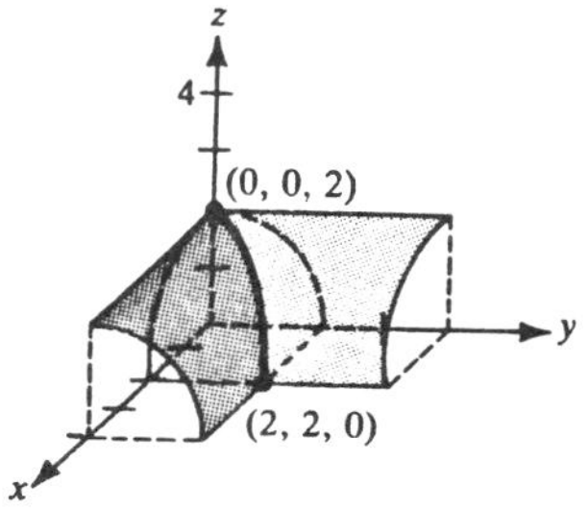

18. $x^2 + y^2 + z^2 = 16$, $xy = 4$ (first octant)

Let $x = t$ then $y = \frac{4}{t}$ and

$x^2 + y^2 + z^2 = t^2 + \frac{16}{t^2} + z^2 = 16$

$z = \frac{1}{t}\sqrt{-t^4 + 16t^2 - 16}$

$(\sqrt{8 - 4\sqrt{3}} \leq t \leq \sqrt{8 + 4\sqrt{3}})$

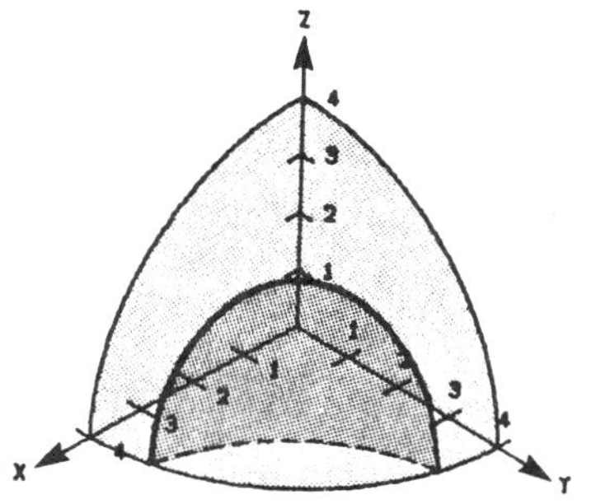

t	$\sqrt{8-4\sqrt{3}}$	1.5	2	2.5	3.0	3.5	$\sqrt{8+4\sqrt{3}}$
x	3.9	1.5	2	2.5	3.0	3.5	3.9
y	1.0	2.7	2	1.6	1.3	1.1	1.0
z	0	2.6	2	2.7	2.3	1.6	0

19. $\lim_{t \to 2} [t\,\mathbf{i} + \frac{t^2 - 4}{t^2 - 2t}\mathbf{j} + \frac{1}{t}\mathbf{k}] = 2\mathbf{i} + 2\mathbf{j} + \frac{1}{2}\mathbf{k}$

Since $\lim_{t \to 2} \frac{t^2 - 4}{t^2 - 2t} = \lim_{t \to 2} \frac{2t}{2t - 2} = 2$ (L'Hopital's Rule)

20. $\lim_{t \to 0} [e^t\,\mathbf{i} + \frac{\sin t}{t}\mathbf{j} + e^{-t}\mathbf{k}] = \mathbf{i} + \mathbf{j} + \mathbf{k}$

Since $\lim_{t \to 0} \frac{\sin t}{t} = \lim_{t \to 0} \frac{\cos t}{1} = 1$ (L'Hopital's Rule)

21. $\lim_{t \to 0} [t^2\,\mathbf{i} + 3t\,\mathbf{j} + \frac{1 - \cos t}{t}\mathbf{k}] = \mathbf{0}$

Since $\lim_{t \to 0} \frac{1 - \cos t}{t} = \lim_{t \to 0} \frac{\sin t}{1} = 0$ (L'Hopital's Rule)

22. $\lim_{t \to 1} [\sqrt{t}\,\mathbf{i} + \frac{\ln t}{t^2 - 1}\mathbf{j} + 2t^2\,\mathbf{k}] = \mathbf{k} + \frac{1}{2}\mathbf{j} + 2\mathbf{k}$

Since $\lim_{t \to 1} \frac{\ln t}{t^2 - 1} = \lim_{t \to 1} \frac{1/t}{2t} = \frac{1}{2}$ (L'Hopital's Rule)

23. $\lim_{t \to 0} [\frac{1}{t}\mathbf{k} + \cos t\,\mathbf{j} + \sin t\,\mathbf{k}]$ does not exist

Since $\lim_{t \to 0} \frac{1}{t}$ does not exist

24. $\lim_{t \to \infty} [e^{-t}\,\mathbf{i} + \frac{1}{t}\mathbf{j} + \frac{t}{t^2 + 1}\mathbf{k}] = \mathbf{0}$

Since $\lim_{t \to \infty} e^{-t} = 0$, $\lim_{t \to \infty} \frac{1}{t} = 0$ and $\lim_{t \to \infty} \frac{t}{t^2 + 1} = 0$

25. $\mathbf{r}(t) = t\,\mathbf{i} + \frac{1}{t}\mathbf{j}$, Discontinuous at $t = 0$

26. $\mathbf{r}(t) = \sqrt{t}\,\mathbf{i} + \sqrt{t - 1}\,\mathbf{j}$, Continuous on $[1, \infty)$

27. $\mathbf{r}(t) = t\,\mathbf{i} + \arcsin t\,\mathbf{j} + (t - 1)\,\mathbf{k}$, Continuous on $[-1, 1]$

28. $\mathbf{r}(t) = \sin t\,\mathbf{i} + \cos t\,\mathbf{j} + \ln t\,\mathbf{k}$, Continuous on $(0, \infty)$

29. $\mathbf{r}(t) = \langle e^{-t}, t^2, \tan t \rangle$, Discontinuous at $t = \frac{\pi}{2} + n\pi$

30. $\mathbf{r}(t) = \langle 8, \sqrt{t}, \sqrt[3]{t} \rangle$, Continuous on $[0, \infty)$

31. Let $\mathbf{r}(t) = x_1(t)\,\mathbf{i} + y_1(t)\,\mathbf{j} + z_1(t)\,\mathbf{k}$ and

$\mathbf{u}(t) = x_2(t)\,\mathbf{i} + y_2(t)\,\mathbf{j} + z_2(t)\,\mathbf{k}$ then

$$\lim_{t \to c} [\mathbf{r}(t) \times \mathbf{u}(t)] = \lim_{t \to c} \{[y_1(t)z_2(t) - y_2(t)z_1(t)]\,\mathbf{i}$$

$$- [x_1(t)z_2(t) - x_2(t)z_1(t)]\,\mathbf{j} + [x_1(t)y_2(t) - x_2(t)y_1(t)]\,\mathbf{k}\}$$

$$= \{[y_1(c)z_2(c) - y_2(c)z_1(c)]\,\mathbf{i} - [x_1(c)z_2(c) - x_2(c)z_1(c)]\,\mathbf{j}$$

$$+ [x_1(c)y_2(c) - x_2(c)y_1(c)]\,\mathbf{k}\}$$

$$= [x_1(c)\,\mathbf{i} + y_1(c)\,\mathbf{j} + z_1(c)\,\mathbf{k}] \times [x_2(c)\,\mathbf{i} + y_2(c)\,\mathbf{j} + z_2(c)\,\mathbf{k}]$$

$$= \lim_{t \to c} \mathbf{r}(t) \times \lim_{t \to c} \mathbf{u}(t)$$

13.2 Differentiation and integration of vector-valued functions

1. $\mathbf{r}(t) = a\cos^3 t\,\mathbf{i} + a\sin^3 t\,\mathbf{j} + \mathbf{k}$

$\mathbf{r}'(t) = -3a\cos^2 t \sin t\,\mathbf{i} + 3a\sin^2 t\cos t\,\mathbf{j}$

2. $\mathbf{r}(t) = \sqrt{t}\,\mathbf{i} + t\sqrt{t}\,\mathbf{j} + \ln t\,\mathbf{k}$

$$\mathbf{r}'(t) = \frac{1}{2\sqrt{t}}\,\mathbf{i} + \frac{3\sqrt{t}}{2}\,\mathbf{j} + \frac{1}{t}\,\mathbf{k}$$

3. $\mathbf{r}(t) = e^{-t}\,\mathbf{i} + 4\mathbf{j}$, $\mathbf{r}'(t) = -e^{-t}\,\mathbf{i}$

4. $\mathbf{r}(t) = \langle \sin t - t\cos t, \cos t + t\sin t, t^2 \rangle$

$\mathbf{r}'(t) = \langle t\sin t, t\cos t, 2t \rangle$

5. $\mathbf{r}(t) = \langle t\sin t, t\cos t, t \rangle$

$\mathbf{r}'(t) = \langle \sin t + t\cos t, \cos t - t\sin t, 1 \rangle$

6. $\mathbf{r}(t) = \langle \arcsin t, \arccos t, 0\rangle$

$\mathbf{r}'(t) = \left\langle \frac{1}{\sqrt{1 - t^2}}, -\frac{1}{\sqrt{1 - t^2}}, 0\right\rangle$

7. $\mathbf{r}(t) = t\,\mathbf{i} + 3t\,\mathbf{j} + t^2\,\mathbf{k}, \qquad \mathbf{u}(t) = 4t\,\mathbf{i} + t^2\,\mathbf{j} + t^3\,\mathbf{k}$

(a) $\mathbf{r}'(t) = \mathbf{i} + 3\,\mathbf{j} + 2t\,\mathbf{k}$

(b) $\mathbf{r}''(t) = 2\,\mathbf{k}$

(c) $\mathbf{r}(t)\cdot\mathbf{u}(t) = 4t^2 + 3t^3 + t^5, \; D_t[\mathbf{r}(t)\cdot\mathbf{u}(t)] = 8t + 9t^2 + 5t^4$

(d) $3\mathbf{r}(t) - \mathbf{u}(t) = -t\,\mathbf{i} + (9t - t^2)\mathbf{j} + (3t^2 - t^3)\,\mathbf{k}$

$D_t[3\mathbf{r}(t) - \mathbf{u}(t)] = -\mathbf{i} + (9 - 2t)\,\mathbf{j} + (6t - 3t^2)\,\mathbf{k}$

(e) $\mathbf{r}(t) \times \mathbf{u}(t) = 2t^4\,\mathbf{i} - (t^4 - 4t^3)\,\mathbf{j} + (t^3 - 12t^2)\,\mathbf{k}$

$D_t[\mathbf{r}(t) \times \mathbf{u}(t)] = 8t^3\,\mathbf{i} + (12t^2 - 4t^3)\,\mathbf{j} + (3t^2 - 24t)\,\mathbf{k}$

(f) $\|\mathbf{r}(t)\| = \sqrt{10t^2 + t^4} = t\sqrt{10 + t^2}$

$D_t[\|\mathbf{r}(t)\|] = \frac{10 + 2t^2}{\sqrt{10 + t^2}}$

8. $\mathbf{r}(t) = t^2\,\mathbf{i} + \sin t\,\mathbf{j} + \cos t\,\mathbf{k}, \qquad \mathbf{u}(t) = \frac{1}{t^2}\,\mathbf{i} + \sin t\,\mathbf{j} + \cos t\,\mathbf{k}$

(a) $\mathbf{r}'(t) = 2t\,\mathbf{i} + \cos t\,\mathbf{j} - \sin t\,\mathbf{k}$

(b) $\mathbf{r}''(t) = 2\mathbf{i} - \sin t\,\mathbf{j} - \cos t\,\mathbf{k}$

(c) $\mathbf{r}(t)\cdot\mathbf{u}(t) = 1 + \sin^2 t + \cos^2 t = 2, \quad D_t[\mathbf{r}(t)\cdot\mathbf{u}(t)] = 0$

(d) $3\mathbf{r}(t) - \mathbf{u}(t) = (3t^2 - \frac{1}{t^2})\,\mathbf{i} + 2\sin t\,\mathbf{j} + 2\cos t\,\mathbf{k}$

$D_t[3\mathbf{r}(t) - \mathbf{u}(t)] = (6t + \frac{2}{t^3})\,\mathbf{i} + 2\cos t\,\mathbf{j} - 2\sin t\,\mathbf{k}$

(e) $\mathbf{r}(t) \times \mathbf{u}(t) = (\frac{1}{t^2}\cos t - t^2\cos t)\,\mathbf{j} + (t^2\sin t - \frac{1}{t^2}\sin t)\,\mathbf{k}$

$D_t[\mathbf{r}(t) \times \mathbf{u}(t)] = (-\frac{1}{t^2}\sin t - \frac{2}{t^3}\cos t + t^2\sin t - 2t\cos t)\,\mathbf{j}$

$+ (t^2\cos t + 2t\sin t - \frac{1}{t^2}\cos t + \frac{2}{t^3}\sin t)\,\mathbf{k}$

(f) $\|\mathbf{r}(t)\| = \sqrt{t^4 + 1}, \qquad D_t[\|\mathbf{r}(t)\|] = \frac{2t^3}{\sqrt{t^4 + 1}}$

9. $\int(2t\,\mathbf{i} + \mathbf{j} + \mathbf{k})\,dt = t^2\,\mathbf{i} + t\,\mathbf{j} + t\,\mathbf{k} + \mathbf{C}$

10. $\int[(2t - 1)\,\mathbf{i} + 4t^3\,\mathbf{j} + 3\sqrt{t}\,\mathbf{k}]\,dt = (t^2 - t)\,\mathbf{i} + t^4\,\mathbf{j} + 2t^{3/2}\,\mathbf{k} + \mathbf{C}$

11. $\int[e^t\,\mathbf{i} + \sin t\,\mathbf{j} + \cos t\,\mathbf{k}]\,dt = e^t\,\mathbf{i} - \cos t\,\mathbf{j} + \sin t\,\mathbf{k} + \mathbf{C}$

12. $\int[\ln t\,\mathbf{i} + \frac{1}{t}\,\mathbf{j} + \mathbf{k}]\,dt = (t\ln t - t)\,\mathbf{i} + \ln t\,\mathbf{j} + t\,\mathbf{k} + \mathbf{C}$

13. $\int[\sec^2 t\,\mathbf{i} + \frac{1}{1 + t^2}\,\mathbf{j}]\,dt = \tan t\,\mathbf{i} + \arctan t\,\mathbf{j} + \mathbf{C}$

14. $\int[e^{-t}\sin t\,\mathbf{i} + e^{-t}\cos t\,\mathbf{j}]\,dt$

$$= \frac{e^{-t}}{2}(-\sin t - \cos t)\,\mathbf{i} + \frac{e^{-t}}{2}(-\cos t + \sin t)\,\mathbf{j} + \mathbf{C}$$

15. $\mathbf{r}(t) = \int(4e^{2t}\,\mathbf{i} + 3e^t\,\mathbf{j})\,dt = 2e^{2t}\,\mathbf{i} + 3e^t\,\mathbf{j} + \mathbf{C}$

$\mathbf{r}(0) = 2\mathbf{i} + 3\mathbf{j} + \mathbf{C} = 2\mathbf{i} \implies \mathbf{C} = -3\mathbf{j}$

$\mathbf{r}(t) = 2e^{2t}\,\mathbf{i} + 3(e^t - 1)\,\mathbf{j}$

16. $\mathbf{r}(t) = \int(2t\,\mathbf{j} + \sqrt{t}\,\mathbf{k})\,dt = t^2\,\mathbf{j} + \frac{2}{3}t^{3/2}\,\mathbf{k} + \mathbf{C}$

$\mathbf{r}(0) = \mathbf{C} = \mathbf{i} + \mathbf{j}, \quad \mathbf{r}(t) = \mathbf{i} + (t^2 + 1)\,\mathbf{j} + \frac{2}{3}t^{3/2}\,\mathbf{k}$

17. $\mathbf{r}'(t) = \int -32\mathbf{j}\,dt = -32t\,\mathbf{j} + \mathbf{C}$

$\mathbf{r}'(0) = \mathbf{C} = 600\sqrt{3}\,\mathbf{i} + 600\mathbf{j}$

$\mathbf{r}'(t) = 600\sqrt{3}\,\mathbf{i} + (600 - 32t)\,\mathbf{j}$

$\mathbf{r}(t) = \int[600\sqrt{3}\,\mathbf{i} + (600 - 32t)\,\mathbf{j}]\,dt = 600\sqrt{3}\,t\,\mathbf{i} + (600t - 16t^2)\,\mathbf{j} + \mathbf{C}$

$\mathbf{r}(0) = \mathbf{C} = \mathbf{0}, \quad \mathbf{r}(t) = 600\sqrt{3}\,\mathbf{i} + (600t - 16t^2)\,\mathbf{j}$

18. $\mathbf{r}'(t) = \int(-4\cos t\,\mathbf{j} - 3\sin 5\,\mathbf{k})\,dt = -4\sin t\,\mathbf{j} - 3(\sin 5)t\,\mathbf{k} + \mathbf{C}$

$\mathbf{r}'(0) = \mathbf{C} = 3\mathbf{k}$

$\mathbf{r}'(t) = -4\sin t\,\mathbf{j} + 3(1 - t\sin 5)\,\mathbf{k}$

$\mathbf{r}'(t) = \int[-4\sin t\,\mathbf{j} + 3(1 - t\sin 5)\,\mathbf{k}]\,dt$

$= 4\cos t\,\mathbf{j} + 3(t - \frac{t^2}{2}\sin 5)\mathbf{k} + \mathbf{C}$

$\mathbf{r}(0) = 4\mathbf{j} + \mathbf{C} = 4\mathbf{j} \Longrightarrow \mathbf{C} = 0$

$\mathbf{r}(t) = 4\cos t\,\mathbf{j} + 3(t - \frac{t^2}{2}\sin 5)\,\mathbf{k}$

19. $\mathbf{r}(t) = \int(te^{-t^2}\,\mathbf{i} - e^{-t}\,\mathbf{j} + \mathbf{k})\,dt = -\frac{1}{2}e^{-t^2}\,\mathbf{i} + e^{-t}\,\mathbf{j} + t\,\mathbf{k} + \mathbf{C}$

$\mathbf{r}(0) = -\frac{1}{2}\mathbf{i} + \mathbf{j} + \mathbf{C} = \frac{1}{2}\mathbf{i} - \mathbf{j} + \mathbf{k} \Longrightarrow \mathbf{C} = \mathbf{i} - 2\mathbf{j} + \mathbf{k}$

$\mathbf{r}(t) = (1 - \frac{1}{2}e^{-t^2})\,\mathbf{i} + (e^{-t} - 2)\,\mathbf{j} + (t + 1)\,\mathbf{k}$

$= (\frac{2 - e^{-t^2}}{2})\,\mathbf{i} + (e^{-t} - 2)\,\mathbf{j} + (t + 1)\,\mathbf{k}$

20. $\mathbf{r}(t) = \int[\frac{1}{1 + t^2}\,\mathbf{i} + \frac{1}{t^2}\,\mathbf{j} + \frac{1}{t}\mathbf{k}]\,dt$

$= \arctan t\,\mathbf{i} - \frac{1}{t}\mathbf{j} + \ln t\,\mathbf{k} + \mathbf{C}$

$\mathbf{r}(1) = \frac{\pi}{4}\mathbf{i} - \mathbf{j} + \mathbf{C} = 2\mathbf{i} \Longrightarrow \mathbf{C} = (2 - \frac{\pi}{4})\,\mathbf{i} + \mathbf{j}$

$\mathbf{r}(t) = [2 - \frac{\pi}{4} + \arctan t]\,\mathbf{i} + (1 - \frac{1}{t})\,\mathbf{j} + \ln t\,\mathbf{k}$

21. $\int_0^1 (8t\,\mathbf{i} + t\,\mathbf{j} - \mathbf{k})\,dt = 4t^2\,\mathbf{i}\Big]_0^1 + \frac{t^2}{2}\mathbf{j}\Big]_0^1 - t\mathbf{k}\Big]_0^1 = 4\mathbf{i} + \frac{1}{2}\mathbf{j} - \mathbf{k}$

22. $\int_{-1}^1 (t\,\mathbf{i} + t^3\,\mathbf{j} + \sqrt[3]{t}\,\mathbf{k})\,dt = \frac{t^2}{2}\mathbf{i}\Big]_{-1}^1 + \frac{t^4}{4}\mathbf{j}\Big]_{-1}^1 + \frac{3}{4}t^{4/3}\,\mathbf{k}\Big]_{-1}^1 = 0$

23. $\int_0^{\pi/2} [(a\cos t)\,\mathbf{i} + (a\sin t)\,\mathbf{j} + \mathbf{k}]\,dt$

$= (a\sin t)\,\mathbf{i}\Big]_0^{\pi/2} - (a\cos t)\mathbf{j}\Big]_0^{\pi/2} + t\,\mathbf{k}\Big]_0^{\pi/2}$

$= a\,\mathbf{i} + a\,\mathbf{j} + \frac{\pi}{2}\mathbf{k}$

24. $\int_0^3 (e^t \mathbf{i} + te^t \mathbf{k})\, dt = e^t \mathbf{i}\Big]_0^3 + (t - 1)\, e^t \mathbf{k}\Big]_0^3 = (e^3 - 1)\,\mathbf{i} + (2e^3 + 1)\,\mathbf{k}$

13.3 Velocity and acceleration

1. $\mathbf{r}(t) = 3t\,\mathbf{i} + (t - 1)\,\mathbf{j}$

$\mathbf{v}(t) = \mathbf{r}'(t) = 3\mathbf{i} + \mathbf{j}$

$\mathbf{a}(t) = \mathbf{r}''(t) = \mathbf{0}$

$x = 3t, \quad y = t - 1$

$y = \dfrac{x}{3} - 1$

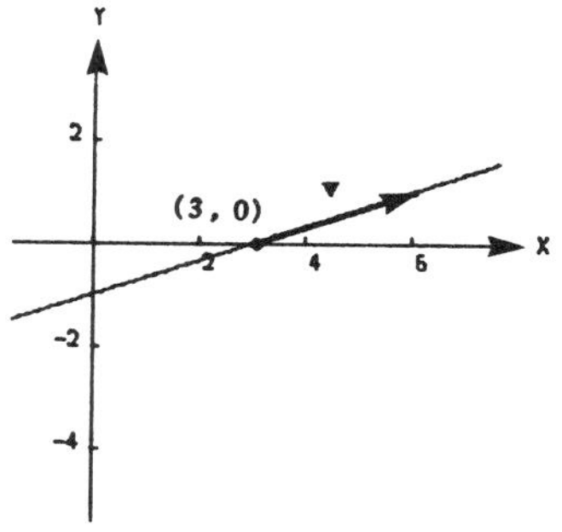

2. $\mathbf{r}(t) = (6 - t)\mathbf{i} + t\,\mathbf{j}$

$\mathbf{v}(t) = \mathbf{r}'(t) = -\mathbf{i} + \mathbf{j}$

$\mathbf{a}(t) = \mathbf{r}''(t) = \mathbf{0}$

$x = 6 - t, \quad y = t$

$y = 6 - x$

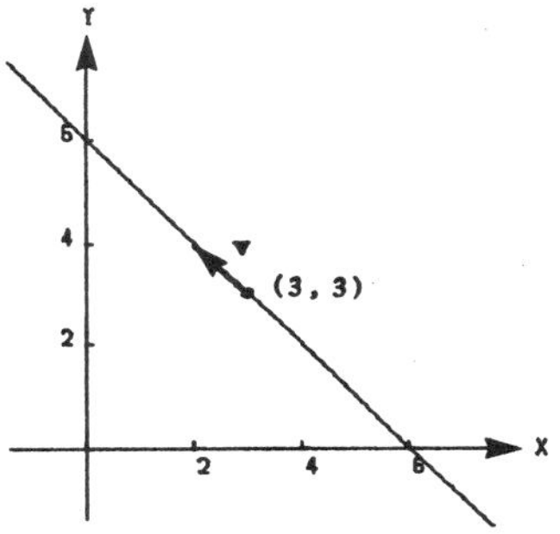

3. $\mathbf{r}(t) = t^2\,\mathbf{i} + t\,\mathbf{j}$

$\mathbf{v}(t) = \mathbf{r}'(t) = 2t\,\mathbf{i} + \mathbf{j}$

$\mathbf{a}(t) = \mathbf{r}''(t) = 2\mathbf{i}$

$x = t^2, \quad y = t, \qquad x = y^2$

At $(4, 2)$, $t = 2$

$\mathbf{v}(2) = 4\mathbf{i} + \mathbf{j}, \quad \mathbf{a}(2) = 2\mathbf{i}$

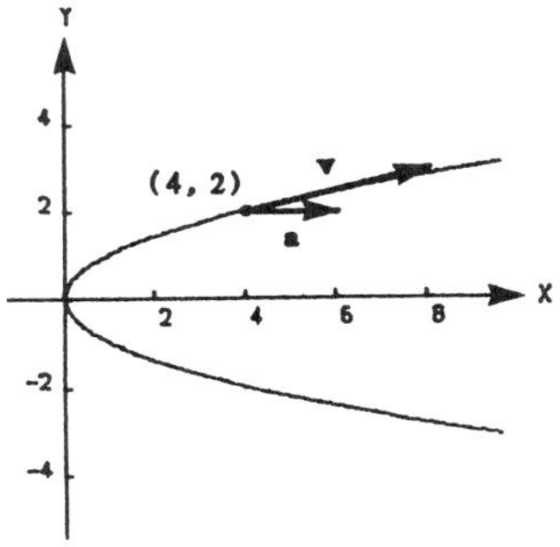

4. $\mathbf{r}(t) = t^3\,\mathbf{i} + t^2\,\mathbf{j}$

$\mathbf{v}(t) = \mathbf{r}'(t) = 3t^2\,\mathbf{i} + 2t\,\mathbf{j}$

$\mathbf{a}(t) = \mathbf{r}''(t) = 6t\,\mathbf{i} + 2\,\mathbf{j}$

$x = t^3, \quad y = t^2, \qquad y = x^{2/3}$

At $(1, 1)$, $t = 1$

$\mathbf{v}(1) = 3\mathbf{i} + 2\mathbf{j}, \quad \mathbf{a}(1) = 6\mathbf{i} + 2\mathbf{j}$

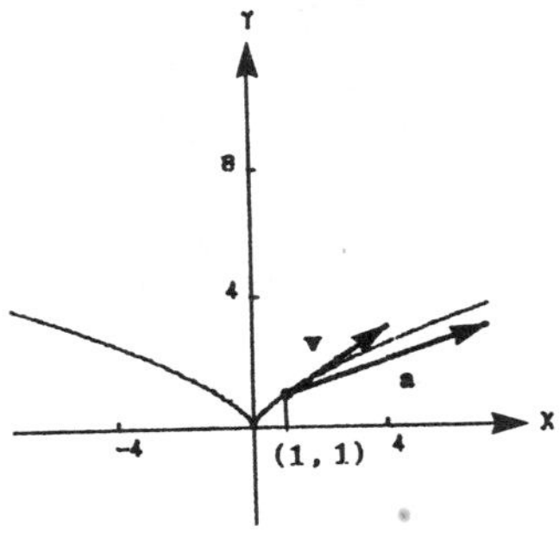

5. $\mathbf{r}(t) = 2\cos t\,\mathbf{i} + 2\sin t\,\mathbf{j}, \quad \mathbf{v}(t) = \mathbf{r}'(t) = -2\sin t\,\mathbf{i} + 2\cos t\,\mathbf{j}$

$\mathbf{a}(t) = \mathbf{r}''(t) = -2\cos t\,\mathbf{i} - 2\sin t\,\mathbf{j}$

$x = 2\cos t, \quad y = 2\sin t$

$x^2 + y^2 = 4$

At $(\sqrt{2}, \sqrt{2})$, $t = \dfrac{\pi}{4}$

$\mathbf{v}(\frac{\pi}{4}) = -\sqrt{2}\,\mathbf{i} + \sqrt{2}\,\mathbf{j}$

$\mathbf{a}(\frac{\pi}{4}) = -\sqrt{2}\,\mathbf{i} - \sqrt{2}\,\mathbf{j}$

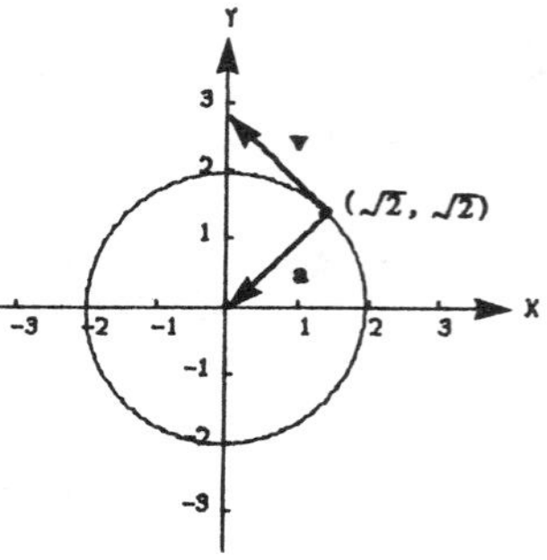

6. $\mathbf{r}(t) = 2\cos t\,\mathbf{i} + 3\sin t\,\mathbf{j}, \quad \mathbf{v}(t) = \mathbf{r}'(t) = -2\sin t\,\mathbf{i} + 3\cos t\,\mathbf{j}$

$\mathbf{a}(t) = \mathbf{r}''(t) = -2\cos t\,\mathbf{i} - 3\sin t\,\mathbf{j}$

$x = 2\cos t, \quad y = 3\sin t$

$\dfrac{x^2}{4} + \dfrac{y^2}{9} = 1$

At $(2, 0)$, $t = 0$

$\mathbf{v}(0) = 3\mathbf{j}, \quad \mathbf{a}(0) = -2\mathbf{i}$

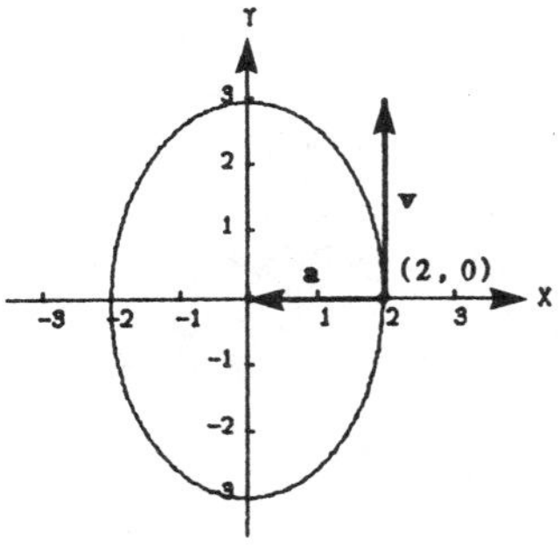

7. $\mathbf{r}(t) = \langle t - \sin t, 1 - \cos t\rangle, \quad \mathbf{v}(t) = \mathbf{r}'(t) = \langle 1 - \cos t, \sin t\rangle$

$\mathbf{a}(t) = \mathbf{r}''(t) = \langle \sin t, \cos t\rangle$

$x = t - \sin t, \quad y = 1 - \cos t$ (cycloid)

At $(\pi, 2)$, $t = \pi$

$\mathbf{v}(\pi) = \langle 2, 0\rangle = 2\mathbf{i}$

$\mathbf{a}(\pi) = \langle 0, -1\rangle = -\mathbf{j}$

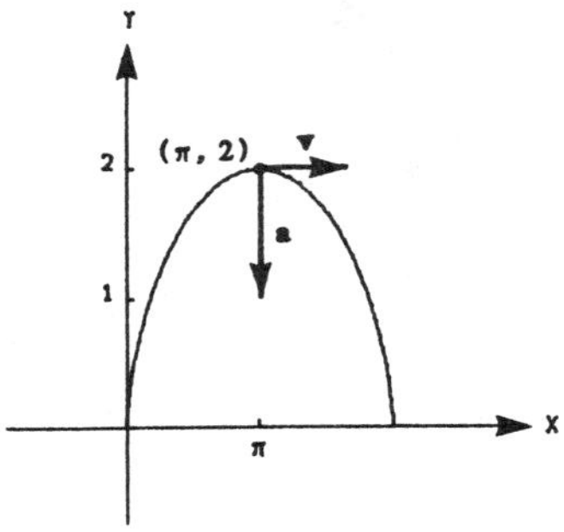

8. $\mathbf{r}(t) = \langle e^{-t}, e^t\rangle, \quad \mathbf{v}(t) = \mathbf{r}'(t) = \langle -e^{-t}, e^t\rangle$

$\mathbf{a}(t) = \mathbf{r}''(t) = \langle e^{-t}, e^t\rangle$

$x = e^{-t} = \frac{1}{e^t}, \quad y = e^t, \quad y = \frac{1}{x}$

At $(1, 1)$, $t = 0$

$\mathbf{v}(0) = \langle -1, 1\rangle = -\mathbf{i} + \mathbf{j}$

$\mathbf{a}(0) = \langle 1, 1\rangle = \mathbf{i} + \mathbf{j}$

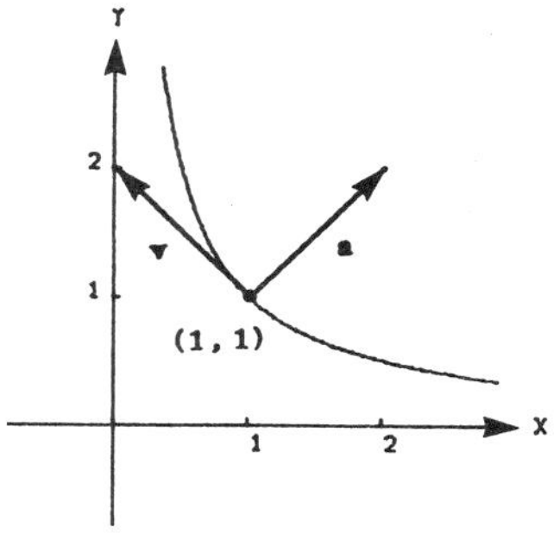

9. $\mathbf{r}(t) = t\,\mathbf{i} + (2t - 5)\,\mathbf{j} + 3t\,\mathbf{k}$

$\mathbf{v}(t) = \mathbf{i} + 2\mathbf{j} + 3\mathbf{k}$

$s(t) = \|\mathbf{v}(t)\| = \sqrt{1 + 4 + 9} = \sqrt{14}$

$\mathbf{a}(t) = \mathbf{0}$

10. $\mathbf{r}(t) = 4t\,\mathbf{i} + 4t\,\mathbf{j} + 2t\,\mathbf{k}$

$\mathbf{v}(t) = 4\mathbf{i} + 4\mathbf{j} + 2\mathbf{k}$

$s(t) = \|\mathbf{v}(t)\| = \sqrt{16 + 16 + 4} = 6$

$\mathbf{a}(t) = \mathbf{0}$

11. $\mathbf{r}(t) = t\,\mathbf{i} + t^2\,\mathbf{j} + \frac{t^2}{2}\mathbf{k}$

$\mathbf{v}(t) = \mathbf{i} + 2t\,\mathbf{j} + t\,\mathbf{k}$

$s(t) = \sqrt{1 + 4t^2 + t^2} = \sqrt{1 + 5t^2}$

$\mathbf{a}(t) = 2\mathbf{j} + \mathbf{k}$

12. $\mathbf{r}(t) = t\,\mathbf{i} + 3t\,\mathbf{j} + \frac{t^2}{2}\mathbf{k}$

$\mathbf{v}(t) = \mathbf{i} + 3\mathbf{j} + t\,\mathbf{k}$

$s(t) = \sqrt{10 + t^2}$

$\mathbf{a}(t) = \mathbf{k}$

13. $\mathbf{r}(t) = t\,\mathbf{i} + t\,\mathbf{j} + \sqrt{9 - t^2}\,\mathbf{k}$

$\mathbf{v}(t) = \mathbf{i} + \mathbf{j} - \frac{t}{\sqrt{9 - t^2}}\mathbf{k}$

$$s(t) = \sqrt{1 + 1 + \frac{t^2}{9 - t^2}} = \sqrt{\frac{18 - t^2}{9 - t^2}}$$

$$\mathbf{a}(t) = -\frac{9}{(9 - t^2)^{3/2}}\mathbf{k}$$

14. $\mathbf{r}(t) = t^2\,\mathbf{i} + t\,\mathbf{j} + 2t^{3/2}\,\mathbf{k}$

$\mathbf{v}(t) = 2t\,\mathbf{i} + \mathbf{j} + 3\sqrt{t}\,\mathbf{k}$

$$s(t) = \sqrt{4t^2 + 1 + 9t} = \sqrt{4t^2 + 9t + 1}$$

$$\mathbf{a}(t) = 2\mathbf{i} + \frac{3}{2\sqrt{t}}\mathbf{k}$$

15. $\mathbf{r}(t) = \langle 4t, 3\cos t, 3\sin t\rangle$

$\mathbf{v}(t) = \langle 4, -3\sin t, 3\cos t\rangle = 4\mathbf{i} - 3\sin t\,\mathbf{j} + 3\cos t\,\mathbf{k}$

$s(t) = \sqrt{16 + 9\sin^2 t + 9\cos^2 t} = 5$

$\mathbf{a}(t) = \langle 0, -3\cos t, -3\sin t\rangle = -3\cos t\,\mathbf{j} - 3\sin t\,\mathbf{k}$

16. $\mathbf{r}(t) = \langle e^t \cos t,\ e^t \sin t,\ e^t \rangle$

$\mathbf{v}(t) = (e^t \cos t - e^t \sin t)\,\mathbf{i} + (e^t \sin t + e^t \cos t)\,\mathbf{j} + e^t\,\mathbf{k}$

$s(t) = \sqrt{e^{2t}(\cos t - \sin t)^2 + e^{2t}(\cos t + \sin t)^2 + e^{2t}} = e^t\sqrt{3}$

$\mathbf{a}(t) = -2e^t \sin t\,\mathbf{i} + 2e^t \cos t\,\mathbf{j} + e^t\,\mathbf{k}$

17. $\mathbf{a}(t) = \mathbf{i} + \mathbf{j} + \mathbf{k}, \quad \mathbf{v}(0) = \mathbf{0}, \quad \mathbf{r}(0) = \mathbf{0}$

$\mathbf{v}(t) = \int(\mathbf{i} + \mathbf{j} + \mathbf{k})\,dt = t\,\mathbf{i} + t\,\mathbf{j} + t\,\mathbf{k} + \mathbf{C}$

$\mathbf{v}(0) = \mathbf{C} = \mathbf{0}, \qquad \mathbf{v}(t) = t\,\mathbf{i} + t\,\mathbf{j} + t\,\mathbf{k}, \qquad \mathbf{v}(t) = t(\mathbf{i} + \mathbf{j} + \mathbf{k})$

$\mathbf{r}(t) = \int(t\,\mathbf{i} + t\,\mathbf{j} + t\,\mathbf{k})\,dt = \frac{t^2}{2}(\mathbf{i} + \mathbf{j} + \mathbf{k}) + \mathbf{C}$

$\mathbf{r}(0) = \mathbf{C} = \mathbf{0}, \qquad \mathbf{r}(t) = \frac{t^2}{2}(\mathbf{i} + \mathbf{j} + \mathbf{k})$

$\mathbf{r}(2) = 2(\mathbf{i} + \mathbf{j} + \mathbf{k}) = 2\mathbf{i} + 2\mathbf{j} + 2\mathbf{k}$

18. $\mathbf{a}(t) = \mathbf{i} + \mathbf{k}, \quad \mathbf{v}(0) = 5\mathbf{j}, \quad \mathbf{r}(0) = \mathbf{0}$

$\mathbf{v}(t) = \int(\mathbf{i} + \mathbf{k})\,dt = t\,\mathbf{i} + t\,\mathbf{k} + \mathbf{C}$

$\mathbf{v}(0) = \mathbf{C} = 5\mathbf{j}, \qquad \mathbf{v}(t) = t\,\mathbf{i} + 5\mathbf{j} + t\,\mathbf{k}$

$\mathbf{r}(t) = \int(t\,\mathbf{i} + 5\mathbf{j} + t\,\mathbf{k})\,dt = \frac{t^2}{2}\mathbf{i} + 5t\,\mathbf{j} + \frac{t^2}{2}\mathbf{k} + \mathbf{C}$

$\mathbf{r}(0) = \mathbf{C} = \mathbf{0}, \qquad \mathbf{r}(t) = \frac{t^2}{2}\mathbf{i} + 5t\,\mathbf{j} + \frac{t^2}{2}\mathbf{k}, \qquad \mathbf{r}(2) = 2\mathbf{i} + 10\mathbf{j} + 2\mathbf{k}$

19. $\mathbf{a}(t) = t\,\mathbf{j} + t\mathbf{k}, \qquad \mathbf{v}(1) = 5\mathbf{j}, \qquad \mathbf{r}(1) = \mathbf{0}$

$\mathbf{v}(t) = \int(t\,\mathbf{j} + t\,\mathbf{k})\,dt = \frac{t^2}{2}\mathbf{j} + \frac{t^2}{2}\mathbf{k} + \mathbf{C}$

$\mathbf{v}(1) = \frac{1}{2}\mathbf{j} + \frac{1}{2}\mathbf{k} + \mathbf{C} = 5\mathbf{j} \implies \mathbf{C} = \frac{9}{2}\mathbf{j} - \frac{1}{2}\mathbf{k}$

$\mathbf{v}(t) = (\frac{t^2}{2} + \frac{9}{2})\,\mathbf{j} + (\frac{t^2}{2} - \frac{1}{2})\,\mathbf{k}$

$\mathbf{r}(t) = \int [(\frac{t^2}{2} + \frac{9}{2})\,\mathbf{j} + (\frac{t^2}{2} - \frac{1}{2})\,\mathbf{k}]\,dt = (\frac{t^3}{6} + \frac{9}{2}t)\,\mathbf{j} + (\frac{t^3}{6} - \frac{1}{2}t)\,\mathbf{k} + \mathbf{C}$

$\mathbf{r}(1) = \frac{14}{3}\mathbf{j} - \frac{1}{3}\mathbf{k} + \mathbf{C} = \mathbf{0} \implies \mathbf{C} = -\frac{14}{3}\mathbf{j} + \frac{1}{3}\mathbf{k}$

$\mathbf{r}(t) = (\frac{t^3}{6} + \frac{9}{2}t - \frac{14}{3})\,\mathbf{j} + (\frac{t^3}{6} - \frac{1}{2}t + \frac{1}{3})\,\mathbf{k}$

$\mathbf{r}(2) = \frac{17}{3}\mathbf{j} + \frac{2}{3}\mathbf{k}$

20. $\mathbf{a}(t) = -\cos t\,\mathbf{i} - \sin t\,\mathbf{j}, \quad \mathbf{v}(0) = \mathbf{j} + \mathbf{k}, \quad \mathbf{r}(0) = \mathbf{i}$

$$\mathbf{v}(t) = \int (-\cos t\,\mathbf{i} - \sin t\,\mathbf{j})\,dt = -\sin t\,\mathbf{i} + \cos t\,\mathbf{j} + \mathbf{C}$$

$$\mathbf{v}(0) = \mathbf{j} + \mathbf{C} = \mathbf{j} + \mathbf{k} \implies \mathbf{C} = \mathbf{k}$$

$$\mathbf{v}(t) = -\sin t\,\mathbf{i} + \cos t\,\mathbf{j} + \mathbf{k}$$

$$\mathbf{r}(t) = \int (-\sin t\,\mathbf{i} + \cos t\,\mathbf{j} + \mathbf{k})\,dt = \cos t\,\mathbf{i} + \sin t\,\mathbf{j} + t\,\mathbf{k} + \mathbf{C}$$

$$\mathbf{r}(0) = \mathbf{i} + \mathbf{C} = \mathbf{i} \implies \mathbf{C} = \mathbf{0}$$

$$\mathbf{r}(t) = \cos t\,\mathbf{i} + \sin t\,\mathbf{j} + t\,\mathbf{k}$$

$$\mathbf{r}(2) = (\cos 2)\,\mathbf{i} + (\sin 2)\,\mathbf{j} + 2\mathbf{k}$$

21. $\mathbf{r}(t) = (v_0 \cos\theta)t\,\mathbf{i} + [h + (v_0 \sin\theta)t - \frac{1}{2}gt^2]\,\mathbf{j}$

$$= \frac{v_0}{\sqrt{2}}t\,\mathbf{i} + (3 + \frac{v_0}{\sqrt{2}}t - 16t^2)\,\mathbf{j}$$

$$\frac{v_0}{\sqrt{2}}t = 300 \text{ when } 3 + \frac{v_0}{\sqrt{2}}t - 16t^2 = 3$$

$$t = \frac{300\sqrt{2}}{v_0}, \qquad \frac{v_0}{\sqrt{2}}\left(\frac{300\sqrt{2}}{v_0}\right) - 16\left(\frac{300\sqrt{2}}{v_0}\right)^2 = 0$$

$$300 - \frac{300^2(32)}{v_0^2} = 0, \quad v_0^2 = 300(32), \qquad v_0 = \sqrt{9600} = 40\sqrt{6}$$

$$v_0 = 40\sqrt{6} \approx 97.98 \text{ ft/sec}$$

The maximum height is reached when the derivative of the vertical component is zero.

$$y(t) = 3 + \frac{t\,v_0}{\sqrt{2}} - 16t^2 = 3 + \frac{40\sqrt{6}}{\sqrt{2}}t - 16t^2 = 3 + 40\sqrt{3}\,t - 16t^2$$

$$y'(t) = 40\sqrt{3} - 32t = 0, \qquad t = \frac{40\sqrt{3}}{32} = \frac{5\sqrt{3}}{4}$$

Maximum height: $y\left(\frac{5\sqrt{3}}{4}\right) = 3 + 40\sqrt{3}\left(\frac{5\sqrt{3}}{4}\right) - 16\left(\frac{5\sqrt{3}}{4}\right)^2 = 78$ ft

22. $\mathbf{r}(t) = (v_0 \cos\theta)t\,\mathbf{i} + [h + (v_0 \sin\theta)t - \frac{1}{2}gt^2]\,\mathbf{j}$

$= (40 \cos 60^\circ)t\,\mathbf{i} + [0 + (40 \sin 60^\circ) - 16t^2]\,\mathbf{j}$

$= 20t\,\mathbf{i} + (20\sqrt{3}\,t - 16t^2)\,\mathbf{j}$

The maximum height is reached when the derivative of the vertical component is zero.

$y(t) = 20\sqrt{3}\,t - 16t^2, \qquad y'(t) = 20\sqrt{3} - 32t = 0, \qquad t = \frac{20\sqrt{3}}{32} = \frac{5\sqrt{3}}{8}$

Maximum height: $y(\frac{5\sqrt{3}}{8}) = 20\sqrt{3}(\frac{5\sqrt{3}}{8}) - 16(\frac{5\sqrt{3}}{8})^2 = \frac{75}{4}$ ft

23. $\mathbf{r}(t) = t(v_0 \cos\theta)\,\mathbf{i} + (t\,v_0 \sin\theta - 16t^2)\,\mathbf{j}$

$\frac{dy}{dt} = v_0 \sin\theta - 32t = 0$ when $t = \frac{v_0 \sin\theta}{32}$

At that time $x(t) = v_0 \cos\theta(\frac{v_0 \sin\theta}{32}) = \frac{v_0}{64}\sin 2\theta = 20$ and

$y(t) = \frac{v_0^2 \sin^2\theta}{32} - 16(\frac{v_0^2 \sin^2\theta}{32^2}) = \frac{v_0^2 \sin^2\theta}{64} = 40$

Solving these equations we have $\tan\theta = 4$ or $\theta \approx 76^\circ$ and

$v_0 = 4\sqrt{170}$ ft/s

24. $\mathbf{r}(t) = t(v_0 \cos\theta)\,\mathbf{i} + (t\,v_0 \sin\theta - 16t^2)\,\mathbf{j}$

$\mathbf{v}(t) = (v_0 \cos\theta)\,\mathbf{i} + (v_0 \sin\theta - 32t)\,\mathbf{j}$

$v_0 \cos\theta = 540$ mph $= 792$ ft/s

$t\,v_0 \sin\theta - 16t^2 = 30{,}000 - 16t^2$

($t = 0$ when bomb released)

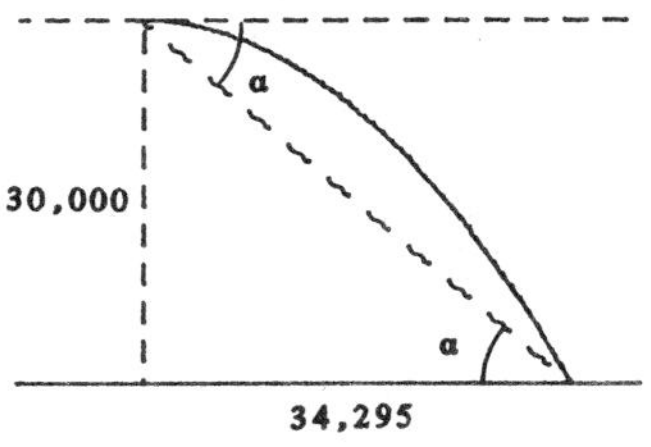

Therefore $\mathbf{r}(t) = (792t)\,\mathbf{i} + (30{,}000 - 16t^2)\,\mathbf{j}$, $\qquad \mathbf{v}(t) = 792\mathbf{i} - (32t)\,\mathbf{j}$

At the time of impact: $30{,}000 - 16t^2 = 0$ or $t = 43.3$ sec

$\mathbf{r}(43.3) = 34295\mathbf{i}, \qquad \mathbf{v}(43.3) = 792\mathbf{i} - 1385.6\mathbf{j}$

Velocity at impact: $\|\mathbf{v}(43.3)\| = 1596$ ft/s ≈ 1088 mph

$\tan\alpha = \frac{30{,}000}{34{,}295} \approx 0.8748$

The angle of depression $= \arctan(0.8748) \approx 47.18^\circ$

25.(a) $\mathbf{r}(t) = t(v_0 \cos\theta)\,\mathbf{i} + (t v_0 \sin\theta - 16t^2)\,\mathbf{j}$

$t(v_0 \sin\theta - 16t) = 0$ when $t = \dfrac{v_0 \sin\theta}{16}$

Range $= x = v_0 \cos\theta\left(\dfrac{v_0 \sin\theta}{16}\right) = \left(\dfrac{v_0^2}{32}\right)\sin 2\theta$

The range will be maximum when: $\dfrac{dx}{dt} = \left(\dfrac{v_0^2}{32}\right)2\cos 2\theta = 0$ or

$2\theta = \dfrac{\pi}{2}$, $\quad \theta = \dfrac{\pi}{4}$ rad

(b) $y(t) = (t v_0 \sin\theta - 16t^2)$

$\dfrac{dy}{dt} = v_0 \sin\theta - 32t = 0$ when $t = \dfrac{v_0 \sin\theta}{32}$

Maximum height $= y\left(\dfrac{v_0 \sin\theta}{32}\right) = \dfrac{v_0^2 \sin^2\theta}{32} - 16\,\dfrac{v_0^{\ 2} \sin^2\theta}{32^2}$

$= \dfrac{v_2 \sin^2\theta}{64}$

Maximum height when $\theta = \dfrac{\pi}{2}$

26. From Exercise 25: Range $= x = \dfrac{v_0^2}{32}\sin 2\theta = \dfrac{1200^2}{32}\sin 2\theta = 3000$

$\sin 2\theta = \dfrac{1}{15}$, $\quad \theta \approx 1.91^\circ$

27. From Exercise 25: Range $= x = \dfrac{v_0^2}{32}\sin 2\theta = \dfrac{v_0^2}{32}\sin 2(10)^\circ = 100$

$v_0^2 = \dfrac{3200}{\sin 20^\circ}$, $\quad v_0 \approx 96.7$ ft/sec

28. $x(t) = t(v_0 \cos\theta)$ or $t = \dfrac{x}{v_0 \cos\theta}$, $\quad y(t) = t(v_0 \sin\theta) - 16t^2 + h$

$y = \dfrac{x}{v_0 \cos\theta}(v_0 \sin\theta) - 16\left(\dfrac{x^2}{v_0^2 \cos^2\theta}\right) + h$

$= (\tan\theta)x - \left(\dfrac{16}{v_0^2}\sec^2\theta\right)x^2 + h$

29. $y = x - 0.005x^2$

From Exercise 28 we know that $\tan\theta$ is the coefficient of x, therefore $\tan\theta = 1$, $\theta = \frac{\pi}{4}$ rad $= 45^\circ$, also

$\frac{16}{v_0^2}\sec^2\theta$ = coefficient of x^2, $\frac{16}{v_0^2}(2) = 0.005$ or $v_0 = 80$ ft/s

$\mathbf{r}(t) = (40\sqrt{2}t)\,\mathbf{i} + (40\sqrt{2}t - 16t^2)\,\mathbf{j}$

when $40\sqrt{2}t = 60$ ft, $t = \frac{60}{40\sqrt{2}} = \frac{3\sqrt{2}}{4}$

$\mathbf{v}(t) = 40\sqrt{2}\,\mathbf{i} + (40\sqrt{2} - 32t)\,\mathbf{j}$

$\mathbf{v}(\frac{3\sqrt{2}}{4}) = 40\sqrt{2}\,\mathbf{i} + (40\sqrt{2} - 24\sqrt{2})\,\mathbf{j} = 8\sqrt{2}(5\mathbf{i} + 2\mathbf{j})$ direction

speed $= \|\mathbf{v}(\frac{3\sqrt{2}}{4})\| = 8\sqrt{2}\sqrt{25 + 4} = 8\sqrt{58}$ ft/s

30. $\mathbf{r}(t) = b(\omega t - \sin\omega t)\,\mathbf{i} + b(1 - \cos\omega t)\,\mathbf{j}$

$\mathbf{v}(t) = b(\omega - \omega\cos\omega t)\,\mathbf{i} + b\omega\sin\omega t\,\mathbf{j} = b\omega(1 - \cos\omega t)\,\mathbf{i} + b\omega\sin\omega t\,\mathbf{j}$

$\mathbf{a}(t) = (b\omega^2\sin\omega t)\,\mathbf{i} + (b\omega^2\cos\omega t)\,\mathbf{j} = b\omega^2[\sin(\omega t)\,\mathbf{i} + \cos(\omega t)\,\mathbf{j}]$

$\|\mathbf{v}(t)\| = \sqrt{2}b\omega\sqrt{1 - \cos(\omega t)}$, $\|\mathbf{a}(t)\| = b\omega^2$

31. $\|\mathbf{v}(t)\| = b\omega\sqrt{(1 - \cos\omega t)^2 + \sin^2\omega t} = \sqrt{2}b\omega\sqrt{1 - \cos\omega t}$

(a) $\|\mathbf{v}(t)\| = 0$ when $\omega t = 0, 2\pi, 4\pi, \ldots.$

(b) $\|\mathbf{v}(t)\|$ is maximum when $\omega t = \pi, 3\pi, \ldots.$ then $\|\mathbf{v}(t)\| = 2b\omega$

32. $\mathbf{r}(t) = b(\omega t - \sin\omega t)\,\mathbf{i} + b(1 - \cos\omega t)\,\mathbf{j}$,

$\mathbf{v}(t) = b\omega[(1 - \cos\omega t)\,\mathbf{i} + (\sin\omega t)\,\mathbf{j}]$

speed $= \|\mathbf{v}(t)\| = \sqrt{2}b\omega\sqrt{1 - \cos\omega t}$ and has a maximum value of $2b\omega$ when $\omega t = \pi, 3\pi, \ldots.$

55 mph = 80.67 ft/s = 80.67 rad/s = ω (since b = 1)

Therefore the maximum speed of a point on the tire is twice the speed of the car: 2(80.67) = 110 mph

33. $\mathbf{v}(t) = -b\omega \sin(\omega t)\,\mathbf{i} + b\omega \cos(\omega t)\,\mathbf{j}$,

$\mathbf{r}(t) \cdot \mathbf{v}(t) = -b^2\omega \sin(\omega t)\cos(\omega t) + b^2\omega \sin(\omega t)\cos(\omega t) = 0$

Therefore $\mathbf{r}(t)$ and $\mathbf{v}(t)$ are orthogonal.

34. Speed $= \|\mathbf{v}\| = \sqrt{b^2\omega^2 \sin^2(\omega t) + b^2\omega^2 \cos^2(\omega t)}$

$= \sqrt{b^2\omega^2[\sin^2(\omega t) + \cos^2(\omega t)]} = b\omega$

35. $\mathbf{a}(t) = -b\omega^2 \cos(\omega t)\,\mathbf{i} - b\omega^2 \sin(\omega t)\,\mathbf{j} = -b\omega^2[\cos(\omega t)\,\mathbf{i} + \sin(\omega t)\,\mathbf{j}]$

$= -\omega^2\mathbf{r}(t)$

$\mathbf{a}(t)$ is a negative multiple of a unit vector from (0, 0) to $(\cos\omega t, \sin\omega t)$ and thus $\mathbf{a}(t)$ is directed toward the origin.

36. $\|\mathbf{a}(t)\| = b\omega^2\|\cos(\omega t)\,\mathbf{i} + \sin(\omega t)\,\mathbf{j}\| = b\omega^2$

37. $\|\mathbf{a}(t)\| = \omega^2 b$, $1 = m(32)$, $F = m(\omega^2 b) = \frac{1}{32}(2\omega^2) = 10$, $\omega = 4\sqrt{10}$ rad/s

$\|\mathbf{v}(t)\| = b\omega = 8\sqrt{10}$ ft/s

38. $\|\mathbf{v}(t)\| = 30$ mph $= 44$ ft/s, and $\omega = \frac{44}{300}$ rad/s

$\|\mathbf{a}(t)\| = b\omega^2$, $F = m(b\omega^2) = \frac{3000}{32}(300)\left(\frac{44}{300}\right)^2 = 605$ lb

39. Let $\mathbf{n}$ be normal to the road.

$\|\mathbf{n}\| \cos\theta = 3000$, $\|\mathbf{n}\| \sin\theta = 605$

Dividing the second equation by the first we have $\tan\theta = \frac{605}{3000}$

$\theta = \arctan\left(\frac{605}{3000}\right) = 11.4^\circ$

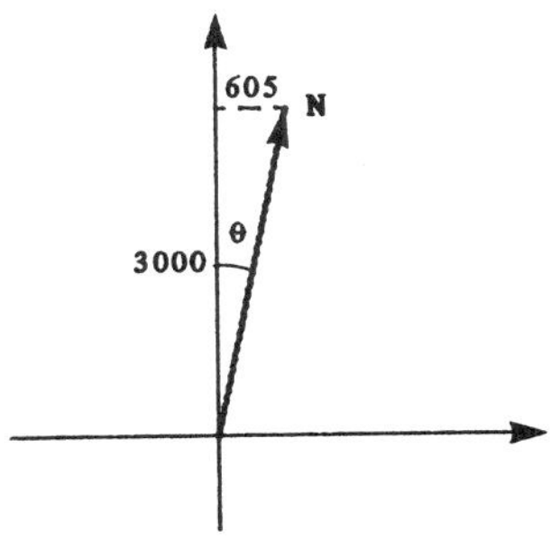

40. $\mathbf{r}(t) = x(t)\,\mathbf{i} + y(t)\,\mathbf{j}$, $\quad \mathbf{v}(t) = x'(t)\,\mathbf{i} + y'(t)\,\mathbf{j}$

$\mathbf{a}(t) = x''(t)\,\mathbf{i} + y''(t)\,\mathbf{j}$

$s(t) = \sqrt{[x'(t)]^2 + [y'(t)]^2} = C$, C a constant

Then $y'(t) = \sqrt{C^2 - [x'(t)]^2}$ and $\mathbf{v}(t) = x'(t)\,\mathbf{i} + \sqrt{C^2 - [x'(t)]^2}\,\mathbf{j}$

$$\mathbf{a}(t) = x''(t)\,\mathbf{i} + \frac{-x'(t)x''(t)}{\sqrt{C^2 - [x'(t)]^2}}\,\mathbf{j}$$

Thus $\mathbf{v}(t) \cdot \mathbf{a}(t) = 0$ and are orthogonal

41. $\mathbf{r}(t) = x(t)\,\mathbf{i} + y(t)\,\mathbf{j}$

$y(t) = m(x(t)) + b$, m, b are constants

$\mathbf{r}(t) = x(t)\,\mathbf{i} + [m(x(t)) + b]\,\mathbf{j}$

$\mathbf{v}(t) = x'(t)\,\mathbf{i} + m\,x'(t)\,\mathbf{j}$

$s(t) = \sqrt{[x'(t)]^2 + [m\,x'(t)]^2} = C$, C a constant

Thus $x'(t) = \dfrac{C}{\sqrt{1 + m^2}}$ and $x''(t) = 0$

$\mathbf{a}(t) = x''(t)\,\mathbf{i} + m\,x''(t)\,\mathbf{j} = \mathbf{0}$

13.4 Tangent vectors and normal vectors

1. $\mathbf{r}(t) = t\,\mathbf{i} + t^2\,\mathbf{j} + t\,\mathbf{k}$, $\quad \mathbf{r}'(t) = \mathbf{i} + 2t\,\mathbf{j} + \mathbf{k}$

When $t = 0$, $\mathbf{r}'(0) = \mathbf{i} + \mathbf{k}$ $\quad$ [$t = 0$ at $(0, 0, 0)$]

Direction numbers: $a = 1$, $b = 0$, $c = 1$

Parametric equations: $x = t$, $y = 0$, $z = t$

2. $\mathbf{r}(t) = t\,\mathbf{i} + t^2\,\mathbf{j} + \frac{2}{3}\mathbf{k}$, $\quad \mathbf{r}'(t) = \mathbf{i} + 2t\,\mathbf{j}$

When $t = 1$, $\mathbf{r}'(1) = \mathbf{i} + 2\mathbf{j}$ $\quad$ [$t = 1$ at $(1, 1, \frac{2}{3})$]

Direction numbers: $a = 1$, $b = 2$, $c = 0$

Parametric equations: $x = t + 1$, $y = 2t + 1$, $z = \frac{2}{3}$

3. $\mathbf{r}(t) = 2\cos t\,\mathbf{i} + 2\sin t\,\mathbf{j} + t\,\mathbf{k}$, $\quad \mathbf{r}'(t) = -2\sin t\,\mathbf{i} + 2\cos t\,\mathbf{j} + \mathbf{k}$

When $t = 0$, $\mathbf{r}'(0) = 2\mathbf{j} + \mathbf{k}$ $\quad$ [$t = 0$ at $(2, 0, 0)$]

Direction numbers: $a = 0$, $b = 2$, $c = 1$

Parametric equations: $x = 2$, $y = 2t$, $z = t$

4. $\mathbf{r}(t) = 3\cos t\,\mathbf{i} + 4\sin t\,\mathbf{j} + \frac{t}{2}\mathbf{k}$

$\mathbf{r}'(t) = -3\sin t\,\mathbf{i} + 4\cos t\,\mathbf{j} + \frac{1}{2}\mathbf{k}$

When $t = \frac{\pi}{2}$, $\mathbf{r}'(\frac{\pi}{2}) = -3\,\mathbf{i} + \frac{1}{2}\mathbf{k}$ $\quad$ [$t = \frac{\pi}{2}$ at $(0, 4, \frac{\pi}{4})$]

Direction numbers: $a = -3$, $b = 0$, $c = \frac{1}{2}$

Parametric equations: $x = -3t$, $y = 4$, $z = \frac{1}{2}t + \frac{\pi}{4}$

5. $\mathbf{r}(t) = \langle t, t^2, \frac{2}{3}t^3\rangle$, $\quad \mathbf{r}'(t) = \langle 1, 2t, 2t^2\rangle$

When $t = 3$, $\mathbf{r}'(3) = \langle 1, 6, 18\rangle$ $\quad$ [$t = 3$ at $(3, 9, 18)$]

Direction numbers: $a = 1$, $b = 6$, $c = 18$

Parametric equations: $x = t + 3$, $y = 6t + 9$, $z = 18t + 18$

6. $\mathbf{r}(t) = \langle t, t, \sqrt{4 - t^2}\rangle$, $\quad \mathbf{r}'(t) = \langle 1, 1, -\frac{t}{\sqrt{4 - t^2}}\rangle$

When $t = 1$, $\mathbf{r}'(1) = \langle 1, 1, -\frac{1}{\sqrt{3}}\rangle$ $\quad$ [$t = 1$ at $(1, 1, \sqrt{3})$]

Direction numbers: $a = 1$, $b = 1$, $c = -\frac{1}{\sqrt{3}}$

Parametric equations: $x = t + 1$, $y = t + 1$, $z = -\frac{1}{\sqrt{3}}t + \sqrt{3}$

7. $\mathbf{r}(t) = \langle 2\cos t, 2\sin t, 4\rangle$, $\quad \mathbf{r}'(t) = \langle -2\sin t, 2\cos t, 0\rangle$

When $t = \frac{\pi}{4}$, $\mathbf{r}'(\frac{\pi}{4}) = \langle -\sqrt{2}, \sqrt{2}, 0\rangle$ $\quad$ [$t = \frac{\pi}{4}$ at $(\sqrt{2}, \sqrt{2}, 4)$]

Direction numbers: $a = -\sqrt{2}$, $b = \sqrt{2}$, $c = 0$

Parametric equations: $x = -\sqrt{2}t + \sqrt{2}$, $y = \sqrt{2}t + \sqrt{2}$, $z = 4$

8. $\mathbf{r}(t) = \langle 2\sin t,\ 2\cos t,\ 4\sin^2 t\rangle$

$\mathbf{r}'(t) = \langle 2\cos t,\ -2\sin t,\ 8\sin t\cos t\rangle$

When $t = \frac{\pi}{6}$, $\mathbf{r}'(\frac{\pi}{6}) = \langle \sqrt{3},\ -1,\ 2\sqrt{3}\rangle$ $\quad [t = \frac{\pi}{6}$ at $(1, \sqrt{3}, 1)]$

Direction numbers: $a = \sqrt{3}$, $b = -1$, $c = 2\sqrt{3}$

Parametric equations: $x = \sqrt{3}t + 1$, $y = -t + \sqrt{3}$, $z = 2\sqrt{3}t + 1$

9. $\mathbf{r}(t) = 4t\,\mathbf{i}$, $\quad \mathbf{v}(t) = 4\,\mathbf{i}$, $\quad \mathbf{v}(2) = 4\mathbf{i}$, $\quad \mathbf{a}(t) = \mathbf{0}$, $\quad \mathbf{a}(2) = \mathbf{0}$

$\mathbf{T}(2) = \dfrac{\mathbf{v}(2)}{\|\mathbf{v}(2)\|} = \mathbf{i}$, $\quad \mathbf{N}(2) = \dfrac{\mathbf{T}'(2)}{\|\mathbf{T}'(2)\|} = \mathbf{0}$

$\mathbf{a}\cdot\mathbf{T} = 0$, $\quad \mathbf{a}\cdot\mathbf{N} = 0$

10. $\mathbf{r}(t) = 4t\,\mathbf{i} - 2t\,\mathbf{j}$, $\quad \mathbf{v}(t) = 4\mathbf{i} - 2\mathbf{j}$, $\quad \mathbf{a}(t) = \mathbf{0}$

$\mathbf{T}(t) = \dfrac{\mathbf{v}(t)}{\|\mathbf{v}(t)\|} = \dfrac{1}{\sqrt{5}}(2\mathbf{i} - \mathbf{j})$, $\quad \mathbf{N}(t) = \dfrac{\mathbf{T}'(t)}{\|\mathbf{T}'(t)\|} = \mathbf{0}$

$\mathbf{a}\cdot\mathbf{T} = 0$, $\quad \mathbf{a}\cdot\mathbf{N} = 0$

11. $\mathbf{r}(t) = 4t^2\,\mathbf{i}$, $\quad \mathbf{v}(t) = 8t\,\mathbf{i}$, $\quad \mathbf{v}(4) = 32\mathbf{i}$, $\quad \mathbf{a}(t) = 8\mathbf{i}$, $\quad \mathbf{a}(4) = 8\mathbf{i}$

$\mathbf{T}(4) = \dfrac{\mathbf{v}(4)}{\|\mathbf{v}(4)\|} = \mathbf{i}$, $\quad \mathbf{N}(t) = \dfrac{\mathbf{T}'(t)}{\|\mathbf{T}'(t)\|} = \mathbf{0}$

$\mathbf{a}\cdot\mathbf{T} = 8$, $\quad \mathbf{a}\cdot\mathbf{N} = 0$

12. $\mathbf{r}(t) = t^2\,\mathbf{j} + \mathbf{k}$, $\quad \mathbf{v}(t) = 2t\,\mathbf{j}$, $\quad \mathbf{v}(0) = \mathbf{0}$, $\quad \mathbf{a}(t) = 2\mathbf{j}$, $\quad \mathbf{a}(0) = 2\mathbf{j}$

$\mathbf{T}(t) = \dfrac{\mathbf{v}(t)}{\|\mathbf{v}(t)\|} = \mathbf{j}$, $\quad \mathbf{T}(0) = \mathbf{0}$

$\mathbf{N}(t) = \dfrac{\mathbf{T}'(t)}{\|\mathbf{T}'(t)\|} = \mathbf{0}$, $\quad \mathbf{N}(0) = \mathbf{0}$

$\mathbf{a}\cdot\mathbf{T} = 0$, $\quad \mathbf{a}\cdot\mathbf{N} = 0$

13. $\mathbf{r}(t) = t\,\mathbf{i} + \frac{1}{t}\mathbf{j}, \quad \mathbf{v}(t) = \mathbf{i} - \frac{1}{t^2}\mathbf{j}, \quad \mathbf{v}(1) = \mathbf{i} - \mathbf{j}$

$\mathbf{a}(t) = \frac{2}{t^3}\mathbf{j}, \quad \mathbf{a}(1) = 2\mathbf{j}$

$$\mathbf{T}(t) = \frac{\mathbf{v}(t)}{\|\mathbf{v}(t)\|} = \frac{t^2}{\sqrt{t^4+1}}\left(\mathbf{i} - \frac{1}{t^2}\mathbf{j}\right) = \frac{1}{\sqrt{t^4+1}}(t^2\,\mathbf{i} - \mathbf{j})$$

$\mathbf{T}(1) = \frac{1}{\sqrt{2}}(\mathbf{i} - \mathbf{j})$

$$\mathbf{N}(t) = \frac{\mathbf{T}'(t)}{\|\mathbf{T}'(t)\|} = \frac{\frac{2t}{(t^4+1)^{3/2}}\mathbf{i} + \frac{2t^3}{(t^4+1)^{3/2}}\mathbf{j}}{\frac{2t}{(t^4+1)}} = \frac{1}{\sqrt{t^4+1}}(\mathbf{i} + t^2\,\mathbf{j})$$

$\mathbf{N}(1) = \frac{1}{\sqrt{2}}(\mathbf{i} + \mathbf{j}), \quad \mathbf{a}\cdot\mathbf{T} = -\sqrt{2}, \quad \mathbf{a}\cdot\mathbf{N} = \sqrt{2}$

14. $\mathbf{r}(t) = t\,\mathbf{i} + t^2\,\mathbf{j}, \quad \mathbf{v}(t) = \mathbf{i} + 2t\,\mathbf{j}, \quad \mathbf{v}(1) = \mathbf{i} + 2\mathbf{j}$

$\mathbf{a}(t) = 2\mathbf{j}, \quad \mathbf{a}(1) = 2\mathbf{j}, \quad \mathbf{T}(t) = \frac{\mathbf{v}(t)}{\|\mathbf{v}(t)\|} = \frac{1}{\sqrt{1+4t^2}}(\mathbf{i} + 2t\,\mathbf{j})$

$\mathbf{T}(1) = \frac{1}{\sqrt{5}}(\mathbf{i} + 2\mathbf{j})$

$$\mathbf{N}(t) = \frac{\mathbf{T}'(t)}{\|\mathbf{T}'(t)\|} = \frac{\frac{-4t}{(1+4t^2)^{3/2}}\mathbf{i} + \frac{2}{(1+4t^2)^{3/2}}\mathbf{j}}{\frac{2}{1+4t^2}}$$

$$= \frac{1}{\sqrt{1+4t^2}}(-2t\,\mathbf{i} + \mathbf{j})$$

$\mathbf{N}(1) = \frac{1}{\sqrt{5}}(-2\mathbf{i} + \mathbf{j}), \quad \mathbf{a}\cdot\mathbf{T} = \frac{4}{\sqrt{5}}, \quad \mathbf{a}\cdot\mathbf{N} = \frac{2}{\sqrt{5}}$

15. $\mathbf{r}(t) = 4\cos(2\pi t)\,\mathbf{i} + 4\sin(2\pi t)\,\mathbf{j}$

$\mathbf{v}(t) = -8\pi\sin(2\pi t)\,\mathbf{i} + 8\pi\cos(2\pi t)\,\mathbf{j}, \quad \mathbf{v}\left(\frac{1}{8}\right) = -4\pi\sqrt{2}\,\mathbf{i} + 4\pi\sqrt{2}\,\mathbf{j}$

$\mathbf{a}(t) = -16\pi^2\cos(2\pi t)\,\mathbf{i} - 16\pi^2\sin(2\pi t)\,\mathbf{j}, \quad \mathbf{a}\left(\frac{1}{8}\right) = -8\pi^2\sqrt{2}\,\mathbf{i} - 8\pi^2\sqrt{2}\,\mathbf{j}$

$\mathbf{T}(t) = \frac{\mathbf{v}(t)}{\|\mathbf{v}(t)\|} = -\sin(2\pi t)\,\mathbf{i} + \cos(2\pi t)\,\mathbf{j}, \quad \mathbf{T}\left(\frac{1}{8}\right) = \frac{1}{\sqrt{2}}(-\mathbf{i} + \mathbf{j})$

$\mathbf{N}(t) = \frac{\mathbf{T}'(t)}{\|\mathbf{T}'(t)\|} = -\cos(2\pi t)\,\mathbf{i} - \sin(2\pi t)\,\mathbf{j}, \quad \mathbf{N}\left(\frac{1}{8}\right) = -\frac{1}{\sqrt{2}}(\mathbf{i} + \mathbf{j})$

$\mathbf{a}\cdot\mathbf{T} = 0, \quad \mathbf{a}\cdot\mathbf{N} = 16\pi^2$

16. $\mathbf{r}(t) = a\cos(\omega t)\,\mathbf{i} + a\sin(\omega t)\,\mathbf{j}$

$\mathbf{v}(t) = -a\omega\sin(\omega t)\,\mathbf{i} + a\omega\cos(\omega t)\,\mathbf{j}$

$\mathbf{v}(t_0) = -a\omega\sin(\omega t_0)\,\mathbf{i} + a\omega\cos(\omega t_0)\,\mathbf{j}$

$\mathbf{a}(t) = -a\omega^2\cos(\omega t)\,\mathbf{i} - a\omega^2\sin(\omega t)\,\mathbf{j}$

$\mathbf{a}(t_0) = -a\omega^2\cos(\omega t_0)\,\mathbf{i} - a\omega^2\sin(\omega t_0)\,\mathbf{j}$

$$\mathbf{T}(t) = \frac{\mathbf{v}(t)}{\|\mathbf{v}(t)\|} = -\sin(\omega t)\,\mathbf{i} + \cos(\omega t)\,\mathbf{j}$$

$\mathbf{T}(t_0) = -\sin(\omega t_0)\,\mathbf{i} + \cos(\omega t_0)\,\mathbf{j}$

$$\mathbf{N}(t) = \frac{\mathbf{T}'(t)}{\|\mathbf{T}'(t)\|} = -\cos(\omega t)\,\mathbf{i} - \sin(\omega t)\,\mathbf{j}$$

$\mathbf{N}(t_0) = -\cos(\omega t_0)\,\mathbf{i} - \sin(\omega t_0)\,\mathbf{j}$

$\mathbf{a}\cdot\mathbf{T} = 0, \qquad \mathbf{a}\cdot\mathbf{N} = a\omega^2$

17. $\mathbf{r}(t) = 2\cos(\pi t)\,\mathbf{i} + 2\sin(\pi t)\,\mathbf{j}$

From Exercise 16 we have

$$\mathbf{T}\left(\frac{1}{3}\right) = -\frac{\sqrt{3}}{2}\,\mathbf{i} + \frac{1}{2}\,\mathbf{j}, \qquad \mathbf{N}\left(\frac{1}{3}\right) = -\frac{1}{2}\,\mathbf{i} - \frac{\sqrt{3}}{2}\,\mathbf{j}$$

$\mathbf{a}\cdot\mathbf{T} = 0, \qquad \mathbf{a}\cdot\mathbf{N} = 2\pi^2$

18. $\mathbf{r}(t) = a\cos(\omega t)\,\mathbf{i} + b\sin(\omega t)\,\mathbf{j}$

$\mathbf{v}(t) = -a\omega\sin(\omega t)\,\mathbf{i} + b\omega\cos(\omega t)\,\mathbf{j}, \qquad \mathbf{v}(0) = b\omega\,\mathbf{j}$

$\mathbf{a}(t) = -a\omega^2\cos(\omega t)\,\mathbf{i} - b\omega^2\sin(\omega t)\,\mathbf{j}, \qquad \mathbf{a}(0) = -a\omega^2\,\mathbf{i}$

$$\mathbf{T}(0) = \frac{\mathbf{v}(0)}{\|\mathbf{v}(0)\|} = \mathbf{j}$$

Motion along $\mathbf{r}(t)$ is counterclockwise therefore $\mathbf{N}(0) = -\mathbf{i}$

$\mathbf{a}\cdot\mathbf{T} = 0, \qquad \mathbf{a}\cdot\mathbf{N} = a\omega^2$

19. $\mathbf{r}(t) = (e^t \cos t)\,\mathbf{i} + (e^t \sin t)\,\mathbf{j}$

$\mathbf{v}(t) = e^t(\cos t - \sin t)\,\mathbf{i} + e^t(\cos t + \sin t)\,\mathbf{j}$

$\mathbf{a}(t) = e^t(-2\sin t)\,\mathbf{i} + e^t(2\cos t)\,\mathbf{j}$

At $t = \frac{\pi}{2}$ $\quad \mathbf{T} = \frac{\mathbf{v}}{\|\mathbf{v}\|} = \frac{1}{\sqrt{2}}(-\mathbf{i} + \mathbf{j})$

Motion along $\mathbf{r}$ is counterclockwise, therefore $\mathbf{N} = \frac{1}{\sqrt{2}}(-\mathbf{i} - \mathbf{j})$

$\mathbf{a} \cdot \mathbf{T} = \sqrt{2}e^{\pi/2}, \qquad \mathbf{a} \cdot \mathbf{N} = \sqrt{2}e^{\pi/2}$

20. $\mathbf{r}(t_0) = (\omega t_0 - \sin \omega t_0)\,\mathbf{i} + (1 - \cos \omega t_0)\,\mathbf{j}$

$\mathbf{v}(t_0) = \omega[(1 - \cos \omega t_0)\,\mathbf{i} + (\sin \omega t_0)\,\mathbf{j}]$

$\mathbf{a}(t_0) = \omega^2[(\sin \omega t_0)\,\mathbf{i} + (\cos \omega t_0)\,\mathbf{j}]$,

$\mathbf{T} = \frac{\mathbf{v}}{\|\mathbf{v}\|} = \frac{(1 - \cos \omega t_0)\,\mathbf{i} + (\sin \omega t_0)\,\mathbf{j}}{\sqrt{2}\sqrt{1 - \cos \omega t_0}}$

Motion along $\mathbf{r}$ is clockwise, therefore

$\mathbf{N} = \frac{(\sin \omega t_0)\,\mathbf{i} - (1 - \cos \omega t_0)\,\mathbf{j}}{\sqrt{2}\sqrt{1 - \cos \omega t_0}}$

$\mathbf{a} \cdot \mathbf{T} = \frac{\omega^2 \sin \omega t_0}{\sqrt{2}\sqrt{1 - \cos \omega t_0}} = \frac{\omega^2}{\sqrt{2}}\sqrt{1 + \cos \omega t_0}$

$\mathbf{a} \cdot \mathbf{N} = \frac{\omega^2}{\sqrt{2}}\sqrt{1 - \cos \omega t_0}$

21. $\mathbf{r}(t_0) = (\cos \omega t_0 + \omega t_0 \sin \omega t_0)\,\mathbf{i} + (\sin \omega t_0 - \omega t_0 \cos \omega t_0)\,\mathbf{j}$

$\mathbf{v}(t_0) = (\omega^2 t_0 \cos \omega t_0)\,\mathbf{i} + (\omega^2 t_0 \sin \omega t_0)\,\mathbf{j}$

$\mathbf{a}(t_0) = \omega^2[(\cos \omega t_0 - \omega t_0 \sin \omega t_0)\,\mathbf{i} + (\omega t_0 \cos \omega t_0 + \sin \omega t_0)\,\mathbf{j}]$

$\mathbf{T}(t_0) = \frac{\mathbf{v}}{\|\mathbf{v}\|} = (\cos \omega t_0)\,\mathbf{i} + (\sin \omega t_0)\,\mathbf{j}$

Motion along $\mathbf{r}$ is counterclockwise therefore

$\mathbf{N}(t_0) = (-\sin \omega t_0)\,\mathbf{i} + (\cos \omega t_0)\,\mathbf{j}$

$\mathbf{a} \cdot \mathbf{T} = \omega^2, \qquad \mathbf{a} \cdot \mathbf{N} = \omega^2(\omega t_0) = \omega^3 t_0$

22. $\mathbf{r}(t) = 4t\,\mathbf{i} - 4t\,\mathbf{j} + 2t\,\mathbf{k}, \qquad \mathbf{v}(t) = 4\mathbf{i} - 4\mathbf{j} + 2\mathbf{k}, \qquad \mathbf{a}(t) = \mathbf{0}$

$\mathbf{T}(t) = \frac{\mathbf{v}}{\|\mathbf{v}\|} = \frac{1}{3}(2\mathbf{i} - 2\mathbf{j} + \mathbf{k}), \qquad \mathbf{N}(t) = \frac{\mathbf{T}'}{\|\mathbf{T}\|} = \mathbf{0}$

$\mathbf{a} \cdot \mathbf{T} = 0, \qquad \mathbf{a} \cdot \mathbf{N} = 0$

23. $\mathbf{r}(t) = t\,\mathbf{i} + t^2\,\mathbf{j} + \dfrac{t^2}{2}\mathbf{k}, \qquad \mathbf{v}(t) = \mathbf{i} + 2t\,\mathbf{j} + t\,\mathbf{k}$

$\mathbf{v}(1) = \mathbf{i} + 2\mathbf{j} + \mathbf{k}, \qquad \mathbf{a}(t) = 2\mathbf{j} + \mathbf{k}$

$$\mathbf{T}(t) = \frac{\mathbf{v}}{\|\mathbf{v}\|} = \frac{1}{\sqrt{1 + 5t^2}}(\mathbf{i} + 2t\,\mathbf{j} + t\mathbf{k})$$

$$\mathbf{T}(1) = \frac{\sqrt{6}}{6}(\mathbf{i} + 2\mathbf{j} + \mathbf{k})$$

$$\mathbf{N}(t) = \frac{\mathbf{T}'}{\|\mathbf{T}'\|} = \frac{\dfrac{-5t\,\mathbf{i} + 2\mathbf{j} + \mathbf{k}}{(1 + 5t^2)^{3/2}}}{\dfrac{\sqrt{5}}{1 + 5t^2}} = \frac{-5t\,\mathbf{i} + 2\mathbf{j} + \mathbf{k}}{\sqrt{5}\sqrt{1 + 5t^2}}$$

$$\mathbf{N}(1) = \frac{\sqrt{30}}{30}(-5\mathbf{i} + 2\mathbf{j} + \mathbf{k}), \qquad \mathbf{a}\cdot\mathbf{T} = \frac{5\sqrt{6}}{6}, \qquad \mathbf{a}\cdot\mathbf{N} = \frac{\sqrt{30}}{6}$$

24. $\mathbf{r}(t) = t\,\mathbf{i} + 3t^2\,\mathbf{j} + \dfrac{t^2}{2}\mathbf{k}, \qquad \mathbf{v}(t) = \mathbf{i} + 6t\,\mathbf{j} + t\,\mathbf{k}$

$\mathbf{v}(2) = \mathbf{i} + 12\mathbf{j} + 2\mathbf{k}, \qquad \mathbf{a}(t) = 6\mathbf{j} + \mathbf{k}$

$$\mathbf{T}(t) = \frac{\mathbf{v}}{\|\mathbf{v}\|} = \frac{1}{\sqrt{1 + 37t^2}}(\mathbf{i} + 6t\,\mathbf{j} + t\,\mathbf{k})$$

$$\mathbf{T}(2) = \frac{1}{\sqrt{149}}(\mathbf{i} + 12\mathbf{j} + 2\mathbf{k})$$

$$\mathbf{N}(t) = \frac{\mathbf{T}'}{\|\mathbf{T}'\|} = \frac{\dfrac{1}{(1 + 37t^2)^{3/2}}[-37t\,\mathbf{i} + 6\mathbf{j} + \mathbf{k}]}{\dfrac{\sqrt{37}}{1 + 37t^2}}$$

$$= \frac{1}{\sqrt{37}\sqrt{1 + 37t^2}}[-37t\,\mathbf{i} + 6\mathbf{j} + \mathbf{k}]$$

$$\mathbf{N}(2) = \frac{1}{\sqrt{37}\sqrt{149}}[-74\mathbf{i} + 6\mathbf{j} + \mathbf{k}] = \frac{1}{\sqrt{5513}}(-74\mathbf{i} + 6\mathbf{j} + \mathbf{k})$$

$$\mathbf{a}\cdot\mathbf{T} = \frac{74}{\sqrt{149}}, \qquad \mathbf{a}\cdot\mathbf{N} = \frac{37}{\sqrt{5513}} = \frac{\sqrt{37}}{\sqrt{149}}$$

25. $\mathbf{r}(t) = 4t\,\mathbf{i} + 3\cos t\,\mathbf{j} + 3\sin t\,\mathbf{k}, \qquad \mathbf{v}(t) = 4\mathbf{i} - 3\sin t\,\mathbf{j} + 3\cos t\,\mathbf{k}$

$\mathbf{v}(\frac{\pi}{2}) = 4\mathbf{i} - 3\mathbf{j}, \qquad \mathbf{a}(t) = -3\cos t\,\mathbf{j} - 3\sin t\,\mathbf{k}, \qquad \mathbf{a}(\frac{\pi}{2}) = -3\mathbf{k}$

$$\mathbf{T}(t) = \frac{\mathbf{v}}{\|\mathbf{v}\|} = \frac{1}{5}(4\mathbf{i} - 3\sin t\,\mathbf{j} + 3\cos t\,\mathbf{k})$$

$$\mathbf{T}(\tfrac{\pi}{2}) = \frac{1}{5}(4\mathbf{i} - 3\mathbf{j}), \qquad \mathbf{N}(t) = \frac{\mathbf{T}'}{\|\mathbf{T}'\|} = -\cos t\,\mathbf{j} - \sin t\,\mathbf{k}$$

$$\mathbf{N}(\tfrac{\pi}{2}) = -\mathbf{k}, \qquad \mathbf{a}\cdot\mathbf{T} = 0, \qquad \mathbf{a}\cdot\mathbf{N} = 3$$

26. $\mathbf{r}(t) = e^t \cos t\, \mathbf{i} + e^t \sin t\, \mathbf{j} + e^t \mathbf{k}$

$\mathbf{v}(t) = (-e^t \sin t + e^t \cos t)\, \mathbf{i} + (e^t \cos t + e^t \sin t)\, \mathbf{j} + e^t \mathbf{k}$

$\mathbf{v}(0) = \mathbf{i} + \mathbf{j} + \mathbf{k}$

$\mathbf{a}(t) = -2e^t \sin t\, \mathbf{i} + 2e^t \cos t\, \mathbf{j} + e^t \mathbf{k}, \qquad \mathbf{a}(0) = 2\mathbf{j} + \mathbf{k}$

$$\mathbf{T}(t) = \frac{\mathbf{v}}{\|\mathbf{v}\|} = \frac{1}{\sqrt{3}}[(-\sin t + \cos t)\, \mathbf{i} + (\cos t + \sin t)\, \mathbf{j} + \mathbf{k}]$$

$$\mathbf{T}(0) = \frac{1}{\sqrt{3}}(\mathbf{i} + \mathbf{j} + \mathbf{k})$$

$$\mathbf{N}(t) = \frac{\mathbf{T}'}{\|\mathbf{T}'\|} = \frac{\frac{1}{\sqrt{3}}[(-\cos t - \sin t)\, \mathbf{i} + (-\sin t + \cos t)\, \mathbf{j}]}{\frac{1}{\sqrt{3}}\sqrt{(-\cos t - \sin t)^2 + (-\sin t + \cos t)^2}}$$

$$= \frac{1}{\sqrt{2}}[(-\cos t - \sin t)\, \mathbf{i} + (-\sin t + \cos t)\, \mathbf{j}]$$

$$\mathbf{N}(0) = \frac{1}{\sqrt{2}}(-\mathbf{i} + \mathbf{j}), \qquad \mathbf{a} \cdot \mathbf{T} = \sqrt{3}, \qquad \mathbf{a} \cdot \mathbf{N} = \sqrt{2}$$

27. From Theorem 13.3 we have $\mathbf{r}(t) = (v_0 t \cos\theta)\, \mathbf{i} + (h + v_0 t \sin\theta - 16t^2)\, \mathbf{j}$

$\mathbf{v}(t) = v_0 \cos\theta\, \mathbf{i} + (v_0 \sin\theta - 32t)\, \mathbf{j}, \qquad \mathbf{a}(t) = -32\mathbf{j}$

$$\mathbf{T}(t) = \frac{(v_0 \cos\theta)\, \mathbf{i} + (v_0 \sin\theta - 32t)\, \mathbf{j}}{\sqrt{v_0^2 \cos^2\theta + (v_0 \sin\theta - 32t)^2}}$$

$$\mathbf{N}(t) = \frac{(v_0 \sin\theta - 32t)\, \mathbf{i} - v_0 \cos\theta\, \mathbf{j}}{\sqrt{v_0 \cos^2\theta + (v_0 \sin\theta - 32t)^2}} \quad \text{(Motion is clockwise)}$$

$$\mathbf{a} \cdot \mathbf{T} = \frac{-32(v_0 \sin\theta - 32t)}{\sqrt{v_0^2 \cos^2\theta + (v_0 \sin\theta - 32t)^2}}$$

$$\mathbf{a} \cdot \mathbf{N} = \frac{32 v_0 \cos\theta}{\sqrt{v_0^2 \cos^2\theta + (v_0 \sin\theta - 32t)^2}}$$

Maximum height when $v_0 \sin\theta - 32t = 0$ (Vertical component of velocity)

At maximum height $\mathbf{a} \cdot \mathbf{T} = 0$ and $\mathbf{a} \cdot \mathbf{N} = 32$

28. Assuming the measurements are in feet and seconds we have

540 mph = 792 ft/s and t = 0 at time of release then

$\mathbf{r}(t) = 792t\,\mathbf{i} + (-16t^2 + 30{,}000)\,\mathbf{j}$, $\mathbf{v}(t) = 792\mathbf{i} - 32t\,\mathbf{j}$, $\mathbf{a}(t) = -32\mathbf{j}$

$$\mathbf{T} = \frac{792\mathbf{i} - 32t\,\mathbf{j}}{8\sqrt{9801 + 16t^2}} = \frac{99\mathbf{i} - 4t\,\mathbf{j}}{\sqrt{9801 + 16t^2}}$$

Motion along $\mathbf{r}$ is clockwise, therefore

$$\mathbf{N} = \frac{-4t\mathbf{i} - 99\mathbf{j}}{\sqrt{9801 + 16t^2}}, \qquad \mathbf{a} \cdot \mathbf{T} = \frac{128t}{\sqrt{9801 + 16t^2}}$$

$$\mathbf{a} \cdot \mathbf{N} = \frac{3168}{\sqrt{9801 + 16t^2}}$$

29. $\mathbf{r}(t) = (a\cos\omega t)\,\mathbf{i} + (a\sin\omega t)\,\mathbf{j}$

From Exercise 16 we know $\mathbf{a} \cdot \mathbf{T} = 0$ and $\mathbf{a} \cdot \mathbf{N} = a\omega^2$

(a) Let $\omega_0 = 2\omega$ then $\mathbf{a} \cdot \mathbf{N} = a\omega_0{}^2 = a(2\omega)^2 = 4a\omega^2$ or the centripetal acceleration is increased by a factor of 4 when the velocity is doubled.

(b) Let $a_0 = \dfrac{a}{2}$, then $\mathbf{a} \cdot \mathbf{N} = a_0\omega^2 = \dfrac{a}{2}\omega^2 = \dfrac{1}{2}a\omega^2$ or the centripetal acceleration is halved when the radius is halved.

30. $\mathbf{r}(t) = (r\cos\omega t)\,\mathbf{i} + (r\sin\omega t)\,\mathbf{j}$, $\mathbf{v}(t) = (-r\omega\sin\omega t)\,\mathbf{i} + (r\omega\cos\omega t)\,\mathbf{j}$

$\|\mathbf{v}(t)\| = r\omega\sqrt{1} = r\omega$, $\mathbf{a}(t) = (-r\omega^2\cos\omega t)\,\mathbf{i} - (r\omega^2\sin\omega t)\,\mathbf{j}$,

$\|\mathbf{a}(t)\| = r\omega^2$

(a) $\mathbf{F} = m[\mathbf{a}(t)] = m(r\omega^2) = \dfrac{m}{r}(r^2\omega^2) = \dfrac{mv^2}{r}$

(b) By Newton's Law: $\dfrac{mv^2}{r} = \dfrac{GMm}{r^2}$, $v^2 = \dfrac{GM}{r}$, $v = \sqrt{(GM/r)}$

31. $v = \sqrt{\dfrac{9.56 \times 10^4}{4100}} \approx 4.83$ mi/sec

32. $v = \sqrt{\dfrac{9.56 \times 10^4}{4200}} \approx 4.77$ mi/sec

33. $v = \sqrt{\dfrac{9.56 \times 10^4}{4385}} \approx 4.67$ mi/sec

34. Let x = distance from the satellite to the center of the earth ($x = r + 4000$). Then

$$v = \frac{2\pi x}{t} = \frac{2\pi x}{24(3600)} = \sqrt{\frac{9.56 \times 10^4}{x}}, \qquad \frac{4\pi^2 r^2}{(24)^2(3600)^2} = \frac{9.56 \times 10^4}{x}$$

$$x^3 = \frac{(9.56 \times 10^4)(24)^2(3600)^2}{4\pi^2} \Longrightarrow x = 26{,}245 \text{ mi}$$

$$v = \frac{2\pi(26{,}245)}{24(3600)} = 1.91 \text{ mi/sec} = 6871 \text{ mph}$$

35. $\|T'\| = K\left(\frac{ds}{dt}\right) = 0$ since $K = 0$ at a point of inflection

36. Let $F_s = mg_s$ be the weight at the surface of the earth and

$F_h = mg_h$ the weight at a height of h miles.

$$F_s = \frac{GMm}{r^2}, \qquad GMm = r^2 F_s, \qquad F_h = \frac{GMm}{(r+h)^2} = \frac{r^2 F_s}{(r+h)^2}$$

$$g_h = \frac{r^2}{(r+h)^2} g_s = \left(\frac{r}{r+h}\right)^2 g_s$$

13.5 Arc length and curvature

1. $\mathbf{r}(t) = t\,\mathbf{i} + 3t\,\mathbf{j}$

$$\frac{dx}{dt} = 1, \quad \frac{dy}{dt} = 3, \quad \frac{dz}{dt} = 0$$

$$s = \int_0^4 \sqrt{1+9}\, dt = \sqrt{10}\int_0^4 dt$$

$$= \sqrt{10}t \Big]_0^4 = 4\sqrt{10}$$

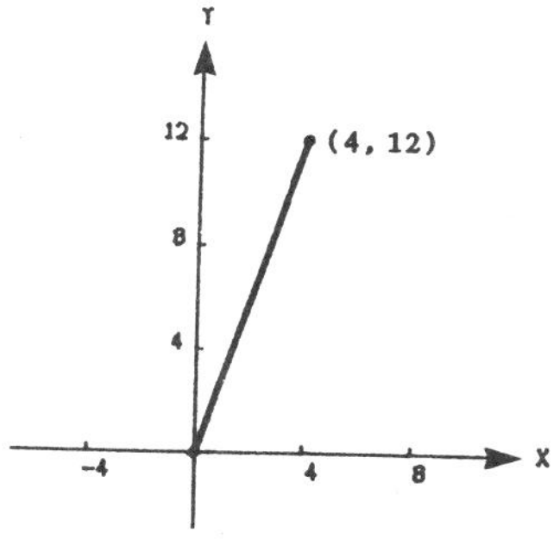

2. $\mathbf{r}(t) = t\,\mathbf{i} + t^2\,\mathbf{k}$

$$\frac{dx}{dt} = 1, \quad \frac{dy}{dt} = 0, \quad \frac{dz}{dt} = 2t$$

$$s = \int_0^4 \sqrt{1 + 4t^2}\, dt$$

$$= \frac{1}{4}\left[2t\sqrt{1+4t^2} + \ln\left|2t + \sqrt{1+4t^2}\right|\right]_0^4$$

$$= \frac{1}{4}[8\sqrt{65} + \ln(8 + \sqrt{65})] \approx 16.818$$

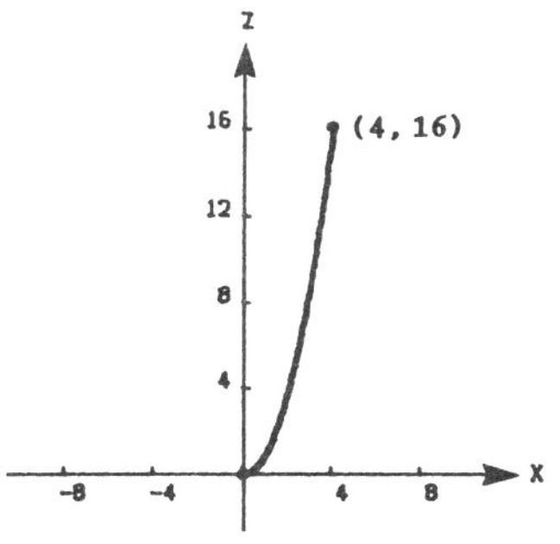

3. $\mathbf{r}(t) = a\cos t\,\mathbf{i} + a\sin t\,\mathbf{j} + bt\,\mathbf{k}$

$$\frac{dx}{dt} = -a\sin t,\quad \frac{dy}{dt} = a\cos t,\quad \frac{dz}{dt} = b$$

$$s = \int_0^{2\pi} \sqrt{a^2\sin^2 t + a^2\cos^2 t + b^2}\,dt$$

$$= \int_0^{2\pi} \sqrt{a^2 + b^2}\,dt = \sqrt{a^2 + b^2}\,t\Big]_0^{2\pi}$$

$$= 2\pi\sqrt{a^2 + b^2}$$

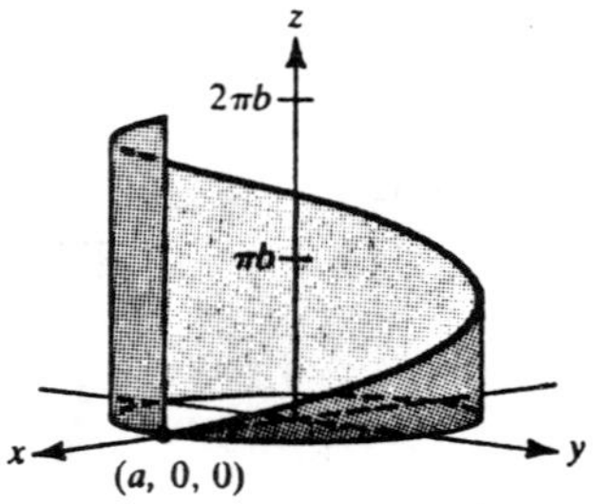

4. $\mathbf{r}(t) = a\cos^3 t\,\mathbf{i} + a\sin^3 t\,\mathbf{j}$

$$\frac{dx}{dt} = -3a\cos^2 t\sin t,\quad \frac{dy}{dt} = 3a\sin^2 t\cos t$$

$$s = 4\int_0^{\pi/2} \sqrt{[3a\cos^2 t(-\sin t)]^2 + [3a\sin^2 t\cos t]^2}\,dt$$

$$= 12a\int_0^{\pi/2} \sin t\cos t\,dt$$

$$= 3a\int_0^{\pi/2} 2\sin 2t\,dt$$

$$= -3a\cos 2t\Big]_0^{\pi/2} = 6a$$

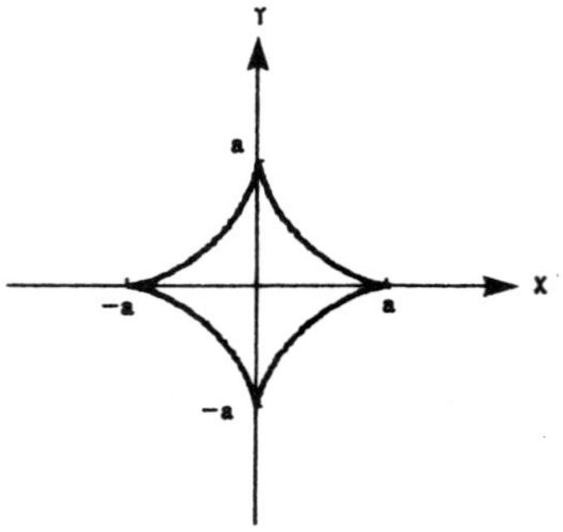

5. $\mathbf{r}(t) = (\sin t - t\cos t)\,\mathbf{i} + (\cos t + t\sin t)\,\mathbf{j} + t^2\,\mathbf{k}$

$$\frac{dx}{dt} = t\sin t,\quad \frac{dy}{dt} = t\cos t,\quad \frac{dz}{dt} = 2t$$

$$s = \int_0^{\pi/2} \sqrt{(t\sin t)^2 + (t\cos t)^2 + (2t)^2}\,dt$$

$$= \int_0^{\pi/2} \sqrt{5t^2}\,dt = \sqrt{5}\int_0^{\pi/2} t\,dt$$

$$= \frac{\sqrt{5}}{2}t^2\Big]_0^{\pi/2} = \frac{\sqrt{5}\pi^2}{8}$$

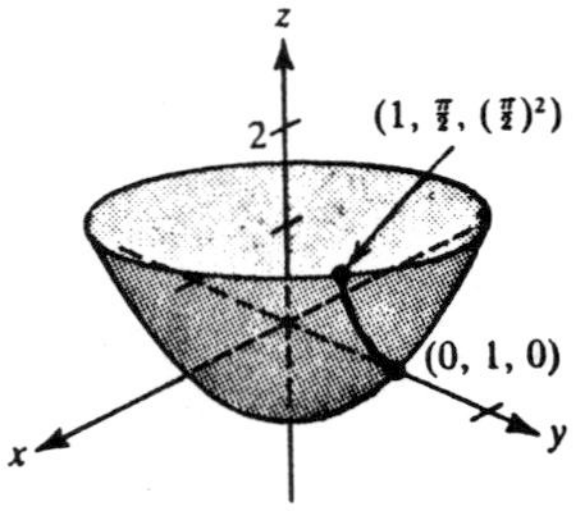

6. $\mathbf{r}(t) = 4t\,\mathbf{i} + 3\cos t\,\mathbf{j} + 3\sin t\,\mathbf{k}$

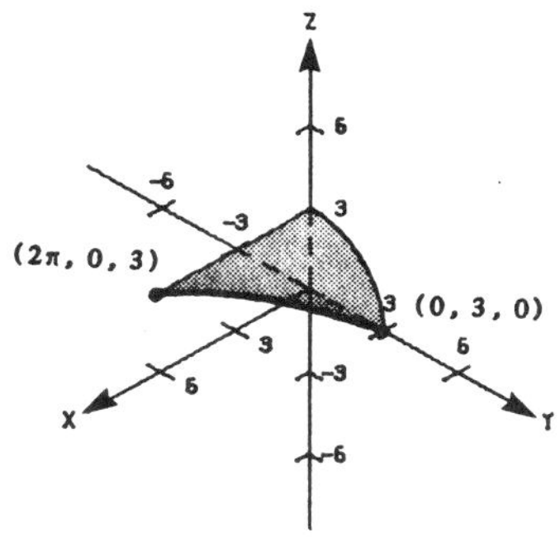

$\dfrac{dx}{dt} = 4, \quad \dfrac{dy}{dt} = -3\sin t, \quad \dfrac{dz}{dt} = 3\cos t$

$$s = \int_0^{\pi/2} \sqrt{(4)^2 + (-3\sin t)^2 + (3\cos t)^2}\,dt$$

$$= \sqrt{25}\int_0^{\pi/2} dt = 5t\Big]_0^{\pi/2} = \frac{5\pi}{2}$$

7. $y = 3x - 2$

Since $y'' = 0$, $K = 0$ and the radius of curvature is undefined.

8. $y = mx + b$

Since $y'' = 0$, $K = 0$ and the radius of curvature is undefined.

9. $y = 2x^2 + 3, \quad y' = 4x, \quad y'' = 4$

$$K = \frac{4}{[1 + (-4)^2]^{3/2}} = \frac{4}{17^{3/2}} \approx 0.057$$

$$\frac{1}{K} = \frac{17^{3/2}}{4} \approx 17.523 \text{ (radius of curvature)}$$

10. $y = x + \dfrac{1}{x}, \quad y' = 1 - \dfrac{1}{x^2}, \quad y'' = \dfrac{2}{x^3}$

$$K = \frac{2}{[(1 + 0^2)]^{3/2}} = 2, \quad \frac{1}{K} = \frac{1}{2} \text{ (radius of curvature)}$$

11. $y = \sqrt{a^2 - x^2}, \quad y' = \dfrac{-x}{\sqrt{a^2 - x^2}}, \quad y'' = \dfrac{2x^2 - a^2}{(a^2 - x^2)^{3/2}}$

At $x = 0$: $y' = 0$ and $y'' = 1/a$

$$K = \frac{1/a}{(1 + 0^2)^{3/2}} = 1/a, \quad \frac{1}{K} = a \text{ (radius of curvature)}$$

12. $y = \dfrac{3}{4}\sqrt{16 - x^2}, \quad y' = \dfrac{-9x}{16y}, \quad y'' = \dfrac{-(9 + 16y'^2)}{16y}$

At $x = 0$: $y' = 0$ and $y'' = -3/16$

$$K = \left|\frac{-3/16}{(1 + 0^2)^{3/2}}\right| = 3/16, \quad \frac{1}{K} = \frac{16}{3} \text{ (radius of curvature)}$$

13. $\mathbf{r}(t) = 4t\,\mathbf{i}$, $\quad \mathbf{v}(t) = 4\mathbf{i}$, $\quad \mathbf{a}(t) = \mathbf{0}$, $\quad K = \dfrac{\mathbf{a}\cdot\mathbf{N}}{\|\mathbf{v}\|^2} = 0$

14. $\mathbf{r}(t) = 4t\,\mathbf{i} - 2t\,\mathbf{j}$, $\quad \mathbf{v}(t) = 4\mathbf{i} - 2\mathbf{j}$, $\quad \mathbf{a}(t) = \mathbf{0}$, $\quad K = \dfrac{\mathbf{a}\cdot\mathbf{N}}{\|\mathbf{v}\|^2} = 0$

15. $\mathbf{r}(t) = 4t^2\mathbf{i}$, $\quad \mathbf{v}(t) = 8t\,\mathbf{i}$, $\quad \mathbf{a}(t) = 8\mathbf{i}$, $\quad \mathbf{T}(t) = \mathbf{i}$, $\quad \mathbf{N}(t) = \mathbf{0}$

$K = \dfrac{\mathbf{a}\cdot\mathbf{N}}{\|\mathbf{v}\|^2} = 0$

16. $\mathbf{r}(t) = t^2\,\mathbf{j} + \mathbf{k}$, $\quad \mathbf{v}(t) = 2t\,\mathbf{j}$, $\quad \mathbf{a}(t) = 2\mathbf{j}$, $\quad \mathbf{T}(t) = \mathbf{j}$

$\mathbf{N}(t) = \mathbf{0}$, $\quad K = \dfrac{\mathbf{a}\cdot\mathbf{N}}{\|\mathbf{v}\|^2} = 0$

17. $\mathbf{r}(t) = t\,\mathbf{i} + \dfrac{1}{t}\mathbf{j}$, $\quad \mathbf{v}(t) = \mathbf{i} - \dfrac{1}{t^2}\mathbf{j}$, $\quad \mathbf{v}(1) = \mathbf{i} - \mathbf{j}$

$\mathbf{a}(t) = \dfrac{2}{t^3}\mathbf{j}$, $\quad \mathbf{a}(1) = 2\mathbf{j}$, $\quad \mathbf{T}(t) = \dfrac{t^2\,\mathbf{i} - \mathbf{j}}{\sqrt{t^4 + 1}}$

$\mathbf{N}(t) = \dfrac{1}{(t^4 + 1)^{1/2}}(2t\,\mathbf{i} + 2t^3\,\mathbf{j})$, $\quad \mathbf{N}(1) = \dfrac{1}{\sqrt{2}}(\mathbf{i} + \mathbf{j})$

$K = \dfrac{\mathbf{a}\cdot\mathbf{N}}{\|\mathbf{v}\|^2} = \dfrac{\sqrt{2}}{2}$

18. $\mathbf{r}(t) = t\,\mathbf{i} + t^2\,\mathbf{j}$, $\quad \mathbf{v}(t) = \mathbf{i} + 2t\,\mathbf{j}$, $\quad \mathbf{v}(1) = \mathbf{i} + 2\mathbf{j}$

$\mathbf{a}(t) = 2\mathbf{j}$, $\quad \mathbf{a}(1) = 2\mathbf{j}$, $\quad \mathbf{T}(t) = \dfrac{\mathbf{i} + 2t\,\mathbf{j}}{\sqrt{1 + 4t^2}}$

$\mathbf{N}(t) = \dfrac{1}{\sqrt{1 + 4t^2}}(-2t\,\mathbf{i} + \mathbf{j})$, $\quad \mathbf{N}(1) = \dfrac{1}{\sqrt{5}}(-2\mathbf{i} + \mathbf{j})$

$K = \dfrac{\mathbf{a}\cdot\mathbf{N}}{\|\mathbf{v}\|^2} = \dfrac{2}{5\sqrt{5}}$

19. $\mathbf{r}(t) = 4\cos(2\pi t)\,\mathbf{i} + 4\sin(2\pi t)\,\mathbf{j}$

$\mathbf{r}'(t) = -8\pi\sin(2\pi t)\,\mathbf{i} + 8\pi\cos(2\pi t)\,\mathbf{j}$

$\mathbf{T}(t) = -\sin(2\pi t)\,\mathbf{i} + \cos(2\pi t)\,\mathbf{j}$

$\mathbf{T}'(t) = -2\pi\cos(2\pi t)\,\mathbf{i} - 2\pi\sin(2\pi t)\,\mathbf{j}$

$K = \dfrac{\|\mathbf{T}'(t)\|}{\|\mathbf{r}'(t)\|} = \dfrac{2\pi}{8\pi} = \dfrac{1}{4}$

20. $\mathbf{r}(t) = a\cos(\omega t)\,\mathbf{i} + a\sin(\omega t)\,\mathbf{j}$

$\mathbf{r}'(t) = -a\omega\sin(\omega t)\,\mathbf{i} + a\omega\cos(\omega t)\,\mathbf{j}$

$\mathbf{T}(t) = -\sin(\omega t)\,\mathbf{i} + \cos(\omega t)\,\mathbf{j}$

$\mathbf{T}'(t) = -\omega\cos(\omega t)\,\mathbf{i} - \omega\sin(\omega t)\,\mathbf{j}$

$$K = \frac{\|\mathbf{T}'(t)\|}{\|\mathbf{r}'(t)\|} = \frac{\omega}{a\omega} = \frac{1}{a}$$

21. $\mathbf{r}(t) = 2\cos(\pi t)\,\mathbf{i} + 2\sin(\pi t)\,\mathbf{j}$, From Exercise 20: $K = \frac{1}{2}$

22. $\mathbf{r}(t) = a\cos(\omega t)\,\mathbf{i} + b\sin(\omega t)\,\mathbf{j}$

$\mathbf{r}'(t) = -a\omega\sin(\omega t)\,\mathbf{i} + b\omega\cos(\omega t)\,\mathbf{j}$

$$\mathbf{T}(t) = \frac{-a\sin(\omega t)\,\mathbf{i} + b\cos(\omega t)\,\mathbf{j}}{\sqrt{a^2\sin^2(\omega t) + b^2\cos^2(\omega t)}}$$

$$\mathbf{T}'(t) = \frac{-ab^2\omega\cos(\omega t)\,\mathbf{i} - a^2b\omega\sin(\omega t)\,\mathbf{j}}{[a^2\sin^2(\omega t) + b^2\cos^2(\omega t)]^{3/2}}$$

$$K = \frac{\|\mathbf{T}'(t)\|}{\|\mathbf{r}'(t)\|} = \frac{\dfrac{ab\omega}{a^2\sin^2(\omega t) + b^2\cos^2(\omega t)}}{\sqrt{a^2\sin^2(\omega t) + b^2\cos^2(\omega t)}}$$

$$= \frac{ab\omega}{[a^2\sin^2(\omega t) + b^2\cos^2(\omega t)]^{3/2}}$$

23. $\mathbf{r}(t) = e^t\cos t\,\mathbf{i} + e^t\sin t\,\mathbf{j}$

$\mathbf{r}'(t) = (-e^t\sin t + e^t\cos t)\mathbf{i} + (e^t\cos t + e^t\sin t)\,\mathbf{j}$

$$\mathbf{T}(t) = \frac{1}{\sqrt{2}}[(-\sin t + \cos t)\,\mathbf{i} + (\cos t + \sin t)\,\mathbf{j}]$$

$$\mathbf{T}'(t) = \frac{1}{\sqrt{2}}[(-\cos t - \sin t)\,\mathbf{i} + (-\sin t + \cos t)\,\mathbf{j}]$$

$$K = \frac{\|\mathbf{T}'(t)\|}{\|\mathbf{r}'(t)\|} = \frac{1}{\sqrt{2}e^t} = \frac{\sqrt{2}}{2}e^{-t}$$

24. $\mathbf{r}(t) = \langle a(\omega t - \sin\omega t),\ a(1 - \cos\omega t)\rangle$

From Exercise 20, Section 13.4 we have: $\mathbf{a}\cdot\mathbf{N} = \frac{a\omega^2}{\sqrt{2}}\cdot\sqrt{1 - \cos\omega t}$

$$K = \frac{\mathbf{a}(t)\cdot\mathbf{N}(t)}{\|\mathbf{v}(t)\|^2} = \frac{(a\omega^2/\sqrt{2})\sqrt{1 - \cos\omega t}}{2a^2\omega^2(1 - \cos\omega t)} = \frac{\sqrt{2}}{4a\sqrt{1 - \cos\omega t}}$$

25. $\mathbf{r}(t) = \langle \cos \omega t + \omega t \sin \omega t, \sin \omega t - \omega t \cos \omega t \rangle$

From Exercise 21, Section 13.4 we have: $\mathbf{a} \cdot \mathbf{N} = \omega^3 t$

$$K = \frac{\mathbf{a}(t) \cdot \mathbf{N}(t)}{\|\mathbf{v}\|^2} = \frac{\omega^3 t}{\omega^4 t^2} = \frac{1}{\omega t}$$

26. $\mathbf{r}(t) = 4t\,\mathbf{i} - 4t\,\mathbf{j} + 2t\,\mathbf{k}, \qquad \mathbf{r}'(t) = 4\mathbf{i} - 4\mathbf{j} + 2\mathbf{k}$

$$\mathbf{T}(t) = \frac{1}{3}(2\mathbf{i} - 2\mathbf{j} + \mathbf{k}), \qquad \mathbf{T}'(t) = \mathbf{0}, \qquad K = \frac{\|\mathbf{T}'(t)\|}{\|\mathbf{r}'(t)\|} = 0$$

27. $\mathbf{r}(t) = t\,\mathbf{i} + t^2\,\mathbf{j} + \frac{t^2}{2}\mathbf{k}, \qquad \mathbf{r}'(t) = \mathbf{i} + 2t\,\mathbf{j} + t\,\mathbf{k}$

$$\mathbf{T}(t) = \frac{\mathbf{i} + 2t\,\mathbf{j} + t\mathbf{k}}{\sqrt{1 + 5t^2}}, \qquad \mathbf{T}'(t) = \frac{-5t\,\mathbf{i} + 2\mathbf{j} + \mathbf{k}}{(1 + 5t^2)^{3/2}}$$

$$K = \frac{\|\mathbf{T}'(t)\|}{\|\mathbf{r}'(t)\|} = \frac{\sqrt{5}/(1 + 5t^2)}{\sqrt{1 + 5t^2}} = \frac{\sqrt{5}}{(1 + 5t^2)^{3/2}}$$

28. $\mathbf{r}(t) = t\,\mathbf{i} + 3t^2\,\mathbf{j} + \frac{t^2}{2}\mathbf{k}, \qquad \mathbf{r}'(t) = \mathbf{i} + 6t\,\mathbf{j} + t\,\mathbf{k}$

$$\mathbf{T}(t) = \frac{\mathbf{i} + 6t\,\mathbf{j} + t\mathbf{k}}{\sqrt{1 + 37t^2}}, \qquad \mathbf{T}'(t) = \frac{-37t\,\mathbf{i} + 6\mathbf{j} + \mathbf{k}}{(1 + 37t^2)^{3/2}}$$

$$K = \frac{\|\mathbf{T}'(t)\|}{\|\mathbf{r}'(t)\|} = \frac{\sqrt{37}/(1 + 37t^2)}{\sqrt{1 + 37t^2}} = \frac{\sqrt{37}}{(1 + 37t^2)^{3/2}}$$

29. $\mathbf{r}(t) = 4t\,\mathbf{i} + 3\cos t\,\mathbf{j} + 3\sin t\,\mathbf{k}, \qquad \mathbf{r}'(t) = 4\mathbf{i} - 3\sin t\,\mathbf{j} + 3\cos t\,\mathbf{k}$

$$\mathbf{T}(t) = \frac{1}{5}[4\mathbf{i} - 3\sin t\,\mathbf{j} + 3\cos t\,\mathbf{k}], \qquad \mathbf{T}'(t) = \frac{1}{5}[-3\cos t\,\mathbf{j} - 3\sin t\,\mathbf{k}]$$

$$K = \frac{\|\mathbf{T}(t)\|}{\|\mathbf{r}'(t)\|} = \frac{3/5}{5} = \frac{3}{25}$$

30. $\mathbf{r}(t) = e^t \cos t\,\mathbf{i} + e^t \sin t\,\mathbf{j} + e^t\,\mathbf{k}$

$$\mathbf{r}'(t) = (-e^t \sin t + e^t \cos t)\,\mathbf{i} + (e^t \cos t + e^t \sin t)\,\mathbf{j} + e^t\,\mathbf{k}$$

$$\mathbf{T}(t) = \frac{1}{\sqrt{3}}[(-\sin t + \cos t)\,\mathbf{i} + (\cos t + \sin t)\,\mathbf{j} + \mathbf{k}]$$

$$\mathbf{T}'(t) = \frac{1}{\sqrt{3}}[(-\cos t - \sin t)\,\mathbf{i} + (-\sin t + \cos t)\,\mathbf{j}]$$

$$K = \frac{\|\mathbf{T}'(t)\|}{\|\mathbf{r}'(t)\|} = \frac{\frac{1}{\sqrt{3}}\sqrt{(-\cos t - \sin t)^2 + (-\sin t + \cos t)^2}}{\sqrt{3}\,e^t} = \frac{\sqrt{2}}{3e^t}$$

31. $y = (x - 1)^2 + 3, \quad y' = 2(x - 1), \quad y'' = 2$

$$K = \frac{2}{(1 + [2(x - 1)]^2)^{3/2}} = \frac{2}{(1 + 4(x - 1)^2)^{3/2}}$$

(a) K is maximum when $x = 1$ or at the vertex $(1, 3)$.

(b) $\lim_{x \to \infty} K = 0$

32. $y = x^3, \quad y' = 3x^2, \quad y'' = 6x, \quad K = \left|\frac{6x}{(1 + 9x^4)^{3/2}}\right|$

(a) K maximum at $\left(\frac{1}{\sqrt[4]{45}}, \frac{1}{\sqrt[4]{45^3}}\right)$ and $\left(\frac{-1}{\sqrt[4]{45}}, \frac{-1}{\sqrt[4]{45^3}}\right)$

(b) $\lim_{x \to \infty} K = 0$

33. $y = x^{2/3}, \quad y' = \frac{2}{3}x^{-1/3}, \quad y'' = -\frac{2}{9}x^{-4/3}$

$$K = \left|\frac{(-2/9)x^{-4/3}}{(1 + (4/9)\,x^{-2/3})^{3/2}}\right| = \left|\frac{6}{x^{1/3}(9x^{2/3} + 4)^{3/2}}\right|$$

(a) $K \to \infty$ as $x \to 0$

(b) $\lim_{x \to \infty} K = 0$

34. $y = \ln x, \quad y' = \frac{1}{x}, \quad y'' = -\frac{1}{x^2}$

$$K = \left|\frac{-1/x^2}{(1 + (1/x)^2)^{3/2}}\right| = \frac{x}{(x^2 + 1)^{3/2}}, \quad \frac{dK}{dx} = \frac{-2x^2 + 1}{(x^2 + 1)^{5/2}}$$

(a) K has a maximum when $x = 1/\sqrt{2}$.

(b) $\lim_{x \to \infty} K = 0$

35. $f(x) = x + \frac{1}{x}, \quad f'(x) = 1 - \frac{1}{x^2} = \frac{x^2 - 1}{x^2}, \quad f''(x) = \frac{2}{x^3}$

$f'(1) = 0$ and $f''(1) = 2$

$K = \left|\frac{2}{(1 + 0^2)^{3/2}}\right| = 2 \qquad \frac{1}{K} = \frac{1}{2}$ (radius of curvature)

Slope at $(1, 2)$ is 0. Normal line is vertical

Center of the closest circular approximation is $(1, \frac{5}{2})$

$$(x - 1)^2 + (y - \frac{5}{2})^2 = (\frac{1}{2})^2$$

35. (continued)

$f(x) = x + \frac{1}{x}, \quad f'(x) = 1 - \frac{1}{x^2}, \quad f''(x) = \frac{2}{x^3}$

$f'(2) = 3/4, \quad f''(2) = 1/4$

$K = \frac{1/4}{[1 + (3/4)^2]^{3/2}} = \frac{16}{125} \quad \frac{1}{K} = \frac{125}{16}$ (radius of curvature)

The equation of the normal line to the curve at (2, 5/2) is:

$$y - \frac{5}{2} = \frac{-4}{3}(x - 2)$$

The center of the circle is on the normal line at a distance of $\frac{125}{16}$ from $(2, \frac{5}{2})$.

$$\sqrt{(2 - x)^2 + (\frac{5}{2} - [\frac{5}{2} - \frac{4}{3}(x - 2)])^2} = \frac{125}{16}$$

$\frac{25}{9}(x - 2)^2 = (\frac{125}{16})^2, \quad x = 2 \pm \frac{75}{16}$

The center of the closest circular approximation is $(\frac{-43}{16}, \frac{35}{4})$

The equation of the circle is $(x + \frac{43}{16})^2 + (y - \frac{35}{4})^2 = (\frac{125}{16})^2$

36. $y = (x + 1)^3 + 3, \quad y' = 3(x + 1)^2, \quad y'' = 6(x + 1)$

$K = \left|\frac{6(x + 1)}{[1 + 9(x + 1)^4]^{3/2}}\right|, \quad K = 0$ when $x = -1$

Point: (-1, 3)

37. Endpoints of the major axis: $(\pm 2, 0)$

Endpoints of the minor axis: $(0, \pm 1)$

$x^2 + 4y^2 = 4, \quad 2x + 8yy' = 0, \quad y' = -x/4y,$

$y'' = \frac{(4y)(-1) - (-x)(4y')}{16y^2} = \frac{-4y - (x^2/y)}{16y^2} = \frac{-(4y^2 + x^2)}{16y^3} = \frac{-1}{4y^3}$

$K = \frac{-1/4y^3}{[1 + (-x/4y)^2]^{3/2}} = \frac{-16}{(16y^2 + x^2)^{3/2}} = \frac{16}{(12y^2 + 4)^{3/2}}$

$= \frac{16}{(16 - 3x^2)^{3/2}}$, Therefore, since $-1 \le y \le 1$,

K is largest when $y = 0$ and smallest when $y = \pm 1$.

38. $y_1 = ax(b - x), \quad y_2 = \dfrac{x}{x + 2}$

We observe that (0, 0) is a solution point to both equations. Therefore, the point P is the origin.

$$y_1 = ax(b - x), \quad y_1' = a(b - 2x), \quad y_1'' = -2a$$

$$y_2 = \frac{x}{x + 2}, \quad y_2' = \frac{2}{(x + 2)^2}, \quad y_2'' = \frac{-4}{(x + 2)^3}$$

At P, $y_1'(0) = ab$ and $y_2'(0) = \dfrac{2}{(0 + 2)^2} = \dfrac{1}{2}$. Since the curves have a common tangent at P, $y_1'(0) = y_2'(0)$ or $ab = \dfrac{1}{2}$

Therefore, $y_1'(0) = \dfrac{1}{2}$

Since the curves have the same curvature at P, $K_1(0) = K_2(0)$

$$K_1(0) = \left|\frac{y_1''(0)}{[1 + (y_1'(0))^2]^{3/2}}\right| = \left|\frac{-2a}{[1 + (1/2)^2]^{3/2}}\right|$$

$$K_2(0) = \left|\frac{y_2''(0)}{[1 + (y_2'(0))^2]^{3/2}}\right| = \left|\frac{-1/2}{[1 + (1/2)^2]^{3/2}}\right|$$

Therefore, $2a = \pm\dfrac{1}{2}$ or $a = \pm\dfrac{1}{4}$

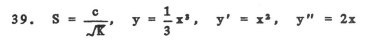

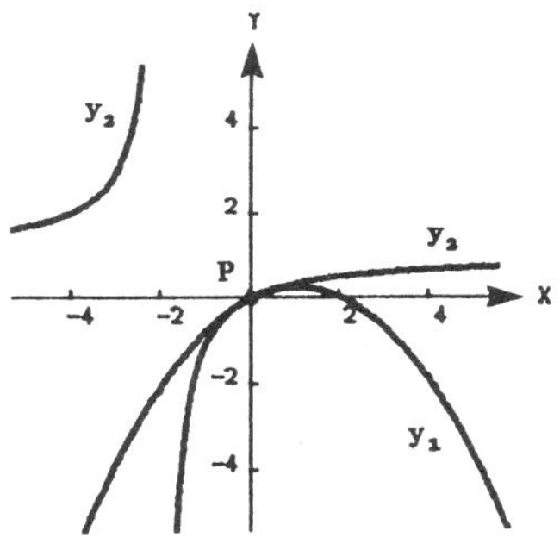

In order that the curves intersect in only one point, the parabola must be concave downward, and thus

$$a = \frac{1}{4} \text{ and } b = \frac{1}{2a} = 2$$

$$y_1 = \frac{1}{4}x(2 - x) \quad \text{and} \quad y_2 = \frac{x}{x + 2}$$

39. $S = \dfrac{c}{\sqrt{K}}, \quad y = \dfrac{1}{3}x^3, \quad y' = x^2, \quad y'' = 2x$

$$K = \left|\frac{2x}{(1 + x^4)^{3/2}}\right|$$

When $x = 1$, $K = \dfrac{1}{\sqrt{2}}$, $S = \dfrac{c}{\sqrt{1/\sqrt{2}}} = \sqrt[4]{2}\,c$, $30 = \sqrt[4]{2}\,c \Rightarrow c = \dfrac{30}{\sqrt[4]{2}}$

$$s\left(\frac{3}{2}\right) = \frac{30}{\sqrt[4]{2}} \frac{1}{\sqrt{\dfrac{3}{[1 + (81/16)]^{3/2}}}} \approx 56.27 \text{ mi/hr}$$

40. $r = 1 + \sin\theta, \quad r' = \cos\theta, \quad r'' = -\sin\theta$

$$K = \frac{|2(r')^2 - rr'' + r^2|}{[(r')^2 + r^2]^{3/2}}$$

$$= \frac{|2\cos^2\theta - (1 + \sin\theta)(-\sin\theta) + (1 + \sin\theta)^2|}{\sqrt[3]{[\cos^2\theta + (1 + \sin\theta)^2]^2}}$$

$$= \frac{3(1 + \sin\theta)}{\sqrt[3]{4(1 + \sin\theta)^2}} = \frac{3\sqrt[3]{1 + \sin\theta}}{\sqrt[3]{4}}$$

41. $r = a\sin\theta, \quad r' = a\cos\theta, \quad r'' = -a\sin\theta$

$$K = \frac{|2(r')^2 - rr'' + r^2|}{[(r')^2 + r^2]^{2/3}}$$

$$= \frac{|2a^2\cos^2\theta + a^2\sin^2\theta + a^2\sin^2\theta|}{\sqrt[3]{[a^2\cos^2\theta + a^2\sin^2\theta]^2}} = \frac{2a^2}{\sqrt[3]{a^4}} = \frac{2a}{\sqrt[3]{a}} = 2\sqrt[3]{a^2}$$

42. $r = \theta, \quad r' = 1, \quad r'' = 0$

$$K = \frac{|2(r')^2 - rr'' + r^2|}{[(r')^2 + r^2]^{2/3}} = \frac{2 + \theta^2}{(1 + \theta^2)^{2/3}}$$

43. $r = e^{\theta}, \quad r' = e^{\theta}, \quad r'' = e^{\theta}$

$$K = \frac{|2(r')^2 - rr'' + r^2|}{[(r')^2 + r^2]^{2/3}} = \frac{2e^{2\theta}}{(2e^{2\theta})^{2/3}} = \sqrt[3]{2e^{2\theta}}$$

Review Exercises for Chapter 13

1. $\mathbf{r}(t) = \mathbf{i} + t\,\mathbf{j} + t^2\,\mathbf{k}$

$x = 1, \quad y = t$

$z = t^2 \Rightarrow z = y^2$

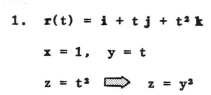

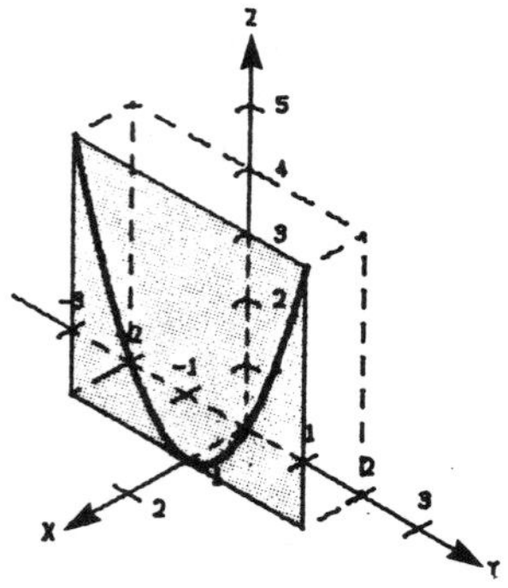

2. $\mathbf{r}(t) = 2t\,\mathbf{i} + t\,\mathbf{j} + t^2\,\mathbf{k}$

$x = 2t, \quad y = t, \quad z = t^2$

$y = \frac{1}{2}x, \quad z = y^2$

t	0	1	−1	2
x	0	2	−2	4
y	0	1	−1	2
z	0	1	1	4

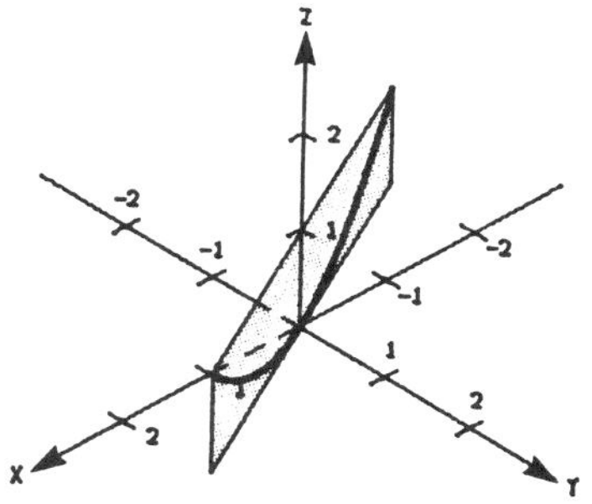

3. $\mathbf{r}(t) = 2\cos t\,\mathbf{i} + t\,\mathbf{j} + 2\sin t\,\mathbf{k}$

$x = 2\cos t, \quad y = t, \quad z = 2\sin t$

$x^2 + z^2 = 4$

t	0	$\pi/2$	π	$3\pi/2$
x	2	0	−2	0
y	0	$\pi/2$	π	$3\pi/2$
z	0	2	0	−2

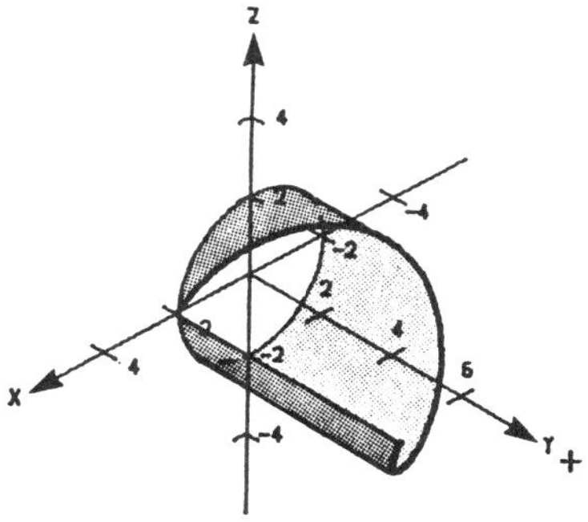

4. $\mathbf{r}(t) = \mathbf{i} + \sin t\,\mathbf{j} + \mathbf{k}$

$x = 1, \quad y = \sin t, \quad z = 1$

t	0	$\pi/2$	π	$3\pi/2$
x	1	1	1	1
y	0	1	0	−1
z	1	1	1	1

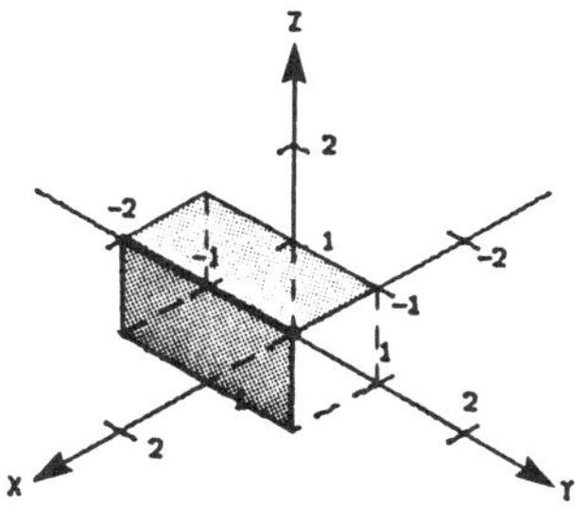

5. $z = x^2 + y^2, \quad x + y = 0, \quad t = x$

$x = t, \quad y = -t, \quad z = 2t^2$

$\mathbf{r}(t) = t\,\mathbf{i} - t\,\mathbf{j} + 2t^2\,\mathbf{k}$

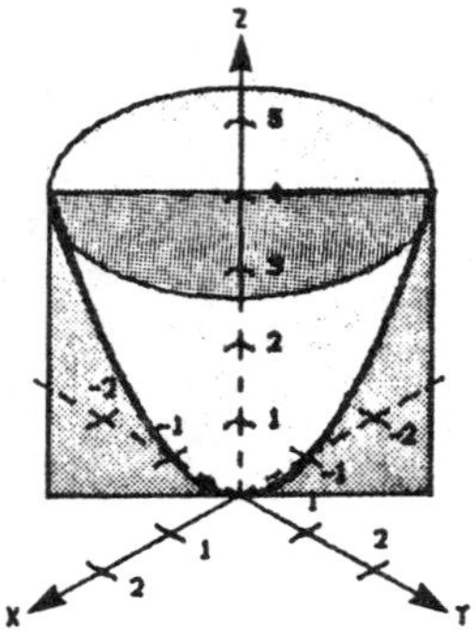

6. $x^2 + z^2 = 4, \quad x - y = 0, \quad t = x$

$x = t, \quad y = t, \quad z = \pm\sqrt{4 - t^2}$

$\mathbf{r}(t) = t\,\mathbf{i} + t\,\mathbf{j} + \sqrt{4 - t^2}\,\mathbf{k}$

$\mathbf{r}(t) = t\,\mathbf{i} + t\,\mathbf{j} - \sqrt{4 - t^2}\,\mathbf{k}$

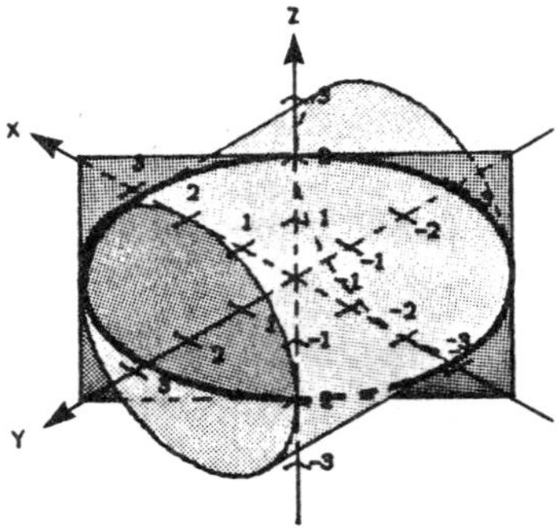

7. $\lim_{t \to 2} (t^2\,\mathbf{i} + \sqrt{4 - t^2}\,\mathbf{j} + \mathbf{k}) = 4\mathbf{i} + \mathbf{k}$

8. $\lim_{t \to 0} \left(\frac{\sin 2t}{t}\,\mathbf{i} + e^{-t}\,\mathbf{j} + e^{t}\,\mathbf{k}\right) = \left(\lim_{t \to 0} \frac{2\cos 2t}{1}\right)\mathbf{i} + \mathbf{j} + \mathbf{k} = 2\mathbf{i} + \mathbf{j} + \mathbf{k}$

9. $\mathbf{r}(t) = t\,\mathbf{i} + \csc t\,\mathbf{k}$

(a) Domain: $t \neq n\pi$, n an integer
(b) Discontinuous at $t = n\pi$, n an integer

10. $\mathbf{r}(t) = \sqrt{t}\,\mathbf{i} + \frac{1}{t - 4}\,\mathbf{j} + \mathbf{k}$

(a) Domain: $[0, 4)$ and $(4, \infty)$
(b) Discontinuous at $t = 4$

11. $\mathbf{r}(t) = \ln t\,\mathbf{i} + t\,\mathbf{j} + t\,\mathbf{k}$

(a) Domain: $(0, \infty)$
(b) Continuous

12. $\mathbf{r}(t) = (2t + 1)\,\mathbf{i} + t^2\,\mathbf{j} + t\,\mathbf{k}$

(a) Domain: $(-\infty, \infty)$
(b) Continuous

13. $\mathbf{r}(t) = 3t\,\mathbf{i} + (t - 1)\,\mathbf{j}, \qquad \mathbf{u}(t) = t\,\mathbf{i} + t^2\,\mathbf{j} + \frac{2}{3}t^3\,\mathbf{k}$

(a) $\mathbf{r}'(t) = 3\mathbf{i} + \mathbf{j}$

(b) $\mathbf{r}''(t) = \mathbf{0}$

(c) $\mathbf{r}(t) \cdot \mathbf{u}(t) = 3t^2 + t^2(t - 1) = t^3 + 2t^2$

$D_t[\mathbf{r}(t) \cdot \mathbf{u}(t)] = 3t^2 + 4t$

(d) $\mathbf{u}(t) - 2\mathbf{r}(t) = -5t\,\mathbf{i} + (t^2 - 2t - 2)\,\mathbf{j} + \frac{2}{3}t^3\,\mathbf{k}$

$D_t[\mathbf{u}(t) - 2\mathbf{r}(t)] = -5\mathbf{i} + (2t - 2)\,\mathbf{j} + 2t^2\,\mathbf{k}$

(e) $\|\mathbf{r}(t)\| = \sqrt{10t^2 - 2t + 1}$

$D_t[\|\mathbf{r}(t)\|] = \dfrac{10t-1}{\sqrt{10t^2 - 2t + 1}}$

(f) $\mathbf{r}(t) \times \mathbf{u}(t) = \frac{2}{3}(t^4 - t^3)\,\mathbf{i} - 2t^4\,\mathbf{j} + (3t^3 - t^2 + t)\,\mathbf{k}$

$D_t[\mathbf{r}(t) \times \mathbf{u}(t)] = (\frac{8}{3}t^3 - 2t^2)\,\mathbf{i} - 8t^3\,\mathbf{j} + (9t^2 - 2t + 1)\,\mathbf{k}$

14. $\mathbf{r}(t) = \sin t\,\mathbf{i} + \cos t\,\mathbf{j} + t\,\mathbf{k}, \qquad \mathbf{u}(t) = \sin t\,\mathbf{i} + \cos t\,\mathbf{j} + \frac{1}{t}\mathbf{k}$

(a) $\mathbf{r}'(t) = \cos t\,\mathbf{i} - \sin t\,\mathbf{j} + \mathbf{k}$

(b) $\mathbf{r}''(t) = -\sin t\,\mathbf{i} - \cos t\,\mathbf{j}$

(c) $\mathbf{r}(t) \cdot \mathbf{u}(t) = 2$

$D_t[\mathbf{r}(t) \cdot \mathbf{u}(t)] = 0$

(d) $\mathbf{u}(t) - 2\mathbf{r}(t) = -\sin t\,\mathbf{i} - \cos t\,\mathbf{j} + (\frac{1}{t} - 2t)\,\mathbf{k}$

$D_t[\mathbf{u}(t) - 2\mathbf{r}(t)] = -\cos t\,\mathbf{i} + \sin t\,\mathbf{j} + (-\frac{1}{t^2} - 2)\,\mathbf{k}$

(e) $\|\mathbf{r}(t)\| = \sqrt{1 + t^2}$

$D_t[\|\mathbf{r}(t)\|] = \dfrac{t}{\sqrt{1 + t^2}}$

(f) $\mathbf{r}(t) \times \mathbf{u}(t) = (\frac{1}{t}\cos t - t\cos t)\,\mathbf{i} - (\frac{1}{t}\sin t - t\sin t)\,\mathbf{j}$

$D_t[\mathbf{r}(t) \times \mathbf{u}(t)] = (-\frac{1}{t}\sin t - \frac{1}{t^2}\cos t + t\sin t - \cos t)\,\mathbf{i}$

$-(\frac{1}{t}\cos t - \frac{1}{t^2}\sin t - t\cos t - \sin t)\,\mathbf{j}$

15. $\int (\cos t\,\mathbf{i} + t\cos t\,\mathbf{j})\,dt = \sin t\,\mathbf{i} + (t\sin t - \sin t)\,\mathbf{j} + \mathbf{C}$

16. $\int (\ln t\,\mathbf{i} + t\ln t\,\mathbf{j} + \mathbf{k})\,dt = (t\ln t - t)\,\mathbf{i} + \frac{t^2}{4}(-1 + 2\ln t)\,\mathbf{j} + t\mathbf{k} + \mathbf{C}$

17. $\int \|\cos t\,\mathbf{i} + \sin t\,\mathbf{j} + t\mathbf{k}\|\,dt = \int \sqrt{1 + t^2}\,dt$

$= \frac{1}{2}[t\sqrt{1 + t^2} + \ln|t + \sqrt{1 + t^2}|] + C$

18. $\int (t\,\mathbf{j} + t^2\mathbf{k}) \times (\mathbf{i} + t\,\mathbf{j} + t\mathbf{k})\,dt$

$= \int [(t^2 - t^3)\,\mathbf{i} + t^2\,\mathbf{j} - t\mathbf{k}]\,dt = (\frac{t^3}{3} - \frac{t^4}{4})\,\mathbf{i} + \frac{t^3}{3}\mathbf{j} - \frac{t^2}{2}\mathbf{k} + \mathbf{C}$

19. $\mathbf{r}(t) = 2\cos t\,\mathbf{i} + 2\sin t\,\mathbf{j} + t\mathbf{k}$

$x = 2\cos t,\quad y = 2\sin t,\quad z = t$

when $t = \frac{3\pi}{4},\quad x = -\sqrt{2},\quad y = \sqrt{2},\quad z = \frac{3\pi}{4}$

$\mathbf{r}'(t) = -2\sin t\,\mathbf{i} + 2\cos t\,\mathbf{j} + \mathbf{k}$

Direction numbers when $t = \frac{3\pi}{4},\quad a = -\sqrt{2},\ b = -\sqrt{2},\quad c = 1$

$x = -\sqrt{2}t - \sqrt{2},\qquad y = -\sqrt{2}t + \sqrt{2},\qquad z = t + \frac{3\pi}{4}$

20. $\mathbf{r}(t) = t\,\mathbf{i} + t^2\,\mathbf{j} + \frac{2}{3}t^3\mathbf{k},\qquad x = t,\quad y = t^2,\quad z = \frac{2}{3}t^3$

When $t = 3,\quad x = 3,\quad y = 9,\quad z = 18$

$\mathbf{r}'(t) = \mathbf{i} + 2t\,\mathbf{j} + 2t^2\mathbf{k}$

Direction numbers when $t = 3,\quad a = 1,\quad b = 6,\quad c = 18$

$x = t + 3,\quad y = 6t + 9,\quad z = 18t + 18$

21. $\mathbf{r}(t) = (1 + 4t)\,\mathbf{i} + (2 - 3t)\,\mathbf{j}$

$\mathbf{v}(t) = 4\mathbf{i} - 3\mathbf{j},\quad s = \|\mathbf{v}\| = 5$

$\mathbf{a}(t) = \mathbf{0},\qquad \mathbf{T}(t) = \frac{1}{5}(4\mathbf{i} - 3\mathbf{j}),\qquad \mathbf{N}(t) = \mathbf{0}$

$\mathbf{a}\cdot\mathbf{T} = 0,\qquad \mathbf{a}\cdot\mathbf{N} = 0,\qquad K = 0$

22. $\mathbf{r}(t) = 5t\,\mathbf{i}, \quad \mathbf{v}(t) = 5\mathbf{i}, \quad \|\mathbf{v}(t)\| = 5, \quad \mathbf{a}(t) = \mathbf{0}$

$\mathbf{T}(t) = \mathbf{i}, \quad \mathbf{N}(t) = \mathbf{0}, \quad \mathbf{a}\cdot\mathbf{T} = 0, \quad \mathbf{a}\cdot\mathbf{N} = 0, \quad K = 0$

23. $\mathbf{r}(t) = 2(t+1)\,\mathbf{i} + \dfrac{2}{t+1}\,\mathbf{j}, \quad \mathbf{v}(t) = 2\mathbf{i} - \dfrac{2}{(t+1)^2}\,\mathbf{j}$

$$\|\mathbf{v}(t)\| = \frac{2\sqrt{(t+1)^4+1}}{(t+1)^2}$$

$$\mathbf{a}(t) = \frac{4}{(t+1)^3}\,\mathbf{j}, \quad \mathbf{T}(t) = \frac{(t+1)^2\,\mathbf{i} - \mathbf{j}}{\sqrt{(t+1)^4+1}}$$

$$\mathbf{N}(t) = \frac{\mathbf{i} + (t+1)^2\,\mathbf{j}}{\sqrt{(t+1)^4+1}}, \quad \mathbf{a}\cdot\mathbf{T} = \frac{-4}{(t+1)^3\sqrt{(t+1)^4+1}}$$

$$\mathbf{a}\cdot\mathbf{N} = \frac{4(t+1)^2}{(t+1)^3\sqrt{(t+1)^4+1}} = \frac{4}{(t+1)\sqrt{(t+1)^4+1}}$$

$$K = \frac{\mathbf{a}\cdot\mathbf{N}}{\|\mathbf{v}\|^2} = \frac{4}{(t+1)\sqrt{(t+1)^4+1}} \cdot \frac{(t+1)^4}{4[(t+1)^4+1]} = \frac{(t+1)^3}{[(t+1)^4+1]^{3/2}}$$

24. $\mathbf{r}(t) = t\,\mathbf{i} + \sqrt{t}\,\mathbf{j}, \quad \mathbf{v}(t) = \mathbf{i} + \dfrac{1}{2\sqrt{t}}\,\mathbf{j}, \quad \|\mathbf{v}(t)\| = \dfrac{\sqrt{4t+1}}{2\sqrt{t}}$

$$\mathbf{a}(t) = -\frac{1}{4t\sqrt{t}}\,\mathbf{j},$$

$$\mathbf{T}(t) = \frac{\mathbf{i} + \frac{1}{2\sqrt{t}}\,\mathbf{j}}{\frac{\sqrt{4t+1}}{2\sqrt{t}}} = \frac{2\sqrt{t}\,\mathbf{i} + \mathbf{j}}{\sqrt{4t+1}}$$

$$\mathbf{N}(t) = \frac{\mathbf{i} - 2\sqrt{t}\,\mathbf{j}}{\sqrt{4t+1}}, \quad \mathbf{a}\cdot\mathbf{T} = \frac{-1}{4t\sqrt{t}\sqrt{4t+1}}, \quad \mathbf{a}\cdot\mathbf{N} = \frac{1}{2t\sqrt{4t+1}}$$

$$K = \frac{2}{(4t+1)^{3/2}}$$

25. $\mathbf{r}(t) = e^t\,\mathbf{i} + e^{-t}\,\mathbf{j}, \quad \mathbf{v}(t) = e^t\,\mathbf{i} - e^{-t}\,\mathbf{j}$

$$\|\mathbf{v}(t)\| = \sqrt{e^{2t} + e^{-2t}}, \quad \mathbf{a}(t) = e^t\,\mathbf{i} + e^{-t}\,\mathbf{j}$$

$$\mathbf{T}(t) = \frac{e^t\,\mathbf{i} - e^{-t}\,\mathbf{j}}{\sqrt{e^{2t}+e^{-2t}}}, \quad \mathbf{N}(t) = \frac{e^{-t}\,\mathbf{i} + e^t\,\mathbf{j}}{\sqrt{e^{2t}+e^{-2t}}}$$

$$\mathbf{a}\cdot\mathbf{T} = \frac{e^{2t} - e^{-2t}}{\sqrt{e^{2t}+e^{-2t}}}, \quad \mathbf{a}\cdot\mathbf{N} = \frac{2}{\sqrt{e^{2t}+e^{-2t}}}$$

$$K = \frac{\mathbf{a}\cdot\mathbf{N}}{\|\mathbf{v}\|^2} = \frac{2}{(e^{2t}+e^{-2t})^{3/2}}$$

26. $\mathbf{r}(t) = t\cos t\,\mathbf{i} + t\sin t\,\mathbf{j}$

$\mathbf{v}(t) = \mathbf{r}'(t) = (-t\sin t + \cos t)\,\mathbf{i} + (t\cos t + \sin t)\,\mathbf{j}$

$\|\mathbf{v}(t)\| = \text{speed} = \sqrt{(-t\sin t + \cos t)^2 + (t\cos t + \sin t)^2} = \sqrt{t^2 + 1}$

$\mathbf{a}(t) = \mathbf{r}''(t) = (-t\cos t - 2\sin t)\,\mathbf{i} + (-t\sin t + 2\cos t)\,\mathbf{j}$

$$\mathbf{T}(t) = \frac{\mathbf{v}(t)}{\|\mathbf{v}(t)\|} = \frac{(-t\sin t + \cos t)\mathbf{i} + (t\cos t + \sin t)\,\mathbf{j}}{\sqrt{t^2 + 1}}$$

$$\mathbf{N}(t) = \frac{-(t\cos t + \sin t)\,\mathbf{i} + (-t\sin t + \cos t)\,\mathbf{j}}{\sqrt{t^2 + 1}}$$

$$\mathbf{a}(t)\cdot\mathbf{T}(t) = \frac{t}{\sqrt{t^2 + 1}}, \qquad \mathbf{a}(t)\cdot\mathbf{N}(t) = \frac{t^2 + 2}{\sqrt{t^2 + 1}}$$

$$K = \frac{\mathbf{a}(t)\cdot\mathbf{N}(t)}{\|\mathbf{v}(t)\|^2} = \frac{t^2 + 2}{(t^2 + 1)^{3/2}}$$

27. $\mathbf{r}(t) = t\,\mathbf{i} + t^2\,\mathbf{j} + \frac{1}{2}t^2\,\mathbf{k}, \qquad \mathbf{v}(t) = \mathbf{i} + 2t\,\mathbf{j} + t\,\mathbf{k}$

$\|\mathbf{v}\| = \sqrt{1 + 5t^2}, \qquad \mathbf{a}(t) = 2\mathbf{j} + \mathbf{k},$

$$\mathbf{T}(t) = \frac{\mathbf{i} + 2t\,\mathbf{j} + t\,\mathbf{k}}{\sqrt{1 + 5t^2}}, \qquad \mathbf{N}(t) = \frac{-5t\,\mathbf{i} + 2\mathbf{j} + \mathbf{k}}{\sqrt{5}\sqrt{1 + 5t^2}}$$

$$\mathbf{a}\cdot\mathbf{T} = \frac{5t}{\sqrt{1 + 5t^2}}, \qquad \mathbf{a}\cdot\mathbf{N} = \frac{5}{\sqrt{5}\sqrt{1 + 5t^2}}$$

$$K = \frac{\mathbf{a}\cdot\mathbf{N}}{\|\mathbf{v}\|^2} = \frac{\sqrt{5}}{(1 + 5t^2)^{3/2}}$$

28. $\mathbf{r}(t) = (t - 1)\,\mathbf{i} + t\,\mathbf{j} + \frac{1}{t}\mathbf{k}, \qquad \mathbf{v}(t) = \mathbf{i} + \mathbf{j} - \frac{1}{t^2}\mathbf{k}$

$$\|\mathbf{v}(t)\| = \frac{\sqrt{2t^4 + 1}}{t^2}, \qquad \mathbf{a}(t) = \frac{2}{t^3}\mathbf{k}$$

$$\mathbf{T}(t) = \frac{t^2\,\mathbf{i} + t^2\,\mathbf{j} - \mathbf{k}}{\sqrt{2t^4 + 1}}, \qquad \mathbf{N}(t) = \frac{\mathbf{i} + \mathbf{j} + 2t^2\,\mathbf{k}}{\sqrt{2}\sqrt{2t^4 + 1}}$$

$$\mathbf{a}\cdot\mathbf{T} = \frac{-2}{t^3\sqrt{2t^4 + 1}}, \qquad \mathbf{a}\cdot\mathbf{N} = \frac{4}{t\sqrt{2}\sqrt{2t^4 + 1}}$$

$$K = \frac{\mathbf{a}\cdot\mathbf{N}}{\|\mathbf{v}\|^2} = \frac{4t^3}{\sqrt{2}(2t^4 + 1)^{3/2}}$$

29. $\mathbf{r}(t) = \frac{1}{2}t\,\mathbf{i} + \sin t\,\mathbf{j} + \cos t\,\mathbf{k}, \quad 0 \le t \le \pi$

$\mathbf{r}'(t) = \frac{1}{2}\mathbf{i} + \cos t\,\mathbf{j} - \sin t\,\mathbf{k}$

$s = \int_0^{\pi} \|\mathbf{r}'(t)\|\,dt = \int_0^{\pi} \sqrt{(1/4) + \cos^2 t + \sin^2 t}\,dt$

$= \frac{\sqrt{5}}{2}\int_0^{\pi} dt = \frac{\sqrt{5}}{2}t\Big]_0^{\pi} = \frac{\sqrt{5}}{2}\pi$

30. $\mathbf{r}(t) = e^t \sin t\,\mathbf{i} + e^t \cos t\,\mathbf{k}, \quad 0 \le t \le \pi$

$\mathbf{r}'(t) = (e^t \cos t + e^t \sin t)\mathbf{i} + (-e^t \sin t + e^t \cos t)\,\mathbf{k}$

$\|\mathbf{r}'(t)\| = \sqrt{(e^t \cos t + e^t \sin t)^2 + (-e^t \sin t + e^t \cos t)^2} = e^t$

$s = \int_0^{\pi} \|\mathbf{r}'(t)\|\,dt = \int_0^{\pi} e^t\,dt = e^t\Big]_0^{\pi} = e^{\pi} - 1$

31. Range $= x = \frac{v_0^2}{32}\sin 2\theta = \frac{(75)^2}{32}\sin 60^\circ \approx 152$ feet

32. Range $= 4 = \frac{v_0^2}{16}\sin\theta\cos\theta$

$= \frac{v_0^2}{16}\cdot\frac{6}{2\sqrt{13}}\cdot\frac{4}{2\sqrt{13}} = \frac{3v_0^2}{104}$

$\frac{416}{3} = v_0^2 \implies v_0 \approx 11.776$ ft/sec

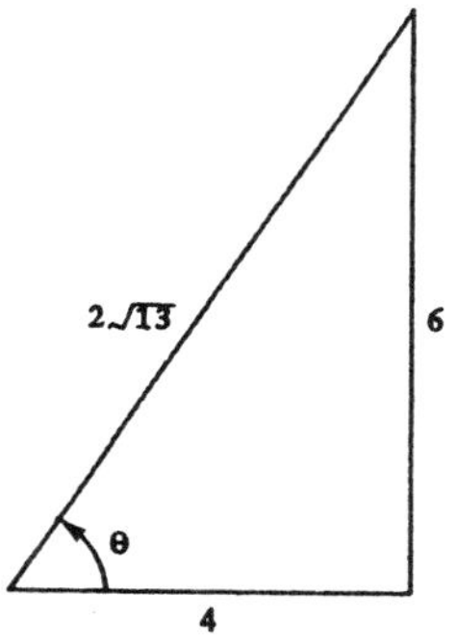

33. $v = \sqrt{\frac{9.56 \times 10^4}{4600}} \approx 4.56$ mi/sec

34. Factor of 4

36. $y = \frac{1}{32}x^{5/2}, \quad y' = \frac{5}{64}x^{3/2}, \quad y'' = \frac{15}{128}x^{1/2}$

$K = \left|\frac{(15/128)\,x^{1/2}}{(1 + (25/4096)\,x^3)^{3/2}}\right|$

At the point (4, 1), $K = \frac{120}{(89)^{3/2}} \implies r = \frac{1}{K} = \frac{(89)^{3/2}}{120} \approx 7$

2 3 4 5 6 7 8 9 10